NEW DEVELOPMENTS IN HIGH-PRESSURE MINERAL PHYSICS AND APPLICATIONS TO THE EARTH'S INTERIOR

DEDICATION

Professor Syun-iti Akimoto (1925–2004)

This volume is dedicated to Professor Syun-iti Akimoto, for his remarkable achievements and pioneering contributions in high-pressure, high-temperature research. Internationally renowned for his leadership over four decades, Professor Syun-iti Akimoto played a major role in elucidation of the deep Earth interior. In 1983, the American Geophysical Union bestowed upon him the prestigious William Bowie Medal for "distinguished attainment and outstanding contribution to the advancement of cooperative research in fundamental geophysics." As the founder of the U.S.-Japan seminar series in High Pressure Research, it is fitting to pay tribute by dedicating this volume to him. His vision, benevolent guidance to his students and colleagues, and interaction with mineral physicists around the world will long be remembered.

Takehiko Yagi
Murli Manghnani
Eiji Ohtani
David C. Rubie
Thomas S. Duffy

ELSEVIER

Physics of the Earth and Planetary Interiors 143–144 (2004) vii–ix

PHYSICS OF THE EARTH AND PLANETARY INTERIORS

www.elsevier.com/locate/pepi

(Abstracts/contents in: Am. Geol. Inst. Bibliogr. Index Geol.; Bull. Signalétique; Curr. Contents; Geo Abstr.; INSPEC; Pascal/CNRS; Phys. Abstr.)

Special Issue: New Developments in High-Pressure Mineral Physics and Applications to the Earth's Interior

doi:10.1016/S0031-9201(04)00176-1

1
2
3
4
5
6
7
8
9
10
11
12
13
14
15
16
17
18
19
20
21
22
23
24
25
26
27
28
29
30
31
32
33
34
35
36
37
38
39
40
41
42
43
44
45
46
47
48
49
50
51
52
53
54
55
56
57
58
59
60
61
62
63
64
65
66
67
68
69
70
71
72
73
74
75
76
77
78
79
80
81
82
83
84
7

Group Photo HPMPS-6, Verbania
(Participants)

1	Raymond Jeanloz
2	Masahiro Osako
3	Mike Walter
4	Tomoaki Kubo
5	Yu Nishihara
6	Astrid Holzheid
7	Takuo Okuchi
8	Hans Joachim Müller
9	Toru Inoue
10	Atsushi Kubo
11	Masaki Akoagi
12	Nobumasa Funamori
13	Peter Ulmer
14	Fred Marton
15	Steve Jacobsen
16	Mario Koch
17	Alan Woodland
18	Tatsuhiko Kawamoto
19	Takaya Nagai
20	Ho-kwang Mao
21	Guillaume Fiquet
22	Jun-ichi Ando
23	Takahiro Kuribayashi
24	Nobuyoshi Miyajima
25	Tadashi Kondo
26	Naotaka Tomioka
27	Norimasa Nishiyama
28	Eiichi Takahashi
29	Satoru Urakawa
30	Yasuhiro Kudoh
31	Yanbin Wang
32	Kiyoshi Fujino
33	Yingwei Fei
34	Patrick Cordier
35	Ulrich Bläss
36	Zizeng Gong
37	Osamu Ohtaka
38	Michael Vaughan
39	Bernie Wood
40	Yoshinori Tange
41	Robert Liebermann
42	Hiroshi Arima
43	Hiroshi Fukui
44	Paul Balog
45	Agnes Mao
46	Wendy Mao

47	Ahmed El Goresy
48	Frank Schilling
49	Christian Holzapfel
50	Eiji Ohtani
51	Guoyin Shen
52	Murli Manghnani
53	Don Weidner
54	Denis Andrault
55	Jay Bass
56	Shun-ichiro Karato
57	Ronald Cohen
58	Petra Ständner
59	Julien Siebert
60	Tetsuo Irifune
61	Dan Frost
62	Surendra Saxena
63	Christian Liebske
64	Thomas J. Ahrens
65	David Rubie
66	Takehiko Yagi
67	Hans Keppler
68	Hélène Couvy
69	Motohiko Murakami
70	Nabil Boctor
71	Eiji Ito
72	Konstantin Litasov
73	Joe Smyth
74	Yuki Asahara
75	Carine Vanpeteghem
76	Thomas Duffy
77	Jiuhua Chen
78	Bettina Schmickler
79	Tomoo Katsura
80	Li Li
81	Maki Okube
82	Alexandre Corgne
83	Valérie Malavergne
84	Stefan Keyssner

Available online at www.sciencedirect.com
SCIENCE @ DIRECT®
ELSEVIER
Physics of the Earth and Planetary Interiors 143–144 (2004) 1–3
PHYSICS OF THE EARTH AND PLANETARY INTERIORS
www.elsevier.com/locate/pepi

Introduction

New Developments in High Pressure Mineral Physics and Applications to the Earth's Interior

Geophysical measurements, such as the lateral variations in seismic wave velocities that are imaged by seismic tomography, provide the strongest constraints on the structure of the Earth's deep interior. However, in order to interpret seismic velocities and the associated density and elastic property estimates in terms of mineralogical/compositional models of the Earth's interior, data from mineral physics are essential. Consequently, considerable effort has been made over the last 30 years to determine the physical and chemical properties of Earth materials (minerals, rocks and melts) under the relevant conditions of high temperatures and pressures. Thermodynamic properties, phase equilibria, crystal chemistry and crystallography at high pressure are examples of areas in which major contributions have been made. Many experimental measurements have been made possible only by a range of technical developments in the quest to achieve high pressures and temperatures in the laboratory. At the same time, analytical methods, including X-ray diffraction, a variety of spectroscopic techniques, electron microscopy, ultrasonic interferometry, and methods for rheological investigations have been developed and greatly improved. In recent years, major progress has been made also in the field of computational mineralogy whereby ab initio simulations are used to investigate the structural and dynamical properties of condensed matter at an atomistic level. These numerical techniques provide an important tool for understanding the fundamental physics behind experimental observations and for probing mineral properties at pressure and temperature conditions not currently accessible by experimental techniques.

Early developments in the field of high-pressure research in Earth Sciences took place mainly in USA and Japan. In the period 1976–1996, scientists in these countries organised and participated in a series of small meetings entitled "US–Japan Seminars on High Pressure–Temperature Research" approximately once every 5 years. The participants (normally around 60) originated almost exclusively from Japan and USA. A notable result of these small meetings was the publication of a series of high quality volumes that beautifully summarized the scientific and technical developments of high-pressure research in mineral physics. Today, these volumes still represent indispensable reference sources. The last three of these volumes were published by the American Geophysical Union, in 1987, 1992 and 1998, as Geophysical Monographs (volumes 39, 67 and 101, respectively).

In the last 15 years, high-pressure research has become more international and leading institutes are now also located in countries such as France, Germany and UK. Because the field is still developing rapidly, both technically and scientifically, a new series of small international meetings has now been established that replace the old "US–Japan" meetings. The new meetings are called "High Pressure Mineral Physics Seminars" (HPMPS) and the first of these, HPMPS-6, took place in Verbania (Northern Italy) on August 26-31, 2002 (following on from the five earlier US–Japan Seminars). More than 90 scientists attended and the number of scientific contributions (talks and posters) was also about 90. This volume has arisen largely from this meeting but also includes a number of contributions from scientists who did not participate.

doi:10.1016/j.pepi.2004.02.001

The aim of this volume is to present an overview of recent developments in high-pressure mineral physics. As can be seen from the contents, the contributions cover a broad range of topics. Although we have attempted to group the 49 papers according to topic, many contributions actually cover two or more aspects.

The first eight papers deal with studies of the elastic properties of mantle minerals—which, as mentioned above, are essential for deriving mineralogical and compositional models of the Earth's mantle from seismic data. As detailed in the various papers (and also in the section on technical developments), elastic properties can now be measured at high pressures using a variety of different techniques, including X-ray diffraction, inelastic X-ray scattering, ultrasonic interferometry at GHz and MHz frequencies, and Brillouin spectroscopy.

In the following section on phase equilibria, contributions deal with mantle mineralogy in realistically complex chemical systems—the results of which are required for developing improved elastic models of the Earth's interior and for interpreting the causes of seismic discontinuities. The effects of oxygen fugacity on mantle mineralogy are discussed as well as mineralogy and crystal chemistry in FeO-bearing systems. It is still challenging to study phase equilibria under conditions of the deep lower mantle and this section includes several state of the art contributions in this research area.

The section on fluids and volatiles in the Earth is the outcome of the realization, which has developed over the past 15 years, that many mantle minerals can accommodate significant quantities of water into their structures. Consequently, water (and other volatiles such as CO_2) are cycled readily between the surface (hydrosphere and atmosphere) and the Earth's deep interior through processes such as subduction and volcanism. At any given time, an enormous volume of water can be stored potentially in the Earth's mantle. Because dissolved water has profound effects on mineral and rock properties (e.g. rheology, diffusion and electrical conductivity), melting processes and the thickness of seismic discontinuities, attempts are made to identify water-rich regions in the mantle, for example using seismic tomography. For this purpose, it is essential to know how dissolved volatiles affect the structure and elastic properties of minerals—which are the topics of several papers in this volume.

In the area of transport properties and rheology, important advances have been made in the last few years, as documented in the five papers of this section. These include measuring heat transport properties (thermal conductivity) in situ at deep mantle conditions and innovative methods for studying the rheology and deformation behavior of mantle materials at high pressures and temperatures. Rheological data are urgently required for understanding the cause of deep earthquakes and modeling mantle convection. Recent progress in this field has been impressive, although the determination of steady-state flow laws during high strain deformation at low strain rates under deep mantle conditions is still elusive.

The volume continues with a set of contributions on melting relations and element partitioning at high pressure. The aim of such studies is to constrain not only melting temperatures deep in terrestrial planets but also to understand the early differentiation of the Earth. Early geochemical evolution during the early accretional history of the Earth and other planets likely involved the crystallisation of deep magma oceans as a consequence of giant impacts.

The properties of iron, particularly with reference to understanding planetary cores, are the subject of several papers. Experiments at conditions of the Earth's core are still extremely challenging and important topics addressed include the crystal structure at ultrahigh pressures, theoretical study of magnetism in iron, and the density and structure of non-crystalline iron. In the case of Mars, equations of state of core forming iron alloys also provide important constraints on the internal structure of the planet.

The final section is concerned with recent technical developments. These papers include reports on new types of multianvil apparatus, the measurement of high pressures and temperatures in multianvil and diamond anvil cell experiments, measurement of elastic properties, recent developments of new diamond-based materials that are potentially valuable for use as anvils in generating high pressures, and some new spectroscopic tools for probing material structures at the atomistic level.

Finally, we would like to thank all the authors for contributing to this volume and especially to the numerous referees who have dedicated their valuable time and assistance in providing detailed and constructive reviews.

D.C. Rubie (Guest Editor) [a,*]
T.S. Duffy (Guest Editor) [b]
E. Ohtani (Guest Editor) [c]

[a] *Universität Bayreuth*
Bayerisches Geoinstitut
Bayreuth 95400, Germany

[b] *Department of Geosciences*
Princeton University, Princeton
NJ 08544, USA

[c] *Institute of Mineralogy*
Petrology and Economic Geology
Faculty of Science, Tohoku University
Sendai 980-8578, Japan

*Corresponding author. Tel.: +49-921-55-3711
fax: +49-921-55-3769
E-mail address: dave.rubie@uni-bayreuth.de
(D.C. Rubie)

1. Elasticity

Available online at www.sciencedirect.com
SCIENCE DIRECT®
ELSEVIER
Physics of the Earth and Planetary Interiors 143–144 (2004) 5–18
PHYSICS OF THE EARTH AND PLANETARY INTERIORS
www.elsevier.com/locate/pepi

Application of inelastic X-ray scattering to the measurements of acoustic wave velocities in geophysical materials at very high pressure

G. Fiquet[a,*], J. Badro[a], F. Guyot[a], Ch. Bellin[a], M. Krisch[b], D. Antonangeli[b], H. Requardt[b], A. Mermet[b], D. Farber[c], C. Aracne-Ruddle[c], J. Zhang[d]

[a] *Laboratoire de Minéralogie Cristallographie, UMR CNRS 7590, Universités Paris VI et VII, Institut de Physique du Globe de Paris, 4 Place Jussieu, 75252 Paris, cedex 05, France*
[b] *European Synchrotron Radiation Facility, BP220, 38043 Grenoble, cedex, France*
[c] *Lawrence Livermore National Laboratory, Energy and Environment Directorate, 4000 East Ave., 94550 Livermore, CA USA*
[d] *CHiPR and Department of Geosciences, SUNY Stony Brook, Stony Brook, NY 11794, USA*

Received 17 January 2003; received in revised form 29 September 2003; accepted 3 October 2003

Abstract

Reference Earth global models used by geophysicists are mostly constrained by analysing seismic waves that travel in the Earth. The interpretation of these seismological models in terms of chemical composition and temperature thus relies on the knowledge of the chemical composition, structure and elastic properties of some candidate materials relevant to geophysics. We will describe some recent advances in experimental mineral physics which yield essential information on the elastic properties of pertinent materials of Earth's mantle and core. In particular, we show that inelastic X-ray scattering (IXS) proves to be a well suited spectroscopic technique for the study of phonon dispersion in materials under high pressure. Inelastic scattering experiments carried out on polycrystalline samples provide measurements of orientationally averaged longitudinal acoustic velocities whereas investigations on single crystals yield elastic moduli. Such a technique is thus highly interesting for mineral physicists and offers a great potential in providing essential elastic data and their pressure and/or temperature dependence on most compounds of geophysical interest.

Keywords: Earth's interior; Mineral physics; High pressure; Physical properties; Elasticity; Techniques

1. Introduction

Knowledge of the elasticity of minerals is essential to the understanding and interpretation of seismic data, in particular radial average sound velocity profiles or three dimensional images from tomographic studies. Successful interpretation of these seismic models in terms of chemical composition, mineralogy, and temperature indeed requires information on the elasticity of deep Earth materials under high pressure and temperature conditions. The experimental determination of elastic properties at extreme conditions is, however, not an easy task. Still, numerous experimental procedures have been developed during these last

* Corresponding author. Tel.: +33-14427-5236; fax: +33-144-273785.
E-mail address: fiquet@lmcp.jussieu.fr (G. Fiquet).

0031-9201/$ – see front matter © 2004 Published by Elsevier B.V.
doi:10.1016/j.pepi.2003.10.005

decades, which couple high-pressure apparatuses to various techniques such as static (at zero frequency), shock-wave compression, or dynamical acoustic vibration (ranging from MHz to GHz frequency range). Among them, we mention the following:

- Shock measurements: measurements of the compressional wave velocities under shock compression along the Hugoniot in iron (Brown and McQueen, 1986), forsterite to 150 GPa (Duffy and Ahrens, 1992), periclase to 27 GPa (Duffy and Ahrens, 1995) or molybdenum to 81 GPa (Duffy and Ahrens, 1994).
- Resonant ultrasound spectroscopy to study the behaviour of minerals at extreme temperature and room pressure, e.g. single crystal periclase or forsterite to 1800 K (Isaak et al., 1989a,b).
- The implementation of ultrasonic interferometric measurements using a 1000 tonnes uniaxial split-cylinder apparatus (see Liebermann and Li, 1998) with applications to polycrystalline olivine to 10 GPa and 1300 K (Li et al., 1998), wadsleyite to 7 GPa and 873 K (Li et al., 2001) or $MgSiO_3$ perovskite to 8 GPa and 800 K (Sinelnikov et al., 1998).
- Optical probes using phonon–photon scattering processes with stimulated Brillouin scattering in a diamond–anvil cell with measurements of single crystal elastic moduli of natural samples of olivine, orthopyroxene and garnet to 20 GPa (Brown et al., 1989; Zaug et al., 1993; Chai et al., 1997a,b; Abramson et al., 1999).
- The extension of pressure range using conventional Brillouin spectroscopy with measurements on periclase to 55 GPa (Zha et al., 2000) or silica glass to 58 GPa (Zha et al., 1994). Data on periclase and synthetic pyrope to 20 GPa (Sinogeikin and Bass, 2000) as well as polycrystalline materials such as natural majorite-pyrope solid solution under pressure (Sinogeikin and Bass, 2002).

Furthermore, first-principles methods have been applied to the study of the high-pressure phases relevant to the Earth's deep interior such as $MgSiO_3$ perovskite (Karki et al., 2000; Oganov et al., 2001), hcp iron (Alfe et al., 2001; Steinle-Neumann et al., 2001) and hcp cobalt (Steinle-Neumann et al., 1999). The reader is invited to refer to the review articles by Stixrude et al. (1998) or Karki et al. (2001).

None of these methods provides unique answers to the standing long geophysical term requirement of obtaining precise measurements of compressional velocities V_P and shear velocities V_S, at extreme pressures and temperatures in chemically and mineralogically complex materials over a large frequency and wavelength range. This is why any technique bringing new information in this field is valuable. Recently, the use of X-ray diffraction (XRD; Mao et al., 1998; Matthies et al., 2001) and of several inelastic scattering techniques such as nuclear resonance inelastic X-ray scattering (NRIXS) (Lubbers et al., 2000; Mao et al., 2001), inelastic neutron (INS) (Klotz and Braden, 2000) and inelastic X-ray scattering (IXS) (Fiquet et al., 2001), have provided new data on geophysically relevant materials. In this paper, we will concentrate on the method of inelastic X-ray scattering with very high energy resolution ($\Delta E/E = 10^{-7}$) which has proven to be well suited for the determination of elastic properties of matter at high pressure (Krisch et al., 1997; Occelli et al., 2001) at THz frequencies. For geophysics, such high frequencies have the interest of getting close to the infinite frequency limit, in complementarity with the low frequency limit provided by radial X-ray diffraction modeling. IXS has more potential in ultra-high pressure geophysics (e.g. lower mantle and core) than INS which requires larger samples. Such INS experiments are therefore limited at present to pressures which do not exceed 10–15 GPa. Furthermore, INS technique is restricted to measurements of relatively low acoustic velocities in disordered or polycrystalline systems. IXS cross section is proportional to $f(Q)^2Q^2\rho/\mu$ where ρ is the specific volume, μ the absorption coefficient and Q the wave vector. The optimal signal for DAC experiments is obtained if the absorption length $t = 1/\mu$ is of the order of 10–40 μm, typically spanning elements with Z between 30 and 50. However, previous experiments have shown that studies on lower Z (Occelli et al., 2001; Krisch et al., 2002) and higher Z materials (Loa et al., 2003) are as well possible. Finally, we note that higher order Brillouin zones can be explored in single crystals, where the signal is enhanced by $f(Q)^2Q^2$. In the following, examples of applications of IXS will be given for polycrystalline iron, iron alloyed with light elements, $MgSiO_3$ perovskite, as well as one example of cobalt single crystal.

2. Experimental setup

The development of IXS with meV energy resolution goes back to work at HASYLAB (Germany) in the 1980s (see Burkel et al. (1987) or Dorner et al., 1987). This method allows to study phonons and phonon-like excitations. The theoretical aspect will not be discussed here and the reader is referred to the articles of Schülke (1991) and Burkel (2000). The basic kinematics of the inelastic scattering process is described in Fig. 1. The incident photon of energy $\hbar\omega_1$ and wave vector $\vec{k}_1$ is scattered by the sample into a photon characterized by energy $\hbar\omega_2$ and wave vector $\vec{k}_2$, a process corresponding to an energy transfer $\hbar\omega = \hbar\omega_1 - \hbar\omega_2$ and the momentum transfer $\hbar\vec{Q} = \hbar\vec{k}_1 - \hbar\vec{k}_2$. In the cases investigated in this study, the energy transfers are of the order of meV, much smaller than the incident photon energy of 15–25 keV, and Q is only controlled by the scattering angle theta, with the following expression:

$$Q = 2k_1 \sin\left(\frac{\theta}{2}\right) \tag{1}$$

At the ESRF (European Synchrotron Radiation Facility), two beamlines (ID16 and ID28) are dedicated to X-ray inelastic scattering spectroscopy. On ID28, the X-ray beam produced by three undulators is monochromatized by a cryogenically cooled silicon (1 1 1) crystal and a high-energy resolution silicon backscattering monochromator (Verbeni et al., 1996). The backscattered beam is focused by a gold-coated toroidal mirror which provides a focal spot at the sample position of 250 μm (horizontal) by 60 μm (vertical) full-width-half-maximum. The scattered photons are analyzed by five analyzers mounted on a Rowland circle at a distance of 650 cm from the sample (Masciovecchio et al., 1996). The energy-analyzed photons are subsequently detected by a Peltier-cooled five element silicon diode detector positioned near the sample position 3 mm above the incident beam axis. According to Eq. (1), the momentum transfer Q is selected by rotating the spectrometer arm in the horizontal scattering plane around a vertical axis passing through the scattering sample (see Fig. 2). The energy scans are performed by varying the monochromator temperature while the temperature of the analyzers is kept fixed. With a linear thermal expansion of silicon at room temperature $\alpha = 2.58 \times 10^{-6}\,\mathrm{K}^{-1}$ and a conversion to the energy scale according to the relation: $\Delta E/E = \alpha\Delta T$, a temperature scan over a few degrees with mK resolution provides the necessary meV resolution in energy needed to probe phonon excitations. The sample orientation and scattering angle are kept fixed whereas the energy of the beam impinging on the sample is changed by varying the monochromator temperature in steps of 0.005–0.020 K over 1–4 K, corresponding to a scanning range of several 10 meV up to a few 100 meV.

In order to illustrate how data processing is made, a typical IXS spectrum is given in Fig. 3 for polycrystalline bcc iron at ambient conditions. On both sides of the elastic line centered at zero-energy transfer, one can notice two excitations which correspond to the scattering of incident X-ray photons by longitudinal acoustic LA phonons in iron. These Stokes (energy loss) and anti-Stokes (energy gain) inelastic interactions, corresponding to phonon creation and annihilation, respectively, are characterized by an energy transfer at a given momentum transfer Q, which corresponds to a particular location in the first Brillouin zone. The energy positions and the widths of the excitations are fitted by using a Lorentzian model function, convoluted with an experimentally determined energy resolution function, by standard χ^2 minimization (shown in Fig. 3).

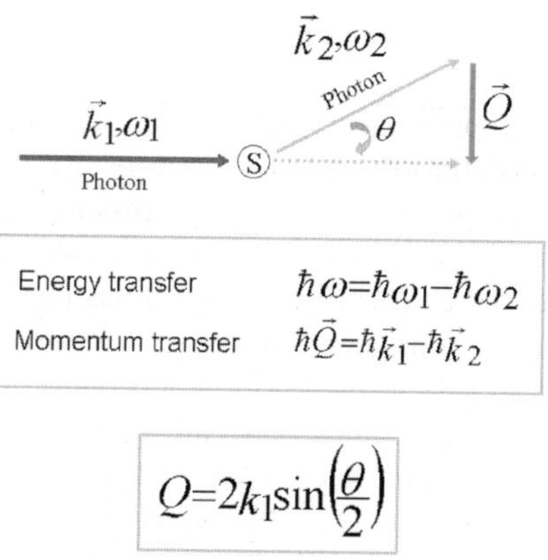

Fig. 1. Scattering geometry and kinematics of the inelastic scattering process. The incident X-ray photon characterized by angular frequency ω_1 and wave vector $\vec{k}_1$, and scattered photon by angular frequency ω_2 and wave vector $\vec{k}_2$. As long as the transferred energy is much smaller than the incident photon energy, the momentum transfer Q is simply connected with the geometry of the experiment, i.e. the scattering angle θ.

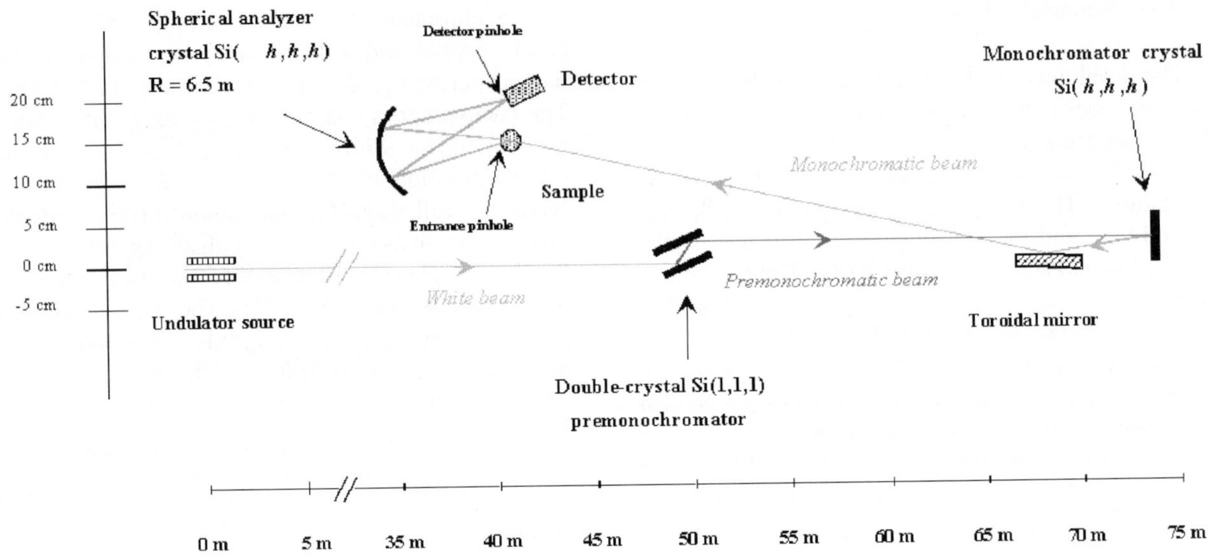

Fig. 2. Optical layout of ID28 beamline at ESRF (European Synchrotron Radiation Facility, Grenoble, France).

2.1. *The example of iron and its implications for the Earth's inner core*

The knowledge of the elastic constants of the phases of iron, which makes up 70–90 wt.% of planetary cores, is essential for comparison with global velocity models of Earth. The hcp (or ε) high-pressure phase is stable to at least 300 GPa at ambient temperature (Mao et al., 1990) and numerous experiments have been carried out in order to characterize the elastic properties of iron phases, such as XRD (Mao et al., 1998; Matthies et al., 2001), NRIXS (Lubbers et al., 2000; Mao et al., 2001), or INS on the low pressure bcc phase (Klotz and Braden, 2000). As shown in Fiquet et al. (2001), IXS provides a way to derive the acoustic velocity from the dispersion of the longitudinal acoustic phonon energy in iron. The results of this study are used here for illustrating how acoustic velocities are obtained from IXS data collected under high pressure.

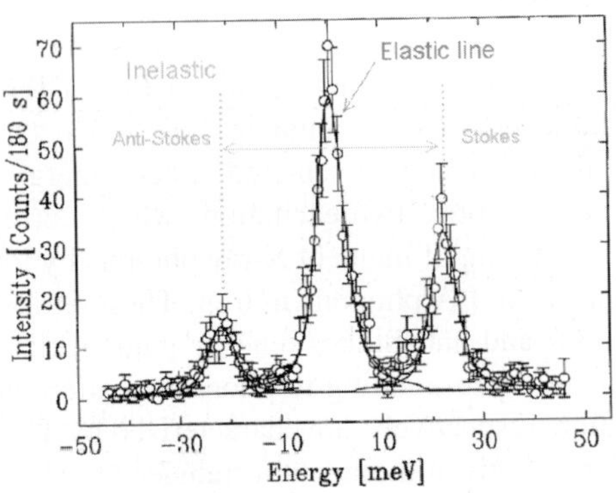

Fig. 3. Example of an inelastic X-ray scattering pattern obtained from a bcc iron foil at ambient conditions using the (8 8 8) monochromator reflection, providing an energy resolution of 5.5 meV. Scanning time is 5 h. $Q = 6.16\,\mathrm{nm}^{-1}$. The experimental data (open circles) are plotted with their error bars along with the corresponding fits.

In this experiment, a polycrystalline sample of iron was loaded into the 120 μm hole of a rhenium gasket and compressed without any pressure transmitting medium between diamond anvils. In the following, only the Stokes part of the spectra has been collected to save time. The peak centered at zero-energy transfer corresponds to the elastic contribution to the signal, whereas two other peaks are visible at higher energy transfer. The knowledge of the phonon dispersion curves of iron (Minkiewicz et al., 1967) and diamond (Warren et al., 1967) allows an unambiguous assignment of these features, which correspond to a LA phonon of iron and a transverse acoustic TA phonon of diamond (see Fig. 4). With the five analyzers located in the horizontal plane at different scattering angles θ, it is possible to map the dispersion curves of acoustic phonons across the first Brillouin zone. Due

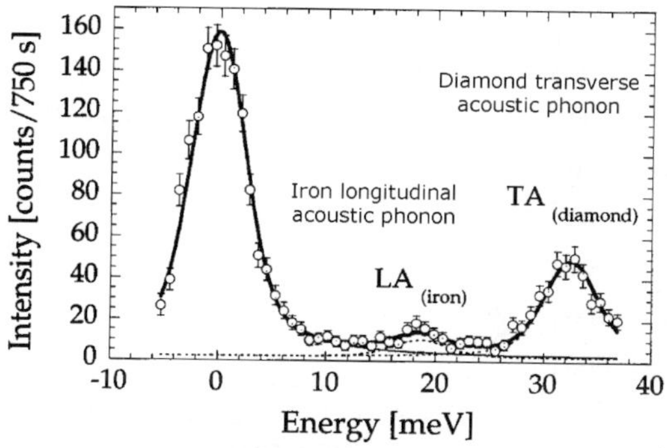

Fig. 4. IXS pattern recorded on a powdered iron sample in the hcp phase, at 28 GPa in a diamond–anvil cell. No pressure transmitting medium is used. The momentum transfer Q is $4\,\text{nm}^{-1}$. Dashed lines represent the inelastic contributions of the LA phonon branch for iron and from a TA branch from diamond, the thin solid line represents the elastic contribution and the thick continuous line shows the summation of individual contributions.

the polycrystalline nature of the sample, only modulus Q is defined, but not its direction. We consequently only record an orientationally averaged LA velocity. Furthermore, the inelastic scattering cross section is proportional to the dot product $(\vec{Q}\vec{e}_q)^2$ where $\vec{Q}$ is the momentum transfer and $\vec{e}_q$ is the phonon eigenvector of mode q (e.g. Ashcroft and Mermin, 1976). Consequently, only modes whose eigenvectors have a component parallel to Q are visible in a coherent inelastic scattering experiment. For momentum transfers within the first Brillouin zone in polycrystalline systems, these modes are predominantly of longitudinal character (Burkel, 2000). However, transverse modes can be detected, though weaker in their intensity, due to the fact that along an arbitrary crystalline direction, there can be a non-zero contribution of the transverse phonon eigenvectors projected onto the momentum transfer direction. Indeed, weak detection of TA mode in iron is evidenced in Fig. 5 for some Q values. Furthermore, in some directions, some LA–TA mixing might occur, but as the intensity contribution of TA modes is weak, this effect is most likely negligible.

As shown in Fig. 5, different excitations corresponding to iron LA or TA phonons can be fitted at different Q values. The TA phonon of diamond visible at $Q = 4\,\text{nm}^{-1}$ disappears at higher Q values in the recorded energy transfer region because of its high sound velocity. The LA iron phonon branch is recorded over the entire momentum and pressure range explored, as

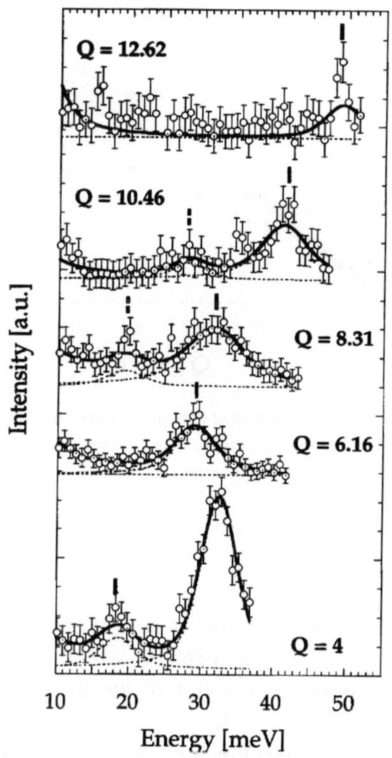

Fig. 5. IXS dispersion of the iron LA phonon recorded in the hcp phase of iron at 28 GPa. These patterns correspond to the signal recorded on the whole analyser set, i.e. at five different Q values. The LA phonons of iron are indicated by ticks. The excitation indicated by broken ticks for Q values at 8.31 and $10.46\,\text{nm}^{-1}$ have been interpreted as a TA phonon in iron sample.

shown in Fig. 6 on the dispersion curves resulting from the fitted energy position of the phonon reported as a function of the momentum transfer Q. We notice that the high-pressure data are at energies well above those at ambient pressure, implying a significant increase of the sound velocity with increasing pressure, in the bcc phase as well as in the hcp phase of iron. The velocity of longitudinal acoustic velocity (V_P) as well as the position of the edge of the first Brillouin zone Q_{MAX} can be estimated by fitting the dispersion relation with a sine law, and by determining its slope in the $Q = 0$ limit with the following expression:

$$E[\text{meV}] = 4.192 \times 10^{-4}\, V_P[\text{ms}^{-1}] \times Q_{\text{MAX}}[\text{ns}^{-1}] \times \sin\left[\frac{\pi}{2}\frac{Q[\text{ns}^{-1}]}{Q_{\text{MAX}}[\text{ns}^{-1}]}\right] \quad (2)$$

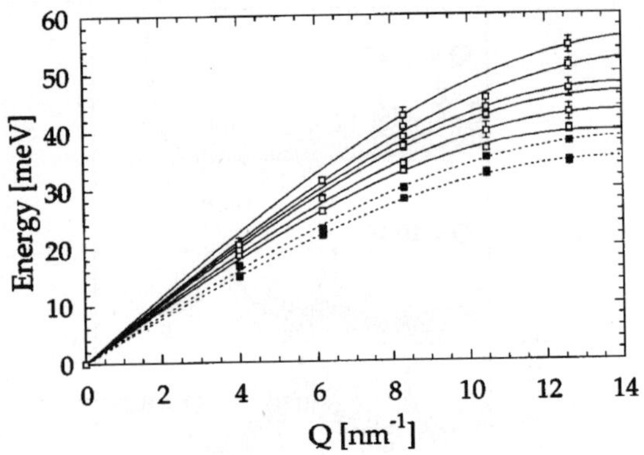

Fig. 6. Orientationally averaged dispersion curve of the longitudinal acoustic phonon branch of iron. The dashed line corresponds to data recorded in the bcc phase at 0.2 and 7 GPa, solid lines stand for fitted dispersion curves in the hcp phase of iron at pressures of 19, 28, 45, 55, 64 and 110 GPa from bottom to top.

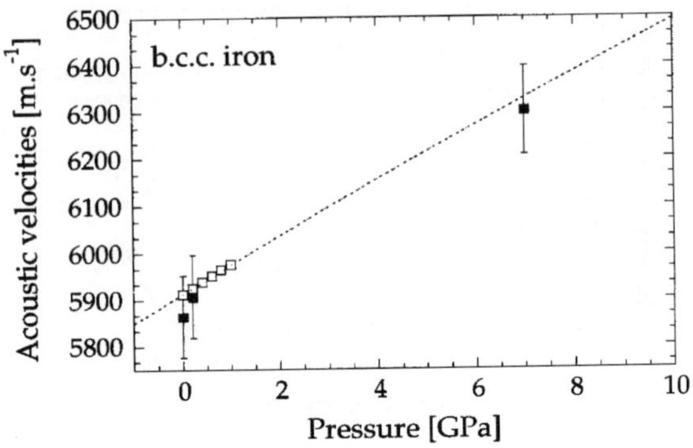

Fig. 7. Acoustic sound velocity as a function of pressure for bcc iron (solid squares) as compared to ultrasonics measurements and extrapolation by Guinan and Beshers (1968).

The first Brillouin zone is approximated here by a sphere defined by radius Q_{MAX}, which has been considered as a free parameter in the fitting procedure. Within uncertainties, the observed values of Q_{MAX} calculated from XRD data by equating the volume of the reciprocal unit-cell to that of a sphere are in agreement with those obtained during the fit. Data recorded at four to five momentum transfers have been used in each dispersion curve to constrain V_P within an estimated error of 3%.

Our results for the low-pressure bcc structure agree within 1% with the ultrasonic data collected up to 1 GPa and extrapolated up to 10 GPa (Guinan and Beshers, 1968), thus confirming the reliability of these measurements (see Fig. 7). Results obtained in the hcp phase are given in Table 1 and reported as a function of pressure in Fig. 8, along with previous measurements obtained with other techniques. At low pressures (below 40 GPa), we find an excellent agreement between our longitudinal acoustic velocity V_P determination and orientationally averaged ultrasonic data (data point at 26.5 GPa in Mao et al., 1998). Up to 100 GPa, our measurements are within error bars in good agreement with radial X-ray diffraction (Mao et al., 1998) or NRIXS data (Mao et al., 2001), whereas theoretical calculations yield much higher velocities (Steinle-Neumann et al., 2001). In Fig. 9, we also show that hcp iron follows a Birch law (see Birch, 1963) for V_P, namely a linear dependence between compressional sound velocity and density, which provides a convenient relationship for extrapolating our measurements to higher pressures. Furthermore, our measurements are in excellent agreement with the Hugoniot shock data of Brown and McQueen (1986) within the frame of the Birch's plot. These shock data have been collected at high temperature whereas our measurements have been carried out at ambient temperature. The important point is that both experiments yield the same observed acoustic velocity at a same given density, which is indeed the Birch assumption (Birch, 1963). It is striking to note that seismic data (Dziewonski and Anderson, 1981) does not fit the linear extrapolation made from the combination of the two sets of data. This could be interpreted as if Earth's

Table 1
Longitudinal acoustic sound velocities (m/s) as a function of specific mass (kg/m^3)

V_P	Specific mass	V_P	Specific mass
Iron		FeO	
7055	9084	6150	5770
7363	9382	6850	6000
7802	9861	7400	6360
8032	10107	7930	6600
8222	10315	8430	7147
8804	11229	8910	7270
		9160	7450
FeS_2		FeSi	
8042	5010	7121	6239
8824	5277	8013	6630
9476	5492	9043	7293
10278	5757	9515	7507

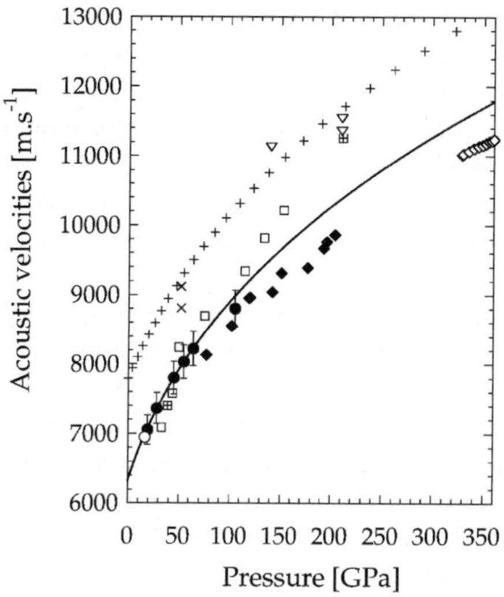

Fig. 8. Longitudinal acoustic sound velocity for the hcp phase of iron as a function of pressure (Fiquet et al., 2001: solid circles). Comparison is made with zero-frequency radial X-ray diffraction measurements (Singh et al., 1998: crosses; Mao et al., 1998: crossed squares), ultrasonics measurements (Mao et al., 1998: empty circle), nuclear resonance inelastic X-ray scattering (Mao et al., 2001: empty squares), Hugoniot sound velocity measurements (Brown and McQueen, 1986: solid diamonds), theoretical calculations (Stixrude and Cohen, 1995; Söderlind et al., 1996; Laio et al., 2000: empty triangles; Steinle-Neumann et al., 2001: small crosses) and PREM (Dziewonski and Anderson, 1981: empty diamonds).

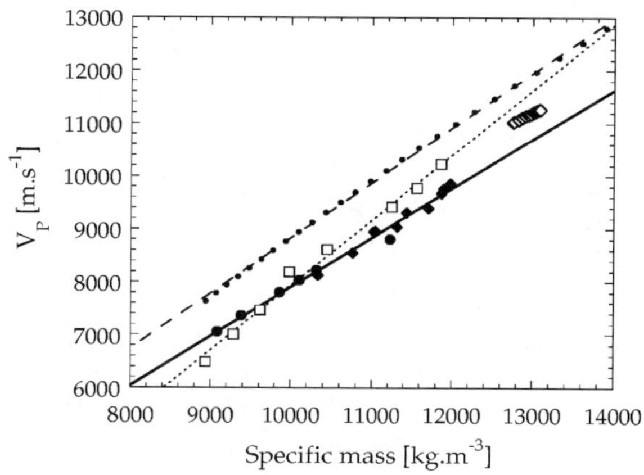

Fig. 9. Birch plot (i.e. the longitudinal acoustic velocity is represented as a function of specific mass) for hcp iron. Our inelastic X-ray scattering data (solid circles) and shock wave data (solid diamonds) closely follow a Birch law, a linear dependence of the longitudinal acoustic sound velocity with the specific mass, depicted by the solid line. Empty squares represent the results obtained by NRIXS (Mao et al., 2001) and the dashed line the associated Birch's fit made from these results only. Experimental observations are plotted along recent theoretical calculations (dots) by Steinle-Neumann et al. (2001).

inner core were 4–5% lighter than hcp iron, as previously suggested by Jephcoat and Olson (1987). Alternatively, this might indicate the presence of another phase of iron or alloy with different physical properties (e.g. Anderson, 1995). The comparison made with NRIXS data by Mao et al. (2001) shows that the extrapolations of the trends outside of the actual measurement ranges yield different values of V_P at inner core conditions, as well as different conclusions on the possible constitution of the inner core. In NRIXS experiments however, the determination of the average LA velocity is less direct than using IXS. The average Debye phonon velocity has to be extracted from a parabolic fit of the low-energy part of the density of states, and has to be converted into V_P and shear wave velocity V_S through an averaging scheme using independent determinations of bulk and shear moduli. Alternatively, such a difference between IXS and NRIXS results might originate in the way preferred orientations that develop at high pressure in diamond–anvil cells in iron are taken into account. On a polycrystalline material, every single crystallite contributes to the total average LA dispersion curve. The average position and the width of LA excitation result from both the intrinsic resolution of the spectrometer and the sample texture. If orientation distribution is not random, the average might be biased toward velocities along that preferred orientation. XRD measurements carried out in parallel to our IXS study on the ID30 ESRF beamline indeed show that the *c* axis of iron is preferentially aligned with the compression axis of the diamond–anvil cell, in accordance with recently reported texture measurements (Wenk et al., 2000). Such measurements proposed in addition a large anisotropy for V_P, with *P*-waves travelling 8% faster at 45° from the *c* axis than either in the *ab* plane or along the *c* axis. According to these texture measurements, our IXS experiment probes mainly LA phonons propagating in the *ab* plane, since the momentum transfer lies approximately perpendicular to the incident X-ray beam and to the *c* axis. When anisotropy curves are considered however (see Wenk et al., 2000), the orientationally averaged V_P values we obtain from IXS are similar to the values corresponding to a predominant but not complete preferential orientation of the *c* axis.

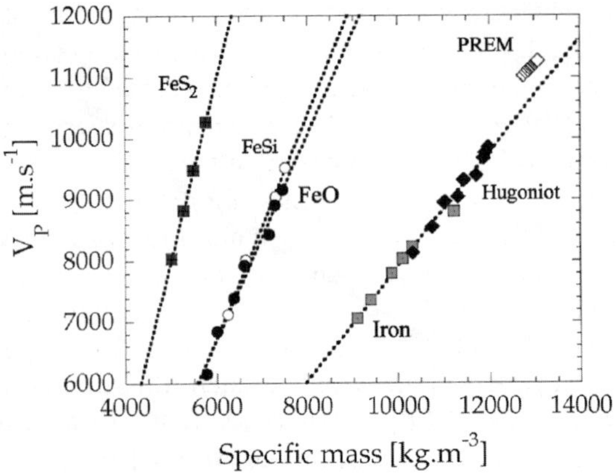

Fig. 10. Longitudinal acoustic sound velocity for iron-bearing compounds at high pressure, as compared to pure iron and the inner core sound velocity (open diamonds). Solid squares: pyrite; hollow circles: iron silicide; solid circles: iron oxide. Dashed lines are Birch's fits made to these observations.

In addition, we performed a series of measurements which show the influence of light elements such as sulfur, silicon or oxygen on acoustic sound velocities of iron alloys at high-pressure. The results are presented in Table 1 and reported as a function of specific mass in Fig. 10. One notices that iron silicide, iron oxide as well as pyrite closely follow a Birch's law, which provides an interesting way to extrapolate data at higher pressure and/or temperature and will therefore bring new constraints to the determination of the light elements content of the Earth's inner core (Badro et al., in preparation).

2.2. *The investigation of the acoustic anisotropy of hcp metals at high-pressure: the example of cobalt*

Beyond studies of the bulk properties of the Earth's core, seismological studies show that the inner core is elastically anisotropic (Woodhouse et al., 1986; Romanowicz et al., 1996; Song, 1997; Garcia and Souriau, 2000). The anisotropy has an axial symmetry and an amplitude of about 3%, with the fast direction oriented parallel to the Earth's rotation axis. Several hypotheses have been proposed to explain this feature: Jeanloz and Wenk (1988) proposed that the anisotropy might be caused by the lattice preferred orientation of the hcp structure of iron due to plastic deformation, whereas Karato (1993) invokes an interaction of the Earth's magnetic field with growing magnetically anisotropic iron crystals. However, it is not clear whether hcp iron is highly anisotropic at very high pressure. Some theoretical results show a rather low intrinsic anisotropy, almost requiring a perfect alignment of iron hcp crystals in order to account for the observed seismic anisotropy (Stixrude and Cohen, 1995). More recent theoretical work shows that intrinsic anisotropy in hcp iron is of the order of 12% at inner core conditions and that perfect texturing is not required by the theoretical results (Steinle-Neumann et al., 2001). Texture X-ray diffraction measurements of iron at very high-pressure (Mao et al., 1998; Wenk et al., 2000) indicate a large compressional-wave anisotropy which relieves the "perfect alignment" textural constraint. This is illustrated in Fig. 11 where the longitudinal acoustic wave velocity is shown as a function of the direction of propagation for hcp iron. The anisotropy proposed by texture measurements (Mao et al., 1998), when compared to calculations (Stixrude and Cohen, 1995), is not only different in magnitude, but as well in direction. Conversely, the direction and magnitude of the anisotropy in another

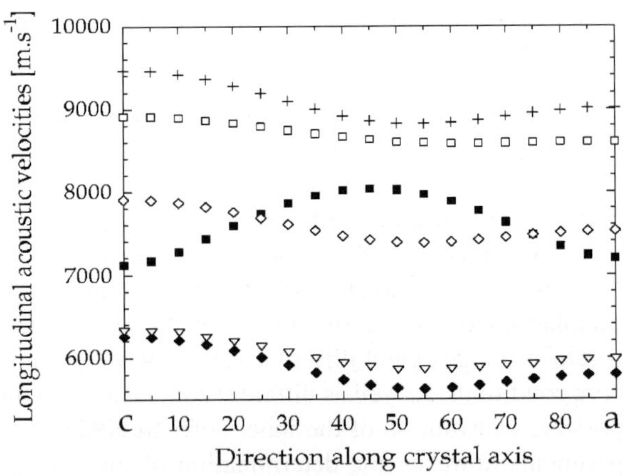

Fig. 11. Calculation of the anisotropy of propagation of longitudinal acoustic velocity for hcp iron and cobalt. Iron: zero-frequency radial X-ray diffraction measurements (solid squares) by Mao et al. (1998) compared at 39 GPa with first principles calculations (open squares) from Stixrude and Cohen (1995) and (crosses) from Steinle-Neumann et al. (2001). Cobalt: room pressure ultrasonics measurements (solid diamonds) by Guinan and Steinberg (1974) plotted along theoretical calculations (open triangles) from Steinle-Neumann et al. (1999). Calculations at 39 GPa are shown for hcp cobalt (open diamonds).

metal with hcp structure such as cobalt compare well (Guinan and Steinberg, 1974) with iron theoretical values.

In order to settle the above outlined discrepancies between the various indirect experimental techniques and theory, a direct experimental determination of the elastic constants of hcp iron and their evolution with pressure and temperature is needed. The key experiments would in fact be the direct comparison of single crystal and polycrystalline data in hcp-Fe. However, obtaining single crystals of hcp-Fe at high pressure is currently not possible. In the next paragraphs, we present preliminary results obtained on cobalt to address the issues of elastic anisotropy. Unlike iron, hcp cobalt is stable at room temperature and ambient pressure to 105 GPa (Yoo et al., 2000). Cobalt is a 3d transition metal located next to iron in the periodic table, and its bulk and thermal properties are close to those of iron. Thus, cobalt makes a good proxy for iron at high-pressure, and single crystals as well as powders can be obtained commercially. The example we present here is focused on the determination of elastic moduli and their pressure derivatives. For the hcp structure there are five independent elastic moduli (C_{11}, C_{33}, C_{44}, C_{12}, C_{13}). These can be derived from the LA and TA phonon dispersion, fitted as shown previously with a sine function where the slope at $Q = 0$ yields acoustic velocities along main directions chosen by orientating the crystal kept at high-pressure in the diamond-anvil cell. With a single crystal with surface normal to [1 1 0], it is possible to get C_{11}, C_{33}, C_{44} and C_{66} which are simply related to LA [1 0 0], LA [0 0 1], TA $[1\,1\,0]_{\langle 0\,0\,1\rangle}$ and TA $[1\,1\,0]_{\langle 1\,0\,0\rangle}$, respectively, whereas C_{12} can be easily obtained from $C_{66} = (1/2)(C_{11} - C_{12})$ and C_{13} is a linear combi-

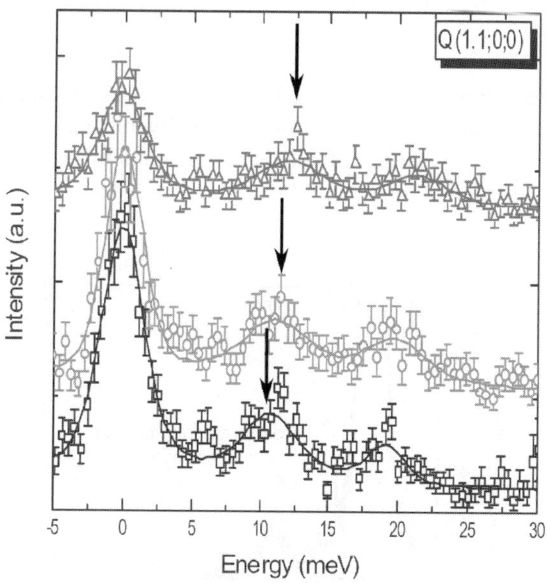

Fig. 12. IXS pattern collected on single crystal hcp cobalt at three different pressures in a diamond-anvil cell. Helium is used as pressure transmitting medium: Squares: 3.3 GPa, circles: 6.5 GPa, triangles: 9.5 GPa. The energy position of the LA phonon is indicated by the arrow. A higher energy phonon can be assigned to an optical mode. Reduced-Q setting is 1.1 along [1 0 0] direction in hcp cobalt. The data were taken with an energy resolution of 3 meV.

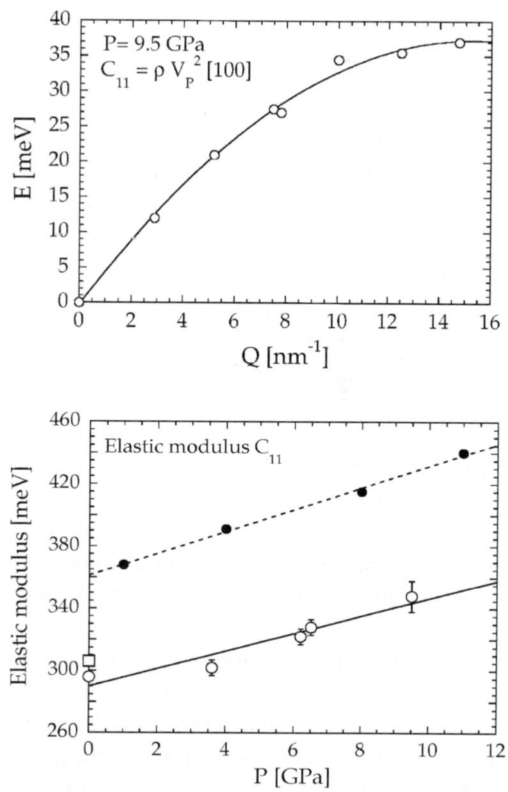

Fig. 13. Top: Dispersion curve of the longitudinal acoustic phonon branch along the [1 0 0] direction in hcp cobalt at 9.5 GPa. Bottom: pressure dependence of C_{11} elastic modulus (open circles). Ultrasound measurements at room pressure (Landolt-Bornstein, 1966) appear as an open diamond whereas theoretical calculations by Steinle-Neumann et al. (1999) are represented by solid dots. Lines are linear fits to the data. The pressure derivative is 5.3 (4) and the extrapolation to room pressure is $C_{11} = 290 \pm 2$ GPa.

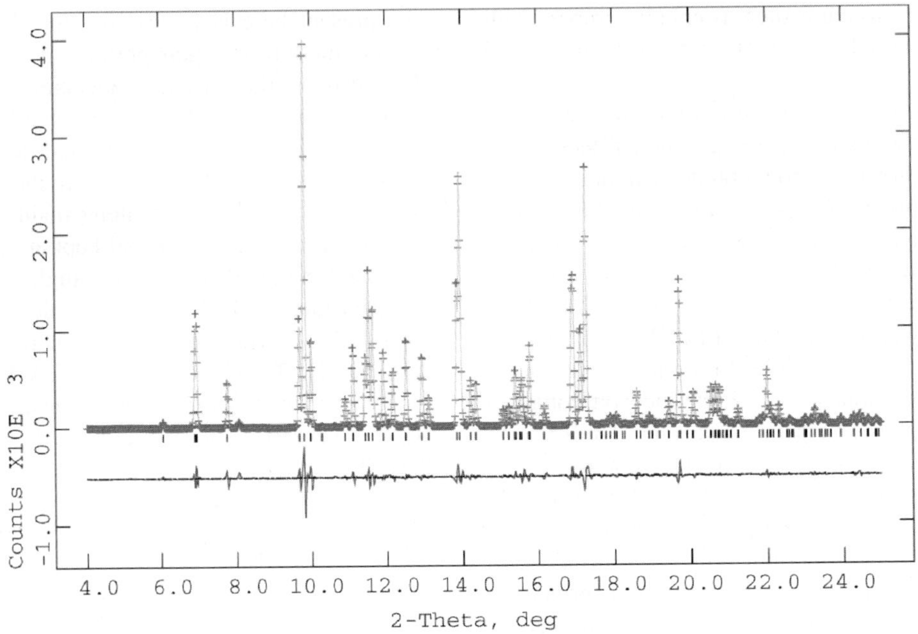

Fig. 14. X-ray diffraction pattern of synthetic $MgSiO_3$ with perovskite structure (space group *Pbnm*) with lattice parameters $a = 4.7765(1)$ Å, $b = 4.9294(1)$ Å, $c = 6.8977(1)$ Å. The pattern has been obtained at a wavelength $\lambda = 0.41584$ Å using an imaging plate located at 363.42 mm from the sample.

nation involving the other four elastic moduli. In order to reach the highest pressure we are aiming at (40–50 GPa), the crystal dimensions compatible with the constraints of the diamond anvil cell are 50 μm in diameter and 15 μm in thickness. These single crystals, which have to be of high quality, have been prepared at the Lawrence Livermore National Laboratory using state-of-the-art polishing and laser cutting techniques, and subsequently loaded in a diamond anvil cell using helium as a pressure transmitting medium. The first set of experiments presented here has been obtained on the ID28 beamline at ESRF using a single crystal with surface normal to [0 0 1]. Typical IXS pattern obtained at pressure up to 9.5 GPa are shown in Fig. 12. It allowed us to obtain the variation of the LA wave velocity along the [1 0 0] direction with increasing pressure. Combining these results with density measurements obtained by XRD, we obtain directly C_{11} and its variation with pressure as shown in Fig. 13, plotted along with recent theoretical determinations (Steinle-Neumann et al., 1999).

These investigations have recently been extended to determine the five elastic constants and their pressure dependences to 42 GPa (Antonangeli et al., in preparation) and will be complemented by measurements on polycrystalline cobalt, in a manner similar to the experiments described above on iron. The sound ve-

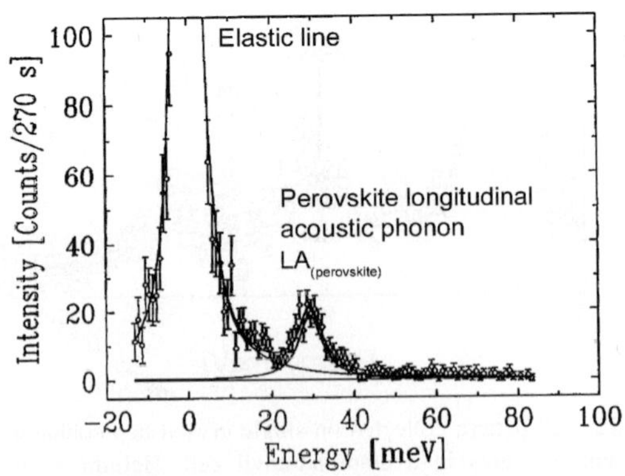

Fig. 15. IXS pattern collected at ambient conditions on a sintered polycrystalline sample of $MgSiO_3$ with perovskite structure with an energy resolution of 3 meV and at a momentum transfer of 4.5 nm^{-1}. The fits of the elastic line and the LA phonon (continuous lines) are plotted along with the experimental data.

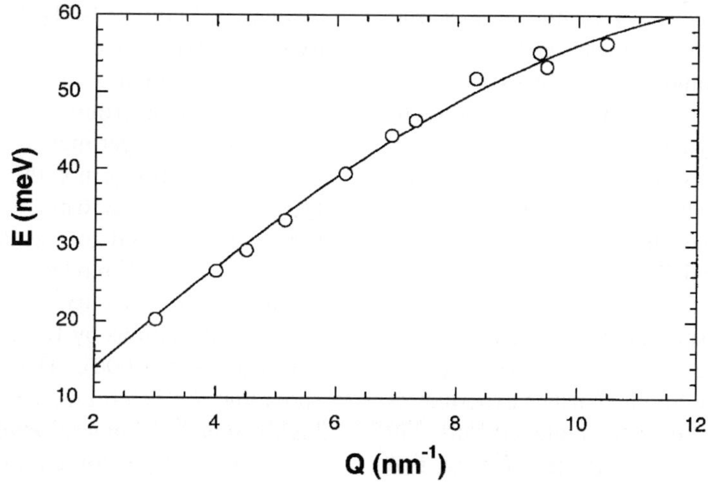

Fig. 16. Polycrystalline longitudinal acoustic phonon dispersion and corresponding sine function fit [Eq. (2)] for $MgSiO_3$ perovskite. The resulting mean longitudinal acoustic sound velocity is 10652 ± 173 m/s. The value of Q_{MAX} is a free parameter in this fit ($Q_{MAX} = 13.92 \pm 0.67\,nm^{-1}$).

locities obtained for the aggregate will be compared to the ones obtained by an orientational averaging of the single crystal data, in order to probe the accuracy of the averaging process. Moreover, we will obtain crucial data on the intrinsic anisotropy of sound velocity propagation in an hcp metal under pressure, and directly use it to estimate the anisotropy of iron at high-pressure.

2.3. Preliminary IXS data on magnesium silicate $MgSiO_3$ perovskite

Physical properties of $MgSiO_3$ with perovskite structure are of interest to better understand both the composition and current state of the Earth's lower mantle (670 to 2900 km in depth). The knowledge of the elastic constants of this compound, which makes

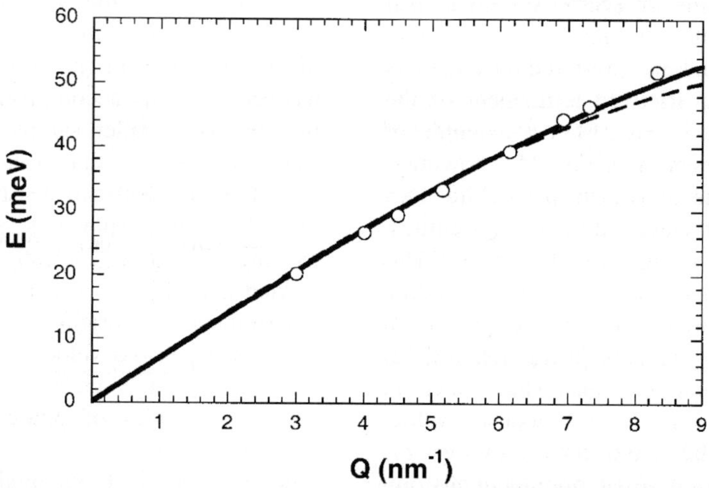

Fig. 17. Experimental measurements plotted along with different simulated dispersion curves calculated using Yeganeh-Haeri et al. (1989) elastic constants (with a Voigt-Reuss-Hill average of LA velocity of 10940 m/s) and different averaging processes. The solid line represents a fit with Q_{MAX} considered as a free parameter; the dashed line is calculated with a spherical first Brillouin zone.

up to 70–100 wt.% of the lower mantle, is essential for comparison with Earth's global seismic models, as well as to translate tomographic images into temperature and composition fields. It is also important to understand the lattice dynamics of this phase, which both determines its thermal properties at high pressure and temperature (e.g. Gillet et al., 2000) and which is related to possible phase transitions via soft-mode mechanisms (Hemley and Cohen, 1992; Wang et al., 1992).

In our preliminary experiment, a sintered sample of polycrystalline magnesium silicate perovskite was measured at ambient conditions. Further experiments will be carried out on $MgSiO_3$ single crystals, but there are some interesting points in proving that working on powders of silicate perovskite is possible. First, it is difficult to keep single crystals at very high pressures whereas powders can be compressed to pressures exceeding the 135 GPa of the core-mantle boundary. Second, IXS on a well sintered powder can still yield information about average acoustic velocities, which can be compared to geophysical models. Our experiments on $MgSiO_3$ perovskite were performed on a polycrystalline sample synthesized at the mineral physics institute of the State University of New York at Stony Brook. The sample was mounted on a glass fiber and checked by X-ray diffraction using a focused monochromatic radiation and imaging plate (MAR345) detector at the ID09 beamline of ESRF. We confirmed that the sample is a pure $MgSiO_3$ compound with perovskite structure, and no preferred orientations are detectable from the Rietveld refinement of the XRD pattern presented in Fig. 14. An example of an IXS spectrum is presented in Fig. 15, showing a well resolved LA mode of silicate perovskite. IXS data were collected at eleven different Q-settings, which allowed us to obtain the first experimental averaged dispersion curve in silicate perovskite with $MgSiO_3$ composition. Results are plotted in Fig. 16 along with the fit made using relation [2] in order to retrieve sound velocities. The results are $V_P = 10652 \pm 173$ m/s for the LA acoustic velocity and $Q_{MAX} = 13.92 \pm 0.67\,nm^{-1}$. In Fig. 17, a simulated dispersion curve of orientationally averaged LA mode of silicate perovskite is also reported. The simulation has been made using elastic constants measured by Yeganeh-Haeri et al. (1989), yielding a Voigt-Reuss-Hill average of (LA) velocity of 10940 m/s, and a spherical first Brillouin zone ($Q_{MAX} = 11.66\,nm^{-1}$). As far as acoustic longitudinal velocity determination is concerned, the agreement between experimental and simulated curves is satisfactory. This is not the case for the parameter Q_{MAX}: above $Q = 6\,nm^{-1}$, however, the IXS pattern is complicated by the dispersion curves of low frequency optical modes crossing the longitudinal acoustic dispersion curve near the zone edge (see theoretical calculations by Karki et al., 2000 or Parlinski and Kawazoe, 2000). The two curves in Fig. 17 show that the determination of sound velocity is not highly sensitive upon a detailed modeling of Q_{MAX}. In order to check for elastic anisotropy, simultaneous measurements of longitudinal phonons with 3D structure of the powder, as well as determination of ab initio calculated dispersion curves are needed for refining such determinations of average V_P in silicate perovskite.

3. Conclusions

Inelastic X-ray scattering is a well suited spectroscopic technique for the study of phonon dispersion of materials under high pressure. On both inelastic scattering beamlines of ESRF (ID16 and ID28), specific emphasis was given to high pressure. Moreover, the focusing optics scheme on ID28 is currently upgraded, and a focal spot size of 30 μm × 60 μm (horizontal × vertical, FWHM) has been recently accomplished, thus enabling investigations on smaller samples, i.e. higher pressures. This work shows that a large variety of samples (metals or oxides) can be studied, either in a polycrystalline form (powders, sintered samples...) or as single crystals. Inelastic scattering experiments carried out on polycrystalline samples provide direct determination of orientationally averaged longitudinal acoustic sound velocities, whereas investigations on single crystals yield elastic moduli and a view on the dynamics of phase transitions. This technique is thus highly interesting for mineral physicists and offers a great potential in providing essential elastic data and their pressure and/or temperature dependence for most compounds of geophysical interest.

Acknowledgements

We thank Bernard Couzinet, Jean Claude Chervin, Bernard Canny for their assistance in diamond-anvil cell helium loading, and Matteo d'Astuto for their help on ID28 inelastic scattering beamline. Mohamed Mezouar and Michael Hanfland are acknowledged for their assistance with unscheduled X-ray diffraction sample characterization on high-pressure beamlines ID30 and ID09 of ESRF. Constructive reviews by L. Stixrude and an anonymous reviewer are appreciated.

References

Abramson, E.H., Brown, J.M., Slutsky, L.J., 1999. Applications of impulsive stimulated scattering in the Earth and planetary sciences. Annu. Rev. Phys. Chem. 50, 279–313.

Alfe, D., Price, G.D., Gillan, M.J., 2001. Thermoynamics of hexagonal-close-packed iron under Earth's core conditions. Phys. Rev. B 64 (45123), 1–16.

Ashcroft, N.W., Mermin, N.D., 1976. Solid State Physics. Saunders College Publishing & Harcourt Brace College Publishers, Orlando, p. 826.

Anderson, O.L., 1995. Mineral physics of iron and of the core. Rev. Geophys. supplement, 429–441.

Birch F., 1963. In: Paul, W., Warschauer, D.M. (Eds.), Solids Under Pressure. McGraw-Hill, New York, pp. 137–162.

Brown, J.M., McQueen, R.G., 1986. Phase transitions, Gruneisen parameter, and elasticity for shocked iron between 77 and 400 GPa. J. Geophys. Res. B 91 (7), 7485–7494.

Brown, J.M., Slutsky, L.J., Nelson, K.A., Cheng, L.T., 1989. single-crystal elastic constants for San Carlos peridot: an application of impulsive stimulated Brillouin scattering. J. Geophys. Res. 95, 9485–9492.

Burkel, E., Peisl, J., Dorner;, B., 1987. Observation of inelastic X-ray scattering from phonons. Europhys. Lett. 3 (8), 957–961.

Burkel, E., 2000. Phonon spectroscopy by inelastic X-ray scattering. Rep. Prog. Phys. 63 (2), 171–232.

Chai, M., Brown, J.M., Slutsky, L.T., 1997a. The elastic constants of pyrope-grossular-almandine garnet to 20 GPa. Geophys. Res. Lett. 24, 523–526.

Chai, M., Brown, J.M., Slutsky, L.T., 1997b. The elastic constants of an aluminous orthopyroxene to 12.5 GPa. J. Geophys. Res. 102, 14779–14785.

Dorner, B., Burkel, E., Illini, T., Peisl, J., 1987. First measurement of a phonon dispersion curve by inelastic X-ray scattering. Z. Phys. B Cond. Matter 69 (2-3), 179–183.

Duffy, T.S., Ahrens, T.J., 1992. Sound velocities at high pressure and temperature and their geophysical implications. J. Geophys. Res. B 97 (4), 4503–4520.

Duffy, T.S., Ahrens, T.J., 1994. The temperature sensitivity of elastic wave velocity at high pressure; new results for molybdenum. Geophys. Res. Lett. 21 (6), 473–476.

Duffy, T.S., Ahrens, T.J., 1995. Compressional sound velocity, equation of state, and constitutive response of shock-compressed magnesium oxide. J. Geophys. Res. 100 (B1), 529–542.

Dziewonski, A.M., Anderson, D.L., 1981. Preliminary reference Earth model. Phys. Earth Planet Int. 25, 297–356.

Fiquet, G., Badro, J., Guyot, F., Requardt, H., Krisch, M., 2001. Sound velocities in iron to 110 GPa. Science 291, 468–471.

Garcia, R., Souriau, A., 2000. Inner core anisotropy and heterogeneity level. Geophys. Res. Lett. 27, 3121–3124.

Gillet, P., Daniel, I., Guyot, F., Matas, J., Chervin, J.C., 2000. A thermodynamic model for $MgSiO_3$-perovskite derived from pressure, temperature and volume dependence of the Raman mode frequencies. Phys. Earth Planet Int. 117, 361–384.

Guinan, M., Steinberg, D., 1974. J. Phys. Chem. Solids 35, 1501.

Guinan, M.W., Beshers, D.N., 1968. Pressure derivatives of the elastic constants of iron to 10 kbs. J. Phys. Chem. Solids 29 (3), 541–549.

Hemley, R.J., Cohen, R.E., 1992. Ann. Rev. Earth Planet. Sci. 20, 553–560.

Isaak, D.G., Anderson, O.L., Goto, T., 1989a. Elasticity of single-crystal forsterite measured to 1700 K. J. Geophys. Res. 94, 5895–5906.

Isaak, D.G., Anderson, O.L., Goto, T., 1989b. Measured elastic moduli of single-crystal MgO up to 1800 K. Phys. Chem. Minerals 16, 704–713.

Jeanloz, R., Wenk, R., 1988. Convection and anisotropy of the inner core. Geophys. Res. Lett. 15, 72–75.

Jephcoat, A., Olson, P., 1987. Is the inner core of the Earth pure iron? Nature 325, 332–335.

Karato, S.-I., 1993. Inner core anisotropy due to the magnetic field-induced preferred orientation of iron. Science 262, 1708–1711.

Karki, B.B., Wentzcovitch, R.M., de Gironcoli, S., Baroni, S., 2000. Ab initio lattice dynamics of $MgSiO_3$ perovskite at high pressure. Phys. Rev. B 62 (22), 14750–14756.

Karki, B.B., Stixrude, L., Wentzcovitch, R.M., 2001. High-pressure elastic properties of major materials of Earth's mantle from first principles. Rev. Geophys. 39 (4), 507–534.

Klotz, S., Braden, M., 2000. Phonon dispersion of bcc iron to 10 Gpa. Phys. Rev. Lett. 85 (15), 3209–3212.

Krisch, M., Mermet, A., San Miguel, A., Sette, F., Masciovecchio, C., Ruocco, G., Verbeni, R., 1997. Acoustic-phonon dispersion in CdTe at 7.5 GPa. Phys. Rev. B 56, 8691–8694.

Laio, A., Bernard, S., Chiarotti, G.L., Scandolo, S., Tosatti, E., 2000. Physics of iron at Earth's conditions. Science 287, 1027–1030.

Landolt-Bornstein, Numerical Data and Functional Relationships in Science and Technology New Series. group III, vol. 1, Springer, Berlin, 1966.

Li, B., Chen, G., Gwanmesia, G.D., Liebermann, R.C., 1998. Sound velocity measurements at mantle transition zone conditions of pressure and temperature using ultrasonic interferometry in a multi-anvil apparatus, Properties of Earth and Planetary Materials at High Pressure and Temperature. Geophysical Monograph 101, AGU, Washington, DC, pp. 41–61.

Li, B., Liebermann, R.C., Weidner, D.J., 2001. *P–V–V_P–V_S–T* measurements on wadsleyite to 7 GPa and 873 K: implications for the 410 km seismic discontinuity. J. Geophys. Res. 106 (B12), 30575–30591.

Liebermann, R.C., Li, B., 1998. In: Hemley, R.J. (Ed.), Elasticity at High Pressures and Temperatures. Reviews in Mineralogy, pp. 459–492.

Loa, I., Kunc, K., Syassen, K., Krisch, M., Mermet, A., Hanfland, M., 2003. Inelastic X-ray scattering in Cs under pressure. High Pressure Res. 23, 1–5.

Lubbers, R., Grünsteudel, H.F., Chumakov, A.I., Wortmann, G., 2000. Density of phonon states in iron at high pressure. Science 287, 1250–1253.

Mao, H.K., Wu, Y., Chen, L.C., Shu, J., Jephcoat, A.P., 1990. Static compression of iron to 300 GPa and $Fe_{0.8}Ni_{0.2}$ alloy to 260 GPa: implications for composition of the core. J. Geophys. Res. 95 (B13), 21737–21742.

Mao, H.K., Shu, J., Shen, G., Hemley, R.J., Li, B., Singh, A.K., 1998. Elasticity and rheology of iron above 220 GPa and the nature of the Earth's inner core. Nature 396, 741–743; erratum: Nature 399, 280.

Mao, H.K., Xu, J., Struzhkin, V.V., Shu, J., Hemley, R.J., Sturhahn, W., Hu, M.Y., Alp, E.E., Vočadlo, L., Alfe, D., Price, G.D., Gillan, M.J., Schwoerer-Böhning, M., Häusermann, D., Eng, P., Shen, G., Giefers, H., Lübbers, R., Wortmann, G., 2001. Phonon density of states of iron up to 153 GPa. Science 292, 914–916.

Masciovecchio, C., et al., 1996. A perfect crystal X-ray analyzer with meV energy resolution. Nucl. Instr. Methods. Phys. Res. B 111, 181.

Minkiewicz, V.J., Shirane, G., Nathans, R., 1967. Phonon dispersion relation for iron. Phys. Rev. 162 (3), 528–531.

Occelli, F., Krisch., M., Loubeyre, P., Sette, F., LeToullec, R., Masciovecchio, C., Rueff; J.P., 2001. Phonon dispersion curves in an argon single crystal at high pressure by inelastic X-ray scattering. Phys. Rev. B 63, 224306, 1–8.

Oganov, A.R., Brodholt, J.P., Price, G.D., 2001. The elastic constants of $MgSiO_3$ perovskite at pressures and temperatures of the Earth's mantle. Nature 411, 934–937.

Parlinski, K., Kawazoe, Y., 2000. Ab initio study of phonons and structural stabilities of the perovskite-type $MgSiO_3$. Eur. Phys. J. B16, 49–58.

Romanowicz, B., Li X.D., Durek, J., 1996. Anisotropy in the inner core: could it be due to low order convection? Science 274, 963–966.

Schülke, W., 1991. In: Brown, G., Moncton, D.E. (Eds.), Handbook on Synchrotron Radiation. North Holland Publishing Company.

Sinelnikov, Y.D, Chen, G., Neuville, D.R., Vaughan, M.T., Liebermann, R.C., 1998. Ultrasonic shear wave velocities of $MgSiO_3$ perovskite at 8 GPa and 800 K and lower mantle composition. Science 281, 677–679.

Singh, A.K., Mao, H.K., Shu, J., Hemley, R.J., 1998. Estimation of single-crystal elastic moduli from polycrystalline X-ray diffraction at high pressure: application to FeO and iron. Phys. Rev. Lett. 80 (10), 2157–2160.

Sinogeikin, S.V., Bass, J.D., 2000. Single-crystal elasticity of pyrope and MgO to 20 GPa by Brillouin scattering in the diamond cell. Phys. Earth Planet Int. 120, 43–62.

Sinogeikin, S.V., Bass, J.D., 2002. Elasticity of Majorite and a Majorite–Pyrope solid solution to high pressure: implications for the transition zone. Geophys. Res. Lett. 29 (2), 52–55.

Söderlind, P., Moriarty, J.A., Wills, J.M., 1996. First-principles theory of iron up to earth-core pressures: structural, vibrational and elastic properties. Phys. Rev. B 53 (21), 14063–14072.

Song, X., 1997. Anisotropy of the Earth's inner core. Rev. Geophys. 35 (3), 297–313.

Steinle-Neumann, G., Stixrude, L., Cohen, R.E., Gülseren, O., 2001. Elasticity of iron at the temperature of the Earth's inner core. Nature 413, 57–60.

Steinle-Neumann, G., Stixrude, L., Cohen, R.E., 1999. First-principles elastic constants for the hcp transition metals Fe, Co, and Re at high pressure. Phys. Rev. B 60 (2), 791–799.

Stixrude, L., Cohen, R.E., 1995. High-pressure elasticity of iron and anisotropy of Earth's inner core. Science 267, 1972–1975.

Stixrude, L., Cohen, R.E., Hemley, R.J., 1998. Theory of minerals at high pressure. In: Hemley, R.J. (Ed.), Reviews in Mineralogy. vol. 37, pp. 639–671.

Verbeni, R., Sette, F., Krisch, M.H., Bergmann, U., Gorges, B., Halcoussis, C., Martel, K., Masciovecchio, C., Ribois, J.F., Ruocco, G., Sinn, H., 1996. X-ray monochromator with 2×10^{-8} energy resolution. J. Synchrotron Radiat. 3, 62.

Wang, Y., Guyot, F., Liebermann, R.C., 1992. Electron microscopy of (Mg,Fe)SiO3 perovskite: evidence for structural phase transitions and implications for the lower mantle. J. Geophys. Res. 97 (B9), 12327–12347.

Warren, J.L., Yarnell, J.K., Dolling, G., Cowley, R.A., 1967. Phys. Rev. 158, 805.

Wenk, H.R., Matthies, S., Hemley, R.J., Mao, H.K., Shu, J., 2000. The plastic deformation of iron at pressures of the Earth's inner core. Nature 405, 1044–1047.

Woodhouse, J.H., Giardini, D., Li, X.D., 1986. Evidence for inner core anisotropy from free oscillations. Geophys. Res. Lett. 13, 1549–1552.

Yeganeh-Haeri, A., Weidner, D.J., Ito, E., 1989. Elasticity of $MgSiO_3$ in the perovskite structure. Science 243, 787–789.

Yoo, C.S., Cynn, H., Soderlind, P., Iota, V., 2000. New beta(fcc)-cobalt to 210 GPa. Phys. Rev. Lett. 84 (18), 4132–4135.

Zaug, J., Abramson, E.H., Brown, J.M., Slutsky, L.J., 1993. Sound velocities in olivine at Earth mantle pressure. Science 260, 1487–1489.

Zha, C.-S., Mao, H.K., Hemley, R.J., Duffy, T.S., Meade, C., 1994. Acoustic velocities and refractive index of SiO_2 glass to 57.5 GPa by Brillouin scattering. Phys. Rev. B 50, 13105–13112.

Zha, C.-S., Mao, H.K., Hemley, R.J., 2000. Elasticity of MgO and a primary pressure scale to 55 GPa. Proc. Natl. Acad. Sci. U.S.A. 97 (25), 13494–13499.

Available online at www.sciencedirect.com

SCIENCE DIRECT

ELSEVIER

Physics of the Earth and Planetary Interiors 143–144 (2004) 19–31

PHYSICS OF THE EARTH AND PLANETARY INTERIORS

www.elsevier.com/locate/pepi

Ultrasonic measurements of the sound velocities in polycrystalline San Carlos olivine in multi-anvil, high-pressure apparatus

Kenneth L. Darling[a], Gabriel D. Gwanmesia[b,c], Jennifer Kung[a,c], Baosheng Li[a,c], Robert C. Liebermann[a,c,d,*]

[a] *Department of Geosciences, Stony Brook University, Stony Brook, NY 11794-2100, USA*
[b] *Department of Physics and Pre-Engineering, Delaware State University, Dover, DE 11901, USA*
[c] *Mineral Physics Institute, Stony Brook University, Stony Brook, NY 11794-2100, USA*
[d] *Laboratoire des Mécanismes de Transfert en Géologie, CNRS UMR 5563, Minéralogie, 39 allées Jules Guesde, 31000 Toulouse, France*

Received 12 February 2003; received in revised form 3 June 2003; accepted 1 July 2003

Abstract

Sound velocities in polycrystalline San Carlos olivine have been measured to pressures of 7 GPa at room temperature in a 1000 t uniaxial, split-cylinder apparatus (USCA-1000) of the Kawai multi-anvil type. These velocities and elastic moduli are in good agreement with those calculated from single-crystal moduli determined by previous investigators. In preliminary experiments in a DIA-type, cubic anvil apparatus (SAM-85), ultrasonic interferometric techniques were used in conjunction with synchrotron X-radiation to determine the elastic properties of San Carlos olivine under simultaneous high pressures and temperatures. We demonstrate the feasibility of such experiments even in specimens which have undergone plastic deformation at elevated pressures and temperatures.

Keywords: Ultrasonic; Elasticity; Pressure; Temperature; Olivine; Polycrystalline

1. Introduction

Many previous investigators have studied the elasticity of olivine $(MgFe)_2SiO_4$, the most abundant mineral in many petrological models of the Earth's upper mantle. However, most of these studies have measured sound wave velocities to high pressures at room temperature (Kumazawa and Anderson, 1969; Webb, 1989; Zaug et al., 1993; Abramson et al., 1997; Zha et al., 1998) or to high temperatures at room pressure (Isaak, 1992).

The purpose of this paper is to report new elasticity data on polycrystalline specimens of San Carlos olivine to high-pressure at room temperature and preliminary results of experiments to simultaneous high pressures and temperatures. These experiments take advantage of several technological developments in our laboratory: (1) hot-pressing of high-quality polycrystalline specimens (Gwanmesia et al., 1993); (2) interfacing of ultrasonic interferometry with multi-anvil, high-pressure apparatus (Li et al., 1996, 1998); and the measurement of sound velocities at simultaneous high pressures and temperatures in conjunction with synchrotron X-radiation (Liebermann et al., 1998; Li

* Corresponding author. Tel.: +1-631-632-8214; fax: +1-631-632-8240.
E-mail address: robert.liebermann@sunysb.edu (R.C. Liebermann).

0031-9201/$ – see front matter
doi:10.1016/j.pepi.2003.07.018

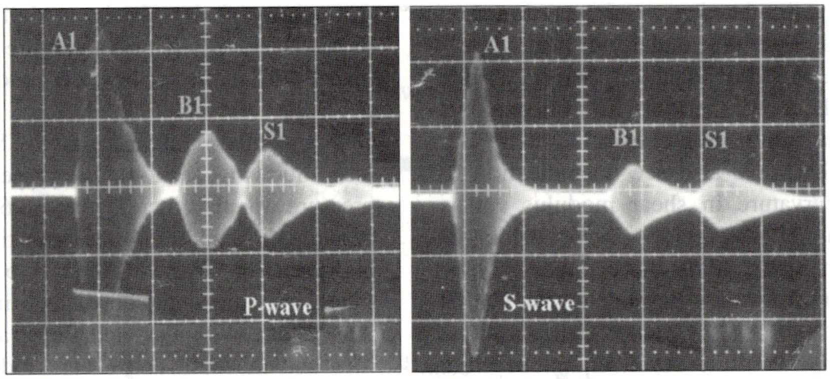

Fig. 6. P and S-waves from polycrystalline specimen S3419 at 10.6 GPa in SAM-85 showing echoes from the WC anvil (A1), the alumina buffer-rod (B1) and specimen (S1). Data for specimen S3419 at 10.6 GPa are as follows: density $= 3.55\ g/cm^3$, $L = 0.928$ mm, for travel times 0.364 μs for S waves and 0.203 μs for P-waves. Elastic properties are 5.10 and 9.14 km/s for shear and compressional velocities, respectively and the bulk and shear moduli are 173.5 and 92.3 GPa, respectively.

pressure derivatives of the elastic moduli are $K_0' = 4.80$ and $G_0' = 1.63$, which brings our determinations of these parameters into better agreement with those of previous investigators (see Table 3), especially those of the previous ultrasonic study by Webb (1989).

4. Velocity measurement at high pressures and high temperatures

Sound velocities in our polycrystalline olivine specimens were determined using dual mode $LiNbO_3$ transducers to simultaneous high pressures and tem-

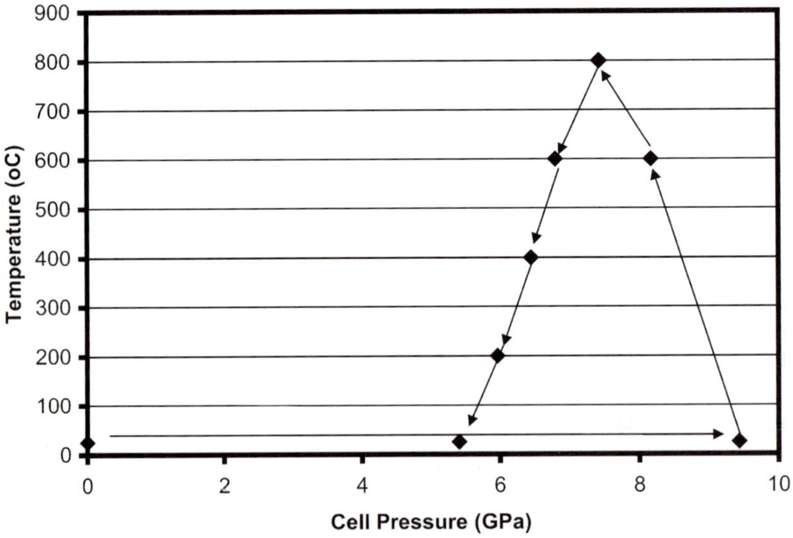

Fig. 7. *P–T* path in SAM-85 experiment for specimen K557 indicating the heating and cooling cycle for the P and S-wave measurements. Cell pressure is first increased at room temperature to 9.4 GPa. On heating from 25 °C under constant oil pressure, the cell pressure decreases to 8.2 GPa at 600 °C and 7.4 GPa at 800 °C due to relaxation of the cell materials. As temperature is subsequently decreased from 800 to 25 °C, the pressure decreases to 5.4 GPa as a consequence of the decrease of thermal pressure. A more typical *P–T* path for such experiments would include additional cooling/heating cycles as the cell pressure is gradually reduced (e.g., Fig. 2 of Li et al., 2004b-this volume).

peratures using a DIA-type cubic anvil (SAM-85) (Weidner et al., 1992) apparatus installed on the X-17B1 beam line of the National Synchotron Light Source (NSLS) of the Brookhaven National Laboratory, using techniques developed in our laboratory over the past 5 years (Liebermann et al., 1998; Liebermann and Li, 1998; Li et al., 2004a,b-this volume). An energy-dispersive X-ray method was employed using white radiation to collect the cell parameters of both the specimen and the NaCl which was used as a reference pressure scale, during the course of the experiment. A $6\,mm^3 \times 6\,mm^3 \times 6\,mm^3$ cell assembly used in this study with a mixture of amorphous boron and epoxy resin used as pressure medium and a graphite furnace to generate the high temperature environment; A W/Re3%–W/Re5% thermocouple located at the center of furnace and in direct contact with the specimen provided the measurement of tem-

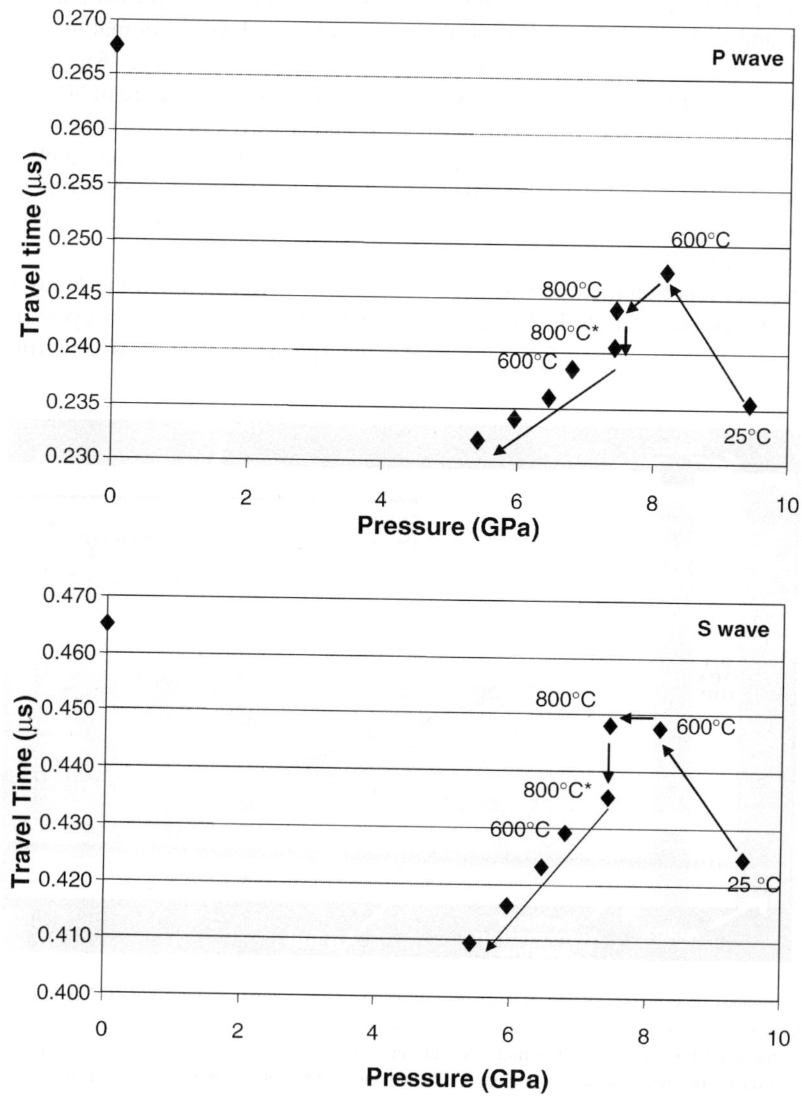

Fig. 8. P and S-wave travel times for sample K557 in SAM-85 experiment. The *P–T* path follows the same pattern of Fig. 7; thin lines and arrows indicate the order of *P–T* points explored. Key temperature points are indicated on the diagram. Note, in particular, the abrupt decrease in travel-times for the data points at 800 °C while the pressure was fixed at 7.4 GPa.

From the X-ray images (Fig. 9), we can obtain directly the specimen length (L_I) along the axis of propagation of the acoustic waves. In Fig. 10, we have plotted the two data points for the compressional and shear velocities in K557 at high-pressure and room temperature on a diagram with the data from the USCA-1000 experiments on specimen S3419. The velocities have been calculated for K557 using the X-ray image lengths calibrated with the specimen length on recovery. We note that the velocities on cold compression at 9.4 GPa agree with the extrapolated fit within 1%; similar agreement is observed for the data at 5.4 GPa after the heating cycle described in Fig. 7. From this comparison, we conclude that it is possible to measure elastic wave velocities at high pressures and temperatures, even in specimens which have undergone plastic deformation during the course of the *P–T* experiment, if direct measurements of specimen length are available.

5. Summary

In this paper, we have presented new data for the sound velocities in polycrystalline San Carlos olivine at high pressures and both room temperature and high temperatures using the new techniques developed in our laboratory over the past few years. These experiments demonstrate the feasibility of performing ultrasonic interferometric measurements even on specimens that have undergone plastic deformation. Recently, Li and Weidner (1999) have proposed to use ultrasonic travel times to monitor deformation as a function of time in specimens held at fixed environmental conditions.

Other papers from our research team describe these technological developments in detail (e. g., Li et al., 1996, 1998, 2004a,b; Liebermann et al., 1998) and report more comprehensive studies of the behavior of ferropericlase (Kung et al., 2002), $MgSiO_3$ ortho- and clino-pyroxenes (Kung et al., 2003), and $CaSiO_3$-perovskite (Li et al., 2004b-this volume).

Acknowledgements

We thank Jiuhua Chen, Michael Vaughan and Donald Weidner for the continued development and operation of the SAM-85 apparatus at the NSLS; and Ian Jackson and Akira Yoneda for their helpful and constructive reviews. This paper was written while the corresponding author (RCL) was a visiting scientist in the CNRS laboratory of Jannick Ingrin and Olivier Jaoul at the Université Paul Sabatier in Toulouse, France; he thanks them for their hospitality and support. This study was supported by the National Science Foundation under grant EAR 99-80491 to RCL. Mineral Physics Institute Publication No. 312. The in situ ultrasonic and X-ray experiments were carried out at the X-17B1 beamline of the National Synchrotron Light Source (NSLS),which is supported by the US Department of Energy, Division of Materials Sciences and Division of Chemical Sciences under Contract No. DE-AC02-76CH00016.

References

Abramson, E.H., Brown, J.M., Slutsky, L.J., Zaug, J., 1997. The elastic constants of San Carlos olivine to 17 GPa. J. Geophys. Res. 102, 12253–12263.

Chen, G., Li, B., Liebermann, R.C., 1996. Selected elastic moduli of single-crystal olivines from ultrasonic experiments to mantle pressures. Science 272, 979–980.

Cook, R.K., 1957. Variation of elastic constants and static strains with hydrostatic pressure: a method for calculating from ultrasonic measurements. J. Acoust. Soc. Am. 29, 445–449.

Darling, K.L., 2002. Ultrasonic measurements of the elastic wave velocities of polycrystalline $(Mg,Fe)_2SiO_4$ San Carlos olivine, M.S. Thesis, Department of Geosciences, Stony Brook University, 95 pp.

Davies, G.F., Dziewonski, A.M., 1975. Homogeneity and constitution of the Earth's lower mantle and outer core. Phys. Earth Planet Interiors 10, 336–343.

Gwanmesia, G.D., Liebermann, R.C., 1992. Polycrystals of high-pressure phases of mantle minerals: hot pressing and characterization of physical properties. In: Syono, Y., Manghnani, M.H. (Eds.), High-Pressure Research: Applications to Earth and Planetary Sciences, Terra Scientific Publishing Co., Tokyo/American Geophysical Union, Washington, DC, pp. 117–135.

Gwanmesia, G.D., Li, B., Liebermann, R.C., 1993. Hot pressing of polycrystals of high-pressure phases of mantle mineral in multi-anvil apparatus. PAGEOPH 141, 467–484.

Isaak, D.G., 1992. High-temperature elasticity of iron-bearing olivines. J. Geophys. Res. 97, 1871–1885.

Jackson, I., Niesler, H., Weidner, D.J., 1981. Explicit correction of ultrasonically determined elastic wave velocities for transducer-bond phase shift. J. Geophys. Res. 86, 3736–3748.

Kawai, N., Endo, S., 1970. The generation of ultrahigh hydrostatic pressures by a split-sphere apparatus. Rev. Sci. Instrum. 41, 1178–1181.

Kawai, N., Togaya, M., Onodera, A., 1973. A new device for pressure vessels. Proc. Jpn. Acad. 49, 623–626.

Kumazawa, M., Anderson, O.L., 1969. Elastic moduli, pressure derivatives, and temperature derivatives of single-crystal olivine and single-crystal forsterite. J. Geophy. Res. 74, 5961–5972.

Kung, J., Li, B., Weidner, D.J., Zhang, J., Liebermann, R.C., 2002. Elasticity of $(Mg_{0.83}, Fe_{0.17})O$ ferropericlase at high-pressure: ultrasonic measurements in conjunction with X-radiation techniques. Earth and Planet Sci. Lett. 203, 557–566.

Kung, J., Li, B., Liebermann, R.C., 2003. Anomalous elasticity behavior in orthoenstatite at high-pressure: onset of phase transition to high-pressure clinoenstatite. EGS-AGU-EUG Joint Assembly, Nice France. Geophys. Res. Abstr. 5, EAE03-a-12376.

Li, B., Weidner, D.J., 1999. Rheological property measurement at high-pressure and high temperature using ultrasonic and X-ray diffraction techniques. In: Proceedings of the International Conference on High-Pressure Science and Technology, AIRAPT-17, Honolulu, 1999, p. 135.

Li, B., Jackson, I., Gasparik, T., Liebermann, R.C., 1996. In situ elastic wave velocity measurement in multi-anvil apparatus to 10 GPa using ultrasonic interferometry. Phys. Earth Planet Interiors 98, 79–91.

Li, B., Chen, G., Gwanmesia, G.D., Liebermann, R.C., 1998. Sound velocity measurements at mantle transition zone conditions of pressure and temperature using ultrasonic interferometry in a multi-anvil apparatus. In: Manghnani, M.H., Yagi, Y.T. (Eds.), Properties of Earth & Planetary Materials at High-Pressure & Temperature. Am Geophys Union, Washington, DC, pp. 41–61.

Li, B., Chen, K., Kung, J., Liebermann, R.C., Weidner, D.J., 2002. Sound velocity measurement using transfer function method, proceedings of AIRAPT conference, Beijing, July 2001. J. Phys. Conden. Matter 14, 11337–11342.

Li, B., Vaughan, M.T., Weidner, D.J., 2004a. Direct length measurement using X-radiography for the determination of acoustic velocities at high-pressure and temperature, Rev. Sci. Instru., submitted for publication.

Li, B., Kung, J., Liebermann, R.C., 2004b. Modern techniques in measuring elasticity of Earth materials at high-pressure and high temperature: Ultrasonic velocity measurements in conjunction with synchrotron X-radiation; Phys. Earth Planet. Interiors 143–144, 559–574.

Liebermann, R.C., Wang, Y., 1992. Characterization of sample environment in a uniaxial split-sphere apparatus. In: In: Syono, Y., Manghnani, M.H. (Eds.), High-Pressure Research: Applications to Earth and Planetary Sciences, Terra Scientific Publishing Co., Tokyo/American Geophysical Union, Washington, DC, pp. 19–31.

Liebermann, R.C., Li, B., 1998. In: Hemley, R.J. (Ed.), Elasticity at high pressures and temperatures. Reviews of Mineralogy, Mineralogical Society of America, Vol. 37, pp. 459–492.

Liebermann, R.C., Chen, G., Li, B., Gwanmesia, G.D., Chen, J., Vaughan, M.T., Weidner, D.J., 1998. Sound velocity measurements in oxides and silicates at simultaneous high pressures and temperatures using ultrasonic techniques in multi-anvil apparatus in conjunction with synchrotron X-radiation determination of equation of state. Rev. High Pressure Sci. Tech. 7, 75–78.

Lloyd, E.C., 1971. Accurate characterization of the high-pressure environment. NBS Spec. Publ. No. 326, Washington, DC, pp. 1–3.

Niesler, H., Jackson, I., 1989. Pressure derivatives of elastic wave velocities from ultrasonic interferometric measurements on jacketed polycrystals. J. Acoust. Soc. Am. 86, 1573–1585.

Rigden, S.M., Gwanmesia, G.D., Jackson, I., Liebermann, R.C., 1992. Pressure dependence of elasticity of high-pressure Mg_2SiO_4 polymorphs and constraints on the composition of the transition zone of the Earth's mantle. In: Syono, Y., Manghnani, M.H. (Eds.), High Pressure Research: Applications to Earth and Planetary Sciences. Terra Scientific Publishing Co., Tokyo/American Geophysical Union, Washington, DC, pp. 167–182.

Webb, S.L., 1989. The elasticity of the upper mantle orthosilicates olivine and garnet to 3 GPa. Phys. Chem. Miner. 16, 684–692.

Weidner, D.J., Vaughan, M.T., Ko, J., Wang, Y., Liu, X., Yeganeh-Haeri, A., Pacalo, R.E., Zhao, Y., 1992. Characterization of stress, pressure and temperature in SAM-85, a DIA type high pressure apparatus. In: Syono, Y., Manghnani, M.H. (Eds.), High Pressure Research: Applications to Earth and Planetary Sciences. Terra Scientific Publishing Co., Tokyo/American Geophysical Union, Washington, DC, pp. 13–17.

Yagi, T., 2001. "KAWAI-type" apparatus. Rev. High Pressure Sci. Tech. 11, 171.

Zaug, J.M., Abramson, E.H., Brown, J.M., Slutsky, L.J., 1993. Sound velocities in olivine at earth mantle pressures. Science 260, 1487–1489.

Zha, C.-S., Duffy, T.S., Mao, H.-k., Downs, R.T., Hemley, R.J., Weidner, D.J., 1998. Brillouin scattering and X-ray diffraction of San Carlos olivine: direct pressure deterlination to 32 GPa. Earth Planet. Sci. Lett. 159, 25–33.

Available online at www.sciencedirect.com

SCIENCE DIRECT®

Physics of the Earth and Planetary Interiors 143–144 (2004) 33–46

PHYSICS OF THE EARTH AND PLANETARY INTERIORS

www.elsevier.com/locate/pepi

Thermal equation of state of $(Mg_{0.91}Fe_{0.09})_2SiO_4$ ringwoodite

Yu Nishihara [a,*], Eiichi Takahashi [a], Kyoko N. Matsukage [b], Tomohiro Iguchi [a], Keisuke Nakayama [a], Ken-ichi Funakoshi [c]

[a] *Magma Factory, Earth and Planetary Sciences, Tokyo Institute of Technology, Ookayama, Meguro-ku, Tokyo 152-8551, Japan*

[b] *Department of Environmental Sciences, Ibaraki University, Bunkyo, Mito, Ibaraki 310-0056, Japan*

[c] *Japan Synchrotron Radiation Research Institute, Kouto, Mikazuki-cho, Sayo-gun, Hyogo 679-5198, Japan*

Received 23 January 2002; received in revised form 20 February 2003; accepted 25 February 2003

Abstract

In situ synchrotron X-ray diffraction (XRD) experiments were conducted using the SPEED-1500 multi-anvil press of SPring-8 on $(Mg_{0.91}Fe_{0.09})_2SiO_4$ ringwoodite (Rw), whose composition is similar to that expected in the Earth's mantle transition zone. Pressure–volume–temperature data were collected using a NaCl or MgO capsule up to 21 GPa and 1273 K. A fit to high-temperature Birch-Murnaghan (HTBM) equation of state (EOS) with fixed values of ambient cell volume $V_0 = 527.83(7)$ Å^3 and isothermal bulk modulus $K_{T0} = 187$ GPa yielded a pressure derivative of isothermal bulk modulus $K'_T = 4.41(1)$, a temperature derivative of bulk modulus $(\partial K_T/\partial T)_P = -0.028(5)$ GPa K^{-1}, and a volumetric thermal expansivity $\alpha = a + bT$ with values of $a = 1.9(2) \times 10^{-5}$ K^{-1} and $b = 1.2(4) \times 10^{-8}$ K^{-2}. These properties are consistent with the analysis using the thermal pressure (Th-P) EOS. The derived K'_T and $(\partial K_T/\partial T)_P$ are consistent with previous studies on Mg_2SiO_4 ringwoodite. However, consistent with measurements at 0 GPa, the present equation of state yields significantly higher thermal expansivity than derived by diamond anvil experiments on Mg_2SiO_4 ringwoodite to 30 GPa and 700 K. On the basis of the equation of state, the density jump around 660 km depth expected for pyrolitic homogeneous mantle (caused by ringwoodite → Mg-perovskite (MgPv) + magnesiowüstite (Mw) and garnet(GRT) → MgPv transitions) was estimated. The density jump calculated for pyrolite (9.2%) is significantly larger than that across the 660 km discontinuity derived by recent seismological studies (4–6%). One possible explanation for this is that the recent seismic data reflect only the sharp transition concerned with the Rw → MgPv + Mw transition.

Keywords: Ringwoodite; Equation of state; High-pressure and high-temperature; X-ray diffraction; Mantle transition zone; 660 km discontinuity

1. Introduction

The pyrolite model has been widely accepted for the chemical composition of the Earth's mantle, and phase relations in this composition at the mantle P–T conditions has been extensively studied by high-pressure (HP) and high-temperature experiments (e.g. Akaogi and Akimoto, 1979; Irifune, 1987, 1994; Irifune and Isshiki, 1998; Wood, 2000; Hirose, 2002). These studies revealed that the pyrolitic mantle is mainly composed of ~60% ringwoodite (Rw, γ-$(Mg,Fe)_2SiO_4$) and ~40% majoritic garnet (Grt) at conditions of the lower part of the mantle transition zone (~520–660 km depths). Although some studies preferred olivine-poorer compositions for the mantle

* Corresponding author. Present address: Department of Geology and Geophysics, Yale University, New Haven, CT 06520, USA. Tel.: +1-203-432-5791; fax: +1-203-432-3134.
E-mail address: yu.nishihara@yale.edu (Y. Nishihara).

0031-9201/$ – see front matter
doi:10.1016/j.pepi.2003.02.001

transition zone (e.g. Bass and Anderson, 1984; Duffy and Anderson, 1989; Nishihara and Takahashi, 2001), ringwoodite is still one of the most abundant minerals in that region. The velocity and density increases across the 660 km discontinuity, which would be caused by decomposition of ringwoodite to Mg-rich silicate perovskite and magnesiowüstite (Mw) are widely used to estimate temperature and mineralogy in this region (e.g. Akaogi et al., 1989; Matsui, 2001). Thus, knowledge of the thermoelastic properties of ringwoodite is very important to understand the nature of the mantle transition zone.

Because of its geophysical importance, thermoelastic properties of ringwoodite have been widely studied: thermal expansion at high-temperatures (Suzuki et al., 1979), room temperature compressibility (Hazen, 1993; Zerr et al., 1993), acoustic velocities at ambient condition (Weidner et al., 1984; Sinogeikin et al., 1997, 1998), at high-temperatures (Jackson et al., 2000; Sinogeikin et al., 2001) and at high-pressures (Rigden and Jackson, 1991; Rigden et al., 1991, 1992; Sinogeikin et al., 2001). However, studies under simultaneous high-pressure and high-temperature conditions are still limited.

Meng et al. (1993) measured unit-cell volumes of ringwoodite with Mg_2SiO_4 composition in the temperature range from 759 to 962 K at 9.8 GPa using synchrotron X-ray diffraction (XRD) in a cubic anvil apparatus, but this pressure is still below the thermodynamically stable conditions of this mineral. Meng et al. (1994) reported the volume measurements on Mg_2SiO_4 ringwoodite to 30 GPa at 300–700 K using an externally heated diamond anvil cell, but the temperatures are far lower than estimated temperatures in the Earth's transition zone.

In the present study, we measured the change of unit-cell volume of $(Mg_{0.91}Fe_{0.09})_2SiO_4$ ringwoodite, whose Fe/(Mg + Fe) ratio is close to that expected for ringwoodite in the Earth's mantle, at pressure and temperature conditions up to 21 GPa and 1273 K with synchrotron radiation, and the thermal equation of state (EOS) was determined. The pressure range essentially overlaps the entire stability field of this mineral, and the temperature range is significantly expanded from previous *P–V–T* studies. By using the present results, the density jump caused by decomposition of ringwoodite to Mg-rich silicate perovskite and magnesiowüstite is calculated, and the expected density jump for pyorlite mantle at Earth's 660 km depth condition is compared with those from recent seismic observations.

2. Experimental procedures

2.1. Samples

$(Mg_{0.91}Fe_{0.09})_2SiO_4$ ringwoodite was synthesized from natural olivine (Mg# = 91), collected from a granular lherzorite xenolith (K-6, from Australia) using the SPI-1000, the 1000 t multi-anvil press at the Magma Factory, Tokyo Institute of Technology (Takahashi et al., 1993). The powdered olivine was enclosed in a Re foil capsule, and experimental conditions were $P = 21.5$ GPa, $T = 1473$ K and $t = 1$ h. The experimental technique and the cell assembly used in the synthesis experiment are similar to those of Wang and Takahashi (2000) and Nishihara and Takahashi (2001). Chemical analyses were performed using a JEOL-8800 microprobe and synthesis of single-phase ringwoodite was confirmed. The chemical composition, including minor elements, of the synthesized ringwoodite is $(Mg_{0.907}Fe_{0.088}Ni_{0.004}Mn_{0.001})_2SiO_4$. According to Frost et al. (2001), the $Fe^{3+}/\Sigma Fe$ content of ringwoodite synthesized using a Re capsule is below 0.03, which is effectively zero. Thus, we ignored the effect of solution of a Fe_2FeO_4 component.

Angle-dispersive X-ray powder diffraction with Cr Kα radiation was carried out at ambient conditions using Si as an external standard to calibrate the diffraction angle. The unit-cell parameters of the ringwoodite were determined to be $a = 8.0816(3)$ Å and $V = 527.83(7)$ Å^3 from the positions of 6 diffraction peaks (220, 311, 400, 422, 511, and 440). Variation of the cell volume of the ringwoodite solid solution in the system Mg_2SiO_4–Fe_2SiO_4 is determined by Marumo et al. (1977) and Akaogi et al. (1989). The linear interpolation of their data predicts $V = 528.0$ Å^3 for $(Mg_{0.91}Fe_{0.09})_2SiO_4$. This volume is in good agreement with our result.

2.2. *P–V–T* experiments

P–V–T experiments were conducted using the SPEED-1500 apparatus installed on beam line

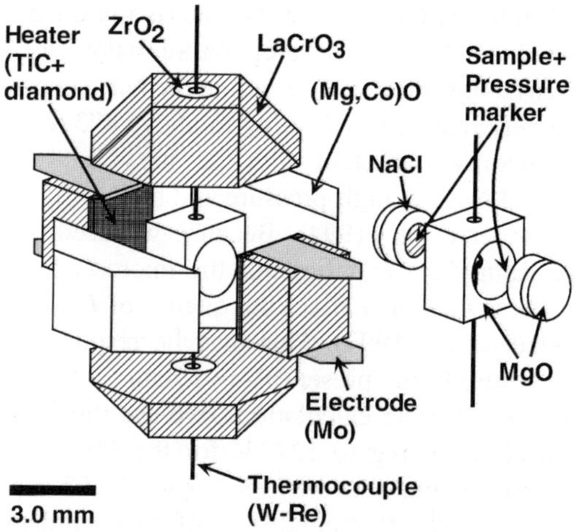

Fig. 1. Schematic illustration of the cell assembly. In the run S602, an MgO capsule was used instead of a NaCl capsule.

BL04B1 at SPring-8, Hyogo, Japan. Detailed information about this system is described by Utsumi et al. (1998). Eight WC anvils (3.0 mm truncation size) were used for high-pressure generation.

Fig. 1 shows schematic illustration of cell assembly for high-pressure and high-temperature X-ray diffraction. Semisintered $LaCrO_3$ is used for the pressure medium as well as thermal insulator and (Mg,Co)O is placed along the X-ray path instead of $LaCrO_3$ for the X-ray window (see Fig. 1). A W5%Re–W26%Re thermocouple wire is inserted vertically, and two sample capsules are placed horizontally at the center of the cell assembly. This enables us to observe two (or more) different samples even when the anvil gap becomes extremely narrow (~0.3 mm) at the highest pressures. The samples are heated with an internal resistance heater made of a mixture of TiC and diamond powders (1:1 by weight). Powdered ringwoodite and a mixture of Au (grain size ~2 μm) and MgO (1:4 by weight, pressure marker) were packed into the NaCl or MgO sample chamber separately. Pressure was determined by EOSs of Au (Anderson et al., 1989; Shim et al., 2002).

A solid-state detector connected to a multi-channel analyzer was used for data collection. The multi-channel analyzer was calibrated with the characteristic X-rays of Cu, Mo, Ag, Ta, Pt, and Pb. The diffracted X-rays from the sample were collected at a fixed 2θ angle of ~5.6° by the energy dispersive method with a horizontal goniometer. The 2θ angle was determined using the unit-cell volume of Au. The incident X-ray beam was collimated to 0.2 and 0.05 mm in the vertical and horizontal dimensions, respectively. Three thin slits between the sample and the detector, 2.0 mm in the vertical dimension and 0.05 and 0.2 mm in the horizontal were used to eliminate diffracted X-rays from the pressure medium. Typical exposure times for collecting diffraction patterns of the sample and pressure marker were 400 and 200 s, respectively. Temperature variation indicated by the thermocouple was within 2 K during this time interval.

Fig. 2 shows an example of the diffraction spectrum of ringwoodite. The unit-cell parameters were calculated from peak positions by least squares technique. In the calculation, at least five diffraction peaks were used from 220, 311, 400, 422, 511, and 440. The pressure was determined from the observed cell volume of Au calculated from four diffraction lines 111, 200, 220, and 311 and the Au pressure scales of Anderson et al. (1989) and Shim et al. (2002).

Anderson's pressure scale has been widely used for in situ X-ray diffraction experiments in the last decade (e.g. Meng et al., 1994; Funamori et al., 1996; Irifune et al., 1998; Fei, 1999). However, Anderson et al. (1989) derived the EOS based on the measured elastic and thermodynamic properties of Au at temperatures only less than 550 K, and this EOS does not reproduce shock compression data for Au (see Shim et al., 2002). It has been pointed out that calculated pressure using Anderson et al.'s scale yields a large discrepancy between the post-ringwoodite phase boundary of Mg_2SiO_4 based on in situ X-ray

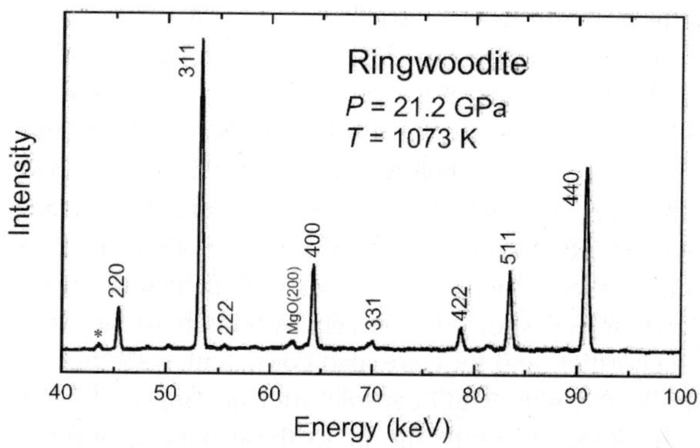

Fig. 2. Example diffraction spectrum of ringwoodite at 21.2 GPa (Anderson et al.'s pressure scale) and 1073 K. Asterick indicates escape peak for 311.

observation (Irifune et al., 1998) and the observed 660 km seismic discontinuity (e.g. Shim et al., 2001; Chudinovskikh and Boehler, 2001).

Very recently, Shim et al. (2002) reported a new *P–V–T* equation of state of Au based on the inversion of recent quasi-hydrostatic compression and shock wave data. However, Shim et al.'s (2002) EOS shows disagreement with the zero-pressure measurements in terms of $(\partial K_S/\partial T)_P$. By using Shim et al.'s EOS of Au, there still exists a ~1.5 GPa discrepancy between the post-ringwoodite boundary and the 660 km discontinuity.

The calculated pressure difference using EOSs by Anderson et al. (1989) and Shim et al. (2001) reaches 0.5 GPa at the maximum in the *P–T* conditions of the present experiments. Because it changes systematically, this difference significantly affects the derived EOS of ringwoodite. However, we cannot judge which EOS is expected to be more accurate. Thus, in our study, the pressure values based on Anderson et al.'s (1989) Au pressure scale are used primarily because it has been used most frequently in EOS measurement. Comparison with Shim et al.'s (2002) is also made.

Three *P–V–T* experiments were conducted. One (experiment no. S602) used an MgO capsule and two (S648 and S649) used NaCl capsules. The samples were compressed to the desired pressure at room temperature and heated to the maximum temperature (873, 1073 or 1273 K) to release non-hydrostatic stress. X-ray diffraction data for both the sample and the pressure marker were collected at fixed temperatures at 200 K intervals under constant load.

The *P–V–T* data collected with the MgO capsule after heating to 873, 1073, and 1273 K showed significant deviation in spite of their relatively sharp diffraction peaks. For example, a fit of the data collected at 300 K to the third-order Birch-Murnaghan EOS (Eq. (1), see below) (with fixed $V_0 = 527.83$ Å^3 and $K_{T0} = 187$ GPa) yields very large root mean square (RMS) misfit 0.77 GPa. On the other hand, the data collected with NaCl capsule after heating to 873 and 1073 K showed relatively good consistency with each other, and they are also consistent with the data collected with MgO capsule after heating to 1273 K. The RMS misfit of the fit to these data at 300 K is reduced to be 0.23 GPa, which is similar in size to the estimated error for pressure data (Table 1). This may mean that the macroscopic differential stress for the sample within the MgO capsule is not effectively released by heating at temperature below 1273 K, whereas that for the sample within the NaCl capsule is effectively released by heating to above 873 K. These are generally consistent with the stress analyses of MgO and NaCl at high-pressure and high-temperature by Weidner et al. (1994). Because we used a powdered sample for experiments, the microscopic deviatric stress may also affect the quality of *P–V–T* data (Weidner et al., 1994). This may be responsible for some scatter in the present *P–V–T* data.

In the following calculations, we use the data collected after heating to 1273 K for the MgO capsule and after heating above 873 K for the NaCl capsule. Fig. 3 shows the *P–T* conditions of the used data and the estimated phase boundary in $(Mg_{0.91}Fe_{0.09})_2SiO_4$ (Akaogi et al., 1989; Ito and Takahashi, 1989; Irifune et al., 1998; Kuroda et al., 2000). In order to avoid back transformation to lower pressure phases, heating was limited to 873 K when outside the stability field of $(Mg_{0.91}Fe_{0.09})_2SiO_4$ (Fig. 3).

3. Results and discussion

The unit-cell parameters of $(Mg_{0.91}Fe_{0.09})_2SiO_4$ ringwoodite at various *P–T* conditions are given in Table 1 and Fig. 4. The slight scatters of the data among each temperature excursion (Fig. 4) may be due to the differences in state of the deviatric stress in the sample, as described above.

3.1. High-temperature Birch-Murnaghan (HTBM) equation of state

The high-temperature Birch-Murnaghan equation of state is often used to fit *P–V–T* data (e.g. Funamori et al., 1996; Wang et al., 1998; Nishihara et al., 2003). The equation of state is given by the following expression:

$$P = \frac{3}{2}K_T\left[\left(\frac{V_{0T}}{V}\right)^{7/3} - \left(\frac{V_{0T}}{V}\right)^{5/3}\right] \times \left\{1 + \frac{3}{4}(K'_T - 4)\left[\left(\frac{V_{0T}}{V}\right)^{2/3} - 1\right]\right\} \quad (1)$$

where K_T, V_{0T}, V, and K'_T are the isothermal bulk modulus, zero-pressure volume, high-pressure vol-

Table 1
Unit-cell parameters of $(Mg_{0.91}Fe_{0.09})_2SiO_4$ ringwoodite at various P–T conditions

P^a (GPa)	P^b (GPa)	T (K)	Ringwoodite, V (Å^3)	Au, V/V_0
S602 (MgO capsule)				
18.17(19)	18.69(19)	1273	494.77(11)	0.9413(8)
17.14(15)	17.52(15)	1073	494.61(14)	0.9402(6)
17.65(18)	18.15(18)	1273	495.50(10)	0.9436(8)
16.74(17)	16.98(17)	873	493.97(11)	0.9365(7)
15.66(17)	15.41(16)	300	491.00(8)	0.9266(6)
20.68(18)	21.24(18)	1273	489.22(15)	0.9309(7)
20.06(25)	20.47(25)	1073	488.96(10)	0.9283(10)
19.48(26)	19.71(26)	873	488.30(12)	0.9257(10)
19.08(33)	19.12(32)	673	487.44(12)	0.9225(12)
18.75(26)	18.59(25)	473	486.37(10)	0.9192(9)
18.71(27)	18.36(26)	300	485.47(11)	0.9156(9)
0.00(16)	0.00(16)	300	529.00(11)	1.0000(9)
S648 (NaCl capsule)				
0.00(13)	0.00(13)	300	528.01(26)	1.0000(8)
9.78(14)	9.94(14)	873	508.95(12)	0.9681(7)
9.13(18)	9.23(18)	673	507.99(13)	0.9646(9)
8.49(19)	8.50(19)	473	507.20(9)	0.9613(9)
8.18(18)	8.11(18)	300	506.38(11)	0.9574(8)
13.56(19)	13.78(19)	873	500.41(11)	0.9502(8)
12.97(15)	13.08(15)	673	499.48(13)	0.9470(7)
12.51(17)	12.48(17)	473	498.62(11)	0.9434(7)
12.19(19)	12.04(18)	300	498.12(14)	0.9401(8)
15.64(20)	15.99(20)	1073	498.12(7)	0.9467(9)
15.09(20)	15.32(20)	873	497.07(12)	0.9434(9)
14.53(16)	14.62(16)	673	496.26(11)	0.9404(7)
14.06(18)	14.00(17)	473	495.54(11)	0.9370(7)
13.88(18)	13.69(18)	300	494.77(11)	0.9334(7)
17.52(13)	17.90(13)	1073	494.23(12)	0.9386(5)
17.08(17)	17.32(17)	873	493.21(15)	0.9351(7)
16.53(15)	16.60(15)	673	492.46(18)	0.9323(6)
16.12(21)	16.02(20)	473	491.59(18)	0.9290(8)
15.82(14)	15.57(13)	300	490.93(18)	0.9260(5)
19.35(18)	19.75(18)	1073	490.47(13)	0.9311(7)
18.83(13)	19.06(13)	873	489.63(19)	0.9282(5)
18.35(15)	18.41(14)	673	488.84(16)	0.9252(5)
17.93(15)	17.78(14)	473	488.13(20)	0.9222(5)
17.82(17)	17.50(17)	300	487.54(24)	0.9187(6)
21.16(14)	21.57(14)	1073	487.67(10)	0.9241(5)
20.51(17)	20.73(17)	873	486.99(7)	0.9218(6)
19.94(16)	19.97(16)	673	486.23(7)	0.9194(6)
19.61(19)	19.43(18)	473	485.45(9)	0.9162(7)
19.49(21)	19.12(21)	300	484.95(10)	0.9129(7)
0.00(9)	0.00(9)	300	528.32(12)	1.0000(5)
S649 (NaCl capsule)				
4.22(14)	4.22(15)	873	524.22(12)	0.9992(9)
3.60(13)	3.61(14)	673	522.93(15)	0.9944(8)
3.08(12)	3.09(12)	473	521.83(18)	0.9892(7)
2.84(14)	2.83(14)	300	521.00(12)	0.9838(7)
0.00(10)	0.00(10)	300	528.43(18)	1.0000(6)

[a] Pressure scale by Anderson et al. (1989).
[b] Pressure scale by Shim et al. (2002).

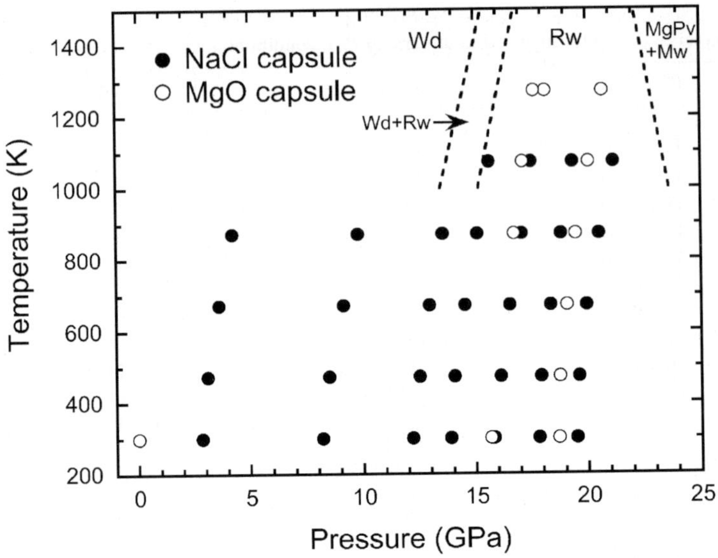

Fig. 3. Pressure–temperature conditions of the present experiments. Solid and open symbols are data collected using NaCl (experiment no. S648 and S649) and MgO (S602) capsules, respectively. The pressures were calculated using the Anderson et al. (1989) EOS for Au. Broken lines are estimated phase boundary in $(Mg_{0.91}Fe_{0.09})_2SiO_4$ composition consistent with Anderson et al.'s Au pressure scale (Akaogi et al., 1989; Ito and Takahashi, 1989; Irifune et al., 1998; Kuroda et al., 2000). Abbreviations: Wd, wadsleyite; Rw, ringwoodite; MgPv, Mg-rich perovskite; Mw, magnesiowüstite.

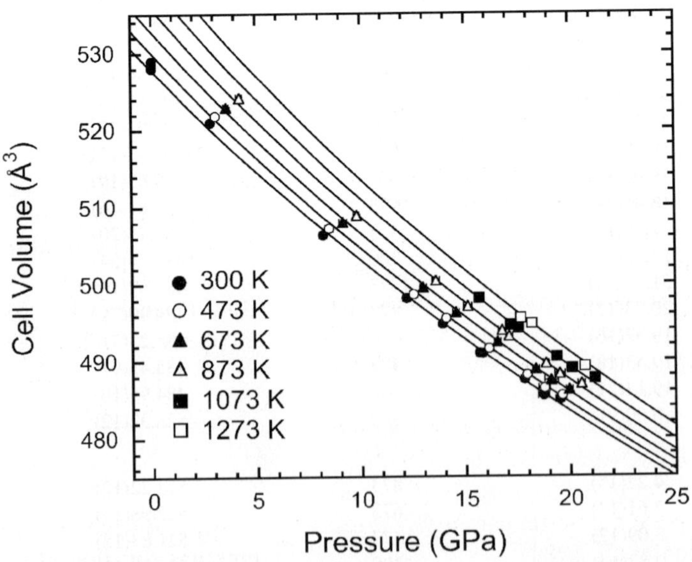

Fig. 4. P–V–T data of $(Mg_{0.91}Fe_{0.09})_2SiO_4$ ringwoodite with calculated isothermal compression curves. Pressures are based on Anderson's scale (Anderson et al., 1989). The isotherms (from lower cell volumes, 300, 473, 673, 873, 1073, and 1273 K) are calculated from thermoelastic parameters derived with fixed value of $K_{T0} = 187$ GPa (Table 2).

ume, and pressure derivative of K_T, respectively. The temperature effects for K_T and V_{0T} are expressed as follows:

$$K_T = K_{T0} + \left(\frac{\partial K_T}{\partial T}\right)_P (T - 300) \quad (2)$$

$$V_{0T} = V_0 \exp\left[\int_{300}^{T} \alpha \mathrm{d}T\right] \quad (3)$$

where $(\partial K_T/\partial T)_P$, V_0 and α are the temperature derivative of the bulk modulus, volume at room condition and the volumetric thermal expansion at atmospheric pressure, respectively. The $(\partial K_T/\partial T)_P$ value is assumed to be constant. V_0 was fixed to the value determined by angle-dispersive X-ray diffraction ($V_0 = 527.83$ Å^3) at ambient condition.

The experimental P–V–T data (Table 1) were fitted to HTBM equation of state. The fits without constraint on K_{T0} and K'_T yielded $K_{T0} = 173(3)$ GPa and $K'_T = 6.5(5)$ (Anderson et al.'s pressure scale). However, this K'_T value is relatively high for a silicate mineral (typically $K'_T = 4 - 5$), and it is well known that there is a trade-off between the values of K_{T0} and K'_T in calculating these properties from P–V data (e.g. Bass et al., 1981). Because of the limited range and precision of the volume data, we could not resolve both K_{T0} and K'_T values simultaneously with confidence.

We also calculated the equation of state by fixing K_{T0}. Sinogeikin et al. (1998, 2001) determined the adiabatic bulk modulus of $(Mg_{0.91}Fe_{0.09})_2SiO_4$ ringwoodite to be $K_{S0} = 188(3)$ GPa by Brillouin scattering (BS). We used an isothermal bulk modulus $K_{T0} = 187$ GPa, which is converted from $K_{S0} = 188(3)$ GPa, to constrain the other thermoelastic parameters accurately. In the conversion from K_S to K_T, the following thermodynamic identity was used: $K_{T0} = K_{S0}/(1+\alpha\gamma_G T)$, where γ_G is the Grüneisen parameter, data for α and γ_G are from Suzuki et al. (1979) and Watanabe (1987), respectively. The fit with fixing $K_{T0} = 187$ GPa yielded $K'_T = 4.4(1)$; $(\partial K_T/\partial T)_P = -0.028(5)$ GPa K^{-1}, and $\alpha = a + bT$ with values of $a = 1.9(2) \times 10^{-5}$ K^{-1} and $b = 1.2(4) \times 10^{-8}$ K^{-2} (Table 2). Compression curves depicted in Fig. 4 are calculated from these thermoelastic parameters (pressure scale is Anderson et al., 1989).

Use of the different Au pressure scales causes significant variation in the derived thermoelastic parameters of $(Mg_{0.91}Fe_{0.09})_2SiO_4$ ringwoodite. Shim et al.'s (2002) scale results in lower absolute value of $(\partial K_T/\partial T)_P$ of ringwoodite ($-0.015(5)$ GPa K^{-1}; Table 2) than the Anderson et al.'s (1989) scales ($-0.028(5)$ GPa K^{-1}; Table 2). This is due to the fact that Shim et al.'s EOS of Au shows extremely small absolute value of $(\partial K_S/\partial T)_P$ compared to zero-pressure measurements (see Fig. 3c of Shim et al., 2002).

Table 2
Thermoelastic parameters of $(Mg_{0.91}Fe_{0.09})_2SiO_4$ ringwoodite using the HTBM equation of state

Parameter		
K_{T0} (GPa)	[187]	[187]
K'_T	4.4(1)	4.0(1)
$(\partial K_T/\partial T)_P$ (GPa K^{-1})	−0.028(5)	−0.015(5)
$\alpha = a + bT$		
a (10^{-5} K^{-1})	1.9(2)	2.1(2)
b (10^{-8} K^{-2})	1.2(4)	0.7(4)
P-Marker[a]	Au-A	Au-S

[a] Pressure was calculated from the equation of state of gold by Anderson et al. (1989) (Au-A) and Shim et al. (2002) (Au-S). V_0 is fixed to be 527.83 Å^3 in the calculation. Square brackets indicate fixed value in the calculation.

Thermal expansion of ringwoodite at 0 and 20 GPa was calculated from the parameters in Table 2, and is depicted in Fig. 5. Thermal expansion at 0 GPa calculated from the present equation of state of $(Mg_{0.91}Fe_{0.09})_2SiO_4$ ringwoodite, which is extrapolated from high-pressures, shows reasonably good agreement with direct measurements at 0 GPa on Mg_2SiO_4 and Fe_2SiO_4 ringwoodites (Suzuki et al., 1979; Yamanaka, 1986; Inoue et al., 2004). As shown in Funamori et al. (1996), the choice of pressure scale significantly affects thermal expansivity at high-pressures. At 20 GPa, Shim et al.'s EOS of Au results in significantly larger thermal expansivity of ringwoodite than that by Anderson et al.'s scale (lower panel of Fig. 5). It is clear that the choice of the pressure scale significantly affects the density estimate of a model mantle composition at high-pressure and high-temperature using thermal equation of state.

3.2. Thermal pressure (Th-P) equation of state

We also analyze the P–V–T data using the thermal pressure equation of state (Anderson, 1984). At any temperature T and for a given volume V, the pressure is expressed as the sum of the static pressure at room temperature $P(V, 300)$ and the thermal pres-

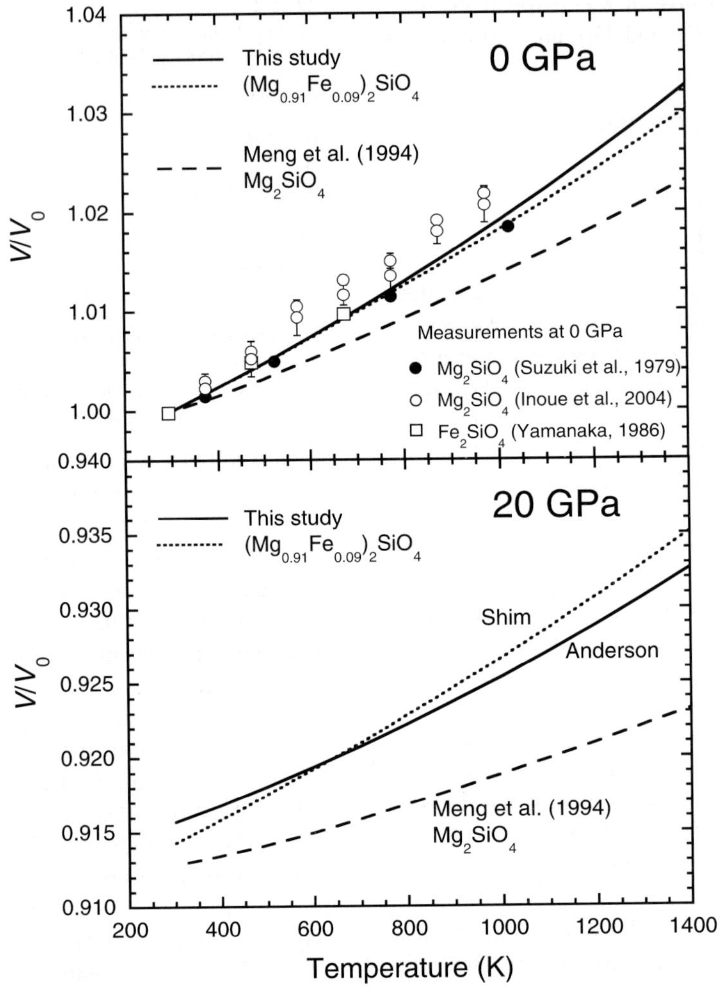

Fig. 5. Normalized volumes (V/V_0) of ringwoodite at 0 GPa (upper panel) and 20 GPa (lower panel) as a function of temperature. Solid and dotted lines are based on Anderson et al.'s (1989) and Shim et al.'s (2002) Au pressure scale, respectively. Broken lines are Meng et al.'s (1994) EOS of Mg_2SiO_4 ringwoodite, which is based on Anderson's pressure scale. Symbols are results of measurements at 0 GPa (1 bar) on Mg_2SiO_4 (solid circles; Suzuki et al., 1979, open circles; Inoue et al., 2004) and Fe_2SiO_4 (open square; Yamanaka, 1986) ringwoodites, respectively.

sure ΔP_{th} (see Jackson and Rigden, 1996). An inspection of Fig. 6 reveals that thermal pressures of $(Mg_{0.91}Fe_{0.09})_2SiO_4$ ringwoodite vary linearly with temperature. In this case, thermal pressure ΔP_{th} is expressed by:

$$\Delta P_{th} = P(V, T) - P(V, 300) = \left[\alpha K_T + \left(\frac{\partial K_T}{\partial T}\right)_V \ln\left(\frac{V_0}{V}\right)\right] \times (T - 300) \quad (4)$$

For expression of $P(V, 300)$, we used Eq. (1).

The results of a fit of the data with fixed $K_{T0} = 187$ GPa is listed in Table 3. The derived values of $K'_T = 4.3(1)$ and $\alpha_0 = 2.5(1) \times 10^{-5}\,K^{-1}$ are reasonably consistent with those derived by the fit to HTBM equation of state ($\alpha_0 = 2.3(2) \times 10^{-5}\,K^{-1}$ from $\alpha_0 = a+b\times300$; Table 2, Anderson et al.'s scale). Within the uncertainty of the fit $(\partial K_T/\partial T)_V$ value is roughly zero, indicating that the thermal pressure is independent of volume. This property is same as other silicate minerals: olivine (Guyot et al., 1996), $CaSiO_3$ perovskite (Wang et al., 1996) and garnet (Wang et al., 1998).

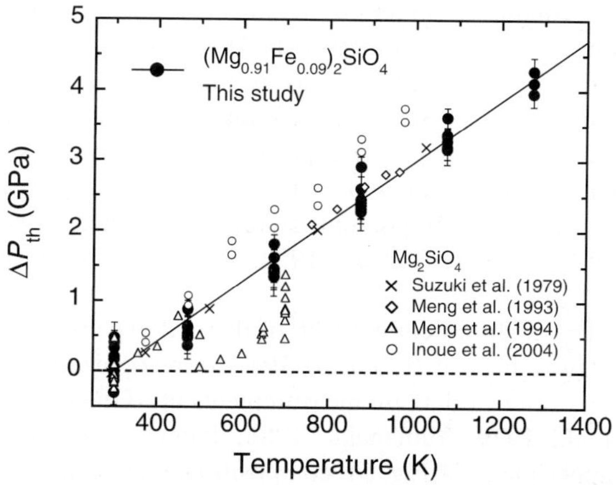

Fig. 6. Thermal pressure of ringwoodite against temperature. Solid circles are $(Mg_{0.91}Fe_{0.09})_2SiO_4$ ringwoodite (this study, pressure scale by Anderson et al., 1989). Data for Mg_2SiO_4 ringwoodite are from Suzuki et al. (1979) (crosses), Meng et al. (1993) (open diamonds), Meng et al. (1994) (open triangles) and Inoue et al. (2004) (open circles). Solid line is calculated from the parameters listed in Table 3 for $V = 498.9\,Å^3$, which is the mean value of data by this study.

Table 3
Thermoelastic parameters of $(Mg_{0.91}Fe_{0.09})_2SiO_4$ ringwoodite from the thermal pressure equation of state

Parameter		
K_{T0} (GPa)	[187]	[187]
K'_T	4.3(1)	3.9(1)
α_0 (10^{-5} K^{-1})	2.5(1)	2.4(1)
$(\partial K_T/\partial T)_V$ (GPa K^{-1})	−0.006(4)	0.006(4)
$(\partial K_T/\partial T)_P$ (GPa K^{-1})	−0.026(4)	−0.011(4)
P-Marker[a]	Au-A	Au-S

[a] Pressure was calculated from the equation of state of gold by Anderson et al. (1989) (Au-A) and Shim et al. (2002) (Au-S). V_0 is fixed to be $527.83\,Å^3$ in the calculation. Square brackets indicate fixed value in the calculation.

By using the thermodynamic relation: $(\partial K_T/\partial T)_P = (\partial K_T/\partial T)_V - \alpha K_T K'_T$, we can obtain $(\partial K_T/\partial T)_P = -0.026(4)$ GPa K^{-1}. This result is in good agreement with previous analysis using HTBM equation of state (Table 2). The results for the data based on Shim et al.'s pressure scale are similarly consistent with those by the HTBM equation of state fit (Tables 2 and 3).

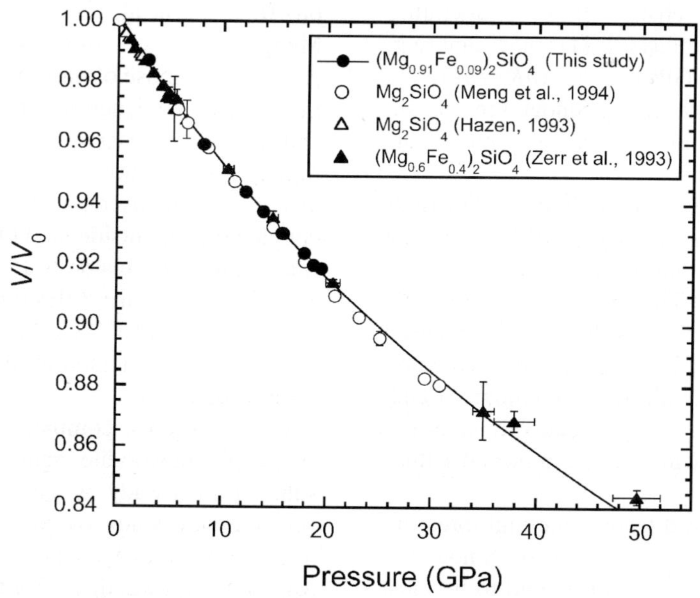

Fig. 7. Comparison of room temperature V/V_0 data of ringwoodites with Mg-rich composition. Solid circles, open circles, open triangles and solid triangles are data from this study ($(Mg_{0.91}Fe_{0.09})_2SiO_4$, Anderson et al.'s scale), Meng et al. (1994) (Mg_2SiO_4), Hazen (1993) (Mg_2SiO_4), and Zerr et al. (1993) ($(Mg_{0.6}Fe_{0.4})_2SiO_4$). Solid line is calculated compression curve with $K_{T0} = 187$ GPa and $K'_T = 4.4$ (Table 2).

3.3. Comparison with previous studies

Fig. 7 shows comparison of room temperature V/V_0 data of ringwoodites with Mg-rich composition. The solid line is a calculated compression curve with $K_{T0} = 187$ GPa and $K'_T = 4.4$ (Table 2). This equation of state of $(Mg_{0.91}Fe_{0.09})_2SiO_4$ ringwoodite is generally consistent with those of previous studies. Slight variation among different studies might be due to a compositional dependence of compressibility.

Elastic properties of ringwoodite have been investigated by many authors. Table 4 summarizes the bulk modulus, its pressure derivative and its temperature derivative of $(Mg,Fe)_2SiO_4$ ringwoodite solid solution determined with various experimental techniques. Adiabatic values (K_{S0}, K'_S, and $(\partial K_S/\partial T)_P$) can be obtained by Brillouin scattering and ultrasonic techniques (UT). These are converted to isothermal values (K_{T0}, K'_T and $(\partial K_T/\partial T)_P$) by using the following equations: $K_{T0} = K_{S0}/(1 + \alpha\gamma_G T)$, $K'_T \approx (1+\alpha\gamma_G T)^{-1}[K'_S - \gamma_G T(\partial K_T/\partial T)_P/K_T]$,; $(\partial K_T/\partial T)_P \approx (\partial K_S/\partial T)_P/(1 + \alpha\gamma_G T) - K_S/(1 + \alpha\gamma_G T)^2[\alpha\gamma_G + (\partial\alpha/\partial T)\gamma_G T]$ (Speziale and Duffy, 2002), respectively. The change in the bulk modulus and its pressure derivative by this conversion are within experimental errors. However, the change of the temperature derivative of the bulk modulus is quite significant (e.g. $(\partial K_S/\partial T)_P = -0.021(2)$ GPa K^{-1} to $(\partial K_T/\partial T)_P = -0.028(2)$ GPa K^{-1}, data from Sinogeikin et al., 2001). Thus, this conversion is necessary for comparison of this property.

The bulk modulus (K_{T0}) of $(Mg,Fe)_2SiO_4$ ringwoodite increases from 184 to ~205 GPa as the Fe content increases from 0 to 100% (Table 4, Sinogeikin et al., 1998). The value $K_{T0} = 187$ GPa for $(Mg_{0.91}Fe_{0.09})_2SiO_4$ ringwoodite, which was used in the above calculation to derive thermal equation of state, is consistent with this systematic trend. For K'_T and $(\partial K_T/\partial T)_P$, no systematic trend is detectable partly due to scatter in the reported values (Table 4).

The K'_T value is derived from some high-pressure studies. Among these K'_T values, that by Sinogeikin et al. (2001) ($K'_T = 4.1(3)$) is considered to be most accurate, because it is from direct measurements of elastic moduli (Brillouin scattering) at relatively wide pressure conditions (0–16 GPa). Although Rigden et al. (1991, 1992) and Rigden and Jackson (1991) also conducted direct measurements of elastic moduli (ultrasonic interferometry) to 3 GPa, their relatively high K'_T (4.7 and 5.6) may result from porosity or grain-boundary scattering (Sinogeikin et al., 2001). The K'_T value from our experiments is consistent with that by Sinogeikin et al. (2001). Our $(\partial K_T/\partial T)_P$ value is consistent with previous studies (Meng et al., 1993, 1994; Sinogeikin et al., 1998, 2001; Jackson et al., 2000).

Fig. 5 shows comparison of thermal expansivity of ringwoodites at 0 and 20 GPa. As recognized from the scatter of data by measurements at 0 GPa (Suzuki et al., 1979; Yamanaka, 1986; Inoue et al., 2004) (upper panel of Fig. 5), compositional dependence of thermal expansivity is not clear. The thermal equation of state of Mg_2SiO_4 ringwoodite by Meng et al. (1994) has been widely used to investigate nature of the 660 km seismic discontinuity (e.g. Weidner and Wang, 1998; Vacher et al., 1998). However, their EOS shows significant difference between our EOS of $(Mg_{0.91}Fe_{0.09})_2SiO_4$ ringwoodite in terms of thermal expansivity, while parameters in Table 4 are consistent each other. Meng et al. (1994) used EOS of Au by Anderson et al. (1989) for pressure standard for their high-temperature experiments, then our results based on Anderson et al.'s scale is comparable. At 0 GPa, Meng et al.'s EOS shows slightly lower thermal expansion than those of ours and zero-pressure measurements. This tendency is preserved at high-pressures, and thermal expansivity of ringwoodite based on Meng et al. is ~60% of ours at 20 GPa (lower panel of Fig. 5). The discrepancy may suggest that thermocouple reading outside the DAC actually does not give true temperature inside in their experiments. The use of different pressure scales for room temperature (ruby fluorescence and EOS of Ne) and high-temperature (EOS of Au, Anderson et al., 1989) is also possibly responsible.

Fig. 6 shows comparison of thermal pressure ΔP_{th} of ringwoodite against temperature. The results of this study is in good agreement with the zero-pressure study by Suzuki et al. (1979) and the measurements at 9.8 GPa by Meng et al. (1993) on Mg_2SiO_4 ringwoodite. Another zero-pressure study by Inoue et al. (2004) shows slightly higher ΔP_{th} than that by this study, and this property is similar to the previous comparison for thermal expansivity (Fig. 5). The results by Meng et al. (1994) show

Table 4
Bulk modulus, its pressure derivative and its temperature derivative of ringwoodite

Composition	K_{T0} (GPa)	K'_T	$(\partial K_T/\partial T)_P$ (GPa K^{-1})	Method[a]	Pressure calibration[b]	References
Mg_2SiO_4	184(2)	[4.8]	–	HP-XRD	Ruby	Hazen (1993)
Mg_2SiO_4	[182.6]	[5.0]	−0.028(3)	HP-XRD	NaCl	Meng et al. (1993)
Mg_2SiO_4	182(3)	4.0(3)	−0.027(5)	HP-XRD	Ruby, Au-A	Meng et al. (1994)
Mg_2SiO_4	183(3)[c]	–	–	BS	–	Weidner et al. (1984)
Mg_2SiO_4	[183][c]	4.7[c]	–	US	RG	Rigden et al. (1991, 1992)
Mg_2SiO_4	184(3)[c]	–	−0.031(3)[c]	BS	–	Jackson et al. (2000)
$(Mg_{0.91}Fe_{0.09})_2SiO_4$	187(3)[c]	4.1(3)[c]	−0.028(2)[c]	BS	Ruby	Sinogeikin et al. (1998, 2001)
$(Mg_{0.91}Fe_{0.09})_2SiO_4$	[187]	4.4(1)	−0.028(5)	HP-XRD	Au-A	This study, HTBM
$(Mg_{0.91}Fe_{0.09})_2SiO_4$	[187]	4.3(1)	−0.026(4)	HP-XRD	Au-A	This study, Th-P
$(Mg_{0.75}Fe_{0.25})_2SiO_4$	192(3)[c]	–	–	BS	–	Sinogeikin et al. (1997)
$(Mg_{0.6}Fe_{0.4})_2SiO_4$	183(2)	5.4(2)	–	HP-XRD	Ruby	Zerr et al. (1993)
$(Mg_{0.4}Fe_{0.6})_2SiO_4$	203(2)	[4.8]	–	HP-XRD	Ruby	Hazen (1993)
$(Mg_{0.2}Fe_{0.8})_2SiO_4$	205(2)	[4.8]	–	HP-XRD	Ruby	Hazen (1993)
Fe_2SiO_4	200[c]	5.6[c]	–	US	RG	Rigden and Jackson (1991)
Fe_2SiO_4	207(3)	[4.8]	–	HP-XRD	Ruby	Hazen (1993)

[a] HP-XRD, X-ray diffraction at high-pressure; BS, Brillouin scattering; US, ultrasonic technique.
[b] Pressure calibration is based on ruby fluorescence (Ruby), EOS of NaCl by Decker (1971) (NaCl), EOS of gold by Anderson et al. (1989) (Au-A) and resistance gauge (RG). The properties derived in this study using high-temperature Birch-Murnghan (HTBM) EOS and thermal pressure (Th-P) EOS are presented. Square brackets indicate fixed value in the calculation.
[c] Converted from adiabatic values.

large scatter and relatively lower ΔP_{th} than those by other studies. Meng et al. (1994) reported a very large absolute value of $(\partial K_T/\partial T)_V = -0.013\,\mathrm{GPa\,K^{-1}}$. However, it is clear from Fig. 6 that the large scatters and limited temperature range of Meng et al.'s (1994) data preclude accurate resolution of this property.

As discussed above, our EOS of $(Mg_{0.91}Fe_{0.09})_2SiO_4$ ringwoodite, based on the data collected over a wide P–T range (up to 21 GPa and 1273 K), is generally consistent with previous reliable zero-pressure (Suzuki et al., 1979; Sinogeikin et al., 1998, 2001) or room temperature (Sinogeikin et al., 2001) measurements. Thus, this EOS of ringwoodite enables highly robust estimation of thermoelasticity of this mineral in the Earth's mantle.

3.4. Implications for the 660 km discontinuity

In order to compare the seismically observed density jump across the 660 km discontinuity ($\Delta\rho_{660}$) with the density jump expected for a model mantle composition (pyrolite), a thermoelastic calculation was carried out on the basis of the present EOS of ringwoodite.

According to recent phase equilibria studies, the density and seismic velocity jumps at around 660 km depth are mainly responsible to two phase transitions: (1) ringwoodite (Rw) → Mg-perovskite (MgPv) + magnesiowüstite and (2) garnet → MgPv (e.g. Weidner and Wang, 1998; Hirose, 2002; Akaogi et al., 2002). At first, we calculated the density jump caused by Rw → MgPv + Mw. The thermoelastic parameters used in the calculation are collected from this study (Rw), Funamori et al. (1996) (MgPv), Fei (1999) (Mw) and Suzuki (1975) (Mw), and all the parameters are based on Anderson et al.'s (1989) pressure scale of Au. The calculated pressure using Anderson et al.'s pressure scale of Au yields a large discrepancy between Rw → MgPv + Mw boundary and the observed 660 km discontinuity (Irifune et al., 1998). However, the calculated density jump (in %) across the phase transition is very insensitive to the choice of pressure scale, because densities of related minerals change similarly by the change of pressure scale. Thus, the use of the parameters based on Anderson et al.'s scale for the calculation of the density jump is not irrelevant.

The zero-pressure densities were corrected to Mg# = 100 Mg/(Mg + Fe) (molar ratio) = 89, 91, and 87 for Rw, MgPv, and Mw, respectively, using the data of Vacher et al. (1998). The calculated density jump at 660 km conditions (23.4 GPa and 1873 K) is 8.9%. If we assume a pyrolitic composition (61 vol.% Rw at the base of transition zone, Hirose, 2002) and chemical homogeneity through the 660 km discontinuity for the mantle, the Rw → MgPv + Mw transition yields a 5.4% density jump.

The Rw → MgPv + Mw transition is very sharp (<~5 km width, Ito and Takahashi, 1989), while the Grt → MgPv transition is relatively broad (~30–50 km width, Hirose, 2002; Akaogi et al., 2002). Estimates of the density jump across the 660 km discontinuity by long-period (>15 s) seismic studies are considered to be the sum of these transitions. By using the estimated contribution of the Grt → MgPv transition for the density jump in pyrolite by Akaogi et al. (2002) (3.8%), the combined density jump around 660 km is 9.2%.

The estimated density jump for a pyrolitic mantle (9.2%) is in good agreement with seismic PREM model ($\Delta\rho_{660} = 9.3\%$; Dziewonski and Anderson, 1981), however, significantly larger than recent seismic models of the 660-km discontinuity derived using long-period data: $\Delta\rho_{660} = 6.2\%$ (Esterbrook and Kind, 1996), 5.2% (Shearer and Flanagan, 1999) and 5.4% (Kato and Kawakatsu, 2001). The reason of this discrepancy is unclear. A recent short-period (2–5 s) study yields similar results ($\Delta\rho_{660} = 4.2\%$, Castle and Creager, 2000) to long-period studies. These results possibly suggest that the recent seismic long-period data sample only the sharp transition concerned with the Rw → MgPv + Mw transition.

Acknowledgements

We thank N. Nishiyama for his advice on experimental techniques, and K. Hirose, T. Komabayashi and M. Shindo for their helpful support for the in situ X-ray diffraction experiments. We also thank T. Inoue for providing *V–T* data for Mg_2SiO_4 ringwoodite before publication. Constructive comments by Y. Wang, T. Katsura, T.S. Duffy and T. Irifune were helpful in improving the manuscript. The in situ X-ray diffraction experiments were performed by using the SPEED-1500 system at BL04B1 in SPring-8 (proposal no. 2000B0548-CD-np and 2001A0272-CD-np). Y.N. is grateful for the Research Fellowships of the Japan Society of the Promotion of Science for Young Scientists.

References

Akaogi, M., Akimoto, S., 1979. High-pressure phase equilibria in a garnet lherzolite, with special reference to Mg^{2+}–Fe^{2+} partitioning among constituent minerals. Phys. Earth Planet. Int. 19, 31–51.

Akaogi, M., Ito, E., Navrotsky, A., 1989. Olivine-modified spinel–spinel transitions in the system Mg_2SiO_4–Fe_2SiO_4: calorimetric measurements, thermochemical calculation, and geophysical application. J. Geophys. Res. 94, 15671–15685.

Akaogi, M., Tanaka, A., Ito, E., 2002. Garnet-ilmenite-perovskite transitions in the system $Mg_4Si_4O_{12}$–$Mg_3Al_2Si_3O_{12}$ at high pressures and high temperatures: phase equilibria, calorimetry and implications for mantle structure. Phys. Earth Planet. Int. 132, 303–324.

Anderson, O.L., 1984. A universal thermal equation of state. J. Geodyn. 1, 185–214.

Anderson, O.L., Isaak, D.G., Yamamoto, S., 1989. Anharmonicity and the equation of state for gold. J. Appl. Phys. 65, 1534–1543.

Bass, J., Anderson, D.L., 1984. Composition of the upper mantle: geophysical tests of two petrological models. Geophys. Res. Lett. 11, 237–240.

Bass, J.D., Liebermann, R.C., Weidner, D.J., Finch, S.J., 1981. Elastic properties from acoustic and volume compression experiments. Phys. Earth Planet. Int. 25, 140–158.

Castle, J.C., Creager, K.C., 2000. Local sharpness and shear wave speed jump across 660-km discontinuity. J. Geophys. Res. 105, 6191–6200.

Chudinovskikh, L., Boehler, R., 2001. High-pressure polymorphs of olivine and the 660-km seismic discontinuity. Nature 411, 574–577.

Decker, D.L., 1971. High-pressure equation of state for NaCl, KCl, and CsCl. J. Appl. Phys. 42, 3239–3244.

Duffy, T.S., Anderson, D.L., 1989. Seismic velocities in mantle minerals and the mineralogy of the upper mantle. J. Geophys. Res. 94, 1895–1912.

Dziewonski, A.M., Anderson, D.L., 1981. Preliminary reference Earth model. Phys. Earth Planet. Int. 25, 297–356.

Esterbrook, C.H., Kind, R., 1996. The nature of the 660-kilometer upper-mantle seismic discontinuity from precursors to the PP phase. Science 274, 1179–1182.

Fei, Y., 1999. Effects of temperature and composition on the bulk modulus of (Mg, Fe)O. Am. Mineral. 84, 272–276.

Frost, D.J., Langenhorst, F., van Aken, P.A., 2001. Fe–Mg partitioning between ringwoodite and magnesiowüstite and the effect of pressure, temperature and oxygen fugacity. Phys. Chem. Mineral. 28, 445–470.

Funamori, N., Yagi, T., Utsumi, W., Kondo, T., Uchida, T., Funamori, M., 1996. Thermoelastic properties of $MgSiO_3$ perovskite determined by in situ X-ray observations up to 30 GPa and 2000 K. J. Geophys. Res. 101, 8257–8269.

Guyot, F., Wang, Y., Gillet, P., Ricard, Y., 1996. Quasi-harmonic computations of thermodynamic parameters of olivines at high-pressure and high-temperature: a comparison with experiment data. Phys. Earth Planet. Int. 98, 17–29.

Hazen, R.M., 1993. Comparative compressibilities of silicate spinels: anomalous behavior of (Mg, Fe)$_2$SiO$_4$. Science 259, 206–209.

Hirose, K., 2002. Phase transitions in pyrolitic mantle around 670-km depth: implications for upwelling of plumes from the lower mantle. J. Geophys. Res. 107, doi: 10.1029/2001JB000597.

Inoue, T., Tanimoto, Y., Irifune, T., Suzuki, T., Fukui, H., Ohtaka, O., 2004. Thermal expansion of wadsleyite, ringwoodite, hydrous wadsleyite and hydrous ringwoodite. Phys. Earth Planet Int. 143–144, 279–290.

Irifune, T., 1987. An experimental investigation of the pyroxene-garnet transformation in a pyrolite composition and its bearing on the constitution of the mantle. Phys. Earth Planet. Int. 45, 324–336.

Irifune, T., 1994. Absence of an aluminous phase in the upper part of the Earth's lower mantle. Nature 370, 131–133.

Irifune, T., Isshiki, M., 1998. Iron partitioning in a pyrolite mantle and the nature of the 410-km seismic discontinuity. Nature 392, 702–705.

Irifune, T., Nishiyama, N., Kuroda, K., Inoue, T., Isshiki, M., Utsumi, W., Funakoshi, K., Urakawa, S., Uchida, T., Katsura, T., Ohtaka, O., 1998. The postspinel phase boundary in Mg$_2$SiO$_4$ determined by in situ X-ray diffraction. Science 279, 1698–1700.

Ito, E., Takahashi, E., 1989. Postspinel transformations the system Mg$_2$SiO$_4$–Fe$_2$SiO$_4$ and some geophysical implications. J. Geophys. Res. 94, 10637–10646.

Jackson, I., Rigden, S.M., 1996. Analysis of *P–V–T* data: constraints on the thermoelastic properties of high-pressure minerals. Phys. Earth Planet. Int. 96, 85–112.

Jackson, J.M., Sinogeikin, S.V., Bass, J.D., 2000. Sound velocities and elastic properties of (γ-Mg$_2$SiO$_4$ to 873 K by Brillouin spectroscopy. Am. Mineral. 85, 296–303.

Kato, M., Kawakatsu, H., 2001. Seismological in situ estimation of density jumps across the transition zone discontinuities beneath Japan. Geophys. Res. Lett. 28, 2541–2544.

Kuroda, K., Irifune, T., Inoue, T., Nishiyama, N., Miyashita, M., Funakoshi, K., Utsumi, W., 2000. Determination of the phase boundary between ilmenite and perovskite in MgSiO$_3$ by in situ X-ray diffraction and quench experiments. Phys. Chem. Miner. 27, 523–532.

Marumo, F., Isobe, M., Akimoto, S., 1977. Electron-density distributions in crystals of γ-Fe$_2$SiO$_4$ and γ-Co$_2$SiO$_4$. Acta Crstallogr. Sect. B 33, 713–716.

Matsui, M., 2001. Density and bulk sound velocity jumps across the 660 km seismic discontinuity. Phys. Earth Planet. Int. 125, 141–146.

Meng, Y., Weidner, D.J., Gwanmesia, G.D., Liebermann, R.C., Vaughan, M.T., Wang, Y., Leinenweber, K., Pacalo, R.E., Yeganeh-Haeri, A., Zhao, Y., 1993. In situ high-*P–T* X-ray diffraction studies on three polymorphs (α, β, γ) of Mg$_2$SiO$_4$. J. Geophys. Res. 98, 22199–22207.

Meng, Y., Fei, Y., Weidner, D.J., Gwanmesia, G.D., Hu, J., 1994. Hydrostatic compression of γ-Mg$_2$SiO$_4$ to mantle pressures and 700 K: thermal equation of state and related thermoelastic properties. Phys. Chem. Miner. 21, 407–412.

Nishihara, Y., Takahashi, E., 2001. Phase relation and physical properties of an Al-depleted komatiite to 23 GPa. Earth Planet. Sci. Lett. 190, 65–77.

Nishihara, Y., Takahashi, E., Matsukage, K., Kikegawa, T., 2003. Thermal equation of state of omphacite. Am. Mineral. 88, 80–86.

Rigden, S.M., Jackson, I., 1991. Elasticity of germanate and silicate spinels at high pressure. J. Geophys. Res. 96, 9999–10006.

Rigden, S.M., Gwanmesia, G.D., Fitz Gerald, J.D., Jackson, I., Liebermann, R.C., 1991. Spinel elasticity and seismic structure of the transition zone of the mantle. Nature 354, 143–145.

Rigden, S.M., Gwanmesia, G.D., Jackson, I., Liebermann, R.C., 1992. Progress in high-pressure ultrasonic interferometry, the pressure dependence of elasticity of Mg$_2$SiO$_4$ polymorphs and constraints on the composition of the transition zone of the Earth's mantle. In: Syono, Y., Manghnani, M.H. (Eds.), High-Pressure Research: Application to Earth and Planetary Sciences. Am. Geophys. Union, Washington, DC, pp. 167–182.

Shearer, P.M., Flanagan, M.P., 1999. Seismic velocity and density jumps across the 410- and 660-kilometer discontinuities. Science 285, 1545–1548.

Shim, S.-H., Duffy, T.S., Shen, G., 2001. The post-spinel transformation in Mg$_2$SiO$_4$ and its relation to the 660-km seismic discontinuity. Nature 411, 571–574.

Shim, S.-H., Duffy, T.S., Takemura, K., 2002. Equation of state of gold and its application to the phase boundaries near 660 km depth in Earth's mantle. Earth Planet. Sci. Lett. 203, 729–739.

Sinogeikin, S.V., Bass, J.D., Kavner, A., Jeanloz, R., 1997. Elasticity of natural majorite and ringwoodite from the Catherwood meteorite. Geophys. Res. Lett. 24, 3265–3268.

Sinogeikin, S.V., Katsura, T., Bass, J.D., 1998. Sound velocities and elastic properties of Fe-bearing wadsleyite and ringwoodite. J. Geophys. Res. 103, 20819–20825.

Sinogeikin, S.V., Bass, J.D., Katsura, T., 2001. Single-crystal elasticity of γ-(Mg$_{0.91}$Fe$_{0.09}$)$_2$SiO$_4$ to high pressures and to high temperatures. Geophys. Res. Lett. 28, 4335–4338.

Speziale, S., Duffy, T.S., 2002. Single-crystal elastic constants of fluorite (CaF$_2$) to 9.3 GPa. Phys. Chem. Miner. 29, 465–472.

Suzuki, I., 1975. Thermal expansion of periclase and olivine and their anharmonic properties. J. Phys. Earth 23, 145–159.

Suzuki, I., Ohtani, E., Kumazawa, M., 1979. Thermal expansion of γ-Mg$_2$SiO$_4$. J. Phys. Earth 27, 53–61.

Takahashi, E., Shimazaki, T., Tsuzaki, Y., Yoshida, H., 1993. Melting study of a peridotite KLB-1 to 6.5 GPa, and the origin of basaltic magmas. Phil. Trans. R. Soc. Lond. A 342, 105–120.

Utsumi, W., Funakoshi, K., Urakawa, S., Yamakata, M., Tsuji, K., Konishi, H., Shimomura, O., 1998. SPring-8 beamlines for high pressure science with multi-anvil apparatus. Rev. High Pres. Sci. Tech. 7, 1484–1486.

Vacher, P., Mocquet, A., Sotin, C., 1998. Computation of seismic profiles from mineral physics: the importance of the non-olivine components for explaining the 660 km depth discontinuity. Phys. Earth Planet. Int. 106, 275–298.

Wang, W., Takahashi, E., 2000. Subsolidus and melting experiments of K-doped peridotite KLB-1 to 27 GPa: its geophysical and geochemical implications. J. Geophys. Res. 105, 2855–2868.

Wang, Y., Weidner, D.J., Guyot, F., 1996. Thermal equation of state of $CaSiO_3$ perovskite. J. Geophys. Res. 101, 661–672.

Wang, Y., Weidner, D.J., Zhang, J., Gwanmesia, G.D., Liebermann, R.C., 1998. Thermal equation of state of garnets along the pyrope-majorite join. Phys. Earth Planet. Int. 105, 59–71.

Watanabe, H., 1987. Physico-chemical properties of olivine and spinel solid solutions in the system Mg_2SiO_4-Fe_2SiO_4. In: Manghnani, M.H., Syono, Y. (Eds.), High-Pressure Research in Mineral Physics. Am. Geophys. Union, Washington, DC, pp. 275–278.

Weidner, D.J., Wang, Y., 1998. Chemical- and Clapeyron-induced buoyancy at the 660 km discontinuity. J. Geophys. Res. 103, 7431–7441.

Weidner, D.J., Sawamoto, H., Sasaki, S., Kumazawa, M., 1984. Single-crystal elastic properties of the spinel phase of Mg_2SiO_4. J. Geophys. Res. 89, 7852–7860.

Weidner, D.J., Wang, Y., Vaughan, M.T., 1994. Yield strength at high pressure and temperature. Geophys. Res. Lett. 21, 753–756.

Wood, B.J., 2000. Phase transformations and partitioning relations in peridotite under lower mantle conditions. Earth Planet. Sci. Lett. 174, 341–354.

Yamanaka, T., 1986. Crystal structures of Ni_2SiO_4 and Fe_2SiO_4 as a function of temperature and heating duration. Phys. Chem. Miner. 13, 227–232.

Zerr, A., Reichmann, H., Euler, H., Boehler, R., 1993. Hydrostatic compression of γ-$(Mg_{0.6}, Fe_{0.4})_2SiO_4$ to 50.0 GPa. Phys. Chem. Miner. 19, 507–509.

Available online at www.sciencedirect.com

SCIENCE DIRECT

Physics of the Earth and Planetary Interiors 143–144 (2004) 47–56

PHYSICS OF THE EARTH AND PLANETARY INTERIORS

www.elsevier.com/locate/pepi

Sound velocities and elastic constants of iron-bearing hydrous ringwoodite

Steven D. Jacobsen[a,*], Joseph R. Smyth[b], Hartmut Spetzler[b,c], Christopher M. Holl[b], Daniel J. Frost[a]

[a] Bayerisches Geoinstitut, Universität Bayreuth, Bayreuth D-95440, Germany
[b] Department of Geological Sciences, University of Colorado, Boulder, CO 80309-0399, USA
[c] Cooperative Institute for Research in Environmental Sciences, University of Colorado, Boulder, CO 80309-0216, USA

Received 4 December 2002; received in revised form 8 May 2003; accepted 24 July 2003

Abstract

The sound velocities and single-crystal elastic constants of Fo_{89} hydrous ringwoodite (γ-$Mg_{1.7}Fe_{0.22}H_{0.16}SiO_4$) containing ~10,000 ppm by weight (1 wt.%) H_2O have been determined from seven separate pure-mode travel-time measurements using gigahertz ultrasonic interferometry. The experiments feature a new Yttrium–Aluminum–Garnet (YAG) P-to-S conversion shear buffer rod (BR), capable of producing pure ultrasonic shear waves with known polarization in the region of 0.5–2.0 GHz. To our knowledge they are the first such single-crystal ultrasonic measurements on a high-pressure phase recovered from the multi-anvil press. The cubic single-crystal elastic constants of hydrous Fo_{89} ringwoodite at ambient conditions are (in GPa): $c_{11} = 298 \pm 13$, $c_{44} = 112 \pm 6$, and $c_{12} = 115 \pm 6$. Hydration of $Fo_{\sim 90}$ ringwoodite to 1 wt.% H_2O reduces the adiabatic bulk ($K_{0S} = 176 \pm 7$ GPa) and shear ($G = 103 \pm 5$ GPa) moduli by about 6 and 13%, roughly equivalent to raising the temperature at room pressure by 600 and 1000 °C, respectively. Assuming a linear trend with hydration, we calculate that P- and S-wave velocities are reduced by about 40 m/s for every 1000 ppm weight (0.1 wt.%) H_2O added to $Fo_{\sim 90}$ ringwoodite. P- and S-wave velocities of the lower transition zone in PREM are consistent with a hydrated ringwoodite-rich composition.

Keywords: Hydrous ringwoodite; Elastic properties; Hydrogen in the transition zone; Gigahertz ultrasonic interferometry

1. Introduction

The hydrogen content of Earth's interior is arguably one of the most poorly constrained geochemical variables, and the total H_2O content of the planet is probably unknown within an order of magnitude. Whether or not there are even small amounts of H (100's ppm) as hydroxyl in the dominant phases of the upper mantle and transition zone has major implications for our understanding of Earth's evolution and chemical budget (e.g. Dixon et al., 2002; Drake and Righter, 2002). Just one-tenth of 1 wt.% (1000 ppm by weight) H_2O in wadsleyite β-$(Mg,Fe)_2SiO_4$ and ringwoodite γ-$(Mg,Fe)_2SiO_4$ in the volume of Earth's transition zone between 410 and 660 km depth is equivalent to about 800 m of liquid water across the surface of the Earth (Smyth, 1994).

Natural olivine from mantle-derived xenoliths normally contains anywhere from zero to about 140 ppm by weight H_2O (Bell and Rossman, 1992; Ingrin and

* Corresponding author. Current address: Geophysical Laboratory Carnegie Institution of Washington, 5251 Broad Branch Rd., NW, Washington, DC 20015, USA. Tel.: +(202) 478-7975; fax: +(202)-478-8901.
E-mail address: s.jacobsen@gl.ciw.edu (S.D. Jacobsen).

0031-9201/$ – see front matter © 2004 Elsevier B.V. All rights reserved.
doi:10.1016/j.pepi.2003.07.019

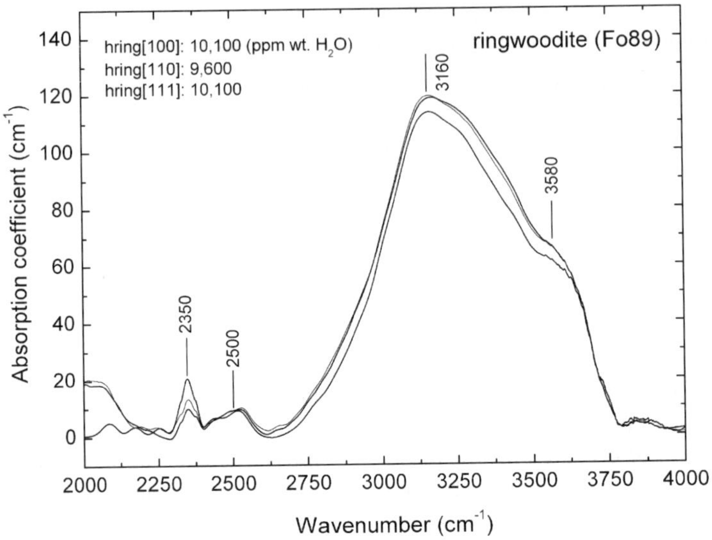

Fig. 1. Unpolarized FTIR spectra from the three Fo_{89} hydrous ringwoodite crystals used in the ultrasonic study.

those at 3120 and 3695 cm^{-1}, respectively, observed by Bolfan-Casanova et al. (2000) in pure-Mg hydrous ringwoodite, with shifts due possibly to the iron-content of our samples. For lack of a specific (ringwoodite) calibration, we used the method of Paterson (1982) to estimate the average water content of the samples, resulting in about 9960 ppm weight (1 wt.%) H_2O, or about 161,000 $H/10^6$ Si.

Combining the microprobe, Mössbauer, and FTIR data results in the formula, $Mg_{1.7}Fe_{0.22}H_{0.16}SiO_4$, with about 10% of the iron the ferric state. Therefore, the calculated density of this hydrous Fo_{89} ringwoodite is 3.651 ± 0.005 g/cm^3.

2.2. Ultrasonics

We recently produced pure ultrasonic shear waves over a broad bandwidth (0.5–2.0 GHz) by P-to-S conversion inside a single-crystal acoustic buffer rod (Jacobsen et al., 2002). The resulting high-frequency shear is of suitable wavelength (~1–10 μm) and has a known polarization direction for elasticity measurements on single-crystal samples as thin as 20–30 μm in length, on the bench or in the diamond anvil cell. The new technique takes advantage of an orthogonal P-to-S conversion inside an oriented YAG single-crystal buffer rod, illustrated in Fig. 2a. Incident P-waves are produced by a ZnO sputtered thin-film transducer. Using Snell's Law, $V_P^{[100]}$, and $V_S^{[100]}$ to calculate the angle incidence (i), we force the conversion to occur at exactly 90°. This allows for both the incident P-wave and reflected S-waves to travel in pure-mode directions of the cubic single-crystal buffer rod, as well as allowing for the symmetric return of the wave (labeled PSSP in Fig. 2) for detection at the source transducer. In this way, delay-line acoustics are carried out in the usual way (Spetzler et al., 1993).

Ultrasonic travel-times are measured using interferometric methods, described in more detail by Spetzler et al. (1993). Briefly, a pulse generator is used to gate a continuous RF signal, resulting short (~100 ns) phase-coherent tone bursts that are introduced to the sample via a buffer rod (Fig. 2). The impedance contrast at the BR-sample interface produces a reflection, which is interfered with an echo produced at the far end of the sample, either by lengthening the width of the input signal, or by carefully delaying two tone bursts such that the first sample echo overlaps with the second BR echo (see Jacobsen et al., 2002 for a standard, thick-sample echo train). In this case, the samples are so thin (and associated travel times so short, 10–20 ns) that the sample echoes cannot be separately resolved in time (see inset of Fig. 3a). Therefore, we measure simply the amplitude of the interfered tone burst where there is only first-order interference

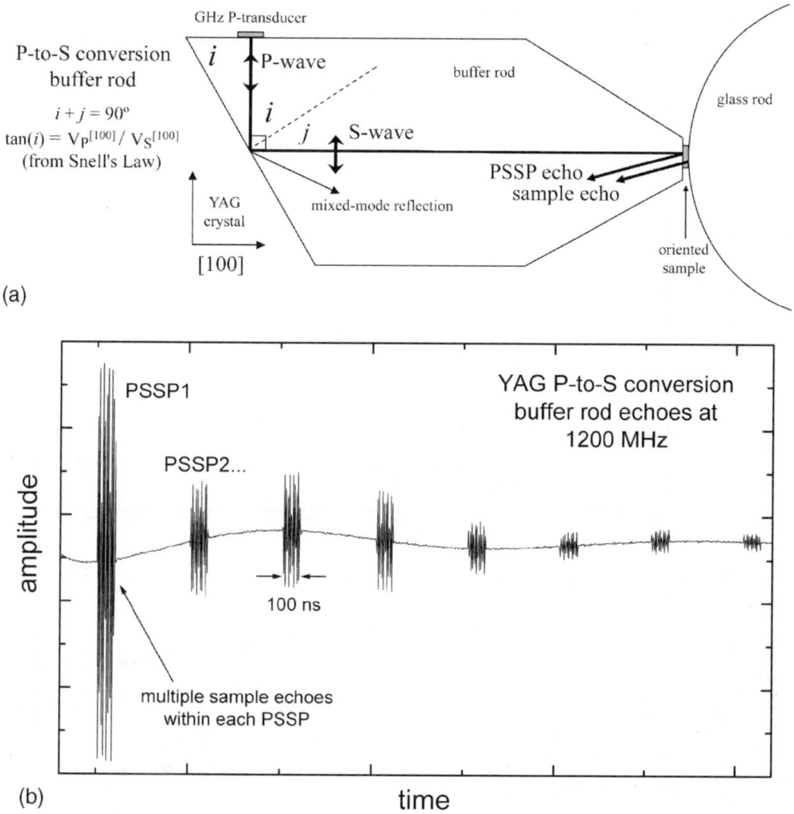

Fig. 2. The new P-to-S conversion ultrasonic shear-wave buffer rod. (a) In schematic, showing the GHz P-wave source transducer, the orthogonal P-to-S conversion, and back reflections from the tip of the buffer rod (PSSP) and from the far end of the sample. The buffer rod is sharpened for placement in the cone of a diamond anvil cell. For bench-top measurements, the sample is gently weighted by an optically aligned glass rod (shown at right). (b) A pulse echo train from the buffer rod resulting from a single sine wave input signal. The echo labeled PSSP1 results from one round trip through the buffer rod. Many PSSP echoes are observed, illustrating the high efficiency of the conversion and precision of the facet orientations. The round trip within the sample is considerably shorter than the width of each burst and can, thus, not be resolved separately. Each echo is about 100 ns in duration and the frequency of the sine wave within them is 1200 MHz. Therefore, there are over 100 cycles per burst. The structure within the bursts (observed here) results from aliasing inherent in the display of a digitizing oscilloscope.

between the sample and the first buffer rod echo. The resulting interference pattern contains maxima (and minima) produced each time there is an integer number (or half-integer for minima) of wavelengths in the round-trip (Fig. 3a). Acoustic coupling between the sample and the buffer rod is achieved by either direct (dry) contact or with wet and cured glue thin-films, but in all cases a weight of a few grams is applied through an optically aligned glass rod (shown in Fig. 2a).

Fig. 3a shows an example of a shear-wave interference pattern measured from 600 to 1200 MHz for the [1 1 0] ringwoodite sample, vertically polarized (T_v), i.e. [0 0 1]. The complex BR and overlapping sample echo are also shown, indicating at what positions the uninterferred and interfered amplitudes were measured. The resulting travel-time data calculated from each f_{max} and f_{min} are shown in Fig. 3b. In this way, we determined seven different pure-mode phase velocities from compression wave travel times in the [1 0 0], [1 1 1], and [1 1 0] directions, and shear-wave travel times in the [1 0 0] and [1 1 1] directions, plus both shear polarizations (T_h and T_v) for the [1 1 0] propagation direction as given by Brugger (1965), and listed in Table 2.

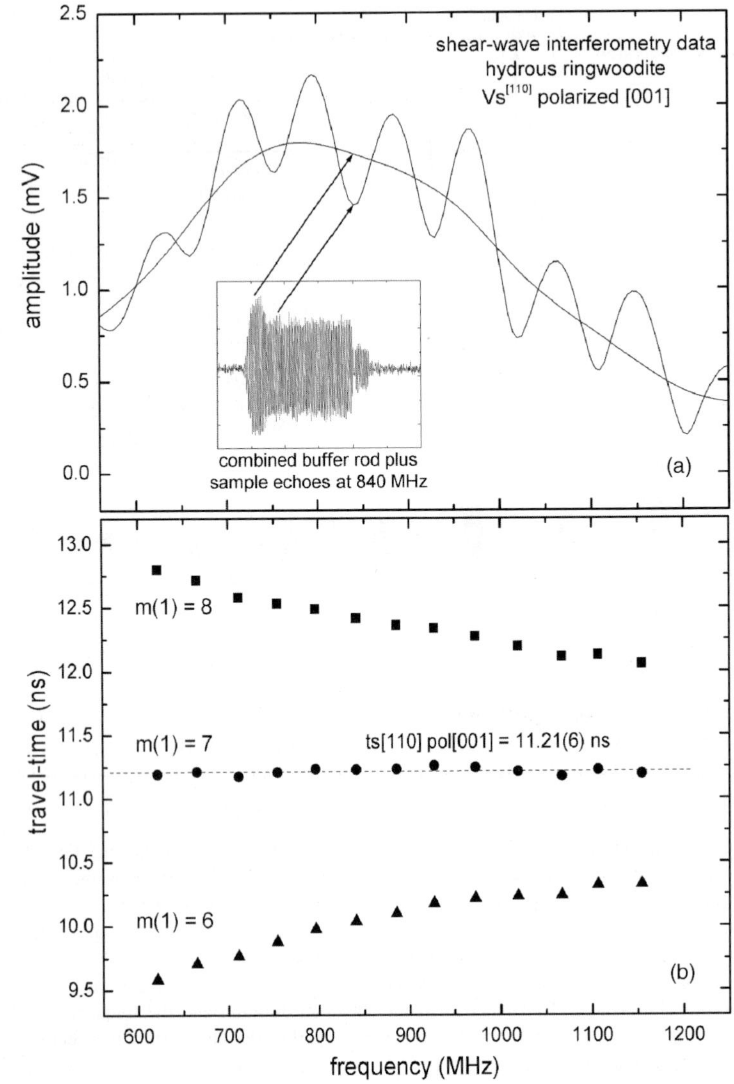

Fig. 3. Micro-crystal shear wave ultrasonic interferometry data from the $V_S^{[1\,1\,0]}$ vertically polarized (T_v) [0 0 1] experiment. (a) Amplitude vs. frequency data from two positions within a 100 ns buffer rod echo (PSSP), shown as inset in the time domain. The smooth curve was measured at a position in the buffer rod echo (PSSP) before the first sample echo arrives. The curve showing ~100 MHz modulation results from interference between the first sample echo and the buffer rod echo. Arrows indicate exactly where in the complex echo each curve was measured. (b) Travel-time data are calculated from the frequency of each maxima and minima, for several values of $m(1)$, the number of whole wavelengths in the round trip ($2l$) for the first maxima. The consequence of choosing the wrong m is illustrated in the travel-time data calculated for $m(1) \pm$ the correct value, in this case, 7.

3. Results and discussion

Measured P- and S-wave travel times were used to calculate ultrasonic wave velocities (Table 2) and the single-crystal elastic constants of Fo_{89} hydrous ringwoodite, compared to other ringwoodite studies in Table 3. Substitution of plane wave solutions into the equations of motion (e.g. Brugger, 1965) gives the c_{11} elastic constant directly from $\rho(V_P^{[1\,0\,0]})^2 = 298 \pm 13$ GPa, where the propagation of error is calculated using standard Gaussian methods. Although travel-times can be routinely measured on

Table 3
Single-crystal and aggregate elastic properties of ringwoodite and hydrous ringwoodite

	Ringwoodite		Hydrous ringwoodite	
	γ-Mg_2SiO_4	γ-$Mg_{1.82}Fe_{0.18}SiO_4$	γ-$Mg_{1.89}Si_{0.97}H_{0.33}O_4$	γ-$Mg_{1.7}Fe_{0.22}SiH_{0.16}O_4$
ρ (g/cm^3)	3.559	3.701 ± 5	3.469	3.651 ± 5
c_{11} (GPa)	327 ± 4	329 ± 3	281 ± 6	298 ± 13
c_{44}	126 ± 2	130 ± 1	117 ± 4	112 ± 6
c_{12}	112 ± 3	118 ± 2	92 ± 5	115 ± 6
A[a]	0.113	0.149	0.160	0.138
K_S (GPa)	184 ± 2	188 ± 3	155 ± 4	176 ± 7
G (GPa)	119 ± 2	119 ± 2	107 ± 3	103 ± 5
V_P (km/s)	9.79	9.69 ± 0.02	9.27	9.26 ± 0.20
V_S (km/s)	5.77	5.68 ± 0.01	5.56	5.31 ± 0.13
Reference	Weidner et al. (1984)	Sinogeikin et al. (1998)	Inoue et al. (1998)	This study

[a] Anisotropy factor $A = [(2c_{44} + c_{12})/c_{11}] - 1$.

micro-crystals to a few parts in 10^3, the rather large uncertainty in the elastic constants are dominated by the much larger uncertainty in the thickness measurement, in this case ±1 μm. There are two ways of obtaining c_{44} directly; from $V_S^{[1\,0\,0]}$ we obtain $c_{44} = 112 \pm 5$ GPa, and from $V_S^{[1\,1\,0]}$ polarized [0 0 1] we obtain $c_{44} = 112 \pm 7$ GPa, in perfect agreement. Given c_{11} and c_{44}, we have four remaining equations for obtaining c_{12} (listed in Table 2), resulting in 115, 118, 110, and 116 GPa, also showing good internal consistency. We use for all subsequent calculations the average value, $c_{12} = 115 \pm 6$ GPa. Despite the rather large reported uncertainties resulting from correct propagation of uncertainty in the length, we note that the internal consistency among the seven independent equations determined from three different crystals is much better. For example the standard deviation in the mean for c_{12} using the four independent equations is only 3 GPa, or about half the reported uncertainty.

Another way of approaching this calculation is to solve all seven equations simultaneously by least-squares. Using a standard over-determined least-squares inversion with data matrix $[\rho V^2]$, we obtain $c_{11} = 297$, $c_{44} = 112$, and $c_{12} = 116$ GPa, in excellent agreement with the (preferred) direct approach.

The single-crystal elastic constants of various ringwoodite samples with and without Fe and water are listed in Table 3. We obtain for the (VRH averaged) aggregate bulk and shear moduli, $K_{0S} = 176 \pm 7$ GPa and $G_0 = 103 \pm 5$ GPa, corresponding to isotropic bulk sound velocities $V_P = 9.26 \pm 0.20$ and $V_S = 5.31 \pm 0.13$ km/s. The bulk sound velocities for hydrous Fo_{90} ringwoodite are about 4.4 and 6.5% slower (for P and S, respectively) than for dry Fo_{90} ringwoodite (Sinogeikin et al., 1998). Thus, hydration of ringwoodite appears to have a larger effect on S-wave velocities than P, so hydration should increase the ratio V_P/V_S in the Earth, an important seismic parameter. We calculate $V_P/V_S = 1.74 \pm 0.04$ for our hydrous sample, compared to $V_P/V_S = 1.706 \pm 0.002$ for dry Fo_{90} ringwoodite (Sinogeikin et al., 1998). Although the ratio V_P/V_S for our sample has a rather large uncertainty due to propagation of the uncertainty in length (around 2%), we note that uncertainties in the ratio V_P/V_S for the various measured crystallographic directions are much lower (around 0.5%) because $V_P/V_S = (2l/t_P)/(2l/t_S) = t_S/t_P$, where l is the sample length and t_S and t_P are the P-wave and S-wave travel-times, respectively. Using travel times and associated uncertainties found in Table 2, we calculate $(V_P/V_S)^{[1\,0\,0]} = 1.64 \pm 0.01$, $(V_P/V_S)^{[1\,1\,1]} = 1.83 \pm 0.01$, $(V_P/V_S)^{[1\,1\,0]} = 1.68 \pm 0.01$ for shear vertically polarized (T_v) and $(V_P/V_S)^{[1\,1\,0]}$ 1.86 ± 0.01 for shear horizontally polarized (T_h).

Compared to dry iron-bearing ringwoodite (Sinogeikin et al., 1998), the c_{11} elastic constant of hydrous $Fo_{\sim 90}$ ringwoodite is reduced from 329 to 298 GPa (or 9.4%), and the c_{44} elastic constant is reduced from 130 to 112 GPa (or 13.8%). In contrast, the c_{12} elastic constant is statistically identical to that measured for nominally dry ringwoodite with or without iron. Consequently, the anisotropy factor

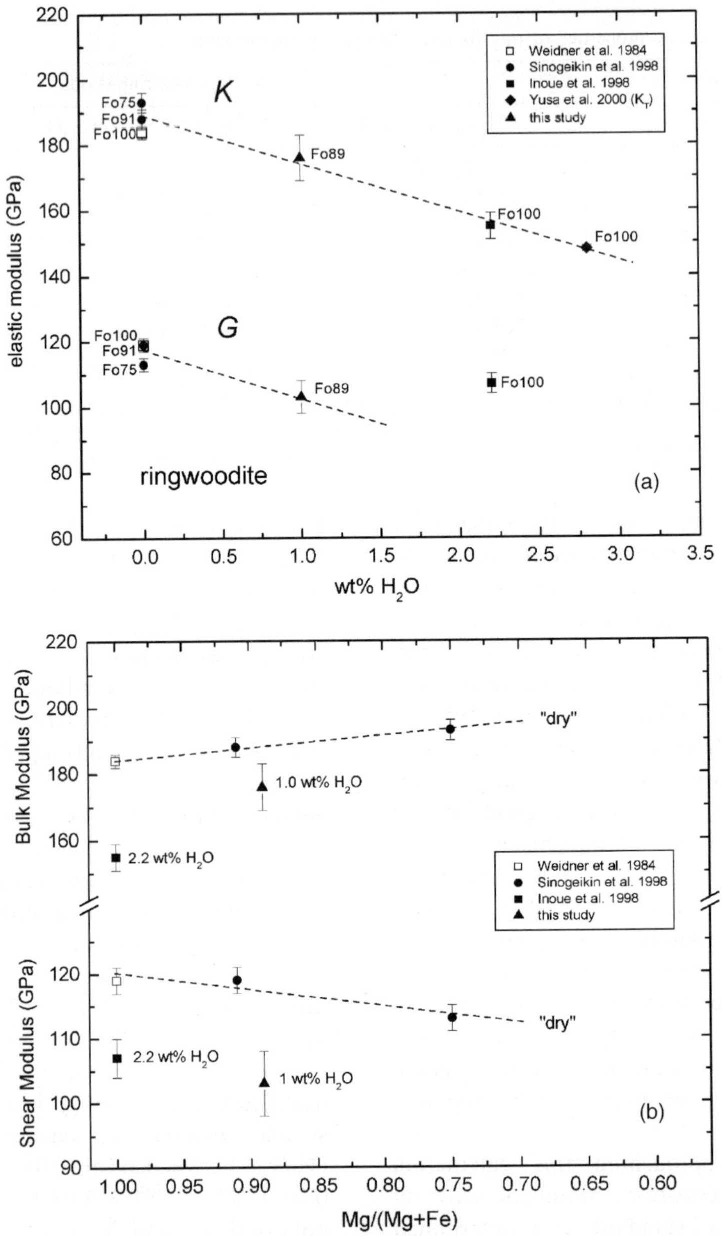

Fig. 4. Variation of the bulk and shear modulus of ringwoodite with (a) water and (b) iron content.

$A = [(2c_{44} + c_{12})/c_{11}] - 1$ is reduced from 0.149 to 0.138 or about 7.4%.

Hydration of $Fo_{\sim 90}$ ringwoodite to 1 wt.% H_2O reduces the bulk ($K_{0S} = 176 \pm 7$ GPa) and shear ($G_0 = 103 \pm 5$ GPa) moduli by about 6 and 13%, respectively, compared to nominally dry iron-bearing ringwoodite (Fig. 4). Our value for the adiabatic bulk modulus is in relatively good agreement with the isothermal value, $K_{0T} = 169 \pm 3$ GPa, measured in a companion paper of this volume (Smyth et al., 2003b). Using the temperature derivatives of the moduli $\partial K_S/\partial T = -0.021$ and $\partial G/\partial T = -0.016$ measured for dry Fo_{91} ringwoodite (Sinogeikin et al., 2001), we estimate that at room pressure hydrating $Fo_{\sim 90}$ ringwoodite to 1 wt.% H_2O is roughly equivalent to raising the temperature by about 600 and 1000 °C. From the adiabatic moduli and density, we calculate bulk sound velocities (in m/s): $V_P = 9260$ and $V_S = 5310$. Anchoring these data to the dry Fo_{91} ringwoodite data from Sinogeikin et al. (1998), we obtain a rough linear estimate for the elastic moduli and velocities $Fo_{\sim 90}$ ringwoodite as a function of water content:

$$K_{S0} = 188 - 0.0012C_{H_2O}\ (\text{GPa})$$

$$G_0 = 119 - 0.0016C_{H_2O}\ (\text{GPa})$$

$$V_P = 9690 - 0.042C_{H_2O}\ (\text{m/s})$$

$$V_S = 5680 - 0.036C_{H_2O}\ (\text{m/s})$$

where C_{H_2O} is the water concentration in ppm H_2O by weight (1000 ppm weight = 0.1 formula wt.% H_2O).

Hydration of $Fo_{\sim 90}$ ringwoodite to 1 wt.% H_2O reduces its density from 3.70 to 3.65 g/cm^3, or about 1.4%. The bulk P- and S-wave velocities are reduced by 4.3 and 6.3%, respectively, thus, reducing the impedance (ρV) by 5.8% for P-waves and 7.6% for S-waves. For comparison, the effect of adding 10% Fe to dry ringwoodite is to increase the impedance by 3.2% for P-waves and 2.4% for shear waves. Without adiabatic moduli for wet wadsleyite, we cannot speculate on the potential impedance contrast between wet–wet versus dry–dry wadsleyite–ringwoodite mantle discontinuities. However, between this study and existing isothermal compression data for hydrous Mg-wadsleyite (Yusa and Inoue, 1997), hydrous Mg-ringwoodite (Yusa et al., 2000), and Fo_{89} hydrous ringwoodite (Smyth et al., 2003b), a larger picture is emerging on the relative magnitude of effects that hydration has on the elasticity of high-pressure silicates. If the effects of relatively small amounts of water (on the order of one-tenth of 1 wt.%) are still large enough at the *P–T* conditions of Earth's transition zone to affect seismic velocities more than the likely range of Fe/Mg ratios or temperature gradients away from subduction zones, then water should be considered a very likely compositional variable able to produce lateral velocity variations and topography on the discontinuities as observed by seismic tomography.

Duffy and Anderson (1989) used the relatively low P- and S-velocities of the transition zone in PREM (Dziewonski and Anderson, 1981) to infer that the mantle in that region could not contain more than about 50% of an olivine-stoichiomery phase, i.e. ringwoodite or wadsleyite. Consideration of the effects of hydration of ringwoodite on velocities removes this constraint so that the P- and S-velocities of the lower transition zone in PREM are consistent with a hydrated ringwoodite-rich composition. Further, it is interesting to note that the current results indicate a strong effect of hydration on V_P/V_S, raising the value from 1.706 for dry to 1.744 for this hydrated ringwoodite at ambient conditions. If these results are confirmed in future studies at elevated pressures and temperatures, it may be possible to distinguish hydration from high temperatures in the transition zone of the mantle.

Acknowledgements

We wish to thank H. Schulze, K. Müller, H. Ohlmeyer, and G. Herrmannsdörfer, S. Linhardt, K. Hain, and K. Klasinski for their technical assistance in construction of the P-to-S conversion buffer rods, high-frequency electronics, and data reduction routines. G. Bromiley assisted with the FTIR measurements. The helpful comments of A. Yoneda and an anonymous reviewer are appreciated. This study was made possible by an Alexander von Humboldt Fellowship to S.D.J. and by the Bayerisches Geoinstitut Visitor Program to J.R.S.

References

Bell, D.R., Rossman, G.R., 1992. Water in Earth's mantle: the role of nominally anhydrous minerals. Science 255, 1391–1397.

Bolfan-Casanova, N., Keppler, H., Rubie, D.C., 2000. Water partitioning between nominally anhydrous minerals in the MgO–SiO_2–H_2O system up to 24 GPa: implications for the distribution of water in the Earth's mantle. Earth Planet. Sci. Lett. 182, 209–221.

Brugger, K., 1965. Pure modes for elastic waves in crystals. J. Appl. Phys. 36, 759–768.

Chen, J., Inoue, T., Yurimoto, H., Weidner, D.J., 2002. Effect of water on the olivine–wadsleyite phase boundary in the $(Mg,Fe)_2SiO_4$ system. Geophys. Res. Lett. 29, 1875 doi: 10.1029/2001GL014429.

Dixon, J.E., Leist, L., Langmuir, C., Schilling, J.-G., 2002. Recycled dehydrated lithosphere observed in plume-influenced mid-ocean-ridge basalt. Nature 420, 385–389.

Drake, M.J., Righter, K., 2002. Determining the composition of the Earth. Nature 416, 39–44.

Duffy, T.S., Anderson, D.L., 1989. Seismic velocities in mantle minerals and the mineralogy of the upper mantle. J. Geophys. Res. 94, 1895–1912.

Dziewonski, A., Anderson, D.L., 1981. Preliminary reference Earth model. Phys. Earth Planet. Int. 25, 297–356.

Ingrin, J., Skogby, H., 2000. Hydrogen in nominally anhydrous upper-mantle minerals: concentration levels and implications. Eur. J. Mineral. 12, 543–570.

Inoue, T., Weidner, D.J., Northrup, P.A., Parise, J.B., 1998. Elastic properties of hydrous ringwoodite (γ-phase) in Mg_2SiO_4. Earth Planet. Sci. Lett. 160, 107–113.

Jackson, J.M., Sinogeikin, S.V., Bass, J.D., 2000. Sound velocities and elastic properties of γ-Mg_2SiO_4 to 873 K by Brillouin spectroscopy. Am. Mineral. 85, 296–303.

Jacobsen, S.D., Reichmann, H.J., Spetzler, H.A., Mackwell, S.J., Smyth, J.R., Angel, R.J., McCammon, C.A., 2002. Structure and elasticity of single-crystal (Mg,Fe)O and a new method of generating shear waves for gigahertz ultrasonic interferometry. J. Geophys. Res. 107 (B2), 2037, doi: 10.1029/2001JB000490.

Kohlstedt, D.L., Keppler, H., Rubie, D.C., 1996. The solubility of water in α, β, and γ phases of $(Mg,Fe)_2SiO_4$. Contrib. Mineral. Petrol. 123, 345–357.

Meng, Y., Fei, Y., Weidner, D.J., Gwanmesia, G.D., Hu, J., 1994. Hydrostatic compression of γ-Mg2SiO_4 to mantle pressures and 700 K: thermal equation of state and related thermoelastic properties. Phys. Chem. Mineral. 21, 407–412.

Paterson, M.S., 1982. The determination of hydroxyl by infrared absorption in quartz, lilicate glasses and similar materials. Bull. Mineral. 105, 20–29.

Sasaki, S., Prewitt, C.T., Sato, Y., 1982. Single-crystal X-ray study of γ-Mg_2SiO_4. J. Geophys. Res. 87, 7829–7832.

Sinogeikin, S.V., Katsura, T., Bass, J.D., 1998. Sound velocities and elastic properties of Fe-bearing wadsleyite and ringwoodite. J. Geophys. Res. 103, 20819–20825.

Sinogeikin, S.V., Bass, J.D., Katsura, T., 2001. Single-crystal elasticity of γ-$(Mg_{0.91}Fe_{0.09})_2$ SiO_4 to high pressures and to high temperatures. Geophys. Res. Lett. 28, 4335–4338.

Smyth, J.R., 1994. A crystallographic model for hydrous wadsleyite (β-Mg_2SiO_4): an ocean in the Earth's interior? Am. Mineral. 79, 1021–1024.

Smyth, J.R., Frost, D.J., 2002. The effect of water on the 410-km discontinuity: an experimental study. Geophys. Res. Lett. 29, 1486 doi: 10.1029/2001GL014418.

Smyth, J.R., Holl, C.M., Frost, D.J., Jacobsen, S.D., Langenhorst, F., McCammon, C.A., 2003a. Structural systematics of hydrous ringwoodite and water in Earth's interior. Am. Mineral. 88, 1402–1407.

Smyth, J.R., Holl, C.M., Frost, D.J., Jacobsen, S.D., 2003b. High pressure crystal chemistry of hydrous ringwoodite and water in the Earth's interior, this volume.

Spetzler, H.A., Chen, G., Whitehead, S., Getting, I.C., 1993. A new ultrasonic interferometer for the determination of equation of state parameters of sub-millimeter single crystals. Pure Appl. Geophys. 141, 341–377.

Weidner, D.J., Sawamoto, H., Sasaki, S., Kumazawa, M., 1984. Elasticity of Mg_2SiO_4 spinel. J. Geophys. Res. 89, 7852–7860.

Yusa, H., Inoue, T., 1997. Compressibility of hydrous wadsleyite (β-phase) in Mg_2SiO_4 by high pressure X-ray diffraction. Geophys. Res. Lett. 24, 1831–1834.

Yusa, H., Inoue, T., Ohishi, Y., 2000. Isothermal compressibility of hydrous ringwoodite and its relation to the mantle discontinuities. Geophys. Res. Lett. 27, 413–416.

Available online at www.sciencedirect.com

SCIENCE DIRECT®

ELSEVIER

Physics of the Earth and Planetary Interiors 143–144 (2004) 57–80

PHYSICS OF THE EARTH AND PLANETARY INTERIORS

www.elsevier.com/locate/pepi

Thermal equation of state of akimotoite $MgSiO_3$ and effects of the akimotoite–garnet transformation on seismic structure near the 660 km discontinuity

Yanbin Wang [a,*], Takeyuki Uchida [a], Jianzhong Zhang [b], Mark L. Rivers [a], Stephen R. Sutton [a]

[a] *Consortium for Advanced Radiation Sources, The University of Chicago, 5640 S. Ellis Ave., Chicago, IL 60637, USA*

[b] *Department of Geosciences, State University of New York at Stony Brook, Stony Brook, NY 11974, USA*

Received 18 February 2003; received in revised form 14 May 2003; accepted 29 August 2003

Abstract

Pressure (P)–volume (V)–temperature (T) relations of akimotoite (ilmenite structured $MgSiO_3$) have been measured up to 19 GPa and 1373 K in a T-Cup multi-anvil apparatus. Polycrystalline specimens synthesized at 21 GPa and 1723 K were loaded in a specially designed cell assembly, together with Au or NaCl pressure standard, for synchrotron X-ray diffraction studies. Two runs were carried out with multiple temperature excursions at various pressures both inside and outside the akimotoite stability field. At conditions outside the stability field, temperature was kept below 873 K, as previous test runs indicated that akimotoite back-transformed to enstatite at about 900 K. In order to minimize effects of non-hydrostatic stress on volume and pressure determination, P–V–T data were collected on cooling cycles from the peak temperature. A total of 134 volume measurements were obtained. Adopting an ambient isothermal bulk modulus (210 GPa) obtained previously using the Brillouin scattering technique, analysis using the third-order Birch–Murnaghan equation of state yielded pressure derivative $K'_{T0} = 5.6(8)$, with zero-pressure volume of 263.9(2) $Å^3$ for $MgSiO_3$ akimotoite. The high-temperature data were fit to several thermal equations of state to cross-examine consistency among the thermoelastic parameters obtained. Using the data with NaCl as the pressure standard combined with previous zero-pressure thermal expansion data of [Phys. Chem. Min. 16 (1988) 239], we obtained $(\partial K_T/\partial T)_P = -0.040(1)$ GPa K^{-1}, with a zero-pressure thermal expansion $\alpha_0 = 2.41(19) \times 10^{-5}$ K^{-1}. The results were applied to model seismic profiles of the transition zone, in order to examine the effects of akimotoite-forming transitions on lateral velocity variations. In depths where no phase transformations and chemical heterogeneities are present, hotter and colder areas relative to a reference geotherm are characterized by similar $\mathrm{dln}V_S/\mathrm{dln}V_P$ and $\mathrm{dln}\rho/\mathrm{dln}V_S$ ratios, which are material properties. The akimotoite–garnet phase transformation, however, skews the ratios because akimotoite has much higher acoustic velocities than garnet. Thus if the mantle geotherm is close to the transition temperature, anomalous apparent $\mathrm{dln}V_S/\mathrm{dln}V_P$ and $\mathrm{dln}\rho/\mathrm{dln}V_S$ ratios may be observed, which are no longer material properties. Aluminum content plays an important role in defining the akimotoite–garnet phase boundary relative to the geotherm and lateral Al variations may cause the transition boundary to fluctuate with depth. Therefore one must be cautious in interpreting seismic tomography observations in the transition zone.

Keywords: Akimotoite; Equation of state; Synchrotron X-ray diffraction; Seismic tomography; Mantle composition models

* Corresponding author. Tel.: +1-630-252-0425; fax: +1-630-252-0436.
E-mail address: wang@cars.uchicago.edu (Y. Wang).

0031-9201/$ – see front matter
doi:10.1016/j.pepi.2003.08.007

1. Introduction

The ilmenite form of $MgSiO_3$, known as akimotoite, is stable at pressures between ca. 18 and 25 GPa and temperatures from ~1100 to 1900 K (Fig. 1) in the pure $MgSiO_3$ system (Gasparik, 1990). The presence of iron is reported to lower slightly the transition pressure from akimotoite to perovskite (Ito and Takahashi, 1989). Aluminum plays an important role in the akimotoite stability field in that pyrope makes complete solid solution with the $MgSiO_3$ end-member majorite, enlarging the garnet field at the expense of that of akimotoite both in pressure and temperature. Fig. 2, after Gasparik (1996), illustrates this effect. Ten mole percent of pyrope in majorite can enlarge the garnet field and push the garnet–akimotoite phase boundary downward by about 100 K and expand the pressure range by more than 1 GPa. Thus, depending on temperature and Al content, akimotoite may be present in certain areas of the transition zone but absent in others. This will cause significant lateral variations in seismic velocities and density and complicate interpretation of seismic tomography observations, as seismically observed lateral variations will not merely reflect temperature derivatives even in a chemically homogeneous mantle. Clearly, information on thermoelastic properties of akimotoite is essential in modeling the effects of the akimotoite–garnet transformation on seismic profiles near the 660 km discontinuity. To this end, we report here the thermal equation of state of akimotoite obtained using synchrotron X-ray diffraction. Based on these new data, we then model the effects of the garnet–akimotoite transition on the seismic structure in the transition zone.

2. Experimental procedure

Pure $MgSiO_3$ akimotoite samples were synthesized using a glass starting material in the 2000 ton uniaxial split-sphere apparatus (USSA-2000) at the Stony Brook High Pressure Laboratory at 21 GPa and 1773 K for 1 h. Initially in this project, a specially designed *P–T* path was followed in order to obtain well-sintered polycrystalline samples free of residual stress. This technique has proven to improve data quality significantly in *P–V–T* measurements (e.g., see Wang et al., 1998). However, all samples recovered were flaky and easily disintegrated during handling, likely due to the narrow metastability field of akimotoite at modest temperatures, and decompression at modestly elevated temperatures caused akimotoite to partially transform

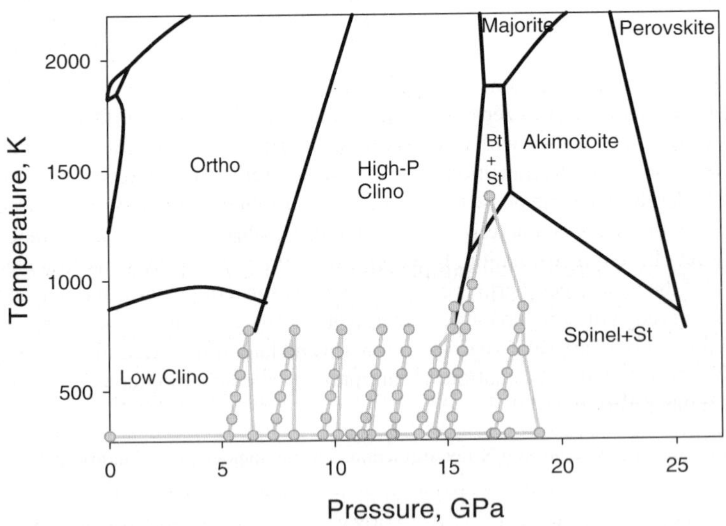

Fig. 1. Phase diagram of the $MgSiO_3$ system, determined from quench experiments (Gasparik, 1990). Superimposed is a *P–T* path in the *P–V–T* measurements (T0150). Due to the high temperature nature of this phase, most *P–V–T* data are collected outside the stability field. Abbreviations: Bt: wadsleyite; St: stishovite; clino: clinoenstatite.

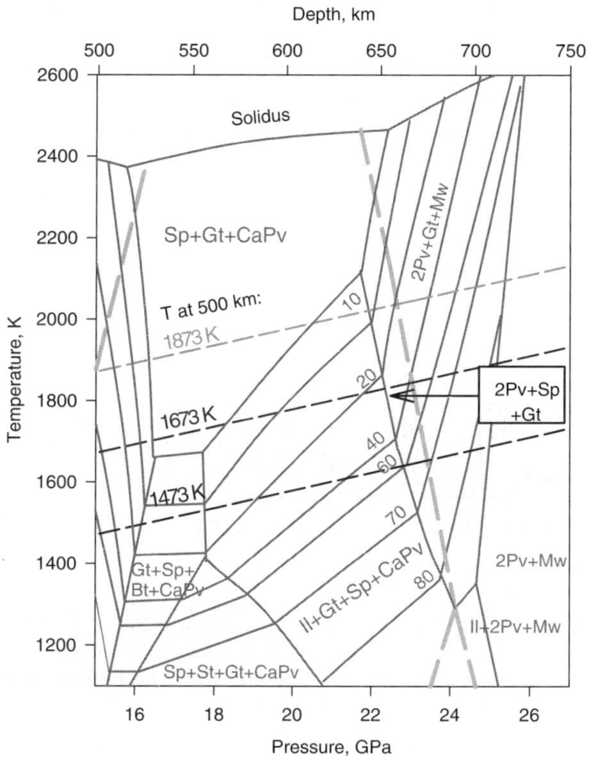

Fig. 2. Stability field of akimotoite and effects of Al on the garnet–akimotoite transformation (after Gasparik, 1996). Phases are identified with abbreviations: Il: akimotite, Gt: garnet; Sp: ringwoodite; Bt: wadsleyite; St: stishovite; CaPv: $CaSiO_3$ perovskite; MgPv: $MgSiO_3$ perovskite; 2Pv: CaPv and MgPv; Mw: magnesiowustite. Thin tie lines indicate pyrope content in garnet (e.g., 40 means 40 mol% pyrope). Longdashed lines indicate boundries between wadsleyite and ringwoodite at low pressures, and ringwoodite and MgPv plus Mw at high pressures. Three mantle geotherms are examined in forward modeling (see text), all with a thermal gradient of $0.35\,K\,km^{-1}$.

into lower pressure phases. Therefore, all samples used in this study were recovered by quenching, rather than decompression at some elevated temperatures.

P–V–T equation of state measurements were carried out using the GeoSoilEnviroCARS large-volume press (LVP) facility at the Advanced Photon Source. All experiments were carried out using the 250 t large-volume press (Wang et al., 2000) in the Bending Magnet station 13-BM-D (see Rivers et al., 1998). We used the T-cup apparatus with 10 mm WC cubic anvils to generate high pressures. Anvil truncation edge length was 2.0 mm and the octahedral sample assembly had an edge length of 7.0 mm. Details of the cell assembly are given in (Uchida et al., 2001). Energy-dispersive diffraction technique was used at a fixed 2θ around 6°. Akimotoite samples were cut into small disks, 1 mm in diameter and 0.5 mm thick, and loaded in one half of the sample chamber, with the other half filled with a homogeneous mixture of pressure standards, either NaCl + BN or Au + MgO, to minimize grain growth at high temperatures. The tungsten–rhenium thermocouple junction was located at the sample-standard interface so that temperature readings represented true conditions in the sample and in the pressure standard. No attempt was made to correct for pressure effects on the thermocouple emf.

Carefully designed *P–T* paths were followed in order to obtain volume measurements at a wide range of pressure and temperature conditions and to minimize effects of non-hydrostatic stress (Wang et al., 1998). Because of limited stability field of akimotoite, temperature must be kept either low enough to keep the sample metastable, or high enough to be inside the stability field. Several initial attempts at measuring equation of state failed because the sample transformed into the low-pressure enstatite phase at temperatures as low as 900 K below 19 GPa. Therefore, when pressures were outside the stability field, the volume measurements were restricted to below 873 K.

Fig. 1 shows a typical *P–T* path in these experiments. Temperature was raised slowly at a constant load (150 t) to 973 K and then quickly increased to 1373 K within a few seconds. This procedure prevented the possibility of breakdown of akimotoite to the two-phase assemblage of ringwoodite and stishovite. As temperature was increased, pressure decreased (see *P–T* path in Fig. 1), so that at 1373 K, a pressure of 17.1 GPa was measured at the same ram load. Akimotoite remained stable at this condition, indicating that it was within the stability field. The diffraction pattern obtained under this condition is shown in Fig. 3. Diffraction lines are well defined and the relative intensities indicate that no significant grain growth occurred. Temperature was then decreased rapidly to 973 K. Volumes of the sample and the pressure standards were determined at 100 K intervals until room temperature was reached, at which point the load was decreased to the next pressure level and the heating cycle repeated. About 8–10 cycles were conducted in each run. Diffraction spectra taken

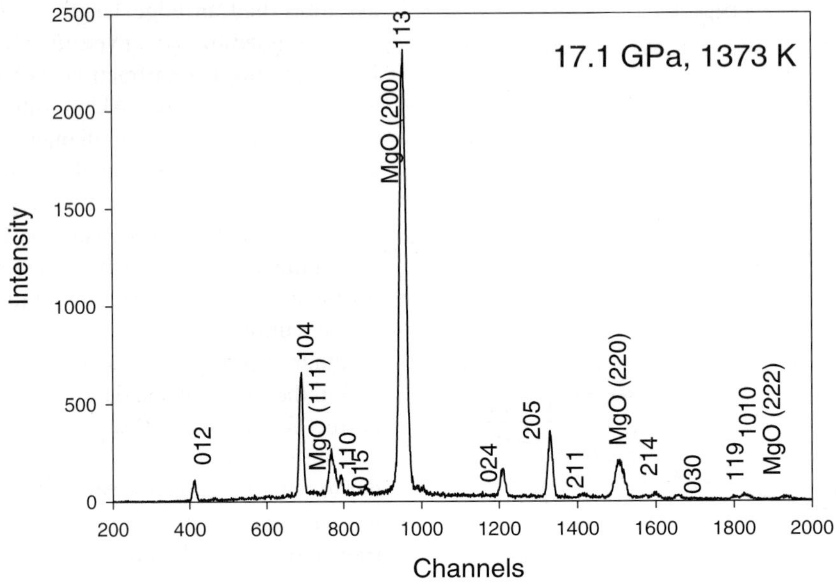

Fig. 3. An X-ray diffraction pattern of akimotoite obtained at 17.1 GPa and 1373 K. All visible peaks are labeled. Peaks of the MgO sample container are also identified. Although the MgO (2 0 0) reflection interferes with the akimotorite (1 1 3), up to 10 sample peaks are resolved. Also note that even at this peak temperature, no significant preferred grain growth has occurred; relative intensities remain unchanged throughout the experiment.

after the samples were recovered at ambient conditions showed no evidence of back transformation and the cell volumes were consistent with those obtained prior to the *P–V–T* cycle. Throughout, volume data collection was performed only on cooling cycles from the maximum temperature, in order to reduce effects of non-hydrostatic stress (Wang et al., 1998).

The low stability pressure for akimotoite at 1373 K is inconsistent with the phase relation data determined in previous quench experiments (e.g., Gasparik, 1990), but in general agreement with in situ X-ray diffraction studies (Irifune et al., 1998). The inconsistency between quench studies using fixed points and in situ experiments with diffraction pressure standards is currently a debated issue. Discrepancies also exist among various diffraction pressure scales, even when the same standard material is used. For example, Brown (1999) proposed to revise the Decker's NaCl pressure scale. There is more disagreement as to which Au pressure scale is the most appropriate (Jamieson et al., 1982; Heinz and Jeanloz, 1984; Anderson et al., 1989; Ono et al., 2001; Shim et al., 2002). At this stage, we prefer to adopt the most widely used equations of state (i.e., Decker, 1971 for NaCl and Anderson et al., 1989 for gold) so that our results can be directly compared with the *P–V–T* datasets accumulated over the past few decades. Realizing the potential inconsistency among different pressure standards, our data were analyzed independently using NaCl and Au.

We also monitored the width of the diffraction peaks and used this parameter as an indication of stress relaxation. Even after heating to 1373 K, diffraction peaks of akimotoite were still about 10% broader then those observed at ambient conditions. The widths of the NaCl and Au diffraction peaks, on the other hand, became comparable to their ambient counterparts. The low temperature range made *P–V–T* data quality control difficult, as unwanted non-hydrostatic stresses could not be completely removed at these temperatures.

3. Results

3.1. *P–V–T* equation of state

A total of about 180 data points were obtained from two runs, one using NaCl, the other Au as the

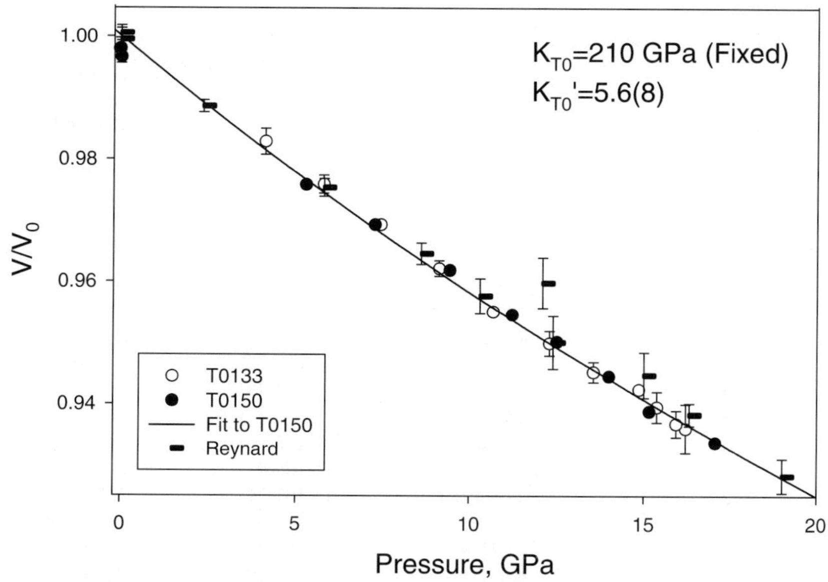

Fig. 4. Room temperature compression data of akimotoite. Open circles data from run T0133 (Au pressure marker) and solid circles from run T0150 (NaCl pressure marker). Solid curve is the third-order Birch–Murnaghan equation of state fit to data from T0150 only, which adopts the bulk modulus from the Brillouin scattering data (Weidner and Ito, 1985). Superimposed are Reynard et al. (1996) data, which, above 12 GPa, show larger errors and overestimate in the cell volume, indicative of the presence of non-hydrostatic stress.

pressure marker. Of those, 134 were obtained after stress relaxation by cooling from high temperatures and therefore could be used in extracting equation of state parameters (Tables 1 and 2). Fig. 4 summarizes the volume data at room temperature. The third-order Birch–Murnaghan equation of state was used to fit these data, with the ambient volume, V_0, treated as a free parameter. Due to the limited pressure range in our dataset, K_{T0} and K'_{T0} could not be simultaneously determined with satisfying accuracy, because of the well-known trade-off among the three parameters K_{T0}, K'_{T0} and V_0. We therefore adopted the adiabatic bulk modulus (K_{S0}) determined from Brillouin scattering measurements (Weidner and Ito, 1985) with an isothermal correction. With K_{T0} fixed at 210 GPa, the fits yielded $K'_{T0} = 4.8(5)$ and $V_0 = 264.2(2)$ Å^3 for run T0133 (Au scale) and $K'_{T0} = 5.6(8)$ and $V_0 = 263.9(2)$ Å^3 for run T0150 (NaCl scale). Although the two fits agreed with each other within the uncertainties, there was some indication that data obtained using Au as the pressure marker were affected by nonhydrostatic stress, especially at low temperatures (see Section 4 on *c/a* ratio that follows). As gold has an exceptionally large ratio of K_T/μ, (μ being the shear modulus), non-hydrostatic stress could significantly affect pressure determinations, as discussed by Weidner et al. (1992). Because of these concerns, we believed that results based on data from T0150 (NaCl) were more reliable and used only those data in subsequent analyses. The K'_{T0} value from run T0150 is also in excellent agreement with a previous roomtemperature diamond-anvil cell determination [5.6(10)] reported by Reynard et al. (1996). Our refined V_0 value, on the other hand, is about 0.6% higher than their results, presumably due to systematic errors in the two experiments.

To analyze the high temperature data using the high temperature Birch–Murnaghan equation of state, we adopted the zero-pressure thermal expansion data from Ashida et al. (1988). A fit to their data yielded a constant ambient thermal expansion $\alpha_0 = 2.48(16) \times 10^{-5}\,K^{-1}$. No information on temperature dependence of thermal expansion could be extracted from their data from 298 to 876 K, because of the relatively large scatter (Fig. 5). High-temperature Birch–Murnaghan fit to data from T0133 (Au scale) using fixed $K_{T0} =$ 210 GPa and $K'_{T0} = 4.8(5)$ gave an unusually large negative $(\partial K_{T0}/\partial T)_P = -0.0553(24)$ GPa K^{-1} and

Table 1
P–V–T data obtained from run T0133. Pressures determined from the Au equation of state (Anderson et al., 1989)

Temperature (K)	Pressure (GPa)	*a* (Å)	*c* (Å)	*V* (Å^3)	*c*/*a*
298	0.00	4.746(5)	13.511(21)	263.59(45)	2.8468(44)
573	17.85	4.651(2)	13.148(12)	246.29(17)	2.8270(26)
773	17.48	4.669(9)	13.136(33)	248.02(57)	2.8132(70)
673	17.06	4.663(12)	13.154(41)	247.67(79)	2.8211(88)
573	16.77	4.663(10)	13.142(34)	247.51(65)	2.8180(72)
473	16.54	4.665(14)	13.120(47)	247.24(92)	2.8124(102)
373	16.18	4.665(18)	13.115(58)	247.18(113)	2.8112(125)
298	16.20	4.665(16)	13.119(54)	247.21(105)	2.8125(117)
573	16.55	4.669(6)	13.135(18)	247.97(36)	2.8133(40)
773	17.05	4.669(11)	13.128(36)	247.88(70)	2.8114(78)
673	16.73	4.668(15)	13.128(49)	247.74(95)	2.8125(105)
573	16.53	4.668(14)	13.131(47)	247.74(91)	2.8132(101)
473	16.26	4.667(13)	13.122(43)	247.51(84)	2.8119(93)
373	16.05	4.669(9)	13.10(29)	247.50(57)	2.8077(62)
298	15.92	4.669(9)	13.107(29)	247.41(57)	2.8074(62)
573	16.08	4.700(11)	13.142(36)	248.17(69)	2.8143(76)
773	16.57	4.671(8)	13.159(26)	248.62(50)	2.8174(56)
673	16.31	4.671(9)	13.158(29)	248.67(56)	2.8166(62)
573	16.06	4.671(10)	13.151(33)	248.484(63)	2.8156(70)
473	15.87	4.670(10)	13.149(34)	248.35(66)	2.8157(73)
373	15.51	4.671(8)	13.140(26)	248.27(50)	2.8131(55)
298	15.37	4.668(10)	13.149(34)	248.13(66)	2.8169(73)
573	15.47	4.679(3)	13.142(11)	249.18(17)	2.8087(24)
773	15.94	4.678(1)	13.161(4)	249.39(6)	2.8135(8)
673	15.66	4.675(4)	13.165(16)	249.18(24)	2.8161(34)
573	15.40	4.6738(4)	13.160(15)	248.96(23)	2.8157(33)
473	15.14	4.674(4)	13.149(14)	248.82(22)	2.8130(31)
373	14.9	4.676(2)	13.143(7)	248.85(10)	2.8107(15)
298	14.85	4.678(1)	13.140(3)	248.89(5)	2.8091(7)
573	14.22	4.678(3)	13.201(11)	250.21(21)	2.8219(24)
773	14.62	4.683(3)	13.195(11)	250.60(22)	2.8175(24)
673	14.40	4.677(6)	13.199(27)	250.06(51)	2.8219(57)
573	14.09	4.676(6)	13.190(28)	249.76(53)	2.8208(59)
473	13.84	4.677(6)	13.182(25)	249.71(49)	2.8186(54)
373	13.64	4.678(5)	13.179(21)	249.75(41)	2.8173(46)
298	13.54	4.676(5)	13.184(23)	249.64(44)	2.8196(49)
573	12.98	4.683(5)	13.226(21)	251.22(41)	2.8240(45)
773	13.33	4.688(5)	13.231(20)	251.78(39)	2.8226(43)
673	13.09	4.688(4)	13.217(19)	251.57(36)	2.8194(40)
573	12.87	4.687(4)	13.212(15)	251.34(30)	2.8190(33)
473	12.64	4.687(5)	13.216(20)	251.42(38)	2.8197(42)
373	12.44	4.683(6)	13.216(28)	251.00(54)	2.8222(60)
298	12.29	4.682(6)	13.214(23)	250.88(53)	2.8221(50)
573	11.76	4.687(2)	13.257(9)	252.19(17)	2.8287(19)
773	12.07	4.691(1)	13.272(2)	252.89(5)	2.8295(5)
673	11.79	4.695(1)	13.257(5)	253.03(9)	2.8238(10)
573	11.46	4.691(1)	13.256(3)	252.66(6)	2.8258(7)
473	11.20	4.689(1)	13.256(2)	252.46(5)	2.8269(5)
373	10.92	4.689(3)	13.250(9)	252.33(18)	2.8257(20)
298	10.67	4.689(1)	13.249(5)	252.24(10)	2.8258(11)
573	10.22	4.699(2)	13.295(9)	254.21(17)	2.8295(18)
773	10.58	4.702(3)	13.311(12)	254.84(28)	2.8312(26)
673	10.30	4.701(4)	13.306(17)	254.63(33)	2.8306(37)

Table 1 (*Continued*)

Temperature (K)	Pressure (GPa)	a (Å)	c (Å)	V (Å^3)	c/a
573	9.99	4.699(4)	13.304(18)	254.47(36)	2.8310(39)
473	9.67	4.698(5)	13.299(20)	254.22(39)	2.8308(43)
373	9.36	4.695(6)	13.305(27)	253.93(52)	2.8341(58)
298	9.12	4.698(4)	13.295(17)	254.11(34)	2.8300(37)
573	8.61	4.708(6)	13.333(20)	255.92(38)	2.8319(42)
773	8.95	4.712(11)	13.351(40)	256.667(66)	2.8336(86)
673	8.71	4.715(1)	13.326(5)	256.61(9)	2.8260(11)
573	8.32	4.7156(4)	13.315(2)	256.412(4)	2.8236(4)
473	8.03	4.716(1)	13.308(6)	256.37(10)	2.8215(12)
373	7.74	4.713(1)	13.310(5)	256.02(9)	2.8241(11)
298	7.44	4.712(1)	13.312(6)	255.99(11)	2.8249(13)
573	7.03	4.720(3)	13.364(15)	257.87(28)	2.8311(31)
773	7.50	4.718(7)	13.400(20)	258.32(48)	2.8401(43)
673	7.10	4.721(4)	13.394(16)	258.50(31)	2.8372(34)
573	6.76	4.723(2)	13.378(9)	258.42(17)	2.8325(19)
473	6.41	4.720(4)	13.378(17)	258.12(34)	2.8344(37)
373	6.07	4.719(4)	13.373(16)	257.99(32)	2.8338(35)
298	5.79	4.716(5)	13.379(20)	257.73(39)	2.8468(44)
573	5.29	4.733(5)	13.406(20)	260.13(46)	2.8270(26)
673	5.42	4.732(6)	13.433(24)	260.34(55)	2.8132(70)
573	5.08	4.7311(6)	13.421(23)	260.16(52)	2.8211(88)
473	4.70	4.732(6)	13.409(22)	260.07(51)	2.8181(72)
373	4.38	4.730(7)	13.406(25)	259.73(58)	2.8124(102)
298	4.12	4.729(7)	13.402(25)	259.58(56)	2.8112(13)
298	0.00	4.744(11)	13.566(49)	264.46(94)	2.8125(12)

Only the quasi-hydrostatic data are given.

$V_0 = 264.26(6)$ Å^3, with $\alpha_0 = 2.48 \times 10^{-5}\,K^{-1}$ as zero-pressure constraint. On the other hand, a fit to the combined dataset of T0150 (NaCl scale) plus Ashida et al. data yielded $(\partial K_{T0}/\partial T)_P = -0.040(1)$ GPa K^{-1} and $\alpha_0 = 2.41(9) \times 10^{-5}\,K^{-1}$, with a fixed $K'_{T0} = 5.6(8)$. Fig. 6 shows the fit to the high temperature data in run T0150. In both fits we scaled the Ashida et al. data according our ambient cell volume to remove systematic errors.

We also examined the data using the approach detailed by Jackson and Rigden (1996), based on the Mie–Gruneisen–Debye (MGD) equation of state. The MGD equation of state fit (Fig. 7) to the *PVT* data using NaCl as the pressure standard, combined with Ashida et al.'s ambient pressure measurements (again scaled according to the fitted ambient volumes), yielded an ambient Gruneisen parameter of 1.54(5), with an rms misfit of 0.33 GPa, for a total of 71 data points.

The room pressure thermal expansivity extracted from the MGD fit increases with temperature from $1.5 \times 10^{-5}\,K^{-1}$ at 300 K to $2.8 \times 10^{-5}\,K^{-1}$ at 2000 K. Over the entire temperature range, the thermal expansion can be fit with the following functional form:

$$\alpha = a + bT - cT^{-2}$$

with $a = 2.047(5) \times 10^{-5}\,K^{-1}$, $b = 4.00(3) \times 10^{-9}\,K^{-2}$, and $c = 0.647(7)$. This fit (Fig. 5) is marginally consistent with the measurements. The extracted $(\partial K_{T0}/\partial T)_P$ values decrease slightly with temperature, from −0.020 at 300 K to −0.026 GP K^{-1} at 2000 K.

Based on their Raman spectroscopy data, Reynard and Rubie (1996) argued that $MgSiO_3$ akimotoite has a quasi-harmonic behavior. Thus above the Debye temperature, the temperature dependence in thermal expansion may be approximated as $(\partial\alpha/\partial T)_P = \delta_T\alpha^2$ (Anderson, 1995). Assuming that $\delta_T = K'_{T0} = 5.6$ (Anderson, 1995), we have, at ambient pressure, $(\partial\alpha/\partial T)_P = 3.3 \times 10^{-9}\,K^{-2}$. Assuming this temperature dependence in the room

Table 2
P–V–T data obtained from run T0150. Pressures are measured using the NaCl equation of state (Decker, 1971)

Temperature (K)	Pressure (GPa)	*a* (Å)	*c* (Å)	*V* (Å^3)	*c*/*a*
298	0	4.736(2)	13.550(13)	263.22(27)	2.8609(43)
673	18.336	4.663(3)	13.110(11)	246.89(18)	2.8114(42)
873	18.332	4.670(2)	13.128(7)	247.59(12)	2.8108(28)
773	18.159	4.666(1)	13.136(5)	247.72(10)	2.8150(17)
673	17.854	4.666(2)	13.129(6)	247.54(9)	2.8139(22)
573	17.604	4.660(3)	13.135(10)	247.00(16)	2.8187(38)
473	17.41	4.661(2)	13.122(6)	246.91(10)	2.8151(23)
373	17.196	4.658(2)	13.123(6)	246.60(10)	2.8171(24)
298	17.044	4.659(1)	13.118(5)	246.60(8)	2.8157(19)
1373	17.112	4.671(2)	13.233(6)	249.99(17)	2.8334(23)
973	16.251	4.6652(8)	13.228(2)	249.32(5)	2.8354(9)
873	16.082	4.6641(7)	13.224(3)	249.13(6)	2.8353(11)
773	15.911	4.662(2)	13.219(6)	248.85(9)	2.8352(22)
673	15.589	4.661(2)	13.218(6)	248.67(10)	2.8359(24)
573	15.452	4.660(1)	13.214(5)	248.47(8)	2.8359(18)
473	15.476	4.658(2)	13.214(6)	248.25(9)	2.8372(22)
373	15.286	4.6564(7)	13.211(3)	248.07(4)	2.8373(10)
298	15.154	4.656(1)	13.207(4)	247.95(6)	2.8365(15)
573	14.953	4.664(2)	13.231(6)	249.21(10)	2.8371(23)
873	15.421	4.669(2)	13.238(6)	249.89(9)	2.8355(22)
773	15.392	4.666(1)	13.237(5)	249.56(8)	2.8370(19)
573	14.199	4.666(2)	13.250(7)	249.86(11)	2.8394(26)
473	14.2	4.6662(8)	13.243(3)	249.70(4)	2.8380(11)
373	14.086	4.6660(4)	13.235(2)	249.54(3)	2.8365(6)
298	13.974	4.6639(9)	13.242(4)	249.46(7)	2.8393(15)
773	13.31	4.678(1)	13.281(4)	251.68(7)	2.8391(16)
673	13.174	4.6764(9)	13.28(3)	251.50(5)	2.8398(13)
573	12.986	4.673(1)	13.279(4)	251.13(7)	2.8416(17)
473	12.776	4.6742(7)	13.271(2)	251.10(4)	2.8393(9)
373	12.667	4.672(2)	13.270(6)	250.89(9)	2.8401(22)
298	12.504	4.675(1)	13.261(4)	250.94(7)	2.8368(17)
773	12.102	4.686(2)	13.311(6)	253.12(10)	2.8406(23)
673	11.992	4.685(1)	13.307(4)	252.92(6)	2.8405(14)
573	11.836	4.6816(6)	13.308(2)	252.60(4)	2.8426(9)
473	11.62	4.680(2)	13.303(6)	252.34(9)	2.8424(21)
373	11.375	4.678(1)	13.304(4)	252.13(6)	2.8440(14)
298	11.213	4.6790(5)	13.298(2)	252.12(3)	2.8420(7)
773	10.312	4.6928(3)	13.355(1)	254.70(2)	2.8458(4)
673	10.122	4.6914(5)	13.351(2)	254.47(3)	2.8458(7)
573	9.94	4.6902(3)	13.348(1)	254.29(2)	2.8459(4)
473	9.762	4.6878(2)	13.3465(9)	254.01(1)	2.8470(3)
373	9.561	4.6878(5)	13.338(2)	253.85(3)	2.8453(7)
298	9.425	4.69185(3)	13.3251(1)	254.03(3)	2.8401(1)
773	8.237	4.702(1)	13.406(5)	256.71(8)	2.8509(20)
673	7.98	4.702(1)	13.397(4)	256.56(7)	2.8489(16)
573	7.825	4.703(1)	13.391(4)	256.50(7)	2.8472(17)
473	7.588	4.6999(9)	13.391(3)	256.16(6)	2.8493(13)
373	7.411	4.6997(8)	13.385(3)	256.04(5)	2.8482(10)
298	7.273	4.7001(10)	13.380(4)	255.99(6)	2.8468(15)
773	6.213	4.7132(11)	13.438(4)	258.57(5)	2.8512(15)
673	5.974	4.7137(5)	13.438(2)	258.42(5)	2.8508(7)

Table 2 (*Continued*)

Temperature (K)	Pressure (GPa)	a (Å)	c (Å)	V (Å^3)	c/a
573	5.836	4.7113(6)	13.429(3)	258.13(5)	2.8503(9)
473	5.612	4.7100(8)	13.425(3)	257.93(6)	2.8503(11)
373	5.435	4.7092(10)	13.419(4)	257.72(8)	2.8496(14)
298	5.283	4.7089(2)	13.4202(8)	257.71(2)	2.8500(3)
298	−0.05	4.7347(15)	13.576(6)	263.57(9)	2.8673(22)

Only the quasi-hydrostatic data are given.

pressure thermal expansion, the Ashida et al. data can be refitted in the form $\alpha = a + bT$, with b fixed as $3.3 \times 10^{-9}\,K^{-2}$, yielding $a = 2.32(8) \times 10^{-5}\,K^{-1}$. This fit is indistinguishable from the constant thermal expansion fit (Fig. 5) over the temperature range where volume measurements are available. Therefore, the overall temperature dependence in the zero-pressure thermal expansion cannot be reliably constrained.

3.2. Anisotropy

Akimotoite is isostructural with ilmenite ($FeTiO_3$), a derivative of the NiAs structure in which the oxygen ions are approximately hexagonally close packed and the cations occupy octahedral sites. The SiO_6 and MgO_6 octahedral sites form layers that lie halfway between the close packed oxygen layers. The cations are completely ordered such that each layer contains all Mg or all Si, alternating so that the closest pairs of ions are always Mg–Si (Horiuchi et al., 1982). These octahedral layers have two thirds of the octahedral sites filled, and each filled octahedron shares a face with another in the layer above or below.

Our data show that linear compressibility along the c-axis is about 1.5 times that along the a-axis for $MgSiO_3$ akimotoite. This property can be understood

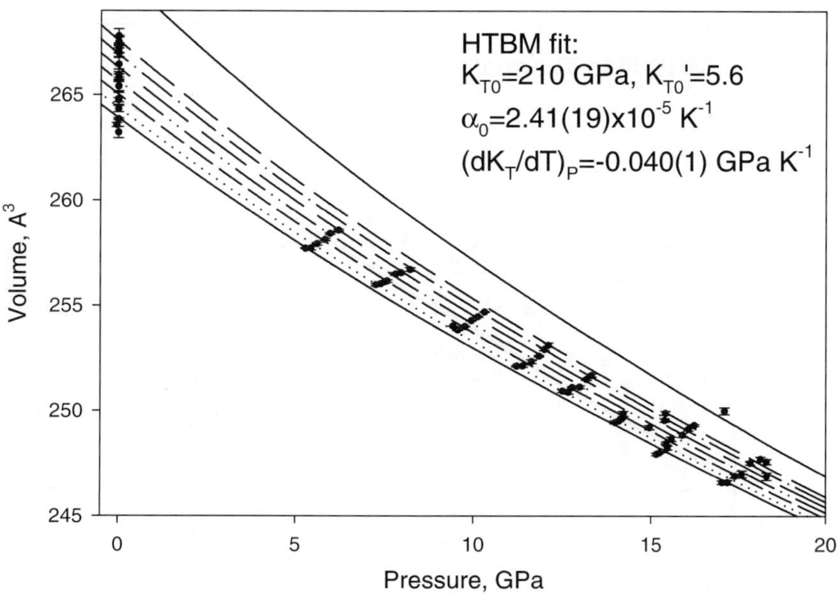

Fig. 5. *P–V–T* data based on NaCl pressure scale (T0150) for akimotoite with the high temperature Birch–Murnaghan (HTBM) equation of state fit. Curves are isothermal volumes at (from bottom upwards) 300, 373, 473, 573, 673, 773, 873, and 1373 K, respectively. Room-pressure volume data are from Ashida et al. (1988), scaled according to zero-pressure volume to remove systematic differences between the two dataset.

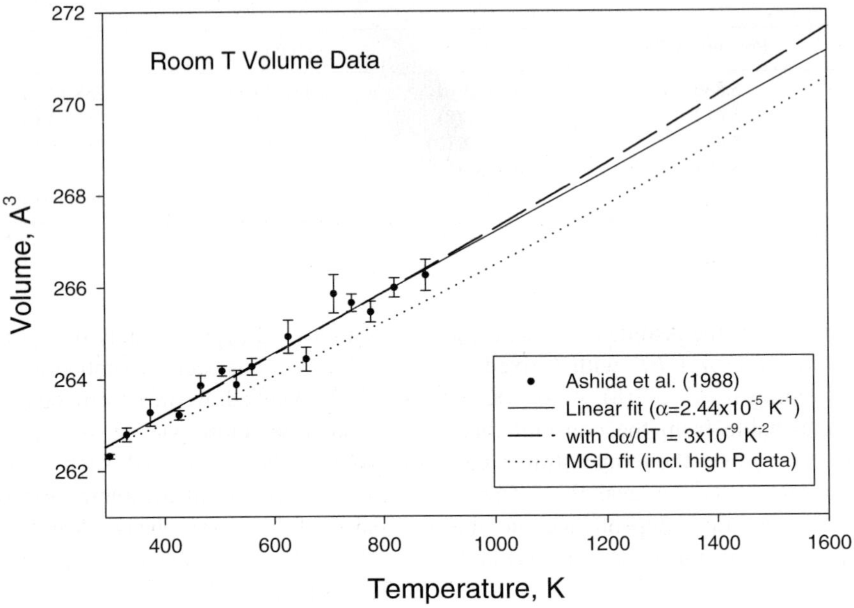

Fig. 6. Room-pressure volume data obtained by Ashida et al. (1988), with various fits as discussed in the text. Solid line is a linear fit to the data, yielding a constant thermal expansion of $2.44 \times 10^{-5}\,K^{-1}$; long dashed curve is a fit by imposing a constant temperature dependence in thermal expansion of $3 \times 10^{-9}\,K^{-2}$. Both fits are solely based on the roompressure data shown in the figure. Dotted curve is from the MGD equation of state fit, which is based on both the room-pressure data of Ashida et al. (1988) and our current *P–V–T* data (NaCl pressure scale only). For direct comparison, all fits are based on the ambient volume V_0 according to Ashida et al.'s data, which is slightly lower than value obtained in the current work.

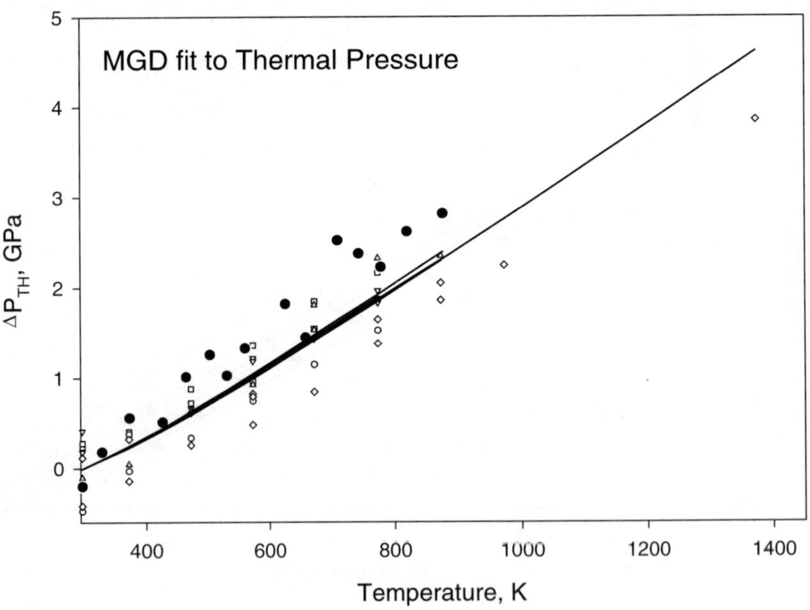

Fig. 7. Mie–Gruneisen–Debye equation of state fit to the combined data set of our *P–V–T* data (open symbols) and Ashida et al.'s room-pressure data (solid circles). The Ashida et al. data show somewhat higher temperature dependence in the thermal pressure. Solid curve is the MGD equation of state fit. No pressure dependence in thermal pressure can be reliably resolved.

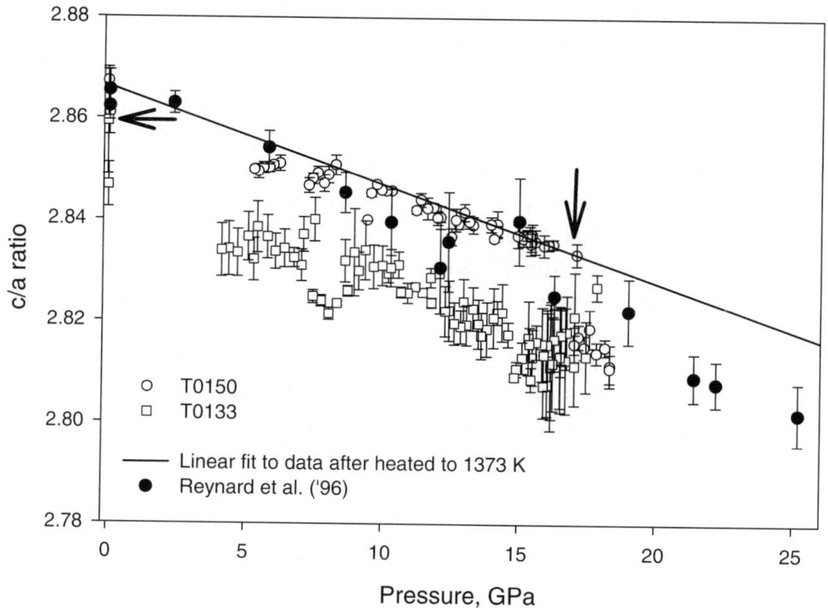

Fig. 8. c/a ratio as a function of pressure for akimotoite. Open symbols data from T0150 (circles; NaCl pressure scale) and T0133 (squares; Au pressure scale). Solid circles data from Reynard et al. (1996) in the diamond cell.

from the stacking of the MgO_6 and SiO_6 octahedral layers in the c direction. Because MgO_6 octahedra are more compressible than SiO_6, compressibility along the c-axis is dominated by that of MgO_6. On the other hand, compression in any direction perpendicular to the c-axis is dominated by that of SiO_6.

Such anisotropy results in a negative dependence on pressure for the c/a ratio, shown in Fig. 8, which includes data at all temperatures. No clear temperature dependence was observed. A discontinuity can be seen in the data from run T0150 near 17 GPa. The P–T paths for both runs T0133 and T0150 were such that data collection started from the highest pressure point, after heating to certain temperatures (Fig. 1). It is likely that at pressures above 17 GPa, non-hydrostatic stresses were not completely removed by heating to a maximum temperature of 873 K. Only when the sample was heated to 1373 K (datum point indicated by the vertical arrow), did we effectively remove most of the non-hydrostatic stress, hence the jump in the c/a ratio. This interpretation is supported by the c/a ratios observed in the dataset using Au as the pressure standard (run T0133): because the temperatures were limited to below 873 K throughout the run, no discontinuity in the c/a ratio was observed, and the high pressure c/a ratios were consistent with those "non-hydrostatic" points in run T0150 above 17 GPa (Fig. 8). Even at pressures below 17 GPa, the c/a ratios remained lower than the hydrostatic data in run T0150, until the pressure was fully released, where the c/a ratio returned to a value (horizontal arrow) consistent with the ambient measurement in run T0150. A linear regression on the T0150 dataset *after* heating to 1373 K (at all temperatures) results in the following pressure dependence:

$$c/a = 2.866 - 1.921 \times 10^{-3}\, P\,(\text{GPa}) \quad (r^2 = 0.84)$$

Clearly, some of the scatter is due to temperature. If only the room temperature data (after annealing) are used, the following very similar fit is obtained:

$$c/a = 2.863 - 1.917 \times 10^{-3}\, P\,(\text{GPa}) \quad (r^2 = 0.90)$$

Reynard et al. (1996) measured the room temperature compression of $MgSiO_3$ akimotoite to 25 GPa, using a diamond-anvil cell with pure H_2O as the pressure medium. These authors claimed that their experimental conditions were quasi-hydrostatic up to 15 GPa, above which the condition degraded considerably in the diamond-anvil cell, as indicated by the broadening of the ruby fluorescence lines. Indeed, their c/a ratios (solid circles, Fig. 8) show excellent agreement with our hydrostatic measurements below 15 GPa, above which the pressure dependence appears to change and becomes more consistent with our non-hydrostatic values. Their *P–V* data (Fig. 4) above ∼12 GPa also have a slight yet systematic shift upwards, with generally larger errors than the low pressure data. All these observations support the notion that, in our experiments, only the *c/a* ratios measured in run T0150, after heating to1373 K, reflect true pressure dependence. These observations also lend support to our approach of using *P–V–T* data in run T0150 alone to obtain thermoelastic parameters for $MgSiO_3$ akimotoite, as Run T0133 was limited to lower temperatures and likely affected by non-hydrostatic stresses.

4. Discussion and conclusions

Table 3 contrasts thermoelastic properties of garnet and akimotoite. $MgSiO_3$ akimotoite has much higher bulk and shear moduli than those of garnet by 31 and 47%, respectively. On the other hand, phase relation data (Fig. 2) show that garnet will be the dominate phase in the transition zone if (1) the mantle temperature is high, or (2) the mantle is rich in Al, or (3) the mantle has a high pyroxene-normative component (i.e., high Si/Mg ratio). Therefore the akimotoite–garnet transformation may have a significant seismic signature. The new data on akimotoite allow us to examine semiquantitatively the effects of the garnet–akimotoite phase transformation on velocity and density structures near the 660 km seismic discontinuity. Here we are not attempting to perform detailed analysis on radial (average) seismic profiles. Many papers have been devoted to this analysis with varying conclusions (e.g., Duffy and Anderson, 1989; Ita and Stixrude, 1992; Zhao et al., 1995; Vacher et al., 1998). Instead, we calculate directly the density and velocity profiles of certain mantle composition models based on mineral physics data alone. Our goal is to examine effects of akimotoite formation under the influence of three variables, mantle temperature, Si/Mg ratio, and Al content, in order to evaluate effects of temperature and Al content on lateral variations in seismic velocities and density.

Table 4
A comparison of thermoelastic properties between garnet and akimotoite

Property	Garnet	Akimotoite	Contrast (%)
ρ (g cc^{-1})	3.6	3.8	5.6
K_{T0} (GPa)	160	210	31.3
K'_{T0}	5	5.1	2.0
μ_0 (GPa)	90	132	46.7
μ'_0	2	1.5 (?)	

Volume-related thermoelastic properties are now available for essentially all major high pressure minerals relevant to the transition zone. The data essential for the modeling presented here are given by Table 1 in Weidner and Wang (1998), except for akimotoite, which are from this study. The phase relation information (Fig. 2) is from Gasparik (1996). Note that Gasparik's results were for the Fe-free, Ca–Mg–Al–Si–O system. Although effects of Fe on the olivine system are relatively well known (and slight modifications have been made in Fig. 2 to reflect these), they are poorly understood for the pyroxene system. Nonetheless, for our purposes, some semi-quantitative conclusions can be drawn without detailed phase equilibrium data. We use an approach similar to that described by Weidner and Wang (1998, 2000) to model the seismic profiles between 500 and 700 km depths.

We examine two commonly referred-to mantle composition models, the pyrolite and the C1 model (e.g., see McDonough and Sun, 1995), whose simplified molar and atomic elemental concentrations have been tabulated and given by Table 3 in Weidner and Wang (1998). We first calculate volume fractions of the phases present under transition zone pressures along the three geotherms shown in Fig. 2, all of which have the same adiabatic thermal gradient of 0.35 K km^{-1}. Throughout this discussion, the "cold" geotherm is anchored at 1473 K at 500 km depth, "warm" at 1673 K and "hot" at 1873 K. The resultant volume fractions of the relevant phases are shown in Fig. 9 for C1 and 10 for pyrolite, respectively.

Along the cold geotherm, the C1 mantle has as much as 40% akimotoite by volume, spreading over a 100 km depth interval between 550 and 650 km, where the garnet volume fraction tumbles from 65 to 25% (Fig. 9A). This is due to the fact that C1 is rich in Si; hence at low temperatures akimotoite dominates the system. In a warm mantle, the maximum akimotoite

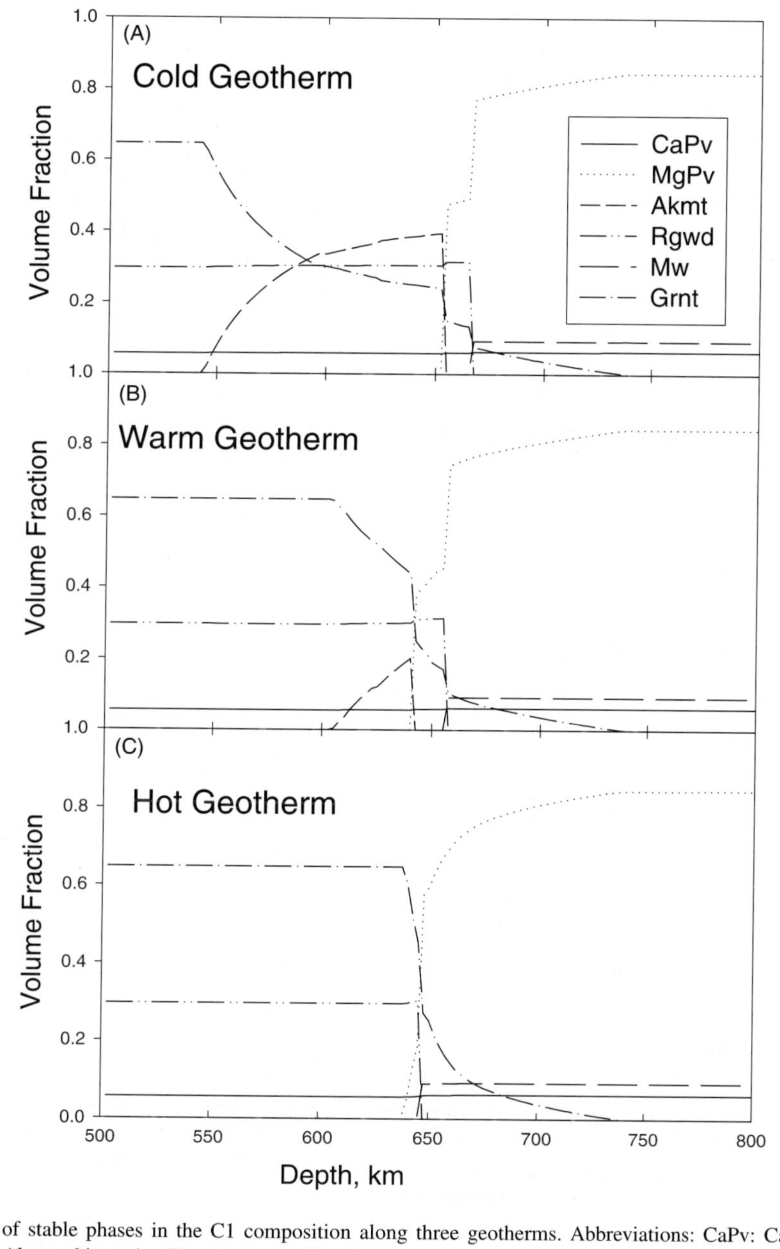

Fig. 9. Volume fractions of stable phases in the C1 composition along three geotherms. Abbreviations: CaPv: $CaSiO_3$ perovskite; MgPv: $(Mg,Fe)SiO_3$ perovskite; Akmt: akimotoite, Rgwd: ringwoodite, Mw: magnesiowustite, Grnt: garnet. Note varying amounts of akimotoite and associated garnet volume fractions along the cold and warm geotherms.

volume fraction is reduced to 20%, and restricted to a much narrower, ~30 km, depth interval (Fig. 9B). No akimotoite is present in a hot mantle (Fig. 9C).

For the pyrolite mantle, which has a lower Si/Mg ratio than that of C1, the maximum akimotoite volume is about 10% in the cold mantle, and spread over about 70 km depths (Fig. 10A). In both warm and hot pyrolite mantles, akimotoite is absent (Fig. 10B and C). Also note that garnet volume fractions increase with depth in the hot mantle (Fig. 10C), in order to satisfy the required equilibrium Al/Si ratios in perovskite and garnet. The large amount of $MgSiO_3$ contributed by

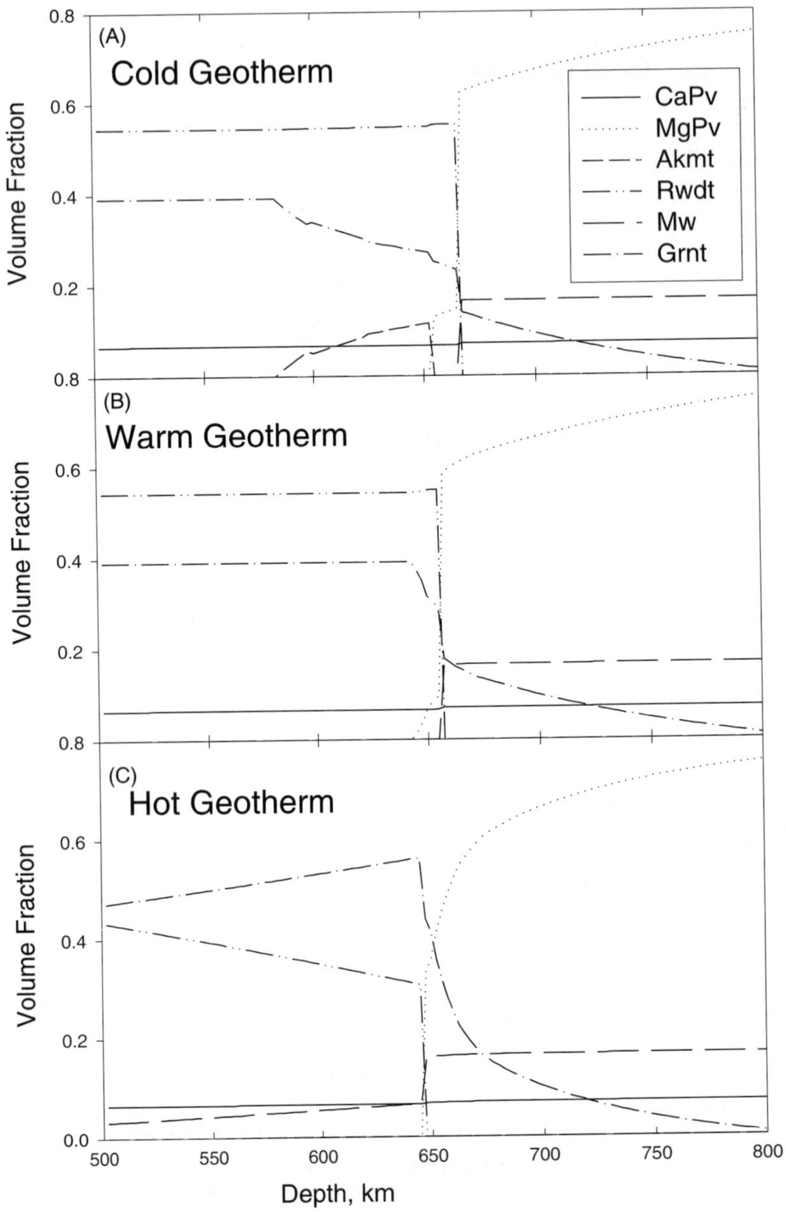

Fig. 10. Volume fractions of stable phases in the pyrolite composition along three geotherms. Same phase abbreviations as in Fig. 9. Note varying amounts of akimotoite and associated garnet volume fractions along the cold geotherm. Also note increasing garnet and decreasing ringwoodite volume fractions along the hot geotherm (see text for discussion).

the ringwoodite breakdown demands that Al be added to perovskite and the only Al source is garnet. Thus, some garnet is forced to transform to perovskite at the ringwoodite breakdown boundary (Weidner and Wang, 1998).

In both mantle composition models, the presence of akimotoite is at the cost of reducing the garnet volume fraction. Because of the large contrast in elastic properties between the two phases, velocity profiles can be dramatically different (note similar Al contents:

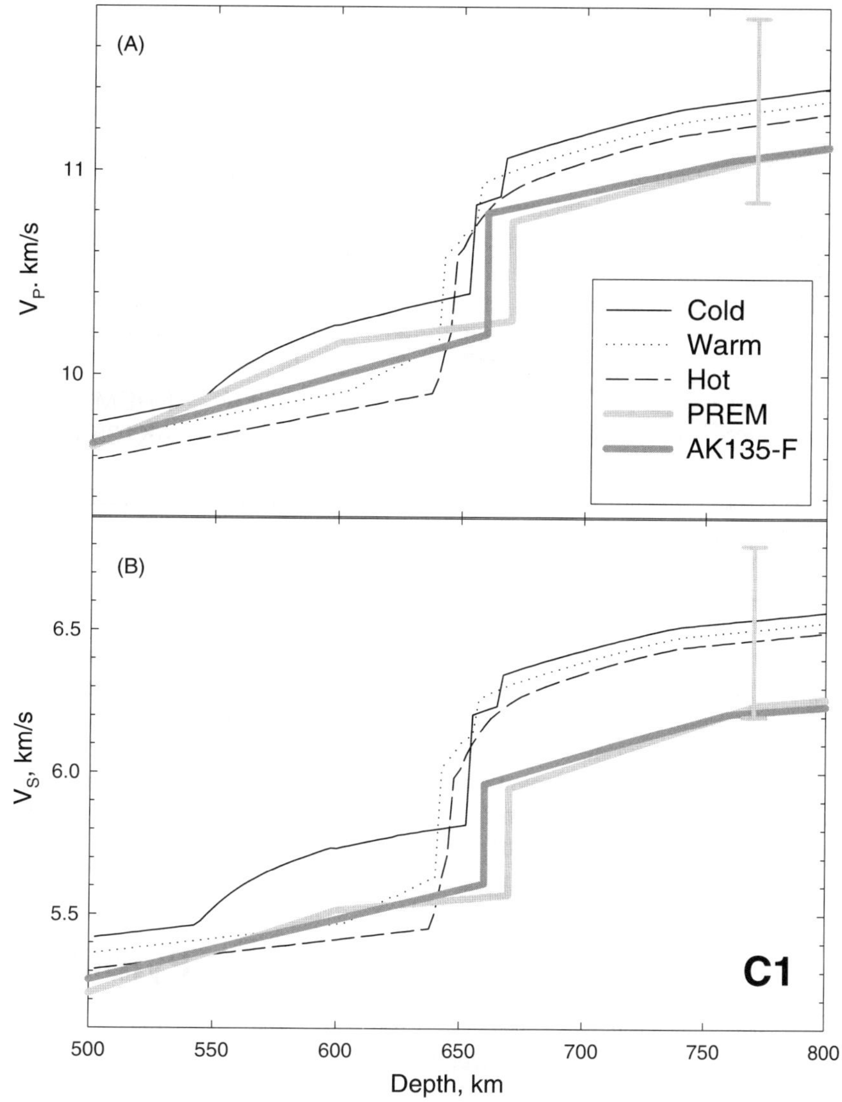

Fig. 11. V_P (A) and V_S (B) profiles of the C1 composition along the three geotherms. Velocities are compared to PREM (Dziewonski and Anderson, 1981) and AK135-F (Kennett et al., 1995). Vertical bars are estimated uncertainties of the forward modeling at 25 GPa and 1800 K (see text).

4.0 atm% in C1 versus 4.5 in pyrolite). Figs. 11 and 12 depict this effect.

For a C1 mantle, large amounts of akimotoite in the cold mantle between 550 and 650 km depths cause a steep velocity rise (Fig. 11A and B), with average velocity gradients about twice that predicted by the radial average seismic earth models PREM (Dziewonski and Anderson, 1981) and AK135-F (Kennett et al., 1995; Montagner and Kennett, 1995). If we take the velocity gradients below 550 km as a baseline, these akimotoite humps correspond to velocity increases by about 3% for V_P and 8% for V_S, respectively. As the mantle temperature increases, garnet becomes more stable and the boundary of the akimotoite–garnet transformation moves to deeper depths, so that in a warm C1 mantle average velocity gradients are comparable with seismic models. On the other hand, velocity gradients in the hot mantle are too low compared

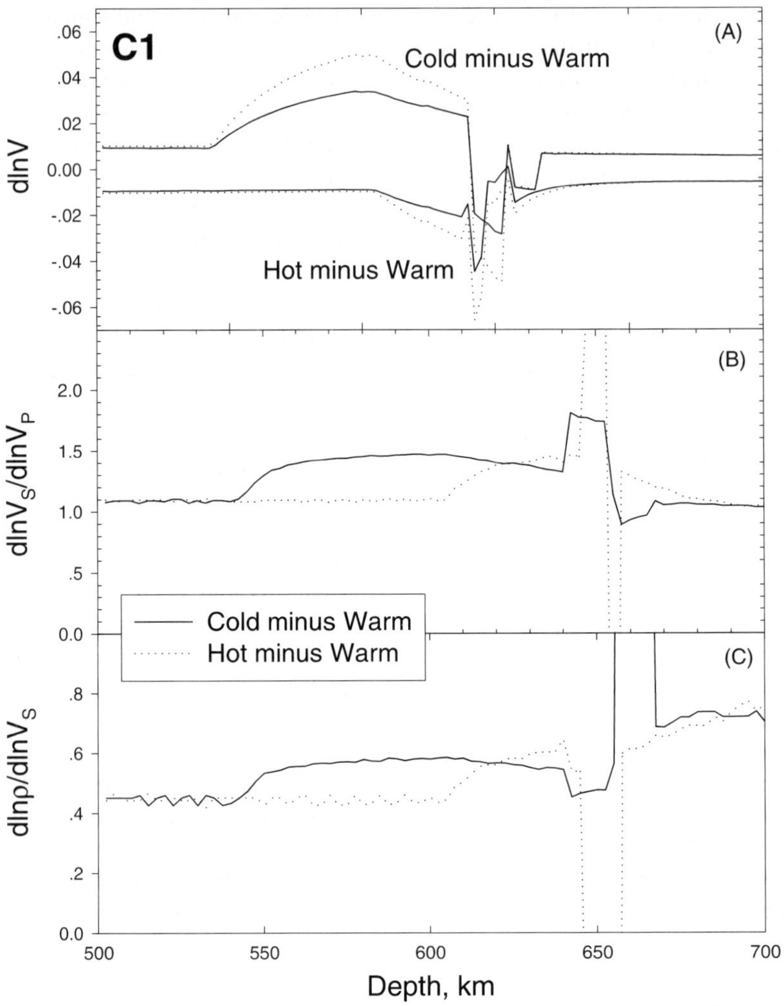

Fig. 13. (A) Tomographic signature in a C1 mantle, plotted as dlnV_S (dotted lines) and dlnV_P (solid lines) as a function of depth. The profiles are computed by assuming a 200 K lateral temperature variation along the warm geotherm, thus "Hot minus Warm" indicates velocity decrease when areas of the mantle are 200 K hotter than the warm mantle, and "Cold minus Warm" shows velocity increase in areas that are 200 K colder. Note asymmetric effects of hotter portions of the mantle vs. colder. The plateaus between 550 and 650 km depths in the "Cold minus Warm" plots are caused by growing akimotoite in colder areas of the mantle. Similarly, velocity decrease between 630–660 km depths in the "Hot minus Warm" plot is due to growing garnet. (B) Ratios of dlnV_S/dlnV_P calculated from (A). Solid curve is Cold minus Warm, and dotted curve Hot minus Warm. Note increase in dlnV_S/dlnV_P due to the negative Clapeyron slope and the akimotoite–garnet transformation in the transition zone. Large variations in the ratio in the vicinity of 650 km are due to interaction of the olivine- and pyroxene-normative components. (C) Ratios of dlnρ/dlnV_S, due to lateral temperature variation in a warm mantle. Same notation as in (B). Note that the ratio is variable and increases gradually from 0.4 near 500 km to 0.7 in the lower mantle.

velocity lows due to the phase transition are more localized (620–640 km), with smaller magnitudes (about 1.5% for V_P and 2.5% for V_S). Across 660, the phase transitions introduce very large velocity drops up to ~7%, which may affect seismic determination on topology of the discontinuity.

For colder regions in a warm C1 mantle, our model predicts a dlnV_S/dlnV_P ratio around 1.1 near 500 km. The ratio rises to about 1.5 near 550 km and falls back to about 1.1 below the 660 km discontinuity, and remains constant in the lower mantle (Fig. 13B). This value is only half that of seismic tomography observations, and is largely due to our assumed temperature derivative ($-0.014\,\mathrm{GPa\,K^{-1}}$) for the shear modulus of perovskite. Wang and Weidner (1994) have argued that based on seismic tomography, shear modulus of perovskite may have stronger temperature dependence than most minerals. Karato (1993), on the other hand, agued that anelastic effects may play an important role. In hotter regions, the rise in $\mathrm{dln}V_S/\mathrm{dln}V_P$ due to the akimotoite–garnet transition is also more localized, only present about 30–40 km above and below the 660 km discontinuity.

Since a scaling ratio between shear wave velocity variation and density is often used in geodynamic models that use seismic tomography as constraints, we also plot the ratio $\mathrm{dln}\rho/\mathrm{dln}V_S$ calculated from the modeling. Fig. 13C shows that this ratio is variable and sensitive to the akimotoite–garnet transition. With colder regions present in a warm C1 mantle, the ratio is about 0.4 near 500 km and increases to about 0.6 between 550 and 640 km, due to the akimotoite–garnet transition. Similar increase in $\mathrm{dln}\rho/\mathrm{dln}V_S$ is more localized in hotter regions.

For the pyrolite model (Fig. 14A), colder regions in a warm mantle (Cold minus Warm) have similar $\mathrm{dln}V_S$ and $\mathrm{dln}V_P$ signatures as in the C1 model, but significantly smaller magnitudes between 580 and 640 km depths (1.8 and 2% for V_P and V_S, respectively), because of smaller amount of akimotoite compared to C1 (Figs. 9A, B, 10A and B). Lower mantle variations are only about 1% (Fig. 14A). However, hotter areas relative to a warm mantle (Hot minus Warm) show large velocity drops throughout the transition zone. V_P drops by as much as 2% and V_S about 4%. This is caused by the increasing garnet fraction in hotter areas in the transition zone (Fig. 10C). In the immediate vicinity of 660, velocity variations revert sign, because of the negative Clapeyron slope and the interaction of the garnet and ringwoodite (Fig. 10).

Colder regions in the warm pyrolite mantle result in $\mathrm{dln}V_S/\mathrm{dln}V_P$ values ranging from 1.0 to 1.3 in the transition zone. Hotter regions give rise to much higher $\mathrm{dln}V_S/\mathrm{dln}V_P$ values, from 1.5 at 500 km to ~1.7 just above the 660 km discontinuity (Fig. 14B). The large spikes at 660 reflect the fact that $\mathrm{dln}V_P$ values are approaching zero.

The pyrolite scaling ratio $\mathrm{dln}\rho/\mathrm{dln}V_S$ varies dramatically and is highly asymmetric relative to temperature fluctuations (Fig. 14C). While colder regions are characterized by values about 0.5 throughout the transition zone, hotter areas exhibit much lower values from 0.17 at 500 km to 0.12 just above the 660, where it cannot be reliably determined due to complex V_S variations across the discontinuity, before adopting a more or less constant value of 0.6 in the lower mantle. The low $\mathrm{dln}\rho/\mathrm{dln}V_S$ values in the transition zone are again the consequence of increasing volume fraction of garnet, whose shear modulus is about 47% lower than that of akimotoite.

Karato (1993) first pointed out that anelastic corrections may be needed when using laboratory shear velocity data to compare with seismic models. Subsequently, many geodynamic studies use a constant correction factor throughout the mantle. Our analysis shows the importance of understanding effects of phase transformations on lateral velocity contrasts. These effects can be more important, especially when the phase boundary is sub-parallel to the geotherm, such as the akimotite-garnet transition. In this case, lateral temperature variation regulates the akimotoite–garnet transformation, resulting in large velocity changes. It is the property contrast between the two phases that dominates the lateral velocity variation; anelastic correction is of secondary importance and may be canceled out when applied to both phases.

Finally, we explore effects of aluminum content on lateral velocity variations. To illustrate this effect, we compare velocity profiles of the "normal" pyrolite (4.5 atm% Al) and C1 (4.0 atm% Al) models with those that have Al contents 2 atm% lower (all other major element proportions remain unchanged), along the same three geotherms. For all three candidate geotherms, Al-poor regions show velocity increase relative to Al-rich regions. This can be easily understood by examining Fig. 2. A decrease in Al content results in expanded akimotoite and perovskite fields to lower pressures. In general, lateral velocity variations are more dramatic than both pure thermal and phase transition effects. Fig. 15A shows $\mathrm{dln}V_S$ and $\mathrm{dln}V_P$ profiles in cold, warm, and hot C1 mantles. In the transition zone, all three mantles have increasingly larger $\mathrm{dln}V_S$ and $\mathrm{dln}V_P$ jumps from 2% and 4%, respectively in cold mantle to 8 and 14%, respectively in hot mantle. The depth span decreases from over 100 km in the cold mantle to only about 30 km in the hot mantle. At shallower depths, no

Reynard, B., Fiquet, G., Itie, J.-P., Rubie, D.C., 1996. High-pressure X-ray diffraction study and equation of state of $MgSiO_3$ ilmenite. Am. Mineral. 81, 45–50.

Reynard, B., Rubie, D.C., 1996. High-pressure, high-temperature Raman spectroscopic study of ilmenite-type $MgSiO_3$. Am. Mineral. 81, 1092–1096.

Rivers, M.L., Duffy, D.S., Wang, Y., Eng, P.J., Sutton, S.R., 1998. A new facility for high-pressure research at the Advanced Photon Source. In: Manghnani, M.H., Yagi, T. (Eds.), Properties of Earth and Planetary Materials at High Pressure and Temperature. AGU, Washington, DC, pp. 79–88.

Shim, S.H., Duffy, T.S., Takemura, K., 2002. Equation of state of gold and its application to the phase boundaries near 660 km depth in Earth's mantle. Earth Planet. Sci. Lett. 203, 729–739.

Uchida, T., Wang, Y., Rivers, M.L., Sutton, S.R., 2001. Stability field and thermal equation of state of ε-iron determined by synchrotron X-ray diffraction in a multianvil apparatus. J. Geophys. Res. 106, 21799–21910.

Vacher, P., Mocquet, A., Sotin, C., 1998. Computation of seismic profiles from mineral physics: the importance of the non-olivine components for explaining the 660 km depth discontinuity. Phys. Earth Planet. Int. 106, 275–298.

Wang, Y., Weidner, D.J., 1994. $(\partial\mu/\partial T)_P$ of the lower mantle. PAGEOPH 146, 533–549.

Wang, Y., Weidner, D.J., Liebermann, R.C., Zhao, Y., 1994. *P–V–T* equation of state of $(Mg,Fe)SiO_3$ perovskite and geophysical implications for the lower mantle. Phys. Earth Planet. Int. 83, 13–40.

Wang, Y., Rivers, M., Uchida, T., Murray, P., Shen, G., Sutton, S., Chen, J., Xu, Y., Weidner, D., 2000. High pressure research using large-volume presses at GeoSoilEnviroCARS, Advanced Photon Source. In: Manghnani, M.H., Nellis, W.J., Nicol, M.F. (Eds.), Science and Technology of High Pressure, Proceedings of AIRAPT-17. Honolulu, Hawaii, 25–30 July 1999, Universities Press (India) Limited, pp. 1047–1052.

Wang, Y., Weidner, D.J., Meng, Y., 1998. Advances in equation of state measurements in SAM-85. In: Syono, Y., Manghnani, M.H. (Eds.), High Pressure Research: Application to Earth and Planetary Sciences. American Geophysical Union, Washington, DC, pp. 365–372.

Weidner, D., Ito, E., 1985. Elasticity of $MgSiO_3$ in the ilmenite phase. Phys. Earth Planet. Int. 40, 65–70.

Weidner, D.J., Wang, Y., 1998. Chemical and Clapeyron induced buoyancy at the 660 km discontinuity. J. Geophys. Res. 103, 7431–7442.

Weidner, D.J., Wang, Y., 2000. Phase transformations: implications for mantle structure. In: Karato, S., Forte, A.M., Liebermann, R.C., Masters, G., Stixrude, L. (Eds.), Earth's Deep Interior: Mineral Physics and Tomography, from the Atomic to the Global Scale, volume 117. Geophysical Monograph Series, AGU, Washington, DC, pp. 215–235.

Weidner, D.J., Vaughan, M.T., Ko, J., Wang, Y., Liu, X., Yeganeh-Haeri, A., Pacalo, R.E., Zhao, Y., 1992. Characterization of stress, pressure, and temperature in SAM85, a DIA type high pressure apparatus. In: Syono, Y., Manghani, M.H. (Eds.), High-Pressure Research: Application to Earth and Planetary Sciences. AGU, Washington, DC, pp. 13–17.

Yeganeh-Haeri, A., 1994. Synthesis and investigation of the elastic properties of singlecrystal magnesium silicate perovskite. Phys. Earth Planet. Int. 87, 111–121.

Zhao, Y., Anderson, D.L., 1994. Mineral physics constraints [miss-spelling] on the chemical composition of the Earth's lower mantle. Phys. Earth Planet. Int. 85, 273–292.

Available online at www.sciencedirect.com
SCIENCE DIRECT
ELSEVIER
Physics of the Earth and Planetary Interiors 143–144 (2004) 81–91
PHYSICS OF THE EARTH AND PLANETARY INTERIORS
www.elsevier.com/locate/pepi

Complicated effects of aluminum on the compressibility of silicate perovskite

Takehiko Yagi [a,*], Kouichi Okabe [a], Norimasa Nishiyama [a,1], Atsushi Kubo [b], Takumi Kikegawa [c]

[a] *Institute for Solid State Physics, University of Tokyo, Kashiwa 277-8581, Japan*
[b] *Institute for Study of the Earth's Interior, Okayama University, Misasa, Tottori-ken 682-0193, Japan*
[c] *Photon Factory, Institute of Materials Structure Science, Tsukuba 305-0801, Japan*

Received 7 February 2003; received in revised form 13 June 2003; accepted 25 July 2003

Abstract

The effect of alumina content on the compressibility of $MgSiO_3$ perovskite was studied using both diamond-anvil and multi-anvil techniques. Hydrostatic compressions were performed on $MgSiO_3$ perovskites using helium as a pressure-transmitting medium up to 33 GPa. Dissolution of up to 10 mol% Al_2O_3 clearly makes perovskite more compressible, and this effect was apparent even when only 1% Al_2O_3 was dissolved. On the other hand, no meaningful effect was observed when the measurement was made using the multi-anvil apparatus. Various results reported so far have shown widely scattered values, and it is difficult to explain this range of results. It is likely that one or more unknown parameters exist that have not yet been accounted for during experimentation. Further careful studies are required to provide an answer this question, which is important to a clear understanding of the lower mantle.

Keywords: Silicate perovskite; Alumina; Bulk modulus; Lower mantle

1. Introduction

Since the discovery of the formation of silicate perovskite under lower mantle conditions (Liu, 1974), intensive studies have been made to clarify its elastic properties such as compressibility and thermal equations of state, which are expected to make a large contribution to our understanding of the nature of the lower mantle. (e.g. Yagi et al., 1982; Kudoh et al., 1987; Knittle and Jeanloz, 1987; Mao et al., 1991; Funamori et al., 1996) Most of these studies, however, were made on the $MgSiO_3$ end member or on its iron-bearing solid solutions. Through these many studies, general consensus was obtained, and until recently the elastic properties of silicate perovskite were considered to be well known. The effect of minor elements such as calcium and aluminum were long regarded to be negligible, based on the fact that in many solid solution systems, a small change in chemical composition does not alter these elastic properties. In 1999, however, Zhang and Weidner (1999) reported that dissolution of Al_2O_3, an important minor component in the lower mantle, drastically changes the compressibility of silicate perovskite. They reported that dissolution of 5 mol% alumina, which is the expected concentration

* Corresponding author.
E-mail address: yagi@issp.u-tokyo.ac.jp (T. Yagi).
[1] Present address: Geodynamic Research Center, Ehime University, Matsuyama 790-8577, Japan.

0031-9201/$ – see front matter
doi:10.1016/j.pepi.2003.07.020

in the lower mantle, decreases the bulk modulus of silicate perovskite by as much as 10%. Since then, several studies have been conducted to check their result. The measurement by Zhang and Weidner (1999) was limited to 10 GPa using a cubic-anvil apparatus with a solid pressure-transmitting medium. In order to avoid ambiguities caused by the possible effect of non-hydrostatic compression, Kubo et al. (2000) made the measurement using a diamond anvil cell (DAC), adopting an ethanol–methanol mixture as a pressure-transmitting medium up to 10 GPa (i.e., in a purely hydrostatic environment), and reported that the bulk modulus decreased, as reported previously by Zhang and Weidner. Subsequently, Daniel et al. (2001) and Andrault et al. (2001) made further studies using diamond anvil cells. The former reported results that were generally in agreement with previous findings, whereas the latter reported findings conflicting with those previously reported—namely, that dissolution of alumina increases the bulk modulus when the pressure range is extended to that of the lower mantle.

In order to understand the effect of alumina on the compressibility of silicate perovskite more clearly, we made measurements using two different techniques. One involves use of a diamond anvil cell with helium as a pressure-transmitting medium up to a pressure of 34 GPa. The other employs a multi-anvil apparatus and allows the unit-cell volumes of two samples, one with and one without alumina, to be measured under the same pressure and temperature conditions. The detail of these results and possible explanations for the complexity of this problem are discussed herein.

2. Experimental

2.1. Diamond-anvil experiment

Three samples with different alumina content (i.e., pure $MgSiO_3$, 1 mol% Al_2O_3, and 10 mol% of Al_2O_3) were prepared in advance and compressed in a DAC at room temperature. Pure $MgSiO_3$ perovskite was synthesized using a laser-heated DAC. $MgSiO_3$ powder prepared from single crystal enstatite was mixed with platinum powder, which works as an absorber of YAG laser light, and compressed to about 35 GPa in a diamond anvil cell without using a pressure medium. The starting material was converted to the perovskite structure by heating it to about 2000 K. Two other samples with alumina were prepared using a Kawai-type multi-anvil apparatus at about 27 GPa and 2300 K. Details of the synthesis have been described previously (Kubo et al., 2000). All samples were confirmed to consist of a single phase with a perovskite structure, and the Al_2O_3 content values measured by EPMA analysis were 1.2 ± 0.6 and 10.3 ± 0.3%, respectively. Unit cell parameters of each sample are summarized in Table 1, together with the parameters observed under various pressures. Samples synthesized by the multi-anvil apparatus were carefully cut into small blocks of about 40 μm × 40 μm × 20 μm and placed in the sample chamber of a diamond anvil cell, together with a small piece of gold foil (15 μm × 30 μm × 2 μm) and several ruby chips to act as pressure markers. The rhenium gasket was pre-indented to a thickness of about 80 μm using diamond anvils with a culet diameter of 0.45 mm. A gasket hole of about 150 μm in diameter was drilled using a Q-switched YAG laser, and the samples were placed in the center of the gasket hole. The sample chamber was then filled with about 200 MPa of helium using a gas loading system that was a modified version of one previously described (Yagi et al., 1996). As shown in Fig. 1, due to the very large compressibility of helium in the lower pressure range, the sample chamber shrinks considerably when the pressure is increased to about 10 GPa. No further shrinkage of the sample chamber was observed above this pressure, and the sample was completely surrounded by the helium pressure-transmitting medium even at the highest pressure studied.

X-ray measurements were made at the synchrotron radiation facility, Photon Factory, KEK, using a system constructed at BL-13A (Yagi et al., 2001). A monochromatized X-ray of about 30 keV from a multi-pole wiggler was focused by a K-B mirror and then passed through a pinhole collimator (30 μm × 30 μm square) to irradiate the sample. Due to the small cross-section of the X-ray beam, it was possible to measure the sample and the gold pressure marker separately. The diffracted X-rays were recorded by an imaging plate (IP) detector located about 300 mm from the sample and equipped with a reader (RIGAGKU R-AXIS IV). The size of each pixel in the IP was 100 μm × 100 μm. Two-dimensional IP

Table 1
Unit cell parameters of $MgSiO_3$ perovskites with various Al_2O_3 contents observed under high pressure and at atmospheric pressure

Composition	Data no.	a (Å)	b (Å)	c (Å)	V (Å^3)	Au a (Å)	Pressure ruby (GPa)	Pressure Au (GPa)
En100	Pv03-02	4.7709 (27)	4.9277 (20)	6.8977 (40)	162.16 (14)		1.00	
En100	Pv03-03	4.7508 (24)	4.9180 (19)	6.8729 (17)	160.58 (11)		3.46	
En100	Pv03-04	4.7380 (18)	4.9031 (14)	6.8513 (13)	159.16 (8)		5.56	
En100	Pv03-05	4.7316 (27)	4.8964 (21)	6.8419 (19)	158.51 (12)		6.82	
En100	Pv03-06	4.7224 (36)	4.8850 (28)	6.8262 (26)	157.47 (16)		8.53	
En100	Pv03-08	4.6437 (14)	4.9035 (11)	6.8521 (10)	159.24 (6)		5.51	
En100	Pv03-09	4.7393 (26)	4.9133 (20)	6.8739 (18)	160.55 (12)		3.59	
En100	Pv03-10	4.7538 (26)	4.9225 (20)	6.8885 (18)	161.62 (12)		1.59	
En100	Pv03-11	4.7663 (20)	4.9244 (15)	6.8909 (14)	161.67 (9)		1.59	
En100	Pv03-12	4.7726 (24)	4.9320 (19)	6.9032 (17)	162.49 (11)		0.00	
En100	Pv09-05	4.7036 (9)	4.8775 (6)	6.8024 (9)	156.06 (4)	3.9987 (1)	11.70	11.68
En100	Pv09-07	4.6881 (8)	4.8560 (5)	6.7790 (8)	154.33 (4)	3.9822 (1)	15.29	14.92
En100	Pv09-09	4.6649 (8)	4.8366 (5)	6.7482 (8)	152.25 (3)	3.9574 (2)	19.94	19.72
En100	Pv09-11	4.6623 (8)	4.8372 (5)	6.7451 (8)	152.12 (4)	3.9573 (2)	20.07	19.74
En100	Pv09-13	4.6877 (8)	4.8583 (5)	6.7797 (7)	154.40 (3)	3.9822 (2)	15.22	14.92
En100	Pv09-15	4.7202 (9)	4.8819 (6)	6.8203 (9)	157.16 (4)	4.0131 (4)	9.80	9.57
En100	Pv09-07	4.7361 (12)	4.8981 (8)	6.8437 (11)	158.76 (5)	4.0314 (5)	6.96	6.69
En100	Pv09-19	4.7620 (9)	4.9171 (6)	6.8751 (9)	160.98 (4)	4.0581 (2)	3.22	2.90
En100	Pv09-21	4.7779 (8)	4.9334 (5)	6.9026 (8)	162.70 (4)	4.0806 (1)	0.00	0.00
En99	Pv06-02	4.6572 (12)	4.8218 (12)	6.7259 (20)	151.04 (7)		21.15	
En99	Pv06-03	4.6243 (10)	4.8001 (11)	6.6922 (13)	148.55 (6)		26.44	
En99	Pv06-04	4.6090 (16)	4.7827 (16)	6.6673 (21)	146.97 (9)		30.11	
En99	Pv06-05	4.5955 (16)	4.7643 (16)	6.6373 (20)	145.32 (8)		34.33	
En99	Pv06-07	4.6196 (14)	4.7952 (15)	6.6819 (19)	148.02 (7)		27.79	
En99	Pv06-08	4.6437 (17)	4.8153 (17)	6.7081 (22)	150.00 (9)		23.28	
En99	Pv06-09	4.6705 (10)	4.8385 (10)	6.7480 (13)	152.41 (5)		18.44	
En99	Pv06-10	4.7040 (10)	4.8677 (10)	6.7965 (13)	155.62 (6)		12.18	
En99	Pv06-11	4.7262 (13)	4.8888 (13)	6.8264 (17)	157.73 (7)		8.55	
En99	Pv06-12	4.7516 (19)	4.9072 (20)	6.8603 (25)	159.96 (11)		4.54	
En99	Pv06-13	4.7810 (9)	4.9337 (9)	6.9047 (12)	162.87 (5)		0.00	
En99	Pv08-02	4.6688 (6)	4.8367 (6)	6.7483 (8)	152.38 (3)	3.9625 (2)	18.68	18.34
En99	Pv08-04	4.6569 (8)	4.8263 (8)	6.7304 (10)	151.27 (4)	3.9519 (2)	20.41	20.49
En99	Pv08-08	4.6418 (8)	4.8114 (8)	6.7099 (10)	149.86 (4)	3.9365 (2)	23.63	23.78
En99	Pv08-10	4.6326 (9)	4.8037 (9)	6.6974 (12)	149.04 (5)	3.9275 (2)	25.58	25.77
En99	Pv08-12	4.6390 (7)	4.8084 (8)	6.7062 (10)	149.59 (4)	3.9324 (2)	24.50	24.68
En99	Pv08-14	4.6634 (7)	4.8612 (8)	6.7351 (10)	151.58 (4)	3.9546 (4)	20.15	19.93
En99	Pv08-16	4.6912 (7)	4.8566 (7)	6.7778 (9)	154.42 (4)	3.9829 (2)	14.75	14.46
En99	Pv08-18	4.7092 (4)	4.8725 (4)	6.8042 (6)	156.12 (2)	4.0005 (1)	11.79	11.35
En99	Pv08-19	4.7265 (4)	4.8865 (5)	6.8279 (6)	157.70 (3)	4.0184 (1)	8.86	8.43
En99	Pv08-23	4.7463 (3)	4.9039 (3)	6.8537 (4)	159.52 (2)	4.0361 (1)	5.99	5.73
En99	Pv08-24	4.7589 (5)	4.9161 (5)	6.8743 (7)	160.83 (3)	4.0532 (1)	3.78	3.32
En99	Pv08-26	4.7806 (5)	4.9349 (5)	6.9049 (6)	162.90 (3)	4.0788 (1)	0.00	0.00
En90	At01-01	4.6715 (8)	4.8535 (8)	6.7851 (11)	153.84 (4)		16.73	
En90	At01-02	4.6521 (6)	4.8353 (6)	6.7584 (7)	152.03 (3)		19.86	
En90	At01-03	4.6398 (10)	4.8246 (10)	6.7384 (13)	150.84 (6)		22.39	
En90	At01-04	4.6174 (10)	4.8059 (10)	6.7070 (13)	148.83 (5)		26.59	
En90	At01-05	4.5915 (10)	4.7877 (10)	6.6719 (13)	146.66 (5)		31.42	
En90	At01-06	4.5844 (12)	4.7782 (13)	6.6611 (16)	145.91 (6)		33.30	
En90	At01-07	4.5910 (11)	4.7829 (11)	6.6711 (14)	146.48 (6)		31.84	
En90	At01-09	4.6043 (12)	4.7962 (13)	6.6942 (16)	147.83 (7)		28.75	

Table 1 (*Continued*)

Composition	Data no.	*a* (Å)	*b* (Å)	*c* (Å)	*V* (Å^3)	Au *a* (Å)	Pressure ruby (GPa)	Pressure Au (GPa)
En90	At01-10	4.6384 (10)	4.8268 (10)	6.7373 (13)	150.54 (5)		23.02	
En90	At01-11	4.6648 (10)	4.8501 (10)	6.7719 (13)	153.21 (5)		17.77	
En90	At01-12	4.6814 (9)	4.8634 (9)	6.7977 (12)	154.77 (5)		14.77	
En90	At01-13	4.7003 (16)	4.8819 (17)	6.8292 (21)	156.71 (9)		11.38	
En90	At01-14	4.7186 (12)	4.8959 (13)	6.8508 (17)	158.27 (7)		8.70	
En90	At01-15	4.7382 (8)	4.9107 (8)	6.8805 (11)	160.10 (5)		5.64	
En90	At01-16	4.7587 (11)	4.9274 (11)	6.9131 (14)	162.10 (6)		2.42	
En90	At01-17	4.7747 (16)	4.9402 (16)	6.9383 (21)	163.66 (9)		0.00	
En90	At02-03	4.6766 (9)	4.8579 (9)	6.7988 (12)	154.46 (5)	3.9777 (2)	15.43	15.45
En90	At02-05	4.6525 (12)	4.8292 (12)	6.7477 (16)	151.61 (7)	3.9507 (2)	20.65	20.78
En90	At02-07	4.6326 (17)	4.8176 (18)	6.7215 (23)	150.01 (9)	3.9324 (2)	23.98	24.71
En90	At02-09	4.6094 (22)	4.7944 (23)	6.6915 (29)	147.88 (12)	3.9120 (3)	28.28	29.47
En90	At02-11	4.5907 (19)	4.7826 (20)	6.6634 (26)	146.30 (10)	3.8965 (5)	31.82	33.33
En90	At02-13	4.6055 (11)	4.7962 (11)	6.6911 (14)	147.80 (6)	3.9100 (5)	28.50	29.94
En90	At02-18	4.6527 (8)	4.8396 (9)	6.7501 (11)	151.99 (5)	3.9540 (1)	19.80	20.10
En90	At02-20	4.6637 (14)	4.8498 (14)	6.7694 (18)	153.11 (7)	3.9669 (1)	17.83	17.51
En90	At02-24	4.6987 (19)	4.8809 (20)	6.8228 (25)	156.47 (11)	3.9985 (1)	11.60	11.73
En90	At02-31	4.7354 (15)	4.9023 (15)	6.8847 (20)	159.82 (9)	4.0363 (1)	6.10	5.74
En90	At02-34	4.7264 (13)	4.9371 (13)	6.9065 (18)	161.16 (8)	4.0516 (1)	3.73	3.55
En90	At02-36	4.7734 (10)	4.9402 (10)	6.9340 (13)	163.51 (6)	4.0789 (1)	0.00	0.00

data were integrated and converted to one-dimensional intensity data. A few minutes of exposure was enough to observe the diffraction from the sample. However, in order to get high quality data, a typical exposure time of 15–30 min was adopted. Because of the small divergence of the X-ray system and the highly hydrostatic nature of the applied pressure, high resolution diffraction data were obtained even at 34 GPa, as shown in Fig. 2. As a result, very precise and reproducible compression curves were obtained for each sample, as will be shown later. Pressures were measured before and after the X-ray exposure using the ruby fluorescence technique (Mao et al., 1986) and were also based on the unit cell parameter of gold using the pressure scale proposed by Anderson et al. (1989). No meaningful difference in pressure was observed between these two pressure scales, and the pressure within the sample chamber, based on observation of the ruby chips, varied by less than 0.5 GPa over the entire pressure range.

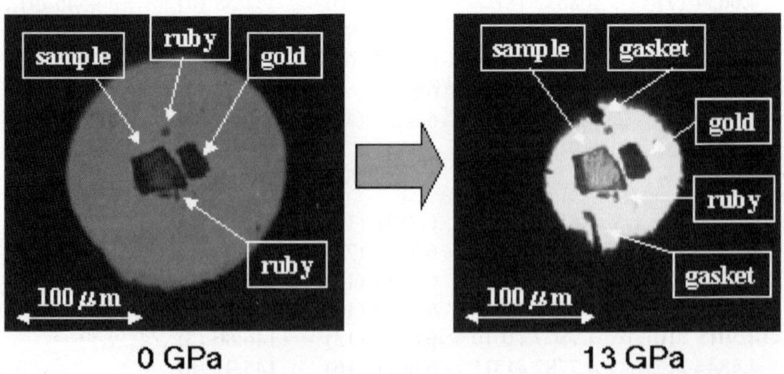

Fig. 1. Sample chamber at 0 and 13 GPa. Due to the large compressibility of the helium pressure-transmitting medium, the sample chamber shrinks considerably at low pressure.

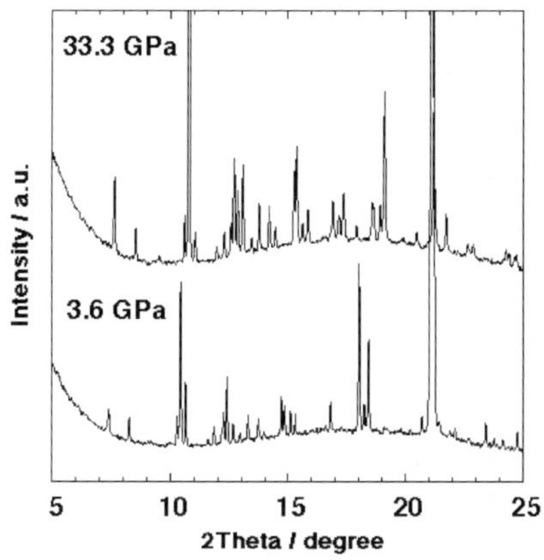

Fig. 2. Examples of diffraction patterns. There was no measurable increase in line width even at 33 GPa.

2.2. Multi-anvil experiment

Two starting materials, pure $MgSiO_3$ and one containing 5 mol% Al_2O_3, respectively, were placed symmetrically above and below the thermocouple in the same sample chamber. The former starting material was a crystalline powder prepared from a single crystal of $MgSiO_3$ enstatite, while the latter was a glass prepared from an oxide mixture. They were put in MgO capsules and heated between two disc-shaped heaters, thus the pressure and temperature conditions of these two samples are expected to be identical. Zirconia plates were used as thermal insulating layers. The sample assembly thus prepared was placed in the middle of eight sintered-diamond anvils (9.5 mm cubes with 2.0 mm truncations) and compressed. A "MAX-III" system (a Kawai-type double stage multi-anvil apparatus combined with an energy-dispersive type X-ray diffraction system) at the BL-14C station of the Photon Factory was used for synthesis and unit cell measurements of the perovskite sample. Details of the experiments are provided elsewhere, together with the results at elevated temperature and those for perovskites with other more complicated compositions (Nishiyama et al., in preparation).

Pressure and temperature conditions of the synthesis and unit cell measurements are summarized in Fig. 3. The samples were first compressed to about 25 GPa at room temperature and then converted to single phases with a perovskite structure by heating to 1500 K. After

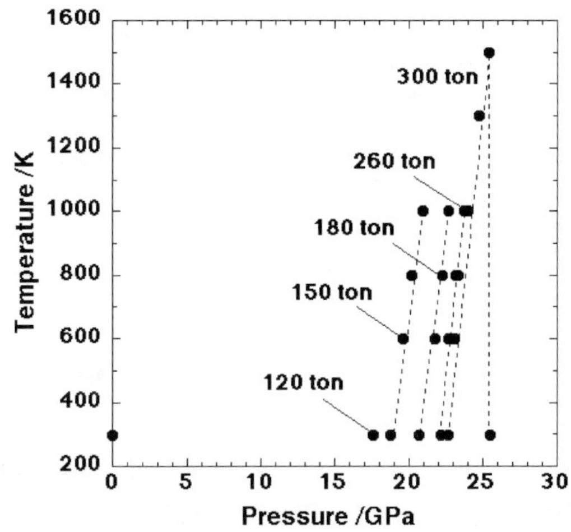

Fig. 3. *P–T* conditions for the multi-anvil experiment. Enstatite was first converted into perovskite at 25.5 GPa and 1500 K, and then the unit cell volume was measured at room temperature. Subsequently, the press load was decreased; the sample was annealed to remove stress inhomogeneity; and the room temperature volume was measured.

the synthesis, the unit cell parameters of two samples were measured separately at room temperature while the press load was kept constant. Then the press load was decreased and the samples were heated to 1000 K to anneal stress and remove the effect of stress inhomogeneity before another set of unit cell measurements was made at room temperature. This procedure was repeated several times until the pressure had been decreased to 19 GPa, then the pressure was reduced to ambient condition without heating to avoid amorphization of the sample. Finally the unit cell parameters of the two recovered samples were measured.

Although the samples were compressed in a solid pressure-transmitting medium, it is known from previous studies (e.g., Funamori et al., 1996) that when the sample is heated to such a high temperature, effects of non-hydrostatic stress become negligibly small and reliable compression curves are obtained.

3. Results

3.1. DAC experiments

Two independent experiments were conducted for each sample with the same chemical composition, and

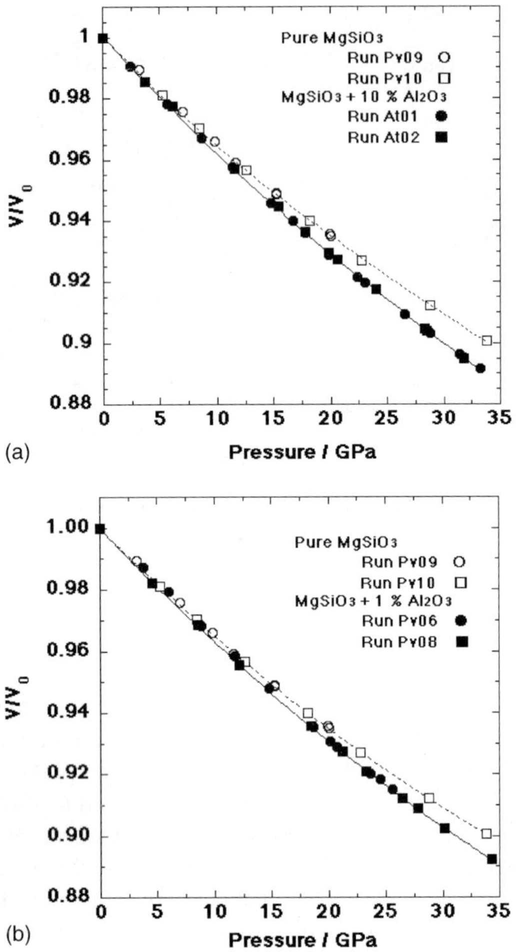

Fig. 4. Isothermal compression curves of perovskites. (a) $MgSiO_3$ perovskite with 10% Al_2O_3 is clearly more compressible than pure $MgSiO_3$ perovskite. (b) The dissolution of only 1% Al_2O_3 resulted in a significant difference.

the resulting compression curves for all the samples are summarized in Table 1 and plotted in Fig. 4a and b. From the results of experiments on samples with the same composition, it is clear that the reproducibility of experiments is high and the effect of alumina content on the compressibility of silicate perovskite can be clearly seen. For the samples with 10% Al_2O_3, a clear decrease of the bulk modulus is observed (Fig. 4a); even for the samples with only 1% Al_2O_3, softening can be clearly seen (Fig. 4b). Birch-Murnaghan equations of state were fitted to each compression curve assuming $dK/dP = 4$, the results of which are shown in Fig. 5, together with other results so far reported.

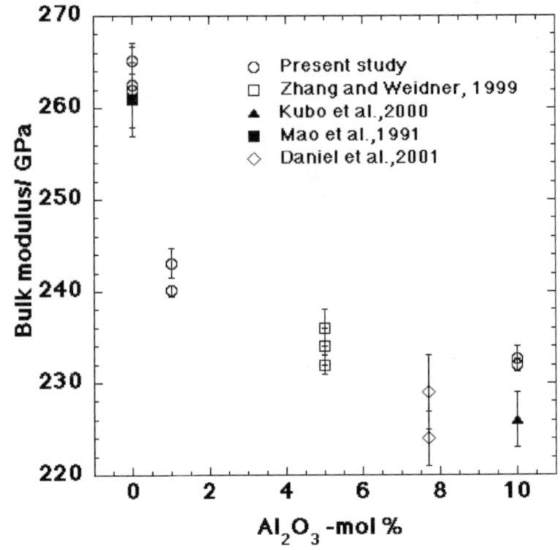

Fig. 5. Bulk modulus vs. Al_2O_3 content for $MgSiO_3$ perovskites.

Present results on the pure $MgSiO_3$ end member are in accord with the well-established bulk modulus of 260 GPa. Results for 1 and 10% Al_2O_3 samples are about 242 and 232 GPa respectively, and are significantly smaller than the bulk modulus of the pure $MgSiO_3$ end member (without Al_2O_3). The results of the present study on the 10% Al_2O_3 samples and those of a previous study by Kubo et al. (2000) using samples from the same batch of synthesis experiments are in reasonable agreement if we take into account the fact that the pressure range was limited to below 10 GPa in the previous study.

3.2. Multi-anvil experiment

Compression curves observed for two samples are plotted in Fig. 6 together with the bulk modulus obtained by fitting a second order Birch-Murnaghan equation of state. It is clear from Fig. 6, in which the absolute values of volume are plotted, that the dissolution of Al_2O_3 into $MgSiO_3$ perovskite clearly expands the unit cell volume. Contrary to the previous results, however, the excess volume caused by the dissolution of Al_2O_3 remained unchanged with pressure, and almost identical values of bulk modulus were obtained for the two perovskites. Bulk modulus obtained for pure $MgSiO_3$ and for perovskite with 5 mol% Al_2O_3 were 258 ± 7 and 257 ± 6 GPa, respectively. Results for end member perovskite are in

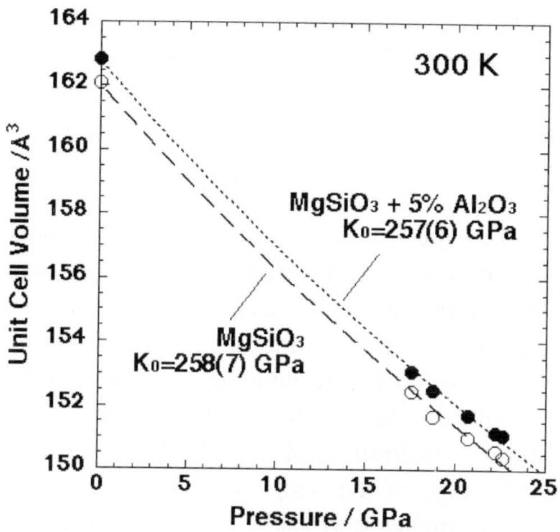

Fig. 6. Compression curves of two perovskites observed in the multi-anvil experiment. No significant difference was observed between these two samples even though the compositions are different.

agreement with those of previous studies, though the result for the 5% Al_2O_3 sample differs significantly from other results reported on aluminous silicate perovskites, shown in Fig. 4a and b.

4. Discussion

For purposes of comparing various results, experimental conditions and parameters of each report are summarized in Table 2. Daniel et al. (2001) studied the compression curve of perovskite with 7.7% Al_2O_3 using DAC. They made two runs using helium and argon, respectively, as pressure-transmitting media. In experiments using helium, perovskite was synthesized in a separate run by compressing the starting material to 35 GPa and heating to about 2100 K using a CO_2 laser. After the synthesis, the sample was recovered at ambient conditions and then loaded in the sample chamber of another DAC together with the helium pressure medium, and compression experiments were carried out. In experiments with argon, the starting material was loaded together with the argon pressure medium and converted to the perovskite phase by heating it at 35 GPa using a CO_2 laser. After the synthesis, the unit-cell volume was measured at room temperature during decompression. The values of bulk modulus reported by Daniel et al. are plotted in Fig. 5, assuming dK/dP of both 2.5 (optimal fit) and 4 (the value used in other studies). Regardless of the assumptions concerning dK/dP, their results are in reasonable agreement with those of other studies.

From Fig. 5, we can conclude that if we only look at the results reported by Zhang and Weidner (1999); Kubo et al. (2000); Daniel et al. (2001), and the present results obtained with the DAC, it is clear that the dissolution of Al_2O_3 into $MgSiO_3$ perovskite decreases the bulk modulus considerably. The effect is highly nonlinear with Al_2O_3 content; the bulk modulus decreases by as much as 7% with only 1% Al_2O_3. It is known that, in general, the bulk modulus is insensitive to small changes in chemical composition. In compounds with the perovskite structure, for example, dissolution of up to 20% iron does not measurably effect the bulk modulus (Mao et al., 1991). Therefore, the large decrease in the bulk modulus by the dissolution of 1% alumina is unusual. One possible explanation for this strong effect is the formation of vacancies associated with the dissolution of alumina. Two different substitution mechanisms

Table 2
Summary of experimental conditions of various experiments

Author	Apparatus	Pressure range (GPa)	Pressure medium	Al_2O_3 (mol%)	Synthesis run (same or separate)	Bulk modulus (GPa)($K'_0 = 4$)
Zhang and Weidner (1999)	Multi anvil	1–10	None	5.0	Separate	234
Kubo et al. (2000)	Diamond anvil	1–9	Alcohol mixture	10.3	Separate	226
Daniel et al. (2001)-1	Diamond anvil	1–32	Argon	7.7	Same	224
Daniel et al. (2001)-2	Diamond anvil	1–32	Helium	7.7	Separate	224
Andrault et al. (2001)-1	Diamond anvil	29–57	Argon	5.0	Same	267
Andrault et al. (2001)-2	Diamond anvil	29–69	Argon	11.0	Same	276
Present study-1	Diamond anvil	4–34	Helium	1.2	Separate	242
Present study-2	Diamond anvil	2–33	Helium	10.3	Separate	232
Present study-3	Multi anvil	17–23	None	5.0	Same	257

are expected when the trivalent cation of Al^{3+} is dissolved into $MgSiO_3$ perovskite, as follows:

$$2Al^{3+} = Mg^{2+} + Si^{4+} \quad (1)$$

$$2Al^{3+} = 2Si^{4+} + O_{Vac} \quad (2)$$

In mechanism (1), aluminum is distributed equally into the Mg and Si sites, and when 25 mol% Al_2O_3 is dissolved, its chemical formula can be written as $(Mg_3, Al)(Al, Si_3)O_{12}$. In the case of (2), aluminum preferentially enters the silicon site and an oxygen vacancy is formed to maintain charge balance. When 33 mol% of Al_2O_3 is dissolved, the hypothetical compound $Mg_2Al_2O_5$ coexisting with SiO_2 is expected to form. In the case of the $CaTiO_3$–Fe_2O_3 join, it is known that a complete solid solution is formed between $CaTiO_3$ and $Ca_2Fe_2O_5$, and the concentration of vacancies increases in proportion to the iron concentration. At a low concentration, the vacancies are randomly distributed; above certain concentrations, they are ordered (Bacerro et al., 1999). $Ca_2Fe_2O_5$, a mineral known as brownmillerite, has a structure similar to that of perovskite, with completely ordered vacancies in one of the six oxygen sites. It is suspected that the formation of vacancies increases the compressibility considerably, and many arguments have been made based on this idea. Ross et al. (2002) studied the compressibility of brownmillerite using high-pressure X-ray diffraction under hydrostatic conditions. They compared the bulk modulus with other calcium compounds having a perovskite structure and concluded that brownmillerite is about 25% more compressible than expected from the systematics of unit cell volume versus bulk modulus relation in perovskites. Brodholt (2000) made ab initio calculations for $(Mg_3, Al)(Al, Si_3)O_{12}$ perovskites and $Mg_2Al_2O_5$ with the brownmillerite structure. He showed that the bulk modulus of $(Mg_3, Al)(Al, Si_3)O_{12}$, which forms by mechanism (1) and contains no vacancies, is about 6% smaller than that of pure $MgSiO_3$ perovskite. On the other hand, in $Mg_2Al_2O_5$, in which 1/6 of the oxygen sites are vacant, the bulk modulus is about 40% lower than that of pure $MgSiO_3$. With regard to perovskites on the $MgSiO_3$–Al_2O_3 join, there has been considerable discussion concerning the dominant substitution mechanism, while experimental results have suggested that at least some vacancies are formed by the dissolution of alumina. On the basis of this idea alone, however, it is difficult to explain the large drop in the bulk modulus for the sample with only 1% Al_2O_3 observed in the present study, because the maximum concentration of vacancies expected from this composition is very small.

Contrary to the results discussed so far, our result obtained in the multi-anvil experiment indicates that the dissolution of Al_2O_3 does not necessary decrease the bulk modulus of perovskite. Based on results shown in Fig. 6, the bulk modulus is identical for samples with and those without alumina, although the unit cell volume is clearly increased with the dissolution of alumina. Andrault et al. (2001) reported that in the pressure range of the deep lower mantle, dissolution of alumina makes perovskite even more incompressible than pure $MgSiO_3$ perovskite. They compressed starting materials with various alumina contents using argon pressure medium, and transformed them to a perovskite structure by heating the sample with a CO_2 laser above 30 GPa. They measured the unit cell volume after each increment of pressure followed by heating and stress relaxation. They observed that the volumes observed were always larger than those of pure $MgSiO_3$ perovskite, and that the difference became even larger with increasing pressure. Their observations were made from 26 GPa up to about 70 GPa, and they concluded that the dissolution of alumina makes the perovskite structure relatively incompressible under lower mantle conditions. Their explanation for these results was based on the fact that the formation of vacancies becomes unfavorable under very high pressure and the concentration of vacancy was re-equilibrated after each heat treatment. If we use the argument of Brodholt (2000), it is reasonable to assert that the bulk modulus increases under very high pressure. However, it cannot exceed the value for pure $MgSiO_3$, in which no vacancies exist. Ab initio calculations by Brodholt (2000) suggest that even for perovskite without any vacancies, the dissolution of alumina slightly decreases the bulk modulus. Therefore, an argument based on vacancies alone does not adequately explain all the experimental results.

Another possible explanation for the large effect of alumina is the partial back transformation of perovskite on release of pressure. It is known that silicate perovskite is marginally metastable at ambient conditions, and that $MgSiO_3$ perovskite transforms into glass when heated to above 400 K at atmospheric pressure. Stability is further reduced when alumina is dissolved, and very often amorphization is observed to result simply in response to grinding the sample by mortar and pestle under atmospheric pressure. This is why we prepared the sample for the DAC experiment without crushing the recovered sample with a mortar and pestle. $CaSiO_3$ perovskite is known to undergo amorphization simply in response to a decrease in pressure at room temperature. Sometimes a small amount of the crystalline phase is recovered at ambient conditions, but the unit cell volume of these samples is slightly different from that derived by extrapolating the decompression curve. A similar possibility exists for metastable aluminous perovskite and the possibility can be tested from the results of Daniel et al. (2001). They conducted volume measurements under pressure in two different ways. First, by increasing the pressure using a sample which had been prepared and recovered at ambient conditions in a previous experiment. Second, by decreasing the pressure on a sample, just after synthesis under high pressure, which had never experienced ambient conditions. They reported finding no difference between the results of these two measurements. To check this point, we also made several measurements, by synthesizing the perovskite phase in a DAC using an argon pressure medium and measuring the volume with decreasing pressure. Argon becomes very soft at low pressure and becomes liquid below 1 GPa. Therefore, very precise volume measurements can be made without heating the sample at all. Still, we could not find any meaningful deviation of volume from that of the compression curve determined using quenched samples.

Another factor possibly affecting the concentration and ordering of vacancies is the synthesis temperature of the sample, which may consequently alter the bulk modulus. In addition, synthesis temperature may affect the unit cell volume measured at ambient conditions. If we compare the unit cell volume of aluminous perovskite reported in the literature (shown in Fig. 7), a general tendency can be seen for the volume of the unit cell to increase with that of increasing alumina content. The measured volumes for a given composition, however, are rather scattered, and it is difficult to see the effect of the synthesis conditions because the expected difference is small and, furthermore, it is

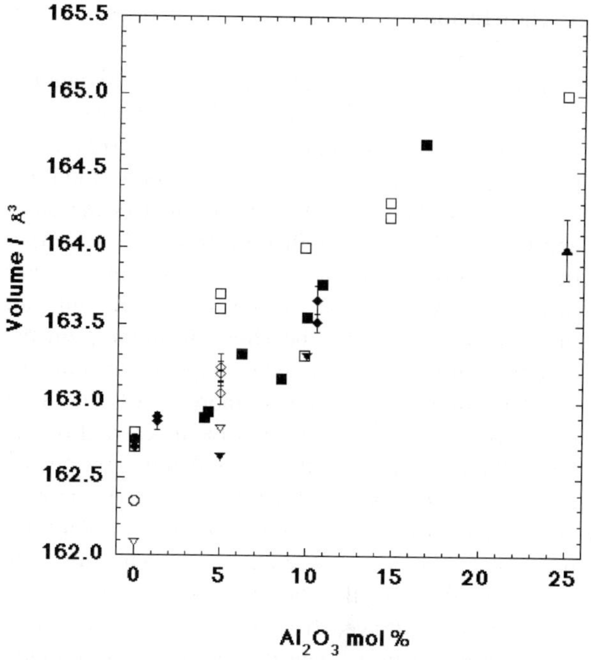

Fig. 7. Unit cell volume vs. Al_2O_3 content of $MgSiO_3$ perovskites. ○: Ito and Matsui, 1978; ●: Yagi et al., 1978; □: Weng et al., 1982; ■: Irifune et al., 1996; ▲: Ito et al., 1998; ◇: Zhang and Weidner; ▼: Kubo and Akaogi, 2000; ▽: Nishiyama, unpublish; ◆: Present study.

difficult to compare the absolute volumes measured in different laboratories with high precision.

In order to clarify the effect of synthesis temperature on the unit cell volume of aluminous perovskite, we heated starting material that had been compressed in argon, using a double-sided laser heating system. The heated area was about 50 μm in diameter and the unit cell volume of the recovered sample was examined by irradiating with a very thin X-ray beam about 15 μm in diameter, using the system at the Photon Factory, which is described above. It was possible to measure the unit cell volume of perovskite at several positions from the center to the margin of the heated sample; thus, the synthesis temperature is expected to be quite different among positions. So far, however, no meaningful difference in unit cell volume has been determined between different positions. Therefore, the synthesis temperature does not appear to explain the discrepancies in the bulk modulus previously reported.

5. Conclusion

In summary, it is difficult at present to estimate the elastic properties of aluminous silicate perovskite in the lower mantle. Dissolution of alumina into $MgSiO_3$ perovskite sometimes decreases the bulk modulus considerably. This is, however, not always the case, and there is a possibility that under deep lower mantle conditions, the bulk modulus becomes even larger than that of pure $MgSiO_3$ perovskite. At this time, all that can be said is that the effect of alumina on the compressibility of silicate perovskite is very complicated, and there must be one or more factors that were not well controlled in the experiments carried out so far. If the large drop in the bulk modulus reported so far in many studies is confirmed, it will considerably affect estimates of chemical composition and/or temperature in the deep lower mantle, which have been made by comparing the elastic data of perovskite with elastic models derived from seismic observations. It is necessary to investigate this problem by conducting further experiments in which various parameters, such as starting composition, synthesis temperature and pressure, method of elasticity measurement, and so on, are carefully controlled.

Acknowledgements

The authors are grateful to M. Akaogi for his help in preparing the starting material used in the multi-anvil experiment and to T. Kondo and N. Miyajima for their helpful discussions. Thanks are also due to I. Jackson, J. Brodholt and D. Rubie for carefully reading the manuscript. Experiments at the Photon Factory were performed at BL 13 (#99S2001) and at BL14 (#99G223).

References

Anderson, O.L., Isaak, D.G., Yamamoto, S., 1989. Anharmonicity and the equation of state for gold. J. Appl. Phys. 65, 1534–1543.

Andrault, D., Bolfan-Casanova, N., Guignot, N., 2001. Equation of state of lower mantle (Al, Fe)-$MgSiO_3$ perovskite. Earth Planet. Sci. Lett. 193, 501–508.

Bacerro, A.I., McCammon, C., Langenhorst, F., Seifert, F., Angel, R.J., 1999. Oxygen-vacancy ordering in $CaTiO_3$-$CaFeO_{2.5}$ perovskite: from isolated defects to infinite sheets. Phase Trans. 69, 133–146.

Brodholt, J.P., 2000. Pressure-induced changes in the compression mechanism of aluminous perovskite in the Earth's mantle, 2000. Nature 407, 620–622.

Daniel, I., Cardon, H., Fiquet, G., Guyot, F., Mezouar, M., 2001. Equation of state of Al-bearing perovskite to lower mantle pressure conditions. Geophys. Res. Lett. 28, 3789–3792.

Funamori, N., Yagi, T., Utsumi, W., Kondo, T., Uchida, T., Funamori, M., 1996. Thermoelastic properties of $MgSiO_3$ perovskite determined by in situ X-ray observations up to 30 GPa and 2000 K. J. Geophys. Res. 101, 8257–8269.

Irifune, T., Koizumi, T., Ando, J., 1996. An experimental study of the garnet-perovskite transformation in the system $MgSiO_3$-$Mg_3Al_2Si_3O_{12}$. Phys. Earth Planet. Inter. 96, 147–157.

Ito, E., Matsui, Y., 1978. Synthesis and crystal-chemical characterization of $MgSiO_3$ perovskite. Earth Planet. Sci. Lett. 38, 443–450.

Ito, E., Kubo, A., Katsura, T., Akaogi, M., Fujita, T., 1998. High-pressure transformation of pyrope ($Mg_3Al_2Si_3O_{12}$) in a sintered diamond cubic anvil assembly. Geophys. Res. Lett. 25, 821–824.

Knittle, E., Jeanloz, R., 1987. Synthesis and equation of state of (Mg, Fe)SiO_3 perovskite to over 100 GPa. Science 235, 668–670.

Kubo, A., Akaogi, M., 2000. Post-garnet transitions in the system $Mg_4Si_4O_{12}$-$Mg_3Al_2Si_3O_{12}$ up to 28 GPa: phase relation of garnet, ilmenite and perovskite. Phys. Earth Planet. Inter. 121, 85–102.

Kubo, A., Yagi, T., Ono, S., Akaogi, M., 2000. Compressibility of $Mg_{0.9}Al_{0.2}Si_{0.9}O_3$ perovskite. Proc. Jpn. Acad. 76B, 103–107.

Kudoh, Y., Ito, E., Takeda, H., 1987. Effect of pressure on the crystal structure of perovskite-type $MgSiO_3$. Phys. Chem. Miner. 14, 350–354.

Liu, L., 1974. Silicate perovskite from phase transformations of pyrope-garnet at high pressure and temperature. Geophys. Res. Lett. 1, 277–280.

Mao, H.K., Xu, J., Bell, P.M., 1986. Calibration of ruby pressure gauge to 800 kbar under quasi-hydrostatic conditions. J. Geophys. Res. 91, 4673–4676.

Mao, H.K., Hemley, R.J., Fei, Y., Chen, L.C., Jephcoat, A.P., Wu, Y., Bassett, W.A., 1991. Effect of pressure, temperature and composition on lattice parameters and density of (Fe, Mg)SiO_3-perovskites to 30 GPa. J. Geophys. Res. 96, 8069–8079.

Ross, N.L., Angel, R.J., Seifert, F., 2002. Compressibility of brownmillerite ($Ca_2Fe_2O_5$): effect of vacancies on the elastic properties of perovskites. Phys. Earth Planet. Inter. 129, 145–151.

Yagi, T., Mao, H.K., Bell, P.M., 1978. Structure and crystal chemistry of perovskite-type $MgSiO_3$. Phys. Chem. Miner. 3, 97–110.

Yagi, T., Mao, H.K., Bell, P.M., 1982. Hydrostatic compression of perovskite-type $MgSiO_3$, vol. 2, in: S.K. Saxena (Ed.), Advances in Physical Chemistry, Springer-Verlag, New York, pp. 317–325.

Yagi, T., Yusa, H., Yamakata, M., 1996. An apparatus to load gaseous materials to the diamond-anvil cell. Rev. Sci. Instrum. 67, 2981–2984.

Yagi, T., Kondo, T., Watanuki, T., Shimomura, O., Kikegawa, T., 2001. Laser heated diamond anvil apparatus at the Photon Factory and SPring-8: problems and improvements. Rev. Sci. Instrum. 72, 1293–1296.

Weng, K., Mao, H.K., Bell, P.M., 1982. Lattice parameters of the perovskite phase in the system $MgSiO_3$-$CaSiO_3$-Al_2O_3. Yearbook Carnegie Institution of Washington 81, 273–277.

Zhang, J., Weidner, D.J., 1999. Thermal equation of state of aluminum-enriched silicate perovskite. Science 284, 782–784.

Available online at www.sciencedirect.com
SCIENCE DIRECT

ELSEVIER

Physics of the Earth and Planetary Interiors 143–144 (2004) 93–105

PHYSICS OF THE EARTH AND PLANETARY INTERIORS

www.elsevier.com/locate/pepi

Elasticity and strength of calcium silicate perovskite at lower mantle pressures

Sean R. Shieh [a,*], Thomas S. Duffy [a], Guoyin Shen [b]

[a] Department of Geosciences, Princeton University, Princeton, NJ 08544, USA
[b] GSECARS, University of Chicago, Chicago, IL 60439, USA

Received 29 November 2002; received in revised form 6 August 2003; accepted 16 October 2003

Abstract

$CaSiO_3$ perovskite was synthesized in a diamond cell and its lattice strain anisotropy was measured under non-hydrostatic compression to conditions corresponding to 61 GPa. Experiments were performed using energy dispersive synchrotron X-ray diffraction in a radial geometry. The equation of state of $CaSiO_3$ perovskite obtained from lattice strains measured at different angles from the loading direction can describe the range of compression curves previously reported under quasi-hydrostatic and non-hydrostatic conditions. The ratio of the differential stress to the shear modulus increases from 0.016(5) to 0.039(4) for $CaSiO_3$ perovskite over pressures from 19 to 61 GPa. In combination with a theoretical prediction for the shear modulus, room-temperature yield strengths are 3–11 GPa for $CaSiO_3$ perovskite over this pressure range. Under the assumption that the effect of the tetragonal distortion is minimal, the elastic constants for $CaSiO_3$ perovskite were recovered. Single-crystal elastic constants of $CaSiO_3$ perovskite are in good agreement with theoretical predictions for the cubic phase. In particular, the elastic anisotropy, S, decreases from 0.0020(7) to 0.0004(2) GPa^{-1} over the 19–61 GPa pressure range. Comparison with theoretical elasticity data provides evidence for possible strength anisotropy.

Keywords: X-ray diffraction; Perovskite; Yield strength; Elasticity; Equation of state

1. Introduction

The single-crystal elastic properties of mantle minerals are essential for interpreting seismic wave velocities and their lateral variations, as well as for understanding seismic anisotropy in the mantle. $CaSiO_3$ perovskite is believed to be an important phase in the Earth's transition zone and lower mantle on the basis of laboratory experiments on expected mantle compositions as well as examination of natural diamond inclusions (Joswig et al., 1995; Fiquet, 2001). Furthermore, it has been suggested as a possible host of large cations such as rare earth elements (REE) and radioactive elements in the deep mantle (Kato et al., 1988). Hence the physical properties of $CaSiO_3$ perovskite are of interest for interpreting seismic observations and understanding the geodynamics and geochemistry of the Earth's deep interior.

The differential stress supported by a sample under loading is a lower bound to the yield strength, and thus, is an important parameter for characterizing mechanical behavior. The low-temperature, high-stress conditions within the non-hydrostatic diamond anvil cell provide insights into rheological behavior in the regime of low-temperature plasticity. The flow law

* Corresponding author. Present address: Department of Earth Sciences, National Cheng Kung University, Taiwan, ROC.
Tel.: +886-6-275-7575x65434; fax: +886-6-274-0285.
E-mail address: sshieh@mail.ncku.edu.tw (S.R. Shieh).

0031-9201/$ – see front matter © 2004 Elsevier B.V. All rights reserved.
doi:10.1016/j.pepi.2003.10.006

under such conditions is governed by the Peierls stress, which is the stress required to move dislocations at 0 K (Evans and Goetze, 1979), and is equivalent to the yield strength. Low-temperature plasticity is important for understanding deformation of the lithosphere (Evans and Goetze, 1979) as well as having technological applications, but relatively little is known about behavior in this regime for silicates. Correlations have been suggested between the room-temperature flow stress and activation energy for high-temperature creep (Karato et al., 1990). In addition to the stress measurements, the texture development observed in radial diffraction experiments can give insights into deformation mechanisms and active slip systems. Perovskites are an important class of compounds and characterizing their mechanical properties across a wide range of conditions is desirable.

The elastic and rheological properties of $CaSiO_3$ perovskite are very poorly understood largely because it is non-quenchable at ambient conditions. Available results on the elastic properties are restricted to analog (Kung et al., 2001) and theoretical studies (Karki and Crain, 1998). Recent studies of the orientation dependence of lattice strains under non-hydrostatic stress in the diamond anvil cell (radial X-ray diffraction method) have shown that constraints on these properties can be obtained for high-symmetry materials (Singh et al., 1998; Kavner and Duffy, 2001a; Merkel et al., 2002; Shieh et al., 2002). Most experimental studies reported that $CaSiO_3$ crystallizes in the cubic structure (Mao et al., 1989; Wang et al., 1996) and first-principles calculations also supported this observation (e.g. Hemley et al., 1987; Sherman, 1993; Wentzcovitch et al., 1995). However, other theoretical calculations favored a tetragonal structure (Chizmeshya et al., 1996; Stixrude et al., 1996) or even lower symmetry (Akber-Knutson et al., 2002; Magyari-Köpe et al., 2002). Calculated energy differences between the proposed phases are sufficiently small that the crystal structure under mantle conditions is uncertain and may be cubic even if a lower symmetry phase is favored at 0 K (Karki and Crain, 1998). Recently, a high-resolution X-ray diffraction study reported the first experimental evidence for the existence of a lower symmetry phase at room temperature through observation of small (<0.4%) splittings in the (2 0 0) and (2 1 1) diffraction lines (Shim et al., 2002). In this paper, we report an in-situ strength and elasticity of $CaSiO_3$ perovskite using energy dispersive radial X-ray diffraction together with lattice strain theory.

2. Experiment

Natural samples from Newburgh, NY were examined by powder X-ray diffraction and confirmed to be wollastonite ($CaSiO_3$). No peaks from any impurity phases were detectable. The powdered sample (grain size ~1–3 μm) was mixed with 10 wt.% Pt, which served as a laser absorber and loaded into a 100-μm Be gasket hole with a pre-indented thickness of ~25 μm in a diamond anvil cell. A small piece of gold foil was placed on the top within 5 μm of the sample center to serve as a positional reference and pressure standard. No pressure-transmitting medium was used. $CaSiO_3$ perovskite was synthesized near 20 GPa by laser heating at temperatures estimated to be less than 1800 K. The sample was heated using the Nd:YLF laser heating set-up at the GSECARS sector of the Advanced Photon Source (Shen et al., 2001). The 20–30 μm diameter laser beam was not scanned across the sample but moved in a step-wise fashion over the whole sample chamber with the laser beam position held fixed for ~2 min at each step. The beam was translated with a step size of ~10 μm in order to ensure overlap of sequential heating spots. The total heating time was more than 30 min. Transformation to the $CaSiO_3$ perovskite phase was confirmed by in situ X-ray diffraction, and no residual diffraction lines attributable to the starting material were observed.

Energy dispersive radial X-ray diffraction experiments were carried out at beamline 13-BM-D of the GSECARS sector of the Advanced Photon Source. All diffraction measurements were made without any further heating after the initial synthesis. The incident X-ray beam was collimated by slits and focused by a pair of Kirkpatrick-Baez mirrors to a size of approximately 12 μm × 11 μm. The beam was directed through the Be gasket, $CaSiO_3$ perovskite, and internal standard. The diffraction signal was detected by a solid-state Ge detector with a fixed angle at $2\theta = 11.998(4)^{\circ}$, which was calibrated with a gold foil outside the diamond cell. Data were collected only after sufficient time (i.e. 1–2 h) elapsed after each compression step to allow for stress relaxation. The change in d-spacing over this time interval was generally less than 0.2% and no further detectable change in d-spacing was observed over the measurement time interval. This was verified by recording diffraction patterns for the same orientation at the beginning and end of the measurements at a given loading step.

Spectra were collected as a function of the angle (ψ) between the diffraction plane normal and the diamond cell loading axis over the range from 0 to 90° at approximately 15° intervals. Hydrostatic pressure was determined from data collected at $\psi = 54.7°$ as described below. Lattice spacings were converted to pressures using the equations of state for platinum (Holmes et al., 1989) and/or gold (Shim et al., 2002). In general, although there is some overlap between gold and platinum peaks especially at lower ψ angles and high pressures, we found that the pressures determined from our available gold peaks are consistent with those from platinum ones, i.e. the pressure difference between gold and platinum is of order 1–3 GPa. This also indicates that errors introduced by any platinum dissolution into gold at high temperature are minimal. Aside from pressure determination, however, the gold and platinum diffraction data were not analyzed in detail.

The measured d-spacings were analyzed using the lattice strain theory described by Singh (1993) and Singh et al. (1998). A sample in the diamond anvil cell is characterized by a stress, σ_3, along the diamond cell loading axis, and a radial stress, σ_1. The tensor describing the state of stress in the sample can be divided into hydrostatic and deviatoric components. The d-spacings obtained from X-ray diffraction are then given by

$$d_m(hkl) = d_p(hkl)[1 + (1 - 3\cos^2\psi)Q(hkl)], \quad (1)$$

where d_m is the measured d-spacing and d_p is the d-spacing under hydrostatic pressure. $Q(hkl)$ is given by

$$Q(hkl) = \left(\frac{t}{3}\right)\{\alpha[2G_R^X(hkl)]^{-1} + (1-\alpha)(2G_V)^{-1}\}, \quad (2)$$

where $G_R^X(hkl)$ is the X-ray shear modulus under the Reuss (isostress) limit and G_V is shear modulus under the Voigt (isostrain) limit (Singh et al., 1998). The parameter, α, which can vary between 0 and 1, is the weighting factor for the relative degree of stress and strain continuity across grain boundaries in the sample. We assumed that the sample was under isostress conditions (Kumazawa, 1969) and hence α is taken to be 1. The effect of varying α will be considered below. According to Eq. (1), when $\psi = 54.7°$, then the quantity $1 - 3\cos^2\psi = 0$ and the measured d-spacing is equivalent to that under hydrostatic stress. The d-spacing at 54.7° is obtained from linear fits of the measured data to Eq. (1). The differential stress, t, can be obtained from the aggregate shear modulus, G, and the average $Q(hkl)$ value obtained from all measured reflections by

$$t = 6G\langle Q(hkl)\rangle. \quad (3)$$

Alternatively, the differential stress corresponding to slip along the plane (hkl) can be obtained if the elastic constants are known through

$$t(hkl) = 6G(hkl)Q(hkl). \quad (4)$$

The differential stress, t, is given by $t = \sigma_3 - \sigma_1 \le Y = 2\tau$ where τ is the shear strength, and Y is the yield strength. The elastic tensor can also be obtained from the measured lattice strain anisotropy (Singh et al., 1998). For a cubic material, a linear fit to $Q(hkl)$ versus $3\Gamma(hkl)$, where $\Gamma(hkl) = (h^2k^2 + k^2l^2 + l^2h^2)/(h^2 + k^2 + l^2)^2$, yields an intercept, m_0, and slope, m_1, that are related to the elastic compliances for the case where $\alpha = 1$ through

$$m_0 = \frac{t}{3}(S_{11} - S_{12}), \quad (5)$$

$$m_1 = -\frac{t}{3}(S_{11} - S_{12} - 0.5S_{44}), \quad (6)$$

where S_{11}, S_{12}, and S_{44} are the three elastic compliances of a cubic crystal (see Singh et al., 1998 for general expressions that apply to any value of α). Additionally, using the relationship of the bulk modulus, K, to the single-crystal compliances

$$\frac{1}{3K} = S_{11} + 2S_{12}, \quad (7)$$

we are able to calculate the three independent elastic compliances or stiffnesses of a cubic crystal (Singh et al., 1998). The bulk modulus can be obtained from the third-order Birch-Manurghan equation of state and a fit to data at $\psi = 54.7°$. The elastic anisotropy of a cubic crystal can be described by the Zener anisotropy ratio

$$A = \frac{2C_{44}}{C_{11} - C_{12}} = \frac{2(S_{11} - S_{12})}{S_{44}}. \quad (8)$$

Alternatively, the anisotropy is also commonly described using

$$S = S_{11} - S_{12} - \frac{S_{44}}{2} = \frac{S_{44}(A-1)}{2}. \quad (9)$$

In the case $\alpha = 1$, the Zener anisotropy can be obtained independent of the values of t and G from (5) and (6)

$$A = \frac{m_0}{m_0 + m_1}. \quad (10)$$

3. Results and discussion

Diffraction spectra of $CaSiO_3$ perovskite were measured to conditions corresponding to a hydrostatic pressure up to 61 GPa. With increasing ψ (or decreasing strain) all $CaSiO_3$ peaks shifted to lower energy (higher d-spacing) (Fig. 1). The total shift of the peaks from $\psi = 0$ to 90° is about 0.5–1 keV at the highest pressure of this study. While there is some variation in peak intensity with ψ, these are not systematic and appear to be primarily a result of small sample position changes. More problematic are peak overlaps which affect mainly $CaSiO_3$ perovskite's (1 1 1) line and the (2 0 0) line to a lesser degree.

The full width at half maximum (FWHM) for the $CaSiO_3$ perovskite peaks increases from 0.3–0.5 keV at low pressure to 0.5–1.0 keV at the highest pressures. Consistent with previous energy dispersive diffraction studies, we are unable to observe any splitting or anomalous broadening associated with the low-symmetry distortion found in the recent high-resolution diffraction study (Shim et al., 2002). The FWHM of the (2 0 0), (2 1 1), and (2 2 0) reflections are all similar and slightly larger than that of the (1 1 0) reflection at all pressures. Furthermore, there is no evidence for any systematic angle dependence of the peak width. As a result, we have analyzed these data using the lattice strain equations for the cubic system. This may not produce major errors if the structure remains pseudo-cubic. Indeed, the effect on the compression curve appears to be minor

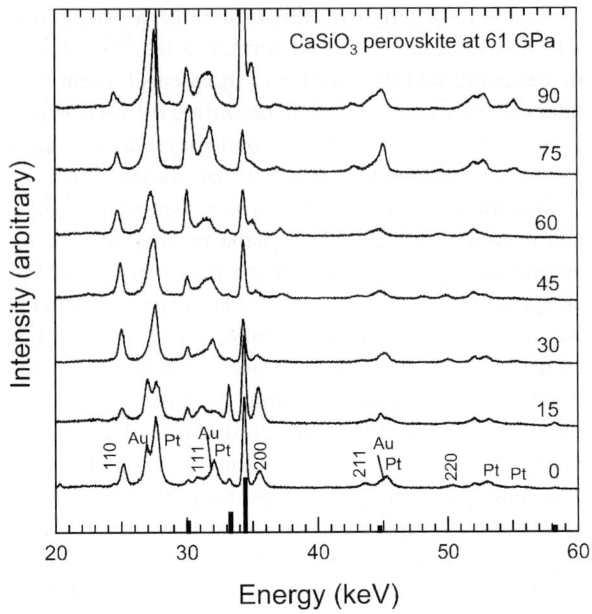

Fig. 1. Energy-dispersive X-ray diffraction spectra of $CaSiO_3$ perovskite obtained at different angles from the loading axis. The value of ψ is indicated next to each spectrum. The pressure, obtained using Eq. (1), is 61 GPa. The sample peaks are labeled with Miller indices. Au is the internal pressure marker, Pt is the laser absorber, and bold vertical lines at the bottom denote Be gasket peaks.

(Shim et al., 2002). However, the effects of the distortion on the elastic tensor and yield strength are difficult to predict: even minor distortions can have major effects near a phase transition boundary (Shieh et al., 2002). Thus, our results for the elastic tensor are only tentative and must be subjected to further examination. We should also note that the symmetry of $CaSiO_3$ perovskite at lower mantle conditions is unknown, and indeed may be cubic (Karki and Crain, 1998). Therefore, the physical properties of cubic $CaSiO_3$ perovskite may well still be of geophysical importance.

In general, our measured d-spacings show a linear relation with $1 - 3\cos^2\psi$, as expected from theory. The value of $Q(hkl)$ was determined from the slope of the linear fit to Eq. (1) and d_p. Fig. 2 shows plots of d-spacings as a function of $1 - 3\cos^2\psi$ at five different pressures. The (1 1 0), (1 1 1), (2 0 0), (2 1 1), and (2 2 0) lines were generally observable through entire range of our measurements and were used for our data analysis. From the linear fit, we obtained $d_p(hkl)$

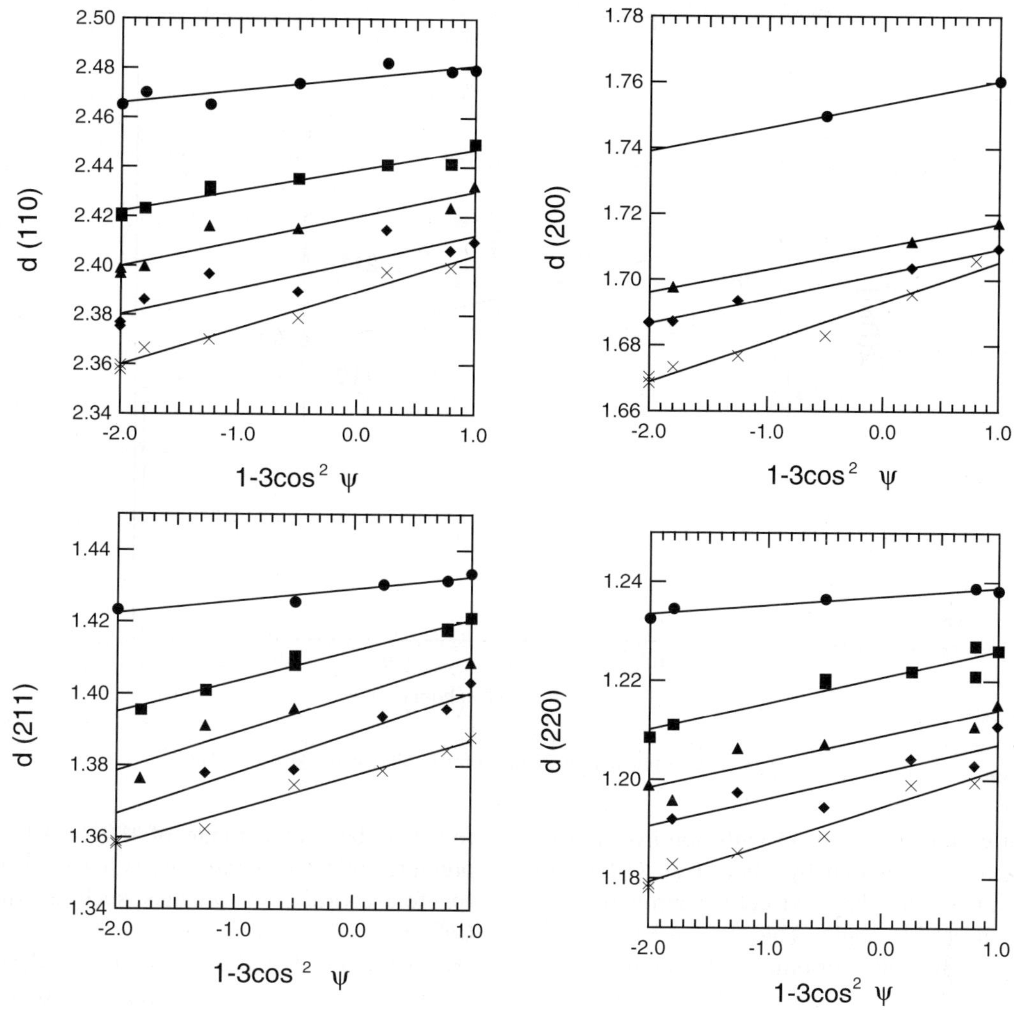

Fig. 2. d-spacings as a function of $1-3\cos^2\psi$ for $CaSiO_3$ perovskite at five different pressures. The solid circles are 18.6 GPa, solid squares are 30.6 GPa, solid triangles are 40.6 GPa, solid diamonds are 52.8 GPa, and crosses are 60.8 GPa. The solid lines are linear fits to all the individual $(h\,k\,l)$ data sets.

values at $\psi = 54.7^\circ$ or $1-3\cos^2\psi = 0$ and used these data to calculate the compression curve of $CaSiO_3$ perovskite. The corresponding pressure was determined from the d-spacing of gold and/or platinum at this angle. Fig. 3 shows a plot of $Q(h\,k\,l)$ as a function of 3Γ. The (2 1 1) line deviates from the linear trend of the data at all pressures. However, there is only a minor effect on the fitted slope and intercept if we removed this line from consideration. The linear fit of $Q(h\,k\,l)$ as function of $3\Gamma(h\,k\,l)$ (Fig. 3) provides two parameters, the intercept, m_0 and the slope, m_1 for resolving single-crystal elastic constants (Eqs. (5) and (6)) as discussed later.

Fig. 4 shows the equation of state determined at angles of 0°, 54.7°, and 90°. In each case, we assumed that the d-spacings of both the pressure marker and sample at this angle correspond to the volume compression under hydrostatic stress. While this is only strictly correct at 54.7°, this method facilitates comparison with literature data which typically uses the same assumption in analyzing axial diffraction data even when the stress is not hydrostatic. Data from $\psi =$

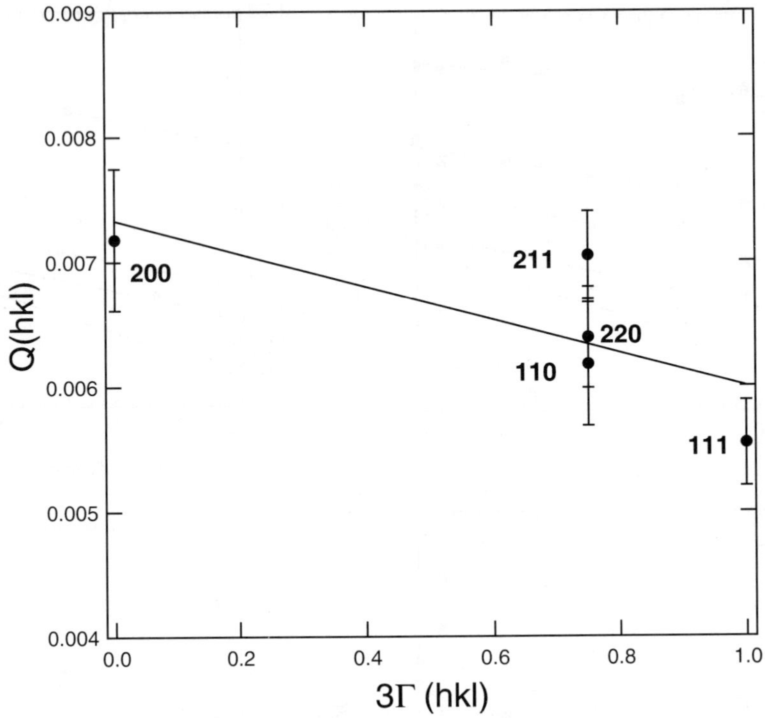

Fig. 3. $Q(hkl)$ as a function of $3\Gamma(hkl)$ for $CaSiO_3$ perovskite at 61 GPa.

90° are in excellent agreement with non-hydrostatic compression data reported by Mao et al. (1989) as expected since axial diamond cell experiments are performed near this angle. This is consistent with radial diffraction results for other materials including SiO_2 (Shieh et al., 2002) and rhenium (Duffy et al., 1999). The data from $\psi = 54.7°$ agree best with quasi-hydrostatic data by Shim et al. (2002) and Wang et al. (1996), but are less compressible than earlier data of Shim et al. (2000). A third-order Birch-Manurghan equation of state fit to the data at 54.7° yields a bulk modulus of 246(18) GPa and pressure derivative of 4.5(1.0), which can be compared with the results of Wang et al. (1996) ($K_o = 232(8)$ GPa, $K'_o = 4.8(3)$), Shim et al. (2000) ($K_o = 236(4)$ GPa, $K'_o = 3.9(2)$) and Shim et al. (2002) ($K_o = 255$ GPa, $K'_o = 4$). We note that Shim et al. (2002)'s equation of state data are based on fitting to a tetragonal cell.

The ratio of differential stress to shear modulus, t/G, is directly obtained from the slope of the d-spacing versus $1 - 3\cos^2\psi$ relationship (Eqs. (1) and (3)). In Fig. 5, t/G of $CaSiO_3$ perovskite is found to be in the range of 0.016(5)–0.039(4) for pressures of 19–61 GPa. This is lower than observed for four-coordinated silicates such as ringwoodite (t/G = ~5–7% in the pressure range of 6–27 GPa) (Kavner and Duffy, 2001a) and olivine (Uchida et al., 1995) but comparable to another six-coordinated silicate, stishovite and its $CaCl_2$-type modification (t/G = ~2–4% in the pressure ranges of 15–61 GPa) (Shieh et al., 2002). Therefore, these results support the previously observation (Shieh et al., 2002) that strength as a fraction of shear modulus for six-coordinated silicates is in general lower than that of four-coordinated silicates at high pressures. Similar results have been reported for silica glass using radial pressure gradient measurements in a diamond anvil cell (Meade and Jeanloz, 1990). It is also worth noting that the t/G values of $CaSiO_3$ perovskite are close to stishovite values, which may indicate a common behavior for octahedral silicates at high pressure. Our observations here can be compared to systematic relations for the normalized flow stress, σ/G, observed for different crystal structures of dense oxides in high-temperature

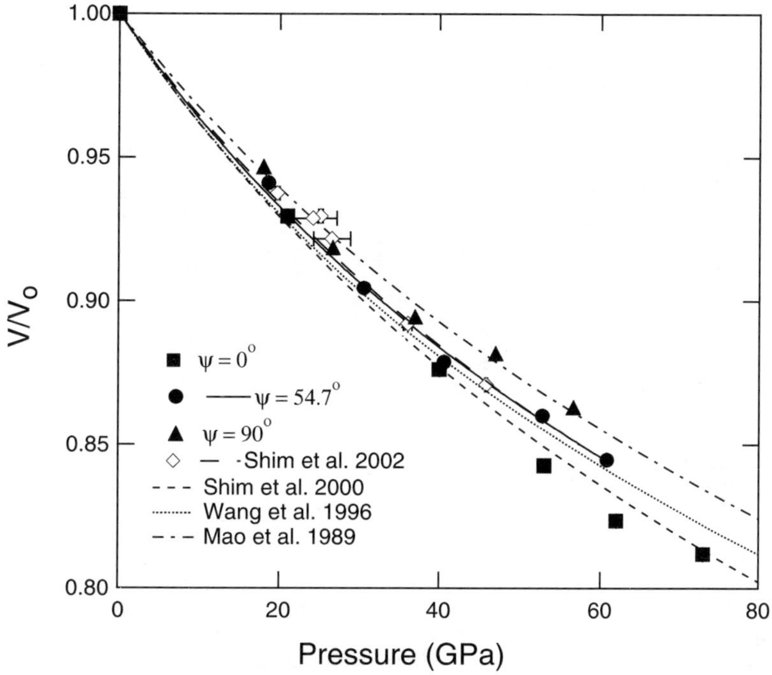

Fig. 4. Equation of state of $CaSiO_3$ perovskite at $\psi = 0$, 54.7, and 90°. Solid symbols are from this study and solid line is a fit to a third-order Birch-Manurghan equation of state. Open diamonds are from Shim et al. (2002) and long dash line is a fit to their data.

deformation experiments (Karato, 1989). In that study, the normalized flow stress was observed to decrease in the order spinel, olivine, and perovskite just as we observed in our low-temperature experiments (Fig. 5). A systematic dependence of yield strength on crystal structure, thus, appears to extend across high- and low-temperature deformation regimes. Further studies are needed, however, to confirm the structural trends observed in diamond cell experiments.

To obtain the differential stress supported by $CaSiO_3$ perovskite, it is necessary to know the shear modulus. While estimates of the shear modulus of $CaSiO_3$ perovskite have been made at ambient pressure based on elasticity systematics (Kung et al., 2001; Sinelnikov et al., 1998), there are no high-pressure experimental data available. However, the shear modulus for cubic $CaSiO_3$ has been determined by density functional theory calculations to pressures above 100 GPa (Karki and Crain, 1998). Using these values (and assuming $t = Y$), the yield strength of $CaSiO_3$ perovskite is found to increase from 3(1) GPa at 19 GPa to 11(1) GPa at 61 GPa (Fig. 6).

We find that six-coordinated silicates (e.g. $CaSiO_3$ perovskite and stishovite) support lower differential stresses than four-coordinated silicates even though the shear moduli of the six-coordinated silicates is larger. There have been several measurements of strength of orthorhombic perovskites with compositions at or near $(Mg_{0.9}Fe_{0.1})SiO_3$ (Meade and Jeanloz, 1990; Chen et al., 2002; Merkel et al., 2003) (Fig. 6). Meade and Jeanloz (1990) used measurements of pressure gradients across the sample to infer shear strengths which we converted to differential stress (yield strength) in Fig. 6. Merkel et al. (2003) used a similar technique as our study, and Chen et al. (2002) inferred the strength at room and high temperatures from measurements of diffraction peak linewidths in a multi-anvil press. For the data of Merkel et al. (2003) and Chen et al. (2002), the reported differential stresses for $(Mg_{0.9}Fe_{0.1})SiO_3$ are 2–3% of the shear modulus, consistent with our results for six-coordinated silicates. The strength of $(Mg_{0.9}Fe_{0.1})SiO_3$ perovskite is greater than $CaSiO_3$ perovskite. However, there remains considerable

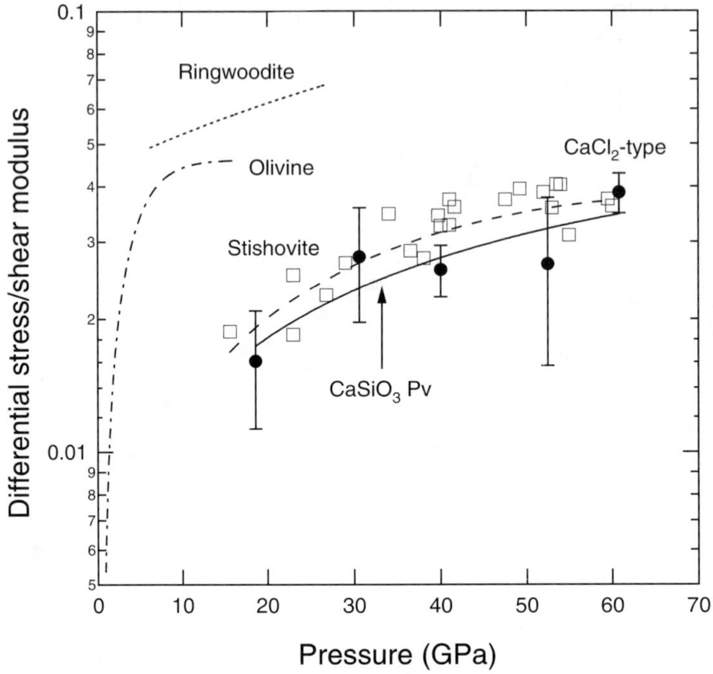

Fig. 5. The ratio of differential stress to shear modulus for $CaSiO_3$ perovskite (solid circles and line). Dotted line—ringwoodite (Kavner and Duffy, 2001a); dot-dashed line—olivine (Uchida et al., 1995); open squares and dashed line—SiO_2 (Shieh et al., 2002). The stishovite–$CaCl_2$-type phase boundary is near 50 GPa.

uncertainty in the strength of $(Mg_{0.9}Fe_{0.1})SiO_3$ as reported values vary by nearly a factor of 3 at 15–20 GPa. In high temperature experiments, it has been noted that the normalized creep strengths of perovskite compositions with orthorhombic symmetry are larger than those of compositions exhibiting cubic symmetry (Wang et al., 1999).

It is notable that lattice strain anisotropy and appreciable deviatoric stresses are observed at our lowest compression step (∼19 GPa) (Figs. 2 and 6) which was measured after laser heating but without further compression. It is generally believed that laser annealing greatly reduces or eliminates differential stresses in diamond cell samples. However, measurements of lattice strain anisotropy now allow the reduction in differential stress to be quantified. Our findings here are consistent with previous results from our laboratory (Shim et al., 2000; Kavner and Duffy, 2001b) which document significant residual differential stresses after laser heating. We note, however, that all these observations follow laser heating restricted to a single heated spot or a series of heated spots, rather than continuous rastering of the laser beam across the sample.

In cases where the elastic tensor is known, Eq. (2) can be solved for individual diffraction lines to investigate possible anisotropy of t (Eq. (4)). For the cubic crystal system, $G_R(h\,k\,l)$ can also be related to the Young's modulus, $E(h\,k\,l)$ in the direction normal to the diffraction plane, and the linear compressibility, β (Uchida et al., 1996)

$$\frac{1}{2G_R^X(hkl)} = \frac{1}{2}\left(\frac{3}{E(hkl)} - \beta\right) = S_{11} - S_{12} - 3S\Gamma(hkl). \quad (11)$$

The results for specific diffraction lines and converting compliances, S_{ij}, into elastic stiffnesses, C_{ij}, are

$$G_R^X(200) = \frac{1}{2}(C_{11} - C_{12}), \quad (12)$$

$$G_R^X(111) = C_{44}, \quad (13)$$

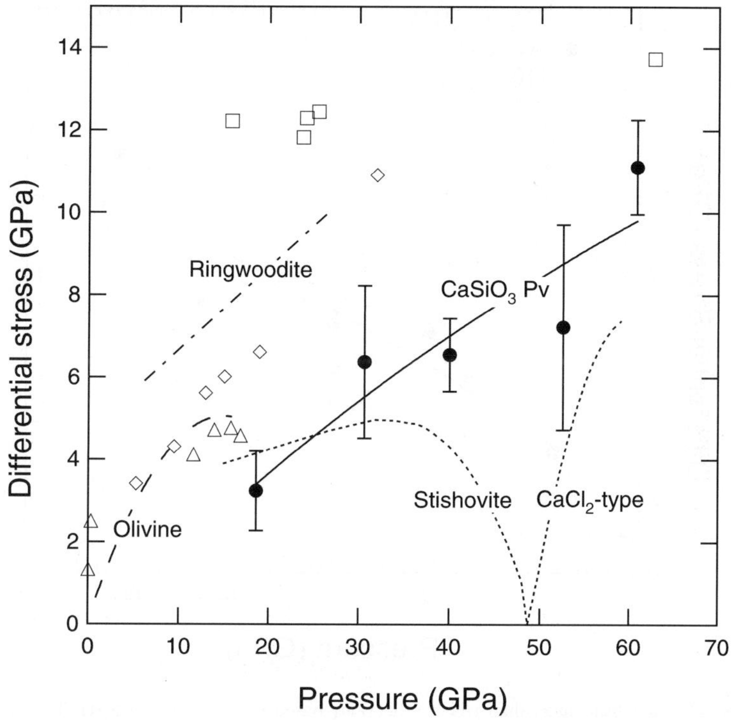

Fig. 6. The supported differential stress for $CaSiO_3$ perovskite using the (cubic) shear modulus of Karki and Crain (1998) (solid circles and line). The dash-dotted line denotes ringwoodite values (Kavner and Duffy, 2001a); dashed line denotes olivine values (Uchida et al., 1995); dotted line denotes SiO_2 values (stishovite to 50 GPa, $CaCl_2$-type structure at >50 GPa) (Shieh et al., 2002). Data for perovskite compositions close to $(Mg_{0.9}Fe_{0.1})SiO_3$ are shown as open squares (Meade and Jeanloz, 1990), open triangles (Chen et al., 2002), and open diamonds (Merkel et al., 2003).

$$G_R^X(211) = G_R^X(110) = \frac{4C_{44}}{A+3}. \quad (14)$$

Using the cubic elastic tensor results of Karki and Crain (1998), we find, for example at 41 GPa, $G_R(2\,0\,0) = 195$ GPa, $G_R(1\,1\,0) = G_R(2\,1\,1) = 266$ GPa, and $G_R(1\,1\,1) = 303$ GPa. Using these results in Eq. (4), we obtain the differential stresses, $t(h\,k\,l)$, shown in Fig. 7. While the (1 1 0), (2 2 0), (2 0 0), and (1 1 1) planes yield similar values, $t(h\,k\,l)$ corresponding to the (2 1 1) planes is larger at high pressure, as expected from Fig. 3. The high value of (2 1 1) relative to (1 1 0) is not affected by uncertainty in Reuss shear modulus for these planes as $G_R(2\,1\,1) = G_R(1\,1\,0)$. Our results suggest that strengths on the (1 0 0), (1 1 0), and (1 1 1) planes are similar in magnitude. This is similar to high-temperature observations on oxide perovskites in which slip is observed on $\langle 0\,0\,1\rangle\{1\,0\,0\}$, $\langle 1\,1\,0\rangle\{1\,\bar{1}\,0\}$, and $\langle 1\,\bar{1}\,0\rangle\{1\,1\,1\}$ systems (Beauchesne and Poirier, 1990). However, deformation experiments on oxide perovskites have revealed a wide range of creep laws, and it is not easy to generalize across compositions (Wright et al., 1992).

An alternative approach, and the way in which strain anisotropy data in the diamond anvil cell have been normally analyzed (e.g. Singh et al., 1998; Duffy et al., 1999; Merkel et al., 2002), is to assume no strength anisotropy and to directly invert for the elastic moduli (Eqs. (3), (5)–(7)). Again, we emphasize that the following results are dependent on the assumption that the slight tetragonal (or lower symmetry) distortion observed by Shim et al. (2002), but undetectable with the techniques of this study, do not affect the results of Fig. 2. This assumption will need to be tested in future work. In general, our single-crystal elastic constants for $CaSiO_3$ perovskite show good agreement with first-principles calculations (Karki

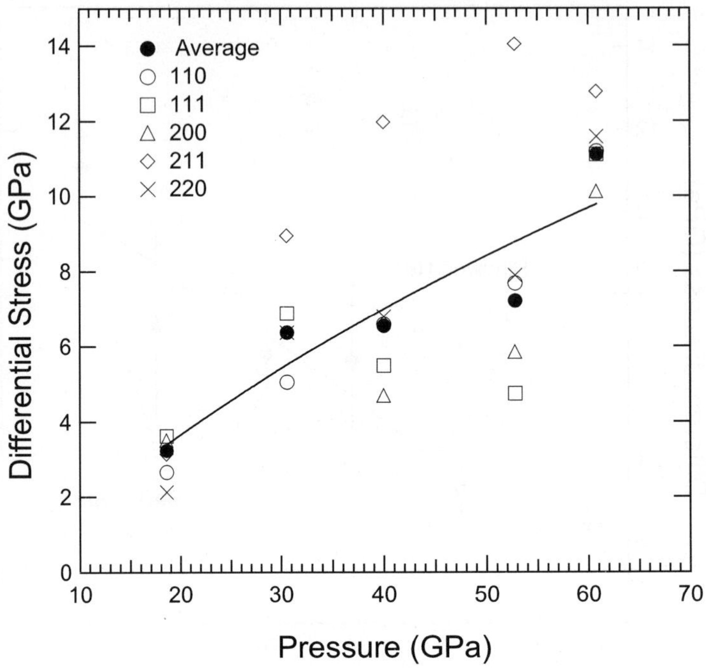

Fig. 7. The differential stress obtained from individual ($h\,k\,l$) reflections based on Eq. (4) (see text). The solid circles are data from the average value of $Q(h\,k\,l)$ based on Eq. (3).

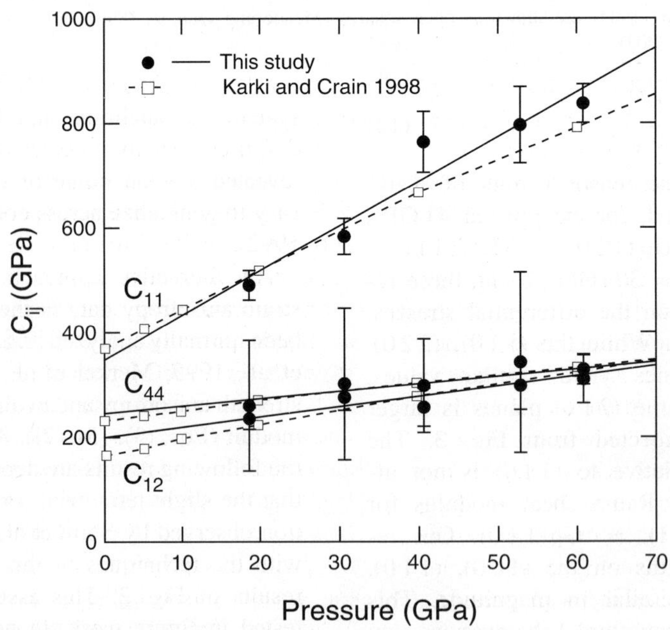

Fig. 8. The second-order cubic elastic constants of cubic $CaSiO_3$ perovskite as a function of pressure. Solid symbols and lines are from this study. Open squares and dashed lines are from Karki and Crain (1998).

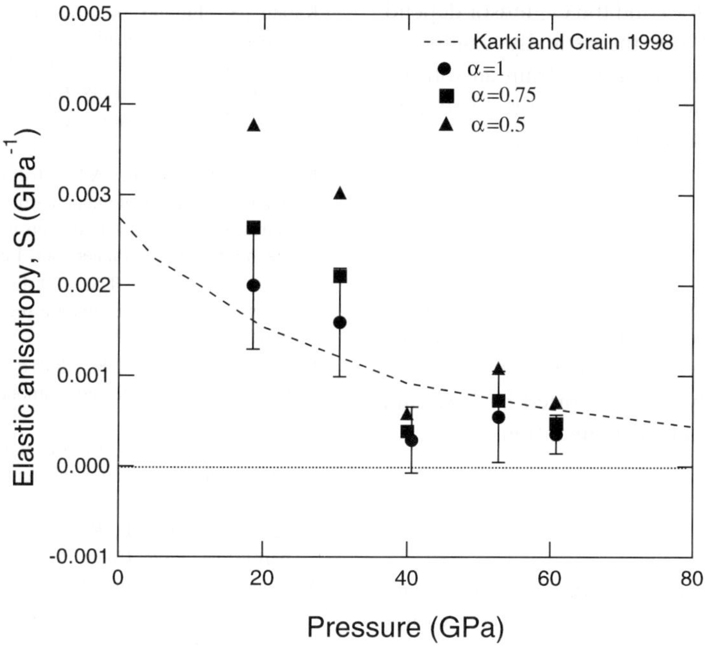

Fig. 9. Elastic anisotropy, S, as a function of pressure. The dashed line is from Karki and Crain (1998). For clarity, error bars are only shown for $\alpha = 1$.

and Crain, 1998) (Fig. 8). When we vary α, the parameter governing stress continuity, from 1 to 0.5, it is found that C_{11} decreases and C_{12} increases. For C_{44}, the values obtained at $\alpha = 0.5$ are much larger at low pressure but becomes close to those of the theoretical predication and values from $\alpha = 1$ at high pressure. The anisotropy factor, $S = S_{11} - S_{12} - 0.5S_{44}$, decreases with increasing pressure, from 0.0020(7) to 0.0004(2) GPa^{-1} when α is assumed to be 1 (Fig. 9). The observed trend is in general good accord with the theoretical predictions (Karki and Crain, 1998). When we reduce the value of α to 0.75 and 0.5, the elastic anisotropy increases by a more pronounced amount at low pressure. Furthermore, the anisotropy deviates increasingly from the theoretical results, suggesting that $\alpha \approx 1$ is appropriate for this experiment. In fact, we continue to observe a strong decrease in elastic anisotropy with pressure in all cases. Thus, a decrease in the elastic anisotropy of $CaSiO_3$ with pressure, consistent with theoretical predictions, is supported by our data, even in the case where α decreases with pressure.

4. Conclusion

Radial X-ray diffraction of $CaSiO_3$ perovskite under non-hydrostatic loading and assuming a pseudo-cubic structure yields a quasi-hydrostatic equation of state in good agreement with previous experiments. The ratio of differential stress to shear modulus, *t/G* varies between 2 and 4% in the pressure range of 19–61 GPa. This value is similar to that of stishovite and suggests a common behavior of six-coordinated silicates at mantle pressures. The differential stress or yield strength for $CaSiO_3$ perovskite increases from 3 to 11 GPa over the pressure range of our study. Direct inversion for the cubic elastic constants and elastic anisotropy of $CaSiO_3$ perovskite from lattice strains yields good agreement with theoretical predictions (Karki and Crain, 1998). However, there is evidence for possible strength anisotropy. Our results indicate the elastic anisotropy decreases with compression. While the recently discovered symmetry-reducing distortion of $CaSiO_3$ perovskite does not appear to measurably affect the angle dependence of lattice

Shim, S.H., Jeanloz, R., Duffy, T.S., 2002. Tetragonal structure of $CaSiO_3$ perovskite above 20 GPa. Geophys. Res. Lett. 29 (24), 10.1029/2002GL016148.
Shukla, Sidorkov, Y.D., Chen, G., Liebermann, R.C., 1998. Elasticity of $CaTiO_3$–$CaSiO_3$ perovskites. Phys. Chem. Miner. 25, 515–520.
Singh, A.K., 1993. The lattice strain in a specimen (cubic system) compressed nonhydrostatically in an opposed anvil device. J. Appl. Phys. 73, 4278–4286.
Singh, A.K., Balasingh, C., Mao, H.K., Hemley, R.J., Shu, J., 1998. Analysis of lattice strains measured under nonhydrostatic pressure. J. Appl. Phys. 83, 7567–7575.
Stixrude, L., Cohen, R.E., Yu, R., Krakauer, H., 1996. Prediction of phase transition in $CaSiO_3$ perovskite and implications for lower mantle structure. Am. Mineral. 81, 1293–1296.
Uchida, T., Funamori, N., Ohtani, T., Yagi, T., 1996. Differential stress of MgO and Mg_2SiO_4 under uniaxial stress field: variation with pressure, temperature, and phase transition. In: Proceedings of the 15th AIRAPT Conference, Warsaw, 1995, pp. 183–185.
Uchida, T., Funamori, N., Yagi, T., 1996. Lattice strains in crystals under uniaxial stress field. J. Appl. Phys. 80, 739–746.
Wang, Y., Weidner, D.J., Guyot, F., 1996. Thermal equation of state of $CaSiO_3$ perovskite. J. Geophys. Res. 101, 661–672.
Wang, Z., Dupas-Bruzek, C., Karato, S., 1999. High temperature creep of an orthorhombic perovskite—$YAlO_3$. Phys. Earth Planet. Inter. 110, 51–69.
Wentzcovitch, R.M., Ross, N.L., Price, G.D., 1995. Ab initio study of $MgSiO_3$ and $CaSiO_3$ perovskites at lower-mantle pressures. Phys. Earth Planet. Inter. 90, 101–112.
Wright, K., Price, G.D., Poirier, J.P., 1992. High-temperature creep of the perovskites $CaTiO_3$ and $NaNbO_3$. Phys. Earth Planet. Inter. 74, 9–22.

Available online at www.sciencedirect.com
SCIENCE DIRECT®

PHYSICS OF THE EARTH AND PLANETARY INTERIORS

ELSEVIER

Physics of the Earth and Planetary Interiors 143–144 (2004) 107–128

www.elsevier.com/locate/pepi

Equations of state of Na–K–Al host phases and implications for MORB density in the lower mantle

Nicolas Guignot*, Denis Andrault

Laboratoire des Géomatériaux, UMR7046, Institut de Physique du Globe de Paris, Université Denis Diderot Paris 7, 4 place Jussieu, 75252 Paris, France

Received 31 January 2003; received in revised form 25 June 2003; accepted 21 September 2003

Abstract

We report equations of state (EoS) of two important silicate phases known to store Na, K, and Al, at pressures and temperatures prevailing in the Earth's lower mantle. Using laser heated diamond anvil cell, we produced the Ca-ferrite type (CF) and the new aluminum rich (NAL) phases using different synthetic and natural starting materials, at pressures ranging from 23 to 65 GPa. Unit cell volumes were investigated by means of angle dispersive X-ray diffraction at the ID30 beamline of the ESRF synchrotron ring (Grenoble, France). CF-phases were synthesized in $NaAlSiO_4$–$MgAl_2O_4$ and $NaAlSiO_4$–$(Mg,Fe)_2SiO_4$ systems, and for a natural mid-ocean ridge basalt (MORB). The insertion of $MgAl_2O_4$ in the $NaAlSiO_4$-based lattice has reduced effects on the CF-phase EoS, whereas that of $(Mg,Fe)_2SiO_4$ is mainly to increase the room pressure volume. CF-phases most relevant to the Earth's mantle show compression behavior comparable to the pure $NaAlSiO_4$-CF-phase, with bulk moduli values close to $K_0 = 190$ GPa and $K'_0 = 4$. We also synthesized (NAL) phases in $NaAlSiO_4$–$MgAl_2O_4$ system, and using CaNa- and K-rich silicate glasses as starting materials. All NAL-compounds investigated show similar compression behavior, except for one anomalously $NaAlSiO_4$-rich composition. For NAL-phase with the most "natural" conditions relevant composition, we refine bulk moduli of $K_0 = 182$ GPa and $K'_0 = 6$.

We then model the density profiles expected for the CF and NAL-phases, using our newly determined EoS and the most accurate chemical compositions reported in the literature. These two phases appear significantly less dense than a pyrolitic type material at all lower mantle pressures and 300 K. However, the density profiles calculated for mineral assemblages corresponding to typical MORB compositions point out to denser MORBs compared to the pyrolite. This is due to the high Fe-content in MORB, yielding a densest Fe-rich perovskite that counterbalances the lower densities of the CF and NAL-phases. The density gap is calculated between 0.4 and 4.0% depending on the starting MORB composition and pressure.

Keywords: Alkaline elements; Aluminum; Lower mantle; Equation of state; MORB

1. Introduction

The behavior of alkaline elements in the deep Earth was already of prime interest at the earliest stage of the high-pressure mineral physics experiments, as attested by the work of Ringwood et al. (1967) on germanate analogous compounds. This report, based on the analysis of the $NaAlGeO_4$ phase diagram, evidenced the importance of the Ca-ferrite (CF) structure in the storage of alkaline elements. Since then, CF, and other derived structures, have been reported in many high-pressure experiments, some being performed on

* Corresponding author. Tel.: +33-144-272488; fax: +33-144-272487.
E-mail address: guignot@ipgp.jussieu.fr (N. Guignot).

0031-9201/$ – see front matter
doi:10.1016/j.pepi.2003.09.014

natural samples at pressures and temperatures characteristic of the Earth's lower mantle. Therefore, these structures should play a major role in the mineralogy and density of the Earth's materials enriched in alkaline elements, such as mid-ocean ridge basalts (MORB). Studying their properties at high pressures and temperatures will therefore help to determine the fate of the subducting slabs into the Earth's mantle (see Irifune and Ringwood, 1993; Kesson et al., 1994b; Hirose et al., 1999; Ono et al., 2001).

There are in fact three different candidate phases to host the alkaline elements in the lower mantle: the CF-phase, the new Al-rich phase, and the $KAlSi_3O_8$ hollandite (Ho) that was shown to be stable up to 95 GPa (Tutti et al., 2001).

It appears that the relative stability of the these three phases is not only dependent on their relative densities, but instead preferentially governed by the composition of the starting material. The presence of the NAL-phase in geochemically relevant systems seems to be correlated to the presence of sufficient amount of K, while the CF-phase appears to be the main host for Na. In the case of an extremely high K content like in sediments, however, hollandite is expected as a K host phase. In MORB-type materials, the Na and K contents can be high enough to produce both CF and NAL-phases. In contrast, the Na and K contents seem too small to require the formation of these phases in the case of a pyrolitic composition (Irifune, 1994; Andrault, 2003).

A large variety of Al-based compounds were already reported to adopt the Ca-ferrite structure, including $MgAl_2O_4$ (Irifune et al., 1991) and $CaAl_2O_4$ (Reid and Ringwood, 1969; Akaogi et al., 1999). It is thus an important candidate to be a main Al host phase in the lower mantle, especially in the case of high-aluminum contents such as in MORB composition. In addition, the CF-structure can host alkaline elements such as Na, as illustrated by the breakdown of jadeite into a mixture of $NaAlSiO_4$-CF and stishovite above 20 GPa (Liu, 1977). In this structure, the Na cations are located in tunnels formed by the linked $(Al,Si)O_6$ octahedra, thus being eightfold coordinated. The structural refinement of the $NaAlSiO_4$-CF-phase results in unit cell parameters of $a = 10.1546(8)$ Å, $b = 8.6642(8)$ Å, $c = 2.7385(4)$ Å, $V = 240.93(3)$ Å^3, the *Pbnm* (or *Pnma*) space group, for a room pressure density of 3.916 g/cm^3 (Yamada et al., 1983). It was shown that this phase can hardly accommodate larger cations such as K (Ringwood et al., 1967).

The phase diagram in the $MgAl_2O_4$–$CaAl_2O_4$ join has been studied at high pressure by Akaogi et al. (1999). Both end-members transform to the same CF-form, but the solid solution appears difficult to realize. As a result, a new Al-rich phase (called NAL-phase by Miyajima et al., 1999, or hexagonal aluminous phase by Akaogi et al., 1999) is obtained for the $CaMg_2Al_6O_{12}$ composition. The cell constants are $a = 8.7616(2)$ Å and $c = 2.7850(1)$ Å, for a hexagonal lattice with *P63/m* space group (Miura et al., 2000). Its local structure is very close to that of the CF-phase, as early suggested by Kesson et al. (1994a). The main difference is the occurrence of two different types of tunnels compared with the only one present in the CF. This characteristic is certainly a key to understand the occurrence of the NAL-phase in various systems, because the insertion of different types of cations, especially large cations like Ca and K, is much facilitated.

Concerning silicates, the occurrence of NAL-phase was reported at several occasions after Irifune and Ringwood (1993) and Kesson et al. (1994a) pioneer studies. For example, the breakdown of a pyrolitic garnet yields a mixture of (Mg,Fe)-perovskite, corundum, and NAL-phase with complex chemical composition (Miyajima et al., 1999). By compiling data from several reports, the general stoichiometric formulae can be expressed as $[K,Na,Ca]_1$-$[Mg,Fe^{II}]_2$-$[Si,Al,Fe^{III},Mg,Fe^{II}]_6O_{12}$ (Akaogi et al., 1999; Miyajima et al., 1999; Gasparik et al., 2000). The largest (K,Na,Ca) cations enter the M3 site (with also possibly some (Mg,Fe^{II})), the M2 smaller site accommodates (Mg,Fe^{II}) almost exclusively, and the M1 small octahedral site accommodates (Si,Al,Fe^{III}), and some (Mg,Fe^{II}).

In this work we investigate the elastic parameters and stability of the CF and NAL-phases under lower mantle pressure and temperature conditions. Our goal is to determine their role in the lower mantle mineralogy, especially in zones where Al-rich phases are abundant such as in the MORB or sediment constituent of subducting slabs. The knowledge of their respective densities can help to determine the density contrast between the different types of material in the deep mantle.

2. Experimental procedure

2.1. Samples

Several starting materials were used (Table 1): natural albite $NaAlSi_3O_8$ from Modane, natural Na-rich MORB (EDUL-DR28 type 2; Humler and Besse, 2002), synthetic glass with composition close to jadeite $NaAlSi_2O_6$, mixture of 75% olivine and 25% albite in mole (AlbOl), mixture of 50% $NaAlSiO_4$ glass, 25% MgO, and 25% Al_2O_3 (MgNa-mix), mixture of $MgAl_2O_4$ and (CaNa)-rich or K-rich glasses to approach the (Ca,Na)-rich or K-rich compositions of the NAL-phase (hereafter named CaNa-mix or K-mix), respectively. Some of these samples, especially those made of mixtures of different materials, were finely ground in an agate mortar for several minutes before the loading in the diamond anvil cell (DAC). Some of the samples were also mixed with 10 wt.% Pt, depending on the IR-laser used for the synthesis at high pressures (see Table 2).

Table 1
Chemical compositions of starting materials (in weight percent)

	Albite	Jadeite[a]	Olivine	Na-rich MORB[d]
SiO_2	68.09	57.65	42.52	50.75
TiO_2	–	–	–	1.5
Al_2O_3	19.61	26.34	–	16.75
FeO	–	–	11.01	8.50
MgO	0.03	–	47.62	7.30
CaO	–	–	0.20	10.36
Na_2O	11.72	16.01	–	4.50
K_2O	0.01	–	–	0.21
Total	99.46	100.00	101.35	100.07
	AlbOl[b]	CaNa-mix[b]	K-mix[b]	MgNa-mix[c]
SiO_2	51.9	38.4	41.4	21.1
Al_2O_3	6.2	31.8	33.7	53.8
FeO	11.4	–	–	–
MgO	27.5	25.3	21.7	14.2
CaO	0.2	4.1	–	–
Na_2O	3.5	1.4	–	10.9
K_2O	–	–	3.2	–
Total	100.7	101.0	100.0	100.0

[a] Synthetic glass.
[b] Powder mixture.
[c] Oxide mixture.
[d] EDUL DR28 type 2 (Humler and Besse, 2002).

2.2. Experimental P and T conditions

We used a membrane-type DAC equipped with diamonds cut to 300 μm culets. Rhenium gaskets were pre-indented to a thickness of 35–40 μm, before electro-erosion drilling to 60–80 μm in diameter. This diameter is ideal to reduce the pressure gradient in the experimental volume. All experiments were performed using Ar pressure medium (loaded cryogenically) to achieve quasi-hydrostatic pressures. The small samples embedded in argon were heated for several minutes using a CO_2 or a YAG laser (Table 2). We used IR-lenses of 25 or 32 mm focal length, yielding hot spot diameter of typically 15–20 μm (or possibly less when using the YAG laser). The hot spot was slowly scanned over the entire samples of about 40–50 μm diameter. The annealing temperature was estimated to 2000(300) K from the visual observation of the color and intensity of the sample thermal emission. The laser power was adjusted at each experimental pressure to achieve a similar annealing temperature. Due to this procedure, we note that the phase composition could eventually evolve along the compression curve in certain cases.

For most experiments, a small ruby chip was deposited on the border of the Re-gasket, close to the sample. Measurement of the pressure, using the fluorescence technique, yields typical error bar of 2–3% of the pressure. Due to the use of IR radiation to relax the deviatoric stresses, we estimate our measurement accurate to 2%, with a minimum value of 0.5 GPa. For some experiments when the ruby fluorescence was not sufficiently intense (or absent), we used the Ar volumes to determine the pressure. We used a Vinet equation of state for solid Ar at room temperature refined from a large set of data recorded in similar experimental conditions than in this study (Guignot et al., in preperation) yielding $K_0 = 1.25$ GPa, $K'_0 = 7.89$ and $V_0 = 177.01$ Å^3. We emphasize that the pressures derived from the Ar EoS and from the ruby fluorescence (when both available) are similar within the experimental error of 2%.

2.3. X-ray diffraction set up

Angle dispersive X-ray diffraction patterns were recorded at 300 K at the ID30 beamline of the

Table 2
Details of each experiment performed in this study

Experiment	Composition	Pt	*P* range (GPa)	Laser	Recovered phases
1	Albite ($NaAlSi_3O_8$)	No	28.2–60.2	CO_2	CF, St
2	Jadeite ($NaAlSi_2O_6$)	No	23.0–65.2	CO_2	CF, St, Co
3	$NaAlSiO_4$–$MgAl_2O_4$ mix	Yes	18.9–45.0	YAG	NAL, Co (traces)
4	$NaAlSiO_4$–$MgAl_2O_4$ mix	Yes	39.2–57.8	YAG	NAL, CF, Co (traces)
5	75% olivine + 25% albite	No	25.9–65.5	CO_2	CF, St, Pv
6	Na-rich natural MORB	No	18.4–48.8	CO_2	CF, St, Pv, Ca-Pv
7	Na-rich natural MORB	No	27.8–62.9	YAG	CF, St, Pv
8	(Ca,Na)-rich oxide mixture	No	30.7–61.9	CO_2	NAL, St, Pv
9	K-rich oxide mixture	No	27.1–48.6	CO_2	NAL, St, Pv

We report the starting composition, the eventual presence of Pt mixed with the sample, the pressure range investigated (GPa), the type of laser used to anneal the sample to about 2000 K, and the recovered phases. Ar pressure medium was used in all experiments. CF: calcium ferrite type phase; NAL: new aluminum rich phase; St: stishovite; Pv: (Mg,Fe) perovskite; Ca-Pv: Ca perovskite; Mw: magnesiowustite; Co: corundum.

European Synchrotron Radiation Facility (ESRF, Grenoble, France). A channel-cut monochromator was used to produce bright, monochromatic X-ray beam at 0.3738 Å wavelength. Two-dimensional images were recorded on an imaging plate in less than 5 min and read online by either the Fastcan detector (Thoms et al., 1998) or the MAR345 detector. Diffraction patterns were integrated using the Fit2d code (Hammersley, 1996) and multi-phases Le Bail and Rietveld refinements were performed using the GSAS program package (Larson and Von Dreele, 1988).

3. Preliminary considerations on the Ca-ferrite and NAL-phases crystal chemistry

In our DAC experiments the run products are about 10^4 μm^3 in size, with a grain size much smaller than 1 μm. Due to this extremely small grain size, we cannot perform accurate chemical analysis of our CF- and NAL-phases (intimately mixed with other phases). We therefore rely on the previous experimental reports, mainly using multi-anvil experiments, to discuss the chemical compositions relevant to the natural CF- and NAL-phases. Various elements are found present, with various types of (partial) solid solutions, and a convenient procedure is thus to describe the CF and NAL-compositions in terms of a sum of isotype end-members. This decomposition is important, since each end-member can affect the elastic parameters of these phases in a different way. We note that another procedure consists in describing the complex chemical composition of the CF and NAL-phases using various types of substitution mechanisms (Kesson et al., 1994b; Miyajima et al., 2001) it yields basically the same results.

3.1. The Ca-ferrite phase

The main end-members reported to date for the CF-structure are $CaFe_2O_4$ (Reid and Ringwood, 1969), $NaAlSiO_4$, $NaTi_2O_4$ (Akimoto and Takei, 1989), $CaAl_2O_4$ (Reid and Ringwood, 1969; Akaogi et al., 1999), $MgAl_2O_4$ (Irifune et al., 1991), $FeAl_2O_4$ and $MnAl_2O_4$ (mentioned by Kesson et al., 1994a). We also note the occurrence of Fe_3O_4, $MgFe_2O_4$, Mn_3O_4, and $CaMn_2O_4$ with the $CaMn_2O_4$-type structure, and of $CaTi_2O_4$ and $MgAl_2O_4$ with the $CaTi_2O_4$-type structure (see Ringwood, 1970; Andrault and Bolfan-Casanova, 2001). Due to strong similarities of local structure in these high-density AB_2O_4 structures, we propose to consider all these compositions as potential end-members for the CF-structure. Furthermore, we find that some CF-phases reported in the literature cannot be decomposed if the $(Mg,Fe)_2SiO_4$ end-member is not considered.

In practical, we can use the Ti, Na, Si, Ca, Al, and (Mg,Fe) contents, in this respective order, to determine the contents in $NaTi_2O_4$, $NaAlSiO_4$, $(Mg,Fe)_2SiO_4$, $CaAl_2O_4$, $(Mg,Fe)Al_2O_4$, and $MgFe_2O_4$ CF end-members. Note that the presence of some amounts of Fe^{III} is considered with the occurrence of the

Table 3
Chemical composition of various CF-phases reported in the literature (in cation per four oxygen atoms)

Reference	A	B1	B2	B3	C	D1	D2	D3	E
P (GPa)	27	30	33	37	25	45	80	100	43.2
Na	0.56	0.58	0.65	0.76	0.89	0.51	0.49	0.43	0.60
Mg	0.41	0.37	0.38	0.27	0.36	0.43	0.32	0.27	0.40
Fe	0.35	0.16	0.15	0.14	0.16	0.16	0.18	0.12	0.15
Al	0.99	1.17	1.12	1.11	1.09	1.31	1.30	1.23	1.14
Si	0.69	0.68	0.70	0.75	0.67	0.58	0.64	0.77	0.68
Ca	0.05	0.02	0.04	0.01	0.02	0.02	0.02	0.02	0.06
Ti	0.02	0.01	0.01	0.01	0.01	–	–	–	0.02
Total	3.07	2.99	3.05	3.05	3.20	3.01	2.95	2.84	3.05
$NaAlSiO_4$	55	58	63	75	–	50	–	–	59
$(Mg,Fe)_2SiO_4$	14	10.5	6	0	–	7	–	–	8
$(Mg,Fe)Al_2O_4$	17	28	20	16	–	37	–	–	21
$MgFe_2O_4$	10	1	7	8	–	2	–	–	5
$CaAl_2O_4$	5	2	3	1	–	2	–	–	5
$NaTi_2O_4$	1	0.5	0.5	0.5	–	–	–	–	1
Total	102	100	99.5	100.5	–	98	–	–	99
V_0 meas. (Å^3)	250.4	248.2	246.6	248.3	244.1	–	–	–	240.3

These compositions are decomposed on percentages of six different CF end-members, as described in the text. CF-compositions are: A from Hirose et al. (1999); B1, B2 and B3 from Ono et al. (2001); C from Ono et al. (2002b); D1, D2 and D3 from Kesson et al. (1994b), and E from Funamori et al. (2000). P and V_0 meas. indicate the synthesis pressure and room pressure volume, respectively.

$MgFe_2O_4$ end-member. This method yields formulae almost perfect with respect to the AB_2O_4 stoichiometry, since we perform a linear combination of various AB_2O_4 end-members (see Table 3). Performing this calculation for all chemical compositions of CF-phases synthesized from materials relevant to MORB composition, we note: (1) the content of the $NaAlSiO_4$ end-member is the highest, accounting for at least 50% in mole of the composition, (2) the $MgAl_2O_4$ end-member is coming second, with variable amounts, (3) the Si/Na ratio is most of the time too high to be simply linked to the presence of the $NaAlSiO_4$ end-member. It points out the presence of up to 10% in mole of the $(Mg,Fe)_2SiO_4$ end-member, (4) the $MgFe_2O_4$ content can be high, accounting for up to 11% in mole of the composition. This information, compared to the occurrence of a remarkably stable iron content at about 0.15 per formula units (pfu), suggests a Fe^{III}/Fe^{II} ratio higher than 1.

For all CF-compositions reported in the literature, we found only two difficulties to use this method: (1) in Ono et al. (2002b) report, the Na content appears to be too high, thus possibly requiring the use of additional end-members accounting for a large substitution of $(Mg,Fe)_2$ by NaAl, (2) in Kesson et al. (1994b) report, Si- and Al-contents seem too high, pointing out the possible presence of cation vacancies.

3.2. The NAL-phase

The NAL-phase shows a remarkable variation of the composition among the several studies (Table 4), to even larger extent than that observed for the CF-phase. It could be partially due to the fact that the starting materials used to produce the NAL-phases have broader ranges of chemical composition compared to the relative uniformity of the starting materials (most often, MORB) used to produce the CF-phases. As stated in introduction, NAL phase is a particular polytype occuring for the sum of CF-type end-members. In order to define proper end-members to decompose the various NAL-phase chemical compositions, we refer to a previous definition of the substitution mechanisms occurring in this phase (Miyajima et al., 2001):

$$Ca^{2+} + Al^{3+} \Leftrightarrow (Na^+, K^+) + Si^{4+} \quad (a)$$

$$2Al^{3+} \Leftrightarrow (Mg^{2+}, Fe^{2+}) + Si^{4+} \quad (b)$$

$$2Al^{3+} \Leftrightarrow 1.5Si^{4+} + 0.5\square \quad \text{(where } \square \text{ is a cationic vacancy)} \quad (c)$$

Table 4
Chemical composition of various NAL-phases reported in the literature (in cation per four oxygen atoms)

Reference	A1	A2	A3	A4	B1	B2	B3	C	D
P (GPa)	27	27	50	23	24	24	22	29.3	26
Na	0.31	0.10	–	–	0.11	0.12	0.13	–	0.55
K	0.04	–	0.06	–	0.19	0.14	0.11	–	0.05
Ca	0.04	0.19	0.03	0.26	0.01	0.04	0.06	0.12	0.04
Mg	0.47	0.62	0.83	0.71	0.87	0.66	0.55	0.76	0.43
Fe	0.35	0.14	0.02	–	0.20	0.10	0.67	0.10	0.29
Al	1.15	1.46[a]	1.28	1.36	0.88	1.58	1.86	1.50	1.04
Si	0.61	0.41	0.59	0.49	0.70	0.32	0.19	0.38	0.66
Total	2.97	2.92	2.81	2.82	2.96	2.96	2.97	2.86	2.84
$NaAlSiO_4$	31	10	–	–	11	12	–	–	54
$KAlSiO_4$	4	–	6	–	19	14	–	–	5
$CaAl_2O_4$	4	19	3	25	1	4	–	12	4
$(Mg,Fe)_2SiO_4$	14	–	–	–	28	0	–	–	6
$(Mg,Fe)Al_2O_4$[b]	44	68	91	75	36	68	–	88	18
$MgFe_2O_4$	4	2	–	–	5	3	–	–	14
Total	101	100	100	100	100	101	–	100	99
V_0 meas. (Å^3)	–	188	–	–	187.7	–	–	187.1	185.9

NAL-compositions are decomposed on percentages of six different end-members, as described in the text. NAL-compositions are: A1, A2, A3, A4 from Miyajima et al. (2001), B1, B2 and B3 from Gasparik et al. (2000), C from Oguri et al. (2000), D from Ono et al. (2002a).
[a] The Cr-content of this compound was added to the Al-content.
[b] Content of $MgSi_{1.5}\square_{0.5}O_4$ end-member is added to that of $MgAl_2O_4$.
P and V_0 meas. indicate the synthesis pressure and room pressure volume, respectively.

From these equations, we can define the $CaAl_2O_4$, $NaAlSiO_4$, $KAlSiO_4$, $(Mg,Fe)Al_2O_4$, $(Mg,Fe)_2SiO_4$, and $(Mg,Fe)Si_{1.5}\square_{0.5}O_4$ end-members. $(Mg,Fe)Si_{1.5}\square_{0.5}O_4$ end-member implies the presence of cationic vacancies in the NAL structure, in agreement with almost all studies showing NAL-compositions with too low cation totals. For questions of simplification, this end-member is finally added to the $(Mg,Fe)Al_2O_4$ content at the end of the calculation (Table 4). As for the case of the CF-phase, we also need to consider the $MgFe_2O_4$ end-member to take into account a relatively high (Mg,Fe) content.

In practical, we can use the Na, K, Ca, Al, Si, and (Mg,Fe) contents, in this respective order, to determine the contents in $NaAlSiO_4$, $KAlSiO_4$, $CaAl_2O_4$, $(Mg,Fe)Al_2O_4$, $(Mg,Fe)Si_{1.5}\square_{0.5}O_4$, $(Mg,Fe)_2SiO_4$, and $MgFe_2O_4$ end-members. In this calculation, we need to adjust the number of cationic vacancies to drive the total cation site content per formula unit as close as possible to 9, according to the NAL-phase stoichiometry (Table 4). The decompositions in end-members evidence that (1) the content of $MgAl_2O_4$ end-member is the highest, accounting for at least 36 % in mole of the composition (except composition D), (2) the $KAlSiO_4$-content can be high, up to 19% in mole. (3) the $NaAlSiO_4$-content is definitely smaller than that observed for the CF-phase. (4) $CaAl_2O_4$, $(Mg,Fe)_2SiO_4$ and $MgFe_2O_4$ contents can be comparable to those of the CF-phase, as evidenced by the A1 composition that corresponds to the only NAL-phase produced from a MORB starting material (Table 4).

For all NAL-compositions reported to date, we found only one difficulty to perform this decomposition into end-members because of a two low Si content compared with those of Na and K (composition B3; Gasparik et al., 2000). Note that the Fe^{3+} content calculated from $MgFe_2O_4$ content is in very good agreement with values reported by Gasparik et al. (2000) (compositions B1 and B2). The composition D (Table 4) reported by Ono et al. (2002a) is also interesting, since it shows a unique amount of (Na,K,Ca) that largely exceeds the M3 site storage capabilities. This implies that some of these large cations have possibly been inserted in the M2 site. It is also the only NAL-composition for which no cation vacancy is required by stoichiometry.

4. Compression behavior of the Ca-ferrite phase

4.1. $NaAlSiO_4$ CF-phase

We first report the experimental results obtained for $NaAlSiO_4$-CF synthesized in situ after CO_2 laser heating of albite or jadeite loaded at 28.2 GPa (Exp. 1) or 23.0 GPa (Exp. 2), respectively. The diffraction patterns of the samples evidence the presence of the Ca-ferrite structure, stishovite and traces of corundum (Fig. 1). The a and b cell parameters of the CF-lattice are well constrained by diffraction peaks at large interplanar distances ($d_{h\,k\,l}$). In contrast, the refinement of the c parameter requires the use of relatively weak diffraction lines at small distances that are only well resolved in high-quality diffraction patterns. In our study, we use in particular reflections (2 0 0), (2 1 0) and (0 2 0) at large $d_{h\,k\,l}$, and (4 1 1), (4 2 1), (0 4 1) and (1 4 1) at smaller $d_{h\,k\,l}$ to refine the lattice parameters accurately (Table 5 and Fig. 1b). No phase transition is observed up to the maximum pressure of 65.5 GPa reached in this study. The two sets of data (Exps. 1 and 2) are highly compatible. After decompression to ambient conditions, we refine $NaAlSiO_4$-CF room pressure volumes V_0 of 241.4(6) Å^3 and 240.9(3) Å^3, in good agreement with values of 240.93(3) Å^3 and 241.1(2) Å^3 reported by Yamada et al. (1983) and Tutti et al. (2000), respectively. Using a third order

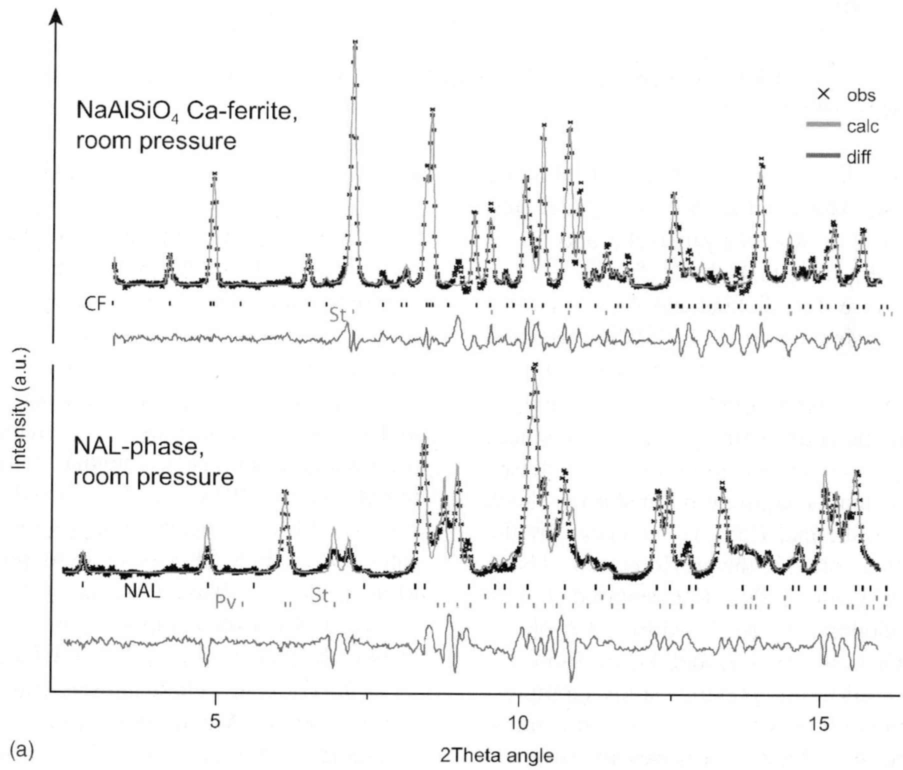

Fig. 1a. Angle dispersive X-ray diffraction patterns of $NaAlSiO_4$-CF (calcium ferrite type structure) and Ca–Na NAL-phase (New Aluminum rich phase) obtained at ambient conditions, after compression to 65.2 GPa and 61.9 GPa, respectively. Background due to Compton diffusion of the diamond anvils was subtracted for clarity. For CF-phase, the refinement is performed in the Rietveld mode. For NAL-phase, the refinement is performed in the Le Bail mode, with no constraint for the diffraction lines intensities.

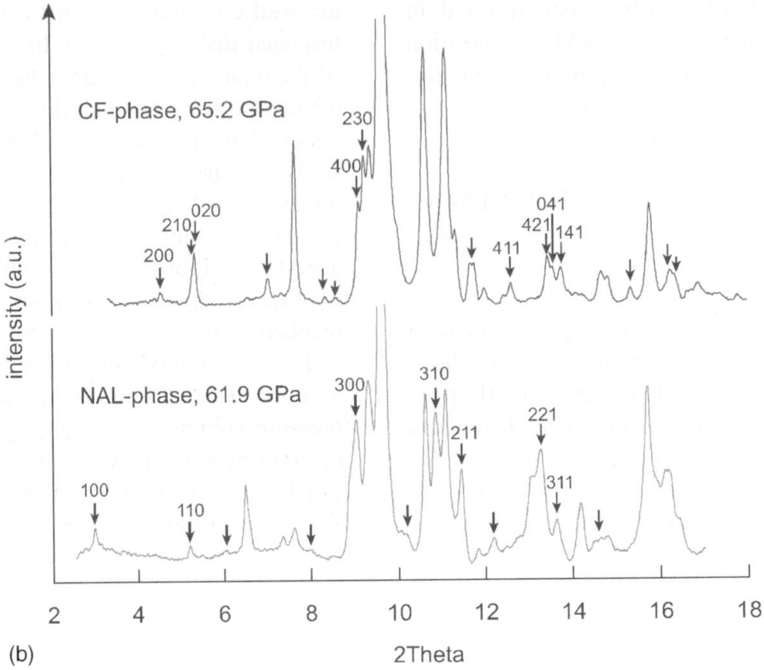

Fig. 1b. Diffraction patterns recorded at the highest pressures performed in this study. The arrows correspond to the major peaks used for the refinements, some of them being indexed here.

Birch–Murnaghan EoS for the whole set of data, we refine $K_0 = 185(5)$ GPa, $K'_0 = 4.6(2)$, and $V_0 = 241.2(3)$ Å^3, or $K_0 = 199(2)$ GPa and $V_0 = 240.6(4)$ Å^3, for K'_0 fixed to 4 (Table 7).

Using shock wave data, Sekine and Ahrens (1992) reported elastic parameters of $K_0 = 275(38)$ GPa and $K'_0 = 1.6(1.5)$ for the $NaAlSiO_4$-CF, using the V_0 value of 245.0(2) Å^3 taken from Liu (1977). These results discard significantly with ours. The differences could be due to a difficult estimation of the Grüneisen and q parameters that is required to transfer the shock wave data into isothermal EoS, as evidenced by the large uncertainty reported in the previous study. Using DAC, Dubrovinsky et al. (2002) also reported slightly harder elastic moduli for the $NaAlSiO_4$ CF-phase; $K_0 = 220(1)$ GPa, $K'_0 = 4.1$, and $V_0 = 243.0$ Å^3. The procedure used in this previous study (compression and decompression without any pressure transmitting medium, no release of stresses by laser annealing) does not seem appropriate to study the EoS of this CF-phase. We also note that the V_0 reported in this previous study seems too large compared with the more reliable measurement of Yamada et al. (1983).

4.2. CF-phases in the $NaAlSiO_4$–$MgAl_2O_4$ join

We now compare our results on the $NaAlSiO_4$-CF with the previous studies on the CF-form of $MgAl_2O_4$. For this compound, Yutani et al. (1997) proposed $K_0 = 241(3)$ GPa with K'_0 fixed to 4, and $V_0 = 240.0(1)$ Å^3. Although this study is limited to approximately 10 GPa the experiments were performed in hydrostatic conditions with reliable volume and pressure measurements. Another study on $MgAl_2O_4$ reports $K_0 = 210$ GPa with K'_0 fixed to 4 (Funamori et al., 1998), but a single high-pressure data point (reported in Fig. 2) was used to refined this EoS. More recently, Irifune et al. (2002) reported a $MgAl_2O_4$-CF EoS compatible with Funamori et al. (1998) data, with $K_0 = 213(3)$ GPa, K'_0 fixed to 4, $V_0 = 240.6(3)$ Å^3. These authors also suggest that the sample used by Yutani et al. (1997) could have been non-stoichiometric, which could lead to inaccurate elastic parameters determination. Finally, Gracia et al. (2002) confirm K_0 value of around 215 GPa by ab initio structure calculations. Comparison between our data for $NaAlSiO_4$ and those of Irifune et al. (2002)

Table 5

Unit cell parameters and volumes of CF-phases refined in this study

P (GPa)	*a* (Å)	*b* (Å)	*c* (Å)	*V* (Å^3)	*P* (GPa)	*a* (Å)	*b* (Å)	*c* (Å)	*V* (Å^3)
$NaAlSiO_4$-CF (Exp. 1, P_{ruby})					$NaAlSiO_4$-CF (Exp. 2, P_{ruby})				
28.2	9.728(30)	8.305(20)	2.662(8)	215.0(8)	23.0	9.798(5)	8.335(3)	2.674(1)	218.4(1)
31.8	9.687(27)	8.250(18)	2.651(8)	211.9(7)	31.3	9.711(3)	8.271(6)	2.649(4)	212.8(3)
42.2	9.579(41)	8.168(30)	2.629(10)	205.7(1.1)	38.0	9.636(2)	8.201(1)	2.639(1)	208.5(1)
48.3	9.515(39)	8.116(29)	2.621(9)	202.4(1.0)	46.1	9.560(4)	8.133(3)	2.618(2)	203.6(2)
54.0	9.480(28)	8.079(19)	2.610(7)	199.9(6)	56.0	9.479(2)	8.044(2)	2.606(1)	198.7(1)
60.2	9.441(29)	8.038(18)	2.600(8)	197.0(6)	65.2	9.392(3)	8.005(2)	2.588(1)	197.8(1)
Quench	10.137(7)	8.643(5)	2.749(2)	240.9(3)	Quench	10.166(21)	8.683(24)	2.735(11)	241.4(6)
$(Na,Mg)(Al,Si)_2O_4$-CF (Exp. 4, P_{Ar})									
51.4	9.512(32)	8.089(22)	2.621(8)	201.6(9)					
51.9	9.504(47)	8.091(32)	2.622(12)	201.6(1.3)					
59.4	9.433(61)	8.031(41)	2.659(18)	197.7(1.8)					
38.8(d)	9.626(55)	8.209(37)	2.647(14)	209.2(1.6)					
Quench	10.101(3)	8.624(2)	2.769(1)	241.2(1)					
Quench	10.131(4)	8.618(2)	2.760(1)	240.9(1)					
$(Na,Mg,Fe)_1(Al,Si,Mg,Fe)_2O_4$-CF (Exp. 5, P_{ruby}), first run					$(Na,Mg,Fe)_1(Al,Si,Mg,Fe)_2O_4$-CF (Exp. 5, P_{ruby}), second run				
25.9	9.751(20)	8.302(17)	2.681(5)	217.0(5)	26.9	9.744(15)	8.299(12)	2.679(3)	216.7(4)
33.5	9.689(22)	8.235(17)	2.657(5)	212.0(6)	36.8	9.607(20)	8.203(20)	2.659(6)	209.6(6)
40.0	9.602(15)	8.186(11)	2.642(4)	207.7(4)	45.5	9.562(31)	8.153(25)	2.628(8)	204.8(8)
46.1	9.542(20)	8.144(16)	2.633(5)	204.6(5)	60.2	9.447(20)	8.024(16)	2.598(5)	196.9(5)
48.4	9.565(23)	8.121(19)	2.619(6)	203.5(6)	65.5	9.371(11)	7.987(09)	2.601(3)	194.7(3)
51.2	9.534(26)	8.083(21)	2.616(6)	201.6(7)	Quench	10.125(16)	8.685(13)	2.765(4)	243.6(4)
Quench	10.175(6)	8.677(5)	2.751(1)	242.9(2)					
Na-rich MORB CF (Exp. 6, P_{Ar})					Na-rich MORB CF (Exp. 7, P_{Ar})				
33.0	9.688(30)	8.226(17)	2.670(6)	212.8(7)	27.7	9.750(19)	8.360(10)	2.706(4)	220.6(4)
40.1	9.617(30)	8.168(18)	2.650(7)	208.2(7)	27.8	9.711(28)	8.354(14)	2.709(5)	219.8(6)
49.6	9.541(29)	8.101(18)	2.627(6)	203.1(7)	38.5	9.646(8)	8.252(4)	2.679(1)	213.2(2)
Quench	10.138(7)	8.644(5)	2.771(2)	242.8(2)	46.2	9.547(21)	8.188(10)	2.658(4)	207.8(5)
					56.4	9.506(16)	8.122(9)	2.627(3)	202.8(4)
					63.4	9.436(10)	8.067(6)	2.616(2)	199.1(2)
					Quench	10.139(12)	8.667(7)	2.782(2)	244.5(3)
					Quench	10.145(17)	8.672(9)	2.786(3)	245.1(4)

(d) data obtained during decompression. P_{ruby}: P derived from ruby fluorescence technique; P_{Ar}: P derived from Ar EoS.

for $MgAl_2O_4$ suggests reduced compositional effect on the CF-phase EoS within the $NaAlSiO_4$–$MgAl_2O_4$ join.

We then report results obtained from a Ca-ferrite phase synthesized using a starting material composed of 50% $NaAlSiO_4$ and 50% $MgAl_2O_4$ in mole (Exp. 4, Table 2). In this experiment, the synthesis was done at 51.4 GPa and a second laser heating was performed at 59.8 GPa. An additional diffraction pattern recorded at 38.8 GPa during decompression completes this limited data set. We find that the CF-phase coexists with corundum and a NAL-phase. We cannot tell the precise chemical composition of this CF-phase in the $NaAlSiO_4$–$MgAl_2O_4$ join. However, as discussed in the previous section, it is expected that coexisting CF and NAL-phases are each one enriched in $NaAlSiO_4$, or $MgAl_2O_4$, respectively. The volumes of this CF-phase are found perfectly bracketed between those of $NaAlSiO_4$-CF and $MgAl_2O_4$-CF at all pressures (Fig. 2). It confirms the reduced compositional effect on the CF compression behavior within the $NaAlSiO_4$–$MgAl_2O_4$ join. The BM3-EoS yields $K_0 = 200(3)$ GPa, with K'_0 fixed to 4, and $V_0 = 241.2(3)$ Å^3 (Table 7).

4.3. CF-phases in the $NaAlSiO_4$–$(Mg,Fe)_2SiO_4$ system

We then investigated the occurrence of a CF-phase from a starting material containing olivine in excess therefore in an Al undersaturated system, as it is likely to be the case in materials relevant to the Earth's mantle. Two different compression runs were performed with a starting material composed of a mixture of 25 mol.% albite + 75 mol.% olivine (Exp. 5). The first result is that a CF-phase is always observed at pressures from 25.9 to 65.5 GPa. This phase coexists with perovskite and traces of stishovite. The experimen-

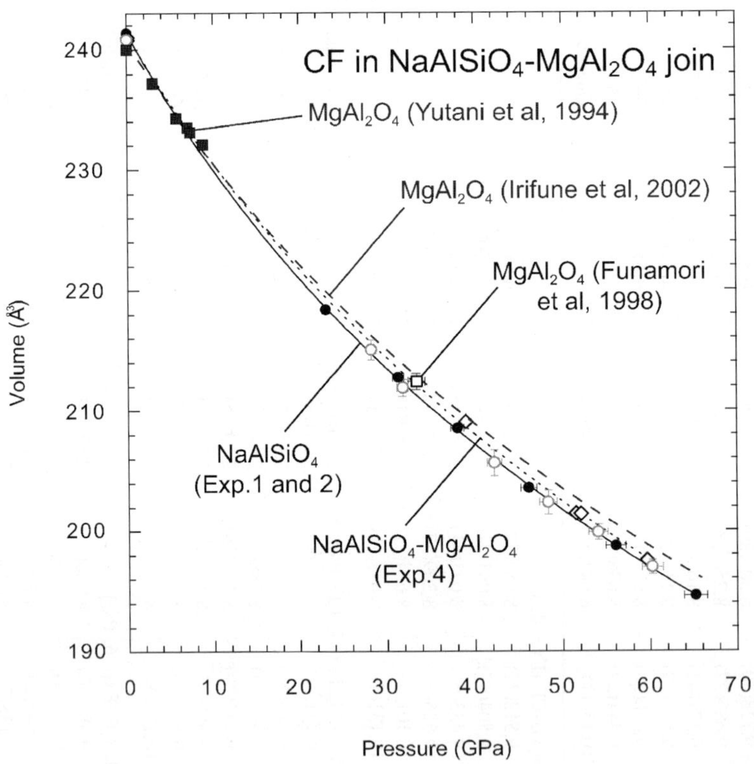

Fig. 2. Compression curves of CF-phases with various compositions in the $NaAlSiO_4$–$MgAl_2O_4$ join. We report previous data set on $MgAl_2O_4$ by Yutani et al. (1997) (solid squares) and Funamori et al. (1998) (open square) and $MgAl_2O_4$-CF EoS by Irifune et al. (2002) (dashed line). Our data concern pure $NaAlSiO_4$-CF (solid and open circles for the phases synthesized from jadeite and albite, respectively), and a $NaAlSiO_4$-rich CF-phase in the $NaAlSiO_4$–$MgAl_2O_4$ system (open diamonds).

tal volumes obtained in the two runs plot in perfect agreement with each other (Fig. 3). As for Exp. 4, the volumes are found at only slightly higher values than those refined for pure $NaAlSiO_4$ CF-phase. This result is compatible with a $NaAlSiO_4$-rich CF-phase synthesized in these two runs, in agreement with all reports using multi-anvil presses (see Table 3). For the complete data set, we calculate $K_0 =$ 184(3) GPa, $K'_0 = 4.2(1)$, and $V_0 = 243.0(2)$ Å^3, or $K_0 = 189(3)$ and $V_0 = 242.8$ Å^3 with K'_0 fixed to 4.

An important information is that our two runs provide CF-phase V_0 of 243. 6(4) Å^3 and 242. 9(2) Å^3, values slightly (but significantly) higher than the ones refined in the $NaAlSiO_4$–$MgAl_2O_4$ join. This effect can be related to the probable insertion of $(Mg,Fe)_2SiO_4$ in the CF-structure, due to the presence of olivine in the starting material. By comparing this V_0 of 243. 0 Å^3 with those of $NaAlSiO_4$ (241. 3 Å^3) and Mg_2SiO_4 with CF structure (estimated to 247. 5 Å^3; Yamada et al., 1983), we estimate that less than 30% in mole of $(Mg,Fe)_2SiO_4$ end-member is integrated in the CF-compound synthesized in the $NaAlSiO_4$–$(Mg,Fe)_2SiO_4$ system. This estimation is compatible with the previous CF chemical analyses showing a maximum $(Mg,Fe)_2SiO_4$-content of 14% (Table 3). Still, even if a relative increase of the V_0 of about 0. 7% is observed, the elastic parameters remain relatively unchanged. With this experiment, we also confirm the occurrence of a Ca-ferrite phase in Al undersaturated systems, as it is expected in the Earth's mantle.

4.4. CF-phase synthesized from a MORB

We now probe the occurrence of a CF-phase using a natural Na-rich MORB composition (Table 1). After CO_2 laser heating at 18.4 and 26.8 GPa, the

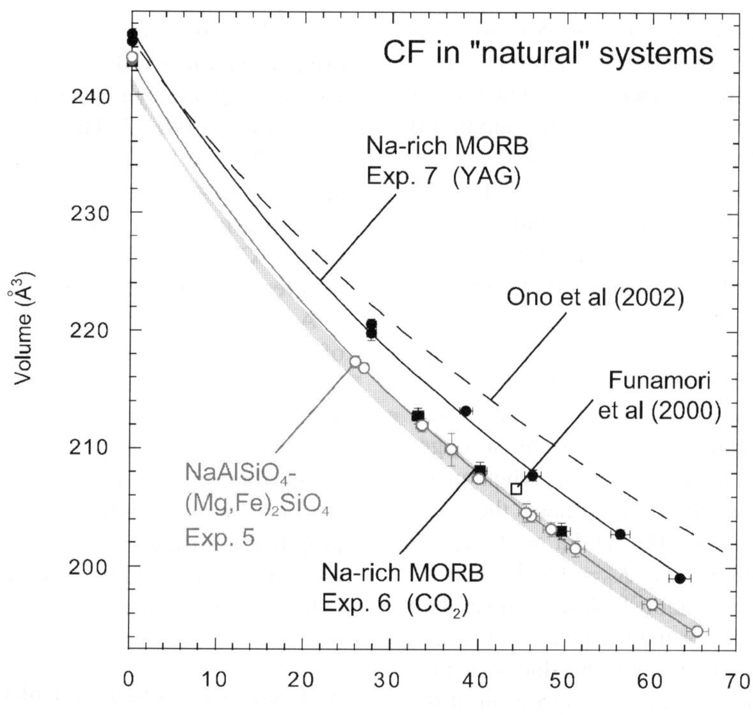

Fig. 3. Compression curves obtained for various CF-phases synthesized in chemically complex systems. We also report the compression curves for the $NaAlSiO_4$ and $MgAl_2O_4$ CF end-members (gray area), and previous reports on MORB-type material by Ono et al. (2002b) (dashed line) and Funamori et al. (1998) (open square). Our results include the CF-phase cell volumes obtained for starting materials made of $NaAlSiO_4$–$(Mg,Fe)_2SiO_4$ mixture (open circles), and of a Na-rich MORB heated using CO_2 (solid squares) or YAG (solid circles) laser. Note that the V_0 seems to be significantly affected by CF-phase chemical composition (Table 3).

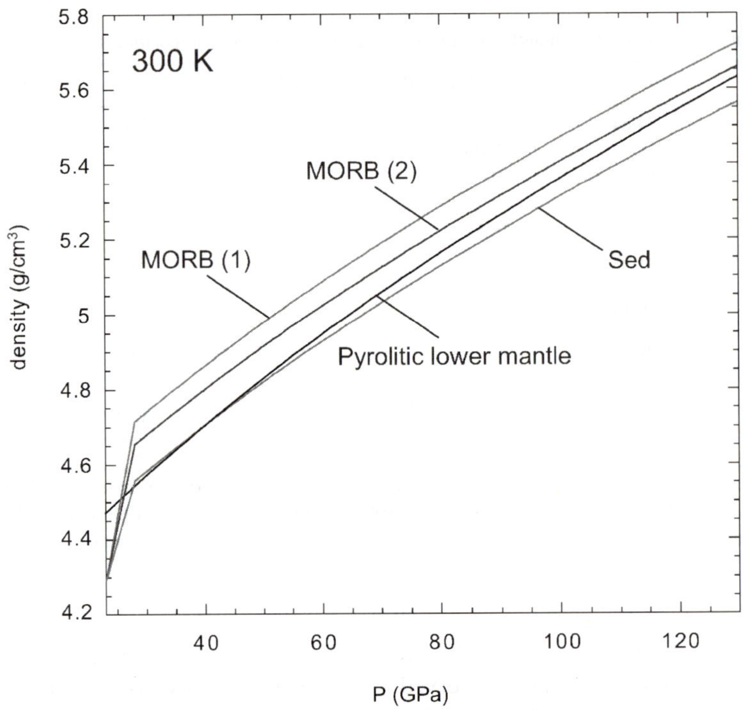

Fig. 6. Density profiles obtained for the different mineralogical assemblages relevant to the lower mantle, derived from pyrolitic, MORB and sediment-type compositions (see Tables 8 and 9). The difference between the two MORB density profiles reported here corresponds to a typical chemical variability that may occur in natural systems plus the error bar of our calculation (see Fig. 5a and text).

dense Fe-rich MORB-Pv. Also, the significant amount of the very incompressible stishovite phase makes the MORB material relatively less compressible than the pyrolite, thus reducing the density gap between the two materials with increasing pressure. The effect of the presence of stishovite is also visible for the density profile of the sediment-type material, but in this case the Al-(Mg,Fe)SiO_3-Pv content is significantly smaller and insufficient to counterbalance the relatively low density of the Na, K, Al and Si-rich phases.

We finally note that there is a flagrant lack of high temperature data to extrapolate these results quantitatively to the high temperatures prevailing in the lower mantle. In any case, as it is believed that a subducting slab (potentially containing significant amount of MORB-type material and maybe some sediments) is colder than the surrounding mantle, the thermal gradient will therefore account for a denser slab, a trend in the same direction than our conclusions based on compressibility and chemical composition considerations. It is thus very likely that the subducted slabs are denser than the environing mantle below 750–800 km, down to the core-mantle boundary. The sink of subducted slabs in the lower mantle is therefore only dependent on their ability to pass through the ~24–28 GPa pressure range when Al-rich majoritic garnet is not fully decomposed.

7. Conclusion

CF and NAL-phases are minerals with particularly complex crystal chemistry. They involve many end-members of different importance, and the chemical composition of these phases can vary severely from a study to another. The occurrence in high-pressure assemblages of only one or two of these "sister" phases relies largely on the chemical composition of the starting material. CF-phase appears to be the ideal phase to host Na, whereas NAL-phase is perfectly adapted to the simultaneous storage of medium and large cations such as Na and K (or Mg and Ca). For very high K-content, the hollandite structure can be involved due to its larger "tunnel" polyhedron.

The compression behavior of CF and NAL-phases can be affected by their varying composition. It is therefore difficult to provide a unique and definite picture of the density profile of the CF and NAL-phases valid for the whole Earth's lower mantle. In this study, we show that the compositional effects on bulk moduli remains limited for the CF-phase, the major effect being the variation of its room pressure volume, whereas it appears more important for room pressure volume and compressibility of the NAL-phase.

The aluminous phases are expected to occur in Al and (Na, K) enriched materials such as MORB and sediments, with typical abundances of about 20% in mole of the whole mineralogical assemblage. Their room pressure densities are found significantly lower than the main other phases like silicate perovskites and magnesiowustite. The CF and NAL-phases are thus significantly less dense than a pyrolitic lower mantle near the 670 km discontinuity. However, due to slightly higher compressibility compared to other lower mantle phases, their density become close to that of the pyrolitic mantle at the highest depth near the core-mantle boundary.

Another important point is the very high stiffness of stishovite. It may prevent SiO_2-enriched materials to sink down to the core-mantle boundary, but our calculation suggests that this effect remains insufficient to stop the descent of MORB materials. Still, further calculations including the effect of temperature on the EoS appear necessary to conclude definitely on this question, regarding the low density difference of 0.4% found between MORB and pyrolitic materials at core-mantle boundary pressures and 300 K. The high temperature data required for this calculation may be difficult to obtain, however, due to complex chemical behavior of the CF and NAL-phases.

Finally, a most fundamental point is the large density difference between Al-(Mg,Fe)SiO_3 perovskites produced from pyrolitic, MORB, and sediment-type materials. Fe content is expected to be significantly higher for MORB and sediment materials, which is remains overall the main reason for the sinking of MORB-type materials down to the highest mantle depth.

Acknowledgements

We thank N. Bolfan-Casanova, T. Le Bihan, M. Mezouar for help in experiments, A. Bezos and E. Humler for providing the MORB sample, and N. Bolfan-Casanova, S. Ono and an anonymous reviewer for useful comments. This work was supported by INSU-CNRS, IPGP, and Paris 7 University.

References

Akaogi, M., Hamada, Y., Suzuki, T., Kobayashi, M., Okada, M., 1999. High pressure transitions in the system $MgAl_2O_4$–$CaAl_2O_4$: a new hexagonal aluminous phase with implication for the lower mantle. Phys. Earth Planet. Inter. 115, 67–77.

Akimoto, S., Takei, H., 1989. Synthesis and crystal structure of $NaTi_2O_4$: a new mixed-valence sodium titanate. J. Sol. St. Chem. 79, 212–217.

Andrault, D., 2001. Evaluation of (Mg, Fe) partitioning between perovskite and magnesiowustite up to 120 GPa. J. Geophys. Res. 106, 2079–2087.

Andrault, D., 2003. Cationic substitution in $MgSiO_3$ perovskite. Phys. Chem. Miner. 4200, 1–12.

Andrault, D., Bolfan-Casanova, N., 2001. High pressure phase transformations in the $MgFe_2SiO_4$ and Fe_2O_3-$MgSiO_3$ systems. Phys. Chem. Miner. 28, 211–217.

Andrault, D., Bolfan-Casanova, N., Guignot, N., 2001. Equation of state of the lower mantle Al-(Mg, Fe)SiO_3 perovskite. Earth Planet. Sci. 193, 501–508.

Andrault, D., Angel, R.J., Mosenfelder, J.L., Le Bihan, T., 2003. Equation of state of the stishovite to lower mantle pressures. Am. Mineral. 88, 301–307.

Dubrovinsky, L.S., Dubrovinskaia, N.A., Prokopenko, V.B., Le Bihan, T., 2002. Equation of state and crystal structure of $NaAlSiO_4$ with calcium-ferrite type structure in the conditions of the lower mantle. High Press. Res. 22, 495–499.

Funamori, N., Jeanloz, R., Nguyen, J.H., Kavner, A., Cadwell, W.A., Fujino, K., Miyajima, N., Shinmei, T., Tomioka, N., 1998. High-pressure transformation in $MgAl_2O_4$. J. Geophys. Res. 103, 20813–20818.

Funamori, N., Jeanloz, R., Miyajima, N., Fujino, K., 2000. Mineral assemblages of basalt in the lower mantle. J. Geophys. Res. 105, 26037–26043.

Gasparik, T., Tripathi, A., Paeise, J.B., 2000. Structure of a new Al-rich phase, $[K,Na]_{0.9}[Mg, Fe]_2[Mg,Fe,Al,Si]_6O_{12}$, synthesized at 24 GPa. Am. Mineral. 85, 613–618.

Gracia, L., Beltran, A., Andrés, J., Franco, R., Recio, J.M., 2002. Quantum-mechanical simulation of $MgAl_2O_4$ under high pressure. Phys. Rev. B 66, 224114.

Hammersley, J., 1996. Fit2d user manual, ESRF, Grenoble, France.

Hirose, K., Fei, Y., 2002. Subsolidus and melting phase relations of basaltic compositions in the uppermost lower mantle. Geochim. Cosmochim. Acta 66 (12), 2099–2108.

Hirose, K., Fei, Y., Ma, Y., Mao, H.K., 1999. The fate of subducted basaltic crust in the Earth's lower mantle. Nature 397, 53–56.

Humler, E., Besse, J., 2002. A correlation between mid-ocean-ridge basalt chemistry and distance to continent. Nature 419, 607–609.

Irifune, T., 1994. Absence of an aluminous phase in the upper part of the Earth's lower mantle. Nature 370, 131–133.

Irifune, T., Ringwood, A.E., 1993. Phase transformations in subducted oceanic crust and buoyancy relationships at depth of 600–800 (km in the mantle. Earth Planet. Sci. 117, 101–110.

Irifune, T., Fujino, K., Ohtani, E., 1991. A new high-pressure form of $MgAl_2O_4$. Nature 349, 409–411.

Irifune, T., Ringwood, A.E., Hibberson, W.O., 1994. Subduction of continental crust and terrigenous and pelagic sediments: an experimental study. Earth Planet. Sci. 126, 351–368.

Irifune, T., Koizumi, T., Ando, J.I., 1996. An experimental study of the garnet-perovskite transformation in the system $MgSiO_3$–$Mg_3Al_2Si_3O_{12}$. Phys. Earth Planet. Inter. 96, 147–157.

Irifune, T., Naka, H., Sanehira, T., Inoue, T., Funakoshi, K., 2002. In situ X-ray observations of phase transitions in $MgAl_2O_4$ spinel to 40 GPa using multianvil apparatus with sintered diamond anvils. Phys. Chem. Miner. 29, 645–654.

Kesson, S.E., Fitz Gerald, J.D., Shelley, J.M., 1994a. Crystal chemistry of calcium-ferrite type structures at lower mantle pressures. EOS trans., Am. Geophys. Union 75, 339.

Kesson, S.E., Fitz Gerald, J.D., Shelley, J.M., 1994b. Mineral chemistry and density of subducted basaltic crust at lower mantle pressures. Nature 372, 767–769.

Kesson, S.E., Fitz Gerald, J.D., Shelley, J.M., Withers, R.L., 1995. Phase relations, structure and crystal chemistry of some aluminous silicate perovskites. Earth Planet. Sci. 134, 187–201.

Kubo, A., Akaogi, M., 2000. Post-garnet transitions in the system $Mg_4Si_4O_{12}$-$Mg_3Al_2Si_3O_{12}$ up to 28 GPa: phase relations of garnet, ilmenite and perovskite. Phys. Earth Planet. Inter. 121, 85–102.

Larson, A.C., Von Dreele, R.B., 1988. GSAS Manual, Los Alamos National Laboratory.

Liu, L.G., 1977. High pressure $NaAlSiO_4$: the first silicate calcium ferrite isotype. Geophys. Res. Lett. 4, 183–186.

Miura, H., et al., 2000. Crystal structure of $CaMg_2Al_6O_{12}$, a new Al-rich pressure form. Am. Mineral. 85, 1799–1803.

Miyajima, N., Fujino, K., Funamori, N., Kondo, T., Yagi, T., 1999. Garnet-perovskite transformation under conditions of the Earth's lower mantle: an analytical transmission electron microscopy study. Phys. Earth Planet. Inter. 116, 117–131.

Miyajima, N., Yagi, T., Hirose, K., Kondo, T., Fujino, K., Miura, H., 2001. Potential host phase of aluminum and potassium in the Earth's lower mantle. Am. Mineral. 86, 740–746.

Oguri, K., Funamori, N., Utchida, T., Miyajima, N., Yagi, T., Fujino, K., 2000. Post-garnet transition in natural pyrope: a multi-anvil study based on in situ X-ray diffraction and transmission electron microscopy. Phys. Earth Planet. Inter. 122, 175–186.

Ono, S., Ito, E., Katsura, T., 2001. Mineralogy of subducted basaltic crust (MORB) from 25 to 37 GPa, and chemical heterogeneity of the lower mantle. Earth Planet. Sci. 190, 57–63.

Ono, S., Hirose, K., Isshiki, M., Mibe, K., Saito, Y., 2002a. Equation of state of hexagonal aluminous phase in basaltic composition to 63 GPa and 300 K. Phys. Chem. Miner. 29, 527–531.

Ono, S., Hirose, K., Kikegawa, T., Saito, Y., 2002b. The compressibility of a natural composition calcium ferrite aluminous phase to 70 GPa. Phys. Earth Planet. Inter. 131, 311–318.

Reid, A.F., Ringwood, A.E., 1969. Newly observed high pressure transformations in Mn_3O_4, $CaAl_2O_4$, and $ZrSiO_4$. Earth Planet. Sci. 6, 205–208.

Ringwood, A.E., 1970. Phase transformations and the constitution of the mantle. Phys. Earth Planet. Inter. 3, 109–155.

Ringwood, A.E., 1979. Origin of the Earth and Moon. Springer, New York.

Ringwood, A.E., Reid, A.F., Wadsley, A.D., 1967. High pressure transformation of alkali aluminosilicates and aluminogermanates. Earth Planet. Sci. 3, 38–40.

Sekine, T., Ahrens, T.J., 1992. Shock-induced transformations in the system $NaAlSiO_4$–SiO_2: a new interpretation. Phys. Chem. Miner. 18, 359–364.

Shim, S.H., Duffy, T.S., Shen, G., 2000. The equation of state of $CaSiO_3$ perovskite to 108 GPa at 300 K. Phys. Earth Planet. Inter. 120, 327–338.

Thoms, R., et al., 1998. An improved detector for use at synchrotrons. Nucl. Instr. Meth. A413, 175–180.

Tutti, F., Dubrovinsky, L.S., Saxena, S.K., 2000. High pressure phase transformation of jadeite and stability of $NaAlSiO_4$ with calcium-ferrite type structure in the lower mantle conditions. Geophys. Res. Lett. 27, 2025–2028.

Tutti, F., Dubrovinsky, L.S., Saxena, S.K., Carlson, S., 2001. Stability of $KAlSi_3O_8$ hollandite-type structure in the Earth's lower mantle conditions. Geophys. Res. Lett. 28, 2735–2738.

Wang, Y., Weidner, D.J., Zhang, J., Gwanrnesia, G.D., 1996. Thermal equation of state of garnets along the pyrope-majorite join. J. Geophys. Res. 101, 661–672.

Yamada, H., Matsui, Y., Ito, E., 1983. Crystal-chemical characterization of $NaAlSiO_4$ with the $CaFe_2O_4$ Structure. Mineral. Mag. 47, 177.

Yutani, M., Yagi, T., Yusa, H., Irifune, T., 1997. Compressibility of calcium ferrite-type $MgAl_2O_4$. Phys. Chem. Miner. 24, 340–344.

Zhang, J., Ko, J., Hazen, R.M., Prewitt, C.T., 1993. High-pressure crystal chemistry of $KAlSi_3O_8$ hollandite. Am. Mineral. 78, 493–499.

2. Mantle mineralogy

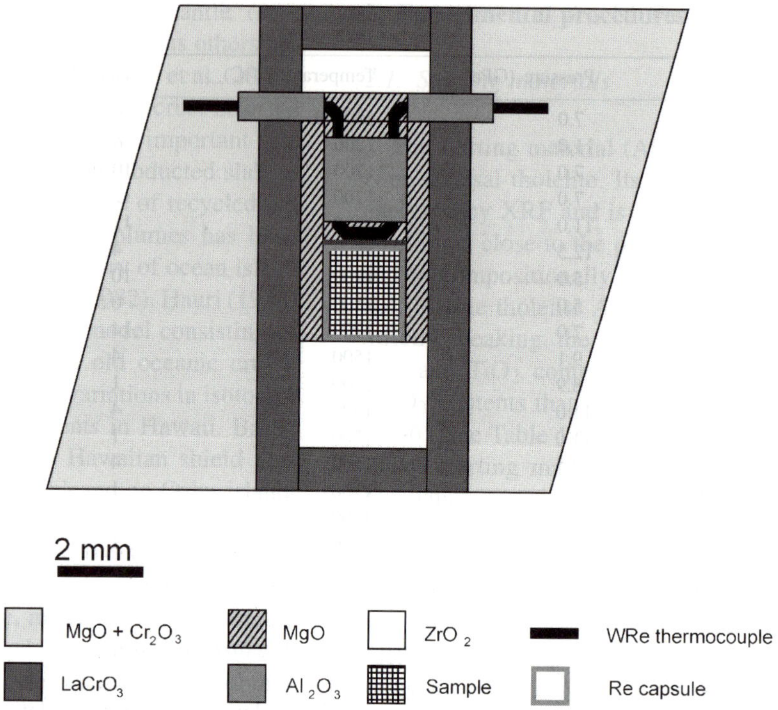

Fig. 1. Cross section of the 14 M multi-anvil assembly (octahedral pressure medium with 14.0 mm edge length) used in this study.

The calibration curves thus constructed are shown in Fig. 2. We found a clear positive temperature effect on pressure generation in our set-up, namely, sample pressure is higher at a given press load at higher furnace temperatures. Frost and Fei (1998) also reported a positive effect on pressure generation with temperature using an 8 mm MgO octahedron (8 M) but the extent of their correction is smaller than ours (cf. Fig. 2 with Fig. 1 of Frost and Fei, 1998). The effect of temperature on pressure generation can be understood in terms of three competing factors: thermal pressure, stress relaxation and plastic deformation of the MgO pressure medium and pyrophyllite gaskets (Meng et al., 1993). In the case of our apparatus using a large pressure assembly (14 M), the efficiency of thermal pressure is greater than that of the stress relaxation and deformation. According to our laboratory experience, the positive thermal effect is most significant when the temperature of the pyrophyllite gasket is relatively cool so that the gaskets are difficult to deform. The thermal effect could be either positive or negative depending on the size of the furnace relative to the pressure medium (e.g., Fig. 2 of Wang and Takahashi, 2000). We, therefore, kept the furnace design of our high-pressure experiments constant throughout this study (Fig. 1). A linear relationship between the heater power and the run temperature was established (1000 °C/270 W) and the runs that drifted more than 15% away from this linear relation during experiments were not used.

Experimental pressures at 1200 and 1700 °C were estimated by interpolating among the three high-temperature calibration curves shown in Fig. 2. The uncertainty in pressure measurements is believed to be <5% in our study and this precision should be higher than usual for multi-anvil experiments using the quenching technique. This is because, in most cases, pressure calibration is made only at room temperature and a single high temperature (e.g., 1000 °C).

3. Analysis and experimental results

3.1. Phase relations and compositional variation

The run conditions and resulting phase assemblages are listed in Table 2. Chemical compositions

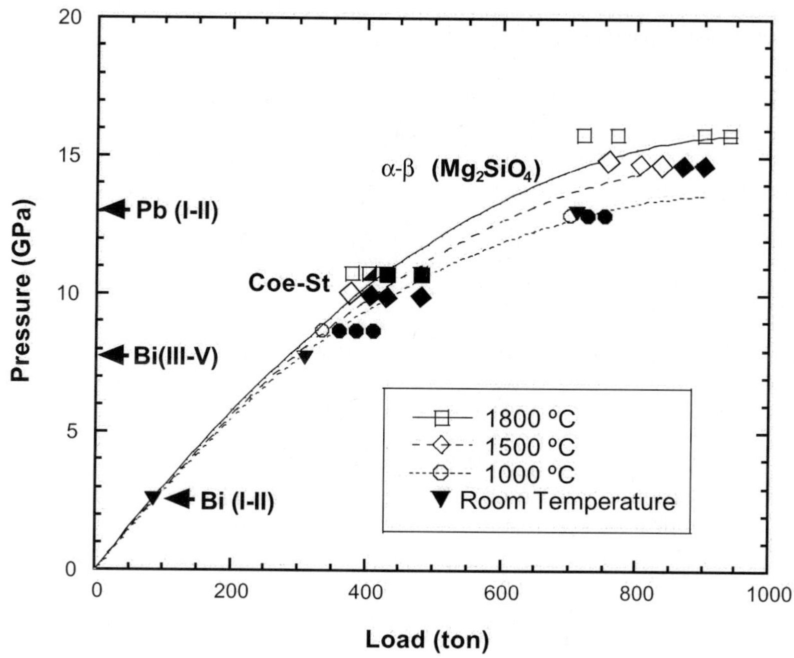

Fig. 2. Pressure calibrations for the 14 M assembly using coesite–stishovite transition (Zhang et al., 1996) and α/β phase transition in Mg_2SiO_4 (Morishima et al., 1994). In order to study the effect of temperature on the pressure generation efficiency, the calibration was carried out at 1000, 1500, and 1800 °C.

of run products were determined by an electron microprobe (JEOL-JXA8800). Representative analyses of chemical compositions of coexisting garnet (Gar) and clinopyroxene (Cpx) are summarized in Table 3. The composition of coesite and stishovite were not determined accurately due to their small crystal sizes (2–4 μm), so their compositions were assumed to be pure SiO_2 in the mass balance calculations. Phases in the run products were identified by an X-ray powder diffractometer (Rigaku-RINT2000). At pressures between 5 and 9 GPa, the charges consist of clinopyroxene (Cpx), garnet (Gar), and coesite (Coe) as shown in Table 2. Above 11 GPa, coesite (Coe) transforms to stishovite (St).

Fig. 3 illustrates the variation in the garnet composition as a function of pressure (at 1500–700 °C). The Si content increases with increasing pressure, while Al and Fe contents decrease in a complementary manner. These systematics are in good agreement with previous experiments at lower temperatures (e.g., Irifune et al., 1986) and can be explained by the dissolution of Cpx into garnet with increasing pressure (see Table 3). This pressure induced reaction requires that the proportion of garnet in run products gradually increases with increasing pressure, whereas that of Cpx decreases in a complementary manner.

The composition of Cpx becomes gradually less aluminous and higher in the jadeite component with increasing pressure as shown in Fig. 4. These variations occur as a result of the pyroxene–majorite phase transition (Akaogi and Akimoto, 1977; Irifune, 1987) in which Na is the last element to dissolve in garnet (Gasparik, 1989, 1992; Ono and Yasuda, 1996).

3.2. Zero-pressure lattice parameters

Run products were examined by a X-ray diffractometer (Rigaku-RINT2000) using Cr $K\alpha_1$, 2.289700 Å (40 kV, 300 mA). The X-ray signals were taken between $2\theta = 40$and 100° at intervals of 0.01° and rotations of 0.3°/min. A representative X-ray diffraction pattern is shown in Fig. 5. Zero-pressure lattice parameters of majorite garnets were calculated from the

Table 3
Representative chemical composition of minerals in run products

	S779		S780		S802		S804	
	5.0 GPa, 1200 °C		5.0 GPa, 1500 °C		7.0 GPa, 1200 °C		8.9 GPa, 1700 °C	
	Gar	Cpx	Gar	Cpx	Gar		Cpx	Cpx
SiO_2	41.59	55.75	42.30	53.92	41.40	56.55	41.64	55.89
TiO_2	0.02	0.28	0.63	0.50	0.80	0.22	1.04	0.41
Al_2O_3	21.34	9.00	22.42	12.53	21.42	8.75	20.63	10.18
FeO	14.55	4.72	13.65	6.03	13.57	4.25	12.71	4.80
MnO	0.18	0.02	0.22	0.06	0.23	0.10	0.27	0.01
MgO	11.42	10.24	12.06	9.33	10.39	10.07	11.44	9.62
CaO	10.17	14.99	10.08	13.58	11.23	14.30	11.69	12.71
Na_2O	0.22	4.18	0.19	3.39	0.27	4.83	0.51	5.03
K_2O	0.49	0.14	–	0.04	0.03	0.35	0.02	0.20
P_2O_5	0.18	0.10	0.05	0.08	0.14	0.01	0.11	0.03
Total	100.15	99.43	101.59	99.46	99.48	99.42	100.05	98.87
No. of oxygens	12	6	12	6	12	6	12	6
Structural formula								
Si	3.079	1.997	3.061	1.928	3.076	2.022	3.075	2.003
Ti	0.001	0.008	0.034	0.013	0.045	0.006	0.058	0.011
Al	1.862	0.380	1.912	0.528	1.876	0.369	1.796	0.430
Fe	0.901	0.141	0.826	0.180	0.843	0.127	0.785	0.144
Mn	0.011	0.001	0.014	0.002	0.014	0.003	0.017	0.000
Mg	1.261	0.547	1.301	0.498	1.150	0.537	1.260	0.514
Ca	0.807	0.575	0.781	0.520	0.894	0.548	0.925	0.488
Na	0.031	0.290	0.027	0.235	0.039	0.335	0.073	0.349
K	0.046	0.006	0.000	0.002	0.003	0.016	0.002	0.009
P	0.022	0.006	0.006	0.005	0.017	0.000	0.013	0.001
Sum	8.021	3.952	7.961	3.911	7.958	3.963	8.003	3.950

	S788		S789		S806		S809	
	11.0 GPa, 1500 °C		11.0 GPa, 1500 °C		14.0 GPa, 1500 °C		14.0 GPa, 1700 °C	
	Gar	Cpx	Gar	Cpx	Gar	Cpx	Gar	Cpx
SiO_2	41.95	56.24	42.84	56.80	44.46	56.83	43.49	55.67
TiO_2	0.99	0.29	0.79	0.27	1.15	0.35	1.24	0.42
Al_2O_3	19.29	8.90	19.71	10.16	18.44	8.76	17.68	9.39
FeO	12.38	4.96	12.13	5.10	10.88	4.69	10.78	4.51
MnO	0.14	0.04	0.19	0.04	0.19	0.00	0.21	0.04
MgO	11.24	10.07	11.43	9.54	11.00	9.79	11.31	9.60
CaO	12.12	13.30	12.31	12.58	11.75	12.33	12.21	12.33
Na_2O	0.93	5.44	0.67	5.64	1.01	6.05	1.32	5.74
K_2O	0.00	0.41	0.00	0.30	0.02	0.26	0.01	0.19
P_2O_5	0.14	0.02	0.10	-	0.10	0.03	0.08	0.00
Total	99.17	99.67	100.16	100.42	98.99	99.10	98.32	97.88
No. of oxygens	12	6	12	6	12	6	12	6
Structural formula								
Si	3.129	2.012	3.152	2.009	3.280	2.036	3.248	2.017
Ti	0.056	0.008	0.043	0.007	0.064	0.009	0.070	0.011
Al	1.696	0.375	1.709	0.423	1.603	0.370	1.556	0.401
Fe	0.772	0.148	0.746	0.151	0.671	0.141	0.673	0.137
Mn	0.009	0.001	0.012	0.001	0.012	0.000	0.013	0.001
Mg	1.250	0.537	1.254	0.503	1.209	0.523	1.259	0.519
Ca	0.968	0.510	0.971	0.477	0.928	0.473	0.977	0.479
Na	0.134	0.377	0.096	0.387	0.144	0.420	0.191	0.403
K	0.000	0.019	0.000	0.014	0.002	0.012	0.001	0.009
P	0.017	0.001	0.012	0.000	0.012	0.002	0.009	0.000
Sum	8.030	3.990	7.995	3.972	7.925	3.985	7.998	3.977

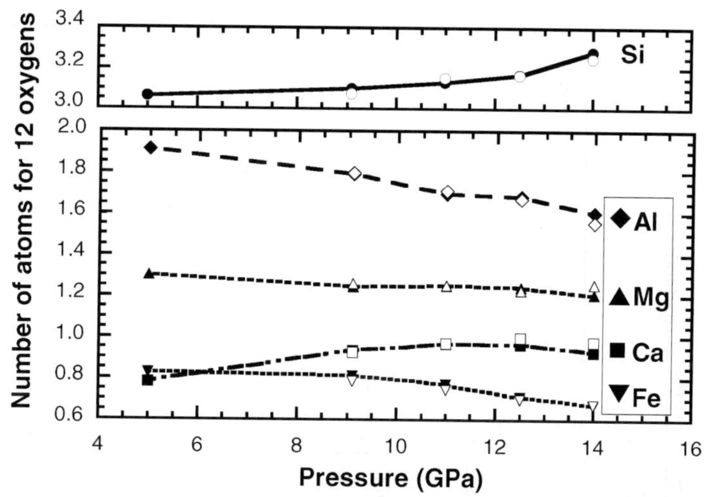

Fig. 3. Composition of garnet in atomic ratio (normalized to 12 oxygens) as a function of pressure. Data obtained at 1500 °C (filled symbols) and those at 1700 °C (open symbols) show no siginificant difference.

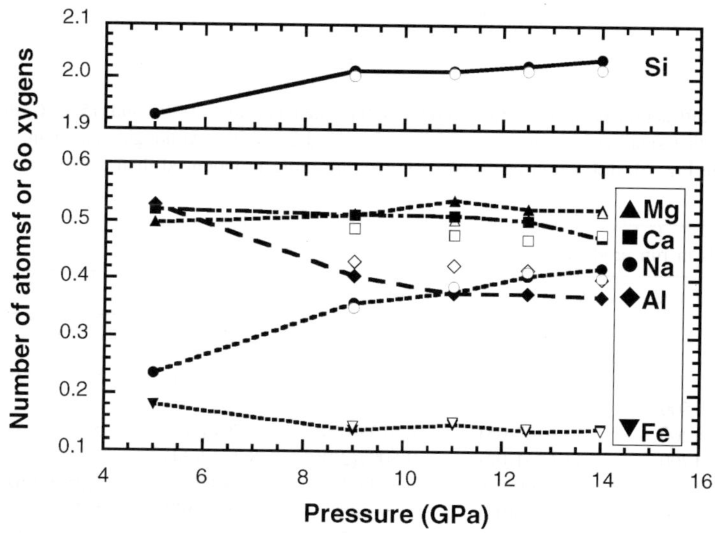

Fig. 4. Composition of clinopyroxene (Cpx) in atomic ratio (normalized to six oxygens) as a function of pressure. Data obtained at 1500 °C (filled symbols) indicate higher Ca contents but lower Al contents than at 1700 °C (open symbols). See text for explanation.

(4 0 0), (4 2 0), (3 3 2), (4 2 2), (4 3 1), (5 2 1), (6 1 1), (4 4 4), (6 4 0), and (6 4 2) reflections. The lattice parameters of Cpx were calculated using 7 to 12 X-ray diffraction lines from the set of (0 2 1), ($\bar{2}$ 2 0), ($\bar{2}$ 2 1), ($\bar{3}$ 1 0), ($\bar{3}$ 1 1), (0 0 2), (2 2 1), (3 1 1), (1 1 2), ($\bar{3}$ 3 0), ($\bar{3}$ 3 1), ($\bar{4}$ 2 1), (0 4 1), ($\bar{4}$ 0 2), and ($\bar{1}$ 3 2). Results for all the run products are summarized in Table 4.

3.3. Zero-pressure densities of individual minerals and mineral proportions

The densities of garnet and Cpx were calculated from the lattice parameters and chemical compositions obtained in this study. The results of the calculations are listed in Table 5 as "XRD". The

where D_i and V_i are the density and the volume fraction of the ith mineral (Cpx, garnet and coesite or sthishovite), respectively. The results of the mass balance calculations are shown in Fig. 6.

4. Discussion

4.1. Comparison with previous work

The most important difference between our results and the data reported by Irifune et al. (1986) is that the Cpx/Gar ratio is much smaller at 1200 °C in our study (see Fig. 6). In our experiments, the Cpx/(Gar + silica phase) ratio increases with increasing temperature at a given pressure (Fig. 6). In other words, garnet and the SiO_2 phase dissolve into Cpx with increasing temperature at a given pressure. This effect can be understood from the phase relations in the system Enstatite–Diopside–Pyrope–Grossular (Boyd, 1970; Yamada and Takahashi, 1984; Irifune, 1987). The MORB composition plots in the Cpx–Gar two-phase region in this system. As the compositional field of Ca-clinopyroxene expands significantly with temperature, whereas that of Pyrope–Grossular garnet changes only slightly, the amount of Cpx relative to garnet increases with temperature (see Fig. 11 of Boyd, 1970).

A small amount of the SiO_2 phase was found in EPMA and XRD analyses in samples from all our experiments (see Table 2 and Fig. 5). The mass balance calculations also confirm that coesite or stishovite appears at pressures below 12 GPa (see Fig. 6). This result, however, is inconsistent with the previous experiments of Irifune et al. (1986) who reported stishovite above 12 GPa as the only SiO_2 phase. The coesite in our experiments is not considered to be a metastable relict mineral, because the present starting materials do not contain silica minerals. The common occurrence of quartz or coesite in natural eclogite may be cited as one of the supporting lines of evidence for stability of the silica minerals in MORB eclogite below 10 GPa. A silica-phase was also found to be stable over a wide pressure and temperature range between 3 GPa at 700 °C and 10 GPa at 1000 °C, in experiments on hydrous MORB (e.g., Schmidt and Poli, 1998; Poli and Schmidt, 1995).

The discrepancies between our study and that of Irifune et al. (1986) may be due to the differences in run durations and starting materials. Run durations in the earlier experiments were between 30 and 90 min, whereas those in this study are greater than 10 h at 1200 °C. Moreover, a glass with primitive MORB composition was used as a starting material in the previous study, while fine-grained natural rock powder of MORB and its hydrated form (synthetic amphibolite) were used in our experiments. Glass as a starting material has a risk that metastable phases tend to appear in the course of raising temperature at high pressure. It is possible that metastable growth of a large amount of Al-rich Cpx in the previous experiments inhibited crystallization of coesite and stishovite, because Cpx (~56 wt.%) is much higher in SiO_2 than Gar (~42 wt.%).

The difference in oxygen fugacity imposed on samples by a Pt capsules (Irifune et al., 1986) and by graphite capsules (this study) might be another cause for the discrepancy. In general, f_{O_2} of the Fe-bearing run products increases in Pt capsules as a function of time (and/or run temperatures) because of the alloying reaction of FeO in silicates with the Pt capsule to form Fe^0. However, because this reaction increases the SiO_2 component in the system by reactions as shown below, it cannot account for the absence of the silica phase in Irifune et al's experiments.

$$FeSiO_3^{Cpx} = Fe^{Pt} + \tfrac{1}{2}O_2 + SiO_2$$

Part of the discrepancy between our results and those by Irifune et al. (1986) may be ascribed to the difference in composition of the starting materials. As the starting material used by Irifune et al. (1986) is higher in CaO than in our study, while SiO_2 is virtually the same (see Table 1), their experimental charges may contain systematically more pyroxene than those in this study. Considering the compositional variation of MORBs along mid oceanic ridges (e.g., Schilling et al., 1983), we believe that our starting material (ARP74) is closer to the average composition of oceanic crust.

4.2. Density of the MORB eclogite at high-pressure and high-temperature

There are several equations of state that are often used to extrapolate densities of individual minerals to high-pressure and high-temperature conditions. We used the third-order Birch–Murnaghan equation of state (EOS) combined with thermal expansivity in the present study. This is because thermodynamic parameters for this EOS in high-pressure phases have been well investigated by in situ X-ray experiments. The EOS is given by the following expression (e.g., Funamori et al., 1996):

$$P = \frac{3}{2}K_{T,0}\left[\left(\frac{V_{T,0}}{V}\right)^{7/3} - \left(\frac{V_{T,0}}{V}\right)^{5/3}\right] \times \left\{1 - \frac{3}{4}(4 - K'_{T,0})\left[\left(\frac{V_{T,0}}{V}\right)^{2/3} - 1\right]\right\}$$

$$K_{T,0} = K_{300,0} + \left(\frac{\partial K_{T,0}}{\partial T}\right)_P (T - 300),$$

$$\left(\frac{\partial K_{T,0}}{\partial T}\right)_P = \text{const.}$$

$$K'_{T,0} = K'_{300,0}$$

$$V_{T,0} = V_0 \exp\int_{300}^{T} \alpha_{T,0}\,\mathrm{d}T$$

$$\alpha_{T,0}=a_0+b_0T - c_0T^{-2}+d_0T^{-1}$$

where $K_{T,0}$, $K_{T,0}'$, $V_{T,0}$ are isothermal bulk modulus, its pressure derivative, and the unit cell volume at temperature T and ambient pressure, respectively.

Necessary parameters for the EOS are a_0, b_0, c_0, d_0, $K_{300,0}$, $K_{300,0}'$, and $(\partial K_{T,0}/\partial T)_P$. The thermoelastic parameters of individual minerals used in the extrapolation calculation are summarized in Table 6. Here, the parameters for diopside (end member in Cpx) and majority of garnet with MORB composition are assumed to be representative of the parameters for Cpx solid solution and garnet solid solution, respectively. Using these parameters, the zero-pressure and room temperature densities listed in Table 5 are calculated to high-pressure and high-temperatures and the results are listed in Table 7.

Table 6
Thermodynamic parameters of minerals

	Majorite	Diopside	Coesite	Stishovite
ρ_0	–	–	2.909	4.287
$K_{300,0}$ (GPa)	160.9 [1]	114.0 [2]	96 [6]	314 [9]
$(\partial K_{T0}/\partial P)_T$	4.0 [1]	4.5 [3]	8.4 [6]	5.1 [10]
$(\partial K_{T0}/\partial T)_P$ (GPa/K)	−0.029 [1]	−0.022 [4]	−0.02 [7]	−0.047 [7]
$\alpha_{T0} = a_0 + b_{0,T} - c_0T^{-2} + d_{0,T}^{-1}$				
a_0 (10^{-5}/K)	2.13 [1]	3.206 [5]	0.543 [8]	1.053 [11]
b_0 (10^{-8}/K^2)	1.71 [1]	0.811 [5]	0.500 [8]	0.903 [11]
c_0 (K)	–	1.8167 [5]	–	−1.22 [11]
d_0 (10^{-2})	–	0.1347 [5]	–	–
	Enstatite	Olivine	β-spinel	
ρ_0	$3.204 + 0.798X_{Fe}$	$3.222 + 1.182X_{Fe}$ ($X_{Fe} = 0.11$)	$3.472 + 1.24X_{Fe}$ ($X_{Fe} = 0.11$)	
$K_{300,0}$ (GPa)	122.8 [12]	127.4 [14]	172.6 [14]	
$(\partial K_{T0}/\partial P)_T$	5.6 [12]	4.8 [14]	4.8 [14]	
$(\partial K_{T0}/\partial T)_P$	−0.020 [4]	−0.021 [14]	−0.027 [14]	
$\alpha T_0 = a_0 + b_0T - c_0T^{-2} + d_0T^{-1}$				
a_0 (10^{-5}/K)	2.86 [13]	2.832 [15]	2.711 [16]	
b_0 (10^{-8}/K^2)	0.72 [13]	0.758 [15]	0.6885 [16]	
c_0 (K)	–	–	0.5767 [16]	
d_0 (10^{-2})	–	–	–	

[1] Aoki (1999), [2] Levien et al. (1979), [3] Levien and Prewitt (1981a), [4] calculating from Saxena et al. (1993), [5] Saxena and Shen (1992), [6] Levien and Prewitt (1981b), [7] Fei et al. (1990), [8] Saxena et al. (1993), [9] Weidner et al. (1982), [10] Rigden et al. (1994), [11] Akaogi et al. (1995), [12] Angel and Hugh-Jones (1994), [13] Zhao et al. (1995), [14] Meng et al. (1993), [15] Bouhifd et al. (1996), [16] Fei et al. (1992).

Table 7
Densities of individual minerals at high-pressures and high-temperatures

Run no.	P (GPa)	T (°C)	Gar (g/cm^3)	Cpx (g/cm^3)	Silica mineral (g/cm^3)	Bulk rock (g/cm^3)
S779	5.0	1200	3.752	3.365	3.038	3.535
S780	5.0	1500	3.690	3.343	3.038	3.454
S802	7.0	1200	3.793	3.420	3.090	3.593
S804	8.9	1700	3.756	3.430	3.144	3.576
S777	9.1	1500	3.791	3.454	3.145	3.615
S788	11.0	1500	3.823	3.517	4.303	3.751
S789	11.0	1700	3.792	3.489	4.303	3.701
S807	12.5	1500	3.845	3.554	4.342	3.795
S808	12.5	1700	3.813	3.501	4.325	3.726
S806	14.0	1500	3.850	3.607	4.363	3.809
S809	14.0	1700	3.848	3.575	4.347	3.821

4.3. Implications for mantle dynamics

Oceanic crust in old subducting slabs (e.g. 100 Ma) would remain at relative low temperatures (<1200 °C) even at depths exceeding 670 km (e.g., Peacock et al., 1994), and the driving force of down going slabs is ascribed at least in part to the density contrast between subducted oceanic crust and surrounding mantle. Fig. 7 compares the density of subducting oceanic crust obtained by this study with the density of ambient mantle peridotite. Densities of subducted oceanic crust at 1200 °C were extrapolated from the results obtained at higher temperatures in this study. Densities of ambient peridotite were calculated from the experimental results of Irifune and Isshiki (1998) using thermoelastic parameters listed in Table 6, along their model geotherm.

We draw two conclusions from Fig. 7. One is that the density of subducted oceanic crust changes discontinuously at around 9 GPa, as a result of the coesite–stishovite phase transition. Subducted oceanic lithosphere (40–80 km thick) consists of three layers, basaltic crust (~7 km thick), harzburgite (~20 km thick) and lherzolite, in which basaltic crust is the only layer that contains a SiO_2 phase. Therefore, the density jump may be less significant in cold slabs consisting of a thick peridotite layer and thin oceanic crust, but may be more effective in hot thinner slabs or slabs with unusually thick crust (e.g. oceanic plateau) where the role of oceanic crust is important. The second conclusion is that the density difference between subducted oceanic crust and the mantle is clearly larger than the difference indicated by previous studies (Irifune et al., 1986; Irifune and Ringwood, 1987), particularly at pressures less than 10 GPa. The inconsistency with previous data results from a discrepancy in the Cpx/(Gar + silica phase) ratio. Accordingly, the role of eclogite as a driving mechanism for subduction processes is larger than previously thought.

The significance of recycled old oceanic crust in basalt magma genesis has been emphasized based on major and trace elements as well as isotopic compositions (e.g., Hofmann and White, 1982; Hauri, 1996; Takahashi et al., 1998; Takahashi and Nakajima, 2002). Fig. 8 demonstrates the density difference between peridotite (pyrolite mantle component) and entrained eclogite blocks (recycled old oceanic crust) in ascending mantle plume at various temperatures. Densities of peridotite in Fig. 8 were calculated from the chemical compositions of minerals synthesized by Irifune and Isshiki (1998), and extrapolated to 1200, 1500 and 1700 °C, respectively. In this calculation, the proportions of minerals in peridotite are assumed to be constant with temperature. This approximation is valid in the shallower part of the mantle because olivine and its high-pressure polymorphs do not interact with other phases, such as Cpx, Gar and orthopyroxene (Opx).

Fig. 8 indicates that the density difference between peridotite and eclogite in the ascending mantle plume becomes small with increasing temperature. As discussed above, this is due to the increase of the Cpx/(Gar + silica phase) ratio in eclogite with increasing temperature. Even at the solidus temperature, however, it is clear that eclogite is denser than peridotite in the studied pressure range ($P < 14$ GPa).

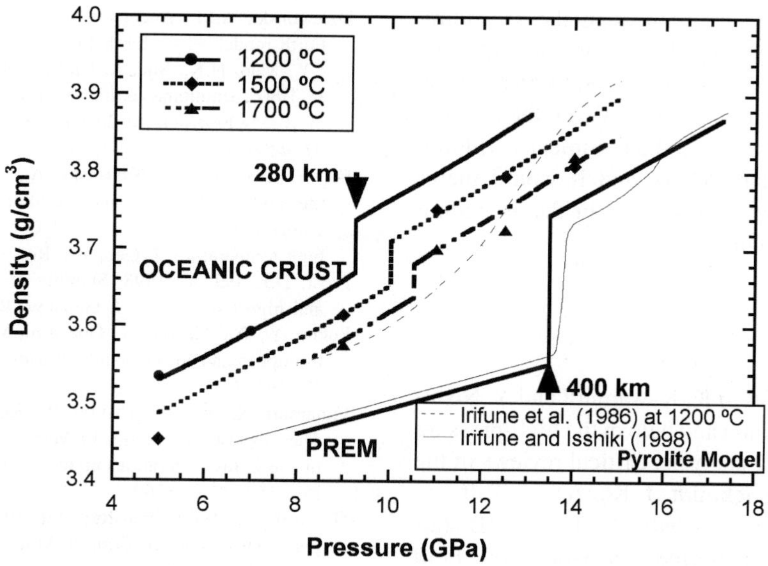

Fig. 7. Density of subducted oceanic crust (MORB eclogite) and ambient mantle (pyrolite). The density profiles of the oceanic crust at 1200, 1500 and 1700 °C were calculated from the mineral proportions (Fig. 6), the zero-pressure densities (Table 5), and the thermoelastic parameters (Table 6). The density profile of the oceanic crust by Irifune and Ringwood (1987) was also recalculated with the same scheme and is shown for comparison (thin broken line). Density profile of the mantle is calculated from the experimental data of Irifune and Isshiki (1998) along their model geotherm (thin solid line). Seismic PREM model (Dziewonski and Anderson, 1981) is shown for comparison. See text for more detail.

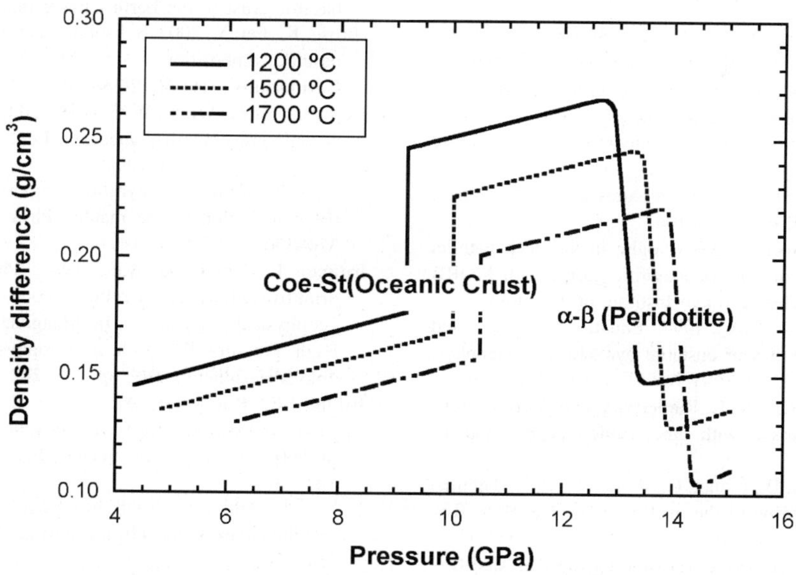

Fig. 8. Density differences (as a function of P and T) between an entrained eclogite block and peridotite in an ascending mantle plume at various temperatures. These profiles were calculated in a similar to those in Fig. 7. Note that the density contrast decreases with increasing temperature.

Therefore, the eclogite component in the ascending plume must be sustained by the upward flow of peridotite in mantle plumes (Yasuda et al., 1998). In the lower part of the transition zone, however, the eclogite may be equally or less dense than ambient peridotite (Ringwood and Irifune, 1988), and it would therefore be easy to become transported in the ascending column of a mantle plume.

Acknowledgements

We are very grateful to K. Kawamura and Y. Nishihara for useful discussion and help in the high-pressure experiments and XRD analysis. Critical reviews of the manuscript by M.W. Schmidt, J. Konzett, J. Tuff and D. Rubie are highly acknowledged. The MORB staring material (ARP74) was kindly provided by T. Fujii (ERI, Tokyo University). Thanks are due to N. Abe for assistance in EPMA analysis. Critical readings of the manuscript by T. Kogiso and Y. Nishihara are highly acknowledged. This work was supported by the grant 12002006 from the Ministry of Education, Culture, Sports, Science, and Technology, Japan.

References

Akaogi, M., Yusa, H., Shiraishi, K., Suzuki, T., 1995. Thermodynamic properties of α-quartz, coesite, and stishovite and equilibrium phase relations at high pressures and high temperatures. J. Geophys. Res. 100, 22337–22347.

Akaogi, M., Akimoto, S., 1977. Pyroxene-garnet solid solution equilibria in the system $Mg_4Si_4O_{12}$–$Mg_3Al_2Si_3O_{12}$ and $Fe_4Si_4O_{12}$–$Fe_3Al_2Si_3O_{12}$ at high pressures and temperatures. Phys. Earth Planet. Inter. 15, 90–106.

Aoki, K., 1999. Density of MORB eclogite in the deep mantle. Chapter II. Equation of state of majoritic garnet with MORB composition. M.Sc. Thesis, Tokyo Institute of Technology.

Angel, R.J., Hugh-Jones, D.A., 1994. Equations of state and thermodynamic properties of enstatite pyroxenes. J. Geophys. Res. 99, 19777–19783.

Baur, W.H., Kahn, A.A., 1971. Rutile-type compounds SiO_2, GeO_2, and a comparison with other rutile-type compounds. Acta Cryst. B27, 2133–2139.

Bouhifd, M.A., Andrault, D., Fiquet, G., Richet, P., 1996. Thermal expansion of forsterite up to the melting point. Geophys. Res. Lett. 23, 1143–1146.

Boyd, F.R., 1970. Garnet peridotites and the system $CaSiO_3$–$MgSiO_3$–Al_2O_3. Mineral. Soc. Amer. Spec. Pap. 3, 63–75.

Cordery, M.J., Davies, G.F., Campbell, I.H., 1997. Genesis of flood basalts from eclogite-bearing mantle plumes. J. Geophys. Res. 102, 20179–20197.

Dziewonski, A.M., Anderson, D.L., 1981. Preliminary reference Earth model. Phys. Earth Planet. Inter. 25, 297–356.

Fei, Y., Mao, H.K., Shu, J., Parthasarathy, G., Basset, W.A., Ko, J., 1992. Simultaneous high-P, high-T X-ray diffraction study of β-$(Mg,Fe)_2SiO_4$ to 26 GPa and 900 K. J. Geophys. Res. 97, 4489–4495.

Fei, Y., Saxena, S.K., Navrotsky, A., 1990. Internally consistent thermodynamic data and equilibrium phase relations for compounds in ther system MgO–SiO_2 at high pressure and high temperature. J. Geophys. Res. 95, 6915–6928.

Frost, D.J., Fei, Y., 1998. Stability of phase D at high pressure and high temperature. J. Geophys. Res. 103, 7463–7474.

Fukao, Y., Widiyantoto, S., Obayashi, M., 2001. Stagnant slabs in the upper and lower mantle transition mregion. Rev. Geophys. 39, 291–323.

Funamori, N., Yagi, T., Utsumi, W., Kondo, T., Uchida, T., 1996. Thermoelastic properties of $MgSiO_3$ perovskite determined by in situ X-ray obervations up to 30 GPa and 2000 K. J. Geophys. Res. 101, 8257–8269.

Gasparik, T., 1989. Transformation of enstatite–diopside–jadeite pyroxenes to garnet. Contrib. Min. Petrol. 102, 389–405.

Gasparik, T., 1992. Enstatite–jadeite join and its role in the Earth's mantle. Contrib. Min. Petrol. 111, 283–298.

Grand, S.P., 1994. Mantle shear structure beneath the Americas and surrounding oceans. J. Geophys. Res. 99, 11591–11621.

Hauri, E.H., 1996. Major-element variability in the Hawaiian mantle plume. Nature 382, 415–419.

Hirose, K., 2002. Phase transitions in pyrolitic mantle around 670-km depth: implication for the upwelling of plumes from the lower mantle. J. Geophys. Res. 107, doi: 10.1029/2001JB000597.

Hirose, K., Fei, Y., Ma, Y., Mao, H.-K., 1999. The fate of subducted basaltic crust in the Earth's lower mantle. Nature 397, 53–56.

Hirose, K., Fei, Y., 2002. Subsolidus and melting phase relations of basaltic composition in the upper most lower mantle. Geochim. Cosmochim. Acta 66, 2099–2108.

Hofmann, A.W., White, W.M., 1982. Mantle plumes from ancient oceanic crust. Earth Planet. Sci. Lett. 57, 421–436.

Irifune, T., 1987. An experimental investigation of pyroxene–garnet transformations in a pyrolite composition and its bearing on the constitution of the mantle. Phys. Earth Planet. Inter. 45, 324–336.

Irifune, T., Ringwood A.E., 1987. Phase transformations in primitive MORB and pyrolite compositions to 25 GPa and some geophysical implications. In: Manghnani, M., Syono, Y. (Eds.), High Pressure Research in Geophysics. TERRAPUB/AGU, Tokyo/Washington, DC, pp. 231–242.

Irifune, T., Ringwood, A.E., 1993. Phase transformations in subducted oceanic crust and buoyancy relationships at depths of 600–800 km in the mantle. Earth Planet. Sci. Lett. 117, 101–110.

Irifune, T., Isshiki, M., 1998. Iron partitioning in a pyrolite mantle and the nature of the 410-km seismic discontinuity. Nature 392, 702–704.

Irifune, T., Sekine, T., Ringwood, A.E., Hibberson, W.O., 1986. The eclogite–garnetite transformations at high pressure and some geophysical implications. Earth Planet. Sci. Lett. 77, 245–256.

Levien, L., Prewitt, C.T., 1981a. High pressure crystal structure and compressibility of coesite. Am. Mineral. 66, 324–333.

Levien, L., Prewitt, C.T., 1981b. High pressure structural study of diopside. Am. Mineral. 66, 315–323.

Levien, L., Weidner, D.J., Prewitt, C.T., 1979. Elasticity of diopside. Phys. Chem. Miner. 4, 105–113.

Lloyd, E.C., 1971. Accurate Characterization of the High Pressure Environment. NBS Spec. Publ. No. 326, Washington, DC, pp. 1–3.

Meng, Y., Weidner, D.J., Gwanmesia, G.D., Liebermann, R.C., Vaughan, M.T., Wang, Y., Leinenweber, K., Pacalo, R.E., Yeganen-Haeri, A., Zhao, Y., 1993. In situ high *P–T* X-ray diffraction studies on three polymorphs (α, β, γ) of Mg_2SiO_4. J. Geophys. Res. 98, 22199–22207.

Morishima, T., Kato, T., Suto, M., Ohtani, E., Urakawa, S., Utsumi, W., Shimomura, O., Kikegawa, T., 1994. The phase boundary between α- and β-Mg_2SiO_4 determined by in situ X-ray observation. Science 265, 1202–1203.

Ono, S., Yasuda, A., 1996. Compositional change of majoritic garnet in a MORB composition from 7 to 17 GPa and 1400 to 1600 °C. Phys. Earth Planet. Inter. 96, 171–179.

Peacock, S.M., Rushmer, T., Thompson, A.B., 1994. Partial melting of subducting oceanic crust. Earth Planet. Sci. Lett. 121, 227–244.

Poli, S., Schmidt, M.W., 1995. H_2O transport and release in subduction zones: experimental constraints on basaltic and andesitic systems. J. Geophys. Res. 100, 22299–22314.

Rigden, S. M., Li B., Liebermann R. C., 1994. Elasticity of stishovite at high pressures, EOS Trans. AGU, 75(44), Fall Meet. Suppl., 596.

Ringwood, A.E., Irifune, T., 1988. Nature of the 650-km seismic discontinuity: implications for mantle dynamics and differentiation. Nature 331, 131–136.

Saxena, S.K., Shen, G., 1992. Assessed data on heat capacity, thermal expansion, and compressibility for some oxides and silicates. J. Geophys. Res. 97, 19813–19825.

Saxena, S. K., Chatterjee N., Fei Y., Shen G., 1993. Thermodynamic Data on Oxides and Silicates. Springer–Verlag, New York, 428 pp.

Schilling, J.-G., Zajac, M., Evans, M., Hhonston, R., White, W., Devine, J.D., Kingsley, R., 1983. Petrologic and geochemical variations along the Mid-Atlantic Ridge from 29 to 73°N. Am. J. Sci. 283, 510–586.

Schmidt, M.W., Poli, S., 1998. The stability of lawsonite and zoisite at high-pressures: experiments in CASH to 92 kbar and implications for the presence of hydrous phases in subducted lithosphere. Earth Planet. Sci. Lett. 124, 105–118.

Smyth, J.R., Artioli, G., Smith, J.V., Kvick, A., 1987. Crystal structure of coesite, a high-pressure form of SiO_2, at 15 and 298 K from single-crystal neutron and X-ray diffraction data: test of bonding models. J. Phys. Chem. 91, 988–992.

Takahashi, T., Mao, H.K., Basset, W.A., 1969. X-ray diffraction study of a high-pressure polymorph. Science 165, 1352–1353.

Takahashi, E., Nakajima, K., 2002. Melting process in the Hawaiian plume: an experimental study. In: Takahashi, E., Garcia, M.O., Lipman, P.W., Naka, J. (Eds.), Hawaiian Volcanoes: Deep Underwater Perspectives, vol. 128. Am. Geophys. Union Monograph, pp. 403–418.

Takahashi, E., Nakajima, K., Wright, T.L., 1998. Origin of the Columbia river basalts: melting model of a heterogeneous plume head. Earth Planet. Sci. Lett. 162, 63–80.

Takahashi, E., Shimazaki, T., Tsuzaki, Y., Yoshida, H., 1993. Melting study of a peridotite KLB-1 to 6.5 GPa and the origin of basaltic magmas. Phil. Trans. R. Soc. London A342, 105–120.

van der Hilst, R., 1995. Complex morphology of subducted lithosphere in the mantle beneath the Tonga trench. Nature 374, 154–157.

Wang, W., Takahashi, E., 2000. Subsolidus and melting experiments of K-doped peridotite KLB-1 to 27 GPa: its geophysical and geochemical implications. J. Geophys. Res. 105, 2855–2868.

Weidner, D.J., Bass, J.D., Ringwood, A.E., Sinclair, W., 1982. The single crystal elastic moduli of stishovite. J. Geophys. Res. 87, 4740–4746.

Yamada, H., Takahashi, E., 1984. Subsolidus phase relations between coexisting garnet and two pyroxenes at 50 to 100 kbar in the system $CaO–MgO–Al_2O_3–SiO_2$. In: Kornprobst, J. (Ed.), Kimberlites. Part II. The Mantle and Crust–Mantle Relations. Elsevier, Amsterdam, pp. 247–255.

Yasuda, A., Fujii, T., Kurita, K., 1998. Ascending subducted oceanic crust entrained within mantle plumes. Geophys. Res. Lett. 25, 1561–1564.

Zhang, J., Li, B., Utsumi, W., Liebermann, R.C., 1996. In situ X-ray observations of the coesite–stishovite transition: reversed phase boundary and kinetics. Phys. Chem. Miner. 23, 1–10.

Zhao, Y., Schiferl, D., Shankland, T.J., 1995. A high *P–T* single-crystal X-ray diffraction study of thermoelasticity of $MgSiO_3$ orthoenstatite. Phys. Chem. Miner. 22, 393–398.

Available online at www.sciencedirect.com
SCIENCE DIRECT
ELSEVIER
Physics of the Earth and Planetary Interiors 143–144 (2004) 145–156
PHYSICS OF THE EARTH AND PLANETARY INTERIORS
www.elsevier.com/locate/pepi

High-pressure transitions of diopside and wollastonite: phase equilibria and thermochemistry of $CaMgSi_2O_6$, $CaSiO_3$ and $CaSi_2O_5$–$CaTiSiO_5$ system

M. Akaogi*, M. Yano, Y. Tejima, M. Iijima, H. Kojitani

Department of Chemistry, Gakushuin University, 1-5-1 Mejiro, Toshima-ku, Tokyo 171-8588, Japan

Received 4 January 2003; received in revised form 23 June 2003; accepted 6 August 2003

Abstract

Phase transitions of $CaMgSi_2O_6$ diopside and $CaSiO_3$ wollastonite were examined at pressures to 23 GPa and temperatures to 2000 °C, using a Kawai-type multiavil apparatus. Enthalpies of high-pressure phases in $CaSiO_3$ and in the $CaSi_2O_5$–$CaTiSiO_5$ system were also measured by high-temperature calorimetry. At 17–18 GPa, diopside dissociates to $CaSiO_3$-rich perovskite + Mg-rich $(Mg,Ca)SiO_3$ tetragonal garnet (Gt) above about 1400 °C. The solubilities of $CaSiO_3$ in garnet and $MgSiO_3$ in perovskite increase with temperature. At 17–18 GPa below about 1400 °C, diopside dissociates to Ca-perovskite + β-Mg_2SiO_4 + stishovite. The Mg, Si-phases coexisting with Ca-perovskite change to γ-Mg_2SiO_4 + stishovite, to ilmenite, and finally to Mg-perovskite with increasing pressure. $CaSiO_3$ wollastonite transforms to the walstromite structure, and further dissociates to Ca_2SiO_4 larnite + $CaSi_2O_5$ titanite. The latter transition occurs at 9–11 GPa with a positive Clapeyron slope. At 1600 °C, larnite + titanite transform to $CaSiO_3$ perovskite at 14.6 ± 0.6 GPa, calibrated against the α–β transition pressure of Mg_2SiO_4. The enthalpies of formation of $CaSiO_3$ walstromite and $CaSi_2O_5$ titanite from the mixture of CaO and SiO_2 quartz at 298 K have been determined as -76.1 ± 2.8, and -27.8 ± 2.1 kJ/mol, respectively. The latter was estimated from enthalpy measurements of titanite solid solutions in the system $CaSi_2O_5$–$CaTiSiO_5$, because $CaSi_2O_5$ titanite transforms to a triclinic phase upon decompression. The enthalpy difference between titanite and the triclinic phase is only 1.5 ± 4.8 kJ/mol. Using these enthalpies of formation and those of larnite and $CaSiO_3$ perovskite, the transition boundaries in $CaSiO_3$ have been calculated. The calculated boundaries for the wollastonite–walstromite–larnite + titanite transitions are consistent with the experimental determinations within the errors. The calculated boundary between larnite + titanite and Ca-perovskite has a slope of 1.3–1.8(±0.4) MPa/K, and is located at a pressure about 2 GPa higher than that determined by [Am. Mineral. 79 (1994) 1219].

Keywords: Diopside; Wollastonite; Titanite; Ca-perovskite; High-pressure transition; Calorimetry

1. Introduction

$CaMgSi_2O_6$ diopside is a major component of clinopyroxene (Cpx) in the crust and upper mantle. $CaSiO_3$ wollastonite is an important ceramic and cement substance, and its high-pressure polymorph with the perovskite structure is accepted as one of major

* Corresponding author. Tel.: +81-3-3986-0221; fax: +81-3-5992-1029.
E-mail address: masaki.akaogi@gakushuin.ac.jp (M. Akaogi).

0031-9201/$ – see front matter
doi:10.1016/j.pepi.2003.08.008

Weidner, D.J., Wang, Y., 2000. Phase transformations: Implications for mantle structure. In: Karato, S., Forte, A.M., Liebermann, R.C., Masters, G., Stixrude, L. (Eds.), Earth's Deep Interior: Mineral Physics and Tomography from Atomic to the Global Scale. Am. Geophys. Union, pp. 215–235.

Warren, M.C., Redfern, S.A.T., Angel, R., 1999. Change from sixfold to fivefold coordination of silicate polyhedra: insight from first-principles calculations of $CaSi_2O_5$. Phys. Rev. B 59, 9149–9154.

Wood, B.J., 2000. Phase transformations and partitioning relations in peridotite under lower mantle conditions. Earth Planet Sci. Lett. 174, 341–354.

Xirouchakis, D., Fritsch, S., Putnam, R.L., Navrotsky, A., Lindsley, D.H., 1997. Thermochemistry and the enthalpy of formation of synthetic end-member ($CaTiSiO_5$) titanite. Am. Mineral. 82, 754–759.

Yagi, T., Uchiyama, Y., Akaogi, M., Ito, E., 1992. Isothermal compression curve of $MgSiO_3$ tetragonal garnet. Phys. Earth Planet Int. 74, 1–7.

Yusa, H., Akaogi, M., Ito, E., 1993. Calorimetric study of $MgSiO_3$ garnet and pyroxene: heat capacities, transition enthalpies, and equilibrium phase relations in $MgSiO_3$ at high pressures and temperatures. J. Geophys. Res. 98, 6453–6460.

Available online at www.sciencedirect.com
SCIENCE DIRECT®
ELSEVIER
Physics of the Earth and Planetary Interiors 143–144 (2004) 157–169
PHYSICS OF THE EARTH AND PLANETARY INTERIORS
www.elsevier.com/locate/pepi

Oxidation state of iron in hydrous mantle phases: implications for subduction and mantle oxygen fugacity

C.A. McCammon[a,*], D.J. Frost[a], J.R. Smyth[b], H.M.S. Laustsen[b], T. Kawamoto[c], N.L. Ross[d], P.A. van Aken[e]

[a] Bayerisches Geoinstitut, Universität Bayreuth, D-95440 Bayreuth, Germany
[b] Department of Geological Sciences, University of Colorado, Boulder, CO 80309, USA
[c] Institute for Geothermal Sciences, Kyoto University, Beppu, Japan
[d] Department of Geological Sciences, Virginia Polytechnic Institute and State University, Blacksburg, VA 24061, USA
[e] Institut für Angewandte Geowissenschaften, Technische Universität Darmstadt, D-64287 Darmstadt, Germany

Received 29 January 2003; received in revised form 8 June 2003; accepted 15 August 2003

Abstract

Samples of olivine, wadsleyite, ringwoodite and majorite with Mg/(Mg + Fe) = 0.85–0.90 were synthesised at high pressure and temperature using a multianvil press under nominally anhydrous conditions as well as hydrous conditions with varying conditions of oxygen fugacity, and examined ex situ using Mössbauer spectroscopy to determine $Fe^{3+}/\Sigma Fe$, and either Fourier Transform Infrared Spectroscopy or Secondary Ion Mass Spectrometry to determine H concentration. Under oxidising conditions, the relative concentration of Fe^{3+} increases with water content in olivine, wadselyite and ringwoodite. Majorite does not incorporate significant water, but shows high $Fe^{3+}/\Sigma Fe$ values when synthesised under oxidising conditions in the presence of water. $Fe^{3+}/\Sigma Fe$ is strongly influenced by oxygen fugacity for all phases studied, and the mechanism of hydration is likely coupled to Fe^{3+} incorporation in the crystal structure. Enhanced $Fe^{3+}/\Sigma Fe$ in mantle-derived olivine and orthopyroxene has a negligible effect on oxygen fugacities estimated using olivine–orthopyroxene–garnet oxybarometers, while increased $Fe^{3+}/\Sigma Fe$ in garnet due to dehydrogenation could lead to an overestimation of oxygen fugacity in exhumed samples. Subduction zones are likely oxidised relative to the rest of the mantle, and hence represent regions where not only water is concentrated, but also Fe^{3+}.

Keywords: Mössbauer spectroscopy; Ferric iron; Water; Subduction

1. Introduction

Mantle hydrous phases have been the focus of numerous studies due to the influence of water on mantle properties, including viscosity, melting behaviour, rheology, elastic properties and the behaviour of fluids. Many studies have focused on the system $MgO–SiO_2–H_2O$, where a large number of stable hydrous phases have been found (reviewed by Angel et al., 2001), demonstrating that iron is not a necessary element for stabilising OH in these phases. Nevertheless, since iron is the most abundant element that adopts multiple valence states, its chemistry is

* Corresponding author. Tel.: +49-921-553709; fax: +49-921-553769.
E-mail address: catherine.mccammon@uni-bayreuth.de (C.A. McCammon).

0031-9201/$ – see front matter © 2004 Elsevier B.V. All rights reserved.
doi:10.1016/j.pepi.2003.08.009

likely to be influenced by the presence of water, particularly since the incorporation of OH may involve the creation of defects in the structure.

Iron is the fourth most abundant element, but due to its nature as a transition element with the ability to exist in multiple valence states, can have an effect on physical and chemical properties greater than its abundance would suggest. Oxygen fugacity and Fe^{3+}/Fe^{2+} equilibria are intimately connected, and while the relation is often complex in mineral systems, the importance of oxygen fugacity to mantle processes has motivated numerous studies. Oxygen fugacity can affect subsolidus and supersolidus phase relations, the presence and speciation of a fluid phase, transport properties such as electrical conductivity and diffusivity, and mechanical behaviour. An ongoing project at Bayerisches Geoinstitut has been to characterise the nature of iron in transition zone and lower mantle phases, particularly with regard to the relation between $Fe^{3+}/\Sigma Fe$ and oxygen fugacity. Previous studies have focused on nominally anhydrous phases, which are currently believed to provide realistic models for bulk mantle mineralogy. The picture is different in regions of subduction, however, where hydrous phases play a major role in determining the physical and chemical properties of downgoing slabs, with obvious implications for mantle dynamics.

The purpose of this paper is to determine the oxidation state of iron in high-pressure mantle phases in the presence of water, and to assess the relation with oxygen fugacity. In particular, we focus on mantle phases that are nominally anhydrous, but have been found to incorporate significant amounts of water: $(Mg, Fe)_2SiO_4$ olivine, β-$(Mg, Fe)_2SiO_4$ wadsleyite, γ-$(Mg, Fe)_2SiO_4$ ringwoodite and $(Mg, Fe)(Si, Al)O_3$ majorite. Many structural details of these phases have been presented elsewhere (see Table 1); here we focus on the measurement of $Fe^{3+}/\Sigma Fe$, OH concentration, and the implications for subduction and mantle oxygen fugacity.

2. Experimental methods

Nominally anhydrous polycrystalline samples of olivine, wadsleyite, ringwoodite and majorite were synthesised from ^{57}Fe-enriched starting materials using the 1000 and 2000 t multianvil presses at Bayerisches Geoinstitut (details in Table 1). The oxygen fugacity imposed on the samples was influenced by the type of capsule used, but with the exception of the run incorporating ReO_2 (S2144), the experiments cannot be considered to have been buffered. The form of the run products was either as a hot-pressed disk (olivine) or polycrystalline aggregates (wadsleyite, ringwoodite and majorite).

Hydrous samples were synthesised from non-^{57}Fe-enriched starting materials using a multianvil press (details in Table 1). In one set of experiments involving the 1000 t multianvil press at Bayerisches Geoinstitut, a double capsule technique was used with a buffer in the outer part and the sample mix in the inner part; these experiments can be considered to have been buffered (H1518; H1505; HZ7). In a different set of experiments involving the 5000 t press at Bayerisches Geoinstitut, sample material was sealed inside large Pt capsules which yielded single crystals in the run products of up to roughly 500 μm in diameter (SZ0004; SZ0005; SZ0006; Z89; SZ0002; SZ9901). Although these experiments were not formally buffered, conditions were likely relatively oxidising due to the high water fugacity and use of oxidised starting materials (magnetite or hematite). In a third set of experiments performed using the Walker-type multianvil press at Arizona State University, starting material was sealed inside $Au_{75}Pd_{25}$ capsules and yielded single crystals in the run products of up to 200 μm in diameter (Wads I; Wads II; Hope; Big Gold). Oxygen fugacity during these experiments was likely influenced by H_2 loss through the walls of the $Au_{75}Pd_{25}$ capsule following dissociation of H_2O according to the reaction

$$H_2O = H_2 + \tfrac{1}{2}O_2, \qquad (1)$$

where O_2 was then consumed by oxidation of Fe^{2+} to Fe^{3+} (Hall, 1999).

The phases present in the run products were identified using powder X-ray diffraction, and chemical compositions were determined using the electron microprobe (Table 1). Structure refinements were performed on some of the single crystals, but results are reported elsewhere (see reference list in Table 1). H concentrations were determined using either Secondary Ion Mass Spectrometry (SIMS) at the Center for Solid State Science, Arizona State University following the procedures of Ihninger et al. (1994) or Fourier Transform Infrared Spectroscopy (FTIR) at

Table 1
Details of sample synthesis and run products

Sample	Phase	T (°C)	P (GPa)	t (h)	Starting materials	Capsule/ buffer	Ø[a] (μm)	Mg #	Si[b]	Mg[b]	Fe[b]	Al[b]	H[b]	H_2O (wt.%)	Method	$Fe^{3+}/\Sigma Fe$	Ref[c]
H636	ol	1200	4	3	ol, opx	Mo	Poly	0.90	1	1.8	0.2		–	–	–	0.00 (1)	[1]
SZ0004	ol	1300	12	5	ol, br, opx, mag	Pt	Poly	0.90	1	1.8	0.2		0.24	0.15	FTIR	0.15 (4)	[2]
S1292	wd	1800	16	4	ol	Re	Poly	0.80	1	1.6	0.4		–	–	–	0.02 (2)	
H1518	wd	1200	15	5	ol, hem, br	Fe/FeO	Poly	0.90	1	1.8	0.2		0.05	0.3	FTIR	0.02 (2)	[3]
H1505	wd	1200	15	5	ol, hem, br	Re/ReO_2	Poly	0.88	1	1.7	0.2		0.05	0.3	FTIR	0.15 (3)	[3]
HZ7	wd	1250	17	3	ol, opx, br	Re/opx	Poly	0.95	1.01	1.75	0.10	0.01	0.24	1.53	FTIR	0.06 (3)	[4]
SZ0006	wd	1400	18	5	ol, opx, hem, qtz, br	Pt	Poly	0.90	0.96	1.72	0.19		0.35	2.2	FTIR	0.28 (3)	[5]
SZ0005	wd	1400	19	5	ol, opx, hem, qtz, br	Pt	Poly	0.90	0.95	1.71	0.19		0.38	2.4	FTIR	0.44 (5)	[5]
Wads II	wd	1350	18	13	KLB1, wüs, br	AuPd	250	0.88	0.97	1.60	0.21	0.01	0.44	2.72	SIMS	0.49 (7)	[6]
Wads I	wd	1400	17	17	KLB1, br	AuPd	250	0.95	0.99	1.73	0.10	0.01	0.35	2.24	SIMS	0.96 (5)	[7]
H594	rg	1200	19	4	ol	Re	Poly	0.85	1	1.7	0.3		–	–	–	0.03 (2)	
Z89	rg	1400	21	5	ol, opx, hem, qtz, br	Pt	300	0.88	0.95	1.63	0.22		0.40	2.2	FTIR	0.15 (4)	[8]
SZ9901	rg	1400	20	5	ol, opx, hem, qtz, br	Pt	250	0.88	0.99	1.70	0.23	0.00	0.15	0.93	FTIR	0.15 (4)	[9]
SZ0002	rg	1400	20	5	ol, opx, hem, qtz, br	Pt	300	0.86	1.00	1.63	0.26	0.01	0.13	0.79	FTIR	0.18 (4)	[9]
S2144	mj	1900	19	1	opx, qtz	Re/ReO_2	Poly	0.86	0.93	0.84	0.14	0.09	–	–	–	0.16 (3)	[10]
Hope	mj	1400	18	22	KLB1, wüs, br	AuPd	200	0.85	0.85	0.78	0.13	0.21	0.00	0.002	FTIR	0.82 (7)	[11]
Big Gold	mj	1400	17	22	KLB1, br	AuPd	200	0.90	0.88	0.81	0.09	0.20	0.00	0.002	FTIR	0.95 (7)	[11]

Notes: ol: olivine; wd: wadsleyite; rg: ringwoodite; mj: majorite; opx: orthopyroxene; br: brucite; mag: magnetite; hem: hematite; qtz: quartz; wüs: wüstite.

[a] Diameter of single crystal for Mössbauer analysis; poly: polycrystalline sample used for Mössbauer analysis

[b] Cations per four oxygen (ol, wd, rg) or per three oxygen (mj). Minor element concentrations (Ti, Cr, Mn, Ca, K, Na) are omitted for brevity, and can be found in the original references.

[c] [1] Poe and Xu (1999); [2] Smyth and Frost (2002); [3] Frost and McCammon (2001); [4] Frost (1999); [5] Smyth et al. (2000); [6] Smyth and Kawamoto (1997); [7] Smyth et al. (1997); [8] Smyth et al. (1999); [9] Smyth et al. (2003); [10] McCammon and Ross (2003); [11] Laustsen, H.M. (née Smith), unpublished data.

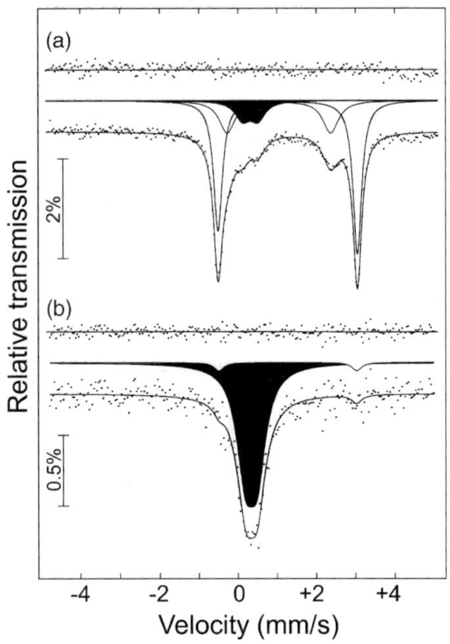

Fig. 5. Room temperature Mössbauer spectra of: (a) nominally anhydrous majorite (S2144); (b) single crystal majorite synthesised under highly oxidising conditions in the presence of water (Big Gold). Doublets corresponding to Fe^{3+} are shaded black, while the low intensity unshaded doublet in the top spectrum corresponds to wadsleyite. Residuals indicating the difference between experimental and calculated data are shown above each spectrum.

taking place across the shared edge. EELS measurements of $Fe^{3+}/\Sigma Fe$ in sample SZ9901 gave the value 0.09 ± 0.05 (Smyth et al., 2003), which is in good agreement with the value of 0.15 ± 0.04 determined by Mössbauer spectroscopy (Table 1).

3.4. Majorite

The Mössbauer spectrum of nominally anhydrous (Mg, Fe)(Si, Al)O_3 majorite (S2144) shows one doublet corresponding to dodecahedral Fe^{2+}, and one doublet corresponding to octahedral Fe^{3+} (Fig. 5, top spectrum). Since iron and aluminium both compete for the octahedral site, there is generally negligible octahedral Fe^{2+}, and $Fe^{3+}/\Sigma Fe$ is less than for (Mg, Fe)SiO_3 majorite synthesised under similar redox conditions (McCammon and Ross, 2003). High-pressure synthesis in the system MgO–FeO–Al_2O_3–SiO_2–H_2O produced majorite with negligible OH (Hope; Big Gold), even though co-existing wadsleyite contained up to 3 wt.% H_2O (Kawamoto et al., 1996). The low hydrogen solubility of majorite has already been reported by Bolfan-Casanova et al. (2000), who found a concentration of 677 ppm H_2O (by weight) in $MgSiO_3$ majorite co-existing with stishovite and melt. The smaller hydrogen content found in the present samples (ca. 2 ppm H_2O) might be a result of their different chemistry. Regardless of their hydrogen content, however, nearly all of the iron in the majorite crystals is Fe^{3+} (Table 1), which is likely a result of the extreme oxidising conditions caused by hydrogen diffusion out of the $Au_{75}Pd_{25}$ capsule. The increase of Fe^{3+} with increasing oxygen fugacity is consistent with previous observations in (Mg, Fe)SiO_3 majorite (McCammon and Ross, 2003).

4. Discussion

4.1. Relation between OH and Fe^{3+}

It is well known that mantle phases in the iron-free system MgO–SiO_2–H_2O can incorporate significant quantities of OH in their crystal structures (reviewed by Angel et al., 2001); hence iron, in particular Fe^{3+}, is not required to stabilise OH. Nevertheless, results from this work show that there is a relation between $Fe^{3+}/\Sigma Fe$ and the incorporation of OH in high-pressure mantle phases. Compared to anhydrous systems, there is a clear trend to higher concentrations of Fe^{3+} for the mantle phases olivine, wadsleyite, ringwoodite and majorite (Fig. 6). The different structures show varying correlations between OH and $Fe^{3+}/\Sigma Fe$, however. Majorite, which incorporates negligible OH in its structure, shows the highest values of $Fe^{3+}/\Sigma Fe$.

Crystal structure refinements have shown that for all of the hydrous phases examined in this study, the principal hydration mechanism is by octahedral site vacancy (Smyth and Frost, 2002; Smyth and Kawamoto, 1997; Smyth et al., 1997; Smyth et al., 2003). An octahedral site vacancy can also be formed in the absence of hydrogen through oxidation of Fe^{2+} to Fe^{3+}:

$$2Fe^{2+} + Mg^{2+} + \tfrac{1}{2}O_2 = 2Fe^{3+} + \square + MgO, \qquad (2)$$

where higher oxygen fugacity would favour a higher concentration of vacancies. The coupling of these two mechanisms was suggested by Kurosawa et al. (1997),

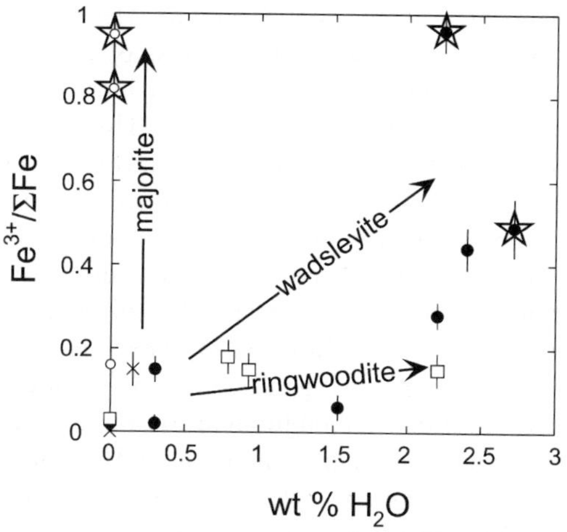

Fig. 6. $Fe^{3+}/\Sigma Fe$ vs. wt.% H_2O for all hydrous high-pressure phases studied. Symbols represent olivine (crosses), wadsleyite (solid circles), ringwoodite (open squares), and majorite (open circles). Stars indicate the runs carried out under extremely oxidising conditions (AuPd capsules), which suggest that $Fe^{3+}/\Sigma Fe$ is strongly related to oxygen fugacity. The trend of $Fe^{3+}/\Sigma Fe$ with OH concentration is indicated by annotated lines for each mineral except olivine, which was omitted due to the limited number of data points.

who found a positive correlation between trivalent and monovalent (including hydrogen) cations in olivine. They suggested a structural model where octahedral vacancies are balanced by protonation of oxygen on one side, and oxidation of Fe^{2+} to Fe^{3+} on the other. Such a model is supported by the Mössbauer data, which indicate Fe^{3+} to occupy sites of octahedral coordination in all phases studied. Hydration under oxidising conditions could therefore stabilise large concentrations of Fe^{3+}.

Such a structural model is consistent with the present wadsleyite results. For samples synthesised under two different conditions of buffered oxygen fugacity, the concentration of Fe^{3+} is nearly ten times higher in the sample buffered at Re-ReO_2 (H1505) compared to the one buffered at Fe–FeO (H1518), even though the amount of incorporated OH is interpreted to be similar for the two samples (Table 1). Under conditions of more extreme oxygen fugacity, nearly all iron is converted to Fe^{3+} (Wads I). This implies that the concentration of Fe^{3+} is determined primarily by oxygen fugacity through the stabilisation of cation vacancies in the structure. The similar concentrations of hydrogen in all of these wadsleyite samples might be related to the preference for protonation to occur at particular oxygen sites, and the limited number of such sites in the structure.

The structure of hydrous ringwoodite contains a single octahedral site, compared to either four (Smyth et al., 1997) or six (Smyth and Kawamoto, 1997) octahedral sites in the different hydrous wadsleyite structures. The smaller concentration of Fe^{3+} in ringwoodite compared to wadsleyite at conditions of similar oxygen fugacity and OH concentration is possibly related to this difference, where the wadsleyite structure has more possibilities for stabilising Fe^{3+} in one of the octahedral sites relative to Fe^{2+}.

All Fe^{3+} in the hydrous ringwoodite structure appears to be involved in rapid electron exchange with Fe^{2+} according to Mössbauer spectroscopy, which might contribute to the stabilisation of a relatively fixed concentration of Fe^{2+} in the structure.

Majorite shows the highest concentrations of Fe^{3+}, despite negligible OH. This is clearly related to the extreme oxidising conditions during the experiment, which stabilised nearly all iron as Fe^{3+}. Charge is balanced through coupled substitution, with essentially no vacancies present in the crystal structure (Table 1).

4.2. Effect on garnet–olivine–orthopyroxene oxygen barometer

The oxygen fugacity in deeper regions of the lithospheric mantle has been determined using the olivine–orthopyroxene–garnet oxybarometer:

$$2Fe_3Fe_2Si_3O_{12}\,(gt) = 8FeSi_{0.5}O_2\,(ol) + 2FeSiO_3\,(opx) + O_2. \quad (3)$$

The equilibrium equation for oxygen fugacity can be written as:

$$\log fO_2 = 9.79 - (20400/T) - (0.045P/T) - 8\log a_{Fa} - 2\log a_{Fs} + 2\log a_{Sk}, \quad (4)$$

where T is in K, P in bars, and a_{Fa}, a_{Fs} and a_{Sk} are the fayalite, ferrosilite and "skiagite" components in olivine, orthopyroxene and garnet, respectively (Gudmundsson and Wood, 1995). Potential influences on oxygen fugacity due to effects related to hydrous phases are: (1) reduction of fayalite and ferrosilite activities due to oxidation of Fe^{2+} to Fe^{3+} at depth;

and (2) increase of "skiagite" activity due to dehydrogenation of garnet during exhumation (e.g. Ingrin and Skogby, 2000).

The activities of fayalite and ferrosilite components are normally determined based on electron microprobe data of olivine and orthopyroxene, which assume that all iron is Fe^{2+}. Normally this assumption is reasonable, according to Mössbauer studies of mantle olivines and orthopyroxenes (e.g. Dyar et al., 1989), but it has not been tested extensively for deep samples. To estimate this effect, we assumed a value of 0.05 for $Fe^{3+}/\Sigma Fe$ in olivine at depth, and using the Gudmundsson and Wood (1995) formulation, calculated that oxygen fugacity would be overestimated by approximately 0.2 log units relative to the fayalite–magnetite–quartz (FMQ) buffer for a typical mantle composition. The overestimation of oxygen fugacity for Fe^{3+} in orthopyroxene would be significantly less, since the contribution from the ferrosilite activity is smaller than the fayalite activity (Eq. (4)). Since oxygen fugacities determined for deep garnet peridotites (ca. 200 km) are relatively low, up to 4 log units below FMQ (Woodland and Peltonen, 1999; Woodland, 2001), our estimation is likely a maximum, since the concentration of Fe^{3+} in olivine would be relatively low under these redox conditions, even in the presence of water. We therefore consider the effect of enhanced Fe^{3+} in olivine or orthopyroxene on the olivine–orthopyroxene–garnet oxybarometer to be negligible.

The second possible effect on the olivine–orthopyroxene–garnet oxybarometer is dehydrogenation during exhumation. The reaction which is generally assumed to control this process,

$$Fe^{2+} + OH^- = Fe^{3+} + O^{2-} + \tfrac{1}{2}H_2, \qquad (5)$$

indicates that $Fe^{3+}/\Sigma Fe$ would increase during dehydrogenation (e.g. Ingrin and Skogby, 2000). Although kinetic studies of garnet have shown that dehydrogenation is likely slower for garnet than for olivine or orthopyroxene (e.g. Wang et al., 1996), timescales of exhumation are sufficiently slow that dehydrogenation could still occur. Another crucial parameter is the degree to which garnet incorporates OH at these depths in the mantle. High-pressure experiments have shown that up to 200 ppm wt. H_2O can be incorporated in pyrope at 10 GPa (Lu and Keppler, 1997), and concentrations measured from an extensive study of mantle samples show up to ca. 100 ppm wt. H_2O (Bell and Rossman, 1992). If we assume a loss of 200 ppm wt. H_2O from garnet, this results in an increase of ca. 0.2 wt.% Fe_2O_3 according to Eq. (4). However, since Fe^{3+} concentrations in garnet are already low at depths of 200 km (e.g. Woodland, 2001), this increase could be a significant proportion of the amount of garnet Fe^{3+}. For example, for a typical garnet peridotite composition with $Fe^{3+}/\Sigma Fe$ equal to 0.05, this value would increase to 0.07 on the loss of 200 ppm wt. H_2O, which would overestimate oxygen fugacity by 0.5 log units.

Such an effect, although small, is possible if garnet contained a moderate amount of H_2O at depth, which would cause an overestimation of oxygen fugacity. One approach to identify this possibility would be to measure $Fe^{3+}/\Sigma Fe$ of co-existing olivine. Based on data reviewed by Ingrin and Skogby (2000), olivine should show a similar increase in Fe^{3+} due to dehydrogenation (perhaps in the form of Fe^{3+}-rich exsolution lamellae), which could be detected by a method such as electron microscopy. An increase in $Fe^{3+}/\Sigma Fe$ of olivine due to dehydrogenation would have no effect on the calculated oxygen fugacity, since the total iron concentration would not have changed.

4.3. Implications for subduction

In order to evaluate the implications of the present work on mantle dynamics, it is important to understand the paradoxical relation between $Fe^{3+}/\Sigma Fe$ and oxygen fugacity that exists within the mantle. In melt or glass systems the relation is simple, where an increase in oxygen fugacity translates directly into an increase in $Fe^{3+}/\Sigma Fe$. In mineral systems, however, oxygen fugacity is monitored by mineral equilibria, which are influenced by the energetics of cation substitution in the relevant phases.

In the upper mantle, oxygen fugacity is relatively high. It has been measured directly from xenoliths and basaltic glasses (e.g. Wood et al., 1990), where even the most reduced values show an average fO_2 of FMQ − 1 (i.e. −1 log units below the FMQ buffer), which is similar to the results obtained for mid-ocean ridge basalt (MORB) glasses. Despite the relatively high oxygen fugacity, however, the bulk $Fe^{3+}/\Sigma Fe$ of the upper mantle is low-estimated to be ca. 3% (O'Neill et al., 1993b). The mineral olivine, which constitutes some 60% of the upper mantle by volume, almost completely excludes Fe^{3+} due to the unfavourable energetics of defect formation in anhydrous olivine, while the mineral spinel, which constitutes only a few percent of the upper mantle by volume, contains significant Fe^{3+} (up to ca. 50% $Fe^{3+}/\Sigma Fe$). Since oxygen fugacity is controlled by equilibria involving the magnetite component in spinel, a high value of $Fe^{3+}/\Sigma Fe$ translates directly to a high oxygen fugacity. At greater depths, laboratory experiments have shown that the increased affinity of high-pressure phases for Fe^{3+} should lower oxygen fugacity (e.g. O'Neill et al., 1993a), and so the general picture is of more reduced deep upper mantle and transition zone.

The relation between $Fe^{3+}/\Sigma Fe$ and oxygen fugacity is reversed in the lower mantle. (Mg, Fe)(Si, Al)O_3 perovskite, which constitutes at least 80% of the lower mantle by volume, incorporates significant Fe^{3+} due to the favourable energetics of trivalent cation substitution in the structure (McCammon, 1997; Lauterbach et al., 2000). For a pyrolite bulk composition, $Fe^{3+}/\Sigma Fe$ of (Mg, Fe)(Si, Al)O_3 perovskite is estimated to be approximately 50%, regardless of oxygen fugacity (during experiments under reduced conditions, charge is balanced through the disproportionation reaction $Fe^{2+} \rightarrow 2/3\,Fe^{3+} + 1/3\,Fe^0$) (Lauterbach et al., 2000). This represents a more than tenfold increase in Fe^{3+} concentration at the top of the lower mantle. Although the oxygen fugacity of the bulk lower mantle is still an open question, a relatively reduced lower mantle would be more compatible with existing evidence than one that were extremely oxidised.

The overall picture of oxygen fugacity within the mantle is therefore a general decrease with depth, and no evidence for a significant discontinuity in oxygen fugacity between the transition zone and lower mantle. In contrast, the bulk concentration of Fe^{3+} is low throughout the upper mantle and (likely) the transition zone, but increases dramatically near the top of the lower mantle as (Mg, Fe)(Si, Al)O_3 perovskite becomes stable. The influence of such an increase in the abundance of Fe^{3+} is greatest for processes that involve an exchange of material through the garnet–perovskite transformation, e.g. subduction and mantle upwelling, where the question of electrostatic charge balance must be addressed.

An estimate of iron oxidation state in subducting slabs can be made based on the results of the present work. Oceanic crust is generally assumed to be relatively oxidised due to sea water alteration, and estimates of oxygen fugacity in the mantle wedge above subduction zones indicate values that are several log units more oxidised than the surrounding mantle (relative to FMQ) (e.g. Parkinson and Arculus, 1999). Results from the present work suggest that a portion of Fe^{2+} in nominally anhydrous high-pressure phases is oxidised to Fe^{3+} during hydration under oxidising conditions, which implies that subducted material could contain higher concentrations of Fe^{3+} than the surrounding mantle. This is supported by Mössbauer data from diamonds believed to have originated in the transition zone (Harte et al., 2002), where inclusions are suggested on the basis of geochemical evidence to represent ancient oceanic crust (e.g. Stachel, 2001; Stachel et al., 2002). The Fe^{3+} concentrations found in such inclusions are higher than those found in high-pressure experiments conducted under highly oxidising, but dry, conditions (McCammon and Ross, 2003), which is consistent with the results of the present work if conditions during such diamond genesis are assumed to be wet and relatively oxidising.

Acknowledgements

PvA would like to thank the Deutsche Forschungsgemeinschaft (Bonn, Germany) for the funding of the upgrade of the Gatan PEELS 666 to the DigiPEELS 766 under Project number AK 26/2-1. The H636 sample was synthesised by Yousheng Xu, and the H_2O concentration of the HZ7 sample was determined by Hans Keppler.

References

Amthauer, G., Annersten, H., Hafner, S.S., 1976. The Mössbauer spectrum of ^{57}Fe in silicate garnets. Z. Krist. 143, 14–55.

Angel, R.J., Frost, D.J., Ross, N.L., Hemley, R., 2001. Stabilities and equations of state of dense hydrous magnesium silicates. Phys. Earth Planet. Int. 127, 181–196.

Bell, D.R., Rossman, G.R., 1992. Water in Earth's mantle: the role of nominally anhydrous minerals. Science 255, 1391–1397.

Bolfan-Casanova, N., Keppler, H., Rubie, D.C., 2000. Water partitioning between nominally anhydrous minerals in the MgO–SiO_2–H_2O system up to 24 GPa: implications for the distribution of water in the Earth's mantle. Phys. Earth Planet. Int. 182, 209–221.

De Grave, E., Van Alboom, A., 1991. Evaluation of ferrous and ferric Mössbauer fractions. Phys. Chem. Minerals 18, 337–342.

Dyar, M.D., McGuire, A.V., Ziegler, R.D., 1989. Redox equilibria and crystal chemistry of coexisting minerals from spinel lherzolite mantle xenoliths. Am. Mineral. 74, 969–980.

Fei, Y., Mao, H.K., Shu, J., Parthasarathy, G., Bassett, W.A., Ko, J., 1992. Simultaneous high-P, high-T X-ray diffraction study of β-$(Mg, Fe)_2SiO_4$ to 26 GPa and 900 K. J. Geophys. Res. 97, 4489–4495.

Frost, D.J., 1999. The stability of dense hydrous magnesium silicates in Earth's transition zone and lower mantle. In: Fei, Y., Bertka, C.M., Mysen, B.O. (Eds.), Mantle Petrology: Field Observations and High-Pressure Experimentation: A Tribute to Francis R. (Joe) Boyd. Geochemical Society, USA, pp. 283–296.

Frost, D., McCammon, C., 2001. The effect of oxygen fugacity on the solubility of water in wadsleyite. J. Conf. Abs. 6 (1), 461.

Gudmundsson, G., Wood, B.J., 1995. Experimental tests of garnet peridotite oxygen barometry. Contrib. Mineral. Petrol. 119, 56–67.

Hall, L., 1999. The effect of water on mantle melting. Ph.D. Thesis, University of Bristol, Bristol, UK.

Harte, B., Harris, J.W., Wilding, M.C., Sautter, V., McCammon, C., 2002. Eclogite-garnetite inclusions in diamonds from the Sao Luiz area, Brazil. In: Proceedings of the 18th General Meeting International Mineralogical Association, abstract vol. 74.

Ihninger, P.D., Hervig, R.L., McMillan, P.F., 1994. Analytical methods for volatiles in glasses. Rev. Mineral. 30, 67–121.

Ingrin, J., Skogby, H., 2000. Hydrogen in nominally anhydrous upper-mantle minerals: concentration levels and implications. Eur. J. Mineral. 12, 543–570.

Kawamoto, T., Hervig, R.L., Holloway, J.R., 1996. Experimental evidence for a hydrous transition zone in the early Earth's mantle. Earth Planet. Sci. Lett. 142, 587–592.

Kurosawa, M., Yurimoto, H., Sueno, S., 1997. Patterns in the hydrogen and trace element compositions of mantle olivines. Phys. Chem. Minerals 24, 385–395.

Lauterbach, S., McCammon, C.A., van Aken, P., Langenhorst, F., Seifert, F., 2000. Mössbauer and ELNES spectroscopy of (Mg, Fe)(Si, $Al)O_3$ perovskite: a highly oxidised component of the lower mantle. Contrib. Mineral. Petrol. 138, 17–26.

Lu, R., Keppler, H., 1997. Water solubility in pyrope to 100 kbar. Contrib. Mineral. Petrol. 129, 35–43.

McCammon, C.A., 1994. A Mössbauer milliprobe: practical considerations. Hyper. Int. 92, 1235–1239.

McCammon, C.A., 1997. Perovskite as a possible sink for ferric iron in the lower mantle. Nature 387, 694–696.

McCammon, C.A., 2000. Insights into phase transformations from Mössbauer spectroscopy. Rev. Mineral. 39, 241–264.

McCammon, C.A., Chaskar, V., Richards, G.G., 1991. A technique for spatially resolved Mössbauer spectroscopy applied to quenched metallurgical slags. Meas. Sci. Technol. 2, 657–662.

McCammon, C.A., Ross, N.L., 2003. Crystal chemistry of ferric iron in (Mg, Fe)(Si, $Al)O_3$ majorite with implications for the transition zone. Phys. Chem. Minerals 30, 206–216.

Nakamura, A., Schmalzried, H., 1983. On the stoichiometry and point defects of olivine. Phys. Chem. Minerals 10, 27–37.

O'Neill, H.S.C., McCammon, C.A., Canil, D.C., Rubie, D.C., Ross II, C.R., Seifert, F., 1993a. Mössbauer spectroscopy of transition zone phases and determination of minimum Fe^{3+} content. Am. Mineral. 78, 456–460.

O'Neill, H.S.C., Rubie, D.C., Canil, D., Geiger, C.A., Ross, II, C.R., Seifert, F., Woodland, A.B., 1993b. Ferric iron in the upper mantle and in transition zone assemblages: implications for relative oxygen fugacities in the mantle. In: Takahashi, T., Jeanloz, R., Rubie, D.C. (Eds.), Evolution of the Earth and Planets. American Geophysical Union, Washington, DC, pp. 73–88.

Parkinson, I.J., Arculus, R.J., 1999. The redox state of subduction zones: insights from arc-peridotites. Chem. Geol. 160, 409–423.

Paterson, M.S., 1982. The determination of hydroxyl by infrared absorption in quartz. Bull. Mineral. 105, 20–29.

Poe, B.T., Xu, Y.S., 1999. In situ complex impedance spectroscopy of mantle minerals measured at 20 GPa and 1400 °C. Phase Trans. 68B, 453–466.

Rancourt, D.G., Mercier, P.H.J., Cherniak, D.J., Desgreniers, S., Kodama, H., Robert, J.-L., Murad, E., 2001. Mechanisms and crystal chemistry of oxidation in annite: resolving the hydrogen-loss and vacancy reactions. Clays Clay Minerals 49, 455–491.

Ross, N.L., 1997. Optical absorption spectra of transition zone minerals and implications for radiative heat transport. Phys. Chem. Earth 22, 113–118.

Shinno, I., 1981. A Mössbauer study of ferric iron in olivine. Phys. Chem. Minerals 7, 91–95.

Smyth, J.R., Kawamoto, T., 1997. Wadsleyite II: a new high pressure hydrous phase in the peridotite–H_2O system. Earth Planet. Sci. Lett. 146, E9–E16.

Smyth, J.R., Frost, D.J., 2002. Effects of hydration on the crystal structure of olivine. J. Conf. Abs. 7, 99.

Smyth, J.R., Kawamoto, T., Jacobsen, S.D., Swope, R.J., Hervig, R.L., Holloway, J.R., 1997. Crystal structure of a monoclinic hydrous wadsleyite [β-$(Mg, Fe)_2SiO_4$]. Am. Mineral. 82, 270–275.

Smyth, J.R., Frost, D.J., McCammon, C.A., 1999. Hydration mechanism in ringwoodite, γ-$Mg_{1.63}Fe_{0.22}H_{0.4}Si_{0.95}O_4$. EOS (Trans. Am. Geophys. Union) 80, F927.

Smyth, J.R., Nguyen, C., Frost, D.J., McCammon, C.A., Langenhorst, F., Bolfan-Casanova, N., 2000. Hydration of Fe-bearing silicate spinels and spinelloids. EOS (Trans. Am. Geophys. Union) 81, F1274.

Smyth, J.R., Holl, C.M., Frost, D.J., Jacobsen, S.D., Langenhorst, F., McCammon, C.A., 2003. Structural systematics of hydrous ringwoodite and water in Earth's interior. Am. Mineral. 88, 1402–1407.

Stachel, T., 2001. Diamonds from the asthenosphere and the transition zone. Eur. J. Mineral. 13, 883–892.

Stachel, T., Harris, J.W., McCammon, C., 2002. Inclusions in ultra-deep diamonds—tracers of ancient slabs? In: Proceedings of the 18th General Meeting on International Mineralogy Association, abstract vol. 74.

van Aken, P.A., Liebscher, B., Styrsa, V.J., 1998. Quantitative determination of iron oxidation states in minerals using Fe $L_{2,3}$-edge electron energy-loss near-edge structure spectroscopy. Phys. Chem. Minerals 25, 323–327.

Wang, L., Zhang, Y., Essene, E., 1996. Diffusion of the hydrous component in pyrope. Am. Mineral. 81, 706–718.

Wood, B.J., Bryndzia, L.T., Johnson, K.E., 1990. Mantle oxidation state and its relationship to tectonic environment and fluid speciation. Science 248, 337–345.

Woodland, A.B., Peltonen, P., 1999. Ferric iron contents of garnet and clinopyroxene and estimated oxygen fugacities of peridotite xenoliths from the Eastern Finland Kimberlite Province. In: Gurney, J.J., Gurney, J.L., Pascoe, M.D., Richardson, S.H. (Eds.), The P.H. Nixon Volume, Proceedings of the VII International Kimberlite Conference. Red Roof Design, Cape Town, pp. 904–911.

Woodland, A.B., 2001. The oxygen fugacity of the Earth's mantle: variations with depth, In: Proceedings of the Eleventh Annual V.M. Goldschmidt Conference. LPI Contribution No. 1088, Lunar and Planetary Institute, Houston, abstract #3148 (CD-ROM).

Available online at www.sciencedirect.com

SCIENCE DIRECT®

Physics of the Earth and Planetary Interiors 143–144 (2004) 171–183

PHYSICS OF THE EARTH AND PLANETARY INTERIORS

www.elsevier.com/locate/pepi

Stability of spinelloid phases in the system Mg_2SiO_4–Fe_2SiO_4–Fe_3O_4 at 1100 °C and up to 10.5 GPa

M. Koch [a,*], A.B. Woodland [b], R.J. Angel [c,1]

[a] *Mineralogisches Institut, Universität Heidelberg, 69120 Heidelberg, Germany*
[b] *Institut für Mineralogie, Universität Frankfurt, 60054 Frankfurt, Germany*
[c] *Bayerisches Geoinstitut, Universität Bayreuth, 95440 Bayreuth, Germany*

Received 28 November 2002; received in revised form 5 May 2003; accepted 25 June 2003

Abstract

Experiments in the system Mg_2SiO_4–Fe_2SiO_4–Fe_3O_4 have produced three intermediate phases that are stable between ~4.0 and 9.0 GPa and 1100 °C. These phases are different spinelloid polytypes and are structural analogues to spinelloids II, III and V in the Ni-aluminosilicate system [J. Solid State Chem. 44 (1982) 257]. The behaviour of spinelloid III is of particular interest since it is isostructural to $(Mg,Fe)_2SiO_4$-wadsleyite. Spinelloid II is the first intermediate phase to appear at low pressures in Mg-poor compositions. With increasing pressure spinelloid II is replaced by an isochemical spinelloid V polytype between 6.0 and 6.5 GPa. Spinelloid III is relatively Si-rich and, depending on pressure, can coexist with either spinelloid II or V. At 1100 °C, spinelloid III gives way to the assemblage olivine + spinelloid V between 7.5 and 8.0 GPa and its maximum Mg content is limited to ~15 mol% Mg_2SiO_4 at 7.5 GPa. Thus, there is a very large gap in composition and pressure separating the stability fields of spinelloid III and wadsleyite in this system. In general, the addition of Mg or an increase in temperature acts to shift the phase stabilities of the spinelloids to higher pressures. At 1100 °C and above 10.0 GPa, no spinelloid phases are stable. We find no compelling evidence for the incorporation of significant Fe^{3+} in olivine, even at very high pressures. In Mg-rich compositions more relevant to the Earth's mantle, the assemblage $olivine_{ss}$ + $spinel_{ss}$ is stable. In such bulk compositions, the spinel is practically Si-free and contains a Mg-ferrite component. Thus there appears to be a mutual incompatibility between Mg and Si in high-pressure spinels. It is conceivable that the presence of Fe^{3+} in a mantle assemblage (e.g. through metasomatism) could stabilise small quantities of either spinelloid V or a magnesioferrite-bearing spinel in equilibrium with olivine. Unlike the spinel that is considered to be present in the lower parts of the transition zone, this phase will contain little or no Si.

Keywords: Phase relations; Spinelloid; Wadsleyite; Ferric iron; Ringwoodite; Earth's mantle

1. Introduction

Wadsleyite and ringwoodite (both $(Mg,Fe)_2SiO_4$) are generally considered to be the major mineralogical constituents of the transition zone of the Earth's mantle. The possible incorporation of Fe^{3+} into these phases will influence their stabilities and this in turn could cause a shift in the depth at which the "410 km"

* Corresponding author. Tel.: +49-6221-544651; fax: +49-6221-544805.
E-mail addresses: mkoch@min.uni-heidelberg.de (M. Koch), woodland@em.uni-frankfurt.de (A.B. Woodland), rangel@vt.edu (R.J. Angel).
[1] Present address: Crystallography Laboratory, Department of Geological Sciences, Virginia Tech, Blacksburg, VA 24060, USA.

0031-9201/$ – see front matter © 2004 Elsevier B.V. All rights reserved.
doi:10.1016/j.pepi.2003.06.001

3. Results and discussion

3.1. Phase relations

The phase relations within the ternary system Fe_2SiO_4–Mg_2SiO_4–Fe_3O_4 are summarised in Fig. 2a–h as a function of pressure at 1100 °C. A couple reconnaissance experiments were also performed at 1200 °C and these are indicated in the Figures (see also Table 1). Spinels in some experimental products have $Si < Mg/2$, indicating a decoupling of Mg from Si and implying the presence of a Mg-ferrite component. Therefore, we present the phase relations in the four-component Mg_2SiO_4–Fe_2SiO_4–Fe_3O_4–$MgFe_2O_4$ system. Compared with the Fe_2SiO_4–Fe_3O_4 subsystem (Woodland and Angel, 2000), the addition of Mg generally acts to shift the stability fields of the various spinelloids to higher pressures. At low pressures, olivine coexists with a spinel that usually incorporates more Si than Mg. However, in very Mg-rich compositions (i.e. olivine with $X_{fo} > 0.80$) the opposite chemical behaviour is observed with the spinel sometimes containing a considerable Mg-ferrite component (Fig. 2a).

The pressure at which an intermediate phase first appears significantly increases in the presence of Mg. Spinelloid II, which is stable at 2.6 GPa in the Mg-free system (Fig. 1), first becomes stable at 4.0 GPa, even though it only contains a maximum of ~5 mol% Mg_2SiO_4 component (Fig. 2b). Between 4.0 and 5.0 GPa, spinelloid II has a limited compositional range of $X_{Fe_3O_4} = 0.50$–0.59, which is similar in extent to the width of the stability field of Mg-free spinelloid II portrayed in Fig. 1. Depending on composition, spinelloid II can coexist with either olivine, a Si-bearing magnetite-rich spinel or both. In Mg-rich bulk compositions that crystallised olivine with $X_{fo} > 0.48$, spinelloid II is no longer stable and olivine coexists with a magnetite-rich spinel (Fig. 2b). The Si content of spinel decreases with progressive incorporation of Mg.

Spinelloid III first appears at 6.0 GPa in Mg-bearing compositions. As with spinelloid II, the stability field of spinelloid III is shifted upwards by ~2.0 GPa due to the incorporation of only small amounts of Mg (Fig. 2c and compare with Fig. 1). Between 6.0 and 6.5 GPa spinelloid II is effectively replaced by spinelloid V, emphasising the polytypic nature of these two phases (Fig. 2d). However, spinelloid II was also produced in a reconnaissance experiment at 7.0 GPa and 1200 °C (8f5h1136) demonstrating that an increase in temperature can stabilise the individual phases to even higher pressures (Fig. 2e). Depending on pressure, spinelloid III can coexist with either spinelloid II or V along with olivine. Spinelloid III can accommodate up to 0.94 cations per formula unit (cpfu) Fe^{3+}, but is always poorer in Fe^{3+} compared to either coexisting polytype. The stability field of spinelloid III is limited in P–T–composition space by the stability of the assemblage olivine + spinelloid V. At 1100 °C, spinelloid III gives way to this assemblage between 7.5 and 8.0 GPa (Fig. 2e and f). In terms of composition, the maximum Mg content of spinelloid III is limited to somewhat more than 10 mol% Mg_2SiO_4 at 7.0 GPa, increasing to ~15 mol% Mg_2SiO_4 at 7.5 GPa. This maximum Mg content is reached in a bulk composition where olivine has a composition between $X_{fo} = 0.34$ and 0.55 (Fig. 2e).

At >8.0 GPa, the only stable intermediate phase is spinelloid V and it spans a large compositional range from $X_{Fe_3O_4} = 0.25$ to 0.62 (Fig. 2f). In an analogous fashion to spinelloid III, the stability field of spinelloid V is limited in terms of Mg content by the appearance of another assemblage, olivine + spinel, which occurs when olivine has a composition of $X_{fo} > 0.58$ or >0.75 at 7.0 and 9.0 GPa, respectively. The maximum Mg content of spinelloid V also increases with pressure, reaching ~30 mol% Mg_2SiO_4 at 9.0 GPa (Fig. 2g). At the same time, the stability field of spinelloid V is considerably reduced at low Mg contents by the increasing extent of solid solution between Fe_2SiO_4-spinel and magnetite. A complete spinel solid solution is stable in the Mg-free system at >7.5 GPa (Fig. 1). At 9.0 GPa, the minimum Mg content of spinelloid V is 5 mol% Mg_2SiO_4. Thus although spinelloid V is stable over a fairly wide range of Fe^{3+}/Si ratio, the compositional range in terms of Mg content is rather limited (Fig. 2f–h).

At 10.5 GPa spinelloid V is no longer stable. In its place the field of spinel solid solution extends into the four-component composition space from the Fe_2SiO_4–Fe_3O_4, Fe_2SiO_4–Mg_2SiO_4 and $MgFe_2O_4$–Fe_3O_4 joins (Fig. 2h). At Mg-rich bulk compositions olivine is still stable and coexists with a Fe^{3+}-rich spinel containing a significant $MgFe_2O_4$ component. This spinel is essentially Si-free.

Table 1
Results of experiments performed at 1100 and 1200 °C

Experiment	T	P	t	Stable phases present	Spinel				Spinelloid				Olivine
	(°C)	(GPa)	(h)		nSi	nMg	nFe^{2+}	nFe^{3+}	nSi	nMg	nFe^{2+}	nFe^{3+}	X_{fo}
5f9b1197	1100	3.0	88	isc + (ol + sp)									0. 11
8f9b1197	1100	3.0	88	ol + isc + (sp)									0. 09
4f8b1200	1100	3.0	68.8	sp + ol + (isc)	0. 06	0.02	1.05	1.87					0. 20
8f8b1200	1100	3.0	68.8	ol + sp	0. 06	0.02	1.04	1.88					0. 20
4f1b1203	1100	3.5	70	Mg-ferrite + ol	0. 02	0.12	0.91	1.95					0. 70
8f1b1203	1100	3.5	70	ol + Mg-ferrite	0. 01	0.38	0.64	1.97					0. 79
4f8b0271	1100	4.0	44	sp + spd II + ol	0. 12	0.04	1.08	1.76	0.42	0.10	1. 32	1.16	0.28
8f8b0271	1100	4.0	44	ol + spd II + (opx + sp)					0. 47	0.03	1.44	1.06	0.08
4f5b0272	1100	4.0	44	ol + sp	0. 06	0.12	0.95	1.87					0. 46
8f5b0272	1100	4.0	44	ol + Mg-ferrite	0. 05	0.12	0.92	1.91					0. 47
5f9b1196	1100	4.0	70.5	spd II + ol + (sp)					0.43	0.07	1. 35	1.15	0.19
8f9b1196	1100	4.0	70.5	spd II + ol + (opx)					0.46	0.04	1. 42	1.08	0.12
4f5b1180	1100	5.0	92	sp + ol	0. 07	0.06	0.98	1.89					0. 49
8f5b1180	1100	5.0	92	ol + sp + (opx)	0. 07	0.06	1.01	1.86					0. 49
5f9b1194	1100	5.0	68	spd II + isc + (ol)					0.45	0.06	1. 40	1.09	0.17
8f9b1194	1100	5.0	68	spd II + ol + (isc)					0. 48	0.04	1.44	1.04	0.11
4f8b1195	1100	5.0	97	spd II + (sp + ol)					0.43	0.09	1. 34	1.14	0.25
8f8b1195	1100	5.0	97	spd II + ol + (sp)					0. 44	0.08	1.37	1.11	0.21
4f1b1199	1100	5.0	68	Mg-ferrite + ol	0. 02	0.13	0.89	1.96					0. 76
8f1b1199	1100	5.0	68	ol + Mg-ferrite	0. 01	0.19	0.83	1.97					0. 83
3f5h1360	1100	6.0	35.7	sp + ol + (Mg-ferrite)	0. 11	0.07	1.03	1.79					0. 51
8f5h1360	1100	6.0	35.7	ol + sp	0. 13	0.07	1.06	1.74					0. 49
4f8v0095	1100	6.0	33.6	spd II + (sp + ol)					0.45	0.17	1. 27	1.11	0.41
8f8v0095	1100	6.0	33.6	spd III + ol + (opx)					0.54	0.11	1. 42	0.93	0.25
3f8v0082	1100	6.5	34.2	sp + (opx)	0. 13	0.04	1.09	1.74					
8f8v0082	1100	6.5	34.2	spd III + ol + opx + (spd V + sp)					0.54	0. 12	1.42	0.92	0.27
4f5v0094	1100	6.5	34.3	sp + ol + (opx)	0. 16	0.10	1.06	1.68					0. 55
8f5v0094	1100	6.5	34.3	ol + spd V + sp + (opx)	0. 18	0.10	1.09	1.63	0.41	0.19	1. 21	1.19	0.53
4f8v0126	1100	6.5	34	spd V + (sp)					0. 42	0.16	1.26	1.16	
8f8v0126	1100	6.5	34	spd III + ol + opx + (spd V + sp)					0.63	0. 15	1.48	0.74	0.29
3f5h1136	1200	7.0	12.7	sp + opx	0. 09	0.14	0.94	1.83					
8f5h1136	1200	7.0	12.7	ol + spd II + sp + opx	0.15	0.10	1.05	1.70	0.45	0. 26	1.18	1.11	0.54
3f8h1137	1200	7.0	12.3	sp + opx	0. 15	0.05	1.10	1.70					
8f8h1137	1200	7.0	12.3	spd III + opx + (spd V + ol + sp)					0.66	0.20	1. 46	0.68	0.34
4f8v0083	1100	7.0	34.5	spd V + sp + (opx)	0.17	0.07	1.10	1.66	0.41	0. 14	1.27	1.18	
8f8v0083	1100	7.0	34.5	spd III + ol + (opx + sp + spd V)		0	0	0	0.64	0.16	1. 47	0.73	0.30
4f5v0084	1100	7.0	34	sp + ol + (opx)	0. 15	0.12	1.04	1.69					0. 60
8f5v0084	1100	7.0	34	ol + spd V + (opx + sp)					0. 41	0.21	1.21	1.17	0.55
4f1v0097	1100	7.0	33.7	ol + Mg-ferrite	0. 06	0.19	0.87	1.88					0. 79
8f1v0097	1100	7.0	33.7	ol + Mg-ferrite	0. 04	0.23	0.81	1.92					0. 84

Table 1 (*Continued*)

Experiment	T (°C)	P (GPa)	t (h)	Stable phases present	Spinel				Spinelloid				Olivine
					nSi	nMg	nFe^{2+}	nFe^{3+}	nSi	nMg	nFe^{2+}	nFe^{3+}	X_{fo}
4f3v0130	1100	7.0	34.3	ol + Mg-ferrite	0. 08	0.21	0.86	1.85					0. 65
8f3v0130	1100	7.0	34.3	ol + Mg-ferrite + (opx)	0.07	0.21	0.86	1.86					0. 63
5f9v0180	1100	7.0	34.5	spd V + (opx + sp)					0.45	0.08	1. 36	1.11	
8f9v0180	1100	7.0	34.5	spd III + opx + (spd V + sp)					0.66	0.11	1. 55	0.68	
4f5v0096	1100	7.5	34.5	sp + ol + (opx)	0. 15	0.13	1.03	1.69					0. 64
5f5v0096	1100	7.5	34.5	sp + ol + (opx)	0. 15	0.13	1.00	1.72					0. 65
8f5v0128	1100	7.5	33	ol + spd II + (opx + sp)					0. 53	0.31	1.22	0.94	0.58
8f8v0128	1100	7.5	33	spd V + (opx + sp)					0.73	0.27	1. 48	0.52	
4f8v0080	1100	8.0	33.3	spd V + sp	0. 23	0.07	1.15	1.55	0.48	0.16	1. 31	1.05	
8f8v0080	1100	8.0	33.3	spd V + ol + Si-sp + (sp)	0.95	0.25	1.69	0.11	0.70	0. 21	1.50	0.59	0.64
4f5v0081	1100	8.0	32.7	sp + ol + spd V + (cpx)	0. 18	0.13	1.06	1.63	0.45	0.28	1. 18	1.09	0.59
8f5v0081	1100	8.0	32.7	ol + spd V + (cpx)					0. 52	0.28	1.24	0.96	0.38
5f8v0125	1100	8.0	34	spd V + sp	0. 28	0.10	1.18	1.44	0.49	0.19	1. 30	1.02	
9f8v0125	1100	8.0	34	Si-sp + (Mg-ferrite)	0. 93	0.35	1.58	0.14					
4f3v0182	1100	8.0	33.5	Mg-ferrite + ol + (sp)	0.11	0.23	0.87	1.79					0. 66
8f3v0182	1100	8.0	33.5	ol + Mg-ferrite	0. 10	0.20	0.89	1.81					0. 63
5f9v0231	1100	8.0	32.9	spd V + (sp + cpx)					0.52	0.08	1. 44	0.96	
8f9v0231	1100	8.0	32.9	Si-sp + spd V + (sp)	0. 87	0.15	1.71	0.27	0.58	0.11	1. 48	0.83	
3f5h0998	1100	9.0	12.5	sp + cpx	0. 10	0.14	0.96	1.80					
8f5h0998	1100	9.0	12.5	ol + spd V + cpx + ? + (Mg-ferrite)					0.57	0.34	1. 23	0.86	0.62
4f8v0053	1100	9.0	36.7	sp + (cpx)	0. 25	0.06	1.20	1.49					
8f8v0053	1100	9.0	36.7	Si-sp + (Mg-ferrite + cpx + ?)	0.73	0.22	1.50	0.55					
4f1v0098	1100	9.0	34.3	Mg-ferrite + ol + (cpx)	0.06	0.26	0.80	1.88					0. 88
8f1v0098	1100	9.0	34.3	ol + Mg-ferrite	0. 08	0.26	0.81	1.85					0. 88
5f8v0099	1100	9.0	33.3	sp + (cpx)	0. 30	0.08	1.22	1.40					
9f8v0099	1100	9.0	33.3	Si-sp + spd V + (cpx)	0. 91	0.32	1.59	0.18	0.66	0.25	1. 41	0.68	
5f8v0124	1100	9.0	23.6	spd V + sp	0. 40	0.13	1.27	1.20	0.55	0.23	1. 32	0.90	
9f8v0124	1100	9.0	23.6	Si-sp + (sp)	0. 93	0.35	1.59	0.13					
4f5v0129	1100	9.0	32	spd V + Mg-ferrite + cpx + ?	0.00	0.13	0.87	2.00	0. 52	0.43	1.09	0.96	
6f5v0129	1100	9.0	32	spd V + ol + (Mg-ferrite)					0. 76	0.56	1.20	0.48	0.68
4f3v0131	1100	9.0	34	spd V + ol + Mg-ferrite + ?	0.00	0.14	0.86	2.00	0. 54	0.45	1.09	0.92	0.72
8f9v0181	1100	9.0	34	ol + (spd V + Mg-ferrite)									0. 66
5f9v0181	1100	9.0	33.7	spd V + sp	0. 37	0.06	1.30	1.27	0.51	0.10	1. 41	0.98	
8fv0181	1100	9.0	33.7	spd V + Si-sp	0. 85	0.15	1.70	0.30	0.67	0.14	1. 53	0.66	
4f5v0055	1100	10.5	37.3	Mg-ferrite + cpx + (sp)	0.00	0.10	0.90	2.00					
8f5v0055	1100	10.5	37.3	Si-sp + (cpx + Mg-ferrite)	0. 99	0.96	1.03	0.02					
4f8v0085	1100	10.5	34	sp + (cpx + Mg-ferrite)	0. 34	0.11	1.22	1.33					
8f8v0085	1100	10.5	34	Si-sp + (cpx + Mg-ferrite)	0. 82	0.29	1.53	0.36					
5f5v0100	1100	10.5	33.7	Mg-ferrite + ? + (cpx)	0.00	0.13	0.88	1.99					
8f5v0100	1100	10.5	33.7	Si-sp + (Mg-ferrite + cpx + ?)	0.90	0.88	1.02	0.20					
4f1v0127	1100	10.5	34.3	Mg-ferrite + ol + ? + (cpx)	0.00	0.32	0.68	2.00					0. 91
8f1v0127	1100	10.5	34.3	ol + Mg-ferrite + ?+ (cpx)	0.00	0.29	0.72	1.99					0. 90
5f9v0232	1100	10.5	32.5	sp + ? + (Mg-ferrite)	0. 56	0.11	1.44	0.89					
9f8v0232	1100	10.5	32.5	Si-sp + (? + Mg-ferrite)	0. 94	0.35	1.58	0.13					

Stable phases produced are listed in the order of abundance. Phases in parentheses are minor constituents in the assemblage (sp: mt-rich spinel; Mg-ferrite: mt-rich spinel but with Si < Mg/2; isc: iscorite; ol: olivine; Si-sp: silica-spinel; spd: spinelloid; II, III and V: spinelloid polytypes; X_{fo} : $X_{Mg_2SiO_4}$, in olivine; ?: unknown phase).

In summary, the sequence of stability of the spinelloid intermediate phases with increasing pressure is the same as that observed in the Mg-free system: spinelloid II → spinelloid III → spinelloid V. These phases tend to become richer in Mg and Si with increasing temperature as well as pressure when coexisting with olivine. However, their stability fields are restricted to quite Fe-rich compositions. At high bulk Mg contents, olivine and a Fe^{3+}-rich spinel coexist.

3.2. Additional phases

Additional phases appeared in some of our experiments that cannot be plotted in composition diagrams such as Fig. 2. The most common extra phase was orthopyroxene or clinopyroxene. Slight oxidation during the experiment shifts the original Fe^{3+}/Fe^{2+} ratio so that excess Si becomes available for producing pyroxene along with spinel, spinelloid or olivine. As such, pyroxene acts as a sink for Si compensating for the presence of excess Fe^{3+} that can be incorporated in spinel or spinelloid. We observed the orthorhombic to high-*P* monoclinic phase transition to occur between 7.5 and 8.0 GPa (Table 1), which is in excellent agreement with experiments of Woodland (1998) on $(Mg,Fe)SiO_3$ pyroxenes. However, due to the unquenchable nature of the high-*P* monoclinic polymorph, the recovered clinopyroxene always had the low-*P* $P2_1/c$ structure.

Several other experiments between 3.0 and 5.0 GPa produced a mixed valence phase informally referred to as "iscorite" (Nell and van den Berg, 1988), in addition to the expected phases. This phase has a nominal composition of Fe_7SiO_{10} and we find only limited substitution of Mg, from 0.65–1.25 wt.% MgO. The low $Fe^{3+}/(Fe^{2+} + Mg)$ ratio indicates that this phase lies to the reduced side of the chemical system that we were investigating (Woodland and Angel, 2000). Thus slight reduction of the starting materials is observed in some experiments while some oxidation occurred in others. Run products from some of the experiments performed at pressures around 9.0 GPa have additional peaks in their X-ray diffraction (XRD) patterns that cannot be indexed by any high-pressure phase known in our chemical system, including that of so-called "wadsleyite-II" (Smyth and Kawamoto, 1997) now demonstrated to be the spinelloid IV structure (Smyth pers. comm.). The positions of the most apparent reflections have *d*-spacings of 2.708 (28), 2.643 (100), 2.444 (24), 2.236 (15), 2.133 (33), 2.046 (54), 1.672 (9), 1.570 (15), 1.512 (11), 1.507 (12) and 1.448 (20) whereby the values in parentheses denote the approximate relative intensities. Unfortunately, investigation of the samples with the microprobe and SEM revealed no additional phase that could be correlated with the XRD data. Additional investigations with Raman spectroscopy are planned to try and shed light on this mysterious phase.

3.3. Equilibrium and Mg–Fe^{2+} partitioning

The aforementioned slight oxidation or reduction in some experiments could lead one to question whether our run products represent equilibrium assemblages or not. However, there are a number of lines of evidence that indicate that local equilibrium was attained in the experiments. Many experiments contained two samples which had the same starting olivine composition, but differed in their magnetite/olivine ratio (i.e. variable bulk Fe^{3+} content). In some cases these samples produced different assemblages, which is not unexpected. However, when the phase relations were such that the same assemblage was produced in the two samples, the mineral compositions were virtually identical to each other, indicating approach to equilibrium from two different bulk Fe^{3+} contents (see Table 1). When olivine was a part of the run product assemblage, its composition was usually different from that of the starting material, demonstrating that this phase recrystallised and adjusted its composition during the experiment.

The partitioning of Fe^{2+} and Mg between phases can also be used as a test of whether chemical equilibrium was attained between two phases in the assemblage. For example, partitioning between orthopyroxene and olivine in our samples is consistent with the experimental data of von Seckendorff and O'Neill (1993), providing evidence of local equilibrium even in the experiments that experienced some oxidation (Fig. 3a). It seems likely that the slight oxidation occurred at the onset of the experiment and that the sample assemblage had adequate time to chemically adjust to this small shift in bulk composition. Pyroxene is always the phase richest in Mg and olivine is always richer in Mg compared to any coexisting spinelloid phase or spinel as can be

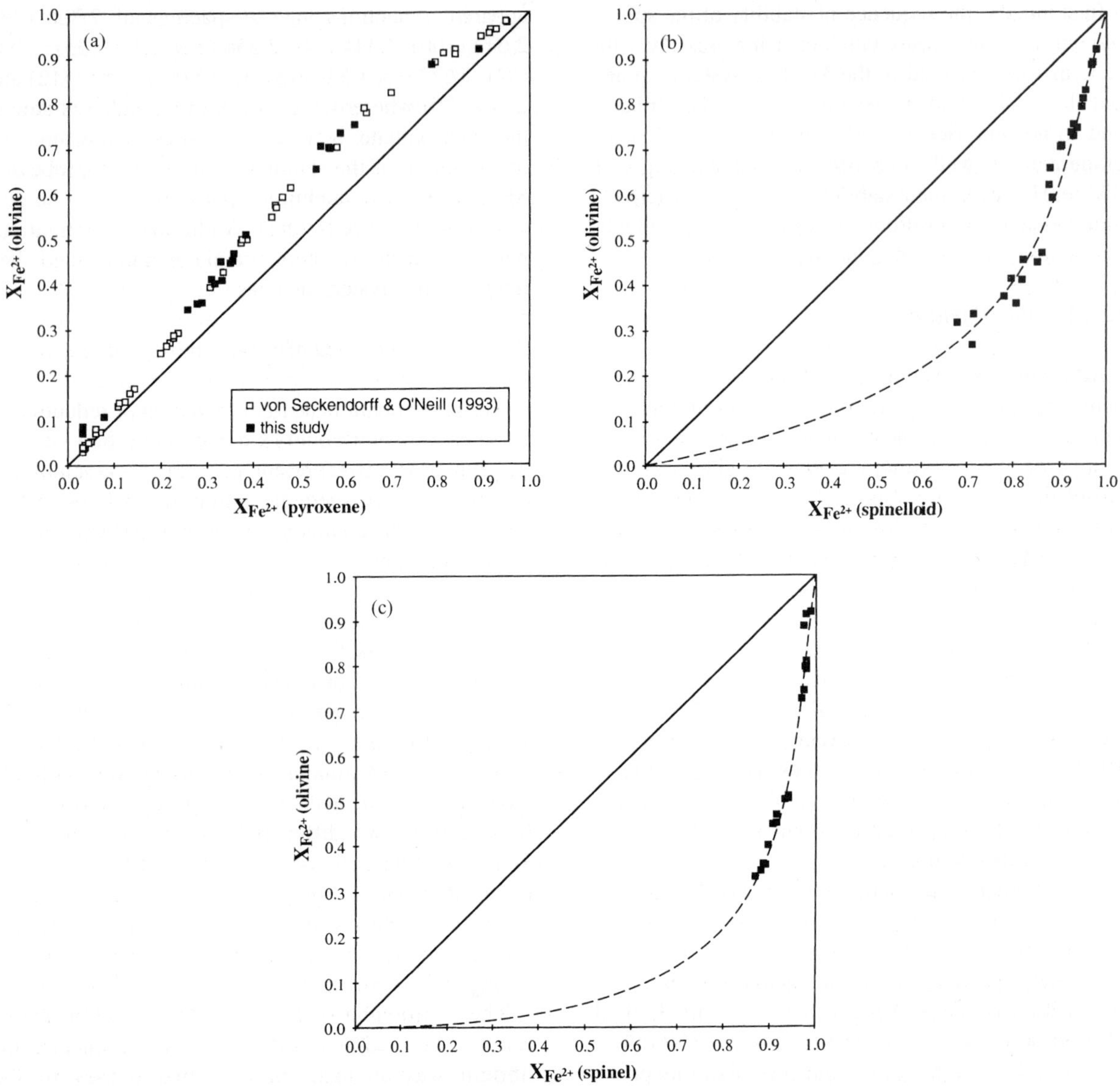

Fig. 3. Roozeboom diagrams of olivines coexisting with (a) pyroxenes; (b) spinelloids and (c) spinels. Similar trends are observed if olivine is substituted by pyroxene in (b) and (c). As reference experimental data from von Seckendorff and O'Neill (1993) have been included in (a). All spinelloid polytypes have been plotted together since their Fe^{2+} values are very similar. Only mt-rich spinels with Si > Mg/2 have been plotted. Dashed lines are drawn by hand.

seen in the Roozeboom diagrams of Fig. 3b and c. The Fe^{2+}–Mg partitioning between olivine and the spinelloid polymorphs and olivine and spinel varies systematically with composition (Fig. 3b and c). Such systematic behaviour also supports our contention that local equilibrium was easily attained during our relatively long-duration experiments. Partitioning between olivine and the spinelloids II, III and V is esssentially identical, thus we plotted all three spinelloids together in Fig. 3b. In all cases there is no noticeable pressure dependence on Fe^{2+}–Mg partitioning between these phases, implying that the spinelloid phases and spinel have a negligible excess volumes of mixing.

3.4. Spinelloid III

The stability field of spinelloid III is restricted to the Fe^{2+}-rich part of the Mg_2SiO_4–Fe_2SiO_4–Fe_3O_4 ternary system. Of the three spinelloid polytypes in this system, spinelloid III contains the highest silicate-bearing component and can coexist with silicate-rich spinel or olivine. Although the incorporation of Mg shifts the stability of spinelloid III to higher pressures this effect is limited. There is clearly a large gap between the stability fields of "true" $(Mg,Fe)_2SiO_4$-wadsleyite (e.g. Katsura and Ito, 1989; Fei et al., 1991) and the isostructural spinelloid III (Fe^{3+}-wadsleyite), both in terms of composition space as well as pressure. It is currrently unclear what thermodynamic relationship there is between these two phases.

3.5. Spinel

Magnetite-rich spinel can incorporate significant amounts of Si and Mg when coexisting with the various spinelloid polytypes (up to 0.42 and 0.17 cpfu, respectively). Such behaviour is particularly apparent in the Mg-free system (Fig. 1). However, in Mg-rich bulk compositions or at pressures >10.0 GPa where olivine coexists directly with spinel solid solutions, Si appears to become excluded from the spinel (e.g. Fig. 2h). This causes a shift in the spinel composition to include a Mg-ferrite component, since Si is no longer present in quantities to balance all the Mg as a Mg_2SiO_4 component. The apparent mutual exclusivity of Mg and Si in magnetite-rich spinels argues against significant solid solution along the Mg_2SiO_4–$MgFe_2O_4$ binary join and suggests complex activity–composition behaviour in $(Mg,Fe^{2+})Fe_2{}^{3+}O_4$–$(Mg,Fe^{2+})SiO_4$ spinel solid solutions.

3.6. Olivine

Olivine solid solutions were produced in experiments over a wide range of pressures from 3.0 to 10.5 GPa (Table 1). The variation of cell volume with composition is consistent with the data of Schwab and Küstner (1977), which were synthesised at atmospheric pressure (Fig. 4). We observe no systematic changes in this relation as a function of pressure, indicating that only very small amounts of Fe^{3+} can be incorporated in olivine at high pressures, even in very Fe^{3+}-rich bulk compositions. In terms of the olivine composition, we find that no matter whether olivine coexists with spinelloid or spinel, its composition measured by the microprobe begins to deviate slightly from ideal stoichiometry in the higher pressure experiments, suggesting the incorporation of a small amount of Fe^{3+} (Fig. 2). Several samples were carefully reanalysed, but didn't reveal any difference in chemical composition. Considering our agreement with the cell volume data of Schwab and Küstner (1977), we consider this small compositional deviation to be an artifact, related either to the presence of finely divided Fe^{3+}-rich inclusions in the olivine or, more likely, to a systematic problem with microprobe data aquisition with Mg-rich compositions. Further investigations of this issue are under investigation by TEM on selected samples.

3.7. Implications for the Earth's mantle

The distinctly separate stability fields of "true" $(Mg,Fe)_2SiO_4$-wadsleyite and the isostructural spinelloid III (Fe^{3+}-wadsleyite) in terms of composition and pressure suggest that spinelloid III, in itself, is of little direct relevance for the Earth's mantle. This is not to say that the incorporation of Fe^{3+} into wadsleyite is unimportant or insignificant. For example, O'Neill et al. (1993) demonstrated that measureable amounts of Fe^{3+} can be incorporated in wadsleyite, even in the presence of Fe-metal. Fei et al. (1992) found $Fe^{3+}/\Sigma Fe = 0.08$ in their wadsleyite sample. The stability of Fe^{3+}-wadsleyite in our experiments suggests a viable substitution mechanism for incorporating small amounts of Fe^{3+} into the "true" wadsleyite through a Fe_3O_4 component.

Aside from $(Mg,Fe)_2SiO_4$-wadsleyite, spinelloid V has the highest pressure stability of any spinelloid found in the Mg_2SiO_4–Fe_2SiO_4–Fe_3O_4 ternary system. Its high-pressure stability limit should also increase somewhat with increasing temperature. In addition it possesses the largest compositional range of any spinelloid and can incorporate more Mg than either spinelloid II or III. The effect of other cations such as Cr or Mn on the stability of spinelloid V is

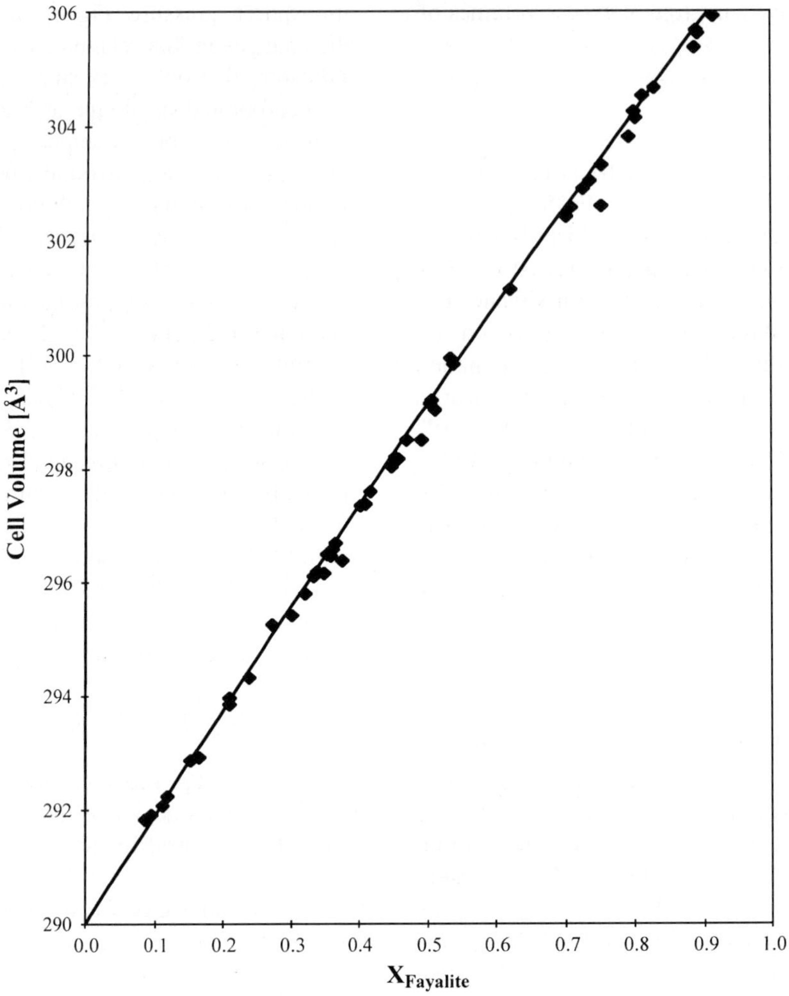

Fig. 4. Variation in the cell volume of olivine solid solutions as a function of composition. The solid curve represents experimental data by Schwab and Küstner (1977). In general all olivines plot on this trend, indicating that olivine is not taking up any Fe^{3+} in its structure.

currently unknown. It is conceivable that provided enough Fe^{3+} is present spinelloid V could be present locally as an additional phase in the mantle assemblage in the deepest parts of the upper mantle and the upper portions of the transition zone. As yet, there is no report of such a phase in diamond inclusions. However, it is likely that this phase could either be mistaken as a spinel solid solution or it could react during decompression to a mixture of olivine and spinel. In the absence of garnet, such as in a depleted harzburgitic composition, the addition of Fe^{3+} through metasomatism could either stabilise spinelloid V or produce a Mg-bearing magnetite-rich spinel that can coexist with olivine. Interestingly, this additional spinel phase would be essentially Si-free in such Mg-rich compositions, which is contrary to the conventional expectation that high-*P* spinel always should contain a significant amount of Si. Magnetite inclusions in diamond have been reported from a number of localities, but most seem to contain only traces of Mg, suggesting another mode of formation for these inclusions (Meyer, 1987; Stachel et al., 1998). In any case, our high-pressure phase relations suggest that even small amounts of added Fe^{3+} could stabilise a magnetite-rich spinel as an accessory phase in equilibrium with olivine. Whether this phase could locally influence the electrical conductivity of the deepest portions of the upper mantle would depend on its physical distribution within the assemblage.

Acknowledgements

This project was supported in part by a grant from the Deutsche Forschungsgemeinschaft (WO 652/4-1 and AL 166/14-2) to Alan Woodland and Rainer Altherr. Additional support for the high-pressure experiments and X-ray diffraction experiments at the Bayerisches Geoinstitut was provided by the EC "Human Capital and Mobility—Access to Large Scale Facilities" programme (contract no. ERBCHGECT940053 to D.C. Rubie). T. Boffa-Ballaran is thanked for access to X-ray diffraction facilities. The thorough reviews of M. Akaogi and J. Smyth are gratefully acknowledged.

References

Akaogi, M., Akimoto, S., Horioka, K., Takahashi, K., Horiuchi, H., 1982. The system $NiAl_2O_4$–Ni_2SiO_4 at high pressures and temperatures: spinelloids and spinel-related structures. J. Solid State Chem. 44, 257–267.

Angel, R.J., Woodland, A.B., 1998. Crystal structure of spinelloid II in the system Fe_3O_4–Fe_2SiO_4. Eur. J. Mineral. 10, 607–610.

Bina, C., Wood, B.J., 1987. The olivine-spinel transition: experimental and thermodynamic constraints and implications for the nature of the 400-km discontinuity. J. Geophys. Res. 92, 4853–4866.

Brey, G.P., Weber, R., Nickel, K.G., 1990. Calibration of a belt apparatus to 1800 °C and 6 GPa. J. Geophys. Res. 95, 15603–15610.

Canil, D., O'Neill, H.S.C., Ross II, C.R., 1991. A preliminary look at phase relations in the system γ-Fe_2SiO_4–Fe_3O_4 at 7 GPa. Terra Abstr. 3, 65.

Fei, Y., Mao, H., Mysen, B.O., 1991. Experimental determination of element partitioning and calculation of phase relations in the MgO–FeO–SiO_2 system at high pressure and high temperature. J. Geophys. Res. 96, 2157–2169.

Fei, Y., Mao, H.-K., Shu, J., Parthasarathy, O., Bassett, W.A., 1992. Simultaneous high-P, high-T X-ray diffraction study of β-$(Mg,Fe)_2SiO_4$ to 20 GPa and 900 K. J. Geophys. Res. 97, 4489–4495.

Horiuchi, H., Horioka, K., Morimoto, N., 1980. Spinelloid: a systematics of spinel-related structures obtained under high-pressure conditions. J. Mineral. Soc. Jpn. 14, 253–264.

Katsura, T., Ito, E., 1989. The system Mg_2SiO_4–Fe_2SiO_4 at high pressures and temperatures: precise determination of stabilities of olivine, modified spinel, and spinel. J. Geophys. Res. 94, 15663–15670.

Larson, A.C., von Dreele, R.B., 1988. GSAS, general structure analysis system manual. Report LAUR 86-748. Los Alamos Natural Laboratory, Los Alamos, New Mexico.

Meyer, H.O., 1987. Inclusions in diamond. In: Nixon, P.H. (Ed.), Mantle Xenoliths. Wiley, Chichester, UK, pp. 501–522.

Nell, J., van den Berg, J.C., 1988. Stability relations of iscorite (Fe_7SiO_{10}) in the system Fe–Si–O. Trans. Inst. Min. Metall. 97, C53–C60.

Ohtaka, O., Tobe, H., Yamanaka, T., 1997. Phase equilibria for the Fe_2SiO_4–Fe_3O_4 system under high pressure. Phys. Chem. Mineral. 24, 555–560.

O'Neill, H.S.C., Pownceby, M.I., 1993. Thermodynamic data from redox reactions at high temperatures. I. An experimental and theoretical assessment of the electrochemical method using stabilised zirconia electrolytes, with revised values for the Fe-"FeO", Co-CoO, Ni-NiO and Cu-Cu_2O oxygen buffers, and new data for the W-WO_2 buffer. Contrib. Mineral. Petrol. 114, 296–314.

O'Neill, H.S.C., McCammon, C.A., Canil, D., Rubie, D.C., Ross II, C.R., Seifert, F., 1993. Mössbauer spectroscopy of mantle transition zone phases and determination of minimum Fe^{3+} content. Am. Mineral. 78, 456–460.

Schwab, R.-G., Küstner, D., 1977. Präzisionsgitterkonstantenbestimmung zur Festlegung röntgenographischer Bestimmungskurven für synthetische Olivine der Mischkristallreihe Forsterit-Fayalit. N. Jb. Mineral., Mh. 5, 205–215.

Smyth, J.R., Kawamoto, T., 1997. Wadsleyite II: A new high pressure hydrous phase in the peridotite–H_2O system. Earth Planet. Sci. Lett. 146, E9–E16.

Stachel, T., Harris, J.W., Brey, G.P., 1998. Rare and unusual mineral inclusions from Mwadui Tanzania. Contrib. Mineral. Petrol. 132, 34–47.

Toby, B.H., 2001. EXPGUI, a graphical user interface for GSAS. J. Appl. Cryst. 34, 210–213.

von Seckendorff, V., O'Neill, H.S.C., 1993. An experimental study of Fe–Mg partitioning between olivine and orthopyroxene at 1173, 1273 and 1423 K and 1.6 GPa. Contrib. Mineral. Petrol. 113, 196–207.

Woodland, A.B., 1998. The orthorhombic to high-P monoclinic phase transition in Mg–Fe pyroxenes: can it produce a seismic discontinuity. Geophys. Res. Lett. 25, 1241–1244.

Woodland, A.B., O'Neill, H.S.C., 1993. Synthesis and stability of $Fe^{2+}{}_3Fe^{3+}{}_2Si_3O_{12}$ garnet and phase relations with $Fe_3Al_2Si_3O_{12}$–$Fe^{2+}{}_3Fe^{3+}{}_2Si_2O_{12}$ solutions. Am. Mineral. 78, 1002–1015.

Woodland, A.B., Angel, R.J., 1998. Crystal structure of a new spinelloid with the wadsleyite structure in the system Fe_2SiO_4–Fe_3O_4 and implications for the Earth's mantle. Am. Mineral. 83, 404–408.

Woodland, A.B., Angel, R.J., 2000. Phase relations in the system fayalite–magnetite at high pressures and temperatures. Contrib. Mineral. Petrol. 139, 734–747.

Available online at www.sciencedirect.com

SCIENCE DIRECT

Physics of the Earth and Planetary Interiors 143–144 (2004) 185–199

PHYSICS OF THE EARTH AND PLANETARY INTERIORS

www.elsevier.com/locate/pepi

Precise determination of phase relations in pyrolite across the 660 km seismic discontinuity by in situ X-ray diffraction and quench experiments

Norimasa Nishiyama [a,*], Tetsuo Irifune [a], Toru Inoue [a], Jun-ichi Ando [b], Ken-ichi Funakoshi [c]

[a] Geodynamics Research Center, Ehime University, 2-5 Bunkyo-cho, Matsuyama 790-8577, Japan
[b] Department of Earth and Planetary Systems Science, Hiroshima University, Higashi-Hiroshima 739-8526, Japan
[c] Japan Synchrotron Radiation Research Institute, Mikazuki, Sayo-gun, Hyogo 679-5198, Japan

Received 5 February 2003; received in revised form 30 June 2003; accepted 29 August 2003

Abstract

Mineral assemblage changes in a pyrolite composition with increasing pressure were observed by in situ X-ray diffraction and quench experiments at pressures near that of the 660 km seismic discontinuity and at a fixed temperature of 1600 °C. According to results obtained by in situ X-ray diffraction experiments, ringwoodite (Rw) was observed with majorite garnet and $CaSiO_3$-rich perovskite at pressures of about 20–22 GPa. Dissociation of ringwoodite to $MgSiO_3$-rich perovskite and magnesiowüstite (Mw) was completed at 22.0 ± 0.2 GPa according to Matsui et al.'s periclase pressure scale, and at 21.7 ± 0.1 GPa according to Shim et al.'s gold pressure scale. Majorite garnet persisted to about 24 GPa where pyrolite transformed to a lower mantle mineral assemblage, i.e. $MgSiO_3$-perovskite, $CaSiO_3$-rich perovskite, and magnesiowüstite. Thus, majorite garnet coexists with the lower mantle assemblage at pressures of about 22–24 GPa. In the quench experiments, an assemblage of $MgSiO_3$-perovskite, magnesiowüstite, $CaSiO_3$-rich perovskite, and majorite garnet was synthesized at 22.5 GPa and 1600 °C, in which Mg-perovskite contained 2.8 wt.% Al_2O_3, and was significantly poorer in Fe than coexisting magnesiowüstite. The Fe–Mg partition coefficient between Mg-perovskite and magnesiowüstite including ferric iron ($K^{app} = 0.27 \pm 0.06$) is very close to that in the Al-free system, which suggests that these P–T conditions are in the vicinity of those of ringwoodite decomposition. Both the results of in situ X-ray diffraction and quench experiments in the present study yield a convergent result that ringwoodite decomposes into Mg-perovskite and magnesiowüstite before the garnet-to-perovskite transition at 1600 °C in pyrolite. The relation between the Al content in Mg-perovskite and K^{app} in pyrolite is non-linear, which is consistent with the Fe–Mg partitioning between Mg-perovskite and magnesiowüstite previously reported for a simpler $MgO–FeO–Al_2O_3–SiO_2$ system.

Keywords: Pyrolite; 660 km seismic discontinuity; Ringwoodite; Majorite garnet; In situ X-ray diffraction experiments; Fe–Mg partitioning

1. Introduction

Pyrolite is a hypothetical rock composition first proposed by Ringwood (1966), and has been considered to be compositionally representative of the upper man-

* Corresponding author. Present address: Building 434 A, Argonne National Laboratory 9700 South Cass Avenue, Argonne, IL 60439, USA. Tel.: +1-630-252-0438; fax: +1-630-252-0436.
E-mail address: nishiyama@cars.uchicago.edu (N. Nishiyama).

0031-9201/$ – see front matter © 2004 Published by Elsevier B.V.
doi:10.1016/j.pepi.2003.08.010

tle. In contrast, there is little direct information about the chemical composition of the lower mantle because of the lack of petrologic evidence. Pyrolite is also considered as a potential candidate of the lower mantle material because the calculated density and seismic velocities of pyrolite under lower mantle conditions using experimentally determined equation-of-state parameters of lower mantle minerals are in good agreement with those obtained from seismic observations (e.g. Yagi and Funamori, 1996; Fiquet et al., 1998), although there are some contradictory opinions (e.g. Jeanloz and Knittle, 1989). Detailed studies of the phase relations in pyrolite under the pressure and temperature conditions of the mantle transition zone and the lower mantle are thus crucial for addressing the composition and dynamics of the entire mantle.

Several attempts have been made to determine the phase relations in pyrolite. Irifune and Ringwood (1987) demonstrated that an assemblage of ringwoodite (Rw), majorite garnet, and $CaSiO_3$-rich perovskite is stable in pyrolite at pressures corresponding to the mantle transition zone (15–22 GPa) at a temperature of 1200 °C using a Kawai-type apparatus. O'Neill and Jeanloz (1990) extended such experiments using pyrolitic material at pressures corresponding to those of the deep lower mantle (−54 GPa) using a laser-heated diamond anvil cell, and observed $MgSiO_3$-rich perovskite, magnesiowüstite (Mw), and $CaSiO_3$-rich perovskite at these conditions. They confirmed the presence of these phases by X-ray diffraction experiments at high pressure after the sample was quenched to room temperature. However, the resolution of the X-ray diffraction using a laboratory X-ray source was relatively poor and the chemical analyses of the run products were not made in that study. Irifune (1994) improved experimental techniques in Kawai-type apparatus to generate higher pressure, and carried out experiments at 28 GPa and 1600 °C. He confirmed the mineral assemblage proposed by O'Neill and Jeanloz (1990) in a pyrolite composition, and determined the chemical compositions of the coexisting phases. He demonstrated that Mg-perovskite can contain most of Al_2O_3 in pyrolite, and that Al-rich phases proposed by earlier studies (e.g. Takahashi and Ito, 1987) are absent in pyrolite at lower mantle conditions. Kesson et al. (1998) further extended the pressure range, and studied the phase transitions in pyrolite at pressures of 70 and 135 GPa using a laser-heated diamond anvil cell. They determined the chemical compositions of the coexisting phases using TEM, and demonstrated that the same phase assemblage as that observed in O'Neill and Jeanloz's and Irifune's studies was present in their run products. Therefore, it has been well established that pyrolite crystallizes into the ringwoodite-bearing and the perovskite-bearing assemblages at conditions corresponding to the mantle transition zone and the lower mantle, respectively.

Detailed phase changes in pyrolite from ringwoodite-bearing to perovskite-bearing assemblages have been under discussion. Irifune (1994) first proposed that ringwoodite decomposes into $MgSiO_3$-rich perovskite and magnesiowüstite before the garnet transition to the perovskite structure is complete, based on quench experiments at about 1600 °C. Thereafter, Wood (2000) confirmed this result by quench experiments (1627 °C) at closely spaced pressure intervals, and determined further detailed phase relation and phase chemistry in pyrolite as follows: ringwoodite dissociates into Al-poor $MgSiO_3$-rich perovskite and Fe-rich magnesiowüstite via a near-univariant transition; above the transition pressure of ringwoodite, garnet starts to transform into aluminous $MgSiO_3$-rich perovskite with Ca-perovskite, the former dissolving into the coexisting Al-poor $MgSiO_3$-rich perovskite that was produced by dissociation of ringwoodite. This transition is estimated to be complete within a pressure interval of about 2 GPa. Nishiyama and Yagi (2003) carried out experiments at temperatures of 1600–2200 °C and pressures of 24 and 30 GPa, the latter being performed using a Kawai-type apparatus with sintered diamond anvils for the second stage anvils. At 24 GPa and 1600 °C, an assemblage of Mg-perovskite, magnesiowüstite, Ca-perovskite, and minor garnet was observed; at 30 GPa and 1600 °C, garnet completely disappeared, and an assemblage of the two perovskites and magnesiowüstite was observed, which supports the previous results of Irifune (1994) and Wood (2000). In contrast, Hirose (2002) proposed different phase relations at 1600 °C based on quench experiments. According to his results, garnet partially transforms into akimotoite before the dissociation pressure of ringwoodite, and the transition of garnet into the perovskite-bearing assemblage is complete before ringwoodite dissociation is complete.

In order to clarify the detailed sequence of changes in the mineral assemblage in pyrolite at conditions corresponding to the mantle transition zone to the uppermost lower mantle, we performed in situ X-ray diffraction experiments on pyrolite at pressures of 20–25 GPa and at a fixed temperature of 1600 °C for a few hours. In addition, quench experiments with longer heating durations (>12 h) were carried out at pressures just above the ringwoodite dissociation pressure in pyrolite, and the chemical compositions of the coexisting phases were analyzed carefully. Using a combination of these experimental data, we examine the detailed phase relations, phase chemistry, and the phase transition pressures in pyrolite under conditions near those of the 660 km discontinuity.

2. Experimental

In situ X-ray diffraction experiments were performed at BL04B1, SPring-8. This beam line has a Kawai-type press (SPEED-1500) and X-ray diffraction system for energy dispersive measurements (Utsumi et al., 1998). Incident X-ray from a bending magnet were collimated to form a thin beam (50 μm in horizontal and 300 μm in vertical dimensions) by tungsten carbide slits and directed to the sample through pyrophyllite gasket and magnesia pressure medium. Diffracted X-ray were detected by a pure Ge solid state detector and a 4096-channel analyzer at a fixed diffraction angle of about 6° through a collimator with a 50 μm horizontal slit and additional tungsten carbide slits (200 μm in horizontal and 500 μm in vertical dimensions). The diffraction angle was calibrated using *d*-values of X-ray diffraction peaks of periclase and gold, which were used as pressure markers, at ambient pressure and temperature. The diffraction angle was determined with uncertainty of less than 0.0005°.

In order to achieve pressures up to about 25 GPa and to perform in situ X-ray diffraction experiments for the samples possessing complex chemical compositions, we adopted a newly designed high pressure cell for use with the second stage tungsten carbide anvils with a truncated edge length (TEL) of 2.5 mm (Fig. 1). Since the anvil gap became narrower in the present cell as compared to our earlier studies with a TEL of 3 mm (Irifune et al., 1998; Kuroda et al., 2000), we placed the sample cylinders horizontally so that the diffraction data of the sample and the pressure markers were able to be easily obtained. We used synthetic pyrolite (SiO_2, 45.2 wt.%; TiO_2, 0.3 wt.%; Al_2O_3, 3.9 wt.%; Cr_2O_3, 0.5 wt.%; FeO, 8.1 wt.%; MgO, 37.5 wt.%; NiO, 0.3 wt.%; CaO,

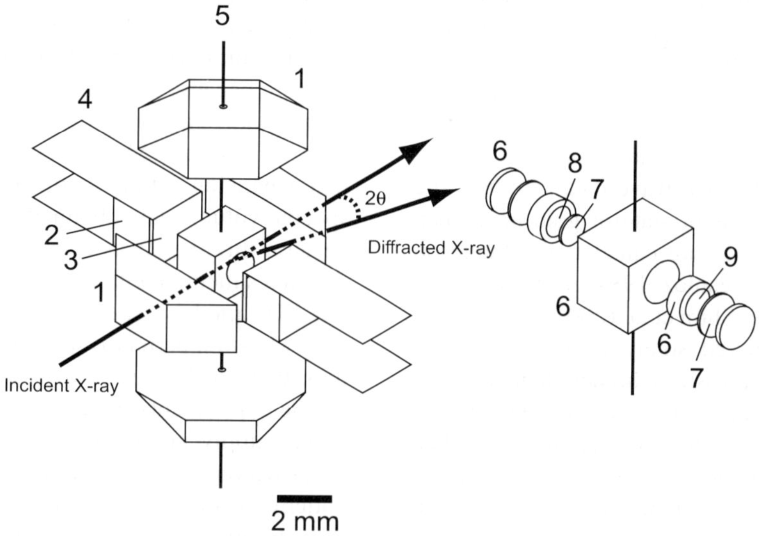

Fig. 1. A schematic illustration of the cell assembly used in the present in situ X-ray diffraction experiments. (1) Pressure medium (CoO-doped MgO); (2) pressure medium (ZrO_2); (3) heater (WC + diamond); (4) electrode (Mo); (5) thermocouple ($W_{97}Re_3$–$W_{75}Re_{25}$); (6) sample capsule (MgO); (7) sample lid (Mo); (8) starting material (pyrolite); (9) pressure marker (MgO + Au).

Available online at www.sciencedirect.com

SCIENCE DIRECT

ELSEVIER

Physics of the Earth and Planetary Interiors 143–144 (2004) 201–213

PHYSICS OF THE EARTH AND PLANETARY INTERIORS

www.elsevier.com/locate/pepi

Phase transitions of (Mg,Fe)O at megabar pressures

Tadashi Kondo[a,*], Eiji Ohtani[a], Naohisa Hirao[a], Takehiko Yagi[b], Takumi Kikegawa[c]

[a] *Institute of Mineralogy, Petrology, and Economic Geology, Graduate School of Science, Tohoku University, Sendai 980-8578, Japan*
[b] *Institute for Solid State Physics, University of Tokyo, Kashiwa 277-8581, Japan*
[c] *Institute of Materials structure Science, High Energy Accelerator Research Organization, Tsukuba 305-0801, Japan*

Received 27 December 2002; received in revised form 11 September 2003; accepted 29 October 2003

Abstract

An in situ X-ray diffraction study of magnesiowüstite at high pressures and high temperatures was performed using a high temperature diamond anvil cell and synchrotron radiation. A rhombohedral distortion starts in $(Mg_{0.20},Fe_{0.80})O$, $(Mg_{0.10},Fe_{0.90})O$, $(Mg_{0.05},Fe_{0.95})O$ and FeO at the same pressure of 16 GPa. The rhombohedral phase was observed even after laser annealing at megabar pressures. At high temperatures above 1500 °C and megabar pressures, a B1 (rock salt) structured phase was observed to be stable in all samples. A B8 (NiAs) structured phase was not found in magnesiowüstite and was only observed in the end member FeO. The diffraction pattern of the B8 phase was consistent with the calculated intensities for the inverse B8 structure. The results indicate that the stable phase of magnesiowüstite would have a B1 structure under lowermost mantle conditions.

Keywords: Lower mantle; Magnesiowüstite; Laser-heated diamond anvil cell; Synchrotron radiation

1. Introduction

It is widely believed that magnesiowüstite (Mg, Fe)O, is one of the major constituents of the Earth's lower mantle. Both wüstite, which has a non-stoichiometric formula $Fe_{1-x}O$ (e.g. McCammon, 1993), and periclase (MgO) have a B1 (rock salt) structure at ambient conditions and can make a complete solid solution to form magnesiowüstite. Many authors have extensively studied high pressure phase transitions of these minerals because of their geophysical importance. While no phase transition has been found in MgO so far in static (Duffy et al., 1995) and shock wave (e.g. Vassiliou and Ahrens, 1982; Duffy and Ahrens, 1995) measurements, the high pressure behavior of FeO is quite different from that of MgO. At about 16 GPa, FeO distorts from the B1 structure into a rhombohedral cell probably due to an antiferromagnetic (AFM) ordering from the paramagnetic (PM) state (Zou et al., 1980; Yagi et al., 1985). The magnetic behavior of FeO under megabar pressures such as magnetic collapse and the Néel transition has been investigated experimentally (Pasternak et al., 1997; Badro et al., 1999) and theoretically (Cohen et al., 1997; Fang et al., 1999), and shows complicated behavior. Further transition to a B8 (NiAs structure) phase at 74 GPa and 900 K has been proposed by Fei and Mao (1994) using an externally heated diamond anvil cell. The B8 phase is expected to be metallic as suggested by Knittle and Jeanloz (1986) based on experiments and also on theoretical grounds (Mazin et al., 1998; Fang et al., 1998a,b). However, available

* Corresponding author. Tel.: +81-22-217-6666; fax: +81-22-217-6675.
E-mail address: tdskondo@mail.tains.tohoku.ac.jp (T. Kondo).

0031-9201/$ – see front matter © 2004 Elsevier B.V. All rights reserved.
doi:10.1016/j.pepi.2003.10.008

data for magnesiowüstite are very limited at present. Recently, the decomposition of ferropericlase with the composition of around $(Mg_{0.5},Fe_{0.5})O$ into an oxide mixture with Mg-rich and Fe-rich components at 1000 K was reported by Dubrovinsky et al. (2000a,b), suggesting the gravitational separation of these phases in the lower mantle. However, the stable phase of (Mg,Fe)O at higher temperatures relevant to the lower mantle is still unclear. In this paper, we report new experimental results of in situ X-ray diffraction study on (Mg,Fe)O with compositions close to the end member FeO in the megabar pressure range.

2. Experimental procedure

Starting materials of $(Mg_{0.20},Fe_{0.80})O$, $(Mg_{0.10},Fe_{0.90})O$, $(Mg_{0.05},Fe_{0.95})O$ and FeO were synthesized from oxide mixtures of Fe_2O_3 and MgO in a gas furnace at 1200 °C for 24 h under controlled oxygen fugacity. The pellets were well sintered after quenching from high temperature and were examined by conventional powder X-ray diffraction technique and electron microprobe analysis. The results are shown in Table 1. We confirmed that each sample was single phase with a rock salt structure and chemically homogenous consistent with the initial composition within the error. The lattice parameter of FeO was $a = 4.307(2)$ Å, corresponding to a composition of $Fe_{0.95}O$ (McCammon and Liu, 1984). The sintered samples were crushed and ground in an agate mortar into fine powders.

Experimental conditions are summarized in Table 2. A Mao-Bell type diamond anvil cell was used with beveled diamond anvils. The inner culet size was 250 μm in diameter. We tried both external heating and laser heating techniques. In the external heating experiments, high temperatures up to 600 °C were generated by a small resistance heater made of Pt13%Rh wire (Fig. 1a). The sample was loaded directly into the hole of a pre-compressed Re gasket with a ruby chip and without any pressure transmitting medium. The initial dimensions of the sample hole were 60 μm in thickness and 100 μm in diameter. We used a ceramic rocker made of well sintered ZrO_2 in the bottom side of the anvil (piston part) for the purpose of thermal insulation to the outer parts. Temperatures were measured by a Pt–Pt13%Rh thermocouple which was attached to the surface of the diamond anvil. The temperature difference between the hot junction of the thermocouple and the sample position in the gasket was measured using fixed melting points of tin (232 °C) and lead (328 °C) at room pressure. The calibrated temperatures at the sample were 10% lower than those measured by the thermocouple. We estimated temperatures up to the maximum temperature of 600 °C by interpolation and extrapolation of the assumed linear correlation of the calibrated points.

Table 1
Starting materials

Sample	Composition	Cell parameter (Å)
$Fe_{0.95}O$	Not analyzed	$a = 4.307(2)$
$(Mg_{0.05},Fe_{0.95})O$	$Mg_{0.055(4)}Fe_{0.945(6)}O$	$a = 4.298(2)$
$(Mg_{0.10},Fe_{0.90})O$	$Mg_{0.106(4)}Fe_{0.894(5)}O$	$a = 4.294(1)$
$(Mg_{0.20},Fe_{0.80})O$	$Mg_{0.202(5)}Fe_{0.798(5)}O$	$a = 4.283(1)$

Table 2
Experimental table

#Run	Starting material	Heating method	Station/wavelength (Å)
M20F801	$(Mg_{0.20},Fe_{0.80})O + Au$	External heating	BL-13B2/0.4422
M20F802	$(Mg_{0.20},Fe_{0.80})O + Au$	External heating	BL-13B2/0.4424
M10F901	$(Mg_{0.10},Fe_{0.90})O + Au$	External heating	BL-13B2/0.4424
M10F902	$(Mg_{0.10},Fe_{0.90})O + Au$	External heating	BL-13B2/0.4428
M10F903	$(Mg_{0.10},Fe_{0.90})O + Au + Al_2O_3$	Laser heating	BL-13A/0.4258
M10F904	$(Mg_{0.10},Fe_{0.90})O + Au + Al_2O_3$	Laser heating	BL-13A/0.4268
M10F905	$(Mg_{0.10},Fe_{0.90})O + Al_2O_3$	Laser heating	BL-13A/0.4258
M05F951	$(Mg_{0.05},Fe_{0.95})O + Al_2O_3$	Laser heating	BL-13A/0.4258
M0F1001	$Fe_{0.95}O + Al_2O_3$	Laser heating	BL-13A/0.4257
M0F1002	$Fe_{0.95}O + Al_2O_3$	Laser heating	BL-13A/0.4258

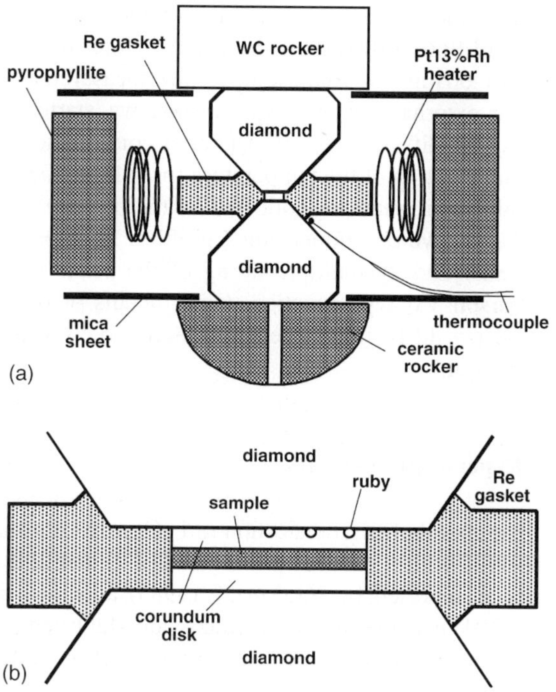

Fig. 1. Schematic illustration of (a) external heating system and (b) sample assembly for laser heating.

A high power multimode Nd:YAG laser (λ = 1.064 μm, LEE LASER, model 8100MQ) was used also to heat the sample to above 1000 °C. The samples were thermally insulated from the diamonds by corundum disks as shown in Fig. 1b. The focus size of the laser spot was about 60 μm in diameter. We measured the temperature distribution across the heated spot with a spatial resolution of a few microns using a method similar to that reported by Watanuki et al. (2000). Each temperature was calculated by fitting the thermal radiation spectra to the Planck radiation function (Boehler, 2000). The central region of the heated spots, showing a flat temperature distribution with a divergence of ±50 K, was almost comparable in size to that sampled by X-rays. The fluctuation of temperature during laser heating was less than ±100 K at constant laser power. In both heating methods, the outer cylinder of the DAC was cooled by a water circulation system to avoid changes in pressure and the position of the X-ray beam due to thermal expansion, and to protect the cell from oxidation.

Pressures at room temperature were measured by the ruby fluorescence method (Mao et al., 1978). The ruby chip was placed in the center of the culet, in the same position as the area sampled by X-rays and the heated spot. In the laser heating experiments, the samples were first compressed to the nominal pressure, and they were then annealed by laser scanning the whole sample at about 1500 K for 10–15 min. After we confirmed that the pressure after annealing coincided with that before heating, the high temperature measurements were carried out at a fixed position to minimize the uncertainty of pressures at high temperature. We heated the samples carefully to keep the temperature below the melting point in order to avoid chemical differentiation of the sample. In all external heating and some laser heating experiments, fine gold powder was mixed with the sample to estimate pressure at high temperatures using the equation of state of gold (Anderson et al., 1989). In the external heating experiments, pressures at high temperatures were calculated using the calibrated temperature from the thermocouple described above. The P–T path in the external heating experiments varies in each run probably due to the difference of the initial stress condition in the sample. The observed thermal pressure in the laser heating experiments was about 6 GPa at 1500 °C, which was nearly reproduced in two runs after the

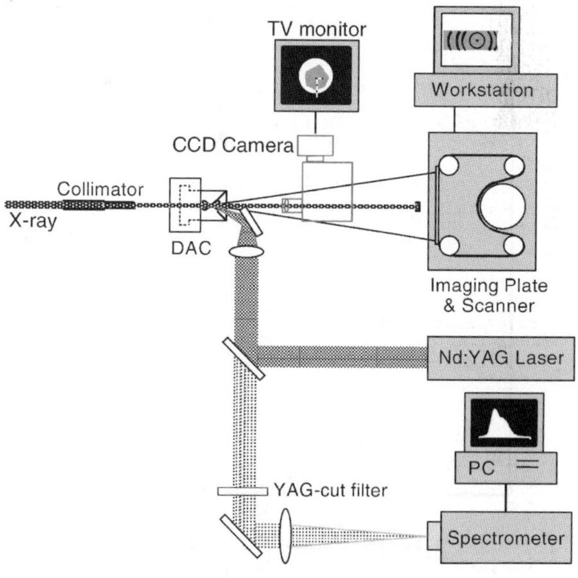

Fig. 2. Schematic diagram of the in situ X-ray diffraction system under high pressure and temperature installed at the Photon Factory, KEK.

sample was annealed enough. Therefore, we added the estimated thermal pressure to the measured value at room temperature assuming a linear relation between the thermal pressure and temperature when the sample did not contain gold. However, the estimated thermal pressure was based on the homogeneous temperature of the standard material. Hence, we were not sure how accurate this estimation was because the actual standard material has a temperature gradient, and adopted this value (6 GPa) as the error in pressure at high temperatures. The differences in pressure between the gold marker and the ruby fluorescence method were within 2 GPa at room temperature.

Angle dispersive X-ray diffraction experiments under high pressure and temperature were performed using synchrotron radiation and an imaging plate (IP) at the BL-13B2 and BL-13A stations in KEK-PF (Tsukuba, Japan). The schematic diagram of the system is shown in Fig. 2. The X-rays were monochromatized to have an energy of about 30 keV and collimated to a 30 μm square beam. The typical exposure time was 5–10 min for high temperature data to avoid Soret diffusion of iron during laser heating, and 20–60 min for room temperature measurements. In the external heating experiments, we started to collect the diffracted X-ray data after confirmation of the stabilization of the temperature for at least 30 min. The diffraction images collected on the IP were integrated to give one dimensional profiles. Each profile was analyzed by multiple fitting of the peaks using a Pearson VII type function. Other details of the in situ X-ray diffraction system are described elsewhere (Yagi et al., 2000).

3. Experimental results

The rhombohedral distortion in FeO at 16 GPa was observed with a sudden splitting of (1 1 1) and (2 2 0) lines as reported by Yagi et al. (1985) as shown in Fig. 3. When the sample contains an MgO component, the distortion is recorded initially by the (2 2 0)

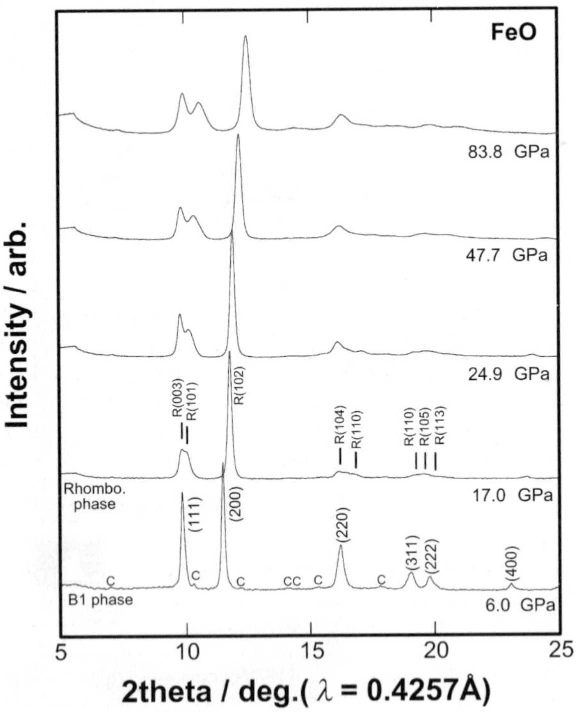

Fig. 3. Compression sequence of FeO at room temperature under non-hydrostatic conditions. Rhombohedral distortion was observed above16 GPa. C: corundum; R: rhombohedaral phase of FeO.

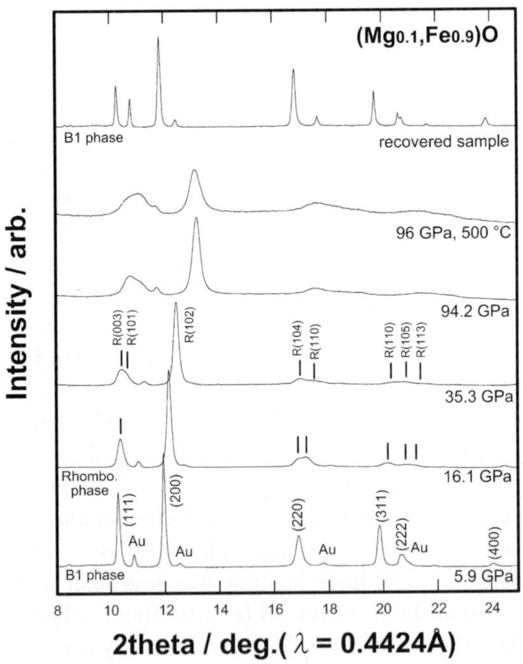

Fig. 4. Rhombohedral distortion observed in $(Mg_{0.10},Fe_{0.90})O$ at room temperature compression under non-hydrostatic conditions. The diffraction lines with higher angle than the (2 2 0) line split at about 16 GPa and broaden gradually with pressure, followed by the splitting of the (1 1 1) line.

line, at the same pressure, followed by the slight splitting of the (1 1 1) line (Fig. 4). In any case, the splitting developed further with increasing pressure in the same manner. However, the internal consistency of the indices for the rhombohedral assignment became worse with increasing pressure, caused by a strong deviation of the (1 0 1) and (1 1 3) lines in both FeO and (Mg,Fe)O. Consequently, the peak identification for the rhombohedral phase at around 100 GPa before heating was difficult because of the low signal to noise ratio of the broadened peaks and the overlapping with other peaks. The effect of non-hydrostatic stress on the sample makes peak identification even more difficult due to a deviation of the *d*-spacings from ideal values (Mao et al., 1996; Dubrovinsky et al., 2000a,b). This problem was serious in the external heating experiments on $(Mg_{0.20},Fe_{0.80})O$ and $(Mg_{0.10},Fe_{0.90})O$, in which the temperatures were not high enough to improve the broad peaks without a phase transformation. Indeed, appearance of new peaks or sharpening of the line widths was not observed up to the maximum temperature of 600 °C at 82 GPa in $(Mg_{0.20},Fe_{0.80})O$ and 500 °C at 96 GPa in $(Mg_{0.10},Fe_{0.90})O$ (Fig. 4), although these conditions were in the stability field of the B8 phase of FeO (Fei and Mao, 1994). On the other hand, the profile after laser annealing of the sample was improved to refine the lattice parameters of the rhombohedral phase in $(Mg_{0.10},Fe_{0.90})O$ (Table 3). We concluded from these observations that the low temperature phase of magnesiowüstite adopts a rhombohedral distortion.

The complex profile at about 10–12° (Figs. 5a and 6a) originates from the overlapping of diffractions of the rhombohedral phase of magnesiowüstite, corundum and its high pressure form. In some cases, gold and/or the quenched B8 phase of FeO were included also in the same angle range. The high pressure phase of corundum is expected to have the Rh_2O_3(II) structure (e.g. Funamori and Jeanloz, 1997) and can ex-

Table 3
Refinement of lattice parameters for the rhombohedral phase (94(6) GPa, room temperature) and B1 phase (110(6) GPa, 1650(200) °C) in $(Mg_{0.10},Fe_{0.90})O$

Index	d_{obs}	d_{cal}	$1 - d_{obs}/d_{cal}$	Intensity
Rhombohedral phase				
0 0 3	2.3854	2.3863	−0.0009	36
1 0 1	2.1959	2.1969	0.0010	17
1 0 2	1.9402	1.9399	0.0003	100
1 0 4	1.4165	1.4144	0.0021	9
1 1 0	1.3330	1.3327	0.0003	4
1 0 5	1.2171	1.2167	0.0004	3
1 1 3	1.1630	1.1635	−0.0005	4
2 0 1	1.1398	1.1394	0.0004	3
B1 phase				
1 1 1	2.2419	2.2401	0.0008	36
2 0 0	1.9394	1.9400	−0.0003	100
2 2 0	1.3717	1.3718	0.0000	18
3 1 1	1.1699	1.1698	0.0000	6
2 2 2	1.1202	1.1200	0.0001	3

Calculation of the rhombohedral phase was based on a hexagonal unit cell of $a = 2.6654(9)$ Å and $c = 7.1588(29)$ Å. Calculation of the B1 phase was based on a cubic unit cell of $a = 3.8799(1)$ Å.

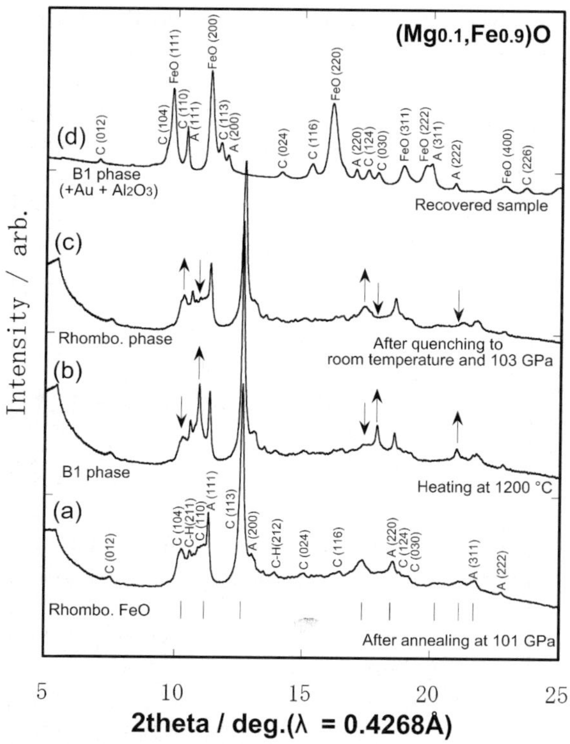

Fig. 5. High temperature phases of $(Mg_{0.10},Fe_{0.90})O$ observed at about 100 GPa. (a) The pattern at room temperature consisted of the rhombohedral phase (indicated by the calculated positions), corundum (C) and its high pressure phase (C-H) and gold (A). A B1 structured phase was observed at 1200 °C (b), which was unquenchable with respect to temperature (c). The recovered sample quenched from about 100 GPa to ambient conditions due to the breakdown of diamond (d). Although the residual stress was found in the spectrum, the pattern was explainable by the starting materials.

plain some other diffraction. Another candidate with the Rh_2O_3(II) structure is the high pressure phase of Fe_2O_3(Rozenberg et al., 2002), which will be stabilized also in the present pressure condition. It is possible that hematite is produced during laser heating by the change of oxygen fugacity in the sample. However, the sample quenched to ambient conditions shows no evidence of the hematite phase (Fig. 5d). Considering the fact that the complex feature was observed only after laser heating above 1000 K as suggested by Funamori and Jeanloz (1997), we interpreted that the profile can be explained not by Fe_2O_3 but by Al_2O_3.

We observed a B1 structured phase in $(Mg_{0.10},Fe_{0.90})O$ and $(Mg_{0.05},Fe_{0.95})O$ in the laser heating experiments. The transformation to the B1 phase happened directly from the rhombohedral phase (Fig. 5b). The observed *d*-spacings were in good agreement with calculated values for a B1 structure. An example of the refinement of lattice parameters is shown in Table 3. Although the (1 1 1) line of the B1 phase overlapped with peaks of the low temperature phase and corundum at the diffraction angle of around 11°, the relative intensities of peaks of the rhombohedral phase decreased at high temperature. The B1 phase observed at high temperatures was unquenchable to room temperature as shown in Fig. 5c. The transformation was completely reversible with respect to temperature and no B8 phase was observed when an MgO component was present.

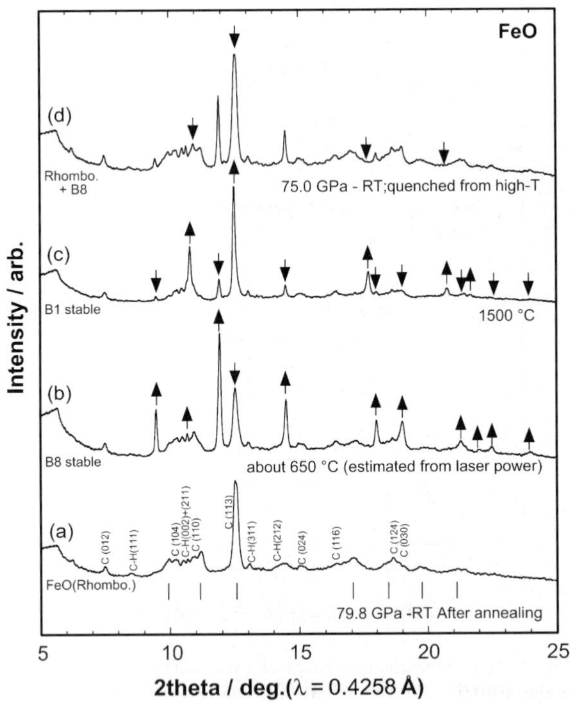

Fig. 6. High temperature phases of FeO observed at about 80 GPa. (a) A pattern at room temperature consisted of diffraction lines from the rhombohedral phase, corundum and its high pressure phase after annealing. Initially, a B8 structured phase appeared below 1000 °C (b), which then transformed to a B1 structured phase at higher temperature (c) The B1 phase was unquenchable with respect to temperature (d).

Fig. 6 shows the results of an in situ X-ray diffraction profile of FeO. The B8 phase, observed only in the end member FeO, exists over a limited temperature range. When the temperature was high enough to give incandescent light from the sample on the CCD monitor, the B8 phase seemed to be unstable because of a sudden reduction of diffraction intensities. Our temperature monitoring system was optimized for visible light. Consequently, a couple of temperature measurements for the B8 phase were unsuccessful due to the low statistics of the photon counting. Therefore, temperatures below about 800 °C were estimated from

Table 4
Refinement of lattice parameters for B8 (84(6) GPa, 750(200) °C) and B1 (87(6) GPa, 1450(200) °C) phases in FeO

Index	d_{obs}	d_{cal}	$d_{obs}/d_{cal} - 1$	Intensity
B8 phase				
0 0 2	2.5755	2.5768	−0.0005	38
1 0 0	2.2321	2.2318	0.0001	–[a]
1 0 1	2.0495	2.0480	0.0007	100
1 0 2	1.6870	1.6870	0.0000	43
1 0 3	1.3610	1.3613	−0.0002	26
1 1 0	1.2886	1.2885	0.0000	25
1 1 2	1.1523	1.1525	−0.0002	10
1 0 4	1.1160	1.1158	0.0002	3
2 0 1	1.0908	0.0906	0.0001	6
2 0 2	1.0239	1.0240	−0.0001	3
B1 phase				
1 1 1	2.2623	2.2622	0.0000	47
2 0 0	1.9576	1.9591	−0.0008	100
2 2 0	1.3853	1.3853	0.0000	16
3 1 1	1.1813	1.1814	−0.0001	8
2 2 2	1.1313	1.1311	0.0002	4

Calculation of the B8 phase was based on the hexagonal unit cell of $a = 2.5771(2)$ Å and $c = 5.1536(9)$ Å. Calculation of the B1 phase was based on a cubic unit cell of $a = 3.9183(4)$ Å.

[a] Not measured because the peak overlapped many other intense lines.

the laser power. The B1 phase was stably formed at higher temperature conditions above the stability field of the B8 phase (Fig. 6c) at pressure and temperature conditions similar to those at which the B1 phase was observed in (Mg,Fe)O. Examples of refinement of lattice parameters for the B8 and B1 phases are shown in Table 4. The B8 phase in FeO was quenchable to room temperature once it was produced, unless pressure was changed after quenching. Therefore, after a cycle of heating and quenching, the quenched B8 phase remained partially in the profile. In the case of quenching from the B1 phase in FeO, the intensities of the B8 phase became weak and sometimes disappeared at room temperature (Fig. 6d). Besides this, the profiles at high temperature sometimes included not only the high temperature phase but also some amount of the low temperature phase because of the temperature gradient around the heated spot. For these reasons,

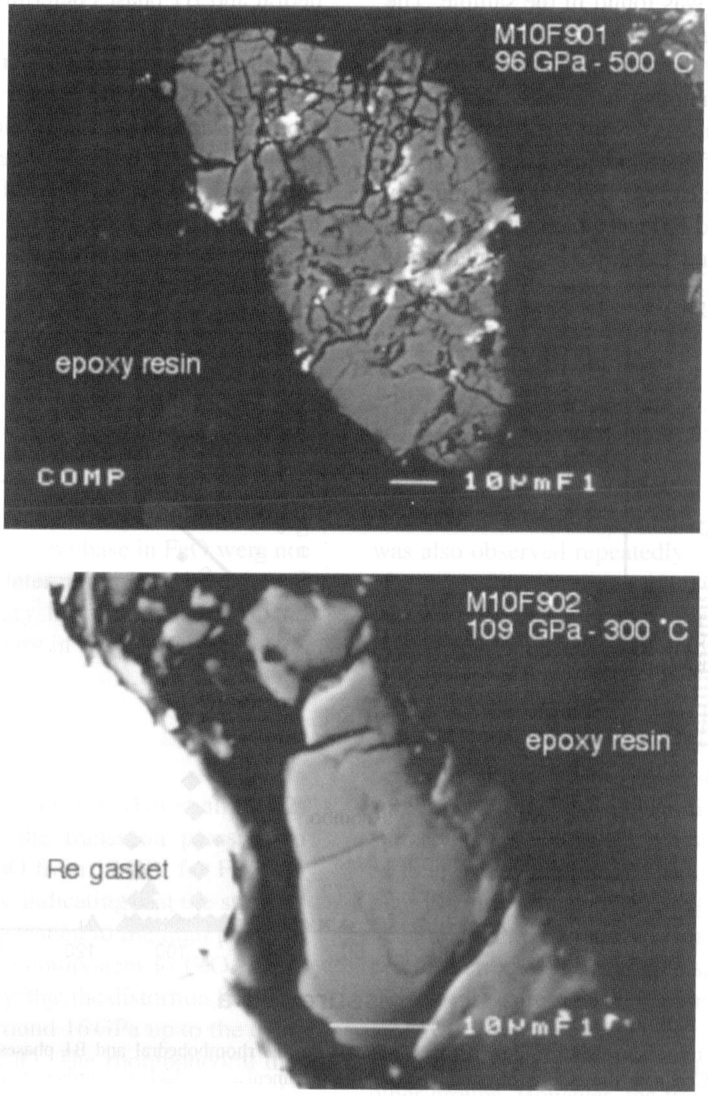

Fig. 7. Back scattered electron images of the recovered samples of M10F901 (top) and M10F902 (bottom). The pressure and temperature values in figures indicate the conditions maintained at the maximum temperature. The white portion in the top figure was due to the charge-up of electrons. Homogeneous compositions identical to that of the starting materials was observed within error.

the advantages of this combined method. The details of the DAC system will be presented elsewhere, and in this paper we mainly report on the use of ATEM in studying ultrahigh pressure materials.

Although ATEM is very powerful for the study of ultrahigh pressure materials, there are several problems that have to be resolved. The first is that the specimens of ultrahigh pressure phases are in general very sensitive to electron beam irradiation and easily become amorphous. In particular, silicate perovskites are very sensitive. We are trying to overcome this problem by the use of a high-sensitivity and high-resolution TV camera system for TEM. With this system, we are expecting to observe TEM images and also electron diffraction patterns of the ultrahigh pressure phases under a very weak electron beam. This is now under way and the results will be presented elsewhere.

The second problem is that in widely used ion-thinning to make ATEM specimens, elements are selectively removed from the surface area of the specimens during ion thinning in some cases, particularly in ultrahigh pressure materials, and analyzed compositions are different from the real values (Fujino et al., 1998). To overcome this problem, we have explored the use of ultramicrotomy for ultrahigh pressure materials because it leaves the chemistry unchanged. In the following, we mainly report the details of using ultramicrotomy on ultrahigh pressure materials, and also report some preliminary results on the mineralogy of the lower mantle.

2. Experimental

2.1. Ultrahigh pressure materials examined

Materials used for the investigation of ultramicrotomy are $Ca(Si,Ti)O_3$, $(Mg,Fe)SiO_3$ and $(Ca,Mg,Fe)SiO_3$ perovskites and the details are summarized in Table 1 together with the high pressure synthesis conditions. Runs 1803 and 1814 were chosen from samples of high pressure and high temperature experiments performed on the system $CaSiO_3$-$CaTiO_3$ at 5.7–13.2 GPa and 1500 °C using a multianvil cell apparatus at the Institute for the Study of the Earth's Interior, Okayama University. Runs A-13, B-02, A-06, A-14 and C-05 are selected samples from the system $CaMgSi_2O_6$(Di)-$CaFeSi_2O_6$(Hd) synthesized at 25–80 GPa and 1700–2000 °C by laser-heated diamond anvil cell apparatus at the Institute for Solid State Physics, University of Tokyo and at Hokkaido University.

In the DAC experiments, diamonds with 450 and 300 μm culets were used for generating pressures of 25–30 and 70–80 GPa, respectively. A powder of a synthetic pyroxene solid solution starting material was loaded into the hole of a stainless steel or rhenium gasket and was sandwiched by NaCl in the diamond cell. Platinum powder was mixed with the $CaMgSi_2O_6$ sample in order to absorb laser radiation for heating. Pressure was measured by the ruby fluorescence technique (Mao et al., 1978). Typical pressure differences between the center and the edge

Table 1
Synthesis conditions of the materials examined

Run no.	Starting material	Pressure (GPa)	Temperature (°C)	Duration (min)	Product phases
$CaSiO_3$–$CaTiO_3$					
1803	$CaSi_{0.8}Ti_{0.2}O_3$	8.3	1500	30	Pv + Wal
1814	$CaSi_{0.8}Ti_{0.2}O_3$	11.3	1500	30	Pv + Tit + Lar
$CaMgSi_2O_6$(Di)–$CaFeSi_2O_6$(Hd)					
A-13	$Di_{100}Hd_0$	29.7	1930	10	Mg-Pv[a] + Ca-Pv[a]
B-02	$Di_{75}Hd_{25}$	31.5	1800	10	Mg-Pv + Mw + St + Ca-Pv[a]
A-06	$Di_{50}Hd_{50}$	29.5	1800	10	Mg-Pv + Mw + St + Ca-Pv[a]
A-14	$Di_{50}Hd_{50}$	26.2	1820	10	Mg-Pv + Mw + St + Ca-Pv[a]
C-05	$Di_{50}Hd_{50}$	78.0	1900	10	Mw[a] + SiO_2[b] + Ca-Pv[a]

Pv: $Ca(Si,Ti)O_3$ perovskite, Wal: walstromite-structured $CaSiO_3$, Tit: titanite solid solution, Lar: larnite (β-Ca_2SiO_4), Mg-Pv: $(Mg,Fe)SiO_3$ perovskite, Ca-Pv: $CaSiO_3$ perovskite, Mw: magnesiowustite, St: stishovite.

[a] Identified by ATEM.

[b] Unknown phase of SiO_2.

of the sample are about 5 GPa. Pressures after heating are listed in Table 1. The sample was heated from both sides by a YAG laser beam focused to 70–80 μm in multimode (A-13, A-06, A-14 and C-05) or a YLF laser beam focused to about 30 μm in TEM_{01} mode (B-02) and temperature of the sample was measured spectroradiometrically.

The experimental details and comprehensive results of high pressure phase relations and structure variations of perovskite phases in both systems will be presented in separate papers.

2.2. Ultramicrotomy

Ultramicrotomy was originally developed in biology to section biological materials into ultrathin foils for TEM observation, and is now applied to high molecular compounds and metallic materials using a diamond knife (Williams and Carter, 1996). In ultramicrotomy the sample, embedded in polymeric resin, is moved across a knife blade while being advanced stepwise (Fig. 1). In our case, the multianvil cell samples were cut into cubes with edges of 100 μm and laser-heated DAC samples were thin disks of 100 μm diameter and 20–30 μm in thickness. The samples were embedded in polymeric resin Quetol 812 (Nisshin EM Co. Ltd.) using micro-molds and hardened at 70 °C for 72 h. Samples were sectioned by an ultramicrotome (Reichert-Nissen ultracuts N) at the Electron Microscope Center of the Faculty of Agriculture, Hokkaido University. A resin block including a sample was cured and trimmed by a glass knife so that the cutting surface becomes 100–300 μm across and was then sectioned by a diamond knife of 45°. The sectioned thin flakes floated off onto water in a trough (Fig. 1), and the flakes which contain the central part of the heated sample were collected on a grid with a carbon film. The sectioned flakes were additionally carbon-coated.

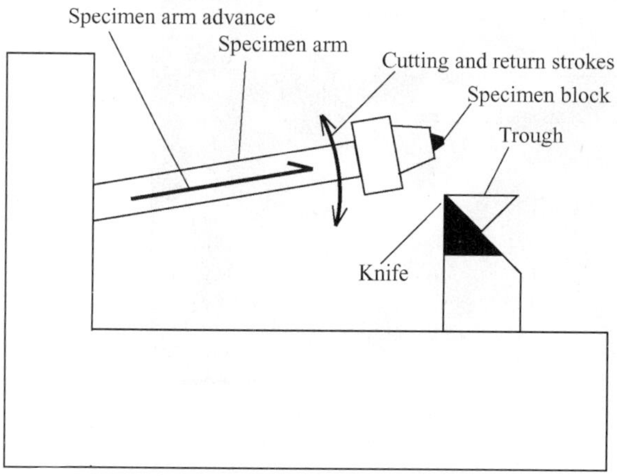

Fig. 1. Schematic drawing of an ultramicrotome.

2.3. TEM observations and quantitative chemical analysis

TEM observations and quantitative chemical analyses were performed using an analytical electron microscope (JEOL, JEM-2010) at Hokkaido University, operated at 200 kV. Quantitative chemical analyses were made with a NORAN Instruments/Voyager II energy-dispersive analytical system attached to the JEM 2010 electron microscope. Silicate perovskites are very sensitive to electron beam irradiation, and in measuring the characteristic X-ray intensities, elements are selectively removed from the specimens when the electron beam density is stronger than some critical value (Fujino et al., 1998). Therefore, TEM observations and chemical analyses were made with a low electron beam current of 4 pA/cm^2 on the small fluorescent screen, and the characteristic X-ray intensities were measured with an electron beam of 100–150 nm diameter and a counting time of 60–120 s. Characteristic X-ray intensities were measured in the central region of the thin foils so that the X-ray Intensities were obtained from the heated area of the sample.

The k-factors (Cliff and Lorimer, 1975) used for the calculation of quantitative chemical analyses were experimentally obtained using the synthetic standard materials $CaSiO_3$ (wollastonite), $CaTiO_3$ (perovskite), Mg_2SiO_4 (forsterite) and Fe_2SiO_4 (fayalite). They were crushed in a pestle and mortar in petroleum ether and were placed on a grid with a holey carbon film. K_{ij}-factors were experimentally obtained from the measured intensity ratios I_i/I_j and known concentration ratios C_i/C_j of the paired elements i and j in the standard materials ($k_{ij} = (C_i/C_j)/(I_i/I_j)$). The k_{ACa}-factors (where A = Si, Ti, Ca) used for $Ca(Si,Ti)O_3$ perovskite were obtained as functions of total X-ray counts of Ca and the k_{ASi}-factors (where A = Mg, Fe, Ca, Si) used for $(Mg,Fe)SiO_3$ and $(Ca,Mg,Fe)SiO_3$ perovskites were obtained as functions of total X-ray counts of Si.

The experimentally obtained k-factors for our ATEM operating at 200 kV with an ultra-thin window and take-off angle of 25° were expressed as the following regression equations.

For $Ca(Si,Ti)O_3$ perovskite

$$k_{TiCa} = 0.540(5) + 9(5) \times 10^{-7} I_{Ca}$$

$$k_{SiCa} = 0.423(16) + 1.2(17) \times 10^{-6} I_{Ca}$$

For $(Mg,Fe)SiO_3$ and $(Ca,Mg,Fe)SiO_3$ perovskites

$$k_{CaSi} = 2.45(18) - 2.6(20) \times 10^{-5} I_{Si}$$

$$k_{MgSi} = 1.14(2) - 5(20) \times 10^{-7} I_{Si}$$

$$k_{FeSi} = 1.80(7) - 9.9(4) \times 10^{-5} I_{Si}$$

where I_{Ca} and I_{Si} are the total counts of characteristic X-rays of Ca and Si, respectively, of the standard materials, and roughly correspond to the foil's thickness. Assignment of k-factors by those intensities includes the correction for X-ray absorption (although standard materials $MgSiO_3$ and $FeSiO_3$ would have been better for this purpose, we could not get those materials with sufficient quality). Numbers in parentheses of the coefficients of the above equations denote one estimated standard deviation (E.S.D.) of the least significant digit. In the above equations, k_{SiCa} is not equal to $1/k_{CaSi}$ because the respective k-factors were obtained from the different data sets.

3. Results and discussion

3.1. Ultramicrotomy of ultrahigh pressure materials

With our ultramicrotome, aggregates of submicron ultrahigh pressure materials could be sectioned into ultrathin foils down to 30 nm thickness. However, this was actually the distance over which the diamond knife was advanced stepwise, and the actual thickness of the foils may be slightly larger, considering that the resin block that contained the sample was not sectioned for every stroke of the diamond knife. Fig. 2 shows the electron micrographs of sectioned foils obtained using the ultramicrotome. Some areas of the

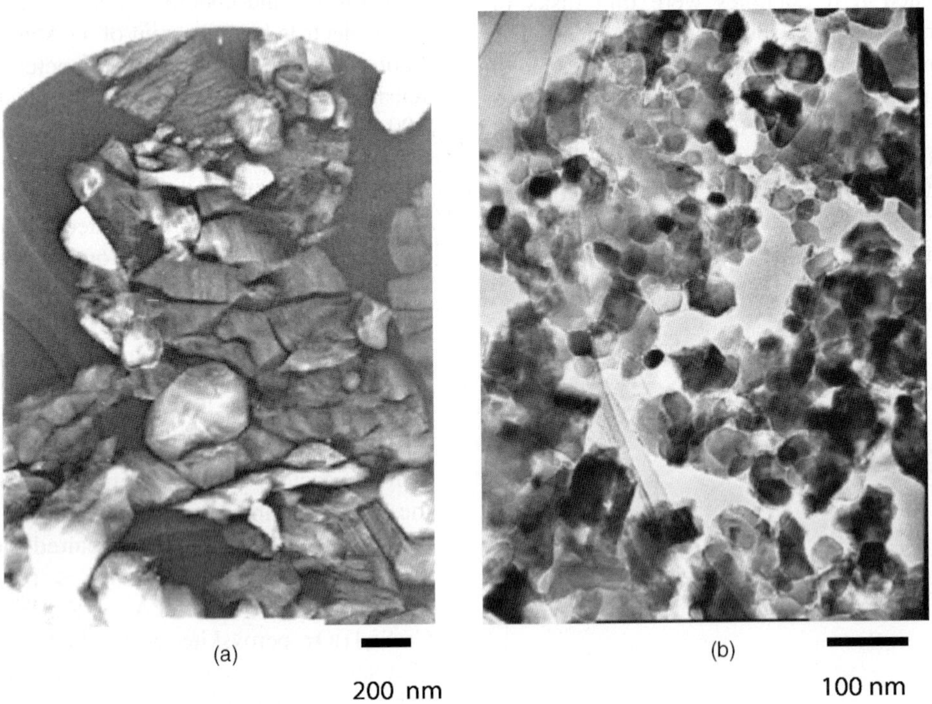

Fig. 2. Electron micrographs of foils prepared by the ultramicrotome. Foils are collected on a grid with a carbon film. (a) Foil of a mutianvil cell sample (sample no. 1814), (b) foil of a laser-heated DAC sample (sample no. A-14).

sectioned foils were often lost. With the multianvil cell samples (Fig. 2(a), No. 1814) the sectioned aggregates are fairly concentrated and coherent, while in the case of the DAC samples (Fig. 2(b), No. A-14) aggregates are rather dispersed. One of the reasons for the dispersion of the DAC samples is that we used NaCl as a pressure medium and the NaCl surrounding the specimen in the sectioned foil dissolved into water in the trough. This problem can be overcome in future by using other materials as pressure medium, which are not soluble in water.

As seen in Fig. 2, the grain sizes of the multianvil cell samples are about 100–500 nm, while those of the DAC samples are much smaller, about 20–100 nm, and both are far below the grain size that can be measured by EPMA. There seem to be some cracks or deformation textures in the multianvil cell samples. Therefore, use of both ion-thinning for structure and texture observations and ultramicrotomy for chemical analyses on the same samples, will be desirable for the preparation of ATEM foils.

3.2. Quantitative chemical analyses of perovskites

For the specimens prepared by ultramicrotomy, quantitative chemical analyses were performed by measuring the characteristic X-ray intensities of the product phases and calculating the compositions using the experimentally obtained *k*-factors. Analyses were quantified in terms of oxides and all iron was treated as Fe^{2+}.

The equations listed in Section 2.3. were used to derive the *k*-factors appropriate for the unknown samples based on the total X-ray counts of Ca or Si of the unknown samples. The estimated E.S.D.s of the obtained *k*-factors, depending on the measured X-ray intensities of Ca or Si, are $\sigma(k_{TiCa}) =\sim 0.005$, $\sigma(k_{SiCa}) =\sim 0.017$ ($I_{Ca} =\sim 2600$), and $\sigma(k_{CaSi}) = 0.19$–0.21, $\sigma(k_{MgSi}) = 0.02$, $\sigma(k_{FeSi}) = 0.15$–0.27 ($I_{Si} = 3000$–6000). Relatively large $\sigma(k_{CaSi})$ and $\sigma(k_{FeSi})$ values seem to originate from the scatter of the measured X-ray intensity ratios of the paired elements in standard materials. Rough estimates of a relative error, $\sigma(C_i/C_j)/(C_i/C_j)$, based on $\sigma(k_{ij})/k_{ij}$, $\sigma(I_i)/I_i$, and $\sigma(I_j)/I_j$ (Williams and Carter, 1996) were 3–5% for $Ca(Si,Ti)O_3$ perovskite (i = Si, Ti, j = Ca) and 3–12% for $(Mg,Fe)SiO_3$ and $(Ca,Mg,Fe)SiO_3$ perovskites (i = Mg, Fe, Ca, j = Si).

Among the experimentally obtained *k*-factors, those for Ca and Fe deviated significantly from the theoretical ones (around 1) installed in the program. Although *k*-factors are not universal but are sensitive to the respective analytical system, so far it is not clear why the *k*-factors of our system deviate so much from the theoretical values.

The calculated compositions, using these *k*-factors, of the perovskite phases in some of the samples of Table 1 are presented in Table 2. These are the averages of several values measured on the same sample, excluding ones that deviated strongly from the stoichiometry of perovskite. The analyzed cation numbers based on three oxygens per formula unit are all close to the ideal chemical formula of perovskite, and the obtained composition of $Ca(Si,Ti)O_3$ perovskite (No. 1803) is nearly the same as that obtained by EPMA (25 mol% of $CaSiO_3$). Therefore, ultramicrotomy,

Table 2
Compositions analyzed by ATEM

Run no.	1803	A-13	B-02	A-13	A-06	C-05
Starting composition	$CaSi_{0.8}Ti_{0.2}O_3$	$Di_{100}Hd_0$	$Di_{75}Hd_{25}$	$Di_{100}Hd_0$	$Di_{50}Hd_{50}$	$Di_{50}Hd_{50}$
Phase	Pv	Mg-Pv	Mg-Pv	Ca-Pv	Ca-Pv	Ca-Pv
Pressure (GPa)	8.3	29.7	31.5	29.7	29.5	78.0
Cation number (O = 3)						
Mg	–	0.958	0.828	0.036	0.092	0.182
Si	0.244	1.021	1.038	1.059	1.040	0.986
Ca	0.998	0.009	0.003	0.847	0.741	0.689
Ti	0.757	–	–	–	–	–
Fe	–	–	0.094	–	0.087	0.158
Total	1.999	1.988	1.963	1.942	1.960	2.015

Pv: $Ca(Si,Ti)O_3$ perovskite, Mg-Pv: $(Mg,Fe)SiO_3$ perovskite, Ca-Pv: $CaSiO_3$ perovskite.

Available online at www.sciencedirect.com

SCIENCE DIRECT®

Physics of the Earth and Planetary Interiors 143–144 (2004) 223–229

PHYSICS OF THE EARTH AND PLANETARY INTERIORS

www.elsevier.com/locate/pepi

Stability of the high-pressure polymorph of zircon ($ZrSiO_4$) in the deep mantle

Yoshinori Tange*, Eiichi Takahashi

Magma Factory, Department of Earth & Planetary Sciences, Tokyo Institute of Technology, 2-12-1 Ookayama Meguroku, Tokyo 152-8551, Japan

Received 26 November 2002; received in revised form 5 July 2003; accepted 16 October 2003

Abstract

The decomposition reaction of $ZrSiO_4$ (scheelite structure) to ZrO_2 (cotunnite structure) and SiO_2 (stishovite) was studied at temperatures between 1500 and 1800 °C in the pressure range 15–25 GPa using a Kawai-type multi-anvil apparatus. In order to ensure the validity of the experimental results, two types of starting materials (synthetic $ZrSiO_4$ and mixture of ZrO_2 and SiO_2) were set in the same furnace assembly and the results were compared in all the experimental runs. Reversed experimental results were obtained for the decomposition reaction. It was revealed that the break down would occur at 20.5 GPa and 1500 °C (equivalent to 600 km deep in the Earth) and the boundary has a positive Clapeyron slope extending to 23.0 GPa at 1800 °C. Our experiments indicate that zircon ($ZrSiO_4$) has limited stability only within the upper mantle.

Keywords: Zircon; $ZrSiO_4$; High-pressure transition; Decomposition; High-pressure experiment

1. Introduction

Although trace in amount, zircon ($ZrSiO_4$) is an important accessory mineral in granite and rhyolite. The mineral is found in virtually all sorts of terrestrial rocks (i.e., basic and intermediate volcanic rocks, metamorphic and sedimentary rocks, and even in mantle peridotites) (e.g., Heaman et al., 1990). The presence of zircon in old Archean rocks is particularly useful because it provides reliable ages based on the U–Pb method (e.g., Bowring et al., 1989). Some of the zircons in old metamorphic rocks show multiple growth histories (often ascribed to different orogenic episodes) when precise ages were measured from core to rim using SIMS (e.g., Vavra et al., 1999). These facts indicate that zircon is stable in a wide range of geologic environments throughout the history of our planet.

Recently, old zircon ages (330–1600 Ma) were reported from the mid-Atlantic Ridge MORBs (Pilot et al., 1998). Considering the young formation ages of MORBs, this zircon must have been derived from old rocks from which the MORB magma was formed. Alternatively, they may have been derived from old continental crust material, which became assimilated in the MORB magma during ascent. Pilot et al. (1998) preferred the former hypothesis and they proposed that zircon survives during global mantle circulation. Accordingly, it is very important to understand the stability of this mineral at very high pressures.

Zircon is known to undergo phase transitions at high-pressure and temperature from experiments using the Bridgman-anvil apparatus (Reid and

* Corresponding author.
E-mail address: tan@geo.titech.ac.jp (Y. Tange).

0031-9201/$ – see front matter © 2004 Elsevier B.V. All rights reserved.
doi:10.1016/j.pepi.2003.10.009

Ringwood, 1969). Reid and Ringwood (1969) reported that zircon transforms to $ZrSiO_4$ (scheelite structure) at ca. 15 GPa and 900 °C. The Natural occurrence of the high-pressure $ZrSiO_4$ polymorph was reported from an upper Eocene impact ejecta (Glass and Liu, 2001). In addition to the phase transition, Liu (1979) found that $ZrSiO_4$ (scheelite structure) decomposed to ZrO_2 (cotunnite structure) plus SiO_2 (stishovite) using a laser-heated diamond anvil cell. Liu (1979) reported that the decomposition occurs at ca. 22 GPa and 1000 °C. However, in a later study, Knittle and Williams (1993) did not observe this decomposition even up to higher pressures of ca. 30 GPa and higher temperatures of ca. 1500 °C using the laser-heated diamond anvil cell. Thus, the high-pressure stability of the $ZrSiO_4$ phase is poorly constrained.

In this report, we determined the decomposition reaction of $ZrSiO_4$ (scheelite structure) using a multi-anvil apparatus. We demonstrate that the phase boundary has a positive Clapeyron slope and lies at 20.5 GPa at 1500 °C and 23.0 GPa at 1800 °C. Our experiments indicate that zircon subducted in sinking oceanic plates becomes unstable and decomposes near the base of the transition zone.

2. Experimental procedures

Synthetic zircon ($ZrSiO_4$) and an oxide mixture (ZrO_2; baddeleyite + SiO_2; quartz) were employed as starting materials in this study. In order to acquire reversed experimental results, these two starting materials were both set into an experimental cell for every run (Fig. 1). The zircon was synthesized from the oxide mixture of ZrO_2 (baddeleyite) and SiO_2 (quartz) at 1 atm and 1400 °C for approximately 250 h. The lattice parameters were calculated from the powder X-ray diffraction data (RINT-2000, Rigaku). The XRD apparatus was calibrated preliminarily using Si powder, but an internal standard was not used in each measurement. The lattice parameters of the synthesized zircon (a = 6.6082 (6) Å, c = 5.9835 (8) Å) are in good agreement with the previous measurement (JCPDS Card No. 6-0224).

High-pressure and high temperature experiments were performed using the Kawai-type multi-anvil apparatus (SPI-1000 and SEDI-1000) installed at the Magma Factory, Tokyo Institute of Technology

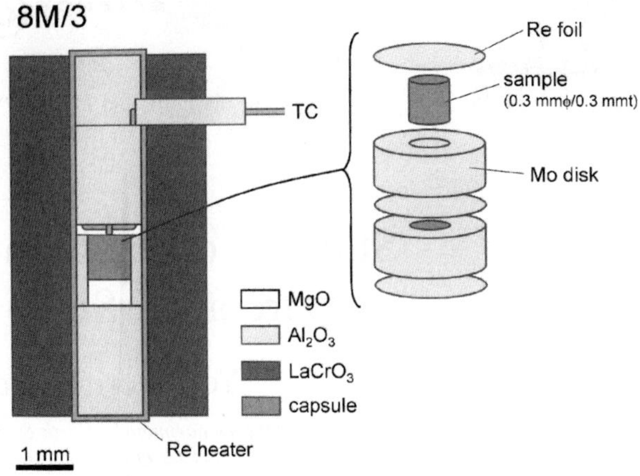

Fig. 1. Schematic cross section of the sample and furnace assembly for the high-pressure and temperature experiments in the 8M/3 type cell-assemblies.

(Takahashi et al., 1993). SPI-1000 was used for the 10M/5 cell-assembly (10.0 mm edge length of the octahedral pressure medium and 5.0 mm truncation edge length of the WC anvil) and SEDI-1000 was used for the 8M/3 cell-assembly. The 10M/5 cell-assembly was adopted for the experiments at 15–20 GPa and the 8M/3 cell-assembly was used in the experiments above 20 GPa. The octahedral pressure medium was semi-sintered MgO containing 5 wt.% Cr_2O_3. The finely powdered samples were packed into the sample chamber (0.3 mm in diameter and 0.3 mm in height) formed by Mo disks and Re foil. The temperature of the sample was monitored using a 0.125 mm diameter W5Re–W26Re thermocouple, the junction of which touched the sample capsule directly. The heater material was $LaCrO_3$ for the 10M/5 and Re for the 8M/3 cell-assembly (see Fig. 1). $LaCrO_3$ was also used as the thermal insulator for the 8M/3 cell-assembly. The pressure calibration by Wang and Takahashi (2000) was applied for the 10M/5 assembly and that by Hirose and Fei (2002) was applied for the 8M/3 assembly.

The heating rate was about 50 °C/min and the experimental temperature was kept at either 1500 or 1800 °C for 2–3 h. These experimental conditions are shown in Table 1. After quenching, the run products were analyzed with a micro-focused X-ray diffractometer (RINT 2000, Rigaku) using Cr Kα radiation collimated to 30 μm diameter.

Table 1
Experimental conditions

Run No.	TEL (mm)	Load (ton)	Pressure (GPa)	Temperature (°C)	Duration (min)	Phase present
S978	5	400	15	1500	180	Sch
S979	5	635	20	1500	180	Sch
C158	3	720	25	1500	180	Cot + St
C160	3	560	22.5	1500	180	Cot + St
C161	3	485	21.3	1500	180	Cot + St
C191	3	410	20	1500	180	Sch
C192	3	560	22.5	1800	120	Sch
C194	3	720	25	1800	120	Cot + St
C196	3	560	22.5	1650	150	Cot + St
C294	3	640	23.8	1800	120	Cot + St

Sch: $ZrSiO_4$ (scheelite structure), Cot: ZrO_2 (cotunnite structure), St: stishovite.

3. Experimental results

Representative X-ray diffraction (XRD) patterns of the low pressure (Run S979; scheelite type $ZrSiO_4$) and that of a high-pressure assemblage (Run C160; consisting of ZrO_2 (cotunnite) and SiO_2 (stishovite)) are shown in Fig. 2a and b. The starting material was synthetic zircon in these experiments. All the X-ray peaks were successfully indexed using JCPDS data. Lattice parameters of present phases are shown in Table 2. As shown in Fig. 2a and b, the low pressure run products and the high-pressure ones are very different and can be readily distinguished. The XRD patterns of selected experiments conducted at 1500 °C are shown as a function of pressure in Fig. 3a and b. The results from experiments using synthetic $ZrSiO_4$ as starting material are shown in Fig. 3a and those using the oxide mixture starting material ($ZrO_2 + SiO_2$) are shown in Fig. 3b.

Table 2
Lattice parameters of synthesized phases

a (Å)	*b* (Å)	*c* (Å)	References
$ZrSiO_4$ (scheelite structure)-tetragonal			
4.734 (2)		10.481 (3)	this study
4.730		1.0480	Reid and Ringwood (1969)
4.712 (2)		10.450 (10)	Liu (1979)
4.734 (1)		10.511 (1)	Kusaba et al. (1985)
4.726 (12)		10.515 (44)	Knittle and Williams (1993)
ZrO_2 (cotunnite structure)-orthorhombic			
3.330 (1)	5.597 (3)	6.473 (5)	this study
3.328 (1)	5.565 (2)	6.503 (3)	Liu (1979)
SiO_2 (stishovite)-tetragonal			
4.173 (5)		2.672 (2)	this study

Our reconnaissance experiments reveal that high-pressure phase transformations of the starting materials (zircon to scheelite structure in $ZrSiO_4$, quartz to stishovite in SiO_2, baddeleyite to cotunnite in ZrO_2) take place in less than 30 min at 1500 °C. Therefore, the starting material, consisting of the low-pressure phases, must have transformed to the high-pressure phases in the beginning of each experiment (with run durations of more than 2 h). We obtained the same products through the different reaction paths in each experiment using the two different starting materials. The experimental results in Figs. 2 and 3, therefore, can be regarded as sort of reversal experiments between the high-pressure phases $ZrSiO_4$ (scheelite structure) and ZrO_2 (cotunnite structure) + SiO_2 (stishovite).

As shown in Fig. 3a and b, all the 1500 °C run products quenched at pressures less than 20 GPa are scheelite-type $ZrSiO_4$ regardless of the starting materials. On the contrary, the run products quenched at pressures higher than 21.3 GPa consist of ZrO_2 (cotunnite structure) + SiO_2 (stishovite). It is thus concluded that the reversed phase boundary for the decomposition reaction lies at 20.6 (6) GPa at 1500 °C. Likewise, the phase boundary is determined to be at 23.2 (5) GPa at 1800 °C (Fig. 4).

Our new experimental result is consistent with the earlier experiments by Liu (1979) but it is contradictory to the study of Knittle and Williams (1993). According to Knittle and Williams (1993), their DAC

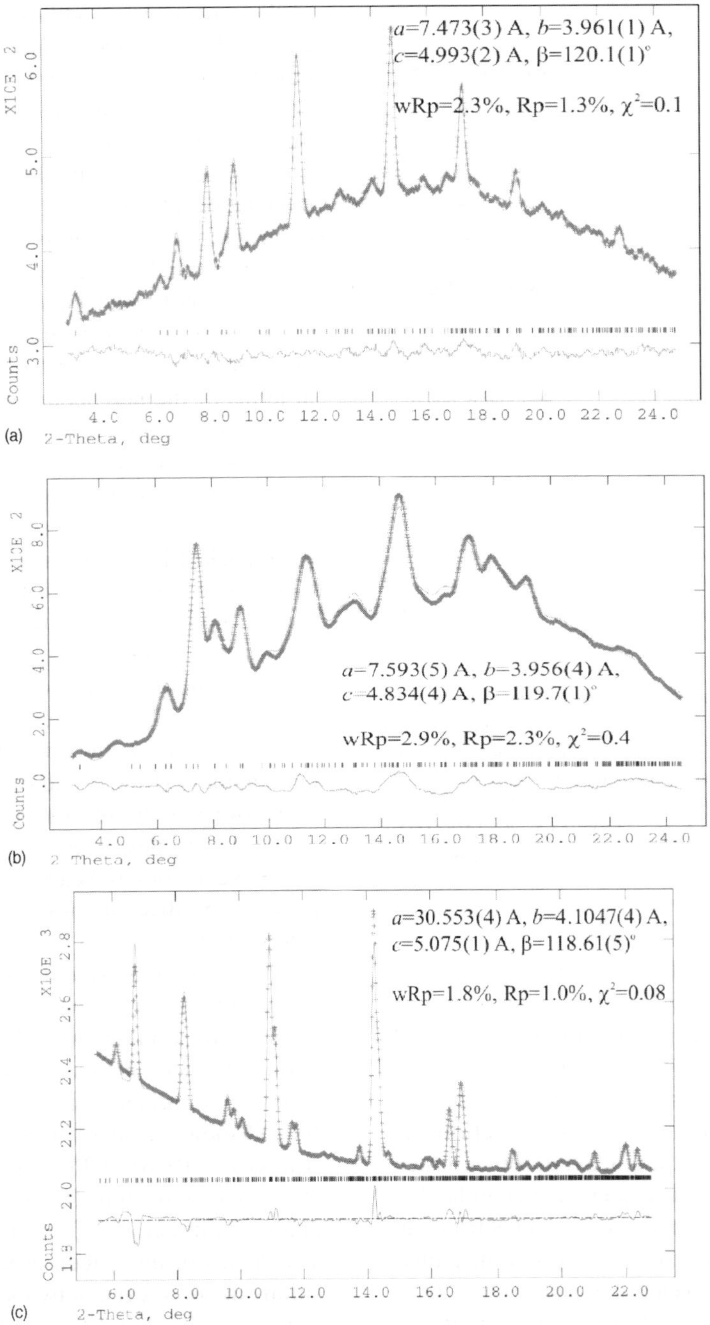

Fig. 6. Profile-fitted X-ray diffraction data of different silica phases obtained: (a) at 34(1) GPa (starting phase α-cristobalite), (b) at 44(2) GPa (starting phase α-quartz), and (c) at ambient conditions after treatment of initial α-cristobalite sample at 112(3) GPa and 1050(30) K during 8 h. The GSAS program package (Larson and Von Dreele, 1994) was used in the LeBail fit. Structural parameters available from authors. Refinement done assuming *P*1 space group. Monoclinic lattice is fixed during the fitting of the atomic coordinates. "Pseudo-monoclinic" symmetry of atomic coordinates (space group $P2_1/c$) derived from lattice dynamics simulations is preserved. Thermal parameters of all atoms are fixed on the value 0.025.

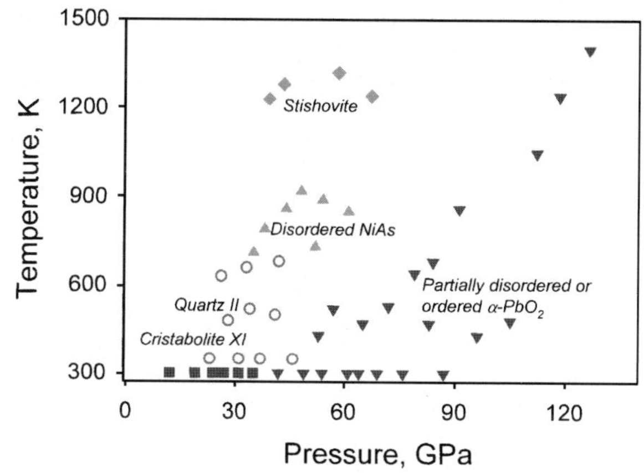

Fig. 7. Phases observed in electrically-heated DAC experiments with α-cristobalite (filled symbols) and α-quartz (open symbols) as starting materials. All partially ordered phases with silicon in tetrahedral and octahedral coordination produced from α-quartz combined under name "Quartz II" (open circles). All phases with silicon atoms partially or fully ordered in octahedral voids in *hcp* oxygen lattice designated as "partially disordered or ordered a-PbO_2" (inverse triangles). Stishovite: diamonds, disordered NiAs structure: triangles, Cristobalite X–I: squares.

temperature range different phases were synthesised (Hemley et al., 1996; Tsuchida and Yagi, 1990; Gratz et al., 1993; Yamakata and Yagi, 1997; Andrault et al., 1998; Dubrovinsky et al., 1997, 2001; Liu et al., 1978; Prokopenko et al., 2001). Silica phases with tetrahedrally co-ordinated silicon demonstrate a significant diversity in densities–from 1.77 g/cm^3 in open framework zeolite-like structures to 2.90 g/cm^3 of coesite (Heaney et al., 1996), while high-pressure modifications of silicon dioxide with hexagonal close packing of oxygen and octahedrally co-ordinated silicon have practically the same densities (~4.28 g/cm^3) at ambient conditions (quenched metastable phases), and very close densities at a megabar pressure range (difference just about 1–1.5% from $CaCl_2$- to α-PbO_2-type and disordered SiO_2 phases).

5. Conclusions

We found that starting from different low-pressure silica polymorphs (quartz, trydimite, cristobalite) depending on history and conditions of treatment in electrically-heated DAC different metastable and stable phases can be synthesed (Fig. 7). All these phases, as well as known high-pressure silica modifications (stishovite, fully disordered phase with niccolite-type structure, α-PbO_2-type, and monoclinic ($P2_1/c$ space group) post-quartz phase) belong to the same class of silica phases with different degree of the ordering of silicon atoms within (distorted) hexagonal close packing of oxygen atoms. The details of the silicon distribution among free positions in the oxygen network strongly depend on the starting material and the conditions of high-pressure high-temperature treatment of silica, but a general trend is that at pressures above 40–45 GPa most of the silicon atoms are placed in octahedral positions.

Acknowledgements

Comments by G. Shen and T. Kondo allowed improve manuscript. Work was supported by Deutsche Forschungsgemeinschaft.

References

Andrault, D., Fiquet, G., Guyot, F., Hanfland, M., 1998. Pressure-induced Landau-type transition in stishovite. Science 282, 720–724.

Dmitriev, V.P., Toledano, P., Torgashev, V.I., Salje, E.K.H., 1998. Theory of reconstructive phase transitions between SiO_2 polymorphs. Phys. Rev. B 58, 11911–11921.

Dubrovinskaia, N.A., Dubrovinsky, L.S., Ahuja, R., Dmitriev, V., Weber, H.P., 2001. Experimental and theoretical identification of new high-pressure TiO_2 polymorph. Phys. Rev. Lett. 87, 5501–5504.

Dubrovinsky, L.S., Saxena, S.K., Lazor, P., Ahuja, R., Eriksson, O., Wills, J.M., Johansson, B., 1997. Experimental and theoretical identification of a new high-pressure phase of silica. Nature 388, 362–365.

Dubrovinsky, L.S., Saxena, S.K., Ahuja, R., Johansson, B., 1998. Theoretical study of the stability of $MgSiO_3$-perovskite in the deep mantle. Geophys. Res. Lett. 25, 4253–4256.

Dubrovinsky, L.S., Saxena, S.K., Tutti, F., Le Bihan, T., 2000. X-ray study of thermal expansion and phase transition of iron at multimegabar pressure. Phys. Rev. Lett. 84, 1720–1723.

Dubrovinsky, L.S., Dubrovinskaia, N.A., Saxena, S.K., Tutti, F., Rekhi, S., Le Bihan, T., Shen, G., Hu, J., 2001. Pressure-induced transformations of cristobalite. Chem. Phys. Lett. 333, 264–270.

El Goresy, A., Dubrovinsky, L., Sharp, T.G., Saxena, S.K., Chen, M., 2000. A new monoclinic post-stishovite polymorph of silica in the SNC meteorite Shergotty. Science 288, 1632–1634.

Giacovazzo, C., 1992. Fundamentals of Crystallography. Oxford University Press, New York, 654 pp.

Gratz, A.J., DeLoach, L.D., Clough, T.M., Nellis, W.J., 1993. Shock amorphization of cristobalite. Science 259, 663–666.

Haines, J., Leger, J.M., Gorelli, F., Hanfland, M., 2001. Crystalline post-quartz phase in silica at high pressure. Phys. Rev. Lett. 87, 155503.

Hemley, R.J., Prewitt, C.T., Kingma, K.J., 1996. High pressure behaviour of silica. Rev. Miner. 29, 41–81.

Heaney, P.J., Prewitt, C.T., Gibbs, G.V. (Eds.), 1996. Silica. Rev. Miner. 29, 606.

Kingma, K.J., Cohen, R.E., Hemley, R.J., Mao, H.K., 1995. Transformation of stishovite to a denser phase at lower-mantle pressures. Nature 374, 243–245.

Larson, A.C., Von Dreele, R.B., 1994. Los Alamos National Laboratory (LAUR), p. 86.

Liu, L., Basset, W.A., Sharry, J., 1978. New high-pressure modifications of GeO_2 and SiO_2. J. Geophys. Res. 83, 2301–2305.

Teter, D.M., Hemley, R.J., Kresse, G., Hafner, J., 1998. High pressure polymorphism in silica. Phys. Rev. Lett. 80, 2145–2148.

Tsuchida, Y., Yagi, T., 1990. New pressure-induced transformations of silica at room temperature. Nature 347, 267–269.

Palmer, D.C., Hemley, L.W., Prewitt, C.T., 1994. Raman spectroscopic study of high-pressure phase transitions in cristobalite. Phys. Chem. Miner. 21, 481–488.

Prokopenko, V.B., Dubrovinsky, L.S., Dmitriev, V., Weber, H.-P., 2001. Raman spectroscopy and X-ray diffraction in situ characterization of phase transitions in cristobalite under high pressure. J. Alloys Comp. 327, 87–95.

Sharp, T.G., El Goresy, A., Wopenka, B., Chen, M., 1999. A post-stishovite SiO_2 polymorph in Shergotty: implication for impact events. Science 284, 1511–1513.

Sekine, T., Akaishi, M., Setaka, N., 1987. Fe_2N-type SiO_2 from shocked quartz. Geochem. Cosmochem. Acta 51, 379–381.

Talyzin, A.V., Dubrovinsky, L.S., Le Bihan, T., Jansson, U., 2002. In situ Raman study of C_{60} polymerization at high pressure high temperature conditions. J. Phys. Chem. 116, 2166–2174.

Vainshtein, B.K., Fridkin, V.M., Indenbom, V.L., 1982. Modern Crystallography, vol. II. Springer, Berlin, 340 pp.

Urusov, V.S., Dubrovinskaia, N.A., Dubrovinsky, L.S., 1990. Construction of Possible Crystal Structures (in Russian). MGU, Moscow, 160 pp.

Yagi, T., Yamakata, M., 2000. Effect of hydrostaticity on the phase transformations of cristobalite. In: Aoki, H., et al. (Eds.), Physics Meets Mineralogy. Cambridge University Press, pp. 242–255.

Yamakata, M., Yagi, T., 1997. New stishovite-like phase of silica formed by hydrostatic compression of cristobalite. Proc. Jpn. Acad. 73B, 6, 85–88.

Available online at www.sciencedirect.com

SCIENCE DIRECT

Physics of the Earth and Planetary Interiors 143–144 (2004) 241–253

PHYSICS OF THE EARTH AND PLANETARY INTERIORS

www.elsevier.com/locate/pepi

Limits to resolution in composition and density in ultra high-pressure experiments on natural mantle-rock samples

Kanani K. M. Lee [a,*], Bridget O'Neill [a,b], Raymond Jeanloz [a]

[a] *Department of Earth and Planetary Science, University of California, Berkeley, 307 McCone Hall, Berkeley, CA 94720-4767, USA*
[b] *Lowell Observatory/NPOI, 1400 W. Mars Hill Road, Flagstaff, AZ 86001, USA*

Received 24 February 2003; received in revised form 12 June 2003; accepted 24 June 2003

Abstract

We compressed and analyzed an undepleted natural peridotite—thought to be representative of the Earth's upper mantle—from ~25 to 110 GPa and heated to 2000–2500 K. Scatter in zero-pressure unit-cell volumes observed for high-pressure phases synthesized from peridotite, of ±0.4% for orthorhombic perovskite and ±0.8% for magnesiowüstite, can be ascribed to compositional heterogeneity in the natural starting material and its transformation products (calcium perovskite was also observed at high pressures, but does not quench to zero pressure). The scatter is larger than the measurement uncertainties, which are as little as ±0.1%, but is less than would be expected from the variation in composition among distinct minerals if the starting material had failed to equilibrate upon transformation to the high-pressure assemblage (e.g., up to ±0.6–0.8% in volume for orthorhombic perovskite). The equilibrium unit-cell volume of orthorhombic perovskite calculated from an average of electron microprobe analyses on the quenched samples, 163.3 (±0.6) $Å^3$, agrees to within 0.4% with the average value measured from the same samples by X-ray diffraction at zero pressure (163.9 ± 0.7 $Å^3$). Similarly, values of zero-pressure density for the whole-rock assemblage of high-pressure phases are identical (4.13 ± 0.03 g/cm^3), whether derived from electron microprobe or X-ray diffraction measurements.

Keywords: Peridotite; Electron microscopy; Chemical analysis; High pressure; Diamond-anvil cells; X-ray diffraction analysis

1. Introduction

There is a great need for experiments that reproduce the conditions of high pressures and temperatures existing deep inside planets. In particular, it is important to characterize the properties of relevant materials at deep-interior conditions in order to determine the constitution, state and evolution of the Earth and planets over geological time. Thus, it is only through high pressure-temperature studies that seismological and other geophysical observations can be interpreted in terms of the identities (e.g., crystal structures), abundances, compositions and other properties of the mineral phases of the Earth's deep interior. It is therefore important to quantify the reliability and resolution of the measurements obtained from high-pressure experiments so that one can determine the significance of models derived from the combined measurements and geophysical observations.

* Corresponding author. Present address: Division of Geological and Planetary Sciences, California Institute of Technology, MC 170-25, Pasadena, CA 91125, USA. Tel.: +1-510-643-8325; fax: +1-510-643-9980.
E-mail addresses: kanani@gps.caltech.edu (K.K.M. Lee), boneill@sextans.lowell.edu (B. O'Neill), jeanloz@uclink.berkeley.edu (R. Jeanloz).

0031-9201/$ – see front matter © 2004 Published by Elsevier B.V.
doi:10.1016/j.pepi.2003.06.002

The quality of such experiments is often described in terms of (i) the range and calibration of the pressures and temperatures that are achieved, and (ii) the quality of the measurements performed at high pressures. With the advent of the laser-heated diamond cell, it is now possible to reproduce essentially the full range of pressures and temperatures existing inside the Earth, and calibration of experimental conditions has received considerable attention to date (Jeanloz and Kavner, 1996; Knittle and Jeanloz, 1987; Mao et al., 1978; Shim et al., 2002). Similarly, the use of synchrotron- and laser-based probes, among other tools, has led to major advances in the quality of measurements that can be performed on samples at high pressures.

The role of the starting material has received relatively little attention, however, as a third source of uncertainty affecting the quality of high-pressure measurements. Indeed, the sample material is often chosen so as to minimize any possible degradation of the measurements. It is for this reason that most ultra high-pressure (~100 GPa) studies to date have focused on the properties of single-phase and, in many cases, synthetic starting materials (e.g., Andrault, 2001; Fiquet et al., 1998; Knittle and Jeanloz, 1987). As the ultimate objective is to determine the properties of planetary materials at deep-interior conditions, however, it also becomes necessary to examine natural samples. This is so if for no other reason than to evaluate the effects of minor elements or the presence of minor (initial) phases on the final results. In addition, certain materials are intrinsically difficult to synthesize as truly homogeneous samples, with the most ideally uniform starting material being amorphous.

Ultramafic rocks, characteristic of the Earth's upper mantle and of chondritic meteorites, offer a good example of such difficulties. Bulk compositions corresponding to peridotite and similar rocks are refractory, and cannot be readily quenched to a glass from the melt. Even if a glass is formed on cooling (e.g., by splat quenching), melting temperatures are sufficiently high that moderately volatile elements such as Na and K are often lost during preparation of the starting sample. Sol–gel techniques can produce an amorphous and relatively uniform starting material of peridotite composition, but invariably lead to contamination, especially by hydrogen. Finally, mechanical comminution (e.g., by ball milling) can produce a uniform, fine-grained and possibly amorphous sample, but with the potential of contamination from the grinding surfaces.

Consequently, it is difficult to obtain pure (uncontaminated) and uniform starting materials having the bulk composition of terrestrial mantles. Sample uniformity is especially critical for the highest-pressure experiments: these rely on small (~μg) samples being compressed inside a diamond-anvil cell, in contrast with the larger (~mg) samples that are compressed (but to lower peak pressures) in multi-anvil devices. For the latter, electron microprobe and other high-quality analytical tools can be used to confirm that individual mineral grains formed at high pressures have uniform and reproducible compositions. In comparison, it is much more difficult to document homogeneity—and therefore a close approach to equilibrium—for individual mineral phases produced in the laser-heated diamond cell.

The objective of the present work is to quantify the resolution in molar volume, composition and density of the high-pressure mineral-phase assemblage formed from a natural sample that is thought to have a bulk composition close to that of the Earth's mantle. In combination with seismological observations, these properties are important for modeling the constitution and state of the deep interior. Our focus here is on zero-pressure measurements, as these can reveal much about the reliability of the data: for example, beginning to quantify the degree to which the sample achieves (at least local) thermodynamic equilibrium, as well as the accuracy of such relevant properties as phase-composition and assemblage-density.

2. Lower-mantle assemblage formed from natural peridotite

We address these issues, including the role of the initial sample in producing equilibrated high-pressure phases, by describing the high-pressure assemblage formed from a natural peridotite at lower-mantle pressures (~25–110 GPa). This is but one example of the rock-types present in the earth's mantle, but is of special interest because it is thought to correspond to the bulk composition of at least the upper mantle (e.g., Ringwood, 1975). The starting material (BN-35), with bulk composition and mineral constituents given in Table 1 (Stern et al., 1986), was transformed in two

Table 1
Composition of BN-35 and its individual phases in wt.% (Stern et al., 1986)

	BN-35 whole-rock composition[a]	Olivine 45 vol.% 43 mol%	Orthopyroxene 25 vol.% 33 mol%	Clinopyroxene 15 vol.% 19 mol%	Garnet 15 vol.% 5 mol%
SiO_2	45.40 (0.26)	34.08	56.89	53.31	43.94
TiO_2	0.25 (0.06)	0.29	0.10	0.45	0.16
Al_2O_3	4.30 (0.69)	4.02	3.49	5.73	23.19
Cr_2O_3	0.30 (0.02)	0	0.35	0.83	1.16
FeO[b]	8.30 (0.53)	10.97	7.63	3.13	8.06
MnO	0.14 (0.01)	0.24	0.08	0.04	0.24
MgO	37.10 (0.93)	52.57	30.99	15.43	19.88
CaO	3.30 (0.13)	0	0.76	20.11	4.97
Na_2O	0.30 (0.03)	0.19	–	1.37	–
K_2O	0.0 (–)	–	–	–	–
NiO	0.24 (–)	–	–	–	–
Total	99.63	102.36	100.29	100.4	101.6
x_{Fe} = Fe/(Fe+Mg)	0.11 (0.02)	0.10	0.12	0.10	0.19
Mg# = Mg/(Mg+Fe)	0.89 (0.02)	0.90	0.88	0.90	0.81

[a] Uncertainties are given in parentheses and are determined by taking the difference in element abundances from an independent measurement of the starting sample (O'Neill, 1994).

[b] Fe^{2+} and Fe^{3+} totals are combined.

sets of experiments using laser-heated diamond cells (diamond culet size ranging from 100 to 350 μm) using pre-compressed (to ~20–40 μm thick) Re or Fe gaskets. In all cases, pressures were determined before and after laser heating using the ruby-fluorescence scale (Mao et al., 1978), with only the values after heating being quoted here. In both sets of experiments, the sample was laser heated by rastering it under the focused beam (hot-spot diameter ~20 μm) of a continuous (cw) TEM_{00}-mode Nd-YAG (λ = 1064 nm) laser, keeping clear of the very edges of the gasket and ruby calibrant to minimize the possibility of contamination. Heating duration at any one spot was short (seconds) at any particular time, in order to minimize the Soret diffusion of Fe out of the hot spot, with total heating of the entire sample taking approximately 10–30 min (both for synthesis and annealing). The laser power was kept sufficiently low to avoid melting, and we observed no evidence of melting either during heating or through analysis of the sample after heating.

Of the two sets of experiments, each run in the first set consisted of compressing a fresh sample of peridotite to a given pressure, laser-heating the sample to temperatures of 2000–2500 K, and then obtaining X-ray diffraction patterns at room temperature before and after decompressing the sample to zero pressure (O'Neill, 1994; O'Neill and Jeanloz, 1990). All of these diffraction patterns were collected at room temperature, using monochromatic Mo Kα (λ = 0.7107 Å) radiation from a rotating-anode source at high pressures and filtered Cu Kα (λ = 1.5418 Å) radiation from a standard tube at zero pressure (after the sample was removed from the diamond cell). In both the cases, the beam size was comparable to the diameter of the sample, ~100–150 μm. The diffraction patterns were collected on film that was digitized and analyzed as described elsewhere (Nguyen and Jeanloz, 1993), and with fitting of individual peaks to determine constituent phase unit-cell volumes (O'Neill, 1994).

In the second set of experiments, the sample was loaded in an Ar medium and laser heated to 2000–2500 K at pressures of 30–52 GPa in order to convert the peridotite to the high-pressure assemblage. X-ray diffraction patterns were then collected at several pressures, on compression to peak pressures of 47–87 GPa and decompression to zero pressure (sample outside the diamond cell), with laser-heating used to further anneal the samples at 41, 47, 66, 78 and 87 GPa. All of the diffraction patterns for the second set of experiments were collected at room temperature using synchrotron radiation and image plates, with the size of the X-ray beam ranging from

60 μm (i.e., much smaller than the sample) for the high-pressure runs to more than 100 μm diameter upon quench (Fig. 1A and B). Data were collected at Stanford Synchrotron Radiation Laboratory (SSRL) beamline 10-2 and Advanced Photon Source (APS) beamline 13-IDD, and were analyzed both through individual fitting of peaks (as in the first set of experiments) and by Rietveld analysis (Larson and Von Dreele, 2000; Toby, 2001) (Fig. 1B).

In both sets of experiments, the samples transformed to assemblages of 64 (±2) mol% (Mg, Fe, Al)SiO_3 orthorhombic perovskite (opv), 31 (±2) mol% (Mg, Fe)O magnesiowüstite (mw) and 5 (±1) mol% $CaSiO_3$ perovskite (cpv), where the relative proportions were determined from the bulk composition of the starting material (see Table 1) and the compositions of the high-pressure mineral phases deduced from the lattice parameters obtained from the X-ray patterns (Jeanloz and Thompson, 1983; O'Neill, 1994; O'Neill and Jeanloz, 1990). These results are compatible with the mineral-phase proportions obtained in the Rietveld fits, so can be considered independently verified.

Complete and essentially smooth powder-diffraction lines were obtained, with some amount of spottiness (from grain growth upon transformation) observed in

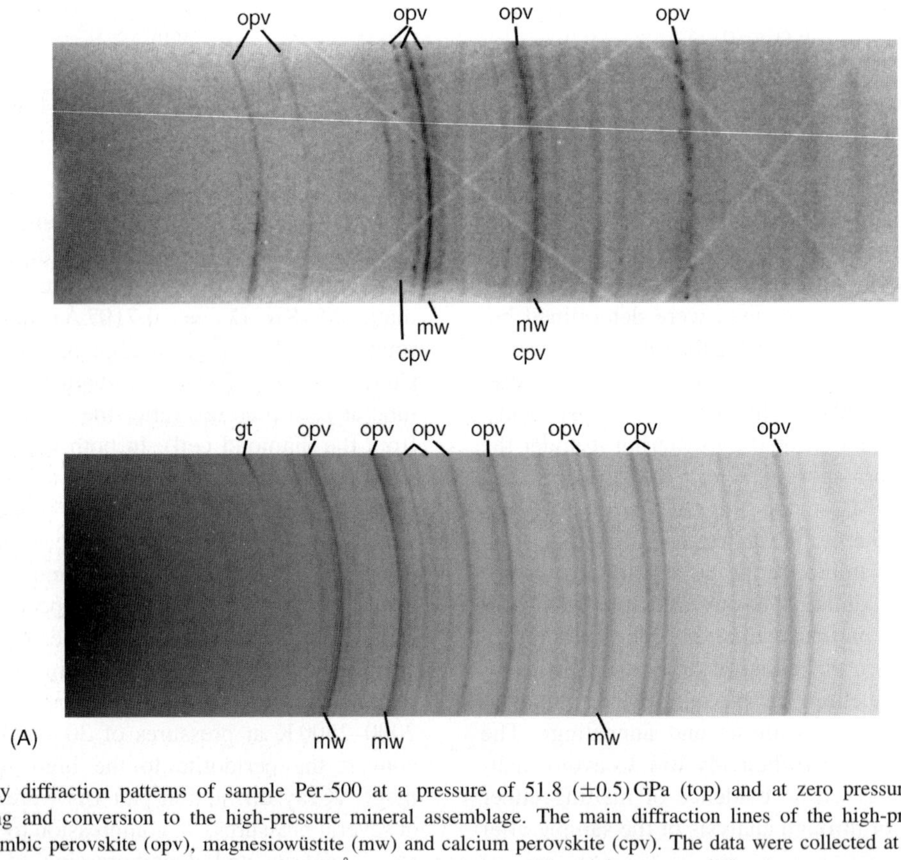

Fig. 1. (A) X-ray diffraction patterns of sample Per_500 at a pressure of 51.8 (±0.5) GPa (top) and at zero pressure (bottom), taken after laser heating and conversion to the high-pressure mineral assemblage. The main diffraction lines of the high-pressure phases are labeled: orthorhombic perovskite (opv), magnesiowüstite (mw) and calcium perovskite (cpv). The data were collected at room temperature at SSRL, using monochromatic radiation ($\lambda = 0.7277$ Å) and image plates. The high-pressure diffraction pattern was obtained using a 60 μm diameter X-ray beam, and the zero-pressure pattern was obtained using a 150 μm diameter beam. Only half the diffraction pattern is shown in each case, and the cross-cutting diagonal lines in the upper pattern are ascribed to Kossel fringes (Lang, 1998) from the diamond anvil (the pattern at zero pressure is of only the sample, without the diamonds). (B) Rietveld refined X-ray diffraction patterns at 51.8 (±0.5) GPa (top) and quenched (bottom) shown in A. Dots are the actual data while the line is the calculated Rietveld fit. Phase diffraction-line positions shown as ticks below each pattern, in descending order: orthorhombic perovskite, magnesiowüstite, and calcium perovskite (or garnet (gt) in quenched pattern). Garnet is seen only in the quenched patterns (and only at the ~0.1 wt.% level) as the X-ray beam probes a larger portion of the sample, including the edges that were not heated and are thus unconverted.

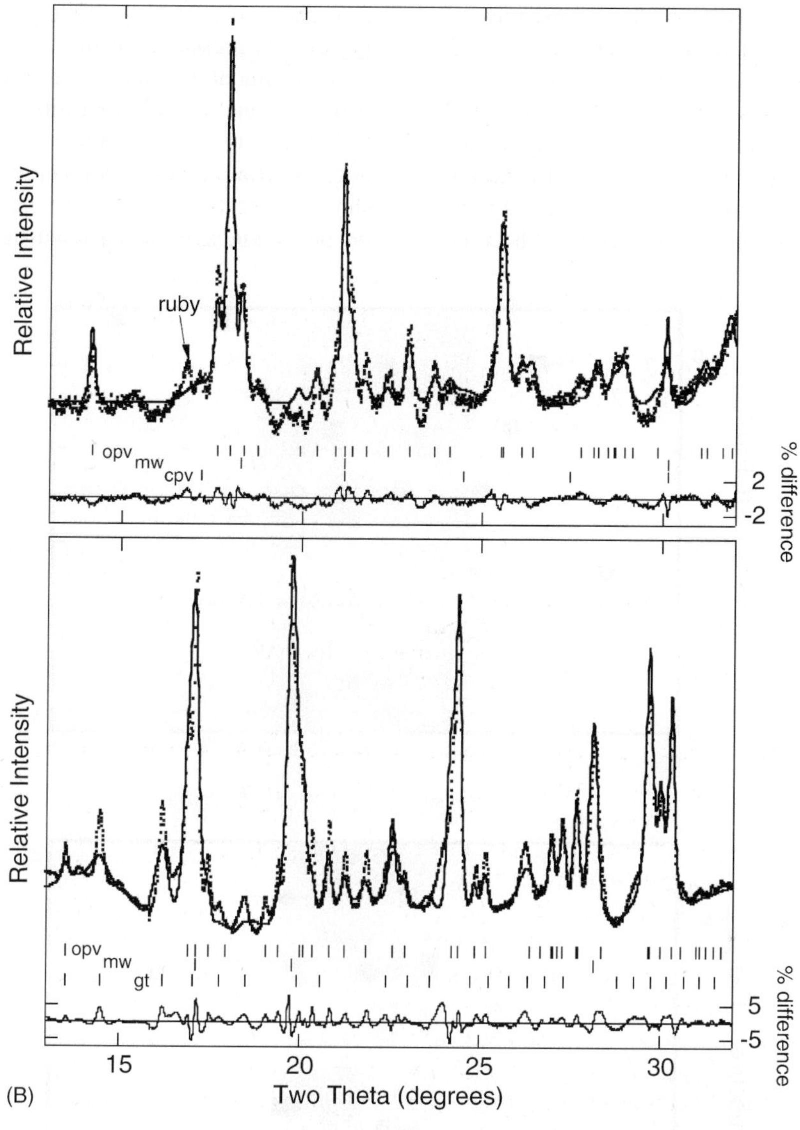

Fig. 1. (*Continued*).

patterns taken with higher spatial resolution (Fig. 1A). The character of the diffraction rings is in line with observations by optical microscopy, which indicate the presence of occasional crystals as large as 2–5 μm but the bulk of the transformed sample having crystal sizes under 1–2 μm. The important point is that the individual crystals were much smaller than the laser-heated spot, both before and after transformation.

Overall, the quality of the diffraction patterns indicates good conversion to the high-pressure assemblage. Except for minor amounts of garnet remaining after heating at pressures below 30–35 GPa, none of the starting material was evident in the diffraction patterns from the heated portions of each sample. Of course, starting material was present in the unheated portions of the samples (e.g., next to the gasket), as we were able to distinguish by comparing diffraction patterns obtained using a small-diameter X-ray beam taken at high pressures with large-diameter beam patterns taken at zero pressure (Fig. 1). The presence of garnet in the heated portions of the lowest-pressure runs is consistent with phase equilibria determined in independent studies (e.g., Kubo and Akaogi, 2000) and with other methods, so does not indicate an inability to transform the entire sample or any other indication of deviation from phase-equilibrium. Because of the small amount of calcium perovskite present in the high-pressure assemblage, and the fact that it becomes amorphous (so has no diffraction lines) on quenching to zero pressure, we limit our present discussion of the diffraction patterns to results for orthorhombic perovskite and magnesiowüstite.

The quality of data obtained from the natural peridotite starting material is quantified in Fig. 2, which shows the scatter in unit-cell volumes for orthorhombic perovskite and magnesiowüstite being about ±1%

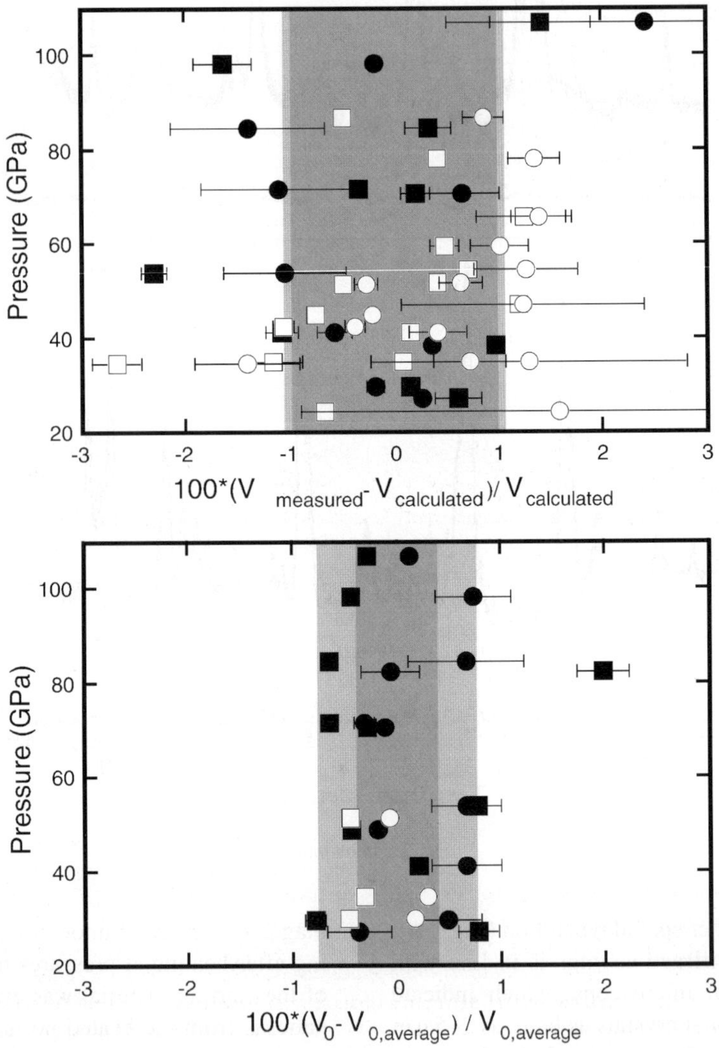

Fig. 2. Percent deviation in unit-cell volumes (V) for orthorhombic perovskite (circles) and magnesiowüstite (squares), shown as a function of pressure at high pressure (top) and as a function of synthesis pressure at zero pressure (bottom). Open and closed symbols designate samples loaded with and without an Ar pressure medium, respectively, and error bars indicate measurement uncertainties when these exceed the size of the symbol. Deviations are calculated from our measured equations of state at high pressures (Lee et al., 2003; O'Neill, 1994), and from weighted-mean values of 163.90 (±0.67) Å^3 for opv and 76.09 (±0.60) Å^3 for mw at zero pressure. Standard deviations reflecting the observed scatter are shown by dark shading for orthorhombic perovskite and light shading for magnesiowüstite.

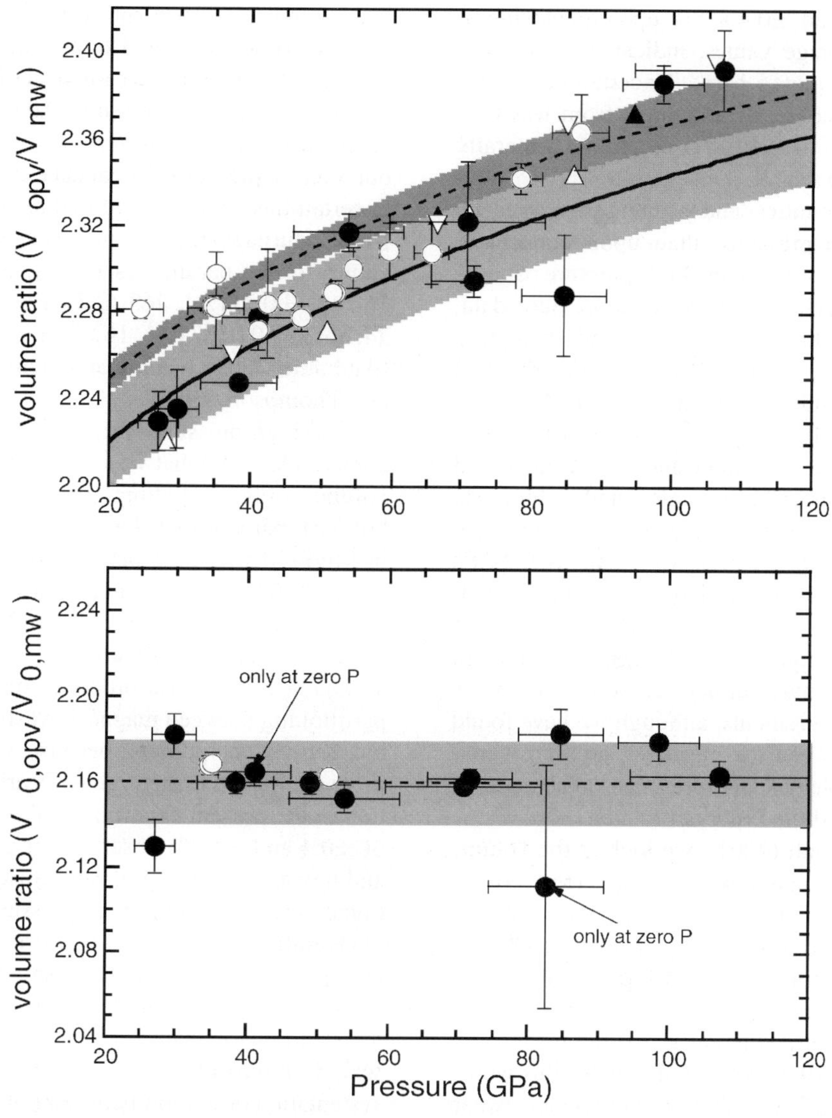

Fig. 3. Volume ratio of orthorhombic perovskite to magnesiowüstite V_{opv}/V_{mw} as a function of pressure at high pressure (top) and as a function of synthesis pressure at zero pressure (bottom). Top: circles are our data, with argon as a pressure medium (open circles) and without argon (closed circles). To compare we also plot data from Table 2 from Andrault's (2001) high-pressure study of San Carlos olivine (open triangles), alumina + San Carlos olivine (closed triangles) and a San Carlos olivine equilibrated at oxygen fugacity f_{O_2} of 8.6×10^{-10} atm (inverted open triangles). The dashed curve shows the predicted volume ratio for pure Mg endmembers, V_{MgSiO_3}/V_{MgO} (dark band indicates uncertainties), whereas the solid curve shows the volume ratio as a function of pressure derived from our equations of state and assuming no effect of pressure on the Mg/Fe distribution ratio between opv and mw (light band indicates uncertainties; at the highest pressures uncertainties are overlaid by the uncertainties of the pure Mg endmembers (dark band)). Bottom: at zero pressure, the average value observed for $V_{0,opv}/V_{0,mw}$ is 2.16 ($\pm$0.02) (dashed line with shaded band indicating uncertainty), and the individual values show no trends with synthesis pressure. Samples for which measurements were only taken at zero pressure (after synthesis at high pressure) are marked. Error bars indicate measurement uncertainties.

at high pressures and ±0.4–0.8% upon quenching to zero pressure (average values, indicated by shading in the figure, are quoted here). Uncertainties in the volumes (individual error bars) range from less than ±0.1% to as much as ±1.0–2.5%, and are generally much less than the scatter.

Both the uncertainties and scatter are larger by ∼25–65% at high pressures than upon quenching. This is partly due to some of the high-pressure patterns being of lower quality than the zero-pressure data, mainly because of the effects of diamond absorption (because the zero-pressure patterns were obtained from samples outside the diamond cell). Additionally, overlapping diffraction lines and uncertainties in the sample pressure also contribute to the increased scatter between observed and calculated volumes at high pressures. In particular, measured variations in pressure across a sample were as small as 0.2 GPa and as large as 12.6 GPa, which corresponds to volume uncertainties of ±0.2–5.0% (calculated using a Birch–Murnaghan equation of state (Birch, 1978)). Non-hydrostaticity may further degrade some of the high-pressure measurements, although we have found no correlation between the measured pressure variations across samples (or the presence or absence of an Ar medium) and the observed scatter.

Following Andrault (2001), we look at the volume ratio between opv and mw, V_{opv}/V_{mw} (Fig. 3). At zero pressure we find an average $V_{0,opv}/V_{0,mw}$ of 2.16 (±0.02) without any noticeable dependence on synthesis pressure, and the results in Fig. 3 (bottom) are merely another representation of the data shown in Fig. 2 (bottom). Following early partitioning experiments (Bell et al., 1979; Yagi et al., 1979), this would suggest that the Mg/Fe distribution coefficient between opv and mw does not depend on pressure, and is otherwise not affected by synthesis conditions over the range covered in the present study.

At high pressure, V_{opv}/V_{mw} is not constant due to the differences in the compressibilities of opv and mw. As Andrault points out (Andrault, 2001), any deviations from the predicted Mg-endmember volume ratio (dashed curve and dark shading in Fig. 3, top) may be attributable to a change in partitioning behavior with pressure. His analysis, however, does not include any effects of Al upon the equation of state of opv nor the uncertainties in the respective equations of state. When we do take these factors into account, we find that within the uncertainties of the known equations of state (solid curve and light shading in Fig. 3, top (Lee et al., 2003)) we cannot confirm any changes of partitioning with pressure. Additionally, the measurements collected at zero pressure appear to rule out such a pressure dependence, at least within our uncertainties. For the Mg-endmembers, we use the Birch–Murnaghan equation of state with the following values for the parameters V_0, K_{0T} and K_{0T}': $MgSiO_3$ 162.47 (±0.31) Å^3, 263 (±3) GPa, 3.9 (±0.5); MgO 74.7243 (±0.0537) Å^3, 162.7 (±2.9) GPa, 4.1 (±0.1) (Andrault, 2001; Jeanloz and Hemley, 1994; Jeanloz and Thompson, 1983).

The high-pressure measurements are therefore compatible with what we infer from the zero-pressure volumes (Fig. 2, bottom) or volume ratios (Fig. 3, bottom), namely that there is no evidence for a shift in Mg/Fe partitioning between opv and mw between 25 and 110 GPa. This is in contrast with inferences recently derived from high-pressure X-ray spectroscopy on magnesiowüstite (Badro et al., 2003) that suggest an increase of several orders of magnitude in the Fe partitioning between magnesiowüstite and orthorhombic perovskite, perhaps because we are studying a more complex (i.e., natural, Al-bearing) bulk composition in the present experiment. The remaining scatter, of ±0.4 and ±0.8% observed for the volumes of opv and mw at zero pressure, can be ascribed to compositional variations in the high-pressure mineral phases.

Overall, as much as 1/2–2/3 of the uncertainty in the opv and mw volumes measured at high pressures is of the same magnitude as—so can be ascribed to the same factors causing—the scatter in volumes observed at zero pressure (Figs. 2 and 3). If there is a systematic correlation between volume deviations for a given sample at high pressures and at zero pressure, however (e.g., Fig. 2, top and bottom), any such correlation is obscured by the added uncertainties in the high-pressure measurements. Consequently, we now focus on the scatter in opv and mw unit-cell volumes measured at zero pressure, and note that it exhibits no correlation with synthesis pressure (Figs. 2 and 3).

3. Electron microprobe analyses

In order to determine whether or not the observed scatter in unit-cell volumes is due to variations in composition, we have performed electron microprobe analyses on several of the quenched samples. The conditions of synthesis, high pressures and temperatures achieved between smooth diamond culet faces, ensure that the transformed peridotite often has a smooth surface suitable for microprobe analysis (Fig. 4). Samples still encased in their gaskets were mounted on glass slides, cleaned and carbon coated. A Cameca SX-51 electron microprobe was used with a 15 keV electron beam focused to a spot size $\leq 5\,\mu m$ diameter at the sample, and pure metal oxides served as standards (Pingitore et al., 1999).

Abundances were determined for Si, Ti, Al, Fe, Mg, Ca, Na and K. Because oxygen was not included in the analyses, the data were reduced assuming stoichiometric compositions. Earlier analyses (Knittle and Jeanloz, 1991), which did include oxygen, provide support for orthorhombic perovskite having

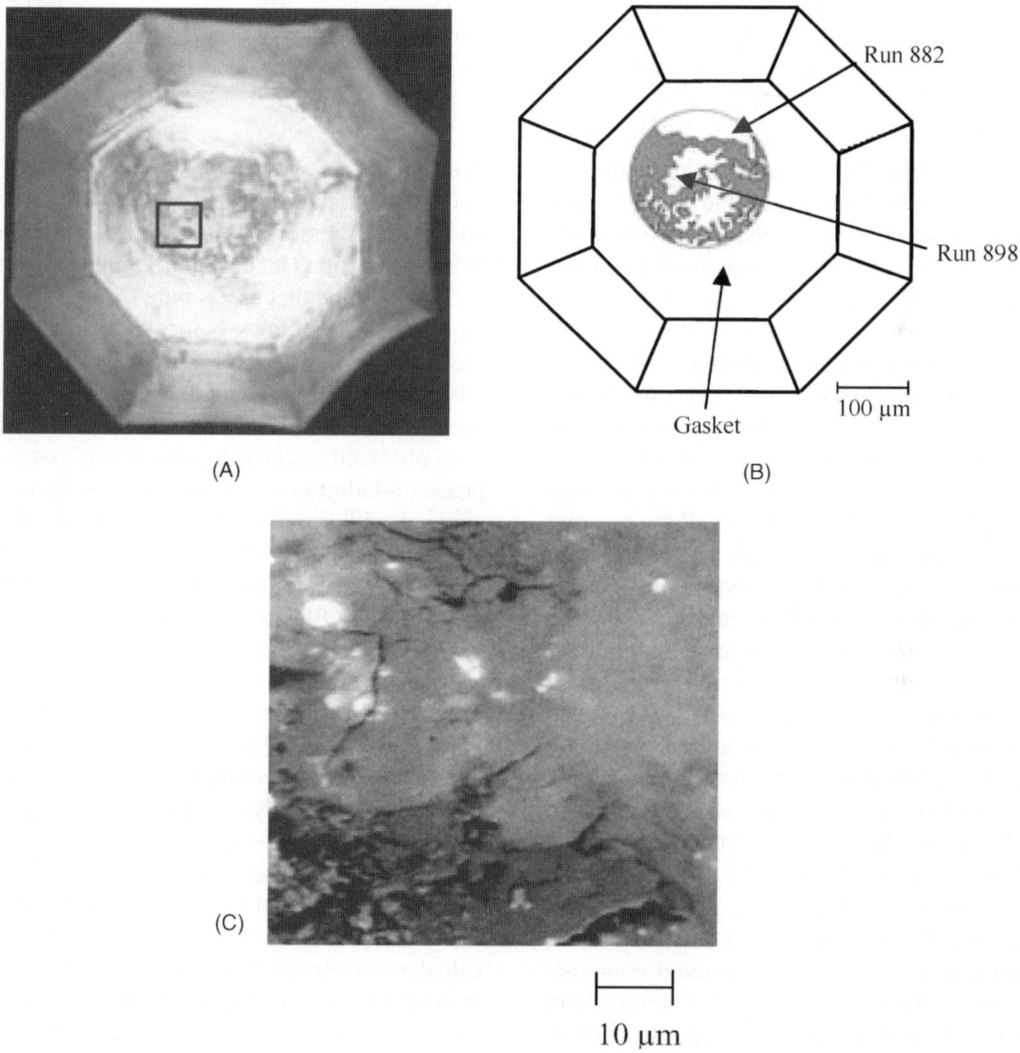

Fig. 4. (A) Photomicrograph in both reflected and transmitted light, and (B) schematic drawing of sample Per_500 quenched to zero pressure after heating, initially at 35 GPa, and compression up to 65 GPa. The locations of microprobe analyses 882 and 898 are marked for reference, and the bright area around the sample is gasket material. Scale bar is 100 μm. (C) Back-scattered electron image (mw appears as bright spots) of region of Per_500 shown in box in A, obtained with an electron beam at 10.00 keV and current of 600 pA. Scale bar is 10 μm. The variations in gray—hence, variations in mean atomic number—in the matrix show the sample's heterogeneity.

Table 2
Electron microprobe analyses of quenched samples in wt.%

Sample analysis	Per_500		Per_601		Average (nine analyses)
	882	898	911	919	
SiO_2	52.34	51.53	44.42	39.98	45.89 (6.32)
TiO_2	0.17	0.12	0.14	0.00	0.08 (0.07)
Al_2O_3	3.87	7.56	1.62	0.20	2.96 (2.94)
FeO	6.47	5.90	9.41	10.19	8.20 (1.96)
MgO	36.32	32.08	43.19	48.86	41.74 (7.57)
CaO	0.74	0.93	1.23	0.14	0.66 (0.37)
Na_2O	0.17	0.13	0.15	0.02	0.11 (0.06)
K_2O	0.02	0.02	0.04	0.01	0.01 (0.01)
Sum	100.10	98.26	100.18	99.39	99.65 (0.81)

stoichiometric composition when formed in the laser-heated diamond cell at comparable pressures and temperatures. However, we note that there is considerable interest in the possibilities of nonstoichiometry and of point defects in opv; this is especially so for more complex (e.g., Al-bearing) compositions, such as may be relevant to the present study (e.g., Navrotsky et al., 2002). Low totals were obtained in some analyses, however these can all be attributed to poor surface conditions or sample orientation in the present experiments. Only those analyses with totals between 98.0 and 102.0% are discussed here (almost all the rejected analyses came from one sample with a poor surface).

Although typical sample dimensions exceed 100 μm in diameter (Fig. 4), an average of only nine analyses is already in good agreement (within mutual uncertainties) with the bulk composition of the rock, except for a deficiency in CaO (Table 2). Because of the fine (mostly sub-μm) grain size of the transformed sample, each analysis is an average over a mixture of the three high-pressure phases, opv, mw and cpv. Indeed, it is impossible to interpret the analyses as being an average over fewer than 2–3 of the high-pressure mineral phases (e.g., an Mg/Si ratio exceeding 1 may be inferred for opv). Four individual analyses illustrating the range of observed compositions are listed in Table 2, and confirm that there can be quite a bit of variability from one spot to another within a sample, as well as from sample to sample.

Interpreting the scatter in individual analyses as being due to varying amounts of opv, mw, cpv and ruby (contamination from the pressure calibrant), we are able to estimate the composition of opv (and relative abundances of each phase) based on two assumptions. First, we assume that $x_{Fe} = 0.20$ for mw, based on the zero pressure lattice-parameter determined by X-ray diffraction (Jeanloz and Thompson, 1983). Second, we assume that cpv is pure $CaSiO_3$, based on the high-pressure volumes being compatible with those previously measured for pure cpv (O'Neill, 1994). Also, there is so little cpv in the high-pressure assemblage that even if the maximum amount of Al that has been observed to enter the structure is assigned to this phase (Takafuji et al., 2002), our conclusions are unaffected within quoted uncertainties. We then calculate the unit-cell volumes for the inferred composition of opv, taking into account aluminum as well as iron substitution (O'Neill and Jeanloz, 1994).

Our calculated opv zero-pressure volumes for the average of nine analyses ($V_0 = 163.3 \pm 0.6$ Å^3) are in excellent agreement with the observed average opv volume ($V_0 = 163.9 \pm 0.7$ Å^3) for those same samples (Table 3). The opv volumes inferred from the individual microprobe analyses generally overlap the values observed for the same samples (e.g., Per_500), but not always Per_601). The difference of up to 0.7% among individual calculated volumes is larger than the estimated uncertainties ($\pm 0.1-0.2\%$), but the average of calculated volumes is still within the uncertainty for the volumes observed by X-ray diffraction (±0.4%). A similar conclusion is reached for the calculated densities of opv, which exhibit differences as large as 4.4% for the individual analyses but an uncertainty less than ±1.5% for the average (Table 3).

These results confirm that the observed scatter in X-ray diffraction measurements at zero pressure can

Table 3
Orthorhombic perovskite compositions inferred from electron microprobe analyses (per 3 O)[a]

Sample analysis	Per_500		Per_601		Average (nine analyses)
	882	898	911	919	
Mg	0.88 ± 0.03	0.87 ± 0.02	0.88 ± 0.07	1.00 ± 0.03	0.93 ± 0.08
Fe	0.07 ± 0.01	0.09 ± 0	0.17 ± 0.02	0 ± 0.02	0.06 ± 0.06
Al	0.08 ± 0.01	0.12 ± 0.04	0.04 ± 0	0.01 ± 0	0.06 ± 0.05
Si	0.97 ± 0.03	0.92 ± 0.02	0.91 ± 0.09	0.99 ± 0.01	0.95 ± 0.05
Calculated opv V_0 (Å^3)	163.43 ± 0.17	163.76 ± 0.21	163.80 ± 0.23	162.64 ± 0.17	163.26 ± 0.62
Observed opv V_0 (Å^3)	164.37 ± 0.19		164.17 ± 0.32		163.90 ± 0.67
Calculated opv ρ_0 (g/cm^3)	4.17 ± 0.02	4.19 ± 0.02	4.29 ± 0.02	4.10 ± 0.04	4.16 ± 0.06

[a] Assumes that the magnesiowüstite is of composition $(Mg_{0.80}Fe_{0.20})O$; all of the Ca is in a pure $CaSiO_3$; and that there might be a finite amount of ruby contamination. Using the volume relation of orthorhombic perovskite with Fe and Al, we calculate the zero-pressure unit-cell volume (V_0) and density (ρ_0) based on the composition determined by electron microprobe analysis (O'Neill and Jeanloz, 1994). Values indicated as '0' mean less than 0.5%.

be completely explained in terms of compositional heterogeneity in the high-pressure assemblage formed from a natural rock sample. Indeed, the X-ray data represent more nearly the average composition of the sample, in comparison with the individual electron microprobe analyses that reflect more of the local heterogeneity within each sample. This is not surprising, given the different sampling volumes characterizing the two methods in the present study: ~5 and ~60 μm diameter for the microprobe and diffraction analyses, respectively.

4. Discussion and conclusions

In order to quantify the range of compositional heterogeneity that might be expected in our samples, we consider the possibility that the individual minerals of the starting material each transforms to the high-pressure phases but without equilibrating amongst each other. Based on the composition and mode of BN-35 (Table 1), we find that the fractional molar abundances of iron and aluminum in the orthorhombic perovskite can vary over the range $x_{Fe} = 0.10$–0.20 and x_{Al} (= Al/(Al + Mg)) = 0.06–0.48 (Table 4). Similarly, the molar abundance of iron in magnesiowüstite is in the range $x_{Fe} = 0.00$–0.10 (mw is derived only from the initial olivine, in this instance).

This calculated variability in composition is larger than we observe, as is the corresponding scatter in calculated zero-pressure volumes: e.g., up to 1.2–1.6% expected volume differences (i.e., up to ±0.6–0.8% scatter) for opv (see Table 4), as compared with ±0.4% observed (Fig. 2, bottom). Therefore, we conclude that whereas compositional variation across the entire sample is reflected in the observed variability in zero-pressure volumes (and electron microprobe analyses), there is at least an approach to equilibration between the high-pressure mineral phases formed from the distinct minerals in the starting material. That is, equilibrium is sufficiently well achieved to yield a reproducible assemblage of phases having relatively well-defined compositions and unit-cell volumes.

The variability in bulk starting composition (Table 1) can explain the scatter in the observed properties of the quenched samples (Figs. 2 and 3). The resulting uncertainty in properties at zero pressure is then generally convolved with additional uncertainties for in-situ measurements at high pressures. In the present case, the average observed uncertainty in unit-cell volumes is roughly 50% greater at high pressures than at zero pressure.

Nevertheless, covariation among the properties of the individual mineral phases can lead to relatively well-constrained properties for the whole-rock assemblage. For example, the whole-rock, zero-pressure densities calculated from analyses 911 and 919 of Tables 2 and 3 are 4.11 and 4.15 g/cm^3, respectively, a difference of only 1% (as opposed to the nearly 5% difference in opv densities given in Table 3). For comparison, the whole-rock density calculated for the average of nine microprobe analyses is 4.13 (±0.03) g/cm^3, and is identical to the density

Table 4
Decomposition reactions of initial minerals to lower-mantle phases[a]

Reaction	Starting composition[b]	Orthorhombic perovskite composition	Magnesiowüstite or calcium perovskite composition
Olivine → orthorhombic perovskite + magnesiowüstite	$(Mg_{1.90}, Fe_{0.22}, Al_{0.06}, Si_{0.82})O_4$	$(Mg_{1.00–0.90}, Fe_{0.12–0.22}, Al_{0.06}, Si_{0.82})O_3$	$(Mg_{0.90–1.00}, Fe_{0.10–0.00})O$
Calculated V_0 (Å^3)[c,d]		163.61–164.20	75.39–74.72
Orthopyroxene → orthorhombic perovskite + calcium perovskite	$(Mg_{0.81}, Fe_{0.11}, Al_{0.07}, Ca_{0.01})\ SiO_3$	$(Mg_{0.82}, Fe_{0.11}, Al_{0.07})\ SiO_3$	$CaSiO_3$
Calculated V_0 (Å^3)[c]		163.61	
Clinopyroxene → orthorhombic perovskite + calcium perovskite	$(Ca_{0.40}, Mg_{0.43}, Fe_{0.05}, Al_{0.13}, Si_{0.99})O_3$	$(Mg_{0.72}, Fe_{0.08}, Al_{0.22}, Si_{0.98})O_3$	$CaSiO_3$
Calculated V_0 (Å^3)[c]		164.24	
Garnet → orthorhombic perovskite + calcium perovskite	$(Mg_{2.10}, Fe_{0.48}, Ca_{0.37}, Al_{1.94}, Si_{3.11})O_{12}$	$(Mg_{0.58}, Fe_{0.13}, Al_{0.53}, Si_{0.76})O_3$	$CaSiO_3$
Calculated V_0 (Å^3)[c]		166.19	

[a] Based on a mass balance, allowing some defects or nonstoichiometry (corresponding to up to −6.3 to +2.0% deviation from charge balance).
[b] Whole-rock and individual phase compositions from Table 1 (Stern et al., 1986).
[c] V_0 calculated for orthorhombic perovskite from (O'Neill and Jeanloz, 1994).
[d] V_0 calculated for magnesiowüstite from (Jeanloz and Thompson, 1983).

determined entirely from the X-ray diffraction measurements (4.13 ± 0.02 g/cm^3).

The reason for this good agreement is that the bulk composition of the rock already determines the mean atomic weight of the sample, so that the density is really determined to within the resolution of the average volumes of the mineral phases. Thus, the zero-pressure density of the assemblage can be obtained by two methods, purely from the X-ray diffraction data (calculated phase compositions and abundances) or purely from the electron microprobe analyses (calculated phase volumes and compositions), and the results agree to well within the ±0.5% uncertainty of either approach. This agreement is important for geophysical applications because it means that the results of high-pressure experiments can be used to quantitatively interpret seismologically observed values of density and elastic moduli for the Earth's deep mantle.

Acknowledgements

We thank M. Feineman for her expertise in and help with collection of the electron microprobe data. We also thank W.R. Panero and L.R. Benedetti for their experimental assistance, E. Girndt for artwork help and S.-H. Shim, L.C. Ming and S.M. Clark for helpful discussions. We thank two anonymous reviewers for a thorough inspection of the manuscript that enabled us to better clarify our presentation. This research was supported by the US National Science Foundation. Portions of this research were carried out at the Stanford Synchrotron Radiation Laboratory, a national user facility operated by Stanford University on behalf of the US Department of Energy, Office of Basic Energy Sciences. Use of the Advanced Photon Source was supported by the US Department of Energy, Office of Basic Energy Sciences, under Contract No. W-31-109-Eng-38.

References

Andrault, D., 2001. Evaluation of (Mg, Fe) partitioning between silicate perovskite and magnesiowüstite up to 120 GPa and 2300 K. J. Geophys. Res. 106 (B2), 2079–2087.

Badro, J., et al., 2003. Iron partitioning in Earth's mantle: toward a deep lower mantle discontinuity. Science 300 (5620), 789–791.

Bell, P.M., Yagi, T., Mao, H.K., 1979. Iron-magnesium distribution coefficients between spinel [(Mg, Fe)$_2$SiO$_4$], Magnesiowüstite [(Mg, Fe)O], and Perovskite [(Mg, Fe)SiO$_3$]. Carnegie Inst. Washington Year Book 78 (1790), 618–621.

Birch, F., 1978. Finite strain isotherm and velocities for single-crystal and polycrystalline NaCl at high pressures and 300 K. J. Geophys. Res. 83, 1258–1267.

Fiquet, G., et al., 1998. P-V-T equation of state of $MgSiO_3$ perovskite. Phys. Earth Planet. Interiors 105 (1/2), 21–31.

Jeanloz, R., Hemley, R.J., 1994. Thermoelasticity of perovskite: an emerging consensus. EOS, Trans., Am. Geophys. Union 75 (41), 476–477.

Jeanloz, R., Kavner, A., 1996. Melting criteria and imaging spectroradiometry in laser-heated diamond-cell experiments. Philos. Trans. R. Soc. London, Ser. A (Math. Phys. Sci.) 354, 1279–1305.

Jeanloz, R., Thompson, A.B., 1983. Phase transitions and mantle discontinuities. Rev. Geophys. Space Phys. 21 (1), 51–74.

Knittle, E., Jeanloz, R., 1987. Synthesis and equation of state of (Mg, Fe)SiO_3 perovskite to over 100 GPa. Science 235 (4789), 668–670.

Knittle, E., Jeanloz, R., 1991. Earth's core–mantle boundary: Results of experiments at high pressures and temperatures. Science 251 (5000), 1438–1443.

Kubo, A., Akaogi, M., 2000. Post-garnet transitions in the system $Mg_4Si_4O_{12}$–$Mg_3Al_2Si_3O_{12}$ up to 28 GPa: Phase relations of garnet, ilmenite and perovskite. Phys. Earth Planet. Interiors 121 (1-2), 85–102.

Lang, A.R., 1998. From Borrmann to super-Borrmann effect: from 2-beam to n-beam diffraction. Cryst. Res. Technol. 33 (4), 613–623.

Larson, A.C., Von Dreele, R.B., 2000. Los Alamos National Laboratory Report No. LAUR 86-748 LANL.

Lee, K.K.M., et al., 2003. Equations of state of the high-pressure phases of a natural peridotite: Implications for the Earth's lower mantle, Earth Planet. Sci. Lett., submitted.

Mao, H.K., Bell, P.M., Shaner, J.W., Steinberg, D.J., 1978. Specific volume measurements of Cu, Mo, Pd, and Ag and calibration of the ruby R1 fluorescence pressure gauge from 0.06 to 1 Mbar. J. Appl. Phys. 49 (6), 3276–3283.

Navrotsky, A., et al., 2002. Aluminum in magnesian silicate perovskite: synthesis and energetics of defect solid solutions. In: Poeppellmeier, K., Navrotsky, A., Wentzcovitch, R.M. (Eds.), Perovskite Materials. Presented at Materials Research Society Symposium. Materials Research Society, Warrendale, PA, pp. 103–108.

Nguyen, J.H., Jeanloz, R., 1993. A computer program to analyze X-ray diffraction films. Rev. Sci. Instrum. 64, 3456–3461.

O'Neill, B., 1994. Experimental petrology of the lower mantle. PhD thesis. University of California, Berkeley, Berkeley, p. 135.

O'Neill, B., Jeanloz, R., 1990. Experimental petrology of the lower mantle: A natural peridotite taken to 54 GPa. Geophys. Res. Lett. 17 (10), 1477–1480.

O'Neill, B., Jeanloz, R., 1994. $MgSiO_3$–$FeSiO_3$–Al_2O_3 in the Earth's lower mantle: Perovskite and garnet at 1200 km depth. J. Geophys. Res. 99 (B10), 19901–19915.

Pingitore Jr., N.E., Donovan, J.J., Jeanloz, R., 1999. Electron microprobe quantification: A new model based on electrons rather than on mass. J. Appl. Phys. 86 (5), 2790–2794.

Ringwood, A.E., 1975. Composition and Petrology of the Earth's Mantle. McGraw-Hill, New York, p. 618.

Shim, S.H., Duffy, T.S., Kenichi, T., 2002. Equation of state of gold and its application to the phase boundaries near 660 km depth in the Earth's mantle. Earth Planet. Sci. Lett. 203 (2), 729–739.

Stern, C.R., Saul, S.L., Skewes, M.A., Futa, K., 1986. Garnet peridotite xenoliths from the Pali-Aike alkali basalts of southernmost South America. In: Ross, J. (Ed.), Proceedings of the Fourth International Kimberlite Conference. Kimberlites and Related Rocks. Blackwell Scientific, Perth, Australia, pp. 735–744.

Takafuji, N., Yagi, T., Miyajima, N., Sumita, T., 2002. Study on Al_2O_3 content and phase stability of aluminous-$CaSiO_3$ perovskite at high pressure and temperature. Phys. Chem. Miner. 29, 532–537.

Toby, B.H., 2001. EXPGUI, a graphical user interface for GSAS. J. Appl. Crystallogr. 34, 210–213.

Yagi, T., Mao, H.K., Bell, P.M., 1979. Lattice parameters and specific volume for the perovskite phase of orthopyroxene composition, (Mg, Fe) SiO_3. Carnegie Inst. Washington Year Book 78 (1790), 612–613.

3. Fluids and volatiles in the Earth

Available online at www.sciencedirect.com
SCIENCE DIRECT®

ELSEVIER

Physics of the Earth and Planetary Interiors 143–144 (2004) 255–269

PHYSICS OF THE EARTH AND PLANETARY INTERIORS

www.elsevier.com/locate/pepi

Water transport into the deep mantle and formation of a hydrous transition zone

Eiji Ohtani*, Konstantin Litasov, Tomofumi Hosoya, Tomoaki Kubo, Tadashi Kondo

Institute of Mineralogy, Petrology, and Economic Geology, Tohoku University, Sendai 980-8578, Japan

Received 14 February 2003; received in revised form 28 August 2003; accepted 3 September 2003

Abstract

The phase relations in the peridotite–water system to the top of the lower mantle reveal that a certain amount of water can be transported and stored in the transition zone by various hydrous minerals resulting in formation of a hydrous transition zone. Results of preliminary experiments to clarify the effect of water on kinetics of the olivine–wadsleyite transformation are given, and show that water enhances the transformation kinetics significantly.

There are potential dehydration sites in the deep mantle; four sites along the descending slabs depending on their temperature profiles, i.e., the mantle wedge of the subduction zone, mantle transition zone, top of the lower mantle where hydrous ringwoodite and superhydrous phase B decompose, and deep in the lower mantle around 1200–1500 km depth where decomposition of phase D(G) occurs. We also expect one site of dehydration melting at the base of the upper mantle where the hot mantle plume hydrated in the transition zone can be dehydrated. The transition zone can serve as a water absorber for the mantle scale water circulation system.

Some of the recent seismic works including the P-wave tomography images reveal the existence of low velocity anomalies in the transition zone suggesting possibly a wet transition zone. Elevation of the 410 km discontinuity and lack of clear evidence for a metastable olivine wedge in cold slabs may be consistent with the equilibrium phase transformation from olivine to wadsleyite in cold and wet slabs on the basis of our kinetic study of the transformation under wet conditions.

Keywords: Water; Subduction; Mantle plume; Dehydration; Transition zone

1. Introduction

Water transport from the surface to the deep earth's interior and its circulation on a global scale are key issues to understand the effect of water on planetary evolution. Water is transported into the deep mantle by hydrous minerals in descending slabs. Various studies have been conducted to clarify the stability of hydrous phases in the peridotite, basalt, and sediment components of slabs at high pressure and temperature. Knowledge of melting relations of hydrous mantle is also essential, since water plays an important role for dynamics of the ascending plumes. Hot plumes under the wet conditions may be partially molten due to depression of the melting temperature. Further, water can play important roles for mantle convection and dynamics of the ascending plumes, since a trace amount of water lowers the viscosity of the mantle in a process

* Corresponding author. Tel.: +81-22-217-6662; fax: +81-22-217-6662.
E-mail address: ohtani@mail.tains.tohoku.ac.jp (E. Ohtani).

0031-9201/$ – see front matter © 2004 Elsevier B.V. All rights reserved.
doi:10.1016/j.pepi.2003.09.015

known as hydrolitic weakening of olivine (e.g., Mei and Kohlstedt, 2000). Water transports elements dissolved in fluids and changes location of phase boundaries (e.g., Wood, 1995). There may be differences in water transport capability in slabs as a results of their thermal states, i.e., hot and young slabs generally transport little water into the deep mantle, whereas colder slabs with a small geothermal gradient may transport significant amounts of water into the deep mantle, although their water transport capabilities are still a matter of great debate.

We have studied so far the stability of hydrous phases, the effect of water on phase transformation kinetics, and the effect of water on melting up to the top of the lower mantle. In this paper, we summarize our results on these topics, and estimate mechanisms for water transport, water storage capacity in the deep mantle, and discuss the possibility of a wet transition zone based on experimental results and recent seismological observations.

2. Phase relation in the peridotite–water system to the top of the lower mantle

2.1. *Stability of high-pressure hydrous phases in slabs*

Hydrous minerals expected to exist in peridotite are summarized in Table 1. The phase relations of hydrous peridotite from 10 to 30 GPa in the $CaO–MgO–Al_2O_3–SiO_2$ (CMAS) system with 2 wt.% water are shown in Fig. 1. The details of the experimental conditions and the chemical compositions of minerals in the run products are given elsewhere (Litasov and Ohtani, 2003a,b,c). The pressure values in Fig. 1 were based on the β–γ phase boundary of Mg_2SiO_4, (Suzuki et al., 2000), the decomposition boundary of ringwoodite into periclase and perovskite based on the ruby scale (Chudinovskih and Boehler, 2001), and Al_2O_3 solubility in $MgSiO_3$ perovskite (Kubo and Akaogi, 2000). The composition of the starting material of the hydrous peridotite shown in this figure is: SiO_2, 45.15 wt.%; Al_2O_3, 5.20 wt.%; MgO, 43.46 wt.%; CaO, 4.19 wt.%; H_2O, 2.00 wt.%. The starting material is a mixture of olivine, pyroxene, anorthite, and brucite.

In peridotite containing 2 wt.% water, hydrous phase E appears coexisting with hydrous wadsleyite at temperatures below 1000 °C, whereas fluid appears at higher temperatures in the upper part of the transition zone. On the other hand, superhydrous phase B coexists with hydrous ringwoodite at temperatures below 1200 °C, whereas fluid coexists above this temperature. Appearance of superhydrous phase B in the transition zone depends on the water content. When the water content is less than ca. 1.2 wt.%, the water content in hydrous ringwoodite is less than ca. 2.0 wt.% and superhydrous phase B does not appear under the transition zone conditions. It appears only as a decomposition product of hydrous ringwoodite at pressures close to the 660 km discontinuity. We observed the mineral assemblage of superhydrous phase B, Mg-perovskite, majorite, and Ca-perovskite without periclase in peridotite with 2 wt.% water under

Table 1
Major hydrous minerals in the peridotite mantle

Minerals	Formula	H_2O (wt.%)	Abbreviations
Chlorite	$Mg_5Al_2Si_3O_{10}(OH)_8$	13	Chl
Serpentine	$Mg_3Si_2O_5(OH)_4$	4 28	Serp
Chondrodite	$Mg_5Si_2O_8(OH)_2$	5.30	Cho
Clinohumite	Mg9Si4O16(OH)2	3	Cli
10 Å Phase	$Mg_3Si_4O_{14}H_6$	13	10 Å
Phase A	$Mg_7Si_2O_8(OH)_6$	12	A
Phase E	$Mg_{2.3}Si_{1.25}H_{2.4}O_6$	11.40	E
Superhydrous phase B/phase C[a]	$Mg_{10}Si_3O_{14}(OH)_4$	5.80	Sb
Phase D(G)[a]	$Mg_{1.14}Si_{1.73}H_{2.81}O_6$	14.5–18	D(G)
Hydrous wadsleyite	Mg_2SiO_4	≤3	Wd
Hydrous ringwoodite	Mg_2SiO_4	≤2.2	Rw

[a] Superhydrous phase B is the same as phase C (Kanzaki, 1993) and phase D (Yang et al., 1997) is also named as phase G and phase F (e.g., Kudoh et al., 1997; Ohtani et al., 2000).

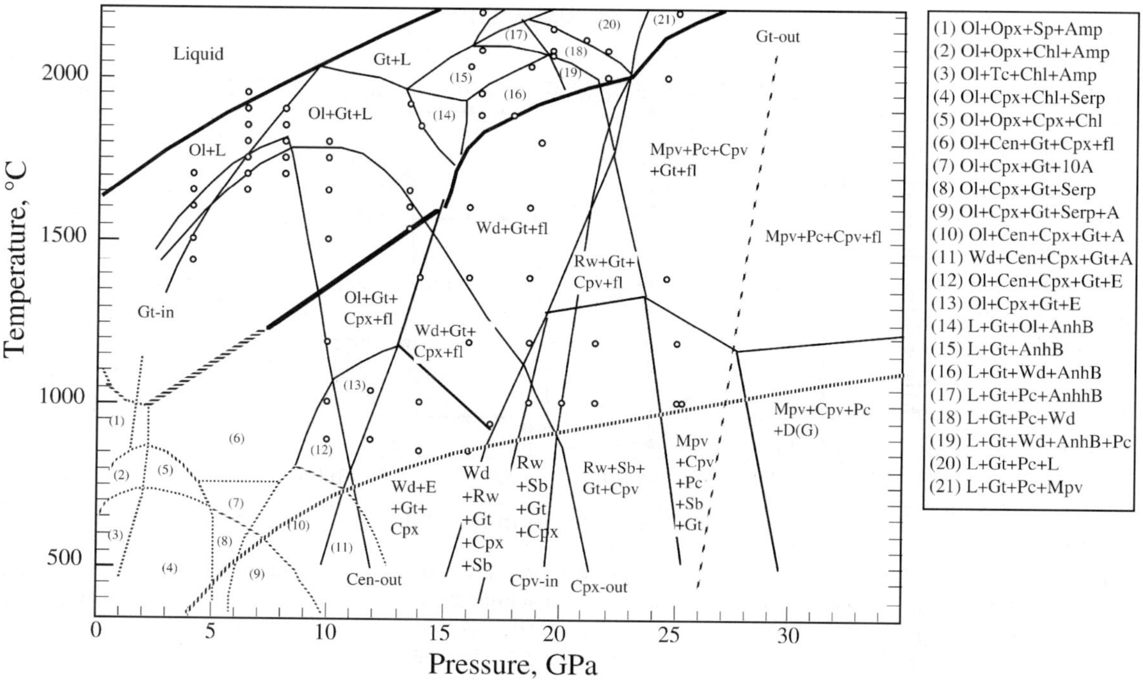

Fig. 1. Phase relation of peridotite–2 wt.% water to 30 GPa. Small circles are the experimental data points by Litasov and Ohtani (2003a). The phase relations below 10 GPa and 800 °C which are shown as dotted lines are estimated based on the phase relation by Schmidt and Poli (1998) with modifications based on the phase rule. The mineral assemblages in the stability fields (1–21) are listed separately. Ol, olivine; Opx, orthopyroxene, Sp, spinel; Amp, amphibole; Chl, chlorite; Tc, talc; Cpx, clinopyroxene; Gt, garnet; Cen, clinoenstatite; fl, fluid; 10 Å, 10 Å phase; Serp, serpentine; A, phase A; E, phase E; L, liquid; AnhB, anhydrous phase B; Wd, wadsleyite; Pc, periclase; Rw, ringwoodite; Sb, superhydrous phase B; Cpv, Ca-perovskite; Mpv, Mg-perovskite; D(G), phase D (phaseG). Gt-out indicates transformation of garnet (majorite) to the lower mantle assemblage with increasing pressure, whereas Gt-in indicates crystallization of garnet from melt. The low temperature slab geotherm (Peacock, 1990) is also given in this figure.

the lower mantle conditions below 1400 °C. Periclase appears in addition to this assemblage for peridotite with lower water contents.

We have observed decomposition of superhydrous phase B into periclase + Mg-perovskite + hydrous phase D(G) in the uppermost part of the lower mantle at around 30 GPa and temperatures below 1200 °C, and it decomposes into periclase + Mg-perovskite + fluid at higher temperature at 30 GPa (Ohtani et al., 2003). Therefore, superhydrous phase B exists up to the depth of 900 km in the upper part of the lower mantle. Fig. 2 summarizes the weight fractions of minerals in peridotite with 2 wt.% water at pressures to 35 GPa along a typical slab geotherm (Peacock, 1990) given in Fig. 1. The figure is drawn based on the experimental results by Litasov and Ohtani (2003a,b) at pressures above 10 GPa, whereas the fractions of mineral assemblage below 10 GPa were based on the previous results by Schmidt and Poli (1998). Since the water storage capacity along the slab geotherm is greater than 2 wt.% due to existence of high-pressure hydrous minerals, fluid is absent along the geotherm as shown in Fig. 2.

Under upper mantle conditions, chlorite and serpentine are the water reservoirs in the cold slabs descending into the base of the upper mantle. Serpentine is replaced by hydrous phase A at the depths greater than 180 km (~6 GPa). The maximum amount of water stored in peridotite as hydrous minerals was estimated by Schmidt and Poli (1998) up to a pressure of 8 GPa.

At pressures below 10 GPa and temperatures below 800 °C, we estimated the phase relations of the peridotite with 2 wt.% water based on the results by Schmidt and Poli (1998). The mineral assemblage of a serpentine peridotite containing more than

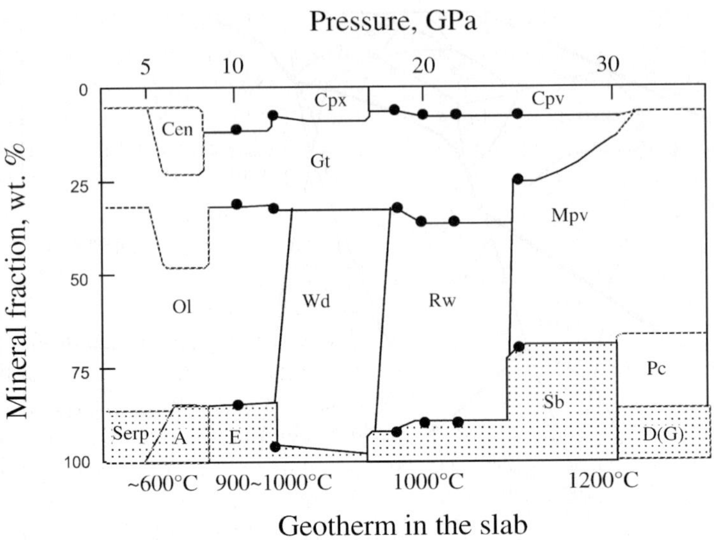

Fig. 2. Modal compositions of mineral assemblages (wt.%) in the peridotite–2 wt.%H_2O bulk composition. The mineral fractions from 10 to 25 GPa given as solid circles are based on the mass balance calculation using the chemical compositions of the coexisting phases given in Litasov and Ohtani (2002). The mineral fractions below 10 GPa are estimated based on the phase relations given by Schmidt and Poli (1998) and Poli and Schmidt (2002), whereas those above 30 GPa are based on Ohtani et al. (2003).

4.6 wt.% water transforms to the assemblage of phase A + Ca-poor clinopyroxene + Ca-rich clinopyroxene + garnet + fluid which is characterized by absence of olivine even in a peridotite bulk composition (Schmidt and Poli, 1998). Since the water content of our starting composition (2.0 wt.%) is lower than that of Schmidt and Poli (1998), we modified their results based on the phase rule of multi-component systems; i.e., olivine is not stable in the water rich system (Schmidt and Poli, 1998), whereas it can coexist with the hydrous phases in our water-poor CMAS peridotite. Hydrous peridotite with 2 wt.% water transforms to the assemblage containing olivine, Ca-poor and Ca-rich clinopyroxenes, garnet, and phase A as shown in Fig. 1.

Although there is no direct information on the amount of water in the peridotite layer of slabs, it should contain water due to serpentinization or chloritization. Serpentine is observed along the fracture zones of the ocean floor. Schmidt and Poli (1998) estimated that the upper 5 km of the peridotite layer suffers about 10% serpentinization, whereas Kesson and Ringwood (1989) suggested more extensive serpentinization.

It has been suggested that water is transported by the slab into the deep upper mantle when the slab temperature is lower than ca. 600 °C at around 6 GPa which is called the choke point by Kawamoto et al. (1996). When the temperature of the slab is lower than this, the stability field of serpentine crosses that of phase A, which is one of the most important dense hydrous magnesium silicate (DHMS) phases stable in the deep upper mantle (Ulmer and Trommsdorff, 1995; Luth, 1995). Although the choke point is important to hold water in the peridotite layer of the slabs, some other mechanisms are also possible for transporting water into the deep upper mantle even at temperatures above the choke point. Details are discussed in the following section.

Recent high-pressure experiments revealed that the major constituent phases in the transition zone, wadsleyite and ringwoodite, can accommodate water up to about 3 and 2 wt.%, respectively, at the temperature conditions of the cold slabs (e.g., Inoue et al., 1995). Thus, the transition zone is a potential water reservoir in the Earth. Wadsleyite and ringwoodite can accommodate water up to about 0.5 wt.% even along the normal mantle geotherm, i.e., the temperature around 1600 °C in the transition zone (Ohtani et al., 2000, 2001). Ilmenite $MgSiO_3$, which is stable in the depleted peridotite mantle, can also accommodate water up to 0.04 wt.% at about

1600–1750 °C near the ilmenite–perovskite phase boundary (Bolfan-Casanova et al., 2000, 2002). Although wadsleyite and ringwoodite are the main water reservoirs under conditions of relatively high temperatures, DHMS phases are stable at lower temperatures corresponding to the descending slabs. The hydrous phases stable in peridotite in the transition zone and the uppermost lower mantle are phase E, superhydrous phase B, and phase D (phase G by Kudoh et al., 1997).

2.2. *Melting relation of the peridotite–water system*

Melting relations of hydrous CMAS peridotite were studied by Asahara and Ohtani (2001) up to 8 GPa, and by Litasov and Ohtani (2003a) up to 25 GPa. The melting relations with 2 wt.% water are given in Fig. 1. Liquidus temperatures of the hydrous CMAS peridotite with 2 wt.% water are about 70–100° lower than those of dry peridotite. The hydrous solidus temperature is difficult to define at pressures above 10 GPa since liquid immiscibility between silicate melt and fluid disappears under such conditions (Bureau and Keppler, 1999). Here we defined the apparent solidus which is the temperature of a rapid increase of the melt fraction.

The apparent solidus is about 1600 °C at 13.5 GPa in the upper mantle, whereas we observed a drastic change of the apparent solidus to 1850 °C at 15.5 GPa at the top of the transition zone. The apparent solidus is 1900 °C at 20 GPa and 2150 °C at 25 GPa, i.e., about 180 °C lower at 20 GPa and 120 °C lower at 25 GPa than the dry solidus temperature, respectively. The drastic change of the apparent solidus temperature may be caused by restricted solubility of water in olivine compared to that of wadsleyite resulting in difference in water activity in the silicate melt. The drastic change of the apparent solidus temperature has important implications for melting processes in the ascending hot mantle plumes as discussed below.

2.3. *Effect of iron on the phase and melting relations*

At pressures below 10 GPa and temperatures below 800 °C, we estimated the phase relations based on the peridotite–water system results of Schmidt and Poli (1998). Although their starting composition contains FeO and we studied FeO-free CMAS peridotite, the stability of the hydrous phases in our system is likely to be similar to the FeO-bearing system with a small (∼1G Pa) shift of the phase boundaries as shown in the location of the choke point and the stability field of phase A in both FeO-free and FeO-bearing systems (Schmidt and Poli, 1998; Kawamoto et al., 1996; Luth, 1995; Ulmer and Trommsdorff, 1995). In the FeO-bearing peridotite, it is known that the olivine–wadsleyite transformation boundary shifts to lower pressures and the stability field of garnet also expands to lower pressures.

Dry melting relations in CMAS peridotite (Litasov and Ohtani, 2003a) are similar to those of KLB-1 peridotite by Zhang and Herzberg (1994); i.e., the differences between liquidus and solidus temperatures of FeO-free and FeO-bearing peridotites are less than 50° in the dry system. Thus, the melting relations between FeO-free and FeO-bearing hydrous peridotites may be similar and the temperature difference of the apparent solidus between the peridotites may be less than 50°.

3. Effect of water on phase transformation kinetics

Water can play an important role in determing rheological properties of the mantle, since a trace amount of water lowers the strength of olivine crystals, i.e., hydrolytic weakening (e.g., Karato, 1986). Water may also affect the phase transformation kinetics of mantle minerals. We have conducted experiments to study the effect of water on kinetics of the olivine–wadsleyite transformation based on both an in situ X-ray diffraction study of polycrystalline olivine and a quenching experiment using olivine single crystals (Kubo et al., 1998a,b).

We performed in situ X-ray diffraction study of kinetics using a mixture of MgO, $Mg(OH)_2$, and SiO_2 with the composition of Mg_2SiO_4−0.5 wt.%H_2O. In situ X-ray observation was made with using the multianvil press, SPEED1500 at beamline BL04B1 of Spring 8 of Japan Synchrotron Research Institute. The mixture of reagents with the above starting composition was enclosed in an AgPd capsule. A pressure standard consisting of NaCl powder was placed outside the capsule. We synthesized the mixture of forsterite and water by heating at 10 GPa and 1000 °C for 60 min, and also annealed the sample to remove the deviatoric stresses and strains at the same time. The grain

Table 2
Experimenal conditions and kinetic parameters for the effect of water on the phase transformation from olivine to wadsleyite

Run no.	Sample	Pressure (Gpa)	Temperature (°C)	Over pressure (Gpa)	Grain size (micron)	Growth rate (m/s)	References
Wet experiments made by in situ X-ray diffraction using magnesia pressure medium							
S826	Mg_2SiO_4 + 0.5 wt.% H_2O	14.2	810	2	5 (3)	$2.4\ (1.5) \times 10^{-9}$	This study
S824	Mg_2SiO_4 + 0.5 wt.% H_2O	15.2	900	2.7	5 (3)	$7.1\ (4)\times 10^{-9}$	This study
Wet experiments made by in situ X-ray diffraction using Boron-epoxy pressure medium							
G1	Mg_2SiO_4	13	900	0.5	2 (1)	$2.4\ (1.4) \times 10^{-10}$	Kubo et al. (1998a)
G2	Mg_2SiO_4	13	900	0.5	2 (1)	$3.3\ (1.7) \times 10^{-10}$	Kubo et al. (1998a)
G3	Mg_2SiO_4	14.1	885	1.6	2 (1)	$7.1\ (3.7) \times 10^{-10}$	Kubo et al. (1998a) wadsleyite, 0.12 wt.% water
G4	Mg_2SiO_4	14.9	875	2.5	2 (1)	$1.5\ (0.8) \times 10^{-9}$	Kubo et al. (1998a)
Wet and quenching experiment							
K1	San Carlos olivine	13.5	1030	0.5	Single crystal	$4.7\ (2.0) \times 10^{-8}$	Kubo et al. (1998b) wadsleyite, 0.21 wt.% water
Dry experiments made by in situ X-ray diffraction using magnesia pressure medium							
S187	Mg_2SiO_4	15.3	930	2.6	5(3)	$8.4\ (3.4) \times 10^{-11}$	This study
S510	Mg_2SiO_4	15.8	890	3.3	5(3)	$2.4\ (1.0) \times 10^{-11}$	This study

size of the recovered sample was measured in a separate run to be an average of 5 microns. The synthesized olivine was further compressed at around 500 °C to the desired pressure in the stability field of wadsleyite and heated to the desired temperatures of around 800–900 °C. X-ray diffraction profiles were collected every 30 s at a fixed press load. The transformation rates were calculated by using the intensities of diffraction lines in olivine and wadsleyite (Table 2 and Fig. 3), and were consistent with those reported by Kubo et al. (1998b).

Quenching experiments using olivine single crystals imply that the transformation is clearly enhanced by the existence of water (Kubo et al., 1998a). FTIR measurements revealed that water content in olivine is about 0.05 wt.% whereas that in wadsleyite is about 0.21 wt.% after quenching of the samples enclosed by the mixture of NaCl:brucite = 10:1 in the capsule. The growth rate was roughly estimated from the thickness of the wadsleyite rim of the olivine single crystals, and is given in Table 2. Kubo et al. (1998b) conducted an in situ X-ray diffraction study on kinetics using a pressure medium made of a mixture of boron and epoxy which supplies water during the run due to decomposition of the epoxy of the pressure medium. These results clearly imply that the transformation was enhanced by water. The drastic effect of water was also

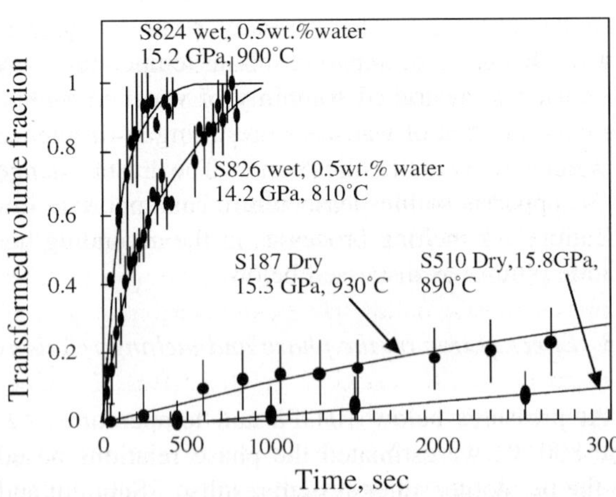

Fig. 3. The time dependence of the volume fraction of wadsleyite during the olivine–wasleyite transformation in forsterite–0.5 wt.% water and dry forsterite.

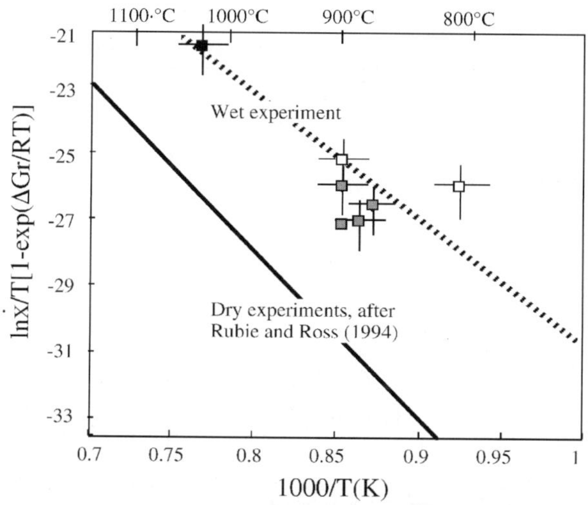

Fig. 4. Growth rates of wadsleyite under wet conditions at various temperatures. Open square are the results of the present in situ X-ray diffraction study of forsterite–0.5 wt.% water.; grey squares are the results of the in situ X-ray diffraction experiments using boron epoxy pressure medium which supplies water at high pressure and temperature (Kubo et al., 1998a). Solid square is the results of a quench experiment using the single crystal San Carlos olivine. The growth rate from the dry runs for the olivine–spinel transformation (Rubie and Ross, 1994) is given as the solid line.

discussed by Mosenfelder et al. (2001) based on the experimental results of Kubo et al. (1998a,b).

Fig. 4 shows the growth rates of wadsleyite as a function of temperature for the dry and wet experiments. In this figure, we plotted the temperature dependence of $\dot{x}/T[1-\exp(\Delta G_r/RT)]$, where $\dot{x}$ is the growth rate, ΔG_r the free energy change between olivine and wadsleyite which can be evaluated by $\Delta V_r \Delta P$. The volume change by the transformation, ΔV_r, is based on Akaogi et al. (1989). ΔP is the pressure of the phase transformation. Fig. 4 and Table 2 clearly imply that a small amount of water about 0.12–0.5 wt.% enhances the olivine–wadsleyite phase transformation kinetics corresponding roughly to a temperature increase of about 150°, although more quantitative studies on the effect of water on the phase transformation kinetics are necessary in the future. We can therefore not expect a metastable olivine wedge in wet slabs, since water enhances the phase transformation from olivine to wadsleyite. Equilibrium can be easily achieved even in the low temperature slabs around 500 °C.

4. Transport and storage of water in the mantle

4.1. Mechanism of water transport into the deep mantle

It has been suggested that water can be transported to the deep mantle when the slab geotherm passes below the choke point at 6 GPa and 600 °C (e.g., Kawamoto et al., 1996). However, several mechanisms have been suggested to transport water into the deeper mantle even at temperatures above the choke point. Ono (1998) shows that lawsonite and phengite are stable in hydrous MORB at temperatures higher than the choke point, and thus the dehydrated water from the peridotite part of the slab can be trapped in the overlying lawsonite or phengite in the MORB layer, and water may be further transported into the earth's deep interior. The dehydrated water at temperatures above the choke point may also be trapped by some hydrous minerals in the sediment component, such as topaz-OH and/or phase Egg ($AlSiO(OH)_3$), and further transported into the transition zone (Schmidt and Poli, 1998). Fumagalli et al. (2001) reported that the 10 Å phase, which is stable at temperatures above the choke point at around 6 GPa in peridotite compositions, is also a candidate for a water carrier in the peridotite layer of slabs. Thus, it is possible that water can be transported even if the slab geotherm passes above the choke point.

4.2. Water storage capacity in the mantle

The hydrous phases existing in mantle peridotite at high pressure and temperature were summarized in the previous section. Many hydrous minerals including DHMS phases are stable only at the low temperatures relevant to the slab conditions, and they are not stable along the normal mantle geotherm. The main water reservoirs in subducting slabs at various pressures are schematically shown in Fig. 2. The phase and melting relations of hydrous peridotite containing 2 wt.% water to 40 GPa are schematically shown in Fig. 5. In the upper mantle, olivine can accommodate only a limited amount of water along the normal mantle geotherm, although the amount of water stored in olivine increases with pressure (Kohlstedt et al., 1996). A small amount of water, however, has a great effect on its rheology (e.g., Karato, 1986) and the transforma-

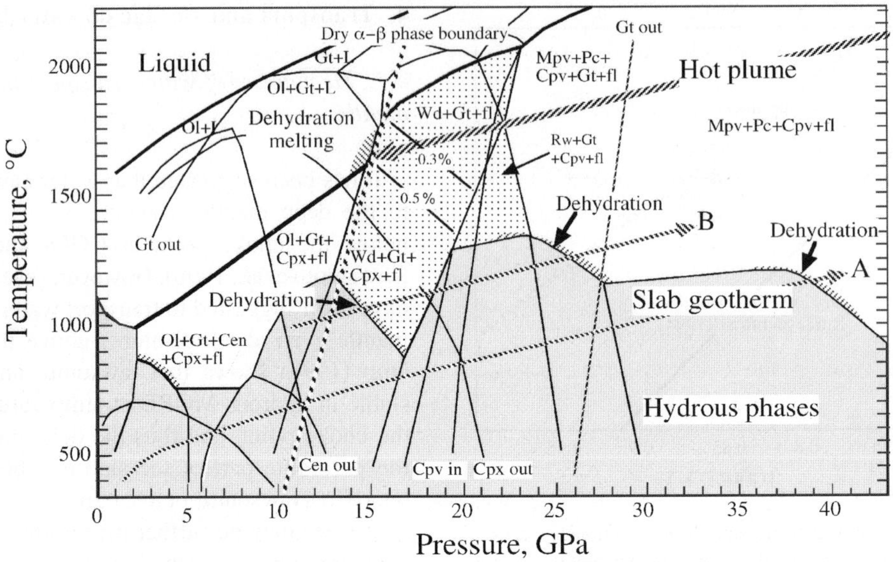

Fig. 5. The phase relations of hydrous peridotite to 40 GPa and the slab and hot plume geotherms in the mantle. Abbreviations are the same as those given in Fig. 1. The low temperature path of the slab (A) the high temperature path (B) and the path of the ascending plume (Peacock, 1990; Litasov and Ohtani, 2003b) are shown in the figure. The water contents (wt.%) in wadsleyite (Litasov and Ohtani, 2003b) are also given in this figure.

tion kinetics (e.g., Kubo et al., 1998a,b) as discussed above.

The major constituent minerals in the transition zone, wadsleyite and ringwoodite, can accommodate a large amount of water (e.g., Inoue et al., 1995; Ohtani et al., 2000). Litasov and Ohtani (2003a) shows that the maximum water solubility in wadsleyite at 15 GPa decreases with increasing temperatures from 2.0 wt.% at 1000 °C to 0.5 wt.% at 1600 °C, whereas it decreases to 0.3 wt.% with increasing pressure to 18 GPa at 1600 °C as shown in Fig. 5. Thus, the shallower region of the mantle transition zone has a larger water storage potential than the deeper region. Water solubility in ringwoodite also decreases with increasing temperature, but it has a weak pressure dependence. The experimental details of the measurement of the water solubility in wadsleyite and ringwoodite are given in Litasov and Ohtani (2003b). The maximum water storage potentials of wadsleyite and ringwoodite are estimated to be about 0.5–1 wt.% along the normal mantle geotherm.

Water storage capacity in the lower mantle is a matter of great debates. At lower temperatures (<1200 °C), superhydrous phase B and phase D(G) can accommodate water in the uppermost part of the lower mantle. The lower mantle is mainly composed of Mg-perovskite, magnesiowustite, and Ca-perovskite, however, the water solubilities of these minerals are poorly known. Meade et al. (1994) reported that water solubility in $MgSiO_3$–perovskite is 60–70 ppm. Whereas, Bolfan-Casanova et al. (2000) showed absence of water (<1 ppm H_2O) in $MgSiO_3$–perovskite. New data revealed that that the water solubility in aluminous perovskite existing in a peridotite bulk composition can accommodate more than 0.1 wt.% H_2O (e.g., Murakami et al., 2002; Litasov et al., 2003). Bolfan-Casanova et al. (2002) and Litasov and Ohtani (2003c) reported magnesiowustite contains only 20–180 ppm H_2O at 25 GPa.

Although there are some ambiguities in the water storage capacity in perovskite, we can generally conclude that there is a layered structure in the water storage capacity of the mantle; i.e., the upper and lower mantles have relatively low water storage capacities, whereas the transition zone has a high water storage capacity. A relatively low water storage capacity in the lower mantle indicates that the slabs subducting into the lower mantle dehydrate at the top of the lower mantle. The water content in the transition zone might be large because the dehydrated water from the subducted slabs together with the primordial water in the lower mantle have been stored in the transition zone during the geological time.

The phase relation of water saturated-peridotite was given in Schmidt and Poli (1998) at pressures below 10 GPa. The equilibrium phase assemblage of the hydrous phases under the water-saturated conditions are clarified experimentally and thermodynamically analyzed by Ohtani et al. (2001, 2003) in the model system of $MgO–SiO_2–H_2O$ in the pressure range from 10 to 30 GPa. These works imply that the stable hydrous phase in the water-saturated peridotite is phase E at the base of the upper mantle and top of the transition zone, superhydrous phase B and phase D(G) in lower part of the transition zone and uppermost lower mantle, and phase D(G) in the lower mantle along the low temperature slab geotherm (curve A in Figs. 5 and 6).

Phase relations reported previously indicate that the hydrous phase under the water-saturated conditions along the high temperature geotherm of the slab (curve B in Figs. 5 and 6) is phase E at the base of the upper mantle, hydrous wadsleyite in the upper part of the transition zone, superhydrous phase B and ringwoodite in the lower part of the transition zone, and superhydrous phase B in the uppermost lower mantle.

A mass balance calculation together with the water content and water partitioning between hydrous minerals (e.g., Ohtani et al., 2000) revealed the maximum water storage capacity in the peridotite layer of slabs as shown in Figs. 7A and B. Along the low temperature geotherm A in Figs. 5 and 6, the mass balance calculation of phase E, garnet, and clinopyroxene revealed existence of 50 wt.% of phase E corresponding to the water content of 5.8 wt.% at the bottom of the upper mantle and the upper part of the transition zone. The similar calculation implies that 52.6% of hydrous phase D(G) and 22.2 wt.% of superhydrous phase B can exist in the lower part of the transition zone and the top of the lower mantle providing the maximum water storage capacity of 10.4 wt.%. Along the low temperature slab geotherm, phase D(G) is the single hydrous phase in the deep lower mantle. Maximum

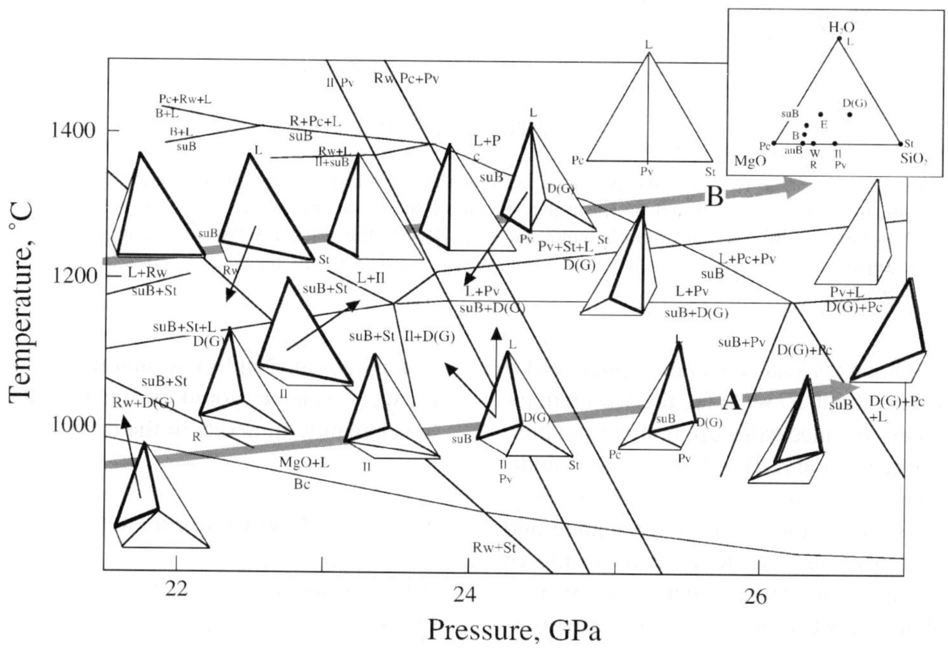

Fig. 6. The phase relations of the $MgO–SiO_2–H_2O$ system at high pressures. The hydrous phases observed in peridotite can be estimated from this system as was shown by Frost (1999) in the peridotite–4.7 wt.% water composition. The phase relations are based on the results of Ohtani et al. (2000, 2001, 2003). The low temperature and high temperature slab geotherms (Peacock, 1990) are also shown as (A and B) in the figure. Both phase D (phase G) and superhydrous phase B (phase C) can coexist along the cold slab geotherm A under water-saturated composition, whereas superhydrous phase B is the water reservoir in the transition zone and the top of the lower mantle along the hotter slab geotherm B.

both dry and wet conditions (e.g., Zhang and Herzberg, 1994; Litasov and Ohtani, 2003a). Although more detailed study is needed to determine the precise chemical composition of the hydrous magma formed at the base of the upper mantle, we may assume that it is close to the partial melt formed by melting of dry PHN1611 peridotite, i.e., IT8720 (Ito and Takahashi, 1987; Suzuki et al., 1995). Using the bulk modulus and its pressure derivative of $K = 25.3$ GPa and $K' = 6.2$ (Ohtani and Maeda, 2001), the density of the dry ultramafic melt IT8720 is estimated to be 3.71 g/cc at 14 GPa and 1600 °C. When the ascending hot plume contains about 0.5 wt.% of water, 10 wt.% partial melting can produce the hydrous ultramafic magma containing about 5 wt.% of water. Assuming that the partial molar volume of water in the melt is the same as the molar volume of the supercritical water estimated by the MD calculation (Belonoshko and Saxena, 1991), we can estimate the density of the hydrous magma to be 3.53 g/cc for the magma with 5 wt.% water at 14 GPa and 1600 °C. This assumption may be reconciled by the data obtained at pressures above 1 GPa (Burnham and Davies, 1971; Ochs and Lange, 1997). The density of the melt is similar to that of the mantle at the base of the upper mantle given by PREM (~3.51 g/cc). Thus, the hydrous melt formed by dehydration melting may be gravitationally stable at the base of the upper mantle. However, the partial molten zone may not last long geologically, since diffusion of hydrogen in mantle olivine is fast and hydrogen in magma tends to rise into the surrounding upper mantle (Mackwell and Kohlstedt, 1990).

6. Possible evidence for hydrated transition zone

We have suggested four dehydration sites related to subducting slabs, i.e., uppermost mantle, transition zone, top of the lower mantle, and deep lower mantle as shown in Fig. 5. One region of dehydration melting in a rising hot plume may also exist at the base of the upper as shown in the same figure. The two dehydration sites provides effective sources for supplying water from subducting slabs to the transition zone; (1) dehydration reaction occurring in the transition zone due to decomposition reaction of phase E and a decrease of water solubility in wadsleyite and ringwoodite with increasing pressure and temperature and (2) dehydration at the bottom of the transition zone and the top of the lower mantle due to the decomposition reactions of superhydrous phase B and ringwoodite to the lower mantle assemblage. Fluids generated at the top of the lower mantle will move upwards and will be trapped in the overlying ringwoodite and wadsleyite with high water storage capacity.

In the previous section, we estimated a high water content in the transition zone based on the simple equilibrium calculation using the partition coefficient of water between wadsleyite and olivine, $D(H_2O)_{\beta-\alpha} = 20$–$40$ (e.g., Kohlstedt et al., 1996). However, the estimation may be too simple, because of the high mobility of hydrogen as suggested in olivine crystals (Mackwell and Kohlstedt, 1990). There are possibilities of heterogeneity in water content in the transition zone even if it has a high water storage capacity, i.e., the transition zone adjacent to the subducting slabs may be enriched in water relative to the normal transition zone. The stratification of the mantle in terms of the water content, i.e., relatively lower water content in the upper and lower mantles and higher water content in the transition zone, may be altered by various processes such as convective circulation, interaction of the ascending plume with the transition zone, and mobility of hydrogen and permeable flow of fluids in the transition zone. The lack of the basic data such as the diffusion coefficients of hydrogen, the dihedral angle, and permeability of water in wadsleyite and ringwoodite makes it difficult to assess the heterogeneity in the water content in the transition zone.

Recent seismic observations suggest existence of a low velocity anomaly at the depths corresponding to the dehydration sites. Revenaugh and Sipkin (1994) suggested a possible existence of a melt at the base of the upper mantle or top of the 410 km discontinuity beneath eastern China. Recent data on P-wave tomography including the data beneath Japan (Zhao, 2001, 2004; Fukao et al., 2001) suggest existence of low velocity anomalies at the base of the upper mantle and the top of the lower mantle. Zhao et al. (1997) reported a low P-velocity anomaly in the transition zone and the base of the upper mantle beneath the Tonga trench which is the coldest among subducted slabs. This low velocity anomaly cannot be assigned to high temperature, but it may originate from existence of water in the transition zone based on the elasticity measurement of hydrous wadsleyite and hydrous ringwoodite (e.g., Inoue et al., 1998). These observations may be evidence for the hydrated mantle transition zone. However, this conclusion is not definitive, since the subduction zones may have complicated geometries at depth with complicated temperature fields, and the observation of low velocity anomalies alone near the Tonga subduction zone may not necessarily imply hydration at the depth (or the absence of hot regions).

Collier et al. (2001) observed a large elevation of the 410 km discontinuity to 60–70 km in the Izu–Bonin slab, and argued that the discontinuity is essentially explained by the equilibrium phase boundary of the olivine–wadsleyite transformation in the cold slab. They observed no evidence for existence of the metastable olivine core in the slab. They argued that a large elevation of the 410 km discontinuity corresponds to the temperature difference nearly 1000° compared to the surrounding normal mantle, which indicates the temperature at the discontinuity to be around 500 °C. The recent studies on phase transition kinetics indicate that under such a cold slab condition existence of metastable olivine is inevitable due to slow transformation kinetics of olivine under the dry condition (e.g., Rubie and Ross, 1994; Kirby et al., 1996), and the depression of the 410 km discontinuity is expected in the cold slabs.

Our results on the effect of water on the olivine–wadsleyite phase transformation kinetics (Fig. 4) imply that a small amount of water about 0.12–0.5 wt.% enhances the olivine–wadsleyite phase transformation kinetics, which corresponds to roughly the temperature increase of about 150°. Further, the olivine–wadsleyite phase boundary moves towards lower pressure by 0.5–1.0 GPa (e.g., Litasov and Ohtani, 2003a). Thus, the elevation of the 410 km discontinuity of the cold Izu–Bonin slab can be explained as the equilibrium phase boundary under the wet conditions. Koper et al. (1998) also conducted a detailed study of the 410 km discontinuity of the Tonga slab, and could not observe evidence for the depression of the discontinuity originating from the metastable olivine wedge in the slab. Thus, current seismic observations of the depth of the 410 km discontinuity are consistent with the wet slabs descending into the transition zone.

The absence of observations supporting the existence of metastable olivine might also imply the existence of wet slabs as discussed above. However, there are other possibilities that dry kinetics determined experimentally might not be applicable to the long time-scales, stress regime, or grain size within slabs; or it might imply that frictional heating during subduction is larger than is often inferred.

Acknowledgements

The authors appreciate Dr. K. Funakoshi for technical support during in situ X-ray diffraction experiments conducted at BL04B1 of Spring-8 (project no. 2001B0415, 2002A0298, and 2002B0567). The manuscript was significantly improved by the constructive reviews by the two anonymous reviewers. This work was partially supported by the grant-in-aid of Scientific Research of Priority Area (B) (no. 12126201) and the grant-in-aid of Scientific Research (S) of the Ministry of Education, Science, Culture, and Technology (no. 14102009) to Eiji Ohtani.

References

Akaogi, M., Ito, E., Navrotsky, A., 1989. Olivine-modified spinel–spinel transition in the system Mg_2SiO_4–Fe_2SiO_4 calorimetric measurements, thermochemical calulation, and geophysical application. J. Geophys. Res. 94, 15671–15685.

Agee, C.B., Walker, D., 1993. Olivine flotation in mantle melt. Earth Planet. Sci. Lett. 90, 144–156.

Asahara, Y., Ohtani, E., 2001. Melting relations of the hydrous primitive mantle in the CMAS-H_2O system at high pressures and temperatures, and implications for generation of komatiites. Phys. Earth Planet. Inter. 125, 31–44.

Burnham, C.W., Davies, N.F., 1971. The role of H_2O in silicate melts: 1. P-V-T relations in the system $NaAlSi_3O_8$–H_2O to 10 kilobars and 1000 °C. Am. J. Sci. 270, 54–79.

Bureau, H., Keppler, H., 1999. Complete miscibility between silicate melts and hydrous fluids in the upper mantle; experimental evidence and geochemical implications. Earth Planet. Sci. Lett. 165, 187–196.

Bolfan-Casanova, N., Keppler, H., Rubie, D.C., 2000. Water partitioning between nominally anhydrous minerals in the MgO–SiO_2–H_2O system up to 24 GPa: implications for the distribution of water in the Earth's mantle. Earth Planet. Sci. Lett. 182, 209–221.

Bolfan-Casanova, N., Mackwell, S., Keppler, H., McCammon, Rubie, D.C., 2002. Pressure dependence of H solubility in magnesiowüstite up to 25 GPa: implications for the storage of water in the Earth's lower mantle. Geophys. Res. Lett., no. 10, 10.1029.

Belonoshko, A., Saxena, S., 1991. A molecular dynamics study of the pressure–volume–temperature properties of super-critical fluids: I. H_2O. Geochim. Cosmochim. Acta. 55, 381–387.

Chudinovskih, L., Boehler, R., 2001. High-pressure polymorphs of olivine and the 660-km seismic discontinuity. Nature 411, 574–577.

Collier, J.D., Helffrich, G.R., Wood, B.J., 2001. Seismic discontinuities and subduction zones. Phys. Earth Planet. Inter. 127, 35–49.

Frost, D.J., 1999. The stability of dense hydrous magnesium silicates in Earth's transition zone and lower mantle. In: Fei, Y., Bertka, C.M., B.O. Mysen (Eds.), Mantle Petrology: Field Observations and High Pressure Experimentation: A tribute to Boyd, F.R., The Geochem. Soc. Special Public, vol. 6, pp. 283–296.

Fukao, Y., Widiyantoro, S., Obayashi, M., 2001. Stagnant slabs in the upper and lower mantle transition region. Rev. Geophys. 39, 291–323.

Fumagalli, P., Stixrude, L., Poli, S., Snyder, L., 2001. The 10Á phase: a high pressure expandable sheet silicate during subduction of hydrated lithosphere. Earth Planet. Sci. Lett. 186, 125–141.

Inoue, T., Yurimoto, H., Kudoh, Y., 1995. $Mg_{1.75}SiH_{0.5}O_4$: a new water reservoir in the mantle transition region. Geophys. Res. Lett. 160, 117–120.

Inoue, T., Weidner, D.J., Northrup, P.A., Parise, J.B., 1998. Elastic properties of hydrous ringwoodite (γ-phase) in Mg_2SiO_4. Earth Planet. Sci. Lett. 160, 107–113.

Ito, E., Takahashi, E., 1987. Melting of peridotite at uppermost lower-mantle conditions. Nature 328, 514–517.

Kanzaki, M., 1993. Calculated powder X-ray pattern of phase B, anhydrous phase B, and superhydrous phase B: re-assessment of previous studies. Mineral. J. 16, 278–285.

Karato, S., 1986. Does partial melting reduces the creep strength of the upper mantle? Nature 319, 309–310.

Kawamoto, T., Hervig, R.L., Holloway, J.R., 1996. Experimental evidence for a hydrous transition zone in the Earth's early mantle. Earth Planet. Sci. Lett. 142, 587–592.

Kesson, S., Ringwood, A.E., 1989. Slab-mantle interactions. 1. Sheared and refertilised garnet peridotite xenoliths-samples of Wadati-Benioff zones? Chemi. Geol. 78, 83–96.

Kirby, S.H., Stein, S., Okal, E.A., Rubie, D.C., 1996. Metastable mantle phase transformations and deep earthquakes in subducting oceanic lithosphere. Rev. Geophys. 34, 261–306.

Kohlstedt, D.L., Keppler, H., Rubie, D.C., 1996. Solubility of water in the α-β-γ phases of $(Mg, Fe)_2SiO_4$. Contrib. Mineral. Petrol. 123, 345–357.

Koper, K.D., Wiens, D.A., Dorman, L.M., Hildebrand, J.A., Webb, S.C., 1998. Modeling the Tonga slab: can travel time data resolve a metastable olivine wedge? J. Geophys. Res. 103, 30079–30100.

Kubo, A., Akaogi, M., 2000. Post-spinel transitions in the system Mg_2SiO_4–$Mg_3Al_2Si_3O_{12}$ up to 28 GPa: phase relations of garnet, ilmenite and perovskite. Phys. Earth Planet. Inter. 121, 85–102.

Kubo, T., Ohtani, E., Kato, T., Morishima, H., Yamazaki, D., Suzuki, A., Mibe, K., Kikegawa, T., Shimomura, O., 1998a. An in situ X-ray diffraction study of the α–β transformation kinetics of Mg_2SiO_4. Geophys. Res. Lett. 25, 695–698.

Kubo, T., Ohtani, E., Kato, T., Shinmei, T., Fujino, K., 1998b. Effect of water on the α–β transformation kinetics in San Carlos Olivine. Science 281, 85–87.

Kudoh, Y., Nagase, T., Mizobata, H., Ohtani, E., Sasaki, S., Tanaka, M., 1997. Structure and crystal chemistry of phase G, a new hydrous magnesium silicate synthesized at 22 GPa and 1050 °C. Geophys. Res. Lett. 24, 1051–1054.

Litasov, K., Ohtani, E., 2003a. Phase relations and melt compositions in CMAS pyrolite–H_2O system up to 25 GPa. Phys. Earth Planet. Inter 134, 105–127.

Litasov, K., Ohtani, E., 2003b. Stability of various hydrous phases in CMAS pyrolite–H_2O system up to 25 GPa. Phys. Chem. Miner. 30, 147–156.

Litasov, K., Ohtani, E., 2003c. Hydrous lower mantle: the water source for wet plumes? In: Proceedings of the 8th International Kimb. Conference, Extended Abstracts, Victoria, Canada, CD-edition.

Litasov, K., Ohtani, E., Langenhorst, F., Yurimoto, H., Kubo, T., Kondo, T., 2003. Water solubility in Mg-perovskites and water storage capacity in the lower mantle. Earth Planet. Sci. Lett. 211, 189–203.

Luth, R.W., 1995. Is phase A relevant to the Earth's mantle? Geochim. Cosmochim. Acta 59, 679–682.

Mackwell, S.J., Kohlstedt, D.L., 1990. Diffusion of hydrogen in olivine: implications for water in the mantle. J. Geophys. Res. 95, 5079–5088.

Meade, C., Reffner, J.A., Ito, E., 1994. Synchrotron infrared absorbance measurements of hydrogen in $MgSiO_3$ perovskite. Science 264, 1558–1560.

Mei, S., Kohlstedt, D.L., 2000. Influence of water on plastic deformation of olivine aggregates. 1. Diffusion creep regime. J. Geophys. Res. 105, 21457–21469.

Mosenfelder, J.L., Marton, F.C., Ross II, C.R., Kerschhofer, L., Rubie, D.C., 2001. Experimental constraints on the depth of olivine metastablity in subducting lithosphere. Phys. Earth Planet. Inter. 127, 165–180.

Murakami, M., Hirose, K., Yurimoto, H., Nakashima, S., Takafuji, N., 2002. Water in Earth's lower mantle. Science 295, 1885–1887.

Ochs III, F.A., Lange, R.A., 1997. The partial molar volume, thermal expansivity and compressibility of H_2O in $NaAlSi_3O_8$ liquid: new measurements and an internally consistent model. Contrib. Mineral. Petrol. 129, 155–165.

Ochs III, F.A., Lange, R.A., 1999. The density of hydrous magmatic liquids. Science 283, 1314–1317.

Ohtani, E., Maeda, M., 2001. Density of basaltic melt at high pressure and stability of the melt at the base of the lower mantle. Earth Planet. Sci. Lett. 193, 69–75.

Ohtani, E., Toma, M., Kubo, T., Kondo, T., Kikegawa, T., 2003. In situ X-ray observation of decomposition of superhydrous phase B at high pressure and temperature. Geophys. Res. Lett. 30 (2), 1029, doi:10.1029/2002GL015549.

Ohtani, E., Toma, M., Litasov, K., Kubo, T., Suzuki, A., 2001. Stability of dense hydrous magnesium silicate phases and water storage capacity in the transition zone and lower mantle. Phys. Earth Planet. Inter. 124, 105–117.

Ohtani, E., Mizobata, H., Yurimoto, H., 2000. Stability of dense hydrous magnesium silicate phases in the systems Mg_2SiO_4–H_2O and $MgSiO_3$–H_2O at pressures up to 27 GPa. Phys. Chem. Miner. 27, 533–544.

Ono, S., 1998. Stability limits of hydrous minerals in sediment and mid-ocean ridge basalt compositions: implications for water transport in subduction zones. J. Geophys. Res. 103, 18253–18267.

Peacock, S.M., 1990. Fluid processes in subduction zone. Science 248, 329–337.

Poli, S., Schmidt, M.W., 2002. Petrology of subducted slabs. Annu. Rev. Earth Planet. Sci. 30, 207–235.

Revenaugh, J., Sipkin, S.A., 1994. Seismic evidence for silicate melt atop the 410-km mantle discontinuity. Nature 369, 474–476.

Rubie, D.C., Ross II, C.R., 1994. Kinetics of olivine-spinel transformation in subducting lithosphere: experimental constraints and implications on deep slab processes. Phys. Earth Planet. Inter. 86, 223–241.

Schmidt, M.W., Poli, S., 1998. Experimental base water budgets for dehydrating slabs and consequences for arc magma generation. Earth Planet. Sci. Lett. 163, 361–379.

Shieh, S.R., Mao, H.K., Ming, J.C., 1998. Decomposition of phase D in the lower mantle and the fate of dense hydrous silicates in subducting slabs. Earth Planet. Sci. Lett. 159, 13–23.

Suzuki, A., Ohtani, E., Kato, T., 1995. Flotation of diamond in mantle melt at high pressure. Science 269, 216–218.

Suzuki, A., Ohtani, E., Morishima, H., Kubo, T., Kanbe, Y., Kondo, T., Okada, T., Terasaki, H., Kubo, T., Kikegawa, T., 2000. In situ determination of the phase boundary between wadsleyite and ringwoodite in Mg_2SiO_4. Geophys. Res. Lett. 27, 803–806.

Ulmer, P., Trommsdorff, V., 1995. Serpentine stability to mantle depths and subduction related magmatism. Science 268, 858–861.

Wallace, P.J., 1998. Water and partial melting in mantle plumes: inferences from the dissolved H_2O concentrations of Hawaiian basaltic magmas. Geophys. Res. Lett. 25, 3639–3642.

Wood, B.J., 1995. The effect of H_2O on the 410-kilometer seismic discontinuity. Science 268, 74–76.

Yang, H., Prewitt, C.T., Frost, D.J., 1997. Crystal structure of the dense hydrous magnesium silicate, phase D. Am. Mineral. 82, 651–654.

Zhang, J., Herzberg, C., 1994. Melting experiments on anhydrous peridotite KLB-1 from 5.0 to 22.5 GPa. J. Geophys. Res. 99, 17729–17742.

Zhao, D., 2004. Global tomographic images of mantle plumes and subducting slabs: insight into deep Earth dynamics. Phys. Earth Planet. Inter., in press.

Zhao, D., 2001. Seismological structure of subduction zones and its implications for arc magmatism and dynamics. Phys. Earth Planet. Inter. 127, 197–214.

Zhao, D., Xu, Y., Wiens, D., Dorman, L., Hildebrand, J., Webb, S., 1997. Depth extent of the Lau back-arc spreading center and its relation to subduction processes. Science 278, 254–257.

Available online at www.sciencedirect.com
SCIENCE DIRECT
ELSEVIER
Physics of the Earth and Planetary Interiors 143–144 (2004) 271–278
PHYSICS OF THE EARTH AND PLANETARY INTERIORS
www.elsevier.com/locate/pepi

High pressure crystal chemistry of hydrous ringwoodite and water in the Earth's interior

Joseph R. Smyth[a,*], Christopher M. Holl[a], Daniel J. Frost[b], Steven D. Jacobsen[b]

[a] *Department of Geological Sciences, University of Colorado, Boulder, CO 80309, USA*
[b] *Bayerisches Geoinstitut, Universität Bayreuth, Bayreuth D95440, Germany*

Received 29 November 2002; received in revised form 12 May 2003; accepted 18 August 2003

Abstract

The crystal chemistry of Fo_{90} hydrous ringwoodite ($Mg_{1.7}Fe_{0.22}H_{0.15}SiO_4$) containing 0.93% H_2O by weight has been studied at pressures up to 11.2 GPa by single-crystal X-ray diffraction in the diamond anvil cell. The unit cell edge and volume have been refined at 20 different pressures in this pressure range. The refined bulk modulus for the unit cell is 169.0 ± 3.4 GPa with a K' of 7.9 ± 0.9. H is accommodated in the structure principally by octahedral cation vacancy. The oxygen position parameter has been refined from X-ray intensity data at 10 pressures over this range. With a K' fixed at 4.0, the bulk modulus of the Si tetrahedron is 245 ± 31 GPa and that of the octahedral site is 150 ± 7 GPa. Consistent with a previous study, we observe a systematic decrease of bulk modulus with H-content. From these data, it appears that hydration of ringwoodite has a larger effect on seismic velocity than temperature within the possible ranges of these parameters under upper mantle conditions. This means that in tomographic images of the Transition Zone in regions distant from subduction zones, blue is more likely to mean 'dry' than it is to mean 'cold'. Observed seismic velocities in the Transition Zone are consistent with significant hydration of the ringwoodite (γ-(Mg, Fe)$_2SiO_4$) structure.

Keywords: Ringwoodite; High pressure; Equation of state; Crystal structure; Water

1. Introduction

Earth is distinguished from its planetary neighbors by the presence of large amounts of liquid water on its surface. Sea level variation studies indicate that sea level has varied relatively little, but perhaps as much as s few hundred meters, through the Phanerozoic (Duval et al., 1998). The presence of quartz-pebble conglomerates of early Archean age indicates that there has been running water (implying the presence of both oceans and land) nearly as far back as we can see in geologic time. Recent geochemical studies of ancient zircons in Archean sediments by Mojzsis et al. (2001) indicate that there may have been liquid water as far back as 4.3 Gy ago. Although the oceans cover 72% of the surface area, they constitute only 0.025% of the planet's mass. Silicate minerals of the crust and mantle can incorporate many times this amount of water, so it is possible that these silicates have played a major role in maintaining Earth's oceans over geologic time.

Approximately 65% of the total mass of the planet is composed of silicate rocks of the mantle and crust in which the only significant anionic species is oxygen. Bulk hydrogen content is perhaps the most poorly

* Corresponding author. Tel.: +1-303-492-5521; fax: +1-303-492-2602.
E-mail address: joseph.smyth@colorado.edu (J.R. Smyth).

0031-9201/$ – see front matter © 2004 Elsevier B.V. All rights reserved.
doi:10.1016/j.pepi.2003.08.011

constrained compositional variable in the planet, and the total water content of the planet is unknown within an order of magnitude (Drake and Righter, 2002). Olivine, generally believed to be the most abundant mineral phase in the upper 400 km, can incorporate up to 0.2 weight percent H_2O (Kohlstedt et al., 1996) at 13 GPa and 1100 °C. The maximum water content of pyroxenes is not known, but natural clinopyroxenes have been reported with up to about 0.18 wt.% H_2O (Rossman and Smyth, 1990; Smyth et al., 1991). In the Transition Zone (410–660 km), wadsleyite (β-Mg_2SiO_4) can incorporate up to about 3.3 wt.% H_2O, (Kohlstedt et al., 1996; Inoue et al., 1995), and ringwoodite (γ-Mg_2SiO_4) has been reported with up to about 2.2 wt.% (Kohlstedt et al., 1996; Kudoh et al., 2000). Thus, if saturated, these nominally anhydrous phases can incorporate at least ten times the amount of water in the planet's oceans and may possibly constitute the planet's largest reservoir of water as H in solid silicates.

In order to test this hypothesis, we have synthesized and characterized single-crystals of hydrous high-pressure silicate minerals for measurement of various physical properties. We have synthesized hydrous ringwoodite of likely mantle compositions, and determined the mechanisms of hydration by means of X-ray single-crystal diffraction, transmission electron microscopy (TEM) and infrared (IR) spectroscopy (Smyth et al., 2003). The principal hydration mechanism in these samples appears to be octahedral cation vacancy, although there is also a small amount of tetrahedral vacancy. In addition to the X-ray structural characterization, Kleppe et al. (2002) have described the Raman spectra to pressures above 50 GPa, and Kavner and Duffy (2001) have studied the deformation strength of several ringwoodites. The crystals are dark blue in color and are too dark for Brillouin spectroscopy, but gigahertz ultrasonic measurements of elastic properties are reported in a companion paper (Jacobsen et al., 2003).

The elastic properties of silicate spinels have been the subject of considerable recent interest. Bulk moduli of natural and synthetic ringwoodites have been measured by powder and single-crystal X-ray diffraction, Brillouin spectroscopy and by ultrasonic methods. From synchrotron X-ray powder diffraction to 50 GPa in the diamond anvil cell, Zerr et al. (1993) report a bulk modulus of 183.1 ± 2.0 GPa with K' of 5.4 ± 0.2 for a sample of $Fe_{0.8}Mg_{1.2}SiO_4$. From single-crystal X-ray diffraction in the diamond cell to more modest pressures (up to 5.5 GPa), Finger et al. (1977) report a bulk moduli for Ni_2SiO_4 and Fe_2SiO_4 of 227 ± 4 and 196 ± 4 GPa, respectively with K' also fixed to 4.0. These authors also were able to refine the crystal structure at pressure and report atom position parameters for oxygen at several pressures. Hazen (1993) measured the compression of several anhydrous ringwoodites to about 5 GPa, (Ni_2SiO_4, Mg_2SiO_4, $Mg_{0.8}Fe_{1.2}SiO_4$, $Mg_{0.4}Fe_{1.6}SiO_4$, and Fe_2SiO_4). He reported bulk moduli of 233 ± 2, 184 ± 2, 203 ± 2, 205 ± 2, 207 ± 3 GPa, respectively with K' fixed at 4.8. These data are in good agreement with values of 184 ± 3 GPa, (Weidner et al., 1986) and 226 ± 2 GPa (Bass et al., 1984) for Mg_2SiO_4 and Fe_2SiO_4, respectively by X-ray powder diffraction and Brillouin spectroscopy.

Yusa et al. (2000) report a very sharply reduced bulk modulus of 148 ± 6 GPa with K' of 7.9 for a hydrous, pure-Mg ringwoodite containing 2.2 wt.% H_2O. So, previous studies indicate an increase of bulk modulus from 184 to 207 for anhydrous compositions from Mg_2SiO_4 to Fe_2SiO_4 and a dramatic decrease of bulk modulus with hydration. Prior to the work of Kohlstedt et al. (1996) it was not realized that ringwoodite could incorporate significant amounts of hydroxyl, so the material was not routinely analyzed for H. Clearly, hydration will have a much larger effect on velocities and velocity contrasts at 525 and 660 km than Fe content (Yusa et al., 2000). In order to quantify and refine the effect of hydration on the bulk modulus and crystal structure of ringwoodite at pressure, we have undertaken a study of a Fo_{90} ringwoodite containing 0.93 wt.% H_2O by single-crystal X-ray diffraction in the diamond anvil cell.

2. Experimental methods

2.1. Synthesis and characterization

Details of the synthesis and room pressure characterization of the material are reported by Smyth et al. (2003) as their sample number SZ9901. Briefly, the sample was synthesized in the 5000 t press at the Bayerisches Geoinstitut (BGI) at 20 GPa and 1400 °C. The starting material was a mixture of San Carlos olivine, brucite, quartz, and FeO. The run (heating) duration was approximately 5 h. The capsule contained crystals of hydrous ringwoodite up to 800 μm in size plus minor amounts of stishovite. The crystals were analyzed by electron microprobe and by IR spectroscopy for H_2O. The composition is approximately Fo_{90} with 0.93 wt.% H_2O.

2.2. X-ray diffraction: unit cell parameters

High pressure X-ray diffraction studies on the ringwoodites were preformed on a Siemens/Bruker P4 automated diffractometer with an 18 kW rotating Mo anode source operating at 50 kV and 250 mA. The diffractometer has a graphite-crystal monochromator diffracting in the horizontal plane. This geometry causes the monochromator to induce small shifts in the proportions of the alpha-one and alpha-two peaks included in low-angle diffractions where the two wavelengths are not effectively separated as is true for all reflections observed in the diamond anvil cell. This requires that the wavelength of the mixed alpha-one-two peak be calibrated frequently against a known standard. The wavelength of the Mo mixed K-alpha was determined from cell refinement of a standard ruby sphere immediately before and after each cell determination. The unit cell parameters of the ringwoodite and quartz were then determined by refinement from the centering angles of five to eight diffraction peaks in each of eight octants in bisecting mode. Cell parameter precisions determined in this manner are generally better than one part in 10,000 and reproducible to about one part in 5000.

A single-crystal approximately 80 μm × 60 μm × 30 μm was loaded in a diamond anvil cell along with a crystal of quartz of similar size and a few small ruby crystals. The pressure medium was a mixture of methanol–ethanol–water at a ratio of 16:3:1. At each pressure the unit cell volumes of the ringwoodite and the quartz were refined as outlined above. The pressure was determined based on the bulk modulus and zero pressure volume of quartz (Angel et al., 1997). The cell parameters of the quartz and ringwoodite at each pressure are reported in Table 1. The error in pressure determination by this method is 0.15–0.5 kbar (Table 1) which is considerably less than could be obtained from the ruby fluorescence system which is about 1.0 kbar. We therefore used the quartz scale throughout this study. The unit cell volume data of ringwoodite was fit to a third order Burch–Murnaghan

Table 1
Unit cell parameters of hydrous ringwoodite and quartz at various pressures

Pressure (GPa)	Quartz			Ringwoodite	
	a (Å)	*c* (Å)	*V* (Å^3)	*a* (Å)	*V* (Å^3)
0.0 (0)				8.0936 (7)	530.18 (14)
0.381 (14)	4.896 (1)	5.389 (2)	111.86 (4)	8.0909 (9)	529.64 (17)
0.546 (18)	4.887 (1)	5.386 (2)	111.40 (5)	8.0868 (11)	528.84 (22)
1.052 (15)	4.865 (1)	5.370 (2)	110.06 (4)	8.0796 (11)	527.45 (22)
1.366 (12)	4.853 (1)	5.358 (1)	109.28 (5)	8.0766 (4)	526.85 (9)
1.449 (21)	4.851 (1)	5.352 (2)	109.08 (5)	8.0752 (11)	526.58 (21)
1.730 (22)	4.839 (1)	5.346 (2)	108.42 (5)	8.0701 (12)	525.57 (25)
1.978 (17)	4.829 (1)	5.338 (2)	107.86 (4)	8.0664 (8)	524.85 (15)
2.640 (24)	4.806 (1)	5.322 (2)	106.45 (5)	8.0564 (10)	522.90 (19)
2.869 (15)	4.797 (1)	5.322 (2)	105.99 (3)	8.0531 (7)	522.27 (14)
3.320 (16)	4.783 (1)	5.306 (2)	105.12 (3)	8.0471 (5)	521.10 (11)
3.530 (16)	4.774 (1)	5.307 (1)	104.73 (3)	8.0435 (7)	520.39 (13)
5.474 (26)	4.7166 (8)	5.2685 (15)	101.50 (4)	8.0189 (9)	515.63 (18)
6.714 (22)	4.685 (1)	5.247 (1)	99.73 (3)	8.0029 (9)	512.56 (17)
7.105 (31)	4.6739 (7)	5.2442 (10)	99.21 (4)	7.9974 (11)	511.49 (22)
7.835 (16)	4.6560 (5)	5.2351 (8)	98.28 (2)	7.9886 (11)	509.82 (21)
8.600 (34)	4.6391 (7)	5.2242 (10)	97.32 (4)	7.9801 (10)	508.19 (20)
9.283 (26)	4.6245 (7)	5.2145 (9)	96.58 (3)	7.9749 (14)	507.19 (27)
9.812 (28)	4.6125 (8)	5.2101 (13)	96.00 (3)	7.9683 (13)	505.94 (25)
10.535 (32)	4.5973 (7)	5.2032 (10)	95.24 (4)	7.9584 (12)	504.05 (24)

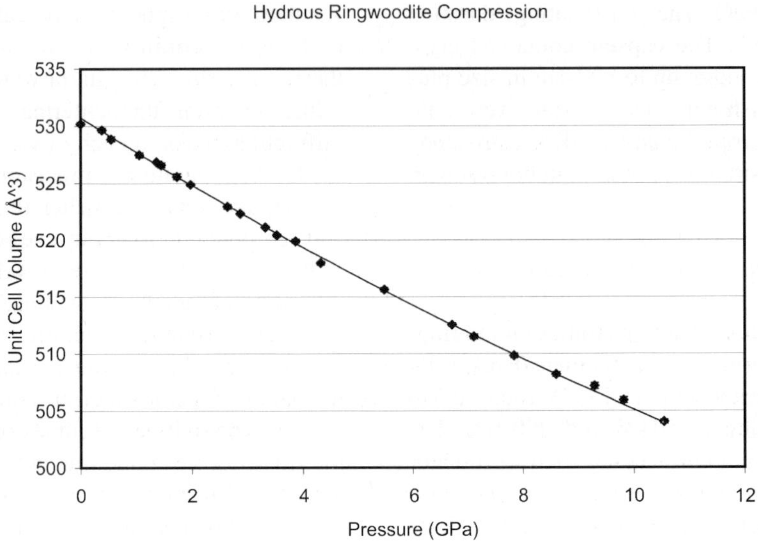

Fig. 1. Unit cell volume of Fo_{90} hydrous ringwoodite as a function of pressure. Symbols contain error bars. Curve is the fit equation of state with $K_0 = 169.0 \pm 3.4$ GPa and $K' = 7.9 \pm 0.9$.

equation of state using the program EOSFIT42 (Angel, 1999 personal communication). The data give a bulk modulus of 169.0 ± 3.4 GPa with a K' of 7.9 ± 0.9. The individual unit cell volumes and the calculated equation of state are shown in Fig. 1.

2.3. X-ray diffraction: intensity data collection

At pressures of 0.44, 0.76, 1.37, 2.64, 7.11, 8.60, 9.81, and 11.22 GPa, the crystal structure of the ringwoodite was refined from intensity measurements of accessible diffractions. Data were corrected for Lorenz and polarization effects and for absorption of the diamonds using the program DACABS (R.T. Downs, personal communication). Refinements of the structure parameters were done using SHELX97 (Sheldrick, 1997) as contained in the package WinGX. Refinement was done by least-squares minimization of F^2. The cation occupancies and atomic displacement parameters were fixed to the room pressure values obtained from a full-octant intensity data set to 90° 2θ. By this method only the oxygen position parameter was refined. The site occupancies and displacement parameters were fixed to those given by Smyth et al. (2003). Final position parameter and $R(F)$-values for structure refinements at each pressure are given in Table 2 along with calculated polyhedral volumes.

Table 2
Crystal structure parameters of hydrous ringwoodite at various pressures

P (GPa)	Total	Obs.	Uniq.	*R* (%)	*a* (Å)	Xox	Tetrahedral volume	Octahedral volume	Non-polyhedral volume
0.437 (2)	352	17	13	3.81	8.0907 (6)	0.2438 (4)	2.368	11.854	321.004
0.763 (2)	184	18	13	3.46	8.0862 (7)	0.2442 (2)	2.389	11.780	321.141
1.37 (2)	184	19	15	3.19	8.0766 (4)	0.2440 (3)	2.369	11.762	319.704
2.64 (2)	267	20	15	3.28	8.0564 (10)	0.2442 (3)	2.364	11.648	317.634
7.11 (3)	301	38	10	3.76	7.9974 (11)	0.2446 (7)	2.334	11.345	311.302
8.60 (3)	406	32	10	3.33	7.9801 (10)	0.2444 (6)	2.308	11.296	308.993
9.81 (3)	408	24	9	2.86	7.9683 (13)	0.2440 (6)	2.296	11.249	307.578
11.22 (5)	315	51	13	2.49	7.9473 (10)	0.2448 (3)	2.300	11.113	305.681

3. Results and discussion

3.1. Polyhedral compression

There are relatively few high-pressure structure studies of silicate spinels in the literature. Finger et al. (1977, 1979) report structure data for Fe_2SiO_4 and Ni_2SiO_4 to 4.0 and 5.5 GPa, respectively. Finger et al. (1979) give oxygen position parameters at pressures to 5.5 GPa for nickel silicate spinel report a polyhedral bulk modulus of 170 ± 10 GPa for the octahedron and >250 GPa for the tetrahedron in Ni_2SiO_4 spinel. However, for the Fe_2SiO_4, they do not observe significant variation of the oxygen position parameter with pressure and do not report polyhedral bulk moduli. Polyhedral bulk moduli for silicate tetrahedra in orthosilicates range from about 190 to over 550 GPa (Smyth et al., 2000). The Mg octahedron bulk moduli in forsterite range from 100 to 140 GPa (Hazen, 1976; Kudoh and Takéuchi, 1985), and Fei (1999) reported a value of 160 ± 2 GPa for periclase (MgO).

The polyhedral volume compression of the tetrahedron, octahedron, and the volume of the unit cell not assigned to either polyhedron are plotted in Fig. 2. As expected, we observe a stiffer tetrahedron than octahedron. The polyhedral volume data were fit to a third order Birch–Murnaghan equation of state. The precision of the data does not permit a significant refinement of K'. However, with K' fixed at 4.0, we obtain bulk moduli of 150 ± 7, and 245 ± 30 GPa for the octahedron and tetrahedron, respectively. With K' fixed to 7.9, the value obtained from the unit cell refinement, we obtain values of 134 ± 6, and 228 ± 30 GPa. So the octahedral site in this ringwoodite is more somewhat compressible than those in MgO or Ni_2SiO_4 spinel and comparable to the octahedral sites in olivine. The bulk modulus of the tetrahedron is comparable to other silicate structures, although there are very few structure refinements in the literature at pressures greater than 6 GPa (Smyth et al., 2000). Surprisingly, the non-polyhedral volume is less compressible than the octahedra. The non-polyhedral volume composes approximately 60% of the total cell volume in this structure. The large bulk modulus (low compressibility) is probably the result of the near cubic-closest-packing arrangement of the oxygen atoms in the spinel structure (Hazen and Yang, 1999).

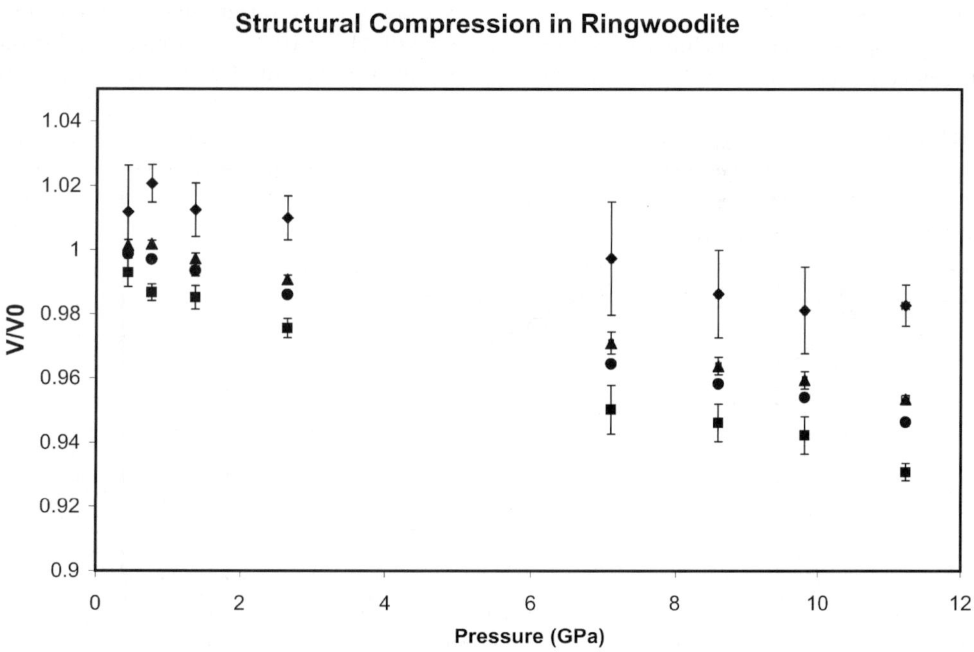

Fig. 2. Relative volume compression of the structural components of hydrous ringwoodite—diamond: tetrahedron; square: octahedron; triangle: non-polyhedral volume, and circle: cell volume. It is clear that most of the compression takes place in the octahedra.

3.2. Equation of state

Plotted in Fig. 1 are the unit cell volumes as a function of pressure. Also plotted here is the fitted third order Birch–Murnaghan equation of state curve with a V_0 of 530.77 Å, K_0 of 169.0 ± 3.4 and K' of 7.9 ± 0.9. The error bars plotted in the figure represent the internal precision in the X-ray data, and do not include systematic errors. The pressures were determined from the equation of state of quartz (Angel et al., 1997). The maximum pressure here is slightly above the maximum in the quartz study and we used a simple extrapolation. The slight flexure in the curve above 9 GPa may be due to the extrapolation. It is also possible that the pressure medium may be causing anisotropic strain above about 9 GPa. To test this hypothesis, we refined the moduli without the data point above 9 GPa and obtained K and K' value not statistically distinct from those reported above. Hazen (1993) refined a value of K' of 6 ± 2 for anhydrous ringwoodite. The presence of H and octahedral cation vacancy in the current structure may account for the relatively large K', but in light of the Hazen (1993) study, the value obtained here is not inconsistent with previous studies. Further study of this material by powder diffraction to higher pressures is in progress.

In Fig. 3 is plotted the bulk modulus as a function of hydrogen content for three hydrous and anhydrous ringwoodites. The bulk modulus of anhydrous pure-Mg ringwoodite reported by Hazen (1993) is 182 ± 2 and for Fe_2SiO_4 is 207 ± 2, so one would expect a value of about 184.5 for Fo_{90} anhydrous ringwoodite. In a powder-diffraction compression study using synchrotron radiation, Yusa et al. (2000) report a K_0 of 148 GPa for a very hydrous ringwoodite with 2.2 wt.% H_2O. The present data are consistent with a near linear trend of bulk modulus with H-content. This variation of bulk modulus with H-content is much larger than the variation with either temperature or Fe content within reasonable ranges of these parameters for the Transition Zone.

3.3. Relative effects of hydration versus temperature

In order to compare the relative effects of temperature and hydration on the bulk modulus we estimated the effective bulk moduli of hydrous and anhydrous ringwoodite at 1200 °C and 20 GPa. To make the estimate we used the V_0 values for Fo_{100} ringwoodite proposed by Smyth et al. (2003) namely 524.1 Å^3 for anhydrous and 525.3 Å^3 for 1% H_2O. We assumed a K' of 6 for anhydrous (Hazen, 1993) and 7.9 for 1% H_2O. We assumed a linear thermal expansion coefficient (α_0) of 24.7×10^{-6} °C^{-1} and an Anderson–Gruneisen parameter, δ_T, of 5.8 (Meng et al., 1993) for all

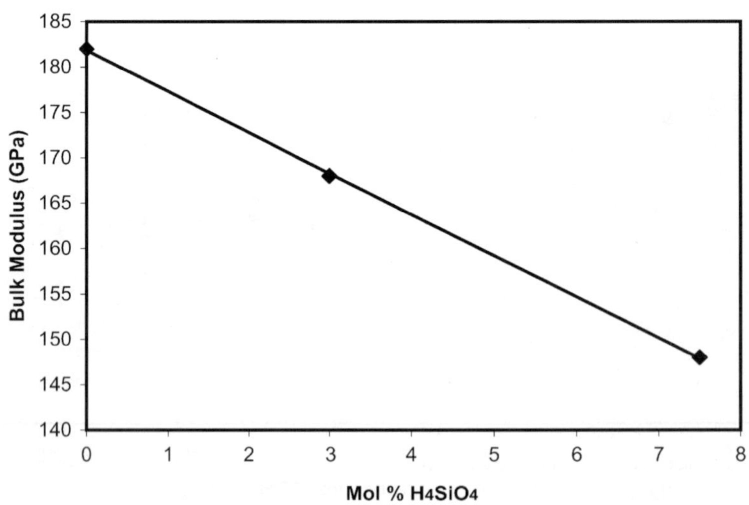

Fig. 3. Bulk modulus vs. H-content of ringwoodite. Data for the anhydrous sample is from Hazen (1993) and that most hydrous sample is from Yusa et al. (2000). The cell bulk modulus indicates a linear trend with H-content.

samples. We then estimated the bulk modulus for the anhydrous sample at 1200 °C to be about 156 GPa, and at 600 °C to be about 170 GPa. From these very rough estimates we can see that the effect of hydrating ringwoodite to 1 wt.% has approximately the same effect on the bulk modulus as raising the temperature by 600 °C. Lateral temperature variation of this magnitude can only be present near subduction zones.

The bulk modulus obtained from static compression measurements is the isothermal bulk modulus, K_T. The adiabatic bulk modulus, K_S, is equal to K_T $(1 + \alpha\gamma T)$ which is equal to ρV_P^2, where α is volumetric thermal expansion and γ, the Grüneisen parameter. The $\alpha\gamma T$ term is relatively small and the K_T is only slightly smaller than K_S, so it is reasonable to assume that hydration will have a similar effect on K_S as on K_T. This means that hydration will have a larger effect than temperature on P-velocity, so that in P-wave tomographic images of the Transition Zone in regions distant from subduction zones, blue is more likely to mean 'dry' than it is to mean 'cold'. Conversely, red is more likely to mean 'wet' than 'hot', and inferring lateral thermal anomalies based on velocity variations is, at best, not well constrained. In a companion paper (Jacobsen et al., 2003), we have measured P and S velocities in this sample at ambient conditions and obtain a similar result for P velocities, but it appears that hydration has a larger effect on shear velocity than on P-velocity, so that V_P/V_S increases with hydration.

Pyrolite models of mantle composition typically contain more than seventy modal percent olivine, and mantle peridotite xenoliths are typically more than ninety modal percent olivine. However, the P and S velocities of ringwoodite are too fast relative to PREM (Dziewonski and Anderson, 1981) to have more than about 50 modal percent ringwoodite in the lower part of the Transition Zone (Duffy and Anderson, 1989; Sinogeikin et al., 1998). Inclusion of the hydration variable places the velocities of a nearly pure ringwoodite composition close to those of PREM and weakens possible constraints on the modal proportions of ringwoodite in the lower Transition Zone (525–660 km depth). However, the possibility of hydration may have a significant effect on the depths of the 410 (Smyth and Frost, 2002) and 660 km depth. Jacobsen et al. (2003) have measured a complete set of elastic constants for this material by ultrasonic methods, and report both P and S velocities at ambient conditions. They have found an (adiabatic) bulk modulus of 168 GPa in close agreement with the present study. Further study of these interesting samples may be able to provide better constraint on the P and S velocities at pressure.

Acknowledgements

This research was supported the US National Science Foundation grants EAR 00-87279 and EAR 02-29315 and by the Bayerisches Geoinstitut Visitor Program. SDJ was supported by an Alexander von Humboldt Foundation fellowship.

References

Angel, R.J., Allan, D.R., Miletich, R., Finger, L.W., 1997. The use of quartz as an internal pressure standard in high pressure crystallography. J. Appl. Cryst. 30, 461–466.

Bass, J.D., Weidner, D.J., Hamaya, N., Ozima, M., Akimoto, S., 1984. Phys. Chem. Miner. 10, 261–266.

Drake, M.J., Righter, K., 2002. Determining the composition of the Earth. Nature 416, 39–44.

Duval, B.C., Cramez, C., Vail, P.R., 1998. Stratigraphic cycles and major marine source rocks. Mesozoic and Cenozoic sequence stratigraphy of European basins. SEPM Spectrosc. Pub. 60, 43–51.

Duffy, T.S., Anderson, D.L., 1989. Seismic velocities in mantle minerals and the mineralogy of the upper mantle. J. Geophys. Res. 94, 1895–1912.

Dziewonski, A., Anderson, D.L., 1981. Preliminary reference Earth model. Phys. Earth. Planet. Int. 25, 297–356.

Fei, Y., 1999. Effects of temperature and composition on the bulk modulus of (Mg, Fe)O. Am. Mineral. 84, 272–276.

Finger, L.W., Hazen, R.M., Yagi, T., 1977. High-pressure crystal structures of spinel polymorphs of Fe_2SiO_4 and Ni_2SiO_4. Carnegie Inst. Wash Yearb. 76, 504–505.

Finger, L.W., Hazen, R.M., Yagi, T., 1979. Crystal structures and electron densities of nickel and iron silicate spinels at elevated temperature or pressure. Am. Mineral. 64, 1002–1009.

Hazen, R.M., 1976. Effects of temperature and pressure on the crystal structure of forsterite. Am. Mineral. 61, 1280–1293.

Hazen, R.M., 1993. Comparative crystal chemistry of silicate spinels: anomalous behavior of $(Mg, Fe)_2SiO_4$. Science 259, 206–209.

Hazen, R.M., Yang, H., 1999. Effects of cation substitution and order disorder on P-V-T equations of state of cubic spinels. Am. Mineral. 84, 1956–1960.

Inoue, T., Yurimoto, H., Kudoh, Y., 1995. Hy, $Mg_{1.75}SiH_{0.5}O_4$: a new water reservoir in the mantle transition region. Geophys. Res. Lett. 22, 117–120.

Jacobsen, S.D., Smyth, J.R., Spetzler, H., 2003. Sound velocities and elasticity of Fe-bearing hydrous ringwoodite. Phys. Earth Planet. Int., this volume.

Kavner, A., Duffy, T.N., 2001. Strength and elasticity of ringwoodite at upper mantle pressures. Geophys. Res. Lett. 28, 2691–2694.

Kleppe, A.K., Jephcoat, A.P., Smyth, J.R., Frost, D.J., 2002. On protons, iron, and the high pressure behaviour of ringwoodite. Geophys. Res. Lett. 29, DOI:10.1029/2002GL015276.

Kohlstedt, D.L., Keppler, H., Rubie, D.C., 1996. The solubility of water in α, β and γ phases of $(Mg, Fe)_2SiO_4$. Contrib. Mineral. Petrol. 123, 345–357.

Kudoh, Y., Takéuchi, T., 1985. The crystal structure of forsterite Mg_2SiO_4 under high pressure up to 149 kbar. Z. Krist. 117, 292–302.

Kudoh, Y., Kuribayashi, T., Mizohata, H., Ohtani, E., 2000. Structure and cation disorder of hydrous ringwoodite, γ-$Mg_{1.89}Si_{0.97}H_{0.34}O_4$. Phys. Chem. Miner. 27, 474–479.

Meng, Y., Weidner, D.J., Gwanmesia, G.D., Liebermann, R.C., Vaughan, M.T., Wang, Y., Leinenweber, K., Pacalo, R.E., Yeganeh-Haeri, A., Zhao, Y., 1993. In Situ high P-T X-ray diffraction studies on three polymorphs (α, β, γ) of Mg_2SiO_4. J. Geophys. Res. 98, 22199–22207.

Mojzsis, S.J., Harrison, T.M., Pidgeon, R.T., 2001. Oxygen isotope evidenc4e from ancient zircons for liquid water at the Earth's surface 4,300 Myr ago. Nature 409, 178–181.

Rossman, G.R., Smyth, J.R., 1990. Hydroxyl and ammonia contents of accessory minerals in mantle eclogites and related rocks. Am. Mineral. 75, 775–780.

Sheldrick, G.M., 1997. SHELXL-97 A program for crystal structure refinement. University of Goettingen, Germany, Release 97-2.

Sinogeikin, S.V., Katsura, T., Bass, J.D., 1998. Sound velocities and elastic properties of Fe-bearing wadsleyite and ringwoodite. J. Geophys. Res. 103 (20), 919–20825.

Smyth, J.R., Bell, D.R., Rossman, G.R., 1991. Hydrous clinopyroxenes from the upper mantle. Nature 351, 732–735.

Smyth, J.R., Jacobsen, S.D., Hazen, R.M., 2000. Comparative crystal chemistry of the orthosilicate minerals. Rev. Mineral. 41, 187–209.

Smyth J.R., Frost, D.J., 2002. The effect of water on the 410-km discontinuity: an experimental study. Geophys. Res. Lett., doi 2002GL014418.

Smyth, J.R., Holl, C.M., Frost, D.J., Jacobsen, S.D., Langenhorst, F., McCAmmon, C.A., 2003. Structural systematics of hydrous ringwoodite and water in the Earth's interior. Am. Mineral. 88, 1402–1407.

Weidner, D.J., Sawamoto, H., Sasaki, S., Kumazawa, M., 1986. Single crystal slastic properties of spinel phase of Mg_2SiO_4. J. Geophys. Res. 89, 7852–7860.

Yusa, H., Inoue, T., Ohishi, Y., 2000. Isothermal compressibility of hydrous ringwoodite and its relation to mantle discontinuities. Geophys. Res. Lett. 27, 413–416.

Zerr, A., Reichmann, H.-J., Euler, H., Boehler, R., 1993. Hydrostatic compression of γ-$(Mg_{0.6}Fe_{0.4})_2SiO_4$ to 50.0 GPa. Phys. Chem. Miner. 19, 507–509.

ELSEVIER

Available online at www.sciencedirect.com

SCIENCE @ DIRECT®

Physics of the Earth and Planetary Interiors 143–144 (2004) 279–290

PHYSICS OF THE EARTH AND PLANETARY INTERIORS

www.elsevier.com/locate/pepi

Thermal expansion of wadsleyite, ringwoodite, hydrous wadsleyite and hydrous ringwoodite

Toru Inoue[a,*], Yasuchika Tanimoto[a], Tetsuo Irifune[a], Taku Suzuki[b], Hiroshi Fukui[c], Osamu Ohtaka[c]

[a] *Geodynamics Research Center, Ehime University, Bunkyo-cho 2-5, Matsuyama 790-8577, Japan*

[b] *Faculty of Environmental Engineering, The University of Kitakyushu, Hibikino 1-1, Wakamatsu-ku, Kitakyushu 808-0135, Japan*

[c] *Department of Space and Earth Sciences, Osaka University, Toyonaka 560-0043, Japan*

Received 7 February 2003; received in revised form 19 July 2003; accepted 21 July 2003

Abstract

The thermal expansions of wadsleyite, ringwoodite, hydrous wadsleyite, and hydrous ringwoodite have been determined by high temperature powder X-ray diffraction at ambient pressure. Wadsleyite and ringwoodite persist up to 700 °C, and their thermal expansion coefficients are measured as $34.0(5) \times 10^{-6}$ and $30.7(6) \times 10^{-6}\,K^{-1}$, respectively, values that are slightly larger than those measured by [J. Phys. Earth 27 (1979) 53; J. Phys. Earth 28 (1980) 273]. Hydrous wadsleyite and hydrous ringwoodite were dehydrated at 450 and 400 °C, respectively. Below their dehydration temperatures, their thermal expansion coefficients are $30.1(14) \times 10^{-6}$ and $27.3(9) \times 10^{-6}\,K^{-1}$, respectively, which are slightly lower than those of their anhydrous forms. Using the present results combined with the previous thermoelastic data, the seismic velocity contrast at the 410 km discontinuity has been calculated for both dry and wet mantle. If a pyrolite composition is assumed, a hydrous mantle transition zone better explains seismic constraints.

Keywords: Wadsleyite; Ringwoodite; Hydrous wadsleyite; Hydrous ringwoodite; Thermal expansion; Mantle transition zone

1. Introduction

Water is one of the main volatile components in the Earth and a number of recent studies have shown that it affects the physical properties and the melting behavior of mantle minerals (e.g. Inoue, 1994; Kawamoto and Holloway, 1997; Yusa and Inoue, 1997; Inoue et al., 1998). Water may be transported into the Earth's mantle transition zone by subducted slabs in the form of dense hydrous magnesium silicates (DHMS) (e.g. Ohtani et al., 2001). Wadsleyite and ringwoodite are believed to be main constituent minerals in the Earth's mantle transition zone, and it has recently been shown that these phases can contain up to ~3 wt.% H_2O in their crystal structures. Thus, it is important to define the physical properties of hydrous wadsleyite and hydrous ringwoodite in order to discuss the mineralogy of and water storage in the Earth's mantle transition zone. The thermal expansivity of hydrous wadsleyite and hydrous ringwoodite, however, have not yet been determined. We have measured the thermal expansions of hydrous wadsleyite and hydrous ringwoodite, and use our results to discuss the possibility of water storage in the transition zone, by comparing

* Corresponding author. Tel.: +81-89-927-9658; fax: +81-89-927-8167.
E-mail address: inoue@sci.ehime-u.ac.jp (T. Inoue).

0031-9201/$ – see front matter
doi:10.1016/j.pepi.2003.07.021

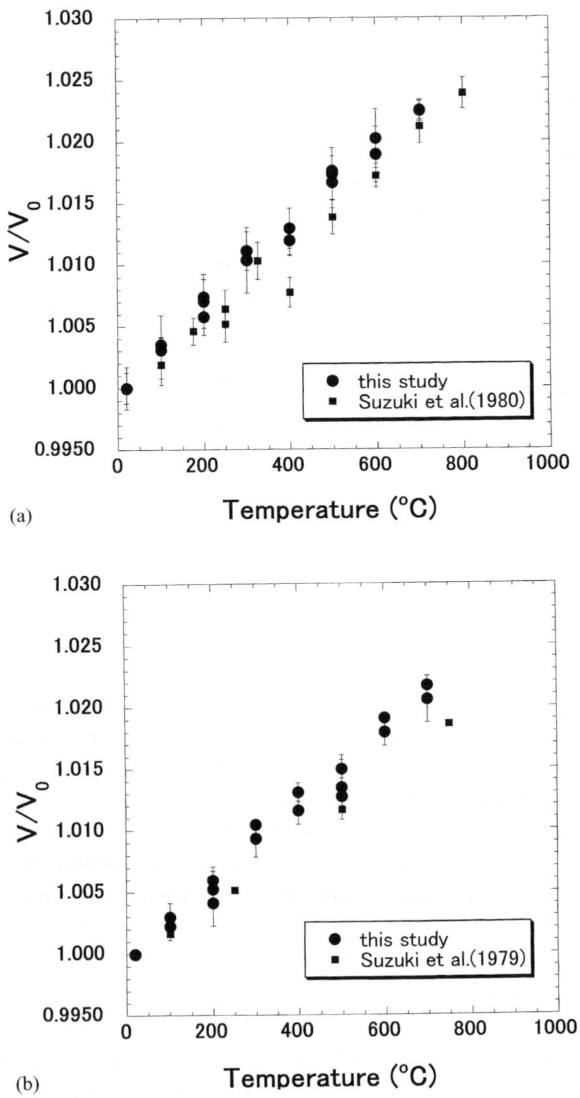

Fig. 4. Thermal expansion of (a) anhydrous wadsleyite and (b) anhydrous ringwoodite. Circles indicate this study and squares indicate the data of Suzuki et al. (1979, 1980).

The dehydration of hydrous ringwoodite was indicated by the abrupt decrease of volume, i.e. the abrupt decrease in the length of the *a*-axis, because the a-axis is larger than that in anhydrous ringwoodite, and should decrease upon dehydration. Liu et al. (2002) reported that hydrous ringwoodite sometimes converted to a defective forsterite structure above 527 °C. The present dehydration temperature was much lower than that reported by Liu et al. (2002).

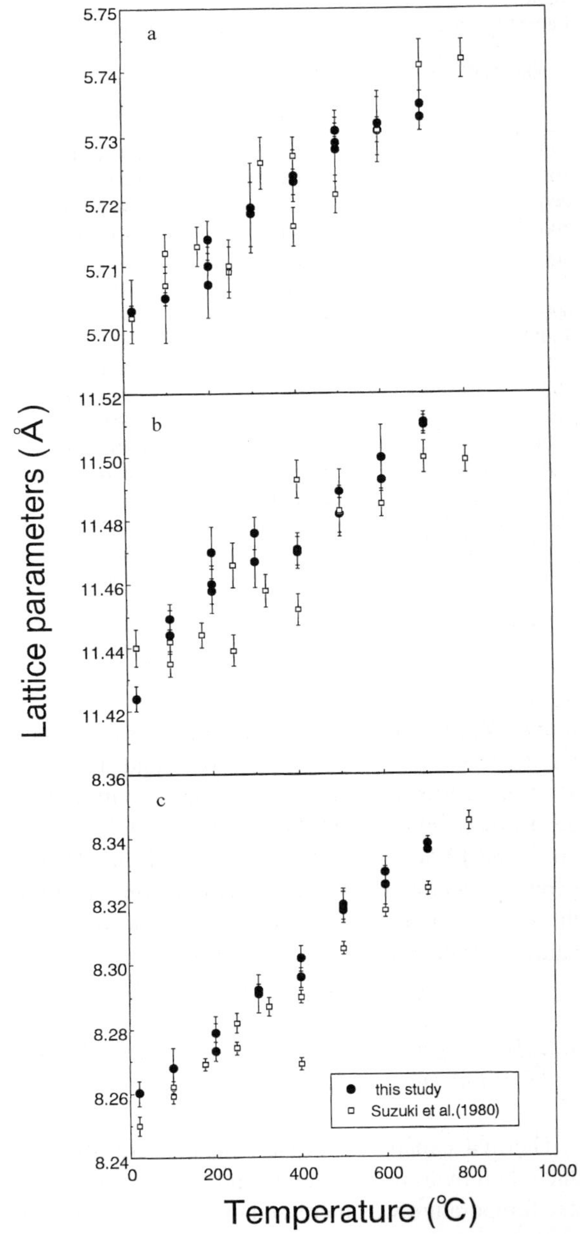

Fig. 5. Thermal expansion of each axis of anhydrous wadsleyite. Circles indicate this study and squares indicate the data of Suzuki et al. (1980).

We used the data up to 350 °C to calculate the thermal expansion coefficients of hydrous wadsleyite and hydrous ringwoodite, which are 30.1(14) × 10^{-6} and 27.3(9) × 10^{-6} K^{-1}, respectively, values that are

Table 2
Lattice parameters and the unit cell volumes of hydrous wadsleyite and hydrous ringwoodite as a function of temperature

Temperature (°C)	Hydrous wadsleyite				Hydrous ringwoodite	
	a (Å)	*b* (Å)	*c* (Å)	*V* (Å^3)	*a* (Å)	*V* (Å^3)
30	5.679 (1)	11.520 (9)	8.252 (3)	539.9 (7)	8.085 (2)	528.5 (4)
30R	5.680 (3)	11.521 (12)	8.255 (3)	540.2 (10)	8.087 (2)	528.9 (4)
30R2	5.681 (2)	11.523 (10)	8.254 (4)	540.3 (9)		
50	5.680 (1)	11.525 (7)	8.254 (2)	540.3 (6)	8.087 (3)	528.9 (6)
100	5.686 (2)	11.521 (13)	8.258 (5)	541.0 (11)	8.091 (1)	529.7 (2)
150	5.686 (2)	11.522 (11)	8.261 (4)	541.2 (10)	8.093 (1)	530.1 (2)
200	5.688 (2)	11.527 (12)	8.266 (4)	542.0 (10)	8.095 (1)	530.5 (2)
250	5.691 (2)	11.538 (10)	8.278 (3)	543.6 (9)	8.101 (2)	531.6 (4)
300	5.694 (1)	11.535 (11)	8.295 (3)	544.8 (8)	8.105 (2)	532.4 (4)
350	5.698 (2)	11.528 (13)	8.299 (4)	545.1 (11)	8.110 (0)	533.4 (0)
400	5.710 (1)	11.530 (10)	8.307 (3)	546.9 (8)	8.092 (1)	529.9 (2)
450	5.717 (1)	11.509 (10)	8.319 (4)	547.4 (8)	8.106 (3)	532.6 (6)
500	5.721 (1)	11.508 (9)	8.327 (4)	548.2 (8)	8.108 (1)	533.0 (2)

30R: measured under decreasing temperature from 350 °C; 30R2: measured under decreasing temperature from 400 °C, standard deviation is given in parenthesis.

slightly smaller than those of the anhydrous forms of these phases.

Liu et al. (1998, 2002) studied hydrous wadsleyite and hydrous ringwoodite at high temperature and ambient pressure using Raman spectroscopy, and inferred that the volume thermal expansivities for both phases are slightly less than those for the anhydrous forms, consistent with our present results. The thermal expansion coefficients of anhydrous wadsleyite, anhydrous ringwoodite, hydrous wadsleyite and hydrous ringwoodite are summarized in Table 3.

3.3. The implication for the possibility of the water storage in the mantle transition zone

We have reported the bulk moduli of hydrous wadsleyite and hydrous ringwoodite previously (Yusa and Inoue, 1997; Inoue et al., 1998; Yusa et al., 2000) (Table 4), and found that the bulk moduli of these phases are much smaller than those of their anhydrous forms. Using the present data in combination with the previous data, we calculated the bulk sound velocity jumps at the 410 km discontinuity in an anhydrous and hydrous mantle. The data used are listed in Table 5.

The bulk sound velocity ($V_\Phi(P, T)$) is defined by the following equation, using the parameters of adiabatic bulk modulus ($\mathrm{Ks}(P, T)$) and density ($\rho(P, T)$).

$$V_\Phi(P, T) = \sqrt{\frac{\mathrm{Ks}(P, T)}{\rho(P, T)}} \tag{3}$$

The bulk modulus at high pressure (P) and high temperature (T) ($\mathrm{Ks}(P, T)$) is defined by the following equation using the bulk modulus at ambient condition (Ks_0), the pressure derivative $(\partial \mathrm{Ks}_0/\partial P)_T$ and the

Table 3
Thermal expansion coefficients of anhydrous wadsleyite, anhydrous ringwoodite, hydrous wadsleyite and hydrous ringwoodite

	H_2O (wt.%)	α ($\times 10^{-6}$ K^{-1})	Temperature range (°C)	
Anhydrous wadsleyite	0	34.0 (5)	20–700	This study
	0	29.4 (8)	20–800	Suzuki et al. (1980)
Anhydrous ringwoodite	0	30.7 (6)	20–700	This study
	0	24.7 (5)	20–750	Suzuki et al. (1979)
Hydrous wadsleyite	2.4 (2)	30.1 (14)	30–350	This study
Hydrous ringwoodite	2.6 (3)	27.3 (9)	30–350	This study

Table 4
Comparison of the thermoelastic properties of anhydrous wadsleyite, hydrous wadsleyite, anhydrous ringwoodite and hydrous ringwoodite

	H_2O (wt.%)	K_0 (GPa)	K_0'	V_0 (Å^3)	ρ_0 (g/cm^3)	dKs/dT ($\times 10^{-2}$ GPa/K)	α ($\times 10^{-6}$ K^{-1})
Anhydrous wadsleyite	–	172[a]	4.2[a]	538.1[b]	3.473[b]	−1.2[a]	34.0[b]
Hydrous wadsleyite	2.4[b]	–	–	–	–	–	30.1[b]
	2.5[c]	155[c]	4.2[c]	539.2[c]	3.363[c]	–	–
Anhydrous ringwoodite	–	184[d]	5.0[e]	524.6[b]	3.562[b]	−2.0[c]	30.7[b]
Hydrous ringwoodite	2.2[f]	155[f]	–	527.2[f]	3.469[f]	–	–
	2.6[b]	–	–	–	–	–	27.3[b]
	2.8[g]	148[g]	5.0[g]	529.1[g]	3.431[g]	–	–

[a] Li et al. (1998).
[b] This study.
[c] Yusa and Inoue (1997).
[d] Weidner et al. (1984).
[e] Meng et al. (1993).
[f] Inoue et al. (1998).
[g] Yusa et al. (2000).

Table 5
Parameters used for density and bulk sound velocity calculation

	H_2O (wt.%)	ρ (g/cm^3)	Ks (GPa)	(dKs/dP)$_T$	(dKs/dT)$_P$ ($\times 10^{-2}$ GPa/K)	α ($\times 10^{-5}$ K^{-1})
Olivine	0	3.227[a]	129[b]	4.2[b]	−1.7[b]	3.752[c]
Anhydrous wadsleyite	0	3.473[d]	172[e]	4.2[e]	−1.2[e]	3.40[d]
Hydrous wadsleyite	2.4	3.367[f]	155.7[f]	4.2[g]	−1.2[g]	3.01[d]

[a] Fugino et al. (1981).
[b] Duffy et al. (1995).
[c] The mean coefficient between 300 and 1673 K from Suzuki (1975).
[d] This study.
[e] Li et al. (1998).
[f] Calculated based on Yusa and Inoue (1997).
[g] Assumed as the same as anhydrous wadsleyite.

Table 6
Density and bulk sound velocity at the 410 km discontinuity (13.5 GPa, 1400 °C)

	ρ (g/cm^3)	$(\rho_\beta-\rho_\alpha)/\rho_\alpha$ (%)	V_ϕ (km/s)	$(V_{\phi\beta}-V_{\phi\alpha})/V_{\phi\alpha}$ (%)
Olivine	3.42	5.3	7.06	9.8
Anhydrous wadsleyite	3.60		7.75	
Olivine	3.42	3.4	7.06	6.5
Hydrous wadsleyite	3.54		7.52	
Velocity contrast of seismological observation (Benz and Vidale, 1993; Nolet et al., 1994)				4.6

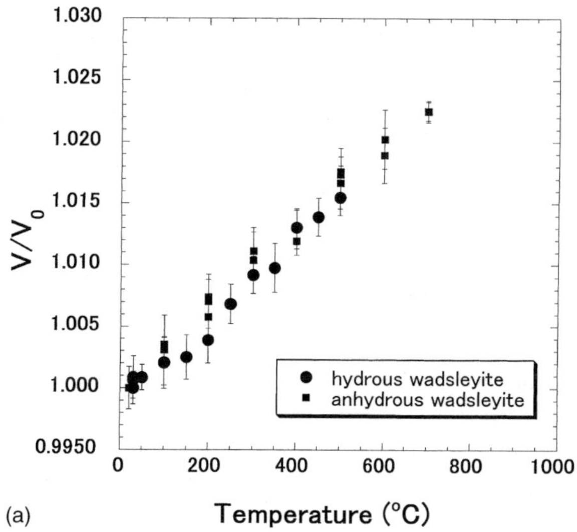

(a)

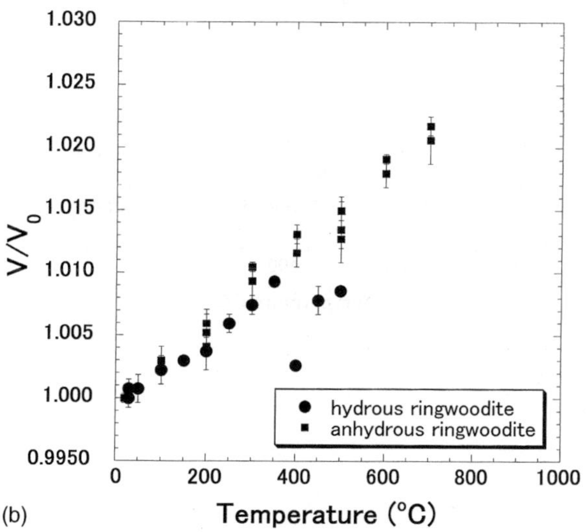

(b)

Fig. 6. Thermal expansion of (a) hydrous wadsleyite and (b) hydrous ringwoodite. Circles and squares indicate the results of hydrous and anhydrous forms, respectively.

temperature derivative $(\partial \mathrm{Ks}_0/\partial T)_P$.

$$\mathrm{Ks}(P,T) = \mathrm{Ks}_0 + \left(\frac{\partial \mathrm{Ks}_0}{\partial P}\right)_T (P - P_0) + \left(\frac{\partial \mathrm{Ks}_0}{\partial T}\right)_P \times (T - T_0) \quad (4)$$

The density at high pressure (P) and high temperature (T) ($\rho(P, T)$) is calculated from the density at ambient condition (ρ_0), the isothermal bulk modulus (K_T) and the thermal expansion coefficient (α).

$$\rho(P,T) = \rho_0 \left[1 + \frac{P - P_0}{K_T} - \alpha(T - T_0)\right] \quad (5)$$

The adiabatic and isothermal bulk moduli can be related by following equation.

$$\mathrm{Ks} = K_T(1 + \alpha\gamma T) \quad (6)$$

where α, γ and T are thermal expansion coefficient, Grüneisen parameter and absolute temperature. At ambient condition, this difference is only ~1%, within experimental error.

The calculated densities and bulk sound velocities of olivine, anhydrous wadsleyite and hydrous wadsleyite at the 410 km discontinuity, corresponding to 13.5 GPa and 1400 °C are listed in Table 6. Olivine can not contain significant H_2O compared to wadsleyite (Chen et al., 2002) in its crystal structure, so we calculated its density and bulk modulus in its anhydrous form. In anhydrous wadsleyite, the density and bulk sound velocity contrasts at the 410 km discontinuity are 5.3 and 9.8%, respectively, whereas in hydrous wadsleyite with 2.4 wt.% of H_2O, those are 3.4 and 6.5%, respectively. Thus, the presence of H_2O in wadsleyite reduces its density and sound velocity, and the contrasts at 410 km discontinuity are reduced correspondingly by the presence of H_2O.

The observed velocity contrast from seismological observation is estimated to be 4.6% from Benz and Vidale (1993) and Nolet et al. (1994). If water does not exist in the mantle, 45–50% olivine is suitable for the mantle transition zone. This is inconsistent with a pyrolite mantle, which contains ~60% olivine. This issue has already been discussed by some authors (e.g. Duffy and Anderson, 1989; Li et al., 1998). If we assume that water is present in the mantle transition zone, this discrepancy will be removed, because H_2O in wadsleyite reduces the velocity. Fig. 8 shows the bulk sound velocity contrast as a function of H_2O and olivine content. If we assume a pyrolite mantle, ~1.5 wt.% of H_2O is needed in wadsleyite. This value is extremely high, although wadsleyite can contain up to 3.3 wt.% water in its crystal structure, and thus from a crystal chemical view, this amount of H_2O (~1.5 wt.%) is possible (Inoue et al., 1995). For further progress, it will be necessary to examine the physical properties of hydrous wadsleyite under high

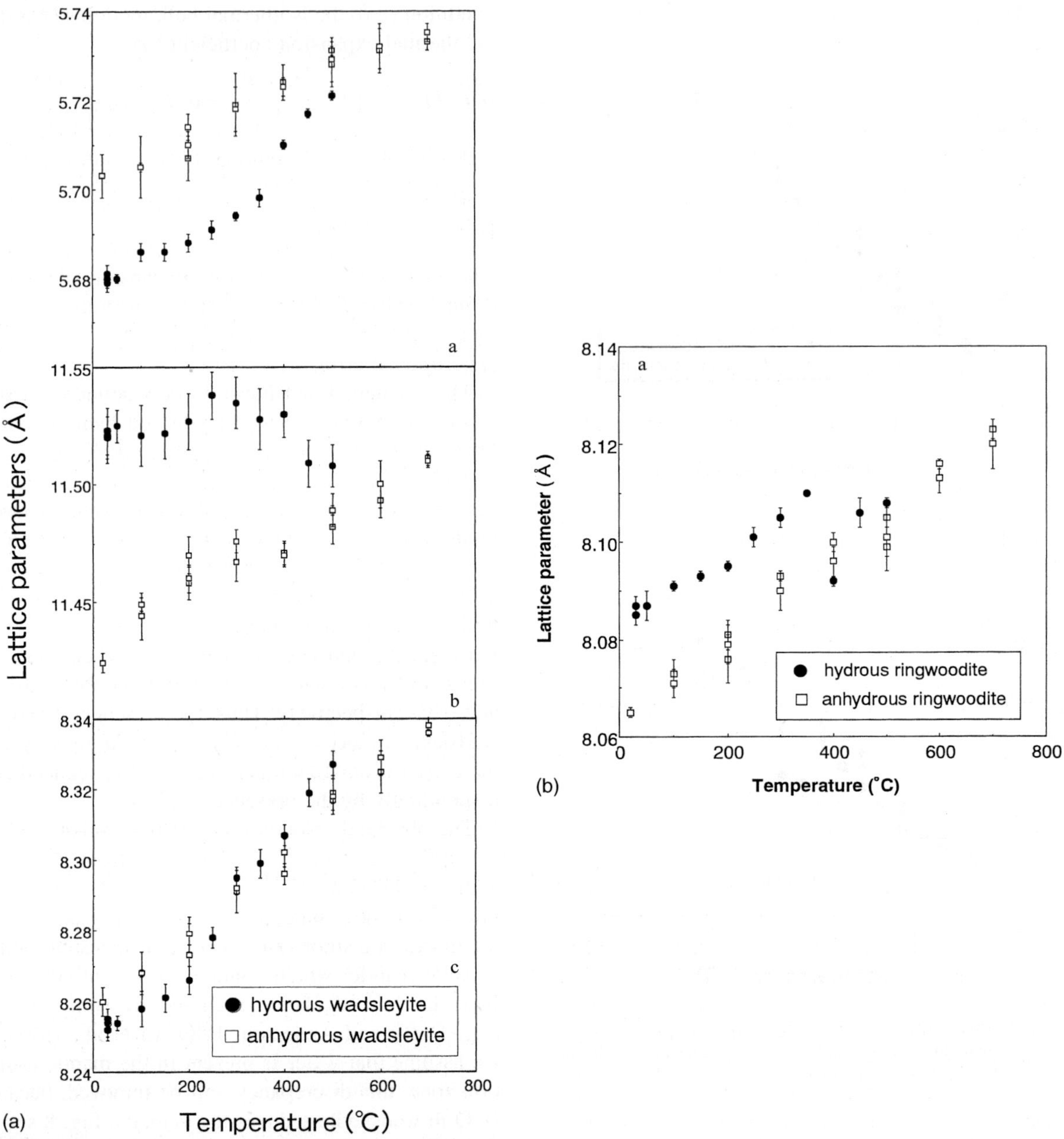

Fig. 7. Thermal expansion of each axis of (a) hydrous wadsleyite and (b) hydrous ringwoodite. Circles indicate the hydrous form and squares indicate the anhydrous form from the present results.

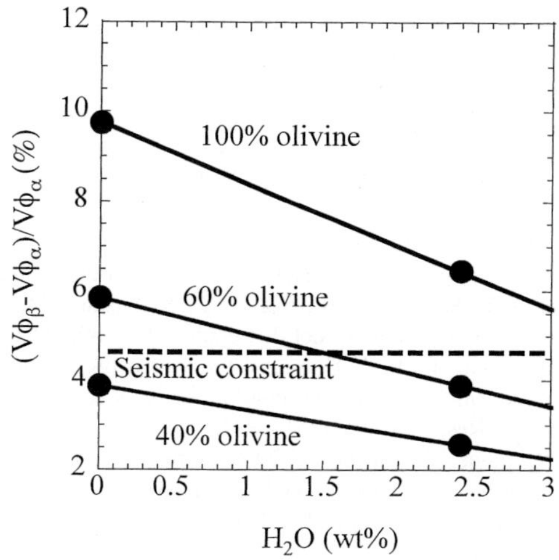

Fig. 8. The bulk sound velocity contrast at 410 km discontinuity corresponding to ~13.5 GPa and ~1400 °C as functions of H_2O and olivine content (100, 60 and 40%). Horizontal broken line indicates the seismic constraint from Benz and Vidale (1993) and Nolet et al. (1994).

pressure and high temperature conditions. Nevertheless, we strongly suggest that the mantle transition zone contains some water, and may be an important water reservoir in the mantle and the Earth's interior.

Acknowledgements

We thank R.P. Rapp and I. Jackson for valuable comments and I. Suzuki, an anonymous reviewer and the associate editor, T. Duffy for their constructive comments and suggestions to improve the manuscript. We are also grateful to H. Yurimoto for supporting the SIMS analysis at Tokyo Institute of Technology. This work was partially supported by the grant-in-aid for Scientific Research of the Ministry of Education, Science, Sport and Culture of the Japanese government (No: 13740270) and the Earthquake Research Institute cooperative research program.

References

Aldebert, P., Traverse, J.P., 1984. α-Al_2O_3: a high-temperature thermal expansion standard. High Temp. High Press. 16, 127–135.

Benz, H.M., Vidale, J.E., 1993. Sharpness of upper-mantle discontinuities determined from high-frequency reflections. Nature 365, 147–150.

Chen, J., Inoue, T., Yurimoto, H., Weidner, D.J., 2002. Effect of water on olivine-wadsleyite phase boundary in the (Mg, $Fe)_2SiO_4$ system. Geophys. Res. Lett. 29 (18), 1875, doi:10.1029/2001GL014429.

Duffy, T.S., Anderson, D.L., 1989. Seismic velocities in mantle minerals and the mineralogy of the upper mantle. J. Geophys. Res. 94, 1895–1912.

Duffy, T.S., Zha, C., Downs, R.T., Mao, H., Hemley, R.J., 1995. Elasticity of forsterite to 16 GPa and the composition of the upper mantle. Nature 378, 170–173.

Fugino, K., Sasaki, S., Takauchi, Y., Sadanaga, R., 1981. X-ray determination of electron distributions in forsterite, fayalite, and tephroite. Acta Cryst. B37, 513–518.

Horiuchi, H., Sawamoto, H., 1981. β-Mg_2SiO_4: single-crystal X-ray diffraction study. Am. Mineral. 66, 568–575.

Inoue, T., 1994. Effect of water on melting phase relation and melt composition in the system Mg_2SiO_4-$MgSiO_3$-H_2O up to 15 GPa. Phys. Earth Planet. Inter. 85, 237–263.

Inoue, T., Yurimoto, H., Kudoh, Y., 1995. Hydrous modified spinel, $Mg_{1.75}SiH_{0.5}O_4$: a new water reservoir in the mantle transition region. Geophys. Res. Lett. 22, 117–120.

Inoue, T., Weidner, D.J., Northrup, P.A., Parise, J.B., 1998. Elastic properties of hydrous ringwoodite (γ-phase) in Mg_2SiO_4. Earth Planet. Sci. Lett. 160, 107–113.

Kawamoto, T., Holloway, J.R., 1997. Melting temperature and partial melt chemistry of H_2O-saturated mantle peridotite to 11 Gigapascals. Science 276, 240–243.

Kudoh, Y., Inoue, T., Arashi, H., 1996. Structure and crystal chemistry of hydrous wadsleyite, $Mg_{1.75}SiH_{0.5}O_4$: possible hydrous magnesium silicate in the mantle transition zone. Phys. Chem. Minerals 23, 461–469.

Kudoh, Y., Inoue, T., 1999. Mg-vacant structural modules and dilution of the symmetry of hydrous wadsleyite, β-$Mg_{2-x}SiH_{2x}O_4$ with $0.00 < x < 0.25$. Phys. Chem. Miner. 26, 382–388.

Li, B., Liebermann, R.C., Weidner, D.J., 1998. Elastic moduli of wadsleyite (β-Mg_2SiO_4) to 7 Gigapascals and 873 Kelvin. Science 281, 675–677.

Liu, L-G., Mernagh, T.P., Lin, C-C., Xu, J-A., Inoue, T., 1998. Raman spectra of hydrous β-Mg_2SiO_4 at various pressure and temperatures. In: Manghnani, M.H., Yagi, T. (Eds.), Properties of Earth and Planetary Materials at High Pressure and Temperature. Am. Geophys. Union, Washington, DC, pp. 523–530.

Liu, L.-G., Lin, C.-C., Mernagh, T.P., Inoue, T., 2002. Raman spectra of hydrous γ-Mg_2SiO_4 at various pressures and temperatures. Phys. Chem. Miner. 29, 181–187.

Meng, Y., Weidner, D.J., Gwanmesia, G.D., Liebermann, R.C., Vaughan, M.T., Wang, Y., Leinenweber, K., Pacalo, R.E., Yeganeh-Haeri, A., Zhao, Y., 1993. In situ high P-T X-ray diffraction studies on three polymorphs (α, β, γ) of Mg_2SiO_4. J. Geophys. Res. 98, 22199–22207.

Miyagi, I., Yurimoto, H., 1995. Water content of melt inclusions in phenocrysts using secondary ion mass spectrometer. Bull. Volcanol. Soc. Jpn. 40, 349–355.

Moore, P.B., Smith, J.V., 1970. Crystal structure of β-Mg_2SiO_4: crystal-chemical and geophysical implications. Phys. Earth Planet. Inter. 3, 166–177.

Nolet, G., Grand, S.P., Kennett, B.L., 1994. Seismic heterogeneity in the upper mantle. J. Geophys. Res. 99, 23753–23766.

Ohtani, E., Toma, M., Kubo, T., Suzuki, A., 2001. Stability of dense hydrous magnesium silicate phases and water storage capacity in the transition zone and lower mantle. Phys. Earth Planet. Inter. 124, 105–117.

Sasaki, S., Prewitt, C.T., Ito, E., 1982. Single-crystal X-ray studies of γ-Mg_2SiO_4. J. Geophys. Res. 87, 7829–7832.

Sawamoto, H., Weidner, D.J., Sasaki, S., Kumazawa, M., 1984. Single-crystal elastic properties of the modified spinel (beta) phase of magnesium orthosilicate. Science 224, 749–751.

Smyth, J.R., Kawamoto, T., Jacobson, S.D., Swope, R.J., Herivig, R.L., Holloway, J.R., 1997. Crystal structure of monoclinic hydrous wadsleyite [β-(Mg, Fe)$_2$SiO$_4$]. Am. Miner. 82, 270–275.

Smyth, J.R., Kawamoto, T., 1997. Wadsleyite II: a new high pressure hydrous phase in the peridotite-H_2O system. Earth Planet. Sci. Lett. 146, E9–E16.

Suzuki, I., 1975. Thermal expansion of periclase and olivine, and their anharmonic properties. J. Phys. Earth 23, 145–159.

Suzuki, I., Ohtani, E., Kumazawa, M., 1979. Thermal expansion of γ-Mg_2SiO_4. J. Phys. Earth 27, 53–61.

Suzuki, I., Ohtani, E., Kumazawa, M., 1980. Thermal expansion of modified spinel, β-Mg_2SiO_4. J. Phys. Earth 28, 273–280.

Weidner, D.J., Sawamoto, H., Sasaki, S., Kumazawa, M., 1984. Elasticity of Mg_2SiO_4 spinel. J. Geophys. Res. 89, 7852–7860.

Yusa, H., Inoue, T., 1997. Compressibility of hydrous wadsleyite (β-phase) in Mg_2SiO_4 by high pressure X-ray diffraction. Geophys. Res. Lett. 24, 1831–1834.

Yusa, H., Inoue, T., Ohishi, Y., 2000. Isothermal compressibility of hydrous ringwoodite and its relation to the mantle discontinuities. Geophys. Res. Lett. 27, 413–416.

Available online at www.sciencedirect.com

SCIENCE DIRECT

Physics of the Earth and Planetary Interiors 143–144 (2004) 291–304

PHYSICS OF THE EARTH AND PLANETARY INTERIORS

www.elsevier.com/locate/pepi

A high-pressure infrared and X-ray study of $FeCO_3$ and $MnCO_3$: comparison with $CaMg(CO_3)_2$-dolomite

J. Santillán*, Q. Williams

Department of Earth Sciences, University of California, Santa Cruz, CA 95064, USA

Received 4 February 2003; received in revised form 24 June 2003; accepted 25 June 2003

Abstract

We have measured the infrared spectra of two carbonates, $FeCO_3$-siderite and $MnCO_3$-rhodochrosite to pressures of 50 GPa at 300 K, and taken X-ray diffraction patterns of these phases before and after in-situ laser heating at pressures to 47 GPa. Our infrared measurements primarily sample the vibrations of the CO_3^{-2} carbonate unit, particularly its asymmetric stretching vibration and out-of-plane and in-plane bends. In particular, we wish to determine whether any high-pressure polymorphism (metastable or otherwise) is present in these phases. Our X-ray results for $FeCO_3$ and $MnCO_3$ demonstrate that the calcite structure is stable for each of these materials to 50 GPa at 300 K, and following heating to 2000 K at pressures to 47 GPa; similarly, within the infrared spectra, all vibrations are observed to shift monotonically with pressure, with no splitting observed. This behavior is in marked contrast to that of dolomite, in which splitting of the infrared vibrations is observed above ~20–30 GPa. Our results demonstrate that carbonates are stable in the calcite structure at lower mantle pressures if the carbonates' divalent cations have zero pressure radii of less than 1 Å. Accordingly, the crystal structure of carbonates within the lower mantle will critically depend on the amount of calcium that enters into carbonates within lower mantle assemblages.

Keywords: Siderite; Rhodochrosite; Dolomite; High-pressure; Infrared-spectroscopy; X-ray diffraction

1. Introduction

The manner and degree to which carbon is stored within the deep earth are the subject of ongoing debate. Possible solid phases that might sequester carbon at depth include carbonates and diamond, with the equilibria between these materials being controlled largely by oxygen fugacity and the identity of the stable carbonate phase. The phases of carbonates stable at high-pressures are known to vary depending on the divalent cation present within the structure. For example, $CaCO_3$-calcite undergoes two metastable phase transitions below 3 GPa, at 1.5 and 2.2 GPa (Merrill and Bassett, 1975), with the stable high-pressure phase being aragonite to a pressure of at least 40 GPa (Kraft et al., 1991). In contrast, calcite-structured $MgCO_3$-magnesite, has been demonstrated to be stable to pressures in excess of 31 GPa at 1200 ± 200 K (Gillet, 1993), and is (at least) metastable to pressures of 70 GPa (e.g. Fiquet and Reynard, 1999; Fiquet et al., 2002).

Indeed, petrologic studies conducted in the (Mg, Fe)O–SiO_2–CO_2 system at deep upper mantle and shallow lower mantle conditions have consistently produced magnesite (Martinez et al., 1998; Katsura and Ito, 1990). These studies have, however, not included calcium or other cations which often occur in

* Corresponding author.
E-mail address: jds@es.ucsc.edu (J. Santillán).

0031-9201/$ – see front matter
doi:10.1016/j.pepi.2003.06.007

3. Infrared results

The ambient pressure infrared spectra of $FeCO_3$-siderite and $MnCO_3$-rhodochrosite have both been extensively studied (e.g. Huang and Kerr, 1960; White, 1974; Dubrawski et al., 1989; Böttcher et al., 1992, 1993; Böttcher and Gehlken, 1995). As with all calcite-structured carbonates, the infrared spectra of both $FeCO_3$-siderite and $MnCO_3$-rhodochrosite between 500 and 1800 cm^{-1} can be divided into three main regions, each produced by a different vibration of the carbonate unit (Table 2): (1) a ν_4-associated band, produced by the in-plane bending vibration of the CO_3^{2-} group, which occurs at 734 cm^{-1} in $FeCO_3$ and 735 cm^{-1} in $MnCO_3$ at zero pressure; (2) a ν_2-associated band, generated by the out-of-plane bending vibration of the CO_3^{2-} group, at 861 and 862 cm^{-1}, respectively; and (3) a broad ν_3 band, associated with the asymmetric stretching vibration of the CO_3^{2-} group at 1412 and 1450 cm^{-1}, respectively.

Under compression, the asymmetric stretching vibration in both $FeCO_3$ and $MnCO_3$ increases linearly to higher frequencies with compression (Figs. 2A and 3A). This band shifts at 3.05 cm^{-1} GPa^{-1} in siderite to a maximum frequency of 1554 cm^{-1} at 49 GPa, and at a somewhat slower rate of 2.45 cm^{-1} GPa^{-1} to a maximum frequency of 1531 cm^{-1} for $MnCO_3$-rhodochrosite at 52 GPa. This band shows minor broadening in each phase upon compression, but the broadening is reversible upon decompression (Figs. 2B and 3B). Indeed, the low-pressure asymmetry of this peak, which is likely generated by a component of TO-LO splitting (e.g. Grzechnik et al., 1999), decreases under compression, implying that these powder measurements sample almost solely the transverse optic vibrations at high-pressure and/or the TO-LO splitting decreases with pressure.

The two lower frequency carbonate bands, the out-of-plane bend and the in-plane bend, shift monotonically with increasing pressure for both phases (Figs. 4 and 5). However, the out-of-plane bend has a negative pressure shift of -0.14 cm^{-1} GPa^{-1} in $FeCO_3$-siderite and -0.03 cm^{-1} GPa^{-1} for $MnCO_3$-rhodochrosite. This behavior is similar to that observed for this vibration in calcite-III, aragonite, and $MgCO_3$ under pressure (Williams et al., 1992; Kraft et al., 1991; Grzechnik et al., 1999), and is likely generated by the increased strength of the divalent cation–oxgen

Table 2
Observed vibrational modes for three calcite-type carbonates

cm^{-1}	Assignment	$d\omega/dP$ (cm^{-1} GPa^{-1})	
Siderite, $\gamma_{K=117}$			
1412	ν_3, Asymmetric stretch of CO_3^{2-}	3.05	0.25
861	ν_2, Out-of-plane bend of CO_3^{2-}	−0.14	−0.02
734	ν_4, In-plane bend of CO_3^{2-}	1.90	0.31
Mean mode Grüneisen parameter for observed vibrational modes:			0.18
Thermodynamic Grüneisen parameter:			1.17
Bulk modulus obtained from Zhang and Reeder (1999). Band assignments obtained from White (1974).			
Rhodochrosite, $\gamma_{K=114}$			
1450	ν_3, Asymmetric stretch of CO_3^{2-}	2.45	0.19
862	ν_2, Out-of-plane bend of CO_3^{2-}	−0.03	−0.003
735	ν_4, In-plane bend of CO_3^{2-}	1.31	0.20
Mean mode Grüneisen parameter for observed vibrational modes:			0.13
Thermodynamic Grüneisen parameter:			1.02
Bulk modulus obtained from Chen et al. (2001). Band assignments obtained from White (1974).			
Magnesite, $\gamma_{K=94}$			
1446	ν_3, Asymmetric stretch of CO_3^{2-}	4.23	0.32
885	ν_2, Out-of-plane bend of CO_3^{2-}	1.16	0.14
749	ν_4, In-plane bend of CO_3^{2-}	2.64	0.39
Mean mode Grüneisen parameter for observed vibrational modes:			0.28
Observed vibrational modes of magnesite from Grzechnik et al. (1999).			

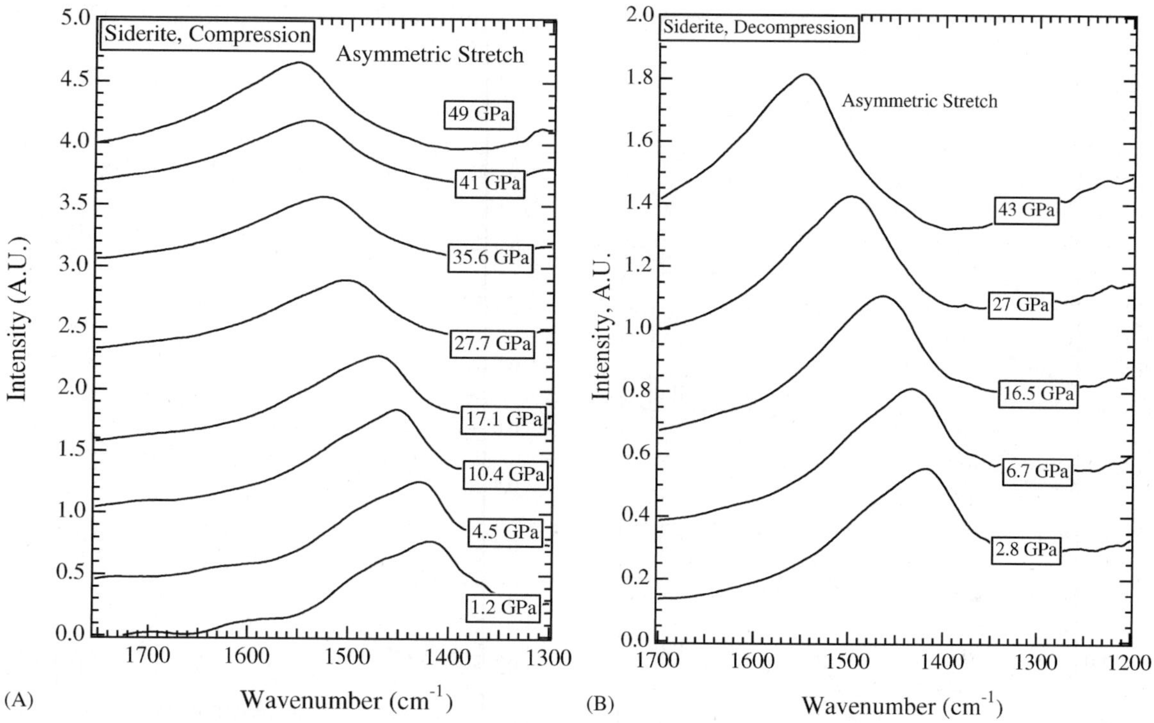

Fig. 2. Representative mid-infrared spectra of the asymmetric stretching (ν_3) vibration of $FeCO_3$ at high-pressures: (A) spectra on compression; (B) spectra on decompression.

bonds under compression (Kraft et al., 1991). This negative shift, which appears to be a generic property of most carbonates, produces a near crossing between the in-plane bend and out-of-plane bend as pressure is increased past 45 GPa (Figs. 4 and 5). An extrapolation of the pressure shifts of these bands indicates that these two peaks should cross at 65 GPa in both $FeCO_3$-siderite and $MnCO_3$-rhodochrosite. Such a crossing is unusual among planar XY_3 ions (Nakamoto, 1986), and implies that the role of oxygen–oxygen repulsion (which modulates the frequency of the in-plane bend) becomes progressively greater under compression, while divalent cation–oxygen attraction contributes to the softening of the out-of-plane bend. The mode shifts of these vibrations are plotted in Figs. 6 and 7 for siderite and rhodochrosite, respectively.

Perhaps the most notable aspect of these results is the stability of the calcite-structured phase in both of these materials to pressures of 50 GPa at 300 K. Indeed, structures similar to the metastable phases of $CaCO_3$, calcite-II and -III, do not appear in either of these compositions. These metastable transformations are characterized by splitting of the ν_3 and ν_4 vibrations in the infrared spectrum of $CaCO_3$ (Adams and Williams, 1980; Williams et al., 1992). The mode shifts and spectra of $FeCO_3$ and $MnCO_3$ at high-pressure are in marked contrast to those of dolomite, which does undergo a pressure-induced transformation to a calcite-III like polymorph (Santillán et al., 2003). Fig. 8 shows the high-pressure mode shifts of dolomite, in which splittings of vibrations initiate near 20 GPa, and become notably extensive above 30 GPa.

4. X-ray results

Our X-ray results confirm the notable high-pressure stability of calcite-structured $FeCO_3$ and $MnCO_3$ documented by our infrared results and supplement these with results derived after samples have been heated

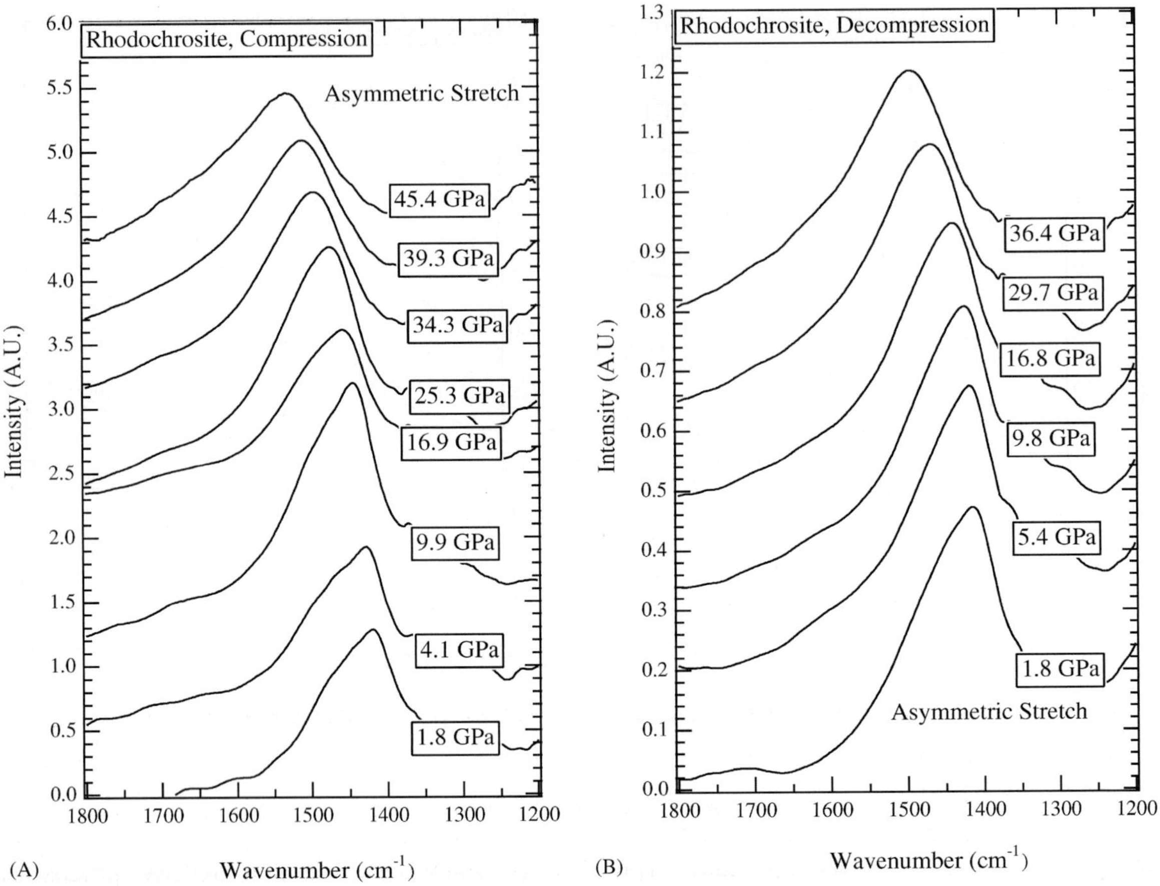

Fig. 3. Representative mid-infrared spectra of the asymmetric stretching (ν_3) vibration of $MnCO_3$ at high-pressures: (A) spectra on compression; (B) spectra on decompression.

at high-pressure conditions. Our X-ray diffraction results for samples held at pressure before and after heating to temperatures estimated at 1900 ± 200 K are shown in Fig. 9. The X-ray reflections become less broad after heating due to relaxation of pressure gradients. The highest intensity (1 0 4) reflection of each carbonate is shown both before and after heating, with a small shift associated with pressure relaxation following heating. Calcite-III-related structures have two peaks in this region of the diffraction pattern (e.g. Santillán et al., 2003), as does aragonite. No new peaks are observed; we expect that any new peaks with amplitudes greater than ~10% of the (1 0 4) peak of siderite and rhodochrosite should be detected. There is thus no compelling evidence for the presence of micro-diamonds or for the associated products of a dissociation reaction in our laser heating charges. For example, in the case of siderite, we find no evidence of Fe_3O_4-magnetite, and in rhodochrosite, we see no evidence of Mn_3O_4-hausmannite below 10 GPa or marokite-structured Mn_3O_4 above 10 GPa (Liu et al., 2001). These results also preclude any aragonite-related transitions from occuring in these materials within our pressure range, and are inconsistent with sub-solidus dissociation of these carbonates under these pressure–temperature conditions. Thus, the known stability ranges of these phases are expanded to 47 GPa and ~1900 K (Fig. 9A and B). Our results are thus not fully consistent with the work of Liu et al. (2001), who observed localized dissociation of rhodochrosite above 6–8 GPa at 2273 K. We speculate that this may be due to our samples being

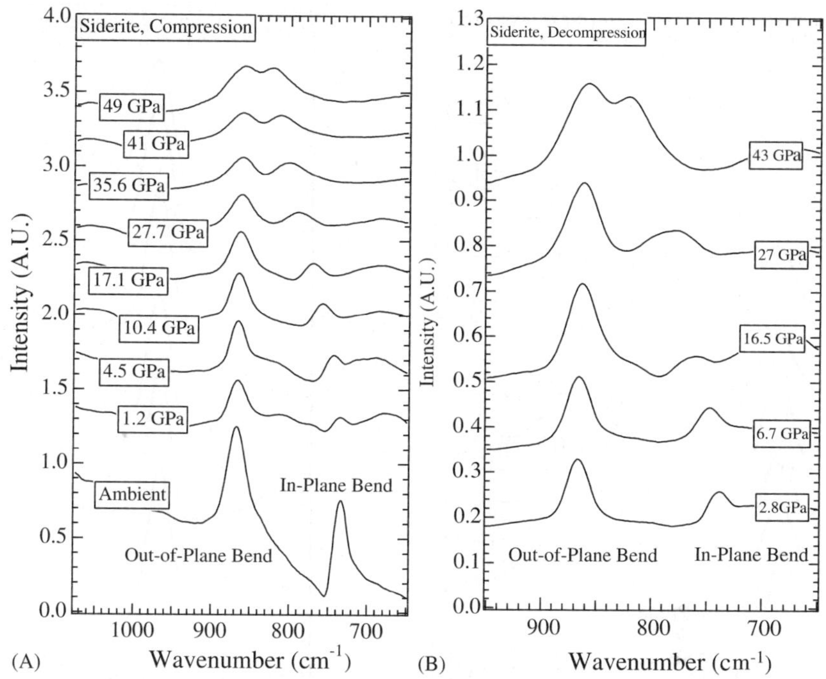

Fig. 4. Representative mid-infrared spectra of $FeCO_3$ out-of-plane (ν_2) and in-plane (ν_4) bending vibrations at high-pressure: (A) spectra on compression; (B) spectra on decompression.

sub-solidus, or the micro-diamonds in their samples being generated by local (and possibly disequilibrium conditions). Extensive work (e.g. Pal'yanov et al., 1999a,b, 2002; Srikanth et al., 1997; Akaishi et al., 1990) has demonstrated that diamonds can be synthesized from carbonate melts, but no indication that diamonds can be synthesized from solid-state carbonates exists. Indeed, Srikanth et al. (1997) demonstrated that solid-state $MnCO_3$ was stable in coexisting carbonate-rich melt to 2173 K and 7.7 GPa: a result also at odds with those of Liu et al. (2001).

5. Discussion

5.1. Systematics of structural stabilities in carbonates under pressure

The reason for the high-pressure stability of the calcite-structured phase in $MnCO_3$ and $FeCO_3$ certainly lies in the smaller ionic radii of the Fe^{2+} (0.78 Å) and Mn^{2+} (0.96 Å) ions relative to that of Ca^{2+} (1.12 Å) (Liu and Bassett, 1986). Indeed, the stability of $MgCO_3$-magnesite to pressures of at least 70 GPa (Fiquet et al., 2002), can be similarly associated with the small radius of Mg^{2+} (0.89 Å). In effect, smaller cations tend to stabilize the calcite structure to high-pressure conditions. Indeed, the calcite-III structure, which both calcite and dolomite convert to under pressure, includes the presence of two different octahedrally-coordinated cation sites, one of which is considerably distorted (Smyth and Ahrens, 1997). It is possible that smaller cations may simply be unable to enter into such a distorted site, thus precluding a transition to the calcite-III structure. Similarly, divalent cations in aragonite are present in an eight-fold coordinate site, and this site is also likely to be energetically difficult for smaller cations to enter.

Fig. 10 illustrates the structure and stability of a number of carbonates in terms of their transition pressures or maximum pressure of known stability as a function of divalent cation radius. The stability field of calcite-I-type structures lies below cation radii of 1 Å at zero pressure. By 20 GPa the calcite-related structure of $CaMg(CO_3)_2$-dolomite, with a mean ionic radius of 0.86 Å, converts to dolomite-II, a strucure

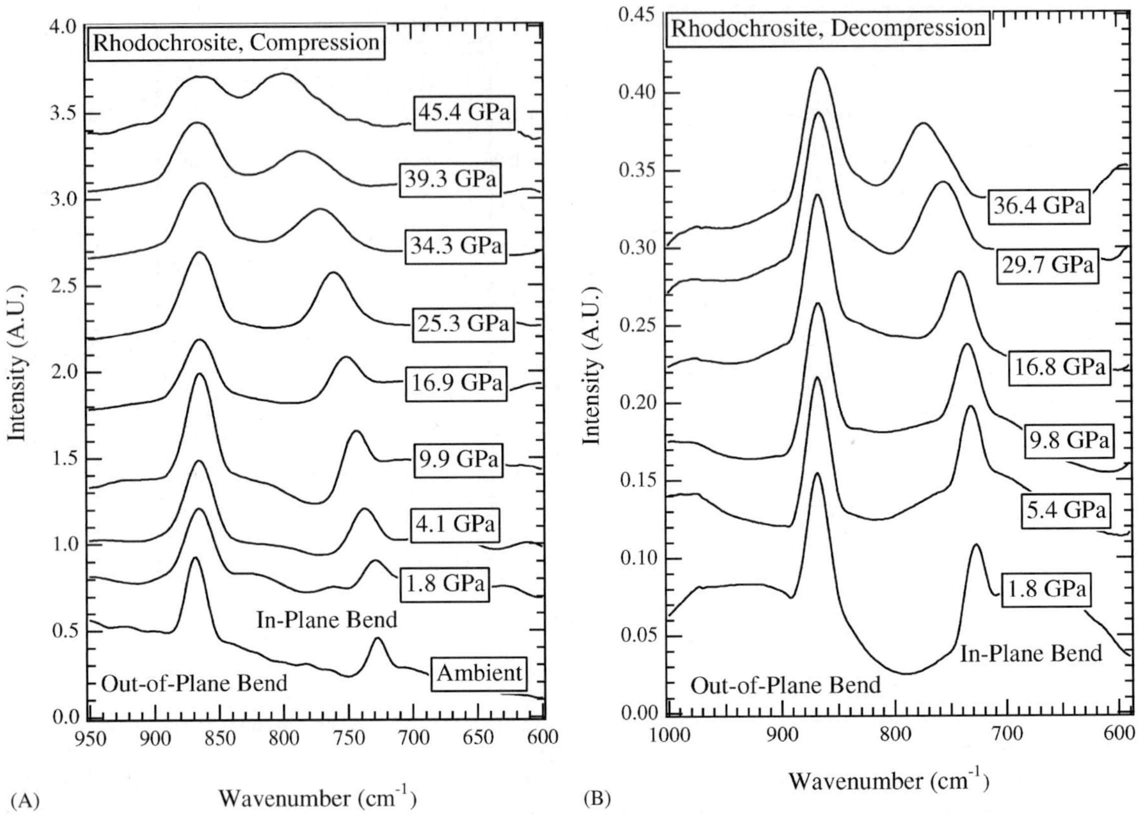

Fig. 5. Representative mid-infrared spectra of the $MnCO_3$ out-of-plane (ν_2) and in-plane (ν_4) bending at high-pressures: (A) shows spectra on compression; (B) on decompression.

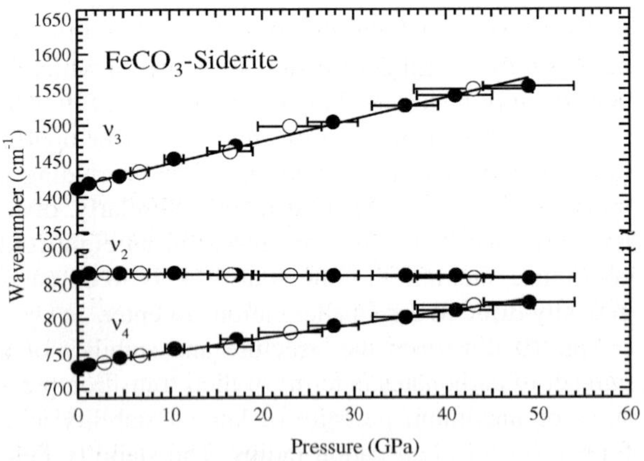

Fig. 6. Pressure-induced mode shifts of the asymmetric stretches (ν_3), out-of-plane (ν_2) and in-plane (ν_4) bends of $FeCO_3$. Closed circles denote data taken on compression and open circles denote data taken upon decompression.

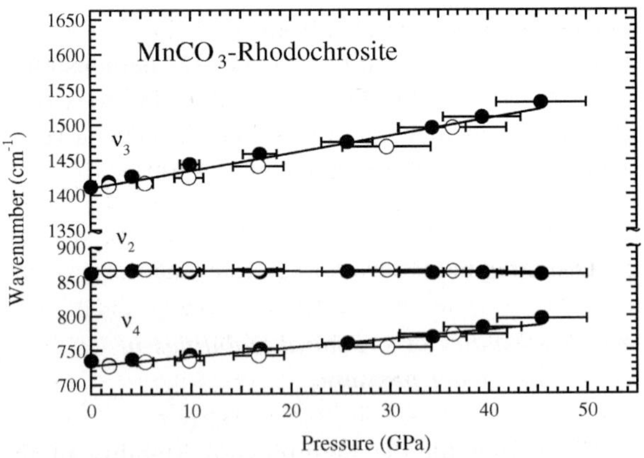

Fig. 7. Pressure-induced mode shifts of the asymmetric stretches (ν_3), out-of-plane (ν_2) and in-plane (ν_4) bends of rhodochrosite. Closed circles denote data taken on compression and open circles denote data taken upon decompression.

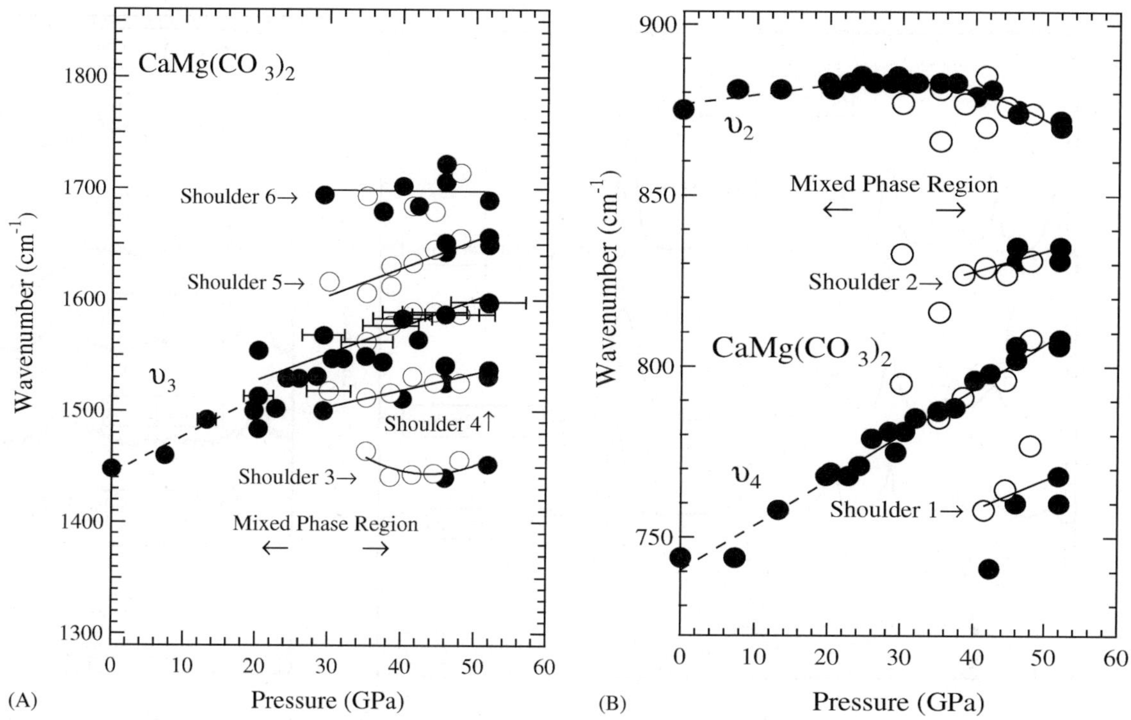

Fig. 8. Mode shifts of dolomite as a function of pressure: (A) shows high frequency modes and (B) shows low-frequency modes. Closed circles denote compression data and open circles denote decompression data. Dashed lines indicate fits to the data below 20 GPa and solid lines indicate fits to the data above 20 GPa. Pronounced splittings of all of the vibrations are clearly visible present above 20 GPa, and a mixed phase region exists above 20 GPa.

related to that of calcite-III (Santillán et al., 2003). The aragonite structure of $CaCO_3$ has an ionic radius of 1.18 Å within the eight-fold coordinate site in this structure, and $CdCO_3$-otavite undergoes a transition from the calcite-I structure to the aragonite-structure at a pressure of ~16 GPa (Liu and Lin, 1997a). The lower bound on the stability of $FeCO_3$ and $MnCO_3$ in the calcite-I structure at high temperatures are shown at 30 and 47 GPa, respectively (Fig. 9). We treat these conditions as a lower bound on the pressure of the calcite-to-aragonite transition in these materials, as we expect this reaction to be kinetically hindered, and our non-observance of this transition to these pressures implies that any transition to an aragonite-structured polymorph must occur at higher pressure conditions. At 300 K, the stability of $FeCO_3$ to 49 GPa, as determined from our infrared results, provides a lower bound on any possible calcite to calcite-III transition, which we do not expect to be kinetically hindered, based on its occurence at 300 K in both $CaCO_3$ and $CaMg(CO_3)_2$ (Fong and Nicol, 1971; Williams et al., 1992; Santillán et al., 2003). Thus, the smaller the zero pressure ionic radius of a carbonate, the higher the pressure required to initiate either the transition to the calcite-III or the aragonite structure. A further implication of Fig. 10 is that the trend for conversion to the aragonite phase may lie at higher pressures (for a given ionic radius) relative to the calcite-III conversion trend. Thus, it is possible that at higher pressure conditions (or for materials with smaller ionic radii), calcite-III related phases might become thermodynamically stable (as opposed to metastable) in carbonate systems (e.g. Santillán et al., 2003).

In order to examine the interplay between cation–oxygen bond strength and the response to pressure of these phases, Fig. 11 shows the pressure-induced mode shifts of the ν_3 and ν_4 peaks of different calcite-structured carbonates as a function of ionic radii, as derived from this study for $MgCO_3$, $FeCO_3$, and $CaMg(CO_3)_2$ and for $MgCO_3$

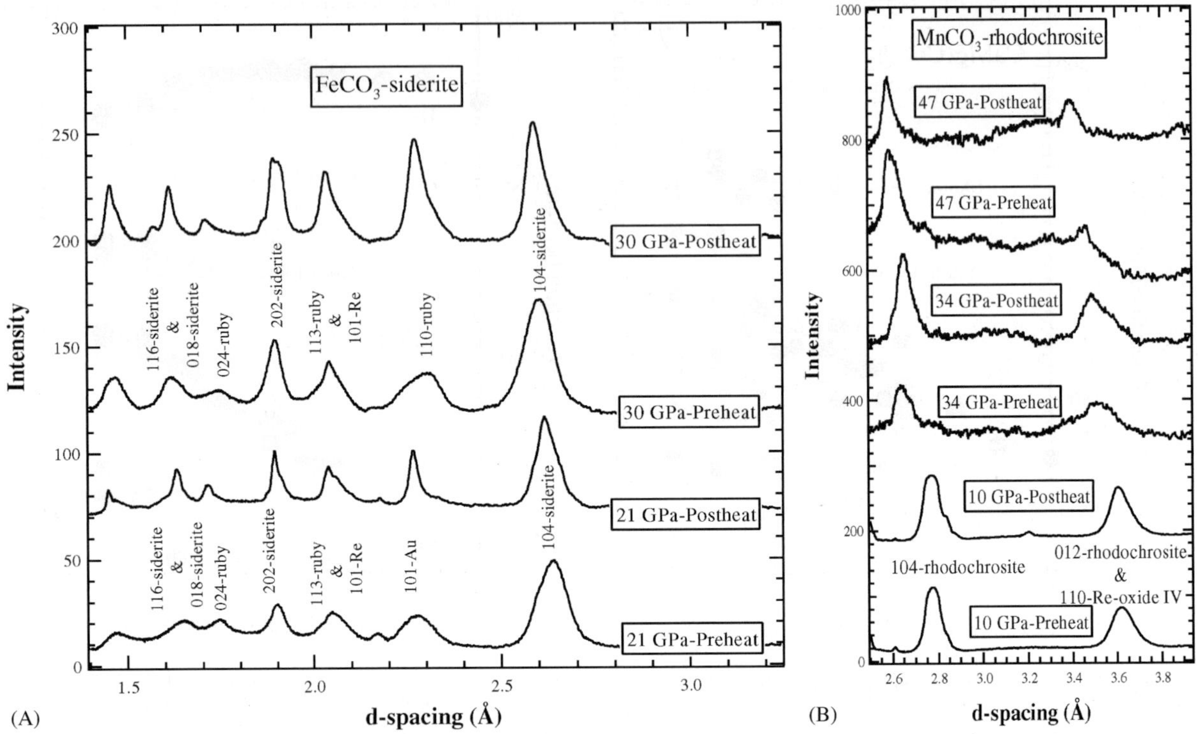

Fig. 9. (A) X-ray diffraction patterns of $FeCO_3$ before and after heating at 1900 K and at 10 and 30 GPa and (B) detailed view of (1 0 4) reflection of rhodochrosite before and after heating at pressures of 10, 34, and 47 GPa.

from Grzechnik et al. (1999). Both of these modes involve oxygen motion that produces deformation and compression of the divalent cation–oxygen bonds, and there is a systematic decrease in the mode shifts of these peaks with increasing ionic radius of those cations. This is due to the progressively greater rigidity of the M–O bonds for smaller cations, and implies that the carbon–oxygen bond is affected by compression substantially more in magnesite than in, for example, rhodochrosite. In effect, the bonding environments surrounding the carbonate ion in these carbonates exhibit a continuum of behavior, from those in which the carbonate behaves more as a rigid molecular-like unit, to magnesite, where the carbonate unit is substantially affected by the compression of the crystal. Such compression-induced effects on the C–O bond in magnesite may contribute to the anomalous behavior of the C–O bond distance under compression, in which the C–O bond has been proposed to initially contract, then expand at pressures above ~50 GPa (Fiquet et al., 2002).

6. Grüneisen parameters

We calculate mode Grüneisen parameters (γ_i) for each of these vibrational modes using Eq. (1) and values of the bulk modulus (K_0) of 117 and 114 GPa for siderite (Zhang and Reeder, 1999) and rhodochrosite (Chen et al., 2001), respectively. Mode Grüneisen parameters for each pressure are calculated for each phase from a linear regression to

$$\gamma_i = \frac{K}{\omega}\left(\frac{d\omega}{dP}\right) \tag{1}$$

Mode Grüneisen parameters for the observed modes are reported in Tables 2 and 3. Notably, the mode Grüneisen parameters for the carbonate vibrations of the two materials are small for each phase, but systematically lower for rhodochrosite relative to siderite. Thus, while the response of the carbonate groups to volumetric compaction is grossly similar between $FeCO_3$ and $MnCO_3$, differences exist in their detailed response to compression.

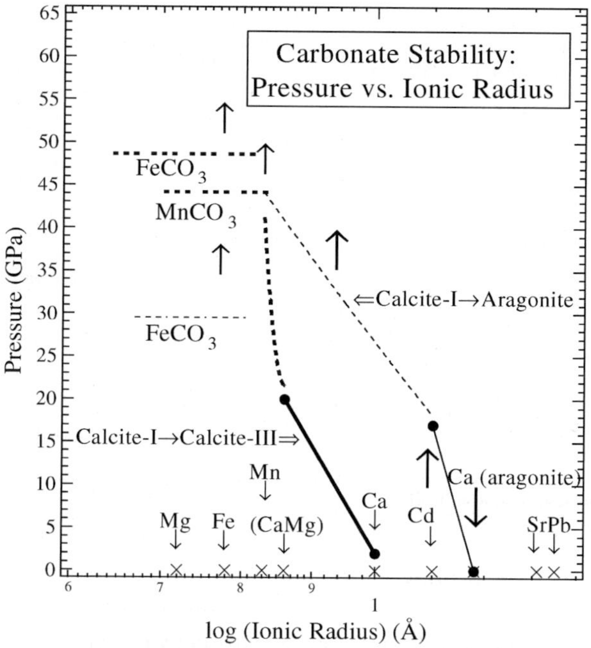

Fig. 10. Phase transition boundaries of the divalent carbonates at pressure as a function of ionic radius. Values for ionic radii are from Liu and Bassett (1986) and the value for dolomite represents an averge value for the Ca^{2+} and Mg^{2+} radii. Closed circles represent known points of phase transitions. The thin line on the right shows the boundary of the calcite-I to aragonite phase transition, as determined from $CaCO_3$ and $CdCO_3$ (Liu and Lin, 1997a). The thin dashed line is the extrapolated transition boundary, which lies at higher pressures than our highest pressure observation of $MnCO_3$ stability, shown as a horizontal dashed line representing the known stability limit we document for $MnCO_3$ at 47 GPa. The highest pressure stability limit for $FeCO_3$ we have established follwing heating at 30 GPa, is also shown by a horizontal dashed line. The bold line represents the known boundary of the calcite-I to calcite-III phase transition, with a lower limit of 2.2 GPa for the calcite-II to calcite-III transition (Merrill and Bassett, 1975), and an upper limit defined by the transition of dolomite to a calcite-III structured polymorph of (Santillán et al., 2003). The bold dashed line is the extrapolated calcite-I to calcite-III boundary, capped by two bold dashed lines, representing the lower limits on calcite-structured $MnCO_3$ and $FeCO_3$ stabilities at 300 K.

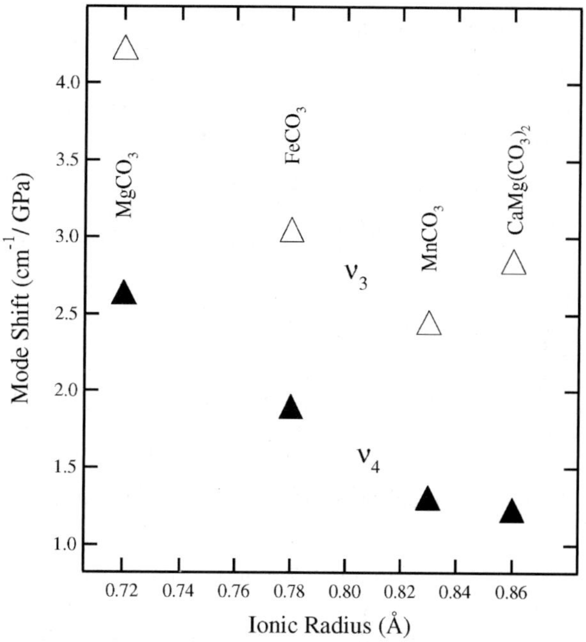

Fig. 11. Dependence of pressure-induced mode shifts on the divalent cation radii for the ν_3 and ν_4 carbonate vibrations of $MgCO_3$ (Grzechnik et al., 1999), $FeCO_3$, $MnCO_3$ (this work), and $CaMg(CO_3)_2$ (Santillán et al., 2003). Notably, the bulk moduli of the different phases are all fairly similar, ranging from 94 to 117 GPa, implying that differences in compressibility of the bulk crystal do not produce these changes in mode shifts.

For comparison, the 300 K thermodynamic Grüneisen parameters for each phase were calculated from

$$\gamma = \frac{\alpha_v K}{\rho_0 C_p} \qquad (2)$$

using values of thermal expansion, α_v of $26.9 \times 10^{-6}\,K^{-1}$ (Pfaff, 1935), a bulk modulus K_s of 117 GPa (estimated from K_0 where $K_s = 1 + \alpha T$, and thus the adiabatic value of the bulk modulus differs from the isothermal bulk modulus by less than 1%) (Zhang and Reeder, 1999), a density, ρ of $3754\,kg\,m^{-3}$ (Christensen, 1972), and a heat capacity, C_p of $82.69\,J\,mol^{-1}\,K^{-1}$ (Robie et al., 1984) for siderite and α_v of $22.8 \times 10^{-6}\,K^{-1}$ (Rao and Murthy, 1970), K_s of 114 GPa (estimated from K_0) (Chen et al., 2001), ρ of $3567\,kg\,m^{-3}$ (Christensen, 1972), and a C_p of $80.87\,J\,mol^{-1}\,K^{-1}$ (Robie et al., 1984) for rhodochrosite (Table 2). For siderite, the thermodynamic Grüneisen parameter is 1.17, compared with 0.95 for rhodochrosite. Mode Grüneisen parameters were also calculated for dolomite (Table 3) using α_v of $22.8 \times 10^{-6}\,K^{-1}$ (Reeder and Markgraf, 1986), K_s of 94 GPa (estimated from K_0) (Ross and Reeder, 1992), ρ of $2860\,kg\,m^{-3}$ (Howie and Broadhurst, 1958), and a C_p of $80.87\,J\,mol^{-1}\,K^{-1}$ (Stout and Robie, 1963). The thermodynamic

Table 3
Observed vibrational modes of dolomite/dolomite-II

cm^{-1}	Assignment	$d\omega/dP$ ($cm^{-1}\ GPa^{-1}$)	$\gamma_{K=94(I)}$
1443	ν_3, Asymmetric Stretch of CO_3^{2-}(I)	3.36	0.22
1478	Vbration replacing ν_3 (II)	2.43	
877	ν_2, Out-of-Plane Bend of CO_3^{2-}(I)	0.27	0.02
865	Vbration replacing ν_2 (II)	1.34	
740	ν_4, In-Plane Bend of CO_3^{2-}(I)	1.31	0.17
740	Vibration replacing ν_4 (II)	1.33	
723	Shoulder 1 (II)	0.88	
875	Shoulder 2 (II)	0.69	
1834	Shoulder 3 (II) (overtone?)	−17.81	
1459	Shoulder 4 (II)	1.50	
1529	Shoulder 5 (II)	2.49	
1700	Shoulder 6 (II) (overtone?)	−0.05	
Mean mode Grüneisen parameter for observed vibrational modes of dolomite (I):			0.14
Thermodynamic Grüneisen parameter of dolomite (I):			0.91

A (I) denotes a mode present in dolomite and a (II) denotes a mode present in dolomite-II.

Grüneisen parameter for dolomite thus yields a value of 0.91.

The markedly low values of the average mode Grüneisen parameter we observe for the carbonate vibrations are 0.18 (siderite), 0.13 (rhodochrosite), and 0.14 (dolomite). In passing, we note that dolomite-II, for which we do not know the bulk modulus or thermal expansion, has an average mode shift, and thus average mode Grüneisen parameter near zero. These low values are compatible with the comparative incompressibility of the carbonate group, and consistent with low-pressure single-crystal X-ray diffraction studies that show that the compaction of the calcite structure is taken up almost entirely by the cation octahedra (e.g. Ross and Reeder, 1992). The high values of the thermodynamic Grüneisen parameters relative to those calculated from the observed mode shifts is certainly due to lower frequency lattice modes (0–500 cm^{-1}), which are expected to have notably higher mode Grüneisen parameters than the higher frequency carbonate group vibrations. The vibrations associated with divalent cation translations are specifically expected to have high mode Grüneisen parameters, as the cation octahedra are the most compressible unit within calcite-structured carbonates. Such high Grüneisen parameters for low-frequency lattice modes associated with cation translations have been previously observed in magnesite (Table 5) (Grzechnik et al., 1999).

7. Conclusions

High-pressure infrared spectroscopy of two calcite-structured minerals, $FeCO_3$-siderite and $MnCO_3$-rhodochrosite, shows that both the CO_3^{2-}-group and the calcite structure is stable in each of these phases to pressures in excess of 49 GPa. No thermally quenchable transition is observed in either phase to pressures of 30–47 GPa following heating, thus precluding the existence of quenchable aragonite-structured polymorphs of either phase in this pressure range. In the infrared spectra, we observe the pressure-induced approach of the out-of-plane and in-plane bends in both phases, which ultimately could generate a Fermi resonance between these modes near 65 GPa. Our infrared and X-ray results are compatible with previous work in terms of the systematics of the stability of carbonate structures at high pressures, and significantly expand the pressure and temperature range over which $FeCO_3$ and $MnCO_3$ have been examined.

The stability of the calcite structure in $FeCO_3$ and $MnCO_3$ appears to be generated by the comparatively small ionic radii of their divalent cations, which likely preclude a transition to either calcite-III-type or aragonite structures. For comparison, $CaMg(CO_3)_2$-dolomite, which has an average cation radius of 0.86 Å, undergoes a metastable phase transition near 20 GPa (Santillán et al., 2003), but does not convert to an aragonite-type phase. In this instance, we believe the presence of calcium ions stabilizes the calcite-III structure, as the larger calcium ions can enter into the distorted octahedral sites in this structure. Taken together, our results imply that the presence of cations smaller than 1 Å could prospectively stabilize the calcite structure in carbonates in the deep interiors of terrestrial planets. Therefore, the crystal structure in which carbonates occur within realistic lower mantle assemblages hinges critically on the ill-constrained divalent cation chemistry (and particularly calcium contents) of carbonates under deep mantle conditions.

Acknowledgements

We thank the NSF and the W.M. Keck Foundation for support, and J. Adler, E. Knittle, C.D. Martin, E. Morris, D. Sampson, W. Schillinger, H. Scott, and L. Whitfield for helpful discussions.

References

Adams, D.M., Williams, A.D., 1980. Vibrational spectroscopy at very high pressures. Part 26. An infrared study of the metastable phases of $Ca[CO_3]$. J. Chem. Soc., Dalton Transactions, 1482–1486.

Akaishi, M., Kanda, H., Yamaoka, S., 1990. Synthesis of diamond from graphite-carbonate systems under very high temperature and pressure. J. Cryst. Growth 104, 578–581.

Böttcher, M.E., Gehlken, P.-L., Usdowski, E., 1992. Infrared spectroscopic investigations of the calcite-rhodochrosite and parts of the calcite-magnesite mineral series. Contrib. Mineral. Petrol. 109, 304–306.

Böttcher, M.E., Gehlken, P.-L., Birch, W.D., Usdowski, E., Hoefs, J., 1993. The rhodochrosite-smithsonite solid-solution series from Broken Hill (NSW), Australia: Geochemical and infrared spectroscopic investigations. N. Jb. Miner. Mh. 1993, 352–362.

Böttcher, M.E., Gehlken, P.L., 1995. Cationic substitution in natural siderite–magnesite ($FeCO_3$–$MgCO_3$) solid solution: a FTIR spectroscopic study, N. Jb. Miner. Mh. 169, 352–362.

Chen, C.C., Lin, C.C., Liu, L.G., Sinogeikin, S.V., Bass, J.D., 2001. Elasticity of single-crystal calcite and rhodochrosite by Brillouin spectroscopy. Am. Miner. 86, 1525–1529.

Christensen, N.I., 1972. Elastic properties of polycrystalline magnesium, iron, and manganese carbonates to 10 kilobars. J. Geophys. Res. 77, 369–372.

Dubrawski, J.V., Channon, A.-L., Warne, S.S., 1989. Examination of siderite-magnesite mineral series by Fourier transform infrared spectroscopy. Am. Miner. 74, 187–190.

Deere, W.A., Howie, R.A., Zussman, J., 1962. Rock Forming Minerals, Non-Silicates, vol. 5. John Wiley and Sons, Inc., New York, pp. 264–265.

Fiquet, G., Guyot, F., Itié, J.P., 1994. High-pressure X-ray diffraction study of carbonates-$MgCO_3$, $CaMg(CO_3)_2$, and $CaCO_3$. Am. Miner. 79, 15–23.

Fiquet, G., Reynard, B., 1999. High-pressure equation of state of magnesite: new data and a reappraisal. Am. Miner. 84, 856–860.

Fiquet, G., Kunz, M., Matas, J., Andrault, D., Hanfland, M., 2002. Structural refinements of magnesite at very high pressure. Am. Miner. 87, 1261–1265.

Fong, M., Nicol, M., 1971. Raman spectrum of calcium carbonate at high pressures. J Chem. Phys. 56, 579–585.

Grzechnik, A., Simon, P., Gillet, P., McMillan, P., 1999. An infrared study of $MgCO_3$ at high pressure. Physica B 262, 67–73.

Gillet, P., 1993. Stability of magnesite ($MgCO_3$) at mantle pressure and temperature conditions: a Raman spectroscopic study. Am. Miner. 78, 1328–1331.

Heinz, D., Jeanloz, R., 1984. The equation of state of the gold calibration standard. J. Appl. Phys. 55, 885–893.

Holl, C.M., Smyth, J.R., Laustsen, H.M.S., Jacobsen, S.D., Downs, R.T., 2000. Compression of witherite to 8 GPa and the crystal structure of $BaCO_3$-II. Phys. Chem. Miner. 27, 467–473.

Holmes, N.C., Moriarty, J.A., Gathers, G.R., Nellis, W.J., 1989. The equation of state of platinum to 660 GPa (6.6 Mbar). J. Appl. Phys. 66, 2962–2967.

Howie, R.A., Broadhurst, F.M., 1958. X-ray data for dolomite and ankerite. Am. Miner. 11–12, 1210–1217.

Huang, C.K., Kerr, P.F., 1960. Infrared study of the carbonate minerals. Am. Miner. 45, 311–324.

Katsura, T., Ito, E., 1990. Melting and subsolidus phase relations in the $MgSiO_3$–$MgCO_3$ system at high-pressures: implications to evolution of the Earth's atmosphere. Earth Planet Sci. Lett. 99, 110–117.

Kraft, S., Knittle, E., Williams, Q., 1991. Carbonate stability in Earth's mantle: a vibrational spectroscopic study of aragonite and dolomite at high pressures and temperatures. J. Geophys. Res. 96, 17,997–18,009.

Liu, L.G., Bassett, W.A., 1986. Elements, Oxides, Silicates: High-Pressure Phases with Implications for the Earth's Interior. Oxford University Press, New York.

Liu, L., Lin, C., 1997a. A calcite-aragonite-type phase transition in $CdCO_3$. Am. Miner. 82, 643–646.

Liu, L., Lin, C., 1997b. High-pressure Raman spectroscopic study of post-aragonite phase transition in witherite ($BaCO_3$). Eur. J. Miner. 9, 785–792.

Liu, L., Lin, C., Yang, Y., 2001. Formation of diamond by decarbonation of $MnCO_3$. Solid State Comm. 188, 195–198.

Magneli, A., 1957. Studies on rhenium oxides. Acta Chem. Scand. 11, 28–33.

Martinez, I., Peréz, E.M.C., Matas, J., Gillet, P., Vidal, G., 1998. Experimental investigation of silicate-carbonate system at high pressure and high temperature. J. Geophys. Res. 103, 5143–5163.

Martinez, I., Zhang, J., Reeder, R.J., 1996. In situ X-ray diffraction of aragonite and dolomite at high pressure and high temperature: Evidence for dolomite breakdown to aragonite and magnesite. Am. Miner. 81, 611–624.

Matas, J., Gillet, P., Ricard, Y., Martinez, I., 2000. Thermodynamic properties of carbonates at high-pressures from vibrational modelling. Eur. J. Miner. 12, 703–720.

Merrill, L., Bassett, W.A., 1975. The crystal structure of $CaCO_3$ (II), a high-pressure metastable phase of calcium carbonate. Acta Cryst. B31, 343–349.

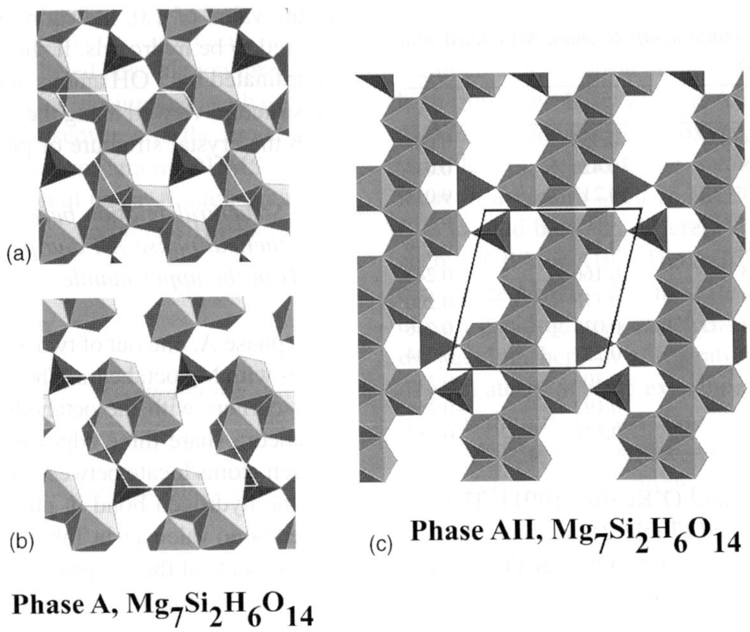

Fig. 1. The crystal structures of phase A (drawn with the atomic coordinates of Horiuchi et al., 1979) and phase AII. **a** phase A viewed along the c axis between $z \approx 0$ and 0.25. **b** phase A viewed along the c axis between $z \approx 0.25$ and 0.50. **c** phase AII viewed along the c axis. The outline of the unit cell is shown.

is a characteristic feature in common for high-pressure dense hydrous magnesium silicates, such as phase A (Horiuchi et al., 1979), phase B (Finger et al., 1991), superhydrous B (Pacalo and Parise, 1992; Kudoh et al., 1994), hydrous wadsleyite (Kudoh et al., 1996) and hydrous ringwoodite (Kudoh et al., 2000).

The present study suggests that phase A can be regarded as one of a homologous series of hypothetical post-humite high-pressure phases derived from ringwoodite as a parent structure from the relation:

$$(m+2n)Mg_2SiO_4 \text{ (ringwoodite)-}nMg_3Si_2O_7 + nH_2O = Mg_{(2m+n)}Si_mH_{2n}O_{2(2m+n)} \text{ (hypothetical post-humite high-pressure phase)}.$$

Although the crystal structure of phase A is made of two types of OT-layer, one OT-layer can change to an O-layer if one Si out of the two Si sites moves along the c axis by about 1.6 Å. The structure of phase A can, therefore, be regarded substantially as derived from the spinel structure which is made of alternating OT-layer and O-layer. In the case of phase A, $m = 2$ and $n = 3$, such as

$$8Mg_2SiO_4 \text{ (ringwoodite)} - 3Mg_3Si_2O_7 + 3H_2O = Mg_7Si_2H_6O_{14} \text{ (phase A)}.$$

This implies possible significance of post-humite high-pressure phases whose stability field might be close to phase A as candidates for the hosts of water in the mantle.

Acknowledgements

The author wishes to thank helpful reviews by Dr. T. Duffy and two anonymous reviewers. The paper benefited from careful reviews and helpful comments by the reviewers.

References

Akimoto, S., Akaogi, M., 1980. The system Mg_2SiO_4–MgO–H_2O at high pressures and temperatures-possible hydrous magnesium silicates in the mantle transition zone. Phys. Earth Planet. Int. 23, 268–275.

Akimoto, S., Yamamoto, K., Aoki, K., 1977. Hydroxyl-clinohumite and hydroxyl-chondrodite: Possible H_2O-bearing minerals in the upper mantle. In: Manghnani, M., Akimoto, S. (Eds.), High-Pressure Research: Application in Geophysics. Academic Press, New York, pp. 163–172.

Aoki, K., Fujino, K., Akaogi, M., 1976. Titanochondrodite and titanoclinohumite derived from the upper mantle in the Buell Park kimberlite, Arizona. USA. Contrib. Mineral. Petrol. 56, 243–253.

Brese, N.E., O'Keeffe, M., 1991. Bond-valence parameters for solids. Acta Crystallogr. B47, 192–197.

Finger, L.W., Hazen, R.M., Prewitt, C.T., 1991. Crystal structures of $Mg_{12}Si_4O_{19}(OH)_2$ (phase B) and $Mg_{14}Si_5O_{24}$ (phase AnhB). Am. Mineral. 76, 1–7.

Horiuchi, H., Morimoto, N., Yamamoto, K., Akimoto, S., 1979. Crystal atructure of $2Mg_2SiO_4 \cdot 3Mg(OH)_2$, a new high-pressure structure type. Am. Mineral. 64, 593–598.

Kitamura, M., Kondoh, S., Morimoto, N., Miller, G.H., Rossman, G.R., Putnis, A., 1987. Planar OH-bearing defects in mantle olivine. Nature 328, 143–145.

Kudoh, Y., Nagase, T., Ohta, S., Sasaki, S., Kanzaki, M., Tanaka, M., 1994. Crystal structure and compressibility of superhydrous phase B, $Mg_{20}Si_6H_8O_{36}$. In: Schmidt, S.C., Shaner, J.W., Samara, G.A., Ross, M. (Eds.), High-Pressure Science and Technology, American Institute of Physics, New York, 1993, pp. 469–472.

Kudoh, Y., Inoue, T., Arashi, H., 1996. $Mg_{1.75}SiH_{0.5}O_4$: possible hydrous magnesium silicate in the mantle transition zone. Phys. Chem. Minerals 23, 461–469.

Kudoh, Y., Kuribatashi, T., Mizobata, H., Ohtani, E., 2000. Structure and cation disorder of hydrous ringwoodite, γ-$Mg_{1.89}Si_{0.98}H_{0.30}O_4$. Phys. Chem. Minerals 27, 474–479.

Kudoh, Y., 2002. Predicted model for hydrous olivine (Hy-α). Phys. Chem. Minerals 29, 387–395.

Lima-de Faria, J., Heller, E., Liebau, F., Makovicky, E., Parthe, E., 1990. Nomenclature of inorganic structure types. Acta Cryst. A46, 1–11.

McGetchin, T.R., Silver, L.T., Chodos, A.A., 1970. Titanoclinohumitr: a possible mineralogical site for water in the upper mantle. J. Geophys. Res. 75, 255–259.

Meier, W.M., Villiger, H., 1969. Die Methode der Abstandsverfeinerung zur Bestimmung der Atomkoordinaten idealisierter Geruestatrukturen. Z. Kristallogr. 129, 411–423.

Pacalo, R.E.G., Parise, J.B., 1992. Crystal structure of superhydrous B, a hydrous magnesium silicate synthesized at 1400 °C and 20 GPa. Am. Mineral. 77, 681–684.

Pauling, L., 1960. The Nature of the Chemical Bond. Cornell University Press, Ithaca, NY, 644 pp.

Ribbe, P.H., Gibbs, G.V., Jones, N.W., 1968. Cation and anion substitutions in the humite minerals. Min. Mag. 36, 966–975.

Taylor, W.H., West, J., 1928. The crystal structure of the chondrodite series. Proc. Roy. Soc. A117, 517–532.

Thompson, A.B., 1992. Water in the Earth's upper mantle. Nature 358, 295–302.

Thompson, J.B., 1978. Biopyriboles and polysomatic series. Am. Min. 63, 239–349.

Wadsley, A.D., 1964. Inorganic non-stoichiometric compounds. In: Mandelcorn, L. (Ed.), Non-Stoichiometric Compounds. Academic Press, New York, London, pp. 98–209.

White, T.J., Hyde, B.G., 1982. Electron microscope study of the humite minerals. Part I. Mg-rich specimens. Phys. Chem. Min. 8, 55–63.

Yamamoto, K., Akimoto, S., 1977. The system MgO-SiO_2-H_2O at high pressures and temperatures-stability field for hydroxyl-chondrodite, hydroxyl-clinohumite and 10 Å phase. Am. J. Sci. 277, 288–312.

4. Transport Properties and Rheology

Available online at www.sciencedirect.com
SCIENCE DIRECT
ELSEVIER
Physics of the Earth and Planetary Interiors 143–144 (2004) 311–320
PHYSICS OF THE EARTH AND PLANETARY INTERIORS
www.elsevier.com/locate/pepi

Simultaneous measurements of thermal conductivity and thermal diffusivity for garnet and olivine under high pressure

Masahiro Osako [a,*], Eiji Ito [b], Akira Yoneda [b]

[a] *Section of Astronomy and Geophysics, National Science Museum, Hyakunin-cho 3-23-1, Shinjuku, Tokyo 169-0073, Japan*
[b] *Institute for Study of the Earth's Interior, Okayama University, Yamada 827, Misasa, Tottori 682-0193, Japan*

Received 14 February 2003; received in revised form 31 October 2003; accepted 31 October 2003

Abstract

Thermal conductivity λ and thermal diffusivity κ have been measured simultaneously for garnet ($Al_{73}Py_{25}Gr_1Sp_1$) and olivine ($Fo_{93}Fa_7$) single crystals up to 8.3 GPa and 1100 K by using a pulse heating method. Anisotropy of thermal conduction is investigated along the three crystallographic axes, [1 0 0], [0 1 0], and [0 0 1] in olivine. The pressure dependence of κ, or $\mathrm{d}\ln\kappa/\mathrm{d}P$, is determined to be 0.03–0.04 GPa^{-1} for the three crystallographic direction in olivine and 0.03 GPa^{-1} for garnet, while $\mathrm{d}\ln\lambda/\mathrm{d}P$ is determined to be ~0.04 GPa^{-1} for olivine and 0.03 GPa^{-1} for garnet. The anisotropy in thermal diffusivity or thermal conductivity of olivine is clearly observed in the present experimental range of pressure and temperature. It is likely that the anisotropy in thermal conduction would be maintained throughout the olivine stability field in the mantle down to 410 km depth. Heat capacities of olivine and garnet are calculated from the present κ and λ. Both the present heat capacities of olivine and garnet are consistent with the previously reported values within 5%. Those obtained for the three crystallographic direction in olivine are also consistent with each other within 6%. Further, the effect of basalt–eclogite transition on heat transportation is discussed based on the present results for garnet. It implies that the eclogite layer of the slab transmits thermal energy more effectively than the upper basaltic layer.

Keywords: Thermal conductivity; Olivine; Garnet; High pressure; Anisotropy; Heat capacity

1. Introduction

Thermal conductivity λ and thermal diffusivity κ of mantle materials are of vital importance for understanding the thermal state and the dynamics nature of the Earth's interior. Several investigations have been published on experimental measurements of λ and κ for mantle phases and rocks. Recently, λ and its pressure derivative for several mantle phases have also been estimated by theoretical approaches (e.g. Manga and Jeanloz, 1997; Hofmeister, 1999; Cohen, 2000). Nevertheless, our knowledge of thermal conduction of the mantle phases is still insufficient to simulate heat transportation in the mantle, in particular under high-pressure and high-temperature conditions.

Thermal conductivity or thermal diffusivity measurements under high pressure were conducted by using cylindrical heat flow or linear heat flow. A cylindrical sample assembly has frequently been used in measurement of λ and κ under high pressure (e.g. Fujisawa et al., 1968; Beck et al., 1978; Katsura, 1995). Due to the requirement of axial symmetric heat flow, this method cannot be applied to investigate anisotropy in thermal conduction. On the other

* Corresponding author. Tel.: +81-3-5332-7174; fax: +81-3-3364-7104.
E-mail address: sako@kahaku.go.jp (M. Osako).

0031-9201/$ – see front matter
doi:10.1016/j.pepi.2003.10.010

hand, thermal conductivity λ of mantle materials were measured under high pressure by means of a one-dimensional thermal conduction experiment. Yukutake and Shimada (1978) measured thermal conductivity of stishovite and periclase up to 4 GPa in a cubic multi-anvil apparatus. Schärmeli (1982) conducted measurement on single crystal olivine up to 2.5 GPa in a piston-cylinder apparatus. It should be noted, however, that relatively large sample is required in those methods to accommodate multiple thermocouples in the sample. Therefore, they are not suitable for high-pressure phases of limited size.

A pulse method of one-dimensional sample configuration was developed by Dzhavadov (1975); he measured λ and κ simultaneously for a relatively large Teflon sample (3.5 mm in thickness and 20 mm in diameter). This method is applicable not only to measurement of material anisotropic in thermal conduction, but also to obtain information on heat capacity under high pressure. We have tried to substantially reduce the sample size so that the technique become applicable to mantle materials and high-pressure phases of limited size. We selected natural garnet and olivine as the samples of the present study. Garnet is accepted as major constituent in the upper mantle and the transition zone and is isotropic in thermal conduction. Measurement along different crystallographic direction in olivine should be a good test for anisotropy of thermal conduction.

2. Experimental

Fig. 1 shows the basic concept of the present experimental technique. The sample consists of three thin identical disks. An impulse heater is placed on one interface between the disks, and a thermocouple is inserted on the other interface. The thermal disturbance caused by impulse heating is monitored by the thermocouple. The temperature variation ΔT at the position of the thermocouple is expressed as

$$\Delta T = A\sum_{n=1}^{\infty}\frac{1}{n^2}\sin\frac{n\pi}{3}\sin\frac{n\pi x}{d}\exp(-n^2Bt) \times[\exp(n^2B\tau)-1] : (t>\tau) \quad (1)$$

where t is time from the onset of heating, x the position measured from the end of the sample, d the total height

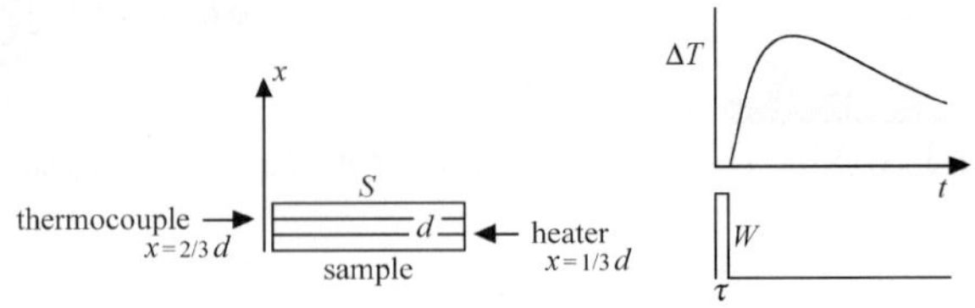

Fig. 1. A pulse heating method for measurement of thermal conductivity and thermal diffusivity after Dzhavadov (1975). Left figure shows basic configuration of the sample. d the total height of the three sample disks, S the area of the heater, x the position from the bottom of the sample (the heater is at $x = d/3$, and the thermocouple is at $x = 2d/3$). Temperature change ΔT caused by impulse heating with a power of W is monitored by the thermocouple (right figure), where τ is the duration of pulse heating, and t the time from onset of the heating.

of the three sample disks, and τ the duration of heating pulse. The quantities A and B are defined as follows:

$$A = \frac{2Wd}{\pi^2\lambda S}, \quad B = \frac{\pi^2\kappa}{d^2} \quad (2)$$

where W is the power of the impulse heating, S the area of the heater. Eq. (1) is derived under the boundary condition of constant temperature at the both sample ends (Dzhavadov, 1975). As the series (1) for ΔT converges rapidly, summation up to $n = 10$ yields very accurate value for the present experimental condition. In the present work, the parameters A and B are determined through the least square fitting of the experimental data by means of Eq. (1) up to $n = 15$.

We prepared natural single crystals samples of garnet and olivine for measurements of κ and λ. The garnet samples from Bahia, Brazil have a composition of 73% almandine and 25% pyrope, 1% grossular and 1% spessartine components from microprobe analysis. The garnet crystals were cut so that the surface was perpendicular to one of the crystallographic axes. The olivine samples from northern Pakistan exhibited a euhedral shape with a composition of 93% forsterite and 7% fayalite from microprobe analysis. The single crystals were cut perpendicularly to the three crystallographic axes, [1 0 0], [0 1 0], and [0 0 1]. The accuracy of the cutting was within 0.5°. A stacking of three identical disks of the sample was used for one measurement. The dimensions of each disk were 4.3 mm in diameter and ~0.35 mm in thickness. The three disks were polished to have same thickness within 0.002 mm.

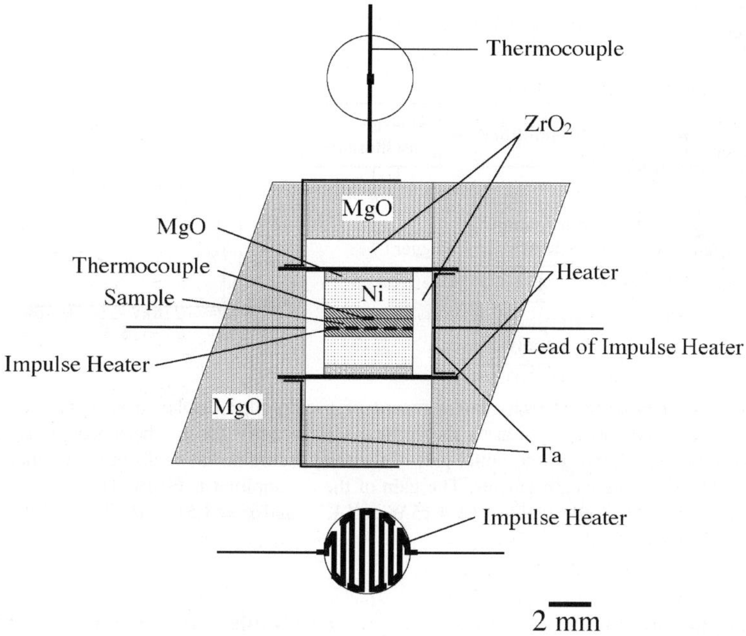

Fig. 2. Cross section of the sample assembly. Top views of thermocouple and impulse heater are also. The flat thermocouple has a thickness of 0.03 mm and a width of 0.3 mm, and the impulse heater has a thickness of 0.03 mm. Although the circuit pattern of impulse heater is shown schematically, the stripe of the impulse heater circuit is folded 20 times.

The high-pressure experiments were performed in a Kawai-type high-pressure apparatus (SHP-1000) at the Institute for Study of the Earth's Interior. A magnesia octahedron with an edge-length of 18 mm was used as a pressure medium in tungsten carbide anvils with a truncation edge length of 11 mm. Calibration of sample pressures to hydraulic oil pressure is based on the Bi I–II (2.54 GPa) and Bi III–IV (7.8 GPa) transitions. The calibration runs were made using a similar setup to that of the actual run.

Fig. 2 shows a view of the whole cell assembly used in this study. The ambient temperature of the sample consisting of three thin disks is controlled by the two plate heaters and the sample temperature was monitored by a thin flat chromel–alumel thermocouple (0.03 mm in thickness and 0.3 mm in width) inserted between the sample disks. A small heater made of nickel–chromium (3.8 mm in diameter and 0.03 mm in thickness) is placed at the other interface of the sample disks to perform impulse heating. The heater which contains photo-etched slots is effective to achieve uniform and nearly one-dimensional heating over a cross section of the sample. Transient heat flow caused by impulse heating passes through the sample disk, and the corresponding transient signal is observe by the thermocouple as a hump on the emf of the ambient temperature. Nickel blocks contacted to the sample serve as heat sink, which is critical to keep the constant temperature boundary condition. A zirconia sleeve and blocks surrounding the assembly serve as thermal insulator for heating of the sample. The sleeve effectively restricts lateral heat flow to realize one-dimensional hear flow in the sample.

Measurements were conducted at pressures up to 8.3 GPa and at temperatures up to 1100 K. The ambient temperature is maintained by the main heater operated by a dc power supply to eliminate inductive noise. This is essential to achieve resolution of $\sim$0.1 μV in the signal (20–100 μV) generated by impulse heating on the thermocouple emf.

Fig. 3 shows the schematic diagram of the measurement and a temperature profile on the storage oscilloscope. Averaging 64, 128, or 256 acquisitions was carried out for the digital records of the oscilloscope to reduce random noise. The direction of the dc impulse current was inverted one after the other on data

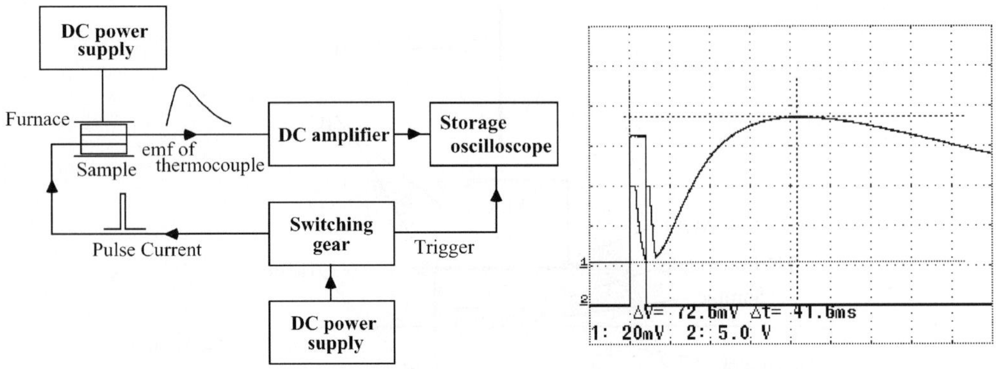

Fig. 3. Schematic diagram of the measurement (left) and a record of oscilloscope display (right) in which an input impulse and the resulted temperature profile are exhibited. One division on the vertical scale is 5 mV for the thermocouple and 5.0 V for the impulse. The horizontal scale is 10 ms per a division. Although the initial part of the temperature profile is disturbed by induction noise from the current of the impulse heater, it does not affect the measurements. The gain of the dc amplifier is 60 dB. The top height of the temperature profile corresponds to ~1 K. The demonstrated profile yields $\lambda = 4.15\,W\,m^{-1}\,K^{-1}$ and $\kappa = 1.51 \times 10^{-6}\,m^2\,s^{-1}$ for impulse heating of 22.1 W input power with a duration of 3.96 ms.

acquisition to cancel the inductive noise caused by the impulse heating current.

3. Results and discussion

3.1. Garnet

Thermal diffusivity κ and conductivity λ of garnet obtained in three runs at room temperature as functions of pressure are shown in Fig. 4. Changes in sample size during compression were corrected by using the compression data of Sato et al. (1978). The κ and λ data obtained in each run are well fitted to the linear line with respect to pressure, and are connected with thin solid lines (Fig. 4). Errors attached to each data point of κ (ca. 2%) are caused mainly from those associated with the uncertainty of sample thickness d and the least-square fitting procedures. On the other hand, errors of λ (ca. 3%) are caused by uncertainty of heating area S and power W besides by those of d and the fitting procedures. Somewhat large systematic variations in κ and λ values among the different runs are inferred to be due to slight movement of the heater and thermocouple relative to the sample disks during the initial compression. Measurements were actually carried out at slightly different temperatures of 293–297 K. The κ and λ values are adjusted to those at 293 K by using the temperature variation of κ in Table 3 (Osako, 1997) and the heat capacity C_P (Watanabe, 1982) at atmospheric pressure. The averaged κ and λ values after the adjustment are shown by the heavy dash-dotted lines (Fig. 4), and tabulated in Table 1. Parameters for linear fitting of κ and λ

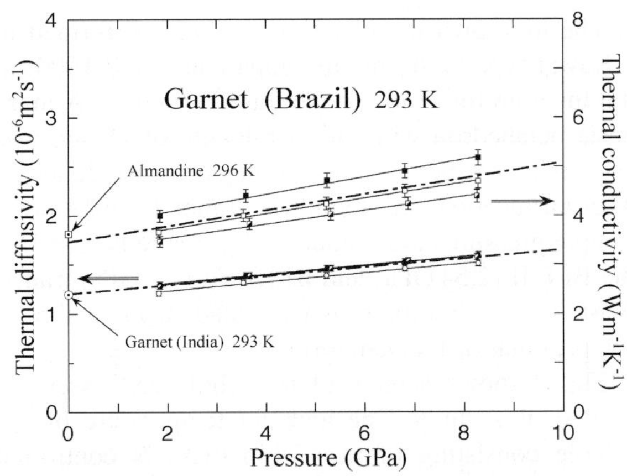

Fig. 4. Thermal diffusivity (circles) and thermal conductivity (squares) of garnet vs. pressure. Measurements were performed at slight different temperature, 293–297 K. Thermal diffusivity κ and thermal conductivity λ were obtained simultaneously for each run. The results for the three separate runs are specified by open, half-closed and closed symbols. Dash-dotted lines are averaged thermal diffusivity and thermal conductivity at 293 K. The dotted circle at zero pressure is a reference thermal diffusivity of a garnet with a similar composition measured using the Ångström method (Osako, 1997), and the dotted square indicates thermal conductivity of natural almandine (Horai, 1971).

Table 1
Thermal diffusivity κ and thermal conductivity λ of olivine and garnet at high pressures P and at various temperatures T

		κ and λ at 293 K, P (GPa) 0	2	4	6	8	10	$d\ln\kappa/dP$, $d\ln\lambda/dP$ (GPa^{-1})	κ and λ at 8.3 GPa, T (K) 400	600	800	1000
Olivine Fo_{93}												
[1 0 0]	κ ($10^{-6}\,m^2\,s^{-1}$)	[2.50 (7)]	2.67 (9)	2.86 (10)	3.05 (12)	3.27 (14)	[3.49 (16)]	0.033 (5)	2.28 (22)	1.50 (18)	1.11 (15)	0.87 (14)
	λ ($W\,m\,K^{-1}$)	[6.61 (29)]	7.12 (35)	7.68 (42)	8.29 (49)	8.94 (57)	[9.64 (66)]	0.038 (5)	7.13 (50)	5.39 (40)	4.52 (34)	4.00 (31)
[0 1 0]	κ ($10^{-6}\,m^2\,s^{-1}$)	[1.53 (8)]	1.66 (10)	1.80 (12)	1.95 (14)	2.11 (16)	[2.28 (20)]	0.040 (7)	1.43 (13)	0.91 (10)	0.65 (9)	0.50 (8)
	λ ($W\,m\,K^{-1}$)	[3.95 (22)]	4.30 (26)	4.67 (30)	5.08 (35)	5.53 (41)	[6.02 (47)]	0.042 (5)	4.28 (49)	3.13 (39)	2.56 (34)	2.21 (31)
[0 0 1]	κ ($10^{-6}\,m^2\,s^{-1}$)	[2.16 (7)]	2.32 (9)	2.48 (11)	2.66 (13)	2.85 (15)	[3.05 (18)]	0.035 (3)	2.05 (23)	1.36 (18)	1.01 (16)	0.80 (15)
	λ ($W\,m\,K^{-1}$)	[5.91 (21)]	6.33 (24)	6.78 (27)	7.26 (31)	7.78 (35)	[8.34 (39)]	0.034 (5)	6.40 (38)	4.96 (30)	4.24 (27)	3.81 (24)
Garnet												
	κ ($10^{-6}\,m^2\,s^{-1}$)	[1.19 (4)]	1.29 (5)	1.38 (6)	1.47 (7)	1.56 (7)	[1.66 (8)]	0.029 (1) (at 8 GPa)	1.23 (16)	0.91 (13)	0.76 (11)	0.66 (10)
	λ ($W\,m\,K^{-1}$)	[3.48 (15)]	3.80 (18)	4.12 (20)	4.44 (23)	4.76 (25)	[5.08 (28)]	0.034 (2) (at 8 GPa)	3.77 (30)	3.19 (25)	2.89 (22)	2.72 (21)

The figures in square brackets are extrapolated values based on the present data. Pressure derivatives are also given.

Table 2
Coefficients of fitting parameters of thermal diffusivity κ and thermal conductivity λ

Garnet				Olivine				
$\kappa = a_0 + a_1 P$		$\kappa = c_0 + c_1/T$			$\kappa = b_0 \exp(b_1 P)$		$\kappa = c_0 + c_1/T$	
a_0 ($10^{-6}\,m^2\,s^{-1}$)	a_1 ($10^{-6}\,m^2\,s^{-1}\,GPa^{-1}$)	c_0 ($10^{-6}\,m^2\,s^{-1}$)	c_1 ($10^{-6}\,m^2\,s^{-1}\,K$)		b_0 ($10^{-6}\,m^2\,s^{-1}$)	b_1 (GPa^{-1})	c_0 ($10^{-6}\,m^2\,s^{-1}$)	c_1 ($10^{-6}\,m^2\,s^{-1}\,K$)
1.19 (6)	0.046 (1)	0.29 (6)	374 (31)	[1 0 0]	2.50 (4)	0.033 (5)	−0.06 (11)	938 (46)
				[0 1 0]	1.53 (6)	0.040 (7)	−0.13 (8)	626 (45)
				[0 0 1]	2.16 (14)	0.035 (3)	−0.03 (17)	832(98)
$\lambda = A_0 + A_1 P$		$\lambda = C_0 + C_1/T$			$\lambda = B_0 \exp(B_1 P)$		$\lambda = C_0 + C_1/T$	
A_0 ($W\,m^{-1}\,K^{-1}$)	A_1 ($W\,m^{-1}\,K^{-1}\,GPa^{-1}$)	C_0 ($W\,m^{-1}\,K^{-1}$)	C_1 ($W\,m^{-1}$)		B_0 ($W\,m^{-1}\,K^{-1}$)	B_1 (GPa^{-1})	C_0 ($W\,m^{-1}\,K^{-1}$)	C_1 ($W\,m^{-1}$)
3.48 (33)	0.160 (26)	2.01 (8)	704 (43)	[1 0 0]	6.61 (13)	0.038 (5)	1.91 (28)	2088 (163)
				[0 1 0]	3.98 (15)	0.042 (5)	0.84 (36)	1377 (157)
				[0 0 1]	5.91 (25)	0.034 (5)	2.08 (38)	1731 (86)

Manga, M., Jeanloz, R., 1997. Thermal conductivity of corundum and periclase and implications for the lower mantle. J. Geophys. Res. 102, 2999–3008.

Osako, M., 1997. Thermal diffusivity of olivine and garnet single crystals. Bul. Natl. Sci. Mus., Tokyo, Ser. E 20, 1–7.

Sato, Y., Akaogi, M., Akimoto, S., 1978. Hydrostatic compression of the synthetic garnets pyrope and almandine. J. Geophys. Res. 83, 335–338.

Schärmeli, G.H., 1982. Anisotropy of olivine thermal conductivity at 2.5 GPa and up to 1500 K measured on optical non-thick samples. In: Schreyer, W. (Ed.), High Pressure Researchs in Geoscience. E. Schweizerbart'sche Verlagsbuchhandlung, Stuttgart, pp. 349–373.

Suzuki, I., 1975. Thermal expansion of periclase and olivine and their anharmonic properties. J. Phys. Earth 23, 145–159.

Suzuki, I., Anderson, O.L., 1983. Elasticity and thermal expansion of a natural garnet up to 1000 K. J. Phys. Earth 31, 125–138.

Yukutake, H., Shimada, M., 1978. Thermal conductivity of NaCl, MgO, coesite and stishovite up to 40 kbar. Phys. Earth Planet. Int. 17, 193–200.

Watanabe, H., 1982. Thermochemical properties of synthetic high-pressure compounds relevant to the Earth's mantle. In: Akimoto, S., Manghnani, M.H. (Eds.), High Pressure Research in Geophysics. Center for Academic Publication, Tokyo, pp. 441–464.

Zha, C.S., Duffy, T.S., Downs, R.T., Mao, H.K., Hemley, R.J., 1998. Brilluin scattering and X-ray diffraction of San Carlos olivine: direct pressure determination to 32 GPa. Earth Planet. Sci. Lett. 159, 25–33.

Available online at www.sciencedirect.com

SCIENCE DIRECT®

ELSEVIER

Physics of the Earth and Planetary Interiors 143–144 (2004) 321–336

PHYSICS OF THE EARTH AND PLANETARY INTERIORS

www.elsevier.com/locate/pepi

Thermal diffusivity and conductivity of olivine, wadsleyite and ringwoodite to 20 GPa and 1373 K

Yousheng Xu [a,*], Thomas J. Shankland [b], Sven Linhardt [a], David C. Rubie [a], Falko Langenhorst [a], Kurt Klasinski [a]

[a] *Bayerisches Geoinstitut, Universität Bayreuth, D-95440 Bayreuth, Germany*

[b] *Earth and Environmental Science Division, Los Alamos National Laboratory, Los Alamos, NM 87545, USA*

Abstract

We present results of lattice thermal diffusivity measurements on $(Mg_{0.9}Fe_{0.1})_2SiO_4$ olivine and its high-pressure polymorphs wadsleyite and ringwoodite under mantle conditions. We used the Ångström method on cylindrical samples in multianvil apparatus at pressures up to 20 GPa and temperatures up to 1373 K. Because of the fine polycrystallinity of the specimens (~30–40 μm or less), there is strong scattering/absorption of light and suppression of radiative transport so that the lattice vibrational component is the dominant heat transfer mode. Lattice thermal conductivities were calculated from the thermal diffusivity results using heat capacity and equation of state data. Olivine thermal conductivities are consistent with previous results obtained at 1 atm [e.g. J. Am. Ceramic Soc. 38 (1954) 107; J. Geophys. Res. 77 (1972) 6966; Science 283 (1999) 1699]. Thermal conductivity increases by approximately 30% at the transition from olivine to wadsleyite (corresponding to the 410 km discontinuity) and a further, but smaller, increase may occur at the transition from wadsleyite to ringwoodite. For each of these phases, lattice conductivity closely follows a $T^{-1/2}$ dependence on temperature T [Phys. Rev. 119 (1960) 507; J. Geophys. Res. 79 (1974) 703; Brown, 2002]. If such a dependence applies to other silicates and complex crystals, there should be a useful way to estimate conductivities at high temperatures from room temperature measurements.

Keywords: Heat transport; High pressure; Earth's mantle; Subducting lithosphere; Multianvil apparatus

1. Introduction

Knowledge of the heat-transport properties of olivine and its high-pressure polymorphs, wadsleyite and ringwoodite, is essential for understanding thermal regimes in the Earth's mantle (e.g. van den Berg and Yuen, 2002; Marton et al., 2002). When thermal conductivity depends on temperature (as opposed to being constant), models of lithospheric geotherms and the thermal structure of subducting lithospheric slabs are affected (e.g. Hofmeister, 1999; Hauck et al., 1999). In addition, jumps in thermal conductivity across the olivine–wadsleyite–ringwoodite phase boundaries in subduction zones can lead to old, fast, steeply-dipping slabs being considerably warmer than previously thought (Hauck et al., 1999; Marton et al., 2002, submitted). Such results have implications for a variety of processes including subduction dynamics and the origin of deep earthquakes (Mosenfelder et al., 2001; Marton et al., submitted).

Because olivine is the principal interconnected phase in the upper mantle, numerous measurements

* Corresponding author. Present address: Department of Geology and Geophysics, Yale University, P.O. Box 208109, New Haven, CT 06520-8109, USA. Tel.: +1-203-432-5791; fax: +1-203-432-3134.
E-mail address: yousheng.xu@yale.edu (Y. Xu).

0031-9201/$ – see front matter
doi:10.1016/j.pepi.2004.03.005

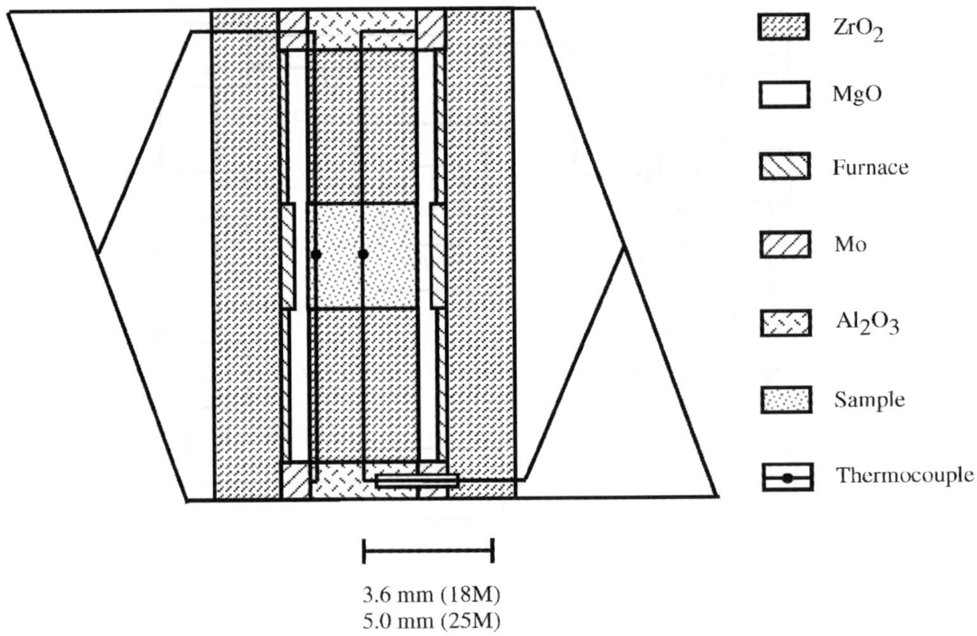

Fig. 2. Schematic cross-section of the high-pressure octahedral cell assemblies used for thermal diffusivity measurements in the multianvil apparatus. Note the different scales for the two sizes used.

Under TEM, wadsleyite shows stacking faults parallel to (010), whereas the large ringwoodite grains are free of stacking faults but contain dislocations in a rather high density of $10^{13}\,m^{-2}$.

The dimensions of the synthesized samples were ~2.5 mm in diameter and ~2.8 mm in length for the 18 M assembly and ~3.5 mm in diameter and ~3.5 mm in length for the 25 M assembly. A hole of 0.3 mm diameter was drilled through the axis of each sample to accommodate the inner thermocouple and a groove was made on the surface of the sample, in the axial direction, to accommodate the outer thermocouple; both thermocouples were fabricated from 0.076 mm diameter wires. The temperatures measured by the two thermocouples were recorded on a computer as a function of time (Figs. 1 and 4).

Measurements on olivine were made using 18 and 25 M assemblies and for those on wadsleyite and ringwoodite the 18 M assembly was used. In each experiment, the sample was first pressurized to the desired value and then heated. Diffusivity measurements were made during heating and for the most part during subsequent cooling, usually at intervals of 100 K. Measurements were performed at five frequencies (0.1, 0.4, 0.7, 1.0, and 1.3 Hz) at each temperature, and the interval between successive measurements was ~10 min in order to allow initial transients to disappear and a steady state to be reached. Data points (2048) were always recorded for 10 periods so that the sampling rate varied with modulation frequency. The precision of recorded temperatures was high as seen in the low noise on the sinusoids of Fig. 4. To obtain precise data for the temperature waves, we fitted 10 steady-state sinusoidal cycles to obtain amplitude and phase at each modulation frequency (Fig. 4). Phase shift was calculated by subtracting the outer phase from the inner (Eq. (2)) and amplitude ratio was calculated by dividing the outer amplitude by the inner amplitude (Eq. (3)) using the fitted parameters for the two temperature waves. Whether the phase shift was larger or smaller than 180° was determined from the magnitude of the phase shift when the modulations started. Although only phase shift was used for determining thermal diffusivity (see below), we kept temperature differences ΔT (and therefore $\Delta T/T$) small, on the order of 5 K, so that any non-linearity introduced by the T-dependence of D would be small.

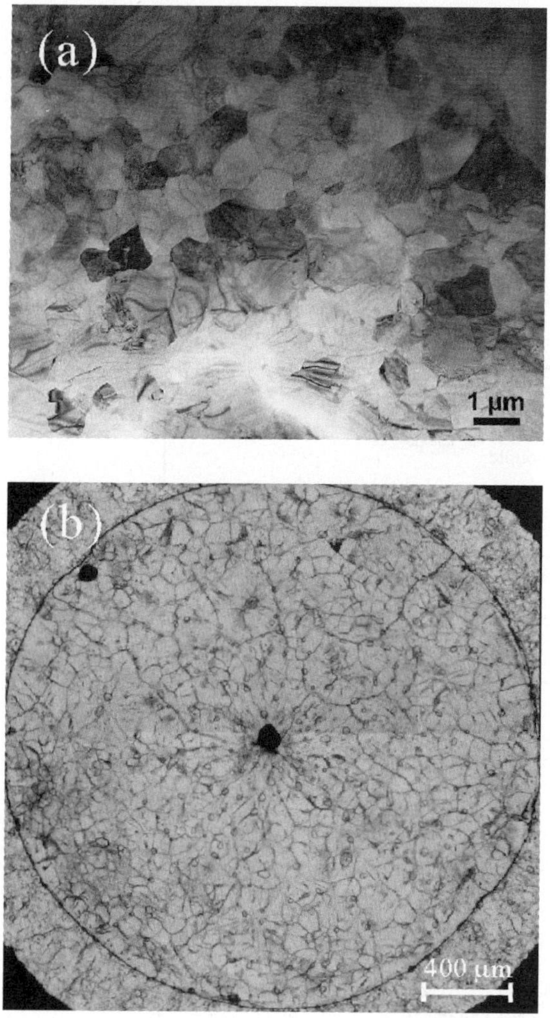

Fig. 3. (a) Bright field TEM image of polycrystalline wadsleyite, revealing grain sizes on the order of 1–2 μm. (b) Optical micrograph of the ringwoodite sample, with a grain size of 20–30 μm, following a thermal diffusivity experiment. The sample is still surrounded by a cylindrical sleeve of MgO (see Fig. 3) and the two thermocouple junctions (opaque) can be clearly seen.

Temperature differences between the outer and inner thermocouples reached a maximum of ~20 K at the highest values of frequency and temperature. For each measurement, the sample temperature was calculated by averaging values recorded by the outer and inner thermocouples. The error of this estimation is insignificant because over a small temperature range (<20 K) the variation in D is negligible.

Following each experiment, the sample was retrieved and was prepared for examination by optical and transmission electron microscopy and X-ray diffraction. The distance d between the two thermocouple junctions (see Fig. 3b) was measured to an accuracy of 5 μm using a microscope fitted with a vernier X–Y stage. The distance d at high pressure and temperature was calculated from the change in density, which is given to first-order by

$$\rho = \rho_0 \left[1 - \alpha(T - T_0) + \frac{P}{K}\right], \quad (4)$$

where ρ is density, α the thermal expansion coefficient, P pressure (GPa), K the bulk modulus (GPa) at high P and T, and the subscript $_0$ donates ambient conditions. High P–T values for K and α are given by

$$K = K_0 + K'P + \left(\frac{\partial K_T}{\partial T}\right)_P (T - T_0), \quad (5)$$

$$\alpha = \alpha_0 + \alpha_1 T + \alpha_2 T^{-2} + \frac{1}{K_0^2}\left(\frac{\partial K_T}{\partial T}\right)_P, \quad (6)$$

where K' is the first pressure derivative of the bulk modulus, $(\partial K_T/\partial T)_P$ the isobaric derivative of the bulk modulus, and α_0, α_1 and α_2 are constants determined by fitting experimental data at ambient-pressure conditions. Finally, the radius at high temperature and pressure is computed as

$$d = d_0(\rho_0/\rho)^{1/3} = d_0[1 - \alpha(T - T_0) + \frac{P}{K}]^{-1/3} \quad (7)$$

using Eqs. (4)–(6), where d_0 is the distance measured at ambient conditions. The parameterization of α is taken from Fei (1995), K_0 and K' are from Ita and Stixrude (1992), and $(\partial K_T/\partial T)_P$ is from Meng et al. (1993).

Uncertainties of the measurements are estimated from the following factors: (1) the errors on the thermocouple positions, measured on the recovered samples, are less than 3%; (2) based on the reproducibility of the 10 collected temperature waves, the errors that originate from fitting these are less than 1%; (3) errors that originate from correcting the frequency-dependence of thermal diffusivity (see below) are <8%. We estimate the total error on the measurements to be ~12%.

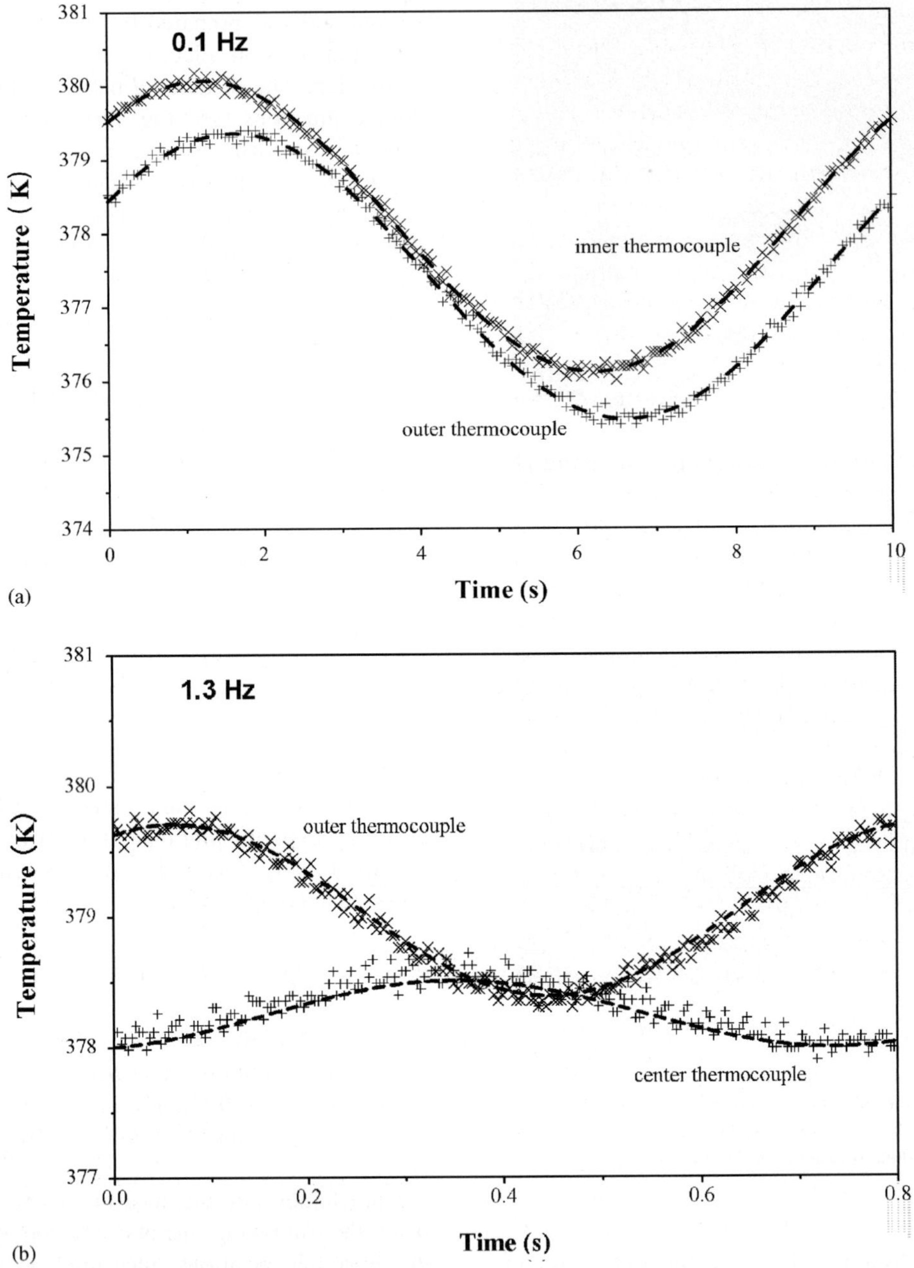

Fig. 4. Examples of recorded temperature waves consisting of 10 cycles measured at frequencies of 0.1 and 1.3 Hz, respectively at 4 GPa and 378 K in run S2354. The 10 cycles of each temperature wave consist of 2048 data points. The symbols represent the measured data and the dash thick lines represent the fitted sine waves.

3. Results

Our measured values of thermal diffusivity often depended on modulation frequency *f*, especially at low temperatures. In these cases, we chose a high-frequency asymptote primarily because both phase and amplitude methods (2) and (3) approach similar values (Fig. 5). The probable reason for disagreement at low frequencies is axial end effects, i.e. departure from the approximation of an infinitely long cylinder when the thermal wave is long at low frequencies (Bhandari and Rowe, 1998). A high-frequency asymptote D_0 was obtained by fitting the data to $D(f) = D_0 + A\exp(-af)$. When no systematic pattern of frequency dependence was observed, diffusivity was obtained by averaging $D(f)$ over the different frequencies.

Previous results show that radiative conduction can contribute significantly to heat transport at high temperature in materials having high transparency, i.e. the case of low optical density or ballistic photon transport (the appendix discusses how absorption and scattering in these fine-grained specimens suppress radiative contributions). However, a further argument derives from the greater effect that radiative heat transport is likely to have on thermal diffusivities determined from amplitude ratios than from phase shifts (Kanamori et al., 1969) when optical density is low. Fig. 6 compares of thermal diffusivities of olivine and ringwoodite obtained using these two approaches. The consistent agreement between the results obtained at high temperature from phase shift and amplitude methods for both olivine and ringwoodite (and also wadsleyite) further supports the conclusion that radiative conduction was not important in these measurements. Considering that thermal diffusivity data obtained from phase shifts are likely to be more reliable than those derived from amplitude ratios (Kanamori et al., 1969), only the former results are emphasized here and used to calculate thermal conductivities.

Obtaining the same result in different experimental configurations satisfies the important criterion of reproducibility and further indicates that radiative heat transport is not significant. Thus, two experiments on olivine (runs S2325 and S2354) were performed at 7 GPa in the smaller 18 M assembly and in the larger 25 M assembly, respectively (Fig. 7). Diffusivities obtained in the two assemblies having significantly different dimensions are identical (within error), which would not be the case if ballistic photon transport were present. In the case of S2325, temperature was increased from 298 to 793 K and then directly to 1473 K,

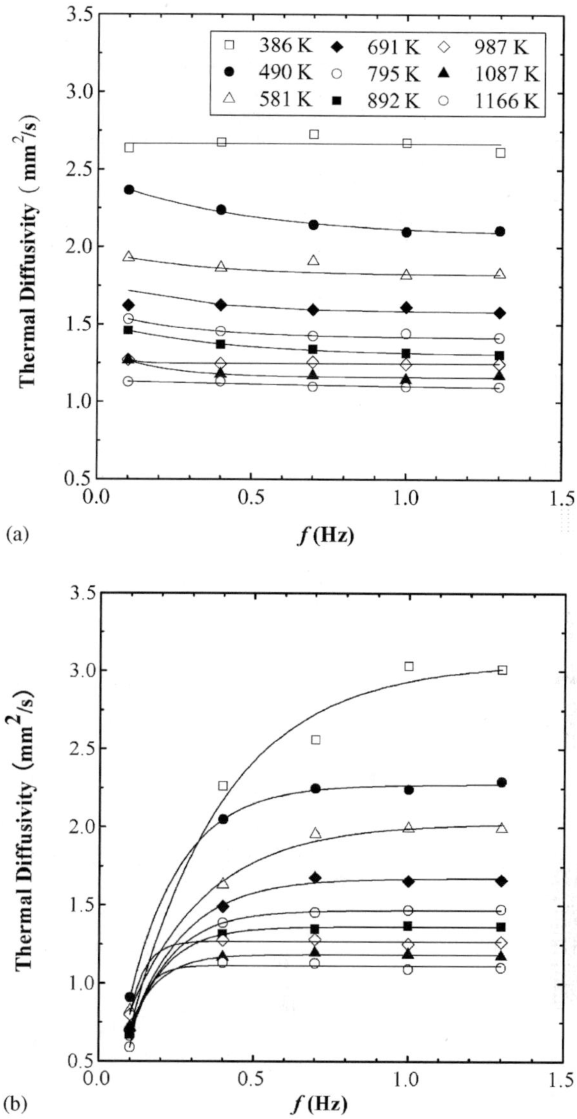

Fig. 5. Thermal diffusivity of ringwoodite vs. frequency at 20 GPa and 386–1166 K. (a) Diffusivity calculated from phase shift; (b) Diffusivity calculated from amplitude ratio. The symbols in Fig. 5b represent the same temperatures as in Fig. 5a. The data were fitted to $D = D_0 + A\exp(-f/f_0)$ except for data at 386 K in Fig. 5a. The high frequency asymptote is D_0.

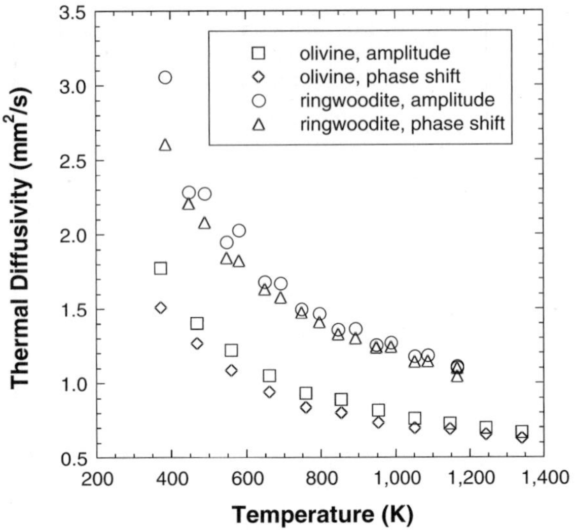

Fig. 6. Comparison of the thermal diffusivities of olivine at 7 GPa and ringwoodite at 20 GPa determined from both phase shift and amplitude.

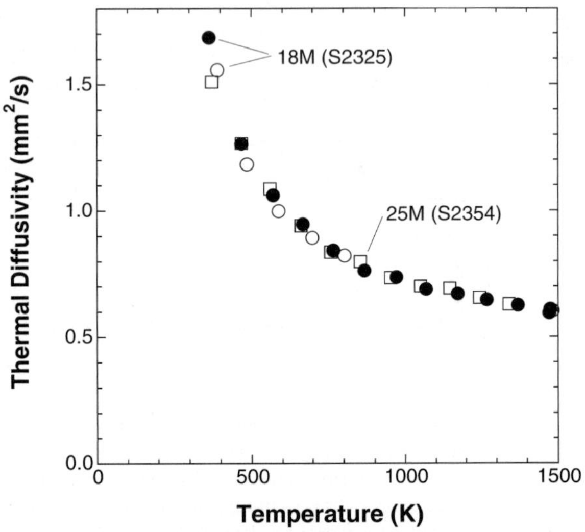

Fig. 7. Comparison of thermal diffusivities of olivine obtained using 18 M (S2325) and 25 M (S2354) multianvil assemblies at 7 GPa to illustrate consistency and reproducibility. Sample diameters were ~2.5 mm in the 18 M assembly and ~3.5 mm in the 25 M assembly. The cycles indicate data obtained using the 18 M assembly and the squares indicate data obtained using the 25 M assembly. Open symbols indicate data taken while increasing the temperature, and solid symbols indicate data obtained during cooling.

and after maintaining 1473 K for 30 min, the sample was cooled back to room temperature. Diffusivity measurements were made during heating between 298 and 793 K and during cooling between 1473 and 298 K at ~100 K intervals. Thermal diffusivity values at 1473 K changed little in 30 min, and the results obtained during heating are consistent with those obtained during cooling (Fig. 7).

Fig. 8 shows thermal diffusivities of olivine (run S2354) obtained in a 25 M assembly as a function of temperature at three different pressures together with results for wadsleyite and ringwoodite. The thermal diffusivity of wadsleyite is significantly higher than that of olivine extrapolated to the same pressure (14 GPa), and diffusivity of ringwoodite at 20 GPa is still higher than that of wadsleyite at 14 GPa. Although ringwoodite diffusivity might be similar to that of wadsleyite at the same pressure, we have not measured pressure derivatives for either of these two phases. It is convenient to use the following power-law formulation for diffusivity as is done for conductivity later,

$$D_{\text{lat}} = D_{298}\left(\frac{298}{T}\right)^n (1 + aP) \tag{8}$$

and results of the fits are as given in Table 1.

Results from Tommasi et al. (2001) from mantle lherzolite and harzburgite rock samples are also shown in Fig. 8. The shaded band represents the range of diffusivities resulting from anisotropic crystal orientations. Their results agree well with our olivine diffusivities extrapolated to low pressure. Our values are a few percent lower than the single-crystal values of Harrell (2002), which is possibly a result of the slight degree of unconsolidation in the olivine samples at 4 GPa and lower temperatures before complete annealing occurs.

4. Thermal conductivities

In order to compute thermal conductivity κ from thermal diffusivity D for comparison with results of other studies, we use

$$\kappa = D\rho C_P, \tag{9}$$

where C_P is isobaric heat capacity. Density as a function of temperature and pressure is calculated from Eqs. (4)–(6) and ρ_0 is from Smyth and McCormick (1995). We do not consider the effect of pressure on

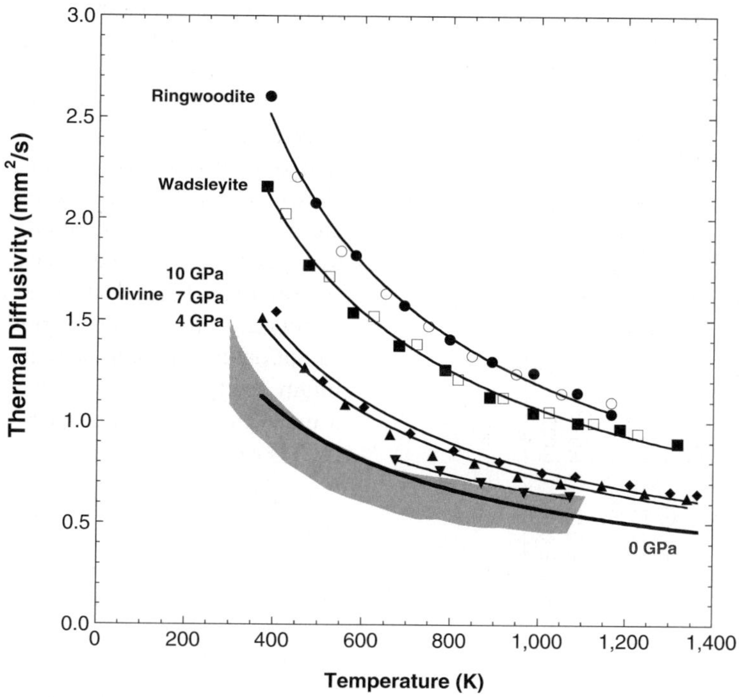

Fig. 8. Thermal diffusivities of olivine (S2354), wadsleyite, and ringwoodite vs. temperature. Solid symbols indicate data taken while increasing temperature (the only points for olivine S2354), and open symbols indicate data obtained during cooling. The lines show the results of fitting a power-law (Eq. (8) and Table 1) and in the case of wadsleyite and ringwoodite are for combined heating and cooling data. The thick solid line shows the extrapolation to zero pressure using Eq. (8), and the shaded area covers the range of diffusivities measured by Tommasi et al. (2001) in anisotropic mantle dunites.

C_P because this parameter is virtually independent of pressure for solids (Navrotsky, 1995). Heat capacity is strongly temperature dependent, and we calculate it using a relation from Saxena (1996)

$$C_P = \alpha + bT + cT^2 + dT^{-2} + eT^{-3} + fT^{-1/2} + \frac{g}{T}, \tag{10}$$

where parameter values also come from Saxena (1996). We assume that density and heat capacity are linearly proportional to the Fe/(Fe + Mg) ratio and calculate values for our samples from data for the Mg- and Fe-end members (Vacher et al., 1998). Because our composition is close to the Mg-end member, errors that result from this assumption are negligible.

Lattice thermal conductivity increases with density and therefore with pressure. For instance, there is the strong dependence given by Roufosse and Jeanloz (1983):

$$\frac{\delta\kappa_{\text{lat}}}{\kappa_{\text{lat}}} \cong 7\frac{\delta\rho}{\rho}.$$

In substances having simple crystal structures, thermal conductivity is theoretically proportional to reciprocal temperature at temperatures above the Debye temperature. As an example, assuming that three-phonon collision processes dominate, lattice thermal conductivity κ_{lat} at temperatures well above the Debye temperature for simple crystal structures is given by Dugdale and MacDonald (1955) as:

$$\kappa_{\text{lat}} = \frac{A_0\rho Cv}{3\alpha\gamma T}, \quad (T > \theta_D),$$

where A_0 is a lattice constant, C the heat capacity per unit mass, ν (acoustic) phonon velocity, γ the Grüneisen parameter, and θ_D the Debye temperature.

However, for complex crystal structures (those having many atoms/unit cell) such as most minerals the reciprocal lattice thermal conductivity is often given in a form that takes account of additional scattering of acoustic phonons through a parameter a, i.e. $\kappa^{-1} = a + bT$. As is frequently the case for such

Table 1
Parameter values derived from fitting the equations $D_{lat} = D_{298}(298/T)^n(1 + aP)$ and $\kappa_{lat} = \kappa_{298}(298/T)^n(1 + aP)$ to the experimental diffusivity data and converted conductivity data respectively

Mineral	P (GPa)	$D_{lat} = D_{298}(298/T)^n(1+aP)$			$\kappa_{lat} = \kappa_{298}(298/T)^n(1+aP)$			$\kappa_{lat} = \kappa_{298}(298/T)^{1/2}(1+aP)$	
		D_{298} (mm^2/s)	n	a (GPa^{-1})	κ_{298} (Wm^{-1}K^{-1})	n	a (GPa^{-1})	κ_{298} (Wm^{-1}K^{-1})	a (GPa^{-1})
Olivine	4–10	1.31 ± 0.05	0.681 ± 0.022	0.036 ± 0.004	4.10 ± 0.14	0.493 ± 0.210	0.032 ± 0.003	4.13 ± 0.11	0.032 ± 0.003
	4	1.29 ± 0.05	0.563 ± 0.035	–	4.08 ± 0.15	0.406 ± 0.035	–	4.49 ± 0.04	–
	7	1.74 ± 0.04	0.720 ± 0.026	–	5.15 ± 0.09	0.491 ± 0.017	–	5.19 ± 0.04	–
	10	1.84 ± 0.05	0.723 ± 0.031	–	5.59 ± 0.10	0.507 ± 0.019	–	5.56 ± 0.04	–
Wadsleyite	14	2.55 ± 0.03	0.721 ± 0.013	–	8.06 ± 0.10	0.494 ± 0.013	0.023*	8.10 ± 0.04	0.023*
Ringwoodite	20	3.09 ± 0.04	0.793 ± 0.017	–	9.83 ± 0.09	0.537 ± 0.011	0.022*	9.54 ± 0.05	0.022*

Pressure coefficients indicated by an asterisk (*) are calculated as in the Note added in proof.

materials (e.g. Kanamori et al., 1968), this approach fails to give a straight line of κ^{-1} versus T for our data, particularly at temperatures well above the Debye temperature (~673 K for olivine, wadsleyite, and ringwoodite).

In other treatments of complex materials, it has been shown both experimentally and theoretically that there are numerous optic vibrational modes that contribute to specific heat but these have low phonon velocities (Hofmeister, 1999; Harrell et al., 2002). Hence, κ_{lat} commonly falls off differently than T^{-1}. On the assumption that phonons at high temperature have a minimum scattering distance that is a small multiple of a mean interatomic distance, Roufosse and Klemens (1974) expressed thermal conductivity as

$$\kappa_{lat} = \frac{B}{T}\left[\frac{2}{3}\left(\frac{T_0}{T}\right)^{1/2} + \frac{1}{3}\left(\frac{T}{T_0}\right)\right]. \tag{11a}$$

Here B and T_0 are constants related to properties of the material. In Eq. (11a) κ_{lat} has a constant asymptotic value at high temperature so that $\kappa_{lat}(T)$ flattens out. Harrell (2002) has successfully used this approach to analyze olivine and pyroxene conductivities and to determine phonon mean free paths that are on the order of twice the mean interatomic distance.

For a phonon scattering regime of short interaction distances given by point defects another approach is possible (Klemens, 1960; Roufosse and Klemens, 1974). Their work indicated that a $T^{-1/2}$ dependence is appropriate to somewhat below the Debye temperature:

$$\kappa_{lat} = AT^{-1/2}. \tag{11b}$$

Brown (2002) has successfully tested Eq. (11b) for a number of materials. Hence, we use a power-law approach for olivine (run S2354), wadsleyite, and ringwoodite (Fig. 9a). Both the room temperature pre-multiplier and the exponent are used as fitting parameters. In this case, results of the fit of the following expression are given in Table 1:

$$\kappa_{lat} = \kappa_{298}\left(\frac{298}{T}\right)^n (1 + aP). \tag{12}$$

It is seen that all the values of the exponent n are close to 1/2. Because of possible effects of unconsolidation and the strong frequency dependence of the lowest-temperature data (below 600 K) for olivine

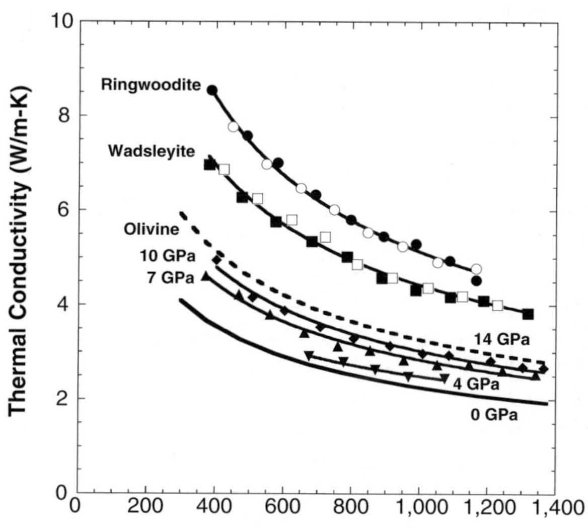

(a)

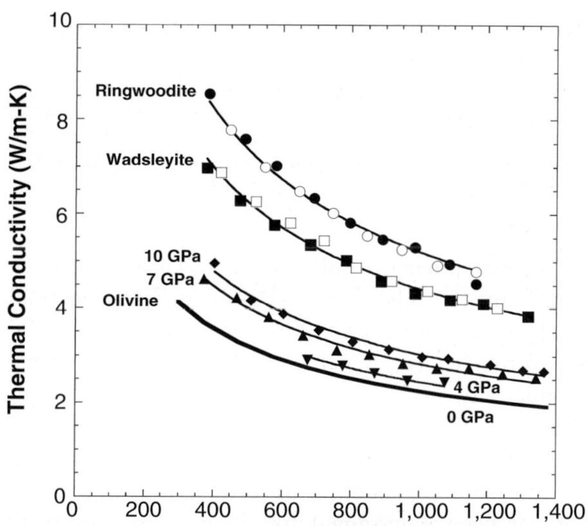

(b)

Fig. 9. Thermal conductivities of olivine, wadsleyite, and ringwoodite indicating a power-law dependence upon temperature. Solid symbols indicate data taken while increasing temperature (the only points for olivine S2354), and open symbols indicate data obtained during cooling. Fitted lines for wadsleyite and ringwoodite are for combined heating and cooling data. (a) Olivine curves are fitted separately according to Eq. (12); the combined fit to Eq. (12) was used to calculate the 0 (solid) and 14 GPa (dashed) extrapolations. (b) Fit to the same data with Eq. (12) in which the only fitting parameter is room temperature conductivity and the exponent is fixed at 1/2.

S2354 at 4 GPa, these points are not displayed or used in the fits. The slope of this curve is still shallower than that of the other curves, and uncertainties in the fitted parameters are larger.

As a test of Eq. (11b) the results of a single-parameter fit to the data (with $n = 1/2$) are shown in Fig. 9b and listed in Table 1. Note that the sole adjustable parameter in Eq. (11b) is the room temperature conductivity κ_{298}, which can be measured at a given pressure more readily than at combined high temperature and pressure. Eq. (11b) was derived for the case of point defects with a nominal atomic fraction of 5% (Klemens, 1960), which are unlikely to exist in such abundance in olivine at least (Hirsch and Shankland, 1993). However, Mg–Fe cation disorder is a likely possibility (Henderson et al., 1996).

Extrapolating olivine Fo_{90} data to ambient conditions yields κ (1 bar, 298 K) = $4.10\pm0.14\,Wm^{-1}\,K^{-1}$ which is slightly lower than the average of previous results of $\sim$4.65 $Wm^{-1}\,K^{-1}$ for Fo_{90} (Birch and Clark, 1940; Chai et al., 1996; Hofmeister, 1999). Our olivine results extrapolated to 1 bar are compared with results of previous studies in Fig. 10a. Our extrapolated results are consistent with previous results for polycrystalline olivine (Kingery et al., 1954; Schatz and Simmons, 1972; Hofmeister, 1999) after correcting for the effect of Fe content, i.e. lowering κ by 0.5 $Wm^{-1}\,K^{-1}$ from the Mg-end member to Fo_{90}, as observed for various silicates (Horai, 1971). However, thermal conductivities of San Carlos olivine obtained from a spatial average of anisotropic lattice conductivities from Schärmeli (1982) are much higher than our extrapolated results at the same pressure of 2.5 GPa (Fig. 10b).

Fig. 11 shows a comparison of our data with those of Hofmeister's (1999) estimations for other Mg_2SiO_4 phases. Hofmeister's estimations were calculated from Eq. (11) and Table 1 in her paper (Hofmeister, 1999). At 14 GPa, our extrapolated olivine results are close to those of Hofmeister's estimations after adjusting for the effect of iron content on conductivity. Our measured $(Mg_{0.9}Fe_{0.1})_2SiO_4$ ringwoodite data are lower (by $\sim$1.2 $Wm^{-1}\,K^{-1}$) than conductivities of Mg_2SiO_4 ringwoodite calculated by Hofmeister (1999). In some models (Benigni and Rogez, 1997; Hofmeister, 1999), the relationship between thermal conductivity and temperature can be represented with both a power-law and an exponential component.

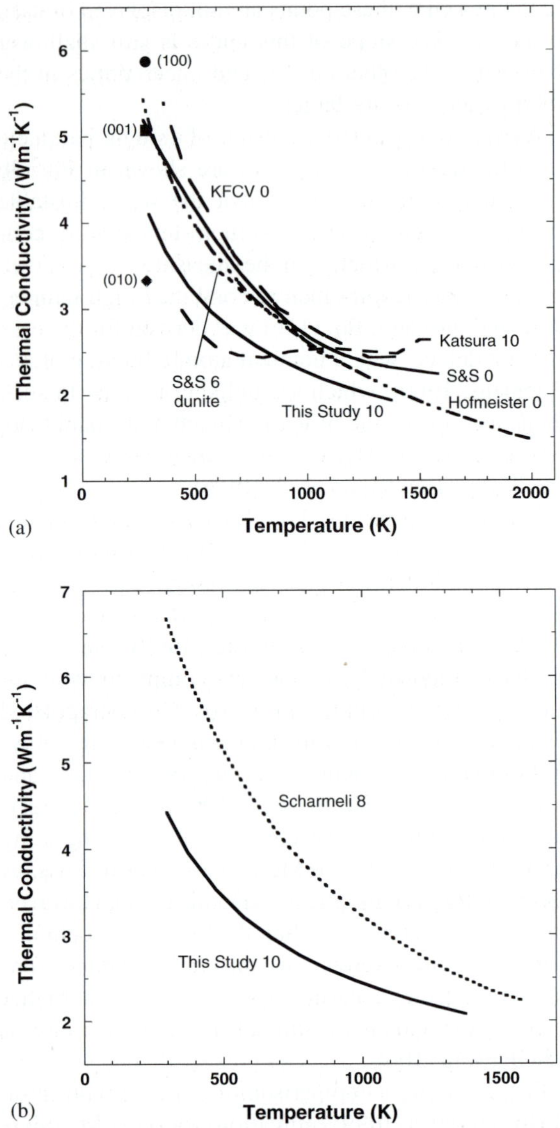

Fig. 10. (a) Comparison of thermal conductivity of olivine with previous results for κ_{lat} at 1 atm. Numbers are fayalite contents (mol%) in olivine. S&S: Schatz and Simmons (1972); Katsura: Katsura (1995); H: Hofmeister (1999), KFCV: Kingery et al. (1954). Data points are room temperature values for different crystallographic directions from Chai et al. (1996). (b) Comparison of calculated thermal conductivity of olivine with Schärmeli's (1982) results at 2.5 GPa.

Hofmeister (1999) has a presentation that contains both power-law and exponential forms and is more complicated than required by our data.

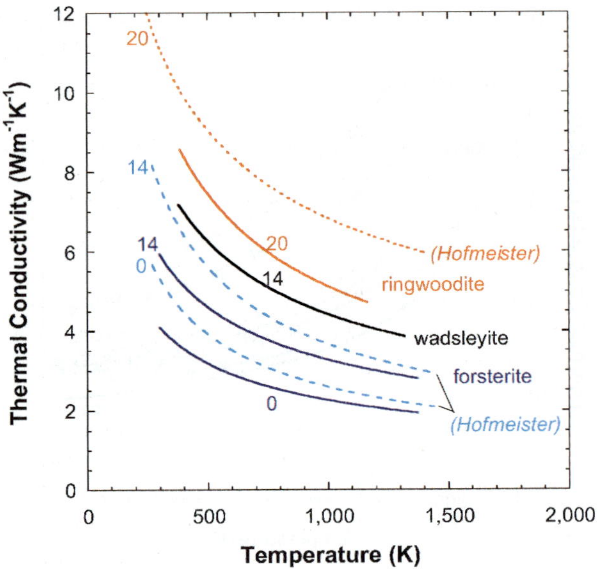

Fig. 11. Comparison of lattice thermal conductivities of $(Mg_{0.9}Fe_{0.1})_2SiO_4$ olivine, wadsleyite, and ringwoodite with Hofmeister's (1999) calculated values for Mg_2SiO_4 with olivine (dashed lines) and ringwoodite (dotted lines) structures. Numbers represent pressures in GPa. The lower two solid lines are for olivine, and the upper two are for wadsleyite and ringwoodite.

Our extrapolated results indicate that when olivine transforms to wadsleyite at 410 km depth, the thermal conductivity increases ~30%, and when wadsleyite transforms to ringwoodite at 520 km, the thermal conductivity might also increase. However, the increase in conductivity should be much smaller at 520 km than at 410 km although this depends on the pressure-dependence of diffusivity, which has not yet been determined for the two high-pressure polymorphs.

An important result to come from Fig. 9 is the realization, following Brown (2002), that for the (Mg, Fe)$_2$SiO$_4$ polymorphs at different pressures there is a good fit to a $T^{-1/2}$ temperature dependence for κ_{lat}. It remains to be seen whether this result can be applied to other silicates or to complex crystals in general; if so, then there is a useful way to extrapolate conductivities measured at ambient temperatures to conditions in the mantle.

Our new thermal conductivity data for olivine, wadsleyite and ringwoodite will enable more refined thermal models of the Earth's interior to be developed and may change some previous conclusions. For example, in models of the thermal structure of subducting slabs a constant value for thermal conductivity was used for many years (e.g. Minear and Toksöz, 1970; Kirby et al., 1996). Such constant thermal conductivity models have been used to predict that metastable olivine would survive in subducting slabs at least to the maximum depths of the deepest earthquakes (e.g. Kirby et al., 1996). Such a conclusion supports the origin of deep earthquakes by transformational faulting, which involves the development of shear instabilities during the transformation of metastable olivine to wadsleyite or ringwoodite (Kirby, 1987; Green and Burnley, 1989). Marton et al. (2002, submitted) have used our new thermal conductivity data for olivine, wadsleyite and ringwoodite in thermo-kinetic subduction models. Including the effects of both variable conductivities and radiative heat transfer results in slabs being considerably warmer than suggested by earlier fixed-conductivity models (see also Hauck et al., 1999). Based on the recent kinetic model of Mosenfelder et al. (2001), the maximum depth of olivine metastability in a rapidly subducting slab of old lithosphere (e.g. Tonga) is reduced by ~50 km compared with fixed-conductivity models. These new models indicate that the depth of olivine metastability in all slabs should be insufficient for explaining deep earthquakes by the mechanism of transformation faulting, even when all uncertainties are considered.

5. Conclusions

Thermal diffusivities of $(Mg_{0.9}Fe_{0.1})_2SiO_4$ olivine, wadsleyite, and ringwoodite have been measured in a multianvil apparatus up to 20 GPa and 1373 K using the Ångström method on cylindrical samples. The major conclusions of this work are as follows:

1. Thermal diffusivities of olivine have been measured at 7 GPa in two different sized multianvil sample assemblies and show consistent and reproducible results.
2. Thermal conductivities calculated from the measured diffusivities show that, for olivine, wadsleyite, and ringwoodite, lattice conductivity can be described well by a $T^{-1/2}$ temperature dependence. This result can be used to obtain thermal conductivities under mantle conditions from room temperature measurements at high pressure, to at least acceptable precision.
3. If we extrapolate the calculated thermal conductivity of olivine to a pressure of 1 atm, our results are consistent with and slightly lower than previous results for olivine. However, for extrapolated conductivities at 2.5 GPa, our results are much lower than those of Schärmeli (1982). Our calculated thermal conductivities of olivine and ringwoodite show that the olivine results are close to Hofmeister's (1999) estimations after we correct for the iron effect, but our ringwoodite results are ~1.2 $W m^{-1} K^{-1}$ lower than those of Hofmeister's calculated values.
4. Extrapolating our data to the 410 km discontinuity indicates that thermal conductivity increases by ~30% at this transition.
5. Incorporating the new data, together with plausible effects of radiative heat transfer, into thermo-kinetic models of subducting slabs reduces the maximum depth of olivine metastability compared with previous estimates. The new results cast doubt on the transformational faulting mechanism as an explanation for deep earthquakes.

Acknowledgements

We gratefully thank H. Schulze, who skillfully prepared samples for the measurements. We also thank D. Frost and B.T. Poe for essential assistance in operating the 5000 t multianvil apparatus and H. Fischer and G. Herrmannsdörfer for technical support. Michael Harrell generously pointed out to us the work of Klemens and Roufosse, and Michael Brown, who has demonstrated this behavior in various silicates, generously provided the suggestion that a $T^{-1/2}$ dependence would apply to a range of materials. TJS thanks the Bayerisches Geoinstitut and the Universität Bayreuth for generous hospitality during part this research, and he gratefully acknowledges the Alexander von Humboldt Foundation for their support. His work at Los Alamos National Laboratory was also supported by a grant from the Geosciences Research Program, Office of Basic Energy Sciences of the Department of Energy.

Appendix A. Evaluation of radiative contributions

So long as a sample is not completely opaque, there should be some measure of a diffusive radiative conductivity κ_R. However, in these experiments this component is small. Because various mechanisms minimize radiative contributions, in these experiments we take the measured diffusivity and conductivity to be the lattice components D_{lat} and κ_{lat}.

The easiest case to treat is that of the wadsleyite and ringwoodite specimens. They simply absorb so strongly that they are effectively opaque. A 30 μm specimen appears darker in the visible range than does 1 cm of olivine; among other reasons, the availability of tetrahedral lattice sites enhances crystal field absorption in Fe ions.

In the case of a coarse-grained olivine of thickness ~1.5 mm one might expect diffusive and ballistic radiative effects because such a material is not optically thick. Intensity of a light beam falls off according to Beer's Law as $\exp(-\alpha_\nu x)$, where optically thickness $\alpha_\nu x$ is the product of absorption coefficient α_ν and distance x in the material. A mean value $\langle\alpha_\nu\rangle$ weighted by the blackbody spectrum is $\sim$0.4 mm^{-1} at 1000 K as obtained from measurements of α_ν at high temperatures (Fukao et al., 1970; Aronson et al., 1970; Shankland et al., 1979). If $\alpha_\nu x$ were $0.4 \times 1.5 = 0.6$, then $\exp(-\alpha_\nu x)$ would not be $\ll 1$ as required for negligible κ_{rad}.

However, our samples are very fine-grained, and grain boundaries in a birefringent material like olivine introduce a scattering coefficient α_s so that total absorption $\alpha_t = \alpha_\nu + \alpha_s$. For $(Mg_{0.9}Fe_{0.1})_2SiO_4$ refractive indices are 1.692, 1.670, and 1.652 (Deer et al., 1966), and a mean difference in refractive index $\langle\Delta n\rangle$ can be of the order of 0.03. We simplify the problem to a two-component system in order to use results from calculations for other birefringent materials. As a kind of worst-case example, we first look at single scattering at interfaces. The problem of birefringent polycrystals is often treated by examining refraction (reflection is negligibly small for such similar refractive indices) by embedded spheres whose refractive indices differ by $\langle\Delta n\rangle$ (e.g. Apetz and van Bruggen, 2003). Although the average scattering angle $\langle\Delta\theta\rangle$ at a single plane interface with $\langle\Delta n\rangle$ ~0.03 is small, less than 2°, scattering from a sphere with the same $\langle\Delta n\rangle$ yields a larger angle. A measure of forward scattering is the quantity $\langle\cos\theta\rangle$, which is ~0.985 for a plane interface with this $\langle\Delta n\rangle$; for a sphere the corresponding angle is ~9° as measured from the direction of an incident beam (Hottel and Serofim, 1967, p. 389). A formula for single scattering is $\alpha_s = \sigma_s(1-\langle\cos\theta\rangle)$ where σ_s is of the order of the reciprocal of grain size G (Pitt and Tozer, 1970; Shankland et al., 1979). For $G = 30\,\mu m = 0.03$ mm, $\alpha_s \approx 0.5$ mm^{-1}, which is comparable to $\langle\alpha_\nu(1000\,K)\rangle \approx 0.4$ mm^{-1} above, even in this single-sphere case.

In the case of multiple scattering the deflection θ can be regarded as the result of a random walk in which light scatters by $\langle\Delta\theta\rangle$ at each of $m = x/G \approx 50$ grains so that $\theta_m = m^{1/2}\langle\Delta\theta\rangle$. Even for $\langle\Delta\theta\rangle \sim 2°, \theta_m \approx 14°$. For multiple scattering, the distribution of scattering angles yields a normal distribution with half-width θ_m (e.g. Dalisa and Seymour, 1973; Julian and Lüty, 1980; Ivey and Bolie, 1991). A beam intensity thus has the angular dependence $I(q) = \text{const.} \times \exp(-\theta^2/2\theta_m^2)$. Julian and Lüty (1980) considered diffraction effects in a birefringent multidomain system to derive the result that $I(\theta = 0)/I_0 = \exp(-\lambda^{-2}\pi^2\langle\Delta n\rangle^2 xG)$ where I_0 = incident intensity and λ = wavelength in vacuum. With the values above and $\lambda \approx 3$ μm at the blackbody peak for 1000 K, $\alpha_s \approx \lambda^{-2}\pi^2\langle\Delta n\rangle^2 G = 30$ mm^{-1}, $\gg \alpha_\nu$. This approximation is sufficient to indicate that multiple scattering can dominate absorption in fine-grained materials.

In the case of direct (ballistic) photon transport across olivine specimens, fine grain size and multiple light scattering events also minimize this effect. In the high-pressure polymorphs, there is the additional strong absorption. Further, if ballistic photon transport were taking place, we would not expect to see the agreement described below for olivine at 7 GPa in specimens having two different dimensions, a radius of 1250 μm in the smaller 18 M assembly and 1750 μm in the larger assembly. Ballistic photons moving at the speed of light would introduce a smaller phase shift in the measurement that would not preserve the proper dependence on sample radius to have these two results agree.

We note three other observations. First, Hofmeister (1999) derived an expression for $\kappa_{rad}(T)$ that is an order of magnitude lower than those based on $\alpha\nu(T)$ cited above and hence would greatly suppress the possibility of a radiative contribution. Because it was not clear how the fourth order polynomial in T in this expression was derived, we have not emphasized this possibility. Second, in Harrell's (2002) experiments olivine samples are optically thin so that there is no radiative contribution, and therefore his $\kappa = \kappa_{lat}$; it is as well described by a $T^{-1/2}$ expression as are the measurements here. Finally, there is a heuristic argument. If radiative transfer were a significant contribution to κ, we would expect a marked turning up of the $\kappa(T)$ curves for olivine in Fig. 8 at high T. Not only is this not seen, the fits to Eq. (12) for all three polymorphs as described above are consistent with each other and with the assumption that our measured $\kappa \propto T^{-1/2}$ represents $\kappa_{\rm lat}$.

Note added in proof

The linear pressure coefficient of Hofmeister (1999) can be used to estimate pressure effect on thermal conductivity. This term is $[1 + (K'/K)P]$ where K is adiabatic bulk modulus and K' is its pressure derivative. We first note that for olivine, for which these numbers are 129.5 GPa (Bass, 1995) and ~4 (assumed), $K'/K = 0.031$. This calculation agrees with the values 0.036 ± 0.04 and 0.032 ± 0.03 in Table 1.

Applying the same approach to wadsleyite and ringwoodite with $K = 174$ and 184 GPa (Bass, 1995) and $K' \approx 4$ (assumed), yields pressure corrections of 0.023 ($P = 14$) and 0.022 ($P = 20$). Pressure effects are then of order 9% and 12% over the respective stability field ranges in the transition zone.

References

Apetz, R., van Bruggen, M.P.B., 2003. Transparent alumina: a light-scattering model. J. Am. Ceram. Soc. 86, 480–486.

Aronson, J.R., Bellotti, L.H., Eckroad, S.W., Emslie, A.G., McConnell, R.K., von Thüna, P.C., 1970. Infrared spectra and radiative thermal conductivity of minerals at high temperatures. J. Geophys. Res. 75, 3443–3456.

Bass, J.B., 1995. Elasticity of minerals, glasses, and melts. In: T.J. Ahrens (Ed.), Mineral Physics and Crystallography: A Handbook of Physical Constants. Am. Geophys. Union, AGU Ref. Shelf 2, 45–63.

Bhandari, C.M., Rowe, D., 1998. Thermal Conduction in Semiconductors. Wiley, New York, pp. 13–34.

Beck, A.E., Darbha, D.M., Schloessin, H.H., 1978. Lattice conductivities of single-crystal and polycrystalline materials at mantle pressures and temperatures. Phys. Earth Planet Inter. 17, 35–53.

Benigni, P., Rogez, J., 1997. High temperature thermal diffusivity measurement by the periodic cylindrical method: The problem of contact thermocouple thermometry. Rev. Sci. Instrum. 68, 2767–2773.

Birch, F., Clark, H., 1940. The thermal conductivity of rocks and its dependence upon temperature and composition, Part I. Am. J. Sci. 238, 529–558.

Brown, J.M., 2002. Private communication.

Chai, M., Brown, J.M., Slutsky, L.J., 1996. Thermal diffusivity of mantle minerals. Phys. Chem. Miner. 23, 470–475.

Dalisa, A.L., Seymour, R.J., 1973. Convolution scattering model for ferroelectric ceramics and other display media. Proc. IEEE 61, 981–991.

Deer, W.A., Howie, R.A., Sussman, J., 1966. An Introduction to the Rock Forming Minerals. Longmans, London.

Dugdale, J.S., MacDonald, D.K.C., 1955. Lattice thermal conductivity. Phys. Rev. 98, 1951–1952.

Fei, Y., 1995. Thermal expansion. In: Ahrens, T.J. (Ed.), Mineral Physics and Crystallography: A Handbook of Physical Constants. American Geophysical Union, Washington, DC, pp. 29–44.

Frost, D.J., Poe, B.T., Trönnes, R.G., Liebske, C., Duba, A., Rubie, D.C., 2004. A new large-volume 6–8 multianvil system. Phys. Earth Planet. Inter., doi: 10.1016/j.pepi.2004.03.003.

Fukao, Y., Mizutani, H., Uyeda, S., 1970. Optical absorption spectra at high temperatures and radiative thermal conductivity of olivines. Phys. Earth Planet. Interiors 1, 57–62.

Fujisawa, H., Fujii, N., Mizutani, H., Kanamori, H., Akimoto, S., 1968. Thermal diffusivity of Mg_2SiO_4, Fe_2SiO_4, and NaCl at high pressures and temperatures. J. Geophys. Res. 73, 4727–4733.

Green II, H.W., Burnley, P.C., 1989. A new self-organizing mechanism for deep-focus earthquakes. Nature 341, 733–737.

Harrell, M.D., 2002 Anisotropic lattice thermal diffusivity in olivines and pyroxenes to high temperatures, Ph.D. Dissertation, University of Washington.

Harrell, M.D., Brown, J.M., Abramson, E.H., 2002. Eos Trans. AGU, 83(47), Fall Meet. Suppl., Abstract MR11A-13.

Hauck II, S.A., Phillips, R.J., Hofmeister, A.M., 1999. Variable conductivity: effects on the thermal structure of subducting slabs. Geophys. Res. Lett. 26, 3257–3260.

Henderson, C.M.B., Knight, K.S., Redfern, S.A.T., Wood, B.J., 1996. High-temperature study of octahedral cation exchange in olivine by neutron powder diffraction. Science 271, 1713–1715.

Hirsch, L.M., Shankland, T.J., 1993. Quantitative olivine-defect chemical model: insights on electrical conduction, diffusion, and the role of Fe content. Geophys. J. Int. 144, 21–35.

Hofmeister, A.M., 1999. Mantle values of thermal conductivity and the Geotherm from phonon lifetimes. Science 283, 1699–1706.

Holt, J.B., 1975. Thermal diffusivity of olivine. Earth Planet. Sci. Lett. 27, 404–408.

Horai, K., 1971. Thermal conductivity of rock-forming minerals. J. Geophys. Res. 76, 1278–1308.

Horai, K., Susaki, J., 1989. The effect of pressure on the thermal conductivity of silicate rocks up to 12 kbar. Phys. Earth Planet. Inter. 55, 292–305.

(Meade and Jeanloz, 1991; Glover et al., 1995; Dobson et al., 2002). Acoustic emission signals are generally detected using miniature piezoelectric transducers operating in the megahertz range. Multiple transducer arrays can be used to locate the source of the signal using travel time residuals.

The multi-anvil-press has proven to be a remarkably successful tool in investigating the phase relations of the earth's upper mantle and transition zone, and pressures and temperatures corresponding to the top 700 km of the mantle are routinely available to experimentalists. Indeed, many of the phase changes implicated in deep-focus earthquake scenarios were first discovered, and their *P–T–X* fields determined, by multi anvil and related techniques. In this paper we describe a technique which combines multi anvil and acoustic emissions technologies to investigate possible seismogenic processes under the conditions of the earth's mantle, and present some sample results from current studies.

2. Methodology

2.1. Acoustic emissions

Acoustic emissions generated by a single increment of crack extension or volume change are detected by the transducers as wave packets, or 'hits', as illustrated in Fig. 1. With modern instrumentation, it is possible to record the full waveform of each hit. Full definition of each waveform, however, requires about 4000 data points, and there may be greater than 10^6 or 10^7 hits per transducer per experiment. This results in the system having to cease recording for the periods necessary to transfer data from data buffers to disk storage, thus risking missing vital data. It is therefore, a common practice to record key parameters that can define the hit. All laboratory environments are acoustically noisy, both from physical and extraneous electrical sources. It is therefore necessary to set a detection threshold (a_t), and only emissions with amplitudes above this threshold trigger the equipment and are recorded as hits. All other parameters are defined from the point at which the first threshold crossing occurs (trigger time, t_t). Note that this is later than the true arrival time of the signal at the transducer (t_a). The rise time (t_r) is the time from t_t to the peak amplitude (a_p), and the duration (d) is the time between the first and last threshold crossings. We consider that the best way of quantitatively describing the amount of AE generated during a process is via the AE energy, because it takes account of both the number of hits and their sizes. The energy is calculated from the envelope of the square of the voltages of individual peaks within the hit duration (Fig. 1). Multiple transducers are required if the sources of hits are to be located. In the present study, two transducers were used to perform a one-dimensional space discrimination. In this case, only hits that are recorded at both transducers within a set time interval (Δt), based on the wave velocity through the various media between source and transducer, are accepted as genuine, and these are called AE 'events'.

Commercially available miniature piezoelectric transducers comprising lead–zirconate–titanate (PZT) ceramic elements of 3.2 mm diameter and 2.1 mm thickness, backed by a half-wave backing plate were glued onto the back truncations of two tungsten carbide anvils. Matched pairs, with a relatively flat high sensitivity band between about 100 kHz and 1 MHz, were used and a transient wave test was performed to check that the travel time from the anvil truncation to the transducer was the same for each pair. Received signals were amplified by 40 dB through close coupled, low noise, pre-amplifiers before transmission to the main recording and analysis instrumentation (PAC Mistras system).

The AE parameters defined above were recorded continuously during the compression and heating cycles of the experiment (Section 2.3). In addition, at discrete stages during the test, the attenuation and travel time of the cell was measured by pulsing alternately on opposite transducers and logging arrivals at both transducers from the induced pulse. These pulser tests were performed at the onset and completion of compression, before the heating stage, at maximum temperature and after temperature quenching. The attenuation between transducers was consistently around 30 dB and maximum variations in travel time due to the effect of pressure and temperature were <0.2 μs. From these tests, we conclude that the maximum variation in travel time across the sample during an experiment is less than 0.05 μs, which is negligible for the Δt filtering of results discussed below. Pulser tests which did not show ~30 dB of attenuation and ~16 μs travel

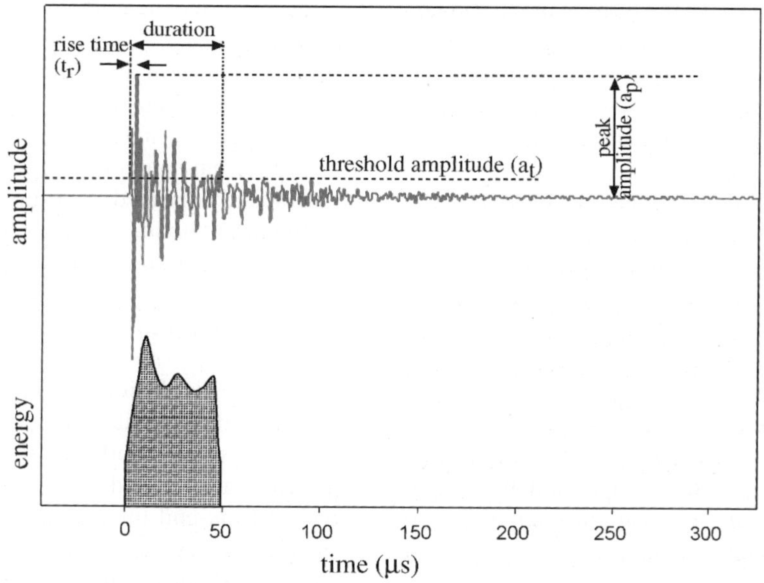

Fig. 1. An acoustic emission (AE) waveform collected during a serpentine dehydration experiment at 6 GPa. The waveform is annotated with the key features of an AE hit. The threshold amplitude (a_t) must be exceeded for an incoming signal to be recorded as a "hit". The rise time (t_r) is the time taken between the first threshold crossing and the peak amplitude (a_p), and the duration (d) is the time from the first to the last threshold crossing of a single hit.

times were indicative of signal transfer problems, usually either failure of the bond between a transducer and anvil, or a poor contact between the waveguide and anvil. Results from these experiments were discarded. Repeat pulser tests showed a reproducibility of 0.1 μs or better, suggesting that source locations would have a precision of approximately 400 μm.

2.2. Multi anvil experiments

Experiments were performed in a 1000 tonne split cylinder multi-anvil-press at the Bayerisches Geoinstitute. Pressure media consisted of Cr-doped MgO, with stepped furnaces and ZrO_2 thermal insulation. Experiments between 2 and 8 GPa used 25 mm edge length octahedra, compressed by 18 mm truncation edge length tungsten carbide anvils (25/18) and with graphite furnaces. Experiments at 16 GPa used a 14/8 octahedron/anvil assembly with a lanthanum chromate furnace. Samples in 25/18 assemblies were 4 mm long and 3 mm diameter, and those for 14/8 assemblies were 2.7 mm long by 1.5 mm diameter. Details of the cell pressure/temperature calibrations have been presented elsewhere (Poli and Schmidt, 1998; Mossenfelder et al., 2000). In principle, modifications to the standard experimental arrangement for AE monitoring are simple, as shown in Fig. 2, and comprise: (a) acoustic waveguides inserted along the furnace axis to couple the sample to the furnace anvils, (b) thermocouples

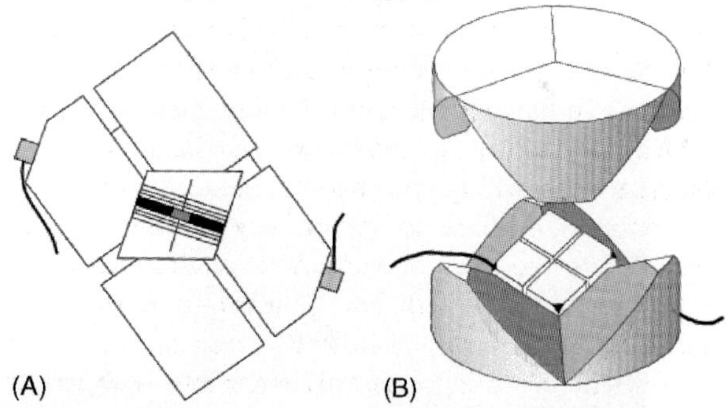

Fig. 2. Schematic of multi anvil AE set-up. (A) Section of cell and anvil set showing acoustic waveguides (black) connecting the sample to the anvils with transducers mounted on the back. (B) Anvil set sitting in first-stage wedges. The top and bottom anvils are not accessible, but the six equatorial anvils can have AE transducers mounted on their back truncations, marked in black.

brought into the cell through the furnace in the equatorial plane of the cell and (c) AE transducers mounted on truncations at the back of the furnace anvils. The success of these experiments, however, depends on careful optimisation of several factors.

1. The choice of waveguide material depends on a trade-off between optimising the transfer function of the cell (low attenuation and impedance intermediate to the sample and anvils) and controlling the deviatoric stress on the sample (relative hardness and compressibility of the waveguides and other pressure media). We have successfully used polycrystalline Al_2O_3 and Fe waveguides. Due to its hardness, Al_2O_3 generates extremely high deviatoric stresses on the sample and the length of the waveguides must be carefully controlled to ensure that the sample is not crushed or faulted (Fig. 3). This, in turn, requires re-calibration of the pressure-load relationship in the cell.
2. Unless the heating circuit is electrically noiseless, the AE transducers must be insulated from the anvils. We used a high-viscosity cyanoacrylate adhesive as the mounting medium to isolate the transducers.
3. A dataset from an un-filtered experiment, with no annealing of the pressure medium can rapidly become unmanageably large ($>10^7$ events). Unwanted noise from sources outside the sample can be significantly reduced through some simple front-end filters and tuning of the pressure–temperature–time (*P*–*T*–*t*) path to the problem being addressed.

2.3. Pressure–temperature–time path

Solid-media devices are inherently noisy, with AE generated during initial compaction due to brittle failure, then subsequently by frictional slipping of the pressure medium and gasket materials. Fig. 4 shows a typical front-end filtered AE trace during a compression/heating cycle. Above about 100 tonnes we find that the AE background in multi anvil cells becomes tolerable. On heating the pressure medium softens and further flows. Based on evidence from in situ pressure measurements from synchrotron multi anvil experiments (e.g. Weidner et al., 1992; Kato et al., 1992)

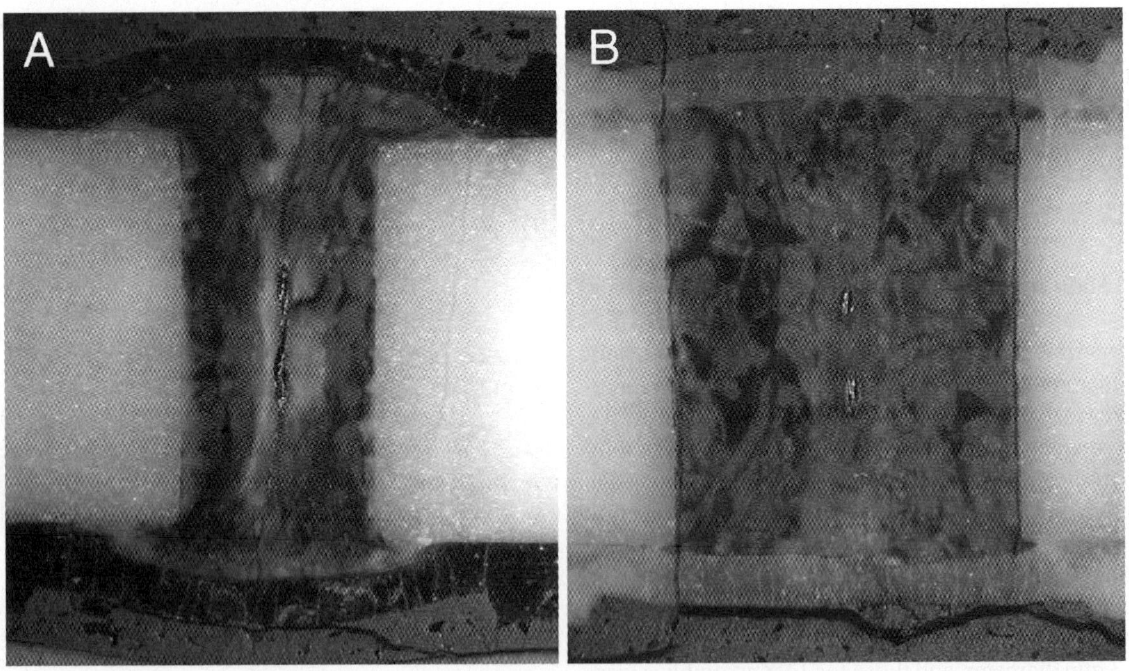

Fig. 3. Serpentine samples with corundum waveguides recovered from 4 GPa. (A) The waveguides were too long causing the sample to suffer >50% strain during compression. (B) The sample suffered ~20% strain when both waveguides were shortened by 700 μm.

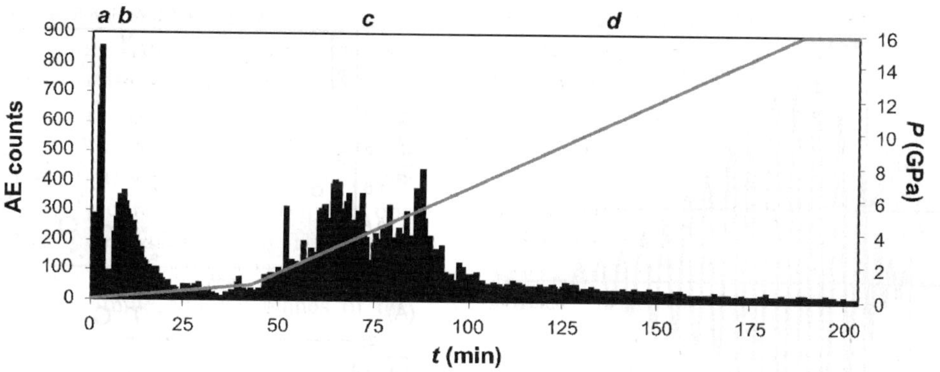

Fig. 4. Typical acoustic emission output during cold compression with front end 70 μs Δt-filter. The black graph shows the number of AE events binned at 30 s intervals and the grey line shows the P–t path. There are four distinct regions identifiable, corresponding to different failure modes within the cell: (a) cracking of the glue and epoxy sheets on initial contact of the first-stage anvils, (b) crushing of the gaskets and pressure medium as the anvils settle-in, (c) brittle failure and frictional sliding within the cell during compression at $P < 6$ GPa and (d) the system becomes substantially quieter at high pressures as plastic and elastic strain dominate. The background event rate at 16 GPa is $\sim 1\,s^{-1}$.

which suggest that temperature induced pressure relaxation reaches a maximum below 600 °C, samples were soaked at between 300 and 500 °C for 1 h, which allows the bulk of the thermal strain to anneal out before the commencement of the test proper (Fig. 4).

2.4. Data filtering

We applied several front-end filters to reduce noise from the experimental system.

1. *Δt Filter*: the travel time of acoustic P-waves between the two transducers was about 16 μs. Events with Δt values longer than this were likely to be either events from outside the anvil assembly or unrelated pairs of hits which happened to arrive at the two transducers within a short duration. We applied a somewhat coarser Δt filter than the 16 μs implied by the travel time to ensure that rise-time effects did not cause us to lose data from within the pressure cell. Even so, the 75 μs Δt filter we used reduced the size of datasets by up to two-orders of magnitude.
2. *Amplitude/threshold crossings*: since the experimental environment is inherently noisy at some background level, we required an AE event to be larger than some pre-determined background amplitude to be counted as genuine. We found that >99% of the background not associated with deformation within the cell was below 25 dB and used this as the threshold value. In addition, however, isolated electrical spikes could have amplitudes significantly larger that 25 dB, but consisted of delta-function-like wave forms. Signals with only one threshold crossing were therefore, also discarded. Both of the above filters vastly reduced the size of the recorded dataset but had very little effect on its total energy.

3. Data analysis

In order to discriminate accurately between those events that originate within the sample and those that originate within the waveguides or pressure medium, we need to correct the measured Δt for the differences between the trigger times and the true arrival times at the two transducers. We note that the wave envelopes are nearly identical in form but the peak amplitudes recorded at the two transducers will be different, since the path lengths will be different except where the source is exactly in the centre of the sample. Fig. 1 shows an example of a waveform collected during a serpentine dehydration experiment and Fig. 6B shows the linear scaling relationship of amplitude against duration indicative of reproducibility of waveform shape for all but the largest events (discussed in Section 4, below). We then simply perform a linear extrapolation from $(a_p, t_t + t_r)$ through (a_t, t_t) to $(0, t_a)$, thus obtaining the true arrival time at each transducer, as shown

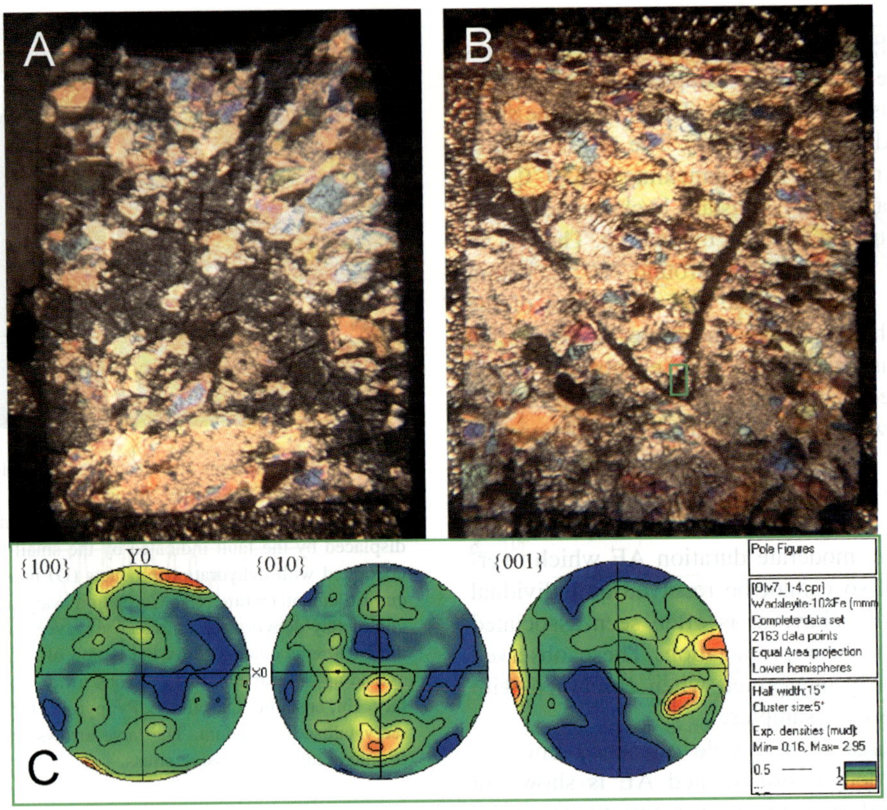

Fig. 8. Thin sections of San Carlos samples recovered from 16 GPa and high temperature. Olivine grains show high second-order birefringence colours and wadsleyite is dull grey. (A) Recovered after AE event at 1190 °C, with ~30% transformation to wadsleyite. (B) Recovered from 1050 °C, before significant AE. Wadsleyite only occurs within the fault set which was introduced during cold compression. (C) Pole figures determined by EBSD for Wadsleyite in the region marked by the green box in (B). {1 0 0} poles are aligned parallel to the direction of shear in the faulted region, indicating that the wadsleyite has deformed by dislocation creep.

Carlos olivine. Samples were cored from a hot-pressed powder of 10–100 μm grain size. During cold compression to 16 GPa samples failed, accompanied by a large AE signal at low Δt. A second set of large AE was observed during heating in about 50% of experiments at temperatures between 1100 and 1200 °C. These AE tended to have Δt values indicating a source just outside the sample.

Samples recovered from 1200 °C contained 30–50% wadsleyite, with a sub-micron grain size, as shown in Fig. 8A. Samples recovered from 1050 °C contained <5% wadsleyite, concentrated along the grain boundaries and completely replacing the fine-grained material filling the low-T fault-set (Fig. 8B). Wadsleyite within the fault zone shows strong lattice preferred orientation by EBSD with {1 0 0} parallel to the direction of shear (Fig. 8C), consistent with previously observed dislocation creep mechanisms in wadsleyite (e.g. Dupas-Bruzek et al., 1998). This suggests that once wadsleyite forms along pre-existing fault planes the sample will deform by aseismic slip on the planes.

We interpret the later AE events to be related to brittle failure of the pressure medium around the reduced volume of the wadsleyite sample. It is impossible to determine the focal mechanism characteristics of such an event with the current arrangement, which uses only two transducers that are aligned along the sample axis. As such, application of these results to deep focus earthquakes would be highly speculative. We note, however, that (1) deformation does continue to be localised along old fault zones, as previously suggested (Silver et al., 1995; Jiao et al., 2000) and (2) although we produce wadsleyite filled faults which look superficially like anti-crack faults generated in olivine-spinel analogue materials (Green et al., 1992), there is no evidence of seismicity associated with this in the current experiments.

5. Conclusions and outlook

The combination of acoustic emission techniques with multi anvil experiments provides a powerful new tool for investigating brittle behaviour in high pressure, high temperature environments. We have developed a methodology which makes this relatively routine and have applied it to the long-standing problem of intermediate- and deep-focus seismicity. While the results demonstrate the power of this new combined technique, they also clearly show that further development is required in order to address some of the key questions regarding the mechanisms of deep-focus faulting. We are currently developing the next generation of AE multi anvil experiments in which improved data collection and manipulation will allow true 3D location of sources and collection of waveforms of AE events. In addition the technique can, in principle, be combined with recently developed synchrotron techniques for in situ stress measurements (e.g. Weidner et al., 1992) to determine the exact P–T–σ regime of brittle failure under extreme conditions. These developments should allow us to determine focal mechanism solutions as well as elucidating the role of relative strength and stress changes in the sample and surrounding matrix during transformation in generating brittle failure.

Acknowledgements

We thank Florian Heidelbach for invaluable assistance in performing SEM analyses. High pressure experiments were performed at the Bayerisches Geoinstitute under the EU "IHP-Access to Research Infrastructures" Programme (Contract No HPRI-1999-CT-004 to D.C. Rubie). D.P. Dobson acknowledges support from the Alexander von Humboldt Foundation and the Royal Society. We thank two anonymous reviewers for their helpful comments.

References

Dobson, D.P., Meredith, P.G., Boon, S.A., 2002. Simulation of subduction zone seismicity by dehydration of serpentine. Science 298, 1407–1410.

Dupas-Bruzek, C., Sharp, T.G., Rubie, D.C., Durham, W.B., 1998. Mechanisms of transformationa and deformation in $Mg_{1.8}Fe_{0.2}SiO_4$ olivina and wadsleyite under non-hydrostatic stress. Phys. Earth Planet. Inter. 108, 33–48.

Frohlich, C., 1989. The nature of deep-focus earthquakes. Annu. Rev. Earth Planet. Sci. 17, 227–254.

Glover, P.W.J., Baud, P., Darot, P., Meredith, P.G., Boon, S.A., Le Ravelec, M., Zoussi, S., Reuschle, T., 1995. The α/β phase transition in quartz monitored using acoustic emissions. Geophys. J. Int. 120, 775–782.

Green, H.W., Burnley, P.C., 1989. A new, self-organising, mechanism for deep earthquakes. Nature 341, 733–737.

Green, H.W., Burnley, P.C., 1990. The failure mechanism for deep-focus earthquakes. In: Knipe, R.J., Rutter, E.H., (Eds.), Deformation Mechanisms, Rheology and Tectonics. Geol. Soc. Lond. Spec. Pub. 54, 133–141.

Green, H.W., Scholz, C.H., Tingle, T.N., Young, T.E., Koczynski, T.A., 1992. Acoustic emissions produced by anticrack faulting during the olivine-spinel transformation. Geophys. Res. Lett. 19, 789–792.

Jiao, W., Silver, P.G., Fei, Y., Prewitt, C.T., 2000. Do intermediate- and deep-focus earthquakes occur on preexisting weak zones? An examination of the Tonga subduction zone. J. Geophys. Res. 105, 28125–28138.

Kato, T., Ohtani, E., Kamays, N., Shimomura, O., Kikegawa, K., 1992. Double stage milti-anvil system with a sintered diamond anvil for X-ray diffraction experiment at high pressures and temperatures. In: Syono, Y., Manghnani, M., (Eds.), High-Pressure Research: Application to Earth and Planetary Science. AGU, Washington, pp. 33–36.

Kirby, S.H., 1987. Localized polymorphic phase transformations in high-pressure faults and applications to the physical mechanism of deep earthquakes. J. Geophys. Res. 92, 13789–13800.

Kirby, S.H., Durham, W.B., Stern, L., 1991. Mantle phase changes and deep-earthquake faulting in subducting lithosphere. Science 252, 216–225.

Lockner, D.A., Byerlee, J.D., Kuksenko, V., Ponomarev, A., Sidorin, A., 1992. Observation of quasistatic fault growth from acoustic emissions. In: Evans, B., Wong, T.-f. (Eds.), Fault Mechanics and Transport Properties of Rocks. Academic Press, San Diego, pp. 1–31.

Meade, C., Jeanloz, R., 1991. Deep-focus earthquakes and recycling of water into the Earth's mantle. Science 252, 68–72.

Meredith, P.G., Knight, K.S., Boon, S.A., Wood, I.G., 2001. The microscopic origin of thermal cracking in rocks: An investigation by simultaneous time-of-flight neutron diffraction and acoustic emission monitoring. Geophys. Res. Lett. 28, 2105–2109.

Mossenfelder, J.L., Connolly, J.A.D., Rubie, D.C., Liu, M., 2000. Strength of $(Mg, Fe)_2SiO_4$ wadsleyite determined by relaxation of transformation stress. Phys. Earth Planet. Inter. 120, 63–78.

Murrell, S.A.F., Ismail, I.A.H., 1976. The effect of decomposition of hydrous minerals on the the mechanical properties of rock. Tectonophysics 31, 207–258.

Poli, S., Schmidt, M.W., 1998. The high-pressure stability of zoisite and phase relationships of zoisite-bearing assemblages. Contrib. Mineral. Petrol. 130, 162–175.

Raleigh, C.B., Patterson, M.S., 1965. Experimental deformation of serpentine and its tectonic implications. J. Geophys. Res. 70, 3965–3985.

Read, M.D., Ayling, M.R., Meredith, P.G., Murrell, S.A.F., 1995. Microcracking during triaxial deformation of porous rocks monitored by changes in rock physical properties: II—pore volumometry and acoustic emission measurements on water saturated rocks. Tectonophysics 245, 223–235.

Silver, P.G., Beck, S.L., Wallace, T.C., Meade, C., Myers, S.C., James, D.E., Kuehl, R., 1995. Rupture characteristics of the deep Bolivian earthquake of 9 June 1994 and the mechanism of deep earthquakes. Science 268, 69–73.

Weidner, D.J., Vaughn, M.T., Ko, J., Wang, Y., Liu, X., Yeganeh-Haeri, A., Pacalo, R.E., Zhao, Y., 1992. Characterisation of stress, pressure, and temperature in SAM85, a DIA type high pressure apparatus. In: Syono, Y., Manghnani, M. (Eds.), High-Pressure Research: Application to Earth and Planetary Science. AGU, Washington, pp. 13–17.

Available online at www.sciencedirect.com
SCIENCE DIRECT®
ELSEVIER
Physics of the Earth and Planetary Interiors 143–144 (2004) 347–356
PHYSICS OF THE EARTH AND PLANETARY INTERIORS
www.elsevier.com/locate/pepi

Deformation experiments using synchrotron X-rays: in situ stress and strain measurements at high pressure and temperature

Jiuhua Chen*, Li Li[1], Donald Weidner[2], Michael Vaughan[3]

Department of Geosciences, Mineral Physics Institute, State University of New York at Stony Brook, Stony Brook, NY 11794-2100, USA

Received 31 January 2003; received in revised form 2 August 2003; accepted 21 September 2003

Abstract

An experimental technique for in situ stress and strain measurements at high pressure and temperature using synchrotron X-rays has been developed at the X17B beamline of the National Synchrotron Light Source. The strain is measured by correlating sample-length marks in recorded X-ray radiographs. The stress is measured by simultaneously collecting energy-dispersive X-ray diffraction patterns of the sample in two perpendicular diffraction planes. Differential stress in the sample is derived from the different lattice strains along the different orientations relative to principal stress direction. Example measurements are conducted to study rheological properties of fayalite. Accuracy of the measurement is about 10–100 MPa for stress and about 10^{-4} for strain.

Keywords: Deformation; High pressure; X-ray diffraction; X-ray imaging; Fayalite; Rheology

1. Introduction

Rheological properties of minerals are very important parameters for understanding Earth's dynamic phenomena such as mantle convection and deep focus earthquakes. As mechanisms of deformation in minerals change with pressure and temperature, rheology studies at mantle conditions are essential to understand the mineral's behavior in the Earth's interior. However, technical difficulties up to now have prevented us from conducting conventional stress–strain rate deformation experiments at pressures above ~3 GPa (Tingle et al., 1993). Recent developments with a multi anvil press have made significant progress in carrying out deformation experiments at higher pressures (Durham and Rubie, 1998; Yamazaki et al., 1996; Bussod et al., 1993; Liebermann et al., 1992; Green et al., 1990). Karato and Rubie (Karato and Rubie, 1997) designed a cell assembly with a thin sample sandwiched between two pistons at 45° angle to their long axis for deformation studies up to 15 GPa. Weidner et al., (Weidner, 1998; Weidner et al., 1998) developed an analytical method to measure the microscopic stress in powder sample by deconvoluting X-ray diffraction peak broadening at the pressure of 20 GPa. However, none of these experimental systems is a direct extension of the traditional macroscopic strain and stress measurement.

* Corresponding author. Tel.: +1-631-632-8058; fax: +1-631-632-8140.
E-mail addresses: jiuhua.chen@sunysb.edu (J. Chen), lilli@ic.sunysb.edu (L. Li), donald.weidner@sunysb.edu (D. Weidner), michael.vaughan@sunysb.edu (M. Vaughan).
[1] Tel.: +1-631-632-8220; fax: +1-631-632-8140.
[2] Tel.: +1-631-632-8211; fax: +1-631-632-8140.
[3] Tel.: +1-631-632-8238; fax: +1-631-632-8140.

0031-9201/$ – see front matter
doi:10.1016/j.pepi.2003.09.021

In many cases, however, the pressure and temperature dependences of the elastic compliances are unknown. Singh et al. (Singh et al., 1998) and Uchida et al. (Uchida et al., 1996) discussed the possibilities of estimating t for different crystal structure systems. For an elastically isotropic cubic material, t can be estimated (Singh et al., 1998) by

$$t = 6\langle Q(hkl)\rangle G \tag{13}$$

where $\langle\ \rangle$ denotes the average value for all lattice reflections, G is the shear modulus at the pressure, and $Q(hkl)$ can be derived by plotting $d(hkl)$ versus $1-3\cos^2\psi$ according to the following equation (Singh et al., 1998),

$$d_m(hkl) = d_p(hkl)[1 + (1 - 3\cos^2\psi)Q(hkl)] \tag{14}$$

In X-ray diffraction experiments, $\varepsilon(hkl)$ is not actually measured at $\psi = 0$. As shown in Fig. 1, energy dispersive diffraction patterns collected in the vertical diffraction has $\psi = \theta$. From Eq. (4), we know $\varepsilon(hkl)$ is linear to $1 - 3\cos^2\psi$. Therefore, $\varepsilon_V(hkl)$ can be derived (Funamori et al., 1994) by

$$\varepsilon_V(hkl) = \frac{\varepsilon_{\psi=\theta}(hkl) - \varepsilon_H(hkl)\sin^2\theta}{\cos^2\theta} \tag{15}$$

If the Bragg angle (θ) is small, e.g. ~3.2° in our experiments, difference between $\varepsilon_V(hkl)$ and $\varepsilon_{\psi=\theta}(hkl)$ is negligible.

3. The experimental setup

The experimental system is installed at the X17B beamline of the National Synchrotron Light Source (NSLS). Experiments are conducted using high energy white X-rays from the superconductor wiggler of X17. Key elements of the experimental system are sketched in Fig. 2. An upstream aperture sets the size of X-rays coming into the experimental hutch. A set of incident slits defines the size of incident X-rays to the sample. The slits are mounted on a motor driven stage, so that they can be moved in and out of the position to change the incident beam size according to the type of measurements. Typically, the upstream aperture sets a 2 mm × 2 mm beam size for sample imaging, and the slits further define a 100 μm × 100 μm beam size for X-ray diffraction.

A 6–8 double-stage multi anvil high pressure module (Tcup), which can generate high pressure beyond 20 GPa (Vaughan et al., 1998), is used for the deformation experiment. With 8 WC second-stage cubes (10 mm × 10 mm × 10 mm), the Tcup module typically leaves only one accessible diffraction-plane which is inclined about 35° from the vertical axis (Fig. 3). To gain the accessibility for diffracted X-rays, we replace 6 of the 8 WC cubes (the top and bottom ones are used for passing the heating current to the cell assembly) with sintered cBN cubes. Because of the low X-ray absorption of boron nitride, diffracted X-rays can be measured in the diffraction plane with any angle to the vertical axis. Fig. 4 shows a radiograph of a sample in the cBN anvils. The maximum intensity reduction

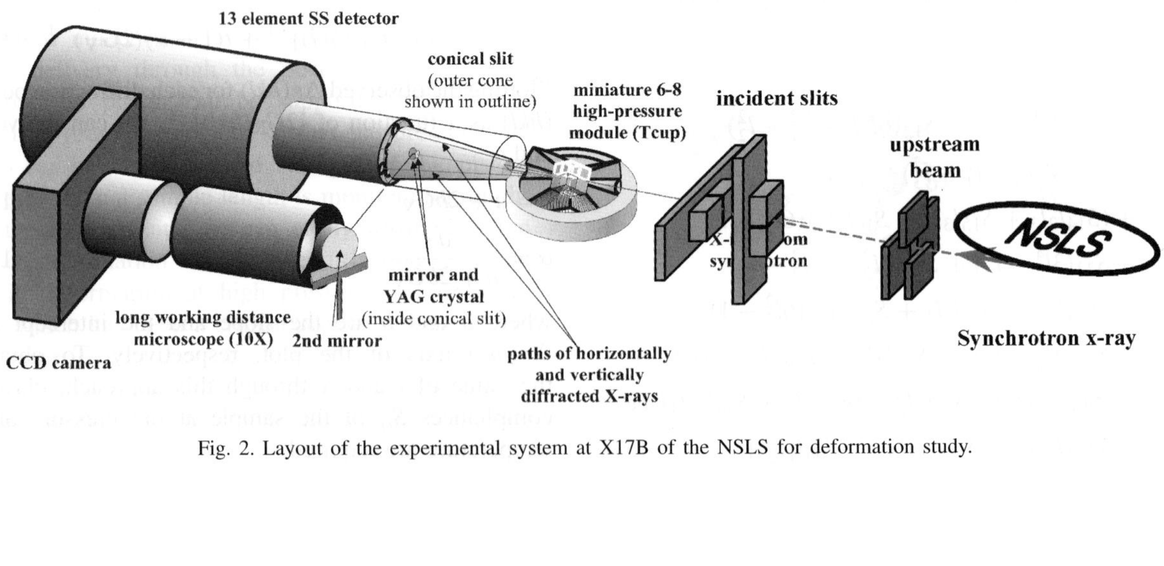

Fig. 2. Layout of the experimental system at X17B of the NSLS for deformation study.

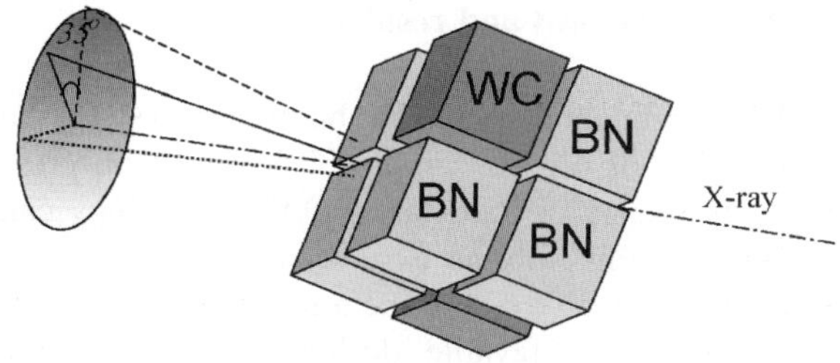

Fig. 3. Diffraction geometry in Tcup module. Solid line indicates the access of diffracted X-rays if all 8 s-stage cubes are made of WC, broken line and dotted line indicate the diffraction in vertical plane and horizontal plane penetrating through BN side cubes.

of the transmitted X-rays by the cBN anvils is about 20%.

A 13-element solid state detector is used to record the energy dispersive X-ray diffraction patterns. The diffracted X-rays are collimated by using a conic slit and detected on four of the 13 detector elements in the vertical and horizontal planes simultaneously (Fig. 5). The conic slit consists of an inner conical block with flat tip and two outer rings. The two rings have different inner diameters and their inner surfaces are tapered to match the conical angle of the inner block. They are attached onto the inner block at about 10 mm apart, as shown in Fig. 5. Spacer shims (typically 50 μm thickness) set a gap between the inner block and outer rings to collimate the diffracted X-rays into the detector. Flux of the diffracted X-rays into the detector is controlled by an aperture placed in between the conic slit and the detector. The aperture consists of two identical four-hole rotational W plates, and the degree of overlapping the holes on the two plates controls the photon flux (see inset of Fig. 5). The mechanical conic

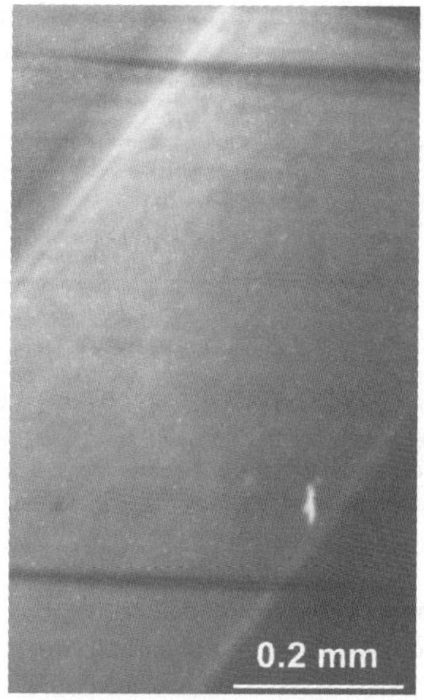

Fig. 4. Radiograph image of fayalite sample and cBN anvils. The bight strip-shape area from top-right to bottom-left indicates the anvil gap. The two horizontal dark lines are the gold foils at each end of the cylindrical sample. Note the dark lines are visible in the anvil areas (grey triangle areas at top-left corner and bottom-right corner).

angle of the inner block is 6.5°, but the actual Bragg angle for each detector depends on the alignment of the slit relative to the incident X-ray. Calibrations of the Bragg angle of each detector are made using X-ray diffraction standard at beginning of the experiment. Fig. 6 shows a typical X-ray diffraction pattern with a data collection time of 95 s.

Both inner conical block and outer rings are made of tungsten-containing brass. The inner conical block

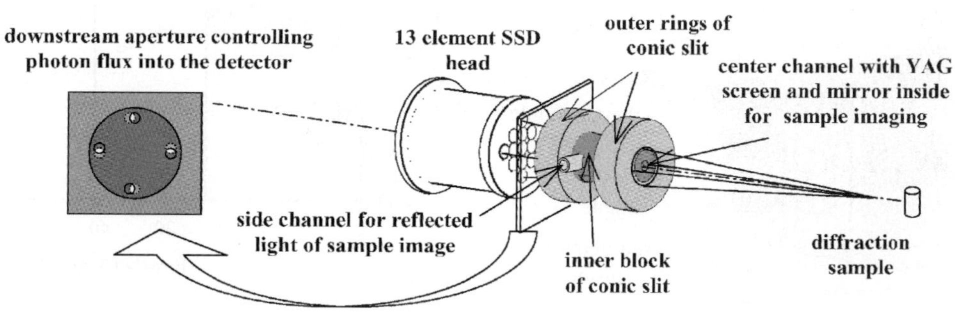

Fig. 5. Conical slit collimating the diffracted X-ray for simultaneous diffraction collection in vertical and horizontal plane with 13-element solid state detector.

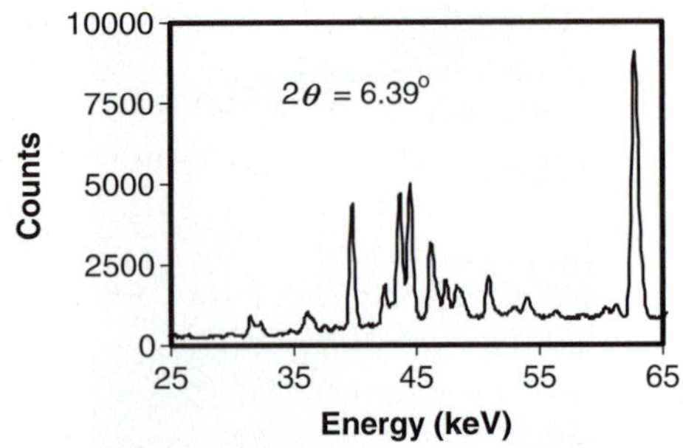

Fig. 6. A typical energy-dispersive X-ray diffraction pattern of fayalite collected at 4.5 GPa and 973 K. The diffracted X-rays transmit through the cBN anvil in the vertical diffraction plane. Data collection time is 95 s.

has two connected holes, one along its axis and one at the side, for accessing of the sample imaging system (see Figs. 2 and 5). A YAG crystal fluorescent screen (6 mm diameter and 100 μm thickness) is mounted in the center hole of the conic slit near the downstream end, perpendicular to the incident X-ray. A mirror is place at the end of the center hole to reflect visible light from the fluorescent screen to the side channel. Transmitted X-rays through the sample impinge on the fluorescent screen where an image of the sample is generated. Visible light from the sample image are reflected by the mirror in the center hole and the outside mirror, into a CCD camera through a long focusing-distance microscope (Fig. 2). Sample strain is derived from the image of two strain marks (6 μm thick gold foils) placed at both ends of the sample (see Fig. 7 for the cell assembly and Fig. 4 for sample image).

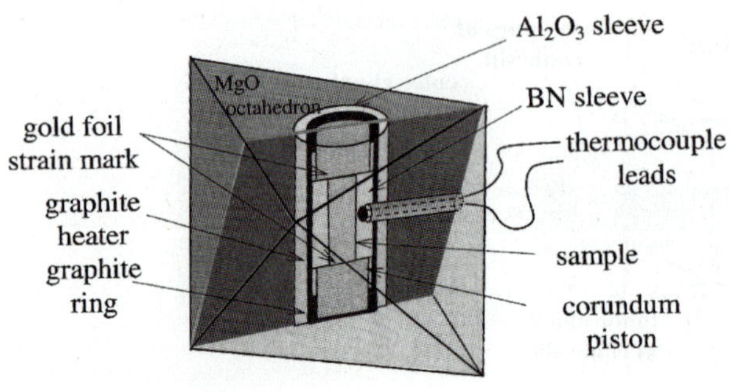

Fig. 7. Cell assembly for high pressure deformation experiments in double-stage Tcup apparatus.

4. The experiment and results

The experimental system has been used to study the deformation of olivine samples. The result on pressure dependence of San Carlos olivine rheology can be found in this volume as a separate report by Li et al. (2004) Here we describe a multiple heating cycle experiment on fayalite, during which the sample transforms from olivine to spinel structure.

Synthetic Fe_2SiO_4 sample is powdered to an average grain size of ~1 μm, and packed into a cylindrical BN sample chamber of 1 mm diameter and 3 mm length. As shown in Fig. 7, the sample is loaded into an 8 mm edge-length octahedral MgO pressure medium, between two hard corundum pistons. A 6 μm thick gold foil is placed between the sample and the piston at each end as strain mark. Edge-length of the truncation on the second stage anvil is 2 mm.

The sample is first compressed at room temperature and then heated to 973 K to sinter the powder sample. After quenching to room temperature, the sample is further compressed to a higher pressure followed by a second heating. A total of four heating cycles are conducted, each at a higher pressure with respect to previous heating. The sample length is measured directly from the distance between two strain marks at each side of the sample on the X-ray radiograph image of the high pressure cell. The change in sample length during the experiment is shown in Fig. 8. Accuracy of the absolute sample length is influenced by the

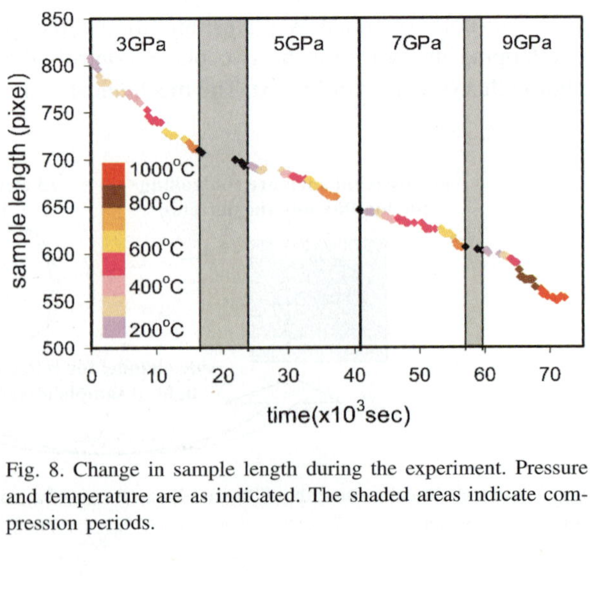

Fig. 8. Change in sample length during the experiment. Pressure and temperature are as indicated. The shaded areas indicate compression periods.

clearness and thickness of the strain mark image (usually in the order of microns). However, a technique of correlating the same strain mark in two consecutive images (Li et al., 2003) increases the accuracy of relative displacement of the strain mark to a fraction of one micron. Therefore, for a sample with about 1 mm length, the strain measurement is accurate to 10^{-4} and strain rate is accurate to 10^{-7}–10^{-6} depending on the time interval between two measurements. As shown in Fig. 8, each heating cycle generates about 10% total plastic strain.

In calculation the stress of olivine phase, we use the elastic constants of single crystal fayalite and their pressure and temperature dependencies reported by Graham et al. (1988). Single crystal elastic constants and their pressure/temperature dependencies of Fe_2SiO_4 spinel phase are not available. We, therefore, use Eq. (13) to derive the stress in the spinel phase using the shear modulus and its pressure dependence values (87.5 GPa and 0.44) suggested by Hofmeister and Mao (Hofmeister and Mao, 2001). In most cases, diffraction peaks (0 3 1), (2 2 0), (1 3 1), (2 1 1), (0 4 1), (3 1 1), (1 5 1) and (2 2 2) are used in the stress calculation for olivine phase, and diffraction peaks (2 2 0), (3 1 1), (2 2 2), (4 0 0), (3 3 1), (5 1 1) and (4 4 0) are used for spinel phase. Fig. 9 shows the calculated apparent pressures based the diffraction in vertical plane ($\psi = 3.2°$) and in horizontal plane ($\psi = 90°$), the derived effective hydrostatic pressure and the differential stress during the first annealing process. Because of the hard pistons in the high pressure cell, the apparent pressure along the piston axis increases dramatically whereas that along the direction perpendicular to this axis barely increases during the initial compression. This indicates that huge differential stress builds up in the sample. Clearly, none of these apparent pressures represent the true pressure in the sample. The effective hydrostatic pressure lies between the two apparent pressures. Upon heating, the apparent pressure along the piston decreases whereas that along the direction perpendicular to the piston increases, indicating a relaxation of the differential stress. At 873 K, the two apparent pressures merge to the value of the effective hydrostatic pressure, and the differential stress is completely relaxed. Further compression after quenching yields a departure of the two apparent pressures from each other indicating rise of the differential stress as well as the effective hydrostatic pressure in the sample.

Table 1 lists the derived stress (t), parameter α and effective hydrostatic pressure (σ_p) of the sample during the second heating cycle. We notice that the α parameter increases with temperature and becomes >1 at 773 K. This indicates that when large plastic strain accumulates in the sample, Reuss and Voigt limits may not be valid in the polycrystal. A similar phenomenon has been also observed in deformation experiments using neutron diffraction due to different elastic and plastic anisotropy (Daymond et al., 1999). So far, there are very limited experimental data on determining the

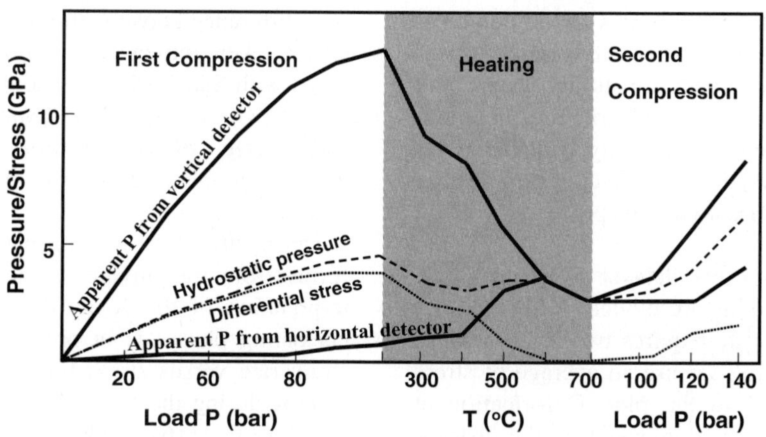

Fig. 9. Apparent pressures calculated from X-ray diffraction at $\psi = 3.2°$ (vertical diffraction) and $\psi = 90°$ (horizontal diffraction), derived hydrostatic pressure and differential stress during first compression, heating at constant load, and continuing compression.

Table 1
Observed stress (t), parameter (α) and effective hydrostatic pressure (σ_P) of fayalite sample during the second heating cycle

T (K)	Pattern #	Time (ks)	t (GPa)	α	σ_P (GPa)
573 (5)	69	25.662	2.7 (1)	0.8 (3)	4.8 (2)
	70	25.877	2.3 (1)	0.7 (3)	4.9 (2)
	71	26.198	2.6 (1)	0.9 (2)	4.9 (2)
	72	28.865	2.2 (1)	0.5 (4)	4.7 (2)
673 (5)	73	29.150	1.9 (1)	0.7 (3)	5.0 (2)
	74	29.426	1.95 (10)	1.0 (2)	4.7 (2)
	75	29.726	2.0 (1)	1.1 (2)	4.5 (2)
	76	30.106	2.0 (1)	1.0 (2)	4.5 (2)
	77	30.340	2.0 (1)	1.0 (2)	4.5 (2)
	78	30.528	1.8 (1)	1.3 (2)	4.6 (2)
773 (5)	79	30.799	1.8 (1)	1.3 (2)	4.4 (2)
	80	30.999	1.7 (1)	1.3 (2)	4.6 (2)
	81	31.199	1.3 (1)	1.3 (2)	4.6 (2)

Note: numbers in parenthesis indicate the experimental uncertainty.

α parameter. Funamori et al. (1994) pioneered such an experiment in a modified Drickamer-type apparatus, and found a decrease of the α parameter from 1 to 0.5 with pressure in NaCl sample. Their experiment was conducted at room temperature. It has not been investigated how the α parameter changes at high temperature when large plastic strain accumulates. In plastic regime, there are also two simplified models, Sachs model (Leffers, 1995; Sachs, 1928) and Taylor model (Bishop and Hill, 1951; Taylor, 1938), describing the coupling between grains in a polycrystal. Similar to Reuss and Voigt models in elastic regime, Sachs and Taylor models assume all the grains in the polycrystal are subjected to the same stress and strain, respectively. However, these models are strongly dependent on the slip systems in each crystal structure, and modeling for olivine phase is not currently available to our knowledge. Data with the α parameter significantly >1 at higher temperature need to be processed with an improved model.

Strain rate versus stress data at 573 and 673 K during the second heating are plotted in Fig. 10. Because deriving strain rate requires two X-ray images taken at a time interval, we use an average of stress over this time period in the plot. Deformation at these low temperatures is dominated by the dislocation glide that is controlled by lattice resistance, and described by Peierls plasticity flow law (Evans and

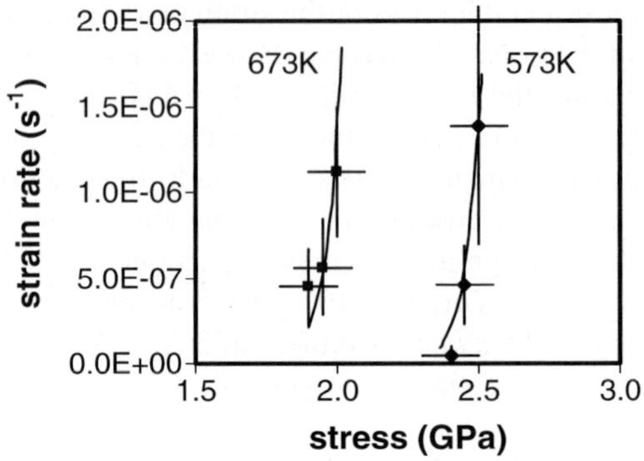

Fig. 10. Strain rate vs. stress of the Fe_2SiO_4 olivine sample at 4.8 GPa, 573 K (solid diamonds) and 4.6 GPa, 673 K (solid squares). Bars attached to the symbols indicate the errors. Solid lines represent the fit lines to the Peierls plasticity flow law (see text) for each data set.

Kohlstedt, 1995)

$$\dot{\varepsilon} = \dot{\varepsilon}_0 \exp\left(-\frac{Q}{RT}\left(1-\frac{\sigma}{\tau}\right)^2\right) \quad (16)$$

where $\dot{\varepsilon}$ is strain rate, $\dot{\varepsilon}_0$ pre-exponential factor (initial strain rate, may be a function of temperature), R the gas constant, T the temperature in Kelvin, Q the activation energy, τ the Peierls stress (lattice friction stress at absolute zero) and σ the stress (the same as t; for consistency with the literature, we use σ to represent stress in the flow law expression). Fitting the experimental data into this equation, we get $Q = 480 \pm 20$ kJ/mol, $\tau = 6.7 \pm 0.5$ GPa at $\dot{\varepsilon}_0 = 1.810^{11}\,s^{-1}$ for $T = 573$ K and $\dot{\varepsilon}_0 = 2.3 \times 10^{12}\,s^{-1}$ for $T = 673$ K. The pressure difference between the two temperature data sets is not taken into account during the fitting. Compared with San Carlos olivine data ($Q = 498$ kJ/mol, $\tau = 9.1$ GPa) reported by Evans and Goetze (Evans and Goetze, 1979), our result indicates a weakening by an increase of iron partitioning in olivine structure.

During the third heating cycle, the sample transforms from the olivine to the spinel structure. Interpretation of the X-ray diffraction data from two coexisting phases is in progress. Fig. 11 shows the strain rate versus stress for spinel phase at 873 and 1073 K during the fourth heating cycle. Fitting these data into Eq. (16) yields $Q = 460 \pm 20$ kJ/mol, $\tau = 6.5 \pm 0.5$ GPa at $\dot{\varepsilon}_0 = 8.1 \times 10^{13}\,s^{-1}$ for $T = 873$ K and $\dot{\varepsilon}_0 = 5.3 \times 10^{15}\,s^{-1}$ for $T = 1073$ K.

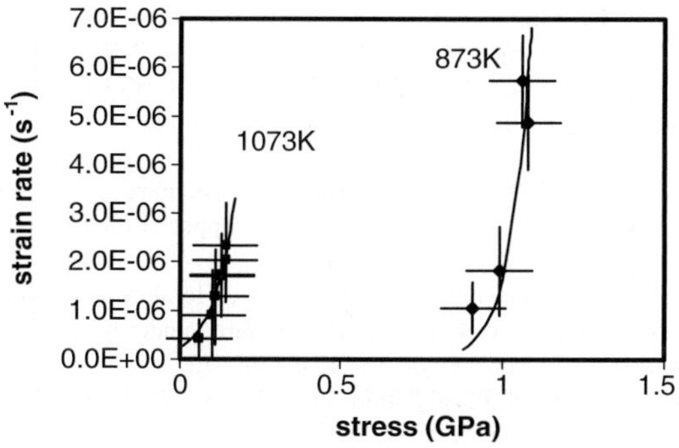

Fig. 11. Strain rate *vs.* stress of the Fe_2SiO_4 spinel sample at 873 K (solid diamonds) and 1073 K (solid squares) under the pressure of 9 GPa. Bars attached to the symbols indicate the errors. Solid lines represent the fit lines to the Peierls plasticity flow law (see text) for each data set.

Uncertainty in the derived stress relies on the resolution of the X-ray diffraction. In general, the resolution of an energy dispersive diffraction is between 10^{-4} and 10^{-3}, which represents the uncertainty of the experimental lattice elastic strain for stress derivation. Therefore, for the materials with elastic modulus in the order of 10^2 GPa, the uncertainty of stress measurement is about 10–100 MPa. However, another source of uncertainty that can not be quantitatively determined is the inadequacy of the model used to describe the coupling between grains in a real polycrystal. Taking this fact into account, the accuracy of the experimental data might be worse than that discussed here. Uncertainties of the derived parameters by fitting experimental data into an empirical equation are estimated through least square fitting process.

5. Remarks

This study demonstrates that the essential components for constructing a deformation flow law, stress and strain rate, can be attained at high pressure up to 10 GPa and temperature beyond 1000 °C by using multi-anvil apparatus in conjunction with synchrotron X-rays. Direct measurements of the strain and stress through sample imaging and X-ray diffraction eliminate the uncertainty introduced in the modeling of indirect measurements at high pressures (e.g. deformation and friction of the pistons that pass the force to the sample). The X-ray techniques are the critical experimental tools which enable the conventional stress–strain rate deformation experiments at pressures beyond the typical pressure limit of 3 GPa. In addition, the multiple X-ray diffractions as a function of ψ supply very important information for understanding the behavior of high pressure cells in terms of stress distribution in the cell and the response of the cell components to the annealing temperatures. In this study, we find that use of hard pistons to sandwich the sample can generate significant differential stress even after an annealing cycle, and therefore the apparent pressure derived from X-ray diffraction at a given ψ angle (i.e. a common high pressure X-ray diffraction setup) is most likely discrepant from the real pressure. The stress and strain measurement reported here is restricted to a relaxation process. Therefore, the flow law for fayalite derived from the experimental data may not perfectly describe a steady-state flow in the sample. Combined with the developing deformation-DIA type apparatus (Wang et al., 2004), the stress and strain can be measured at a controlled strain rate. One should be able to develop the deformation flow law at a near steady-state.

Acknowledgements

Research carried out (in part) at the National Synchrotron Light Source, Brookhaven National Laboratory, which is supported by the US Department of Energy, Division of Materials Sciences and Division of Chemical Sciences, under Contract No. DE-AC02-98CH10886. We would like to thank Dr. L. Wang and Dr. Z. Zhong for their technical support at the X17 beamline. MPI pub. 314.

References

Bishop, J.F.W., Hill, R., 1951. A theory of the plastic distortion of a polycrystalline aggregate under combined stresses. Phil. Mag. 42, 414.

Bussod, G.Y., Katsura, T., Rubie, D.C., 1993. The large volume multi-anvil press as a high *P–T* deformation apparatus. Pure Appl. Geophys. 141, 579–599.

Daymond, M.R., Bourke, M.A.M., Von Dreele, R.B., 1999. Use of Rietveld refinement to fit a hexagonal crystal structure in the presence of elastic and plastic anisotropy. J. Appl. Phys. 85 (2), 739–747.

Durham, W.B., Rubie, D.C., 1998. Can the multianvil apparatus really be used for high-pressure deformation experiments. In: Manghnani, M.H., Yagi, T. (Eds.), Properties of Earth and Planetary Materials at High Pressure and Temperature. AGU Geophys. Monogr. 101, 63–70.

Evans, B., Kohlstedt, D.L., 1995. Rheology of Rocks. Rock Physics and Phase Relations, A Handbook of Physical Constants. Washington, DC, AUG, pp. 148–165.

Evans, B., Goetze, C., 1979. The temperature variation of hardness of olivine and its implication for polycrystalline yield stress. J. Geophys. Res. 84, 5505–5524.

Funamori, N., Yagi, T., Uchida, T., 1994. Deviatoric stress measurement under uniaxial compression by a powder X-ray diffraction method. J. Appl. Phys. 75 (9), 4327–4331.

Graham, E.K., Schwab, J.A., Sopkin, S.M., Takei, H., 1988. The pressure and temperature dependence of the elastic properties of single-crystal fayalite Fe_2SiO_4. Phys. Chem. Mineral. 16, 186–198.

Green, H.W., Young, T.E., Walker, D., Scholz, C.H., 1990. Anticrack-associated faulting at very high pressure in natural olivine. Nature 348, 720–722.

Hearmon, R.F.S., 1956. The elastic constants of anisotropic materials—II. Adv. Phys. 5, 323–382.

Hofmeister, A.M., Mao, H.K., 2001. Evaluation of shear moduli and other properties of silicates with the spinel structure form IR spectroscopy. Am. Mineral. 86, 622–639.

Karato, S.I., Rubie, D.C., 1997. Toward an experimental study of deep mantle rheology: a new multianvil sample assembly for deformation studies under high pressures and temperatures. J. Geophys. Res. 102, 20,111–20,122.

Kavner, A., Duffy, T.S., 2001. Strength and elasticity of ringwoodite at upper mantle pressures. Geophys. Res. Lett. 28 (14), 2691–2694.

Leffers, T., 1995. Long-range stresses associated with boundaries in deformed materials. Phys. Stat. Solut. 149, 69.

Li, L., Raterron, P., Weidner, D., Chen, J., 2003. Olivine flow mechanisms at 8 GPa. Phys. Earth Planet. Int. 138 (2), 113–129.

Li, L., Weidner, D., Raterron, P., Chen, J., Vaughan, M., 2004. Stress Measurements of Deforming Olivine at High Pressure. Phys. Earth Planet Int. 143–144, 357–367.

Liebermann, R.C., Wang, Y., 1992. Characterization of sample environment in a multianvil split-sphere apparatus. In: Syono, Y., Manghnani, M.H. (Eds.), High Pressure Research: Application to Earth and Planetary Sciences. AGU Geophys. Monogr. 67, pp. 19–31.

Mao, H.-K., Shu, J., Shen, G., Hemley, R.J., Li, B., Singh, A.K., 1998. Elasticity and rheology of iron above 220 GPa and the nature of the Earth's inner core. Nature 396 (6713), 741–743.

Sachs, G., 1928. Zur Ableitung einer Fliebbedingung. Z. Ver. Deu. Ing. 72 (22), 734.

Singh, A.K., 1993. The lattice strain in a specimen (cubic system) compressed nonhydrostatically in an opposed anvil device. J. Appl. Phys. 73 (9), 4278–4286.

Singh, A.K., Balasingh, C., Mao, H.K., Hemley, R.J., Shu, J., 1998. Analysis of lattice strains measured under nonhydrostatic pressure. J. Appl. Phys. 83 (12), 7567–7575.

Taylor, G.I., 1938. Plastic strain in metals. J. Inst. Metals 62, 307.

Tingle, T.N., Green II, H.W., Young, T.E., Koczynski, T.A., 1993. Improvements to Griggs-type apparatus for mechanical testing at high pressures and temperatures. Pure Appl. Geophys. 141, 523–543.

Uchida, T., Funamori, N., Yagi, T., 1996. Lattice strains in crystals under uniaxial stress field. J. Appl. Phys. 80 (2), 739–746.

Uchida, T., Yagi, T., Funamori, N., Anonymous, 1997. Differential stress in Mg_2SiO_4 under uniaxial stress field; variation with pressure, temperature, and phase transformation. American Geophysical Union, 1997 fall meeting 78 (Suppl. 46) 722.

Vaughan, M.T., Weidner, D.J., Wang, Y.B., Chen, J.H., Koleda, C.C., Getting, I.C., 1998. T-cup: A new high-pressure apparatus for X-ray studies. Rev. High Pressure Sci. Technol. 7, 1520–1522.

Wang, Y., Durham, W.B., Getting, I.C., 2004. D-DIA: A new apparatus for tiaxial deformation at pressures up to 15 GPa, this volume.

Weidner, D.J., 1998. Rheological studies at high pressure. In: Hemley, R.J. (Ed.), Ultrahigh-Pressure Mineralogy: Physics and Chemistry of the Earth's Deep Interior, vol. 37. Mineralogical Society of America, Washington, DC, pp. 493–524.

Weidner, D.J., Wang, Y., Chen, G., Ando, J., Vaughan, M.T., 1998. Rheology measurements at high pressure and temperature. In: Manghnani, M.H., Yagi, T. (Eds.), Properties of earth and planetary materials at high pressure and temperature. AGU Geophys. Monogr. 101, 473–482.

Yamazaki, D., Kato, T., Ohtani, E., 1996. Preferred orientation of olivine in stress field under ultra-high pressures. J. Seismol. Soc. Jpn. 49, 39–53.

Yamazaki, D., Karato, S., 2001. High-pressure rotational deformation apparatus to 15 GPa. Rev. Sci. Instrum. 72 (11), 4207–4211.

Available online at www.sciencedirect.com
SCIENCE DIRECT
ELSEVIER
Physics of the Earth and Planetary Interiors 143–144 (2004) 357–367
PHYSICS OF THE EARTH AND PLANETARY INTERIORS
www.elsevier.com/locate/pepi

Stress measurements of deforming olivine at high pressure

Li Li[a,*], Donald Weidner[a], Paul Raterron[b], Jiuhua Chen[a], Michael Vaughan[a]

[a] *Department of Geosciences, Mineral Physics Institute, State University of New York at Stony Brook, Stony Brook, NY 11794-2100, USA*
[b] *Laboratoire de Structure et Propriétés de l'État Solide (Associated to CNRS), Université des Sciences et Technologies de Lille, F-59655 Villeneuve d'Ascq Cedex, France*

Received 17 April 2003; received in revised form 16 July 2003; accepted 2 September 2003

Abstract

Rheological properties of mantle minerals are critical for understanding the dynamics of the Earth's deep interior. Due to limitations in experimental technique, previous quantitative studies of the rheological properties of mantle minerals are limited to either low pressure or low temperature. The present understanding of mantle flow is mostly inferred from the extrapolation of relatively low-pressure data to mantle high-pressure conditions. However, the effect of pressure (represented by activation volume) on the rheological properties of olivine is still controversial. Therefore, deformation experiments, carried out at mantle pressures, are necessary to understand and model mantle flow. Here we report an experimental study of plastic deformation of San Carlos olivine $(Mg, Fe)_2SiO_4$ under upper mantle conditions. Macroscopic differential stress and strain rates have been measured in situ in a large-volume high-pressure apparatus using newly developed techniques. The differential stress at high temperature and high pressure that we measured is significantly lower than that estimated by many currently accepted olivine flow laws. We document the first in situ experimental differential stress results in a multi-anvil press. Our results give direct evidence for a relatively small activation volume (less than 10^{-5} m^3 mol^{-1}). This shows that the effect of pressure on dislocation creep is small.

Keywords: Olivine; Stress; Strain; Dislocation; Pressure; Grain size

1. Introduction

The effect of pressure (P) on viscosity is important in defining the rheological properties of mantle minerals. It is, however, poorly constrained. The structure and dynamics of the Earth are directly linked to the variation of flow properties with depth. Some seismic observations, such as regions of low velocity and high attenuation at the top of asthenosphere, are explained by the effects of water and/or partial melting (Weidner, 1974). These agents are further implicated in creating the low viscosity aesthenosphere. An increase in creep rate by a factor of 2–5 at a given temperature (T) and stress can result from 3 to 10% melt (Kohlstedt and Chopra, 1994). The maximum effect of adding water to the system, that has been observed, is an order of magnitude increase in strain rate for a given stress and temperature (Hirth and Kohlstedt, 1996). However, for an activation volume (V^*) of 20 cm^3 mol^{-1}, pressure will decrease the strain rate by a factor of 1000 by a depth of 100 km. Questions, such as the variation of the viscosity of the mantle with depth, can only be addressed by studying the pressure dependence of the rheology of

* Corresponding author. Tel.: +1-631-632-8220; fax: +1-631-632-8140.
E-mail address: lilli@notes.cc.sunysb.edu (L. Li).

0031-9201/$ – see front matter © 2004 Elsevier B.V. All rights reserved.
doi:10.1016/j.pepi.2003.09.022

mantle minerals (Poirier, 1985). In addition, the pressure dependence of olivine rheology is also relevant to the mechanism of deep earthquakes since a small pressure strengthening is required for the observed Wadati–Benioff zone seismicity to be bounded by a constant limiting strength (Brodholt and Stein, 1988).

The effect of pressure on the rheological properties of olivine has been a controversial and ambiguous issue for a few decades even with the realization of its importance. The various experimental attempts to measure the activation volume of olivine dislocation creep at high temperature give bimodal results, yielding values of 0–6 $cm^3 mol^{-1}$ (Béjina et al., 1997; Jaoul et al., 1983), or 15–27 $cm^3 mol^{-1}$ (Green and Borch, 1987; Borch and Green, 1987; Ross et al., 1979; Kohlstedt and Wang, 2001; Karato and Wu, 1993). The lower values of activation volume for high-temperature creep of olivine come from studies of silicon or oxygen diffusion (Béjina et al., 1997). High-pressure dislocation recovery experiments suggest an activation volume between 6 and 11 $cm^3 mol^{-1}$ (Karato et al., 1993; Kohlstedt et al., 1980). Deformation reports of olivine at high temperature and low pressure (<300 MPa) concluded that activation volumes range from 13 to 27 $cm^3 mol^{-1}$ (Green and Borch, 1987; Ross et al., 1979; Kohlstedt and Wang, 2001; Karato and Wu, 1993) while Borch and Green (1987) pointed out that activation volume may decrease with further increase of pressure.

Here we report the results of a relaxation-style deformation experiment, in which differential stresses and strain rates are measured using synchrotron X-radiation in a multi-anvil high-pressure apparatus. Together with our TEM investigations of the run products and our previous studies (Li et al., 2003), we conclude that the observed small effect of pressure on the rheology of olivine is consistent with the lower range of activation volume (<10 $cm^3 mol^{-1}$) at high temperature and high pressure.

2. Flow mechanism

In our previous studies (Li et al., 2003; referred to as Li1), we defined the high-pressure flow mechanism of olivine under the same experimental conditions as the current study. In Li1, olivine samples with two grain-sizes were symmetrically placed in the high-pressure cell assembly and deformed at 8 GPa and $T \leq 1473$ K. The strain rates of the two samples were found to be the same indicating a grain-size independent flow under the assumption that

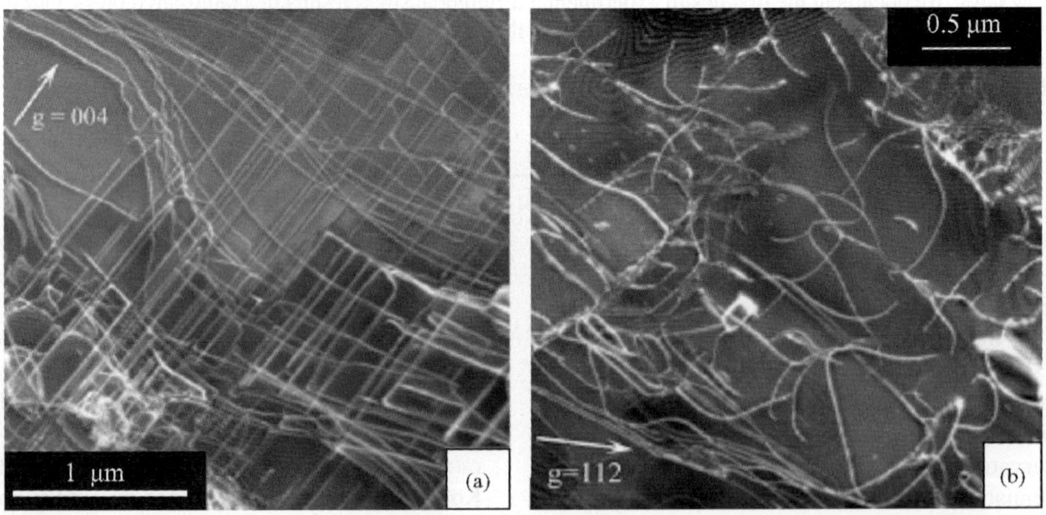

Fig. 1. Dark field weak beam TEM micrographs of olivine dislocation microstructures, comparable in both coarse- and fine-grain materials, here shown in the coarse-grain samples. (a) At 973 K, olivine grains exhibit c dislocations well confined in their glide plane (here (0 1 0) plane) with straight screw segments extending throughout the grains and (mobile) curved edge segments. (b) At 1273 K, c dislocations still dominate the deformation although a dislocations are occasionally observed. At that temperature, all dislocation lines are curly and not always confined in their glide plane, which shows that dislocation climb is now an active process.

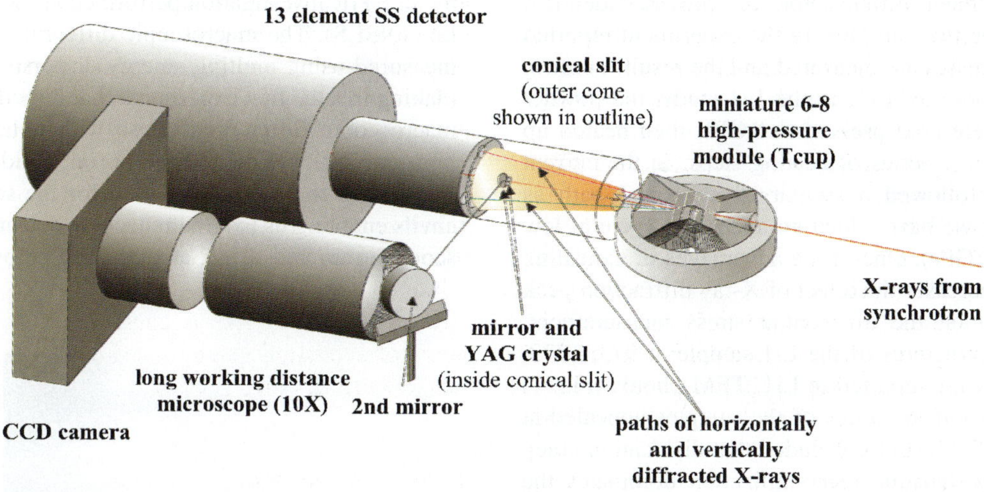

Fig. 2. Multi-anvil high-pressure system. White X-rays from the synchrotron illuminate the sample placed in the T-cup. A conical slit system defines the two theta angle as 6.5°. The collimated diffracted X-ray beams are detected by four elements of a 13-element solid state detector which converts the signals into energy dispersive patterns. A mirror and a fluorescent YAG crystal, mounted inside the conical slit, convert transmitted X-rays into visible light that are recorded by the CCD camera after being magnified by a long working distance microscope.

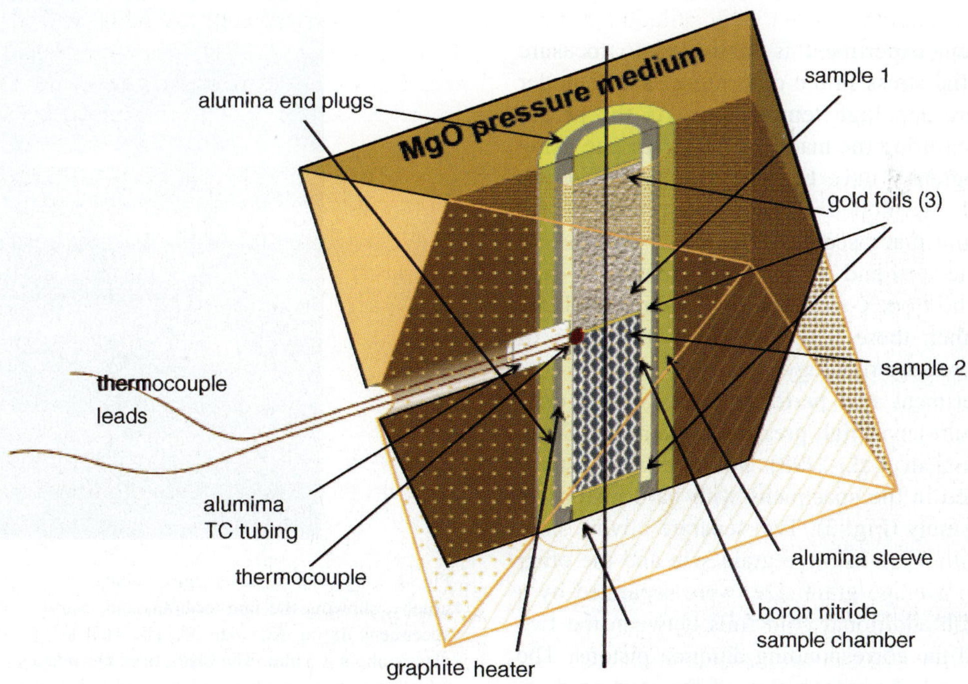

Fig. 3. Schematic cross section of T-cup apparatus cell assembly. An MgO octahedron is used as the pressure medium. A cylindrical furnace is placed inside the MgO octahedron and contains alumina end plugs and two samples symmetrically arranged. The end plugs and the two samples are separated by gold foils, which is used as strain markers. The temperature is measured with a W3%Re–W25%Re thermocouple. Temperature gradients within specimens are less than $10\,K\,mm^{-1}$, as calibrated previously in similar cell assemblies.

the arrangement of the sample-cell provides identical stress in the two samples. In the experiment reported here the stresses are measured and the results confirm the assumption of Li1. In the Li1 study, the powder samples were cold pressed to 8 GPa, then heated up to 1473 K in a series of heating steps. In the current study, we followed a similar experimental path as Li1 except we have added an annealing cycle at low pressure (3 GPa). One of the advantages of annealing is that it decreases the effect of X-ray diffraction peak broadening on the differential stress measurement. The microstructures of the Li1 sample at 973, 1273, and 1473 K are reported in Li1. TEM photos (Fig. 1) show the microstructures of the samples annealed at 973 and 1273 K. Li1 concluded that dislocation creep assisted by dynamic recrystallization dominates the flow at T of 1273 and 1473 K at high pressure while low-temperature dislocation glide dominates olivine flow at T of 973 K.

3. Experimental protocol

The current experiment is motivated to measure the differential stress in the deforming sample under high pressure and high temperature. As the procedures for measuring the macroscopic strain rate using X-ray radiography have become routine, we report here a newly developed experimental procedure for measuring the macroscopic differential stress using multiple detectors and anvils transparent to X-rays with a synchrotron X-radiation light source (NSLS). Taken together, these new tools form the basis of these high-pressure rheology studies.

The experiment was performed with a 6–8 double stage multi-anvil high-pressure system, the T-cup (Fig. 2) (Vaughan et al., 1998). Non-hydrostatic stress was produced in the specimens with hard pistons in the cell assembly (Fig. 3). Two separate olivine samples, one with 5 μm average grain size and the other with 0.5 μm average grain size, were separated by a gold foil with additional gold foils between the two samples and the corresponding alumina pistons. The preparation and characterization of the starting samples as well as the evolution of the grain sizes during deformation are described in detail elsewhere (Li1). The water content in the starting material and recovered run product is less than 1 wt. ppm as indicated by an FTIR investigation performed at the U2A beam line (NSLS). The macroscopic differential stress was measured using multiple energy dispersive detectors, making this the first experiment that directly measures macroscopic differential stress with a large-volume high-pressure system. Cubic boron nitride, which is transparent to X-rays, was used for the second-stage anvils enabling us to simultaneously detect the macroscopic strain and stress, as discussed below. Combin-

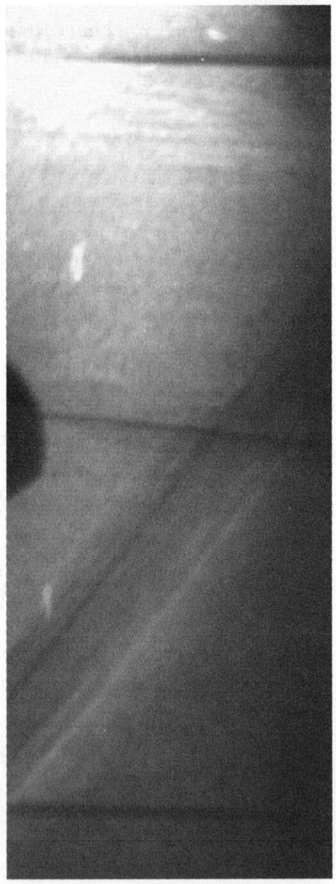

Fig. 4. Magnified digital images, obtained in situ by X-ray radiography, showing the fine (bottom) and coarse (top) grain olivine specimens during Run San_36. The total length of the two samples is about 1.5 mm. The black lines are images of the gold foils which are strain markers. Above and below the samples are the images of the hard-alumina rods used as the pistons in this study. The diagonal features are the shadows of nearly transparent cubic boron nitride anvils. The black blob at the left of the figure is the tip of the thermocouple. The few bright areas in the image are flaws in the fluoresence screen.

ing these techniques, we are able to measure the stress and strain inside the specimen in situ at high P and T as a function of time.

The data reported here are from our experiments performed with SAM85 (Weidner et al., 1992). A white X-ray beam passes through polycrystalline San Carlos olivine samples in the radial direction, and probes the axial and radial lattice strains of the samples as a function of time. At the same time, downstream X-ray radiographic images were captured and magnified (Fig. 4). Gold foils, used as strain markers, are imaged; the resulting images allow the determination of strain to a precision of 10^{-4} (as described in Li1). This yields a precision for strain rate of $10^{-6}\,s^{-1}$ for data taken at 100 s intervals or longer. In this experiment, the powder samples were annealed at 3 GPa and 1273 K during the first heating cycle, and were then deformed at 8 GPa at temperatures up to 1473 K. The stress relaxed during the annealing cycle. However, further compression created sufficient differential stress to deform the sample again at 1130 K.

4. Strain analysis

Strain is derived from the comparison of two images created by X-ray radiographs. While the measurement of the absolute length of the sample is accurate to a few pixels, changes in length are measurable to fractions of pixels. The method used here is described in detail in the appendix of Li1.

The results of strain versus time are illustrated in Fig. 5 and documented in Table 1. Strains evolve similarly in the two samples except at temperature equal to 1129 K where the coarse grain sample exhibits higher strains. At temperature of 1504 K or higher, both samples shorten by ~3%. The time-resolved strain rates are shown in Fig. 6 and also documented in Table 1. In general, the strain rates of the two samples are the same (except at 1129 K) and decrease with time at each temperature. The data at 1129 K and 8 GPa indicated the onset of measurable flow. At this point, the strength of the material is low enough that stress initiates plastic flow. Strain rate measurements in this study are sparser than in Li1 since experimental time

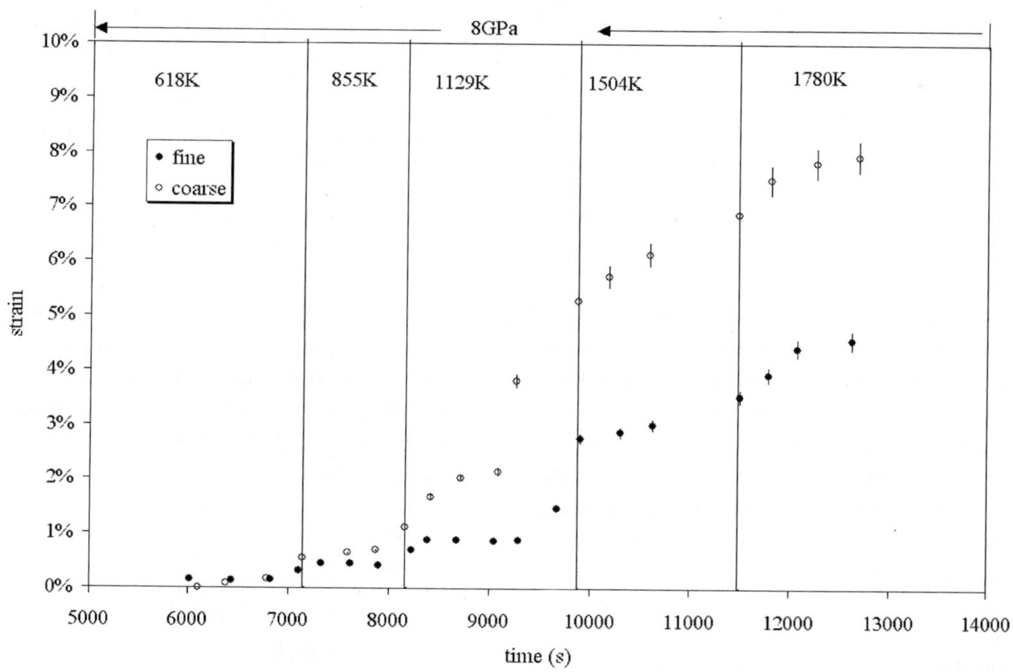

Fig. 5. Time-resolved strain results from Run San_36. The open symbols represent coarse grain sample, the solid symbols represent fine grain sample. Coarse grain sample experienced higher total strain than fine grain sample during the initial yielding period ($T = 1129$ K). At higher temperatures, both samples shorten by ~3%.

Table 1
Run San_36[a] condition, and strain and strain-rate results

P (±0.5 GPa)	T^{b} (±10 K)	Time (±1 s)	Fine[c] (%)	Fine[c] ($10^{-6}\,s^{-1}$)
8	618	6019	0.1 (0.01)	3.4 (0.4)
8	618	6826	0.2 (0.01)	0.5 (0.7)
8	855	7332	0.5 (0.02)	4.4 (0.6)
8	855	7615	0.5 (0.02)	0.4 (0.6)
8	1129	8386	0.9 (0.03)	6.4 (0.6)
8	1129	8671	0.9 (0.03)	0.2 (0.5)
8	1129	9282	0.9 (0.03)	0.8 (0.5)
8	1504	9901	2.8 (0.1)	33 (0.8)
8	1504	10297	2.9 (0.1)	3.4 (0.6)
8	1504	10623	3.0 (0.1)	1.6 (0.2)
8	1780	11782	3.9 (0.2)	14 (0.7)
8	1780	12068	4.4 (0.2)	8.9 (0.4)
8	1780	12620	4.6 (0.2)	3.1 (0.4)
P (±0.5 GPa)	T^{b} (±10 K)	Time (±1 s)	Coarse[c] (%)	Coarse[c] ($10^{-6}\,s^{-1}$)
8	618	6786	0.1 (0.03)	2.3 (0.5)
8	855	7590	0.6 (0.02)	3.7 (0.6)
8	855	7867	0.7 (0.02)	1.8 (0.6)
8	1129	8410	1.1 (0.04)	19 (0.7)
8	1129	8706	1.7 (0.06)	9.2 (0.5)
8	1129	9082	2.0 (0.07)	6.6 (1)
8	1504	9871	3.8 (0.1)	48 (1)
8	1504	10176	5.3 (0.2)	11 (0.5)
8	1504	10587	5.7 (0.2)	4.6 (0.3)
8	1780	11805	6.9 (0.3)	14 (0.5)
8	1780	12259	7.5 (0.3)	7.2 (0.5)
8	1780	12689	7.8 (0.3)	2.6 (0.4)

Note. Numbers in parenthesis indicate the one standard deviation.
[a] Images of fine and coarse specimens are taken at different times, numbers in the time column are simply converted from hour, minute, and second in the recorded time, then normalized to the starting point of the experiment.
[b] W3%Re–W25%Re thermocouple, no pressure correction was applied.
[c] Fine and coarse stand for fine- and coarse-powder specimens, respectively. About half of the strains were generated in the first 1000 s at each temperature.

was divided between strain measurements and stress measurements. In the relaxation experiment, strain rates of $10^{-4}\,s^{-1}$ or higher occur in the first 1000 s after each increase in temperature. The measurements in these experiments can be repeated in every 500 s. High strain rate happened during the first 1000 s may not be measured in time. As a result, this approach is limited to investigating deformation experiments with strain rates greater than $10^{-4}\,s^{-1}$.

5. Stress analysis

Macroscopic differential stress manifests, as a uniform lattice strain, when an orientation constrained by the geometry of our uniaxial deformation cell assembly. Analysis of the lattice strain for an orthorhombic system under non-hydrostatic pressure is described in detail elsewhere (Singh et al., 1998). In summary, lattice strain ε_{ij} and differential stress σ_{ij} is given by Weidner (1998):

$$\varepsilon_{ij} = \sum_{k=1}^{3}\sum_{l=1}^{3} S_{ijkl}\sigma_{kl}$$
$$\varepsilon_{hkl} = \frac{[d(h\,k\,l) - d_{\mathrm{P}}(h\,k\,l)]}{d_{\mathrm{P}}(h\,k\,l)} \tag{1}$$

where $d(h\,k\,l)$ is the d spacing for each $h\,k\,l$, and $d_{\mathrm{P}}(h\,k\,l)$ is the d spacing under hydrostatic pressure alone. The elastic compliance tensor, S, is rotated to

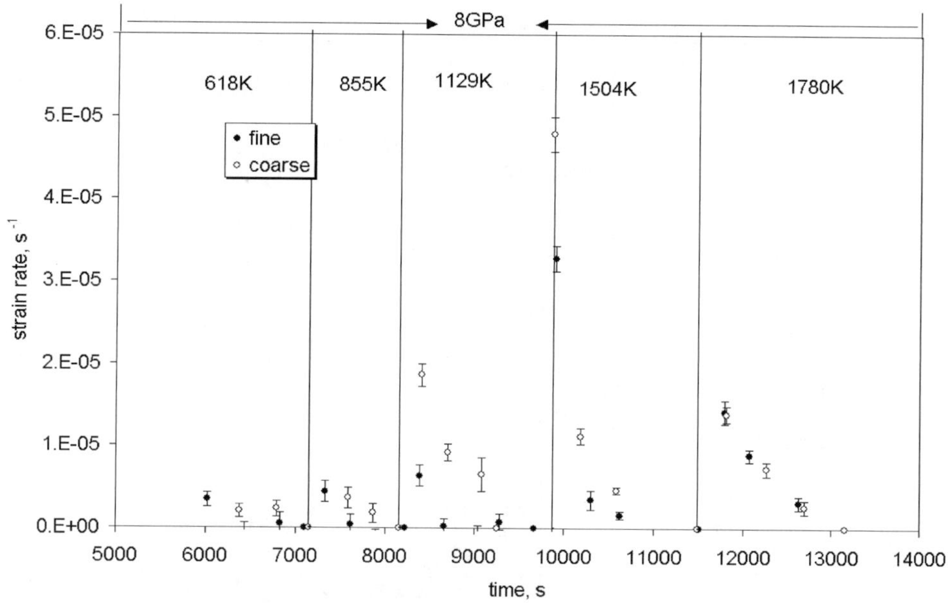

Fig. 6. Time-resolved strain rate results from Run San_36. Open symbols represent coarse grain sample, close symbols represent fine grain sample. Two samples demonstrate very similar flow behavior. At $T < 1100$ K, no significant flow was observed. Significant flow occurs at 1129 K and continues at 1504 and 1780 K. When the temperature is increased, the flow rate is initially high and decays with time as expected in a relaxation style experiment.

the coordinate system of the stress system for each hkl. S in our calculations is from experiment data from Isaak (1992) and Zha et al. (1996).

6. Experimental results

We have measured the macroscopic differential stress as a function of temperature at $P = 8(\pm 0.5)$ GPa. Stress results are illustrated in Fig. 7 and documented in Table 2. Pressure is calculated using the olivine sample itself as the pressure standard with the third-order Birch–Murnaghan equation of state with parameters from Isaak (1992) and Zha et al. (1996). The volume measured after the experiment (quenched and open press) is used as the reference V_0 (=292.5769 Å^3). Stress is determined using Eq. (1) for four different diffraction peaks {(1 1 2), (1 3 0), (1 3 1), (0 2 1)}. The plotted value is for the average of these determinations for the four diffraction peaks, while the range of the individual values is indicated by the error bars. An analysis of the expected error of any one measurement is dominated by the uncertainty in the d spacing resolution and the single-crystal elastic constants at the P, T of the sample and is expected to be less than 0.1 GPa, which is much less than the indicated error bar. On the other hand, the stress defined by each diffraction peak represents the average stress experienced by the subset of grains that contribute to the specific diffraction peak. The grains in any one of these subsets share elastic and plastic properties that may contribute to a unique stress field for the specific subset. As a result the distribution of stress among the different subsets represents the aggregation/stress-distribution process and contributes as the main uncertainty in defining the aggregate stress.

The stresses and the strain rates for the two samples are very similar once plastic flow is initiated ($T > 1129$ K), supporting the assertion that the flow process is grain-size independent. The stress initially increases with heating and reaches the maximum of 1.7 GPa at 875 K and decays rapidly after about 1100 K. The increase in stress with temperature is probably due to the relaxation of the cell assembly on heating.

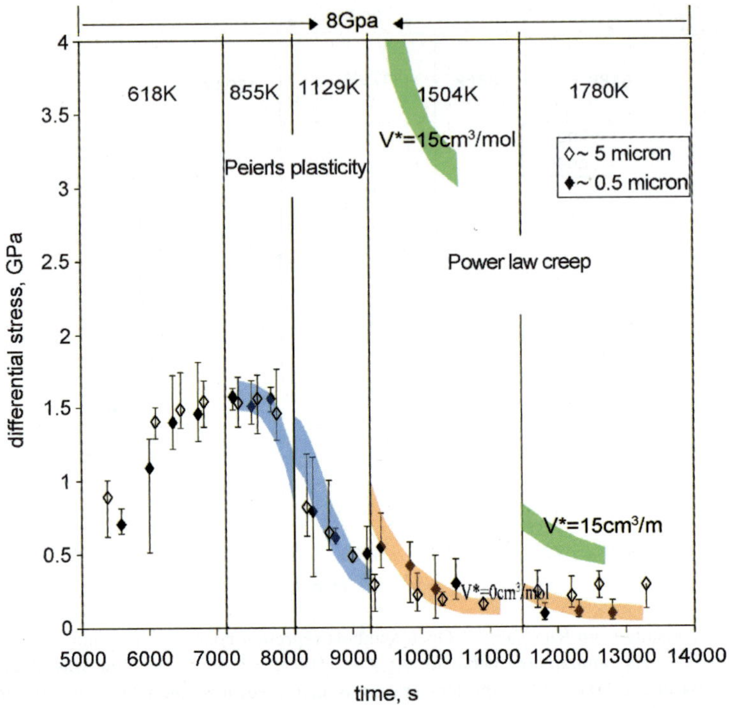

Fig. 7. Macroscopic differential stresses for Run San_36. Measured stresses are represented by either open symbols (for the 5 μm grains size) or solid symbols (for the 0.5 μm grain size). The plotted values are the average of the values of measured stress from diffraction peaks of (1 1 2), (1 3 1), (1 3 0), and (0 2 1). The error bar represents the range of stress values from these four peaks. The colored fields represent the range of stress that is calculated from different flow models at different temperatures as described in Tables 1 and 2 using the strain rate that is given in Fig. 5. The brown region is based on the power law creep flow law parameters given in Table 2 for $V^* = 0\,cm^3\,mol^{-1}$. The green region is based on the power law creep flow law parameters given in Table 2 for $V^* = 15\,cm^3\,mol^{-1}$. The blue region is based on the Peierls plasticity flow law parameters given in Table 1. The data indicate no grain size sensitivity in either strain rate or stress and are consistent with the literature models only if V^* is small.

7. Discussion and conclusion

Grain-size independent flow dominates the experiments in Li1 and in this study. With the combination of stress and strain rate measurements that are reported here, we have the opportunity to test dry olivine flow laws that have been reported in the literature for a range of activation volumes. In this discussion, we interpret our measurements as steady state data even though they are gathered during a relaxation process. Phenomena such as transient creep are superimposed on and difficult to separate from the steady state process. However, we expect that cold compression will introduce a high dislocation density into the sample and will thus work-harden the sample. This will make the stress field higher than the steady state value.

The Peierls plasticity flow law (Evans and Goetze, 1979) is used for fitting the data at $T = 855$ and 1129 K. Power-law creep parameters from the literature (Kirby, 1983; Chopra and Paterson, 1984; Karato and Wu, 1993) are used to fit the stress–strain rate data at $T = 1504$ and 1780 K with V^* as 0 and $15\,cm^3\,mol^{-1}$. The parameters for the Peierls plasticity and power-law creep flow laws are listed in Table 3. The stress is calculated using the measured strain rate and the flow laws with the parameters given in Table 4. Fig. 7 illustrates the entire range of calculated stresses for the three listed flow models assuming $V^* = 0\,cm^3\,mol^{-1}$ by the brown-colored band. The calculated stress with $V^* = 15\,cm^3\,mol^{-1}$ at $T = 1504$ and 1780 K is indicated by the light green-colored band. Our data are clearly inconsistent

Table 2
Run San_36[a] condition, and stress results

P (±0.5 GPa)	T (±10 K)	Time (±1 s)	σ fine[b] (GPa)	$-\Delta\sigma$ fine[c] (GPa)	$+\Delta\sigma$ fine[c] (GPa)
8	25	5590	0.71	0.11	0.07
8	618	6010	1.09	0.20	0.58
8	618	6370	1.40	0.32	0.18
8	618	6740	1.46	0.35	0.19
8	855	7250	1.57	0.06	0.09
8	855	7530	1.51	0.17	0.12
8	855	7820	1.56	0.08	0.09
8	1129	8420	0.79	0.37	0.45
8	1129	8750	0.61	0.06	0.04
8	1129	9210	0.50	0.18	0.17
8	1129	9410	0.55	0.23	0.14
8	1504	9830	0.42	0.16	0.25
8	1504	10200	0.26	0.23	0.20
8	1504	10500	0.29	0.17	0.11
8	1780	11800	0.09	0.07	0.04
8	1780	12300	0.10	0.08	0.04
8	1780	12800	0.09	0.09	0.05
P (±0.5 GPa)	T (±10 K)	Time (±1 s)	σ coarse[b] (GPa)	$-\Delta\sigma$ coarse[c] (GPa)	$+\Delta\sigma$ coarse[c] (GPa)
8	25	5390	0.90	0.12	0.27
8	618	6110	1.40	0.10	0.11
8	618	6480	1.48	0.26	0.12
8	618	6830	1.54	0.14	0.17
8	855	7620	1.56	0.16	0.24
8	855	7910	1.46	0.30	0.18
8	1129	8330	0.81	0.37	0.19
8	1129	8650	0.64	0.36	0.12
8	1129	9000	0.49	0.06	0.04
8	1129	9320	0.29	0.07	0.18
8	1504	9930	0.22	0.15	0.10
8	1504	10300	0.19	0.05	0.04
8	1504	10900	0.15	0.03	0.04
8	1780	11700	0.24	0.14	0.10
8	1780	12200	0.21	0.14	0.08
8	1780	12600	0.28	0.09	0.09

Note. Numbers in parenthesis indicate the one standard deviation.

[a] X-ray diffraction spectra of specimens are taken at different times, that is, the stresses are measured at different time for the two specimens. Fine and coarse stand for fine- and coarse-powder specimens, respectively.

[b] Stresses represent the average of stresses measured from diffraction peaks of (1 1 2), (1 3 1), (1 3 0), and (0 2 1).

[c] Errors of stresses represent the range of stresses for the four diffraction peaks. The error for the stress measured from an individual peak is expected to be <0.1 GPa.

with the higher activation volume. Rather, a relatively small activation volume (0–5 $cm^3\,mol^{-1}$), as derived from the silicon diffusion experiments of Béjina et al. (1997), is consistent with the grain-size independent flow observation at the condition of 8 GPa, differential stress below 2 GPa and grain size between 0.5 and 5 μm. This activation volume is the average activation volume over 8 GPa of pressure and could be compatible with the larger values that have been reported for low pressure.

We have documented the first in situ experimental strain rate and differential stress measurements in a multi-anvil press. Our results give the direct evidence for a relatively small activation volume for dislocation creep in dry olivine. These data must be taken as qualitative until the transient phenomena associated

Table 3
Flow laws

Peierls plasticity: $\dot{\varepsilon} = \dot{\varepsilon}_0 \exp\left(-\frac{Q_d}{RT}(1-\frac{\sigma}{\sigma_0})^2\right)$	
San Carlos Peridotite (Evans and Goetze, 1979)	
$\sigma_0 = 9.1\,\text{GPa}$	Peierls stress
$\dot{\varepsilon}_0 = 1.3 \times 10^{12}\,\text{s}^{-1}$	Reference strain rate
$Q_d = 498\,\text{kJ mol}^{-1}$	Activation energy
σ	Stress
$\dot{\varepsilon}$	Strain rate
Power law creep: $\dot{\varepsilon} = A\sigma^n \exp - \left(\frac{E^* + PV^*}{RT}\right)$	
A	Material constant
E^*	Activation energy
V^*	Activation volume
n	Stress exponent
R	Gas constant, $8.3144\,\text{J mol}^{-1}\,\text{K}^{-1}$

Table 4
Parameters for power law creep

Dry olivine	Flow law 1	Flow law 2	Flow law 3
A	6.31E−17[a] ($\text{MPa}^{-n}\,\text{s}^{-1}$)	3.50E+22[b] (s^{-1})	4.85E+04[c] ($\text{MPa}^{-n}\,\text{s}^{-1}$)
n	3.5[a]	3.5[b]	3.5[c]
E* (J mol^{-1})	5.33E+05[a]	5.40E+05[b]	5.35E+05[c]
V^* ($\text{cm}^3\,\text{mol}^{-1}$)	0 and 15	0 and 15	0 and 15

[a] Kirby (1983).
[b] Karato and Wu (1993), the value for A is obtained by normalizing stress to shear modulus in the flow law.
[c] Chopra and Paterson (1984).

with the relaxation style experiments are fully understood. The present study demonstrates that the in situ experimental studies of plastic deformation of mantle minerals at mantle pressures are possible. Our new techniques should be able to contribute to a better understanding of the dynamics of the Earth's deep interior.

Acknowledgements

The authors thank SAM group (especially Liping Wang) for providing technical support, Zhenxian Liu for his support at the beam line U2A of the National Synchrotron Light Source (NSLS), and Zhong Zhong for his support at the NSLS beam line. This research was carried out at the NSLS, which is supported by the U.S. Department of Energy, Division of Material Sciences and Division of Chemical Sciences under contract no. DE-AC02_98CH10886. Operations of the beam lines X17 and U2A are supported by COMPRES (EAR0135554). This research was supported by the NSF Grant EAR-9909266, EAR0135551, and EAR0229260 and the Centre National de la Recherche Scientifique (CNRS). MPI publication no. 313.

References

Béjina, F., Raterron, P., Zhang, J., Jaoul, O., Liebermann, R.C., 1997. Activation volume of silicon diffusion in San Carlos olivine. Geophys. Res. Lett. 24, 2597–2600.
Borch, R.S., Green II, H.W., 1987. Dependence of creep in olivine on homologous temperature and its implications for flow in the mantle. Nature (Lond.) 330, 345–348.
Brodholt, J., Stein, S., 1988. Rheological control of Wadati–Benioff zone seismicity. Geophys. Res. Lett. 15 (10), 1081–1084.
Chopra, P.N., Paterson, M.S., 1984. The role of water in the deformation of dunite. J. Geophys. Res. 89, 7861–7976.
Evans, B., Goetze, C., 1979. The temperature variation of hardness of olivine and its implication for polycrystalline yield stress. J. Geophys. Res. 84, 5505–5524.
Green, H.W., Borch, R.S., 1987. The pressure dependence of creep. Acta Metall. 35, 1301–1315.
Hirth, G., Kohlstedt, D.L., 1996. Water in the oceanic upper mantle implications for rheology, melt extraction and the evolution

of the lithosphere. Earth Planet. Sci. Lett. 144 (1–2), 93–108.

Isaak, D.G., 1992. High-temperature elasticity of iron-bearing olivines. J. Geophys. Res. B: Solid Earth and Planets 97 (2), 1871–1885.

Jaoul, O., Houlier, B., Abel, F., 1983. Study of (super 18) O diffusion in magnesium orthosilicate by nuclear microanalysis. J. Geophys. Res. B 88 (1), 613–624.

Karato, S.-I., Rubie, D.C., Yan, H., 1993. Dislocation recovery in olivine under deep upper mantle conditions implications for creep and diffusion. J. Geophys. Res. B: Solid Earth Planets 98, 9761–9768.

Karato, S.-I., Wu, P., 1993. Rheology of the upper mantle a synthesis. Science 260, 771–778.

Kirby, S.H., 1983. Rheology of the lithosphere. Rev. Geophys. Space Phys. 21, 1458–1487.

Kohlstedt, D.L., Chopra, P.N., 1994. Influence of basaltic melt on the creep of polycrystalline olivine under hydrous conditions magmatic systems. Int. Geophys. Ser. 57, 37–53.

Kohlstedt, D.L., Nichols, H.P.K., Hornack, P., 1980. The effect of pressure on the rate of dislocation recovery in olivine. J. Geophys. Res. 85 (B6), 3122–3130.

Kohlstedt, D.L., Wang, Z., 2001. Grain-boundary sliding accommodated dislocation creep in dunite. Eos. Trans. AGU 82 (47), F1137.

Li, L., Raterron, P., Weidner, D., Chen, J., 2003. Olivine flow mechanisms at 8 GPa. Phys. Earth Planet. Interior 138 (2), 113–129.

Poirier, J.P., 1985. Creep of Crystals, High-Temperature Deformation Processed in Metals, Ceramics and Minerals. Cambridge University Press, Cambridge, pp. 63–75.

Ross, J.V., Ave Lallemant, H.G., Carter, N.L., 1979. Activation volume for creep in the upper mantle. Science 203 (4377), 261–263.

Singh, A.K., Balasingh, C., Mao, H.-K., Hemley, R., Shu, J., 1998. Analysis of lattice strains measured under nonhydrostatic pressure. J. Appl. Phys. 83 (12), 7567–7575.

Vaughan, M.T., et al., 1998. T-cup: a new high-pressure apparatus for X-ray studies. Rev. High Pressure Sci. Technol. 7, 1520–1522.

Weidner, D.J., 1974. Rayleigh wave phase velocities in the Atlantic Ocean. The Geophys. J. Royal Astronom. Soc. 1 (1), 105–139.

Weidner, D.J., 1998. Rheological studies at high pressure. Rev. Mineral. Ultrahigh-Pressure Mineral. 37, 493–522.

Weidner, D.J., et al., 1992. Characterization of stress, pressure, and temperature in SAM85, A DIA type high pressure apparatus. In: Syono, Y., Manghnani, M.H. (Eds.), High-Pressure Research: Application to Earth and Planetary Sciences. Terra Scientific Publishing Company and American Geophysical Union, Tokyo and Washington, DC, pp. 13.

Zha, C.-S., Duffy, T.S., Downs, R.T., Mao, H.-K., Hemley, R.J., 1996. Sound velocity and elasticity of single-crystal forsterite to 16 GPa. J. Geophys. Res. B: Solid Earth Planets 101 (8), 17535–17545.

5. Melting and early Earth history

Available online at www.sciencedirect.com
SCIENCE @ DIRECT®
ELSEVIER
Physics of the Earth and Planetary Interiors 143–144 (2004) 369–386
PHYSICS OF THE EARTH AND PLANETARY INTERIORS
www.elsevier.com/locate/pepi

Shock-induced superheating and melting curves of geophysically important minerals

Sheng-Nian Luo*, Thomas J. Ahrens

Lindhurst Laboratory of Experimental Geophysics, Seismological Laboratory, California Institute of Technology, Pasadena, CA 91125, USA

Received 5 February 2003; received in revised form 22 April 2003; accepted 23 April 2003

Abstract

Shock-state temperature and sound-speed measurements on crystalline materials, demonstrate superheating-melting behavior distinct from equilibrium melting. Shocked solid can be superheated to the maximum temperature, $T_{c'}$. At slightly higher pressure, P_c, shock melting occurs, and induces a lower shock temperature, T_c. The Hugoniot state, (P_c, T_c), is inferred to lie along the equilibrium melting curve. The amount of superheating achieved on Hugoniot is, $\Theta_H^+ = T_{c'}/T_c - 1$. Shock-induced superheating for a number of silicates, alkali halides and metals agrees closely with the predictions of a systematic framework describing superheating at various heating rates [Appl. Phys. Lett. 82 (12) (2003) 1836]. High-pressure melting curves are constructed by integration from (P_c, T_c) based on the Lindemann law. We calculate the volume and entropy changes upon melting at (P_c, T_c) assuming the $R \ln 2$ rule (R is the gas constant) for the disordering entropy of melting [J. Chem. Phys. 19 (1951) 93; Sov. Phys. Usp. 117 (1975) 625; Poirier, J.P., 1991. Introduction to the Physics of the Earth's Interior. Cambridge University Press, Cambridge, 102 pp.].

(P_c, T_c) and the Lindemann melting curves are in excellent accord with diamond-anvil cell (DAC) results for NaCl, KBr and stishovite. But significant discrepancies exist for transition metals. If we extrapolate the DAC melting data [Phys. Rev. B 63 (2001) 132104] for transition metals (Fe, V, Mo, W and Ta) to 200–400 GPa where shock melting occurs, shock temperature measurement and calculation would indicate $\Theta_H^+ \sim 0.7$–2.0. These large values of superheating are not consistent with the superheating systematics. The discrepancies could be reconciled by possible solid–solid phase transitions at high pressures. In particular, this work suggests that Fe undergoes a possible solid–solid phase transition at ~200 GPa and melts at ~270 GPa upon shock wave loading, and the melting temperature is ~6300 K at 330 GPa.

Keywords: Superheating; Melting curve; Shock waves; Sound-speed; Molecular dynamics

1. Introduction

Since the initial studies of dynamic compression of gabbro and dunite (Hughes and McQueen, 1958), it has been recognized that polymorphic, melting and freezing transitions in minerals often require excess pressure and temperature relative to the equilibrium phase boundary. In practice hysteresis has long been observed in many phase transformations (Taylor et al., 1991). Of particular interest is the superheating-melting behavior of shock-loaded crystalline materials. Shock melting experiments are useful for studying high-pressure melting of

* Corresponding author. Present address: P-24 Plasma Physics, Los Alamos National Lab, MS E526, Los Alamos, NM 87545, USA. Tel.: +1-505-664-0037; fax: +1-505-665-3552.
E-mail address: sluo@lanl.gov (S.-N. Luo).

0031-9201/$ – see front matter © 2004 Elsevier B.V. All rights reserved.
doi:10.1016/j.pepi.2003.04.001

geophysically and physically important materials. Shock-state sound-speed and temperature measurements with planar shock wave loading on silicates, alkali halides and metals have demonstrated the existence of superheating-melting behavior which is distinct from equilibrium melting. Superheating concept has been applied to interpret the sharp drops in temperature along Hugoniots (Lyzenga et al., 1983; Boness and Brown, 1993; Holland and Ahrens, 1997; Boness, 1999). Superheating has also been observed in intense laser irradiation (Williamson et al., 1984; Fabricius et al., 1986; Herman and Elsayed-Ali, 1992; Murphy et al., 1993). We attempt here to reconcile the issues related to the interpretation of temperature measurements in shock melting experiments, and whether the large discrepancies between the melting temperatures obtained from shock wave loading and recent diamond-anvil cell (DAC) (Errandonea et al., 2001) results for transition metals such as Fe, V, Mo, W, and Ta, can be simply attributed to the effect of superheating. Of special geophysical interest is the discrepancy between the shock melting data (Ahrens et al., 2002) and extrapolated DAC melting curve (Boehler, 1993) for iron.

Previously, there were no systematic theoretical investigations (either phenomenological or first-principles) directly addressing shock-induced superheating. The lack of a practical superheating theory is partly due to the difficulty in achieving superheating in low heating-rate experiments, where heterogeneous nucleation dominates. Heterogeneous nucleation is favored at free surfaces, grain boundaries and impurities, etc., which significantly lower the energy barriers for nucleation. Homogeneous nucleation of crystals by undercooling a liquid is more manageable experimentally and a body of data have been documented (Kelton, 1991), while appreciable superheating at low heating rates has been observed in very few cases using special experimental designs (Uhlmann, 1980).

Ultrafast heating (10^9 to 10^{13} K/s) is characteristic of dynamic melting experiments and is responsible for the observed superheating, as the temperature can rise faster than the rate of rearrangement of atoms required for melting. To quantify the amount of superheating and relate superheating to melting at high pressures, it is important to predict the maximum superheating achievable at heating rates comparable to those in shock wave loading. Previous theories proffered involve the catastrophes of entropy (Fecht and Johnson, 1988), rigidity and volume (Tallon, 1989) at melting. Other efforts to describe superheating utilized kinetic nucleation theory (Lu and Li, 1998; Rethfeld et al., 2002). These theories are not applicable because they either predict too high an upper bound to superheating or do not take heating rate into account.

Superheating is internally limited by the nucleation energy barrier and externally by the heating process. Recently we developed a framework of systematics to relate superheating to undercooling parameters, systematics of energy barriers for nucleation, and heating rates (Luo and Ahrens, 2003; Luo et al., 2003). In this paper we review our theory, survey the existing shock compression superheating data and discuss their implications to high-pressure melting of silicates, alkali halides and transition metals.

2. Superheating systematics

Our previous work on superheating and undercooling systematics (Luo and Ahrens, 2003; Luo et al., 2003) is based on homogeneous nucleation theory, undercooling experiments, and molecular dynamics simulations of melting and refreezing.

2.1. Homogeneous nucleation theory and systematics of energy barrier for nucleation

The nucleation rate per unit volume, I, for steady-state homogeneous nucleation, can be written as (Turnbull and Fisher, 1949; Christian, 1965; Walton, 1969; Motorin and Musher, 1984; Kelton, 1991)

$$I = g(m, T)\exp\left(-\frac{E_a}{kT}\right)\exp\left(-\frac{\Delta G_c}{kT}\right) \quad (1)$$

where $g(m, T)$ is a function of material properties (m) and temperature (T), E_a the activation energy, ΔG_c the critical Gibbs free energy for nucleation, and k the Boltzmann constant. Or, more simply (Rethfeld et al., 2002; Kelton, 1991)

$$I = M(m, T)\exp\left(-\frac{\Delta G_c}{kT}\right) \quad (2)$$

where M is a function of material properties m, such as the heat of fusion per unit volume ΔH_m, melting temperature T_m, solid–liquid interfacial energy

γ_{sl}, viscosity, heat capacity and critical size, etc. The T-dependence of $M(m, T)$ is negligible compared to the exponential ΔG_c term, thus $M(m, T)$ can be regarded as a constant, I_0 (Walton, 1969; Porter, 1981; Kelton, 1991). Consider a spherical liquid nucleus of critical radius r_c (the minimum size beyond which the energy of formation of nuclei decreases) inside a perfect crystalline lattice, the critical nucleation energy ΔG_c (Christian, 1965) is

$$\Delta G_c = \frac{16\pi\gamma_{sl}^3}{3(\Delta G_{sl} - E_\epsilon)^2} \quad (3)$$

where ΔG_{sl} is the Gibbs free energy difference per unit volume between solid and liquid state, approximated as

$$\Delta G_{sl} \approx \left(\frac{T}{T_m} - 1\right) \Delta H_m \quad (4)$$

assuming that the heat capacities of solid and liquid are equal (Turnbull, 1952; Fecht and Johnson, 1988; Kelton, 1991). E_ϵ is the strain energy due to the volume change between solid and liquid. We assume $E_\epsilon = 0$ (the effect of E_ϵ is small). I is often reduced to the form (Walton, 1969; Porter, 1981; Kelton, 1991)

$$I = I_0 \exp\left\{-\frac{16\pi\gamma_{sl}^3 T_m^2}{3\Delta H_m^2 kT(T - T_m)^2}\right\} \quad (5)$$

We define the energy barrier for nucleation, β, as a dimensionless quantity (Luo and Ahrens, 2003)

$$\beta(\gamma_{sl}, \Delta H_m, T_m) = \frac{16\pi\gamma_{sl}^3}{3\Delta H_m^2 kT_m} \quad (6)$$

and introduce reduced temperature, $\theta = T/T_m$. Thus $I = I_0 f(\beta, \theta)$ with

$$f(\beta, \theta) = \exp\left\{-\frac{\beta}{\theta(\theta - 1)^2}\right\} \quad (7)$$

The prefactor I_0 can be obtained experimentally or theoretically. Nucleation is essentially controlled by $f(\beta, \theta)$, namely, by the dimensionless quantity β at given temperature. The form of $f(\beta, \theta)$ is simple, but it reflects the physics of nucleation. As shown in Fig. 1, during superheating of solids, f (i.e., I/I_0) increases with temperature monotonically, as the mobility of atoms and the chemical driving force for melt nucleation both increase with T. Upon undercooling of liquids, the chemical driving force induced by undercooling is partly offset by the decrease in mobility, thus there is a maximum for f at $\theta = 1/3$. Although diffusion processes are not explicitly included in f, they were accounted for, by the functional form of f.

The dimensionless quantity β is characteristic of a material and depends on γ_{sl}, ΔH_m and T_m. An estimate the value of β can be obtained from hard sphere

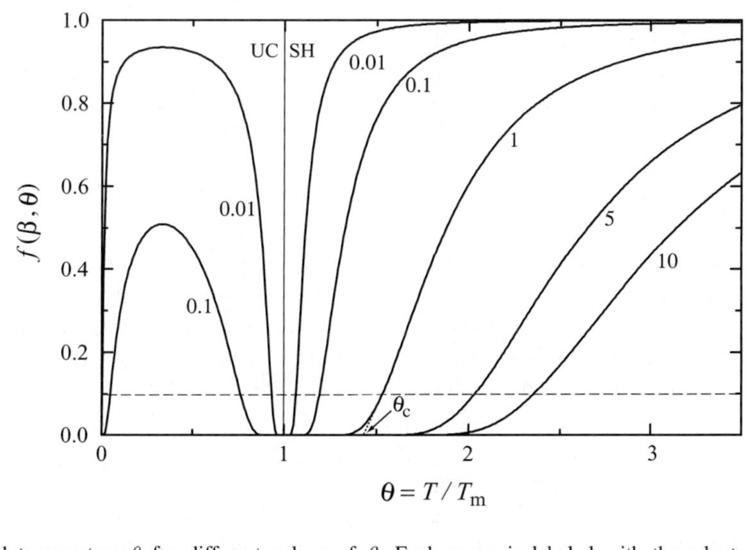

Fig. 1. $f(\beta, \theta)$ vs. reduced temperature θ for different values of β. Each curve is labeled with the adopted value of β. UC denotes undercooling and SH superheating. θ_c is the critical temperature obtained from the tangent method.

Table 1
Normalized energy barrier β for elements, alkali halides and silicates at ambient pressure (Luo et al., 2003)

Ag	1.67	Ni	2.63	Al	1.47	LiF	2.05	CsF	1.27
Au	1.48	Pd	1.66	Bi	5.83	LiCl	2.05	CsCl	1.41
Cd	1.64	Pt	1.73	Ga	8.15	LiBr	0.63	CsBr	1.56
Co	1.69	Rh	1.99	Ge	4.87	NaF	2.22	CsI	2.40
Cu	1.51	Ru	0.86	In	3.04	NaCl	1.27	Albite	1.52
Fe	3.11	Ta	1.96	Pb	5.87	NaBr	1.27	Quartz	5.01
Hf	1.62	Ti	1.45	Sb	2.48	KCl	1.27		
Hg	6.30	V	1.95	Se	0.20	KBr	1.27		
Mn	2.12	W	1.55	Sn	5.34	KI	1.27		
Mo	1.31	Zr	2.00	Te	4.43	RbCl	1.27		

systems (hss) which are simple and yet have theoretical significance. Let the diameter of a hard sphere be σ. The solid–liquid interfacial energy for a hard sphere (γ_{sl}^{hss}), is shown to be $0.61kT/\sigma^2$ (Davidchack and Laird, 2000). Heat of fusion per unit volume for hss, is $\Delta H_m^{hss} = 1.16kT/V^{hss}$, where $V^{hss} = \pi\sigma^3/6$ (Hoover and Ree, 1968). Thus, $\beta \sim 0.77$ for a hss. We may expect for a real system, the value of β is of a similar magnitude. For materials of interest, $\gamma_{sl} \sim 0.1\,J/m^2$, $T_m \sim 10^3\,K$ and $\Delta H_m \sim 10^9\,J/m^3$ (Kelton, 1991) yields $\beta \sim 1.2$. The elements studied in undercooling experiments (Kelton, 1991; Shao and Spaepen, 1996) include transition metals, simple metals and semiconductors in the third to sixth row of the periodic table. The systematics for elements may be valid for materials in general. The values of β for these elements are calculated (Table 1 and Fig. 2). Transition metals (Group IVB–IIB) have relatively constant value of β with an average close to 2, except for Hg ($\beta = 6.30$). Group IIA–VIA elements have significantly higher values of β, with maximum at Ga ($\beta = 8.15$) except for Al ($\beta = 1.47$) and Se which has the lowest $\beta = 0.20$. These variations largely reflect the periodicity of electronic structure of elements. As γ_{sl} increases with T_m and ΔH_m (Luo et al., 2003), the variations in β are relatively small from material to material. It is demonstrated that the quantity $\gamma_g/\Delta H_m V$ is nearly a constant for a variety of elements and compounds

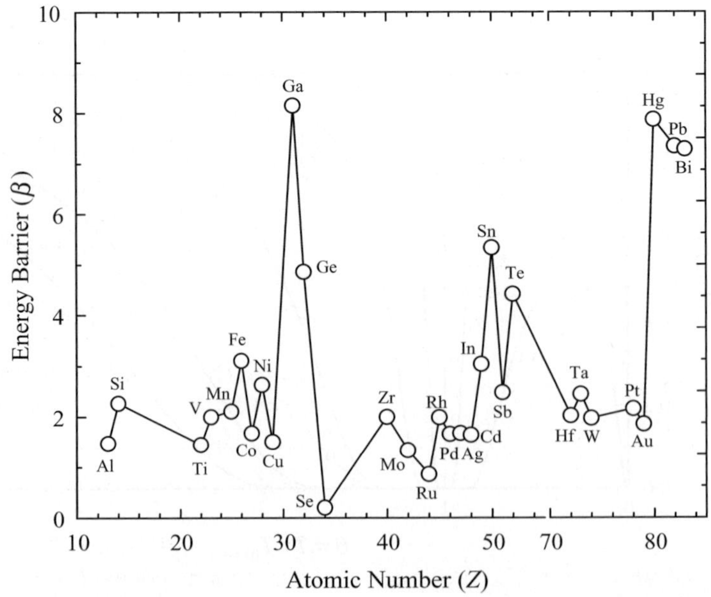

Fig. 2. Systematics of β for elements (β vs. Z) (Luo and Ahrens, 2003).

(Kelton, 1991; Uhlmann, 1980). The gram-interfacial energy is $\gamma_g = \gamma_{sl} V^{2/3} N_A^{1/3}$, where V is the molar volume and N_A the Avogadro's number. As $\Delta H_m = T_m \Delta S_m$, where ΔS_m is the entropy change upon melting, thus from its definition β is proportional to $\Delta S_m V$, i.e., to the entropy of melting per mole atoms ($\sim R$). Therefore β should vary over a narrow range for different materials. The same result can be obtained from the argument that $\gamma_{sl} \propto \Delta H_m V/a^2$, where a^2 is the effective atomic surface area (Walton, 1969). As the entropy of melting (per mole atoms) at high pressures is not significantly different from that at ambient pressure (see Section 4), we expect that β for most solids under high pressures (e.g., megabar pressure under shock compression) should be of the same order as at ambient pressure. Moreover, molecular dynamics simulations of superheating and undercooling of Al at 0–100 GPa demonstrate a weak pressure-dependence of β (Luo et al., 2003). Values of β for alkali halides and silicates are also listed in Table 1.

2.2. The maximum amount of superheating

The limit of superheating was investigated by assuming an absolute value of $I = 1\,\text{cm}^{-3}\,\text{s}^{-1}$ (Lu and Li, 1998) for some metals. Similarly, assuming the nucleation time $t_M = (IV_c)^{-1}$, where $V_c = (4/3)\pi r_c^3$, the critical volume of a nucleus with critical radius r_c, the maximum superheating can be estimated at given nucleation time scale (referred to as the V_c method) (Rethfeld et al., 2002). Consider the normalized nucleation rate, $I/I_0 = f(\theta, \beta)$ (Fig. 1). For nucleation of melt in a superheated solid, virtually no nuclei are formed until a critical temperature T_c (or $\theta_c = T_c/T_m$) is reached at which point a catastrophic nucleation occurs (Porter, 1981). It is similar for nucleation of crystal in an undercooled liquid. θ_c^+ or $\Theta^+ = \theta_c^+ - 1$ refers to the maximum superheating, and θ_c^- or $\Theta^- = 1 - \theta_c^-$ to the maximum undercooling. (Superscript + denotes superheating and − undercooling.) First we construct a tangent of $f(\beta, \theta)$ at a certain f for a given β, say, $f_0 = 0.1$ (the dashed line in Fig. 1). The tangent line with a slope

$$\left.\frac{\partial f}{\partial \theta}\right|_{\theta=\theta(f_0,\beta)} \tag{8}$$

intersects the θ axis at θ_c (dotted line in Fig. 1). This then yields a simple treatment of the maximum superheating (or undercooling). This method (referred to as tangent method) is easy to visualize, but its difficulty lies in the arbitrariness of choosing f_0 which is not defined from first-principles yet.

Given the β systematics, next we develop a Q-dependent (Q is the heating or cooling rate) scheme to probe the systematics of superheating characteristic of elements and compounds. The undercooling experiments on various elements and compounds were documented (Kelton, 1991), and serve the basis for our calculating the superheating at various heating rates. The parameters for undercooling such as γ_{sl}, ΔH_m, T_m and molar volume V can be regarded as equal to those for superheating. In the case of superheating, the probability (Kelton, 1991) x of v moles of parent phase containing no new phase (liquid) under heating rate Q^+ is

$$x = \exp\left\{-\frac{vT_m I_0 V}{Q^+}\int_1^{\theta_c^+} f(\beta, \theta)\,\mathrm{d}\theta\right\} \tag{9}$$

and the expression is similar for the undercooling case. If we assume I_0, v, and x for superheating are similar to those for undercooling, the maximum superheating temperature θ_c^+ can be estimated from the documented θ_c^- using Eq. (9). Given undercooling θ_c^- (Kelton, 1991) under typical cooling rate $Q^- = 1$ K/s, θ_c^+ under certain heating rate Q^+ can be obtained by solving

$$\frac{1}{Q^-}\int_{\theta_c^-}^{1} f(\beta, \theta)\,\mathrm{d}\theta = \frac{1}{Q^+}\int_1^{\theta_c^+} f(\beta, \theta)\,\mathrm{d}\theta \tag{10}$$

For superheating, we consider heating rates between $Q^+ = 1$ K/s (corresponding superheating denoted as Θ_s^+) and 10^{12} K/s (Θ_{ns}^+). These heating rates should be regarded as typical but not exact, because a factor of 10^2 difference in Q would yield a negligible difference in θ given a reasonable value of $f(\beta, \theta)$. θ_c^+ at typical heating rates is calculated as shown in Fig. 3 for elements along with experimental values for θ_c^-. The numerical relationship between β and θ_c at given Q can be fitted with a simple analytical form (Luo et al., 2003)

$$\beta = (A_0 - b\log Q)\theta_c(\theta_c - 1)^2 \tag{11}$$

where $A_0 = 59.39$, $b = 2.33$ and Q is normalized by 1 K/s. This functional form applies to both superheating and undercooling. Thus the material property β,

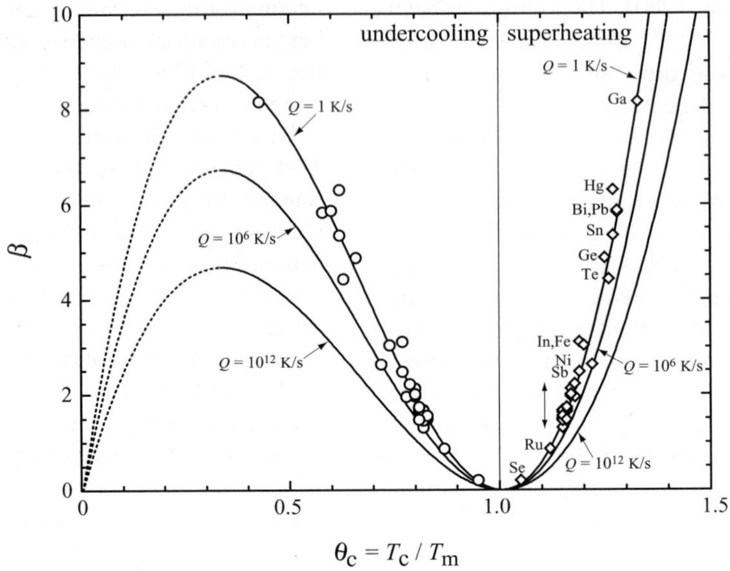

Fig. 3. Maximum undercooling and superheating (β vs. θ_c) of elements under various heating (cooling) rates Q. θ_c in superheating regime is calculated with Eq. (10). Note that there are a maximum for β and a minimum for θ_c (maximum undercooling) in the undercooling regime for each Q. Both superheating and undercooling under certain Q are described in Eq. (11). Solid curves are plotted with different values of Q, and dotted curves denote the non-physical portion of the undercooling. The elements within the double-headed arrow are Ti, Al, Au, Cu, Hf, Cd, Pd, Ag, Co, Pt, Ta, Rh, Zr, Mn, Si, Sb, Ni, In and Fe in β-increasing order. After Luo et al. (2003).

the maximum superheating (undercooling), and heating (cooling) process are related via Eq. (11). The similarity between Eqs. (7) and (11) simply indicates a catastrophic behavior of nucleation near θ_c.

The limit for superheating Θ^+ is inherently limited by the material property β, and increases monotonically with β and Q (Fig. 3). In general, the energy barrier β is in the range of 0.2–8.2 (Table 1). For $Q = 1$ to 10^{12} K/s, Θ^+ varies in a narrow range of 0.05–0.43 for all elements considered, with maximum at Ga and minimum at Se. A change in heating rate from 1 to 10^{12} K/s induces an increase of less than 10% in the maximum superheating (θ_c). Thus, the effect of Q is more important in achieving superheating experimentally (Section 3) than achieving larger superheating. Group IIIA–VIA elements (except Se) can be superheated more than transition elements (except Hg). Transition metals such as Fe, V, Ta, Mo and W can be superheated by $0.30T_m$ at most (we expect that β at high pressures, does not vary much as argued before). At similar heating and cooling rates, the maximum undercooling (Θ^-) is larger than maximum superheating (Θ^+).

A partial verification of the above superheating systematics can be achieved through molecular dynamics simulations. The heating (and cooling) rate in typical MD simulations is $\sim 10^{12}$ K/s. For a perfect supercell with three-dimensional periodic boundaries, superheating has long been observed in MD simulations of melting (Allen and Tildesley, 1987). There are few systematic studies addressing superheating previous to Luo et al. (2003). MD simulations were conducted for f.c.c. metals and Be (h.c.p.) with quantum-corrected Sutton–Chen many-body potentials and single- and two-phase simulation techniques (Luo et al., 2003). The crystal melts at superheated temperature $T_{1,m}$ in single-phase simulations, and the equilibrium melting temperature is obtained from solid–liquid coexisting phase (two-phase) simulations as $T_{2,m}$. Thus, superheating achieved in MD is $\Theta^+_{md} = T_{1,m}/T_{2,m} - 1$, and listed in Table 2. The simulations agree closely with the prediction (Θ^+_{ns}) of the superheating systematics. It is also worth mentioning that both superheating and undercooling achieved in these MD simulations (Luo et al., 2003), are in accord with the superheating–undercooling systematics

Table 2
Molecular dynamics simulations (Luo et al., 2003) of closed-packed metals (except Ta) with quantum-corrected Sutton–Chen potential at $P = 0$

	$T_{1,m}$ (K)	$T_{2,m}$ (K)	T_m^{exp} (K)	Θ^+_{md}	Θ^+_{ns}
Be	1600	1350	1560	0.19	–
Al	1100	925	933	0.19	0.20
Ni	1700	1375	1728	0.24	0.26
Cu	1350	1070	1356	0.21	0.19
Rh	2700	2125	2239	0.27	0.23
Pd	1850	1475	1825	0.25	0.21
Ag	1200	1000	1234	0.20	0.21
Ir	3400	2740	2683	0.24	–
Pt	2450	1925	2042	0.27	0.21
Au	1400	1075	1336	0.30	0.20
Pb	700	575	601	0.22	0.37
Ta	3650	3150	3253	0.15	0.23

T_m^{exp} is the experimental value of T_m. Ta (b.c.c.) results are from Strachan et al. (2001).

(Eq. (11)). In MD simulations of high-pressure melting for stishovite, ϵ-Fe and the Lennard-Jones system with different potentials, similar values of maximum superheating were observed (Belonoshko et al., 2000; Luo et al., 2002a, 2003, 2004a).

Given the above methods of estimating maximum superheating, we compare their predictions. The melting theories based on catastrophe of entropy (Fecht and Johnson, 1988), catastrope of rigidity and volume (Tallon, 1989) predict a wide range of superheating of $\Theta^+ = 0.2$–2.0 for crystalline solids (Lu and Li, 1998). In particular for Al, Θ^+ is 0.38, 0.29 and 0.24, for entropy, rigidity and volume catastrophe respectively, while by assuming $I = 1\,\mathrm{cm^{-3}\,s^{-1}}$ and homogeneous nucleation, $\Theta^+ = 0.21$ (Lu and Li, 1998). For Al ($\beta = 1.47$), the tangent method predicts $\Theta^+ = 0.51$ with $f_0 = 0.1$ and $\Theta^+ = 0.16$ with $f_0 = 10^{-20}$. Note that these methods do not involve heating rates. The V_c method predicts $\Theta^+ = 0.34$ for Al with $t_M = 1$ ps, and the superheating systematics predict $\Theta^+ = 0.16$ and 0.23 at $Q = 1$ and 10^{12} K/s, respectively. The present systematics based on undercooling experiments and β-systematics define tightest upper bounds for superheating and incorporate effects of heating rates.

3. Superheating achieved in ultrafast experiments

Both diamond-anvil cell experiments and shock wave loading can be useful tools to study the melting behavior under extreme conditions. The temperature achievable and measurable in high-pressure DAC melting experiments, is limited by laser-heating techniques. Heating in DAC experiments is essentially steady-state. In contrast, pressure and temperature are raised simultaneously on ultrashort time scales upon shock compression. The typical risetime of a strong shock front for planar impact is of the order of nanosecond (ns), and the temperature increase is $\sim 10^3$ K, thus the effective heating rates Q is $\sim 10^{12}$ K/s. Q is similar in intense laser irradiation experiments, depending on energy deposited, irradiation time and material properties. We note that in shock wave experiments, solids can be heated at rates higher than those required for rearrangement of atoms upon melting. During shock wave loading, solids are heated inside as the shock front advances, thus surface melting can be suppressed. Kinetics inherent in melting may play an important role at these time scales, thus inducing superheating.

Fig. 4 shows the schematic diagram of the two distinct types of melting behavior upon shock wave loading: equilibrium and non-equilibrium melting. For equilibrium melting, the $T-P$ Hugoniot states lie on

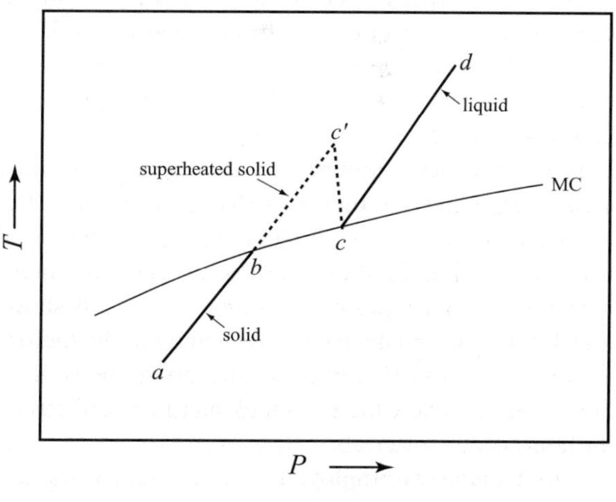

Fig. 4. Schematic of equilibrium and non-equilibrium melting upon shock wave loading. MC is the equilibrium melting curve and assumed to have positive dT_m/dP slope. For equilibrium melting, the $T-P$ Hugoniot states lie on abcd, where bc coincides with MC. The Hugoniot represented by abc′cd indicates that shocked solid is superheated above MC. As P increases, T in successive Hugoniot states reaches a maximum ($T_{c'}$) and then drops to T_c upon melting near P_c. The maximum superheating on the Hugoniot with non-equilibrium melting, is $\Theta^+_H = T_{c'}/T_c - 1$.

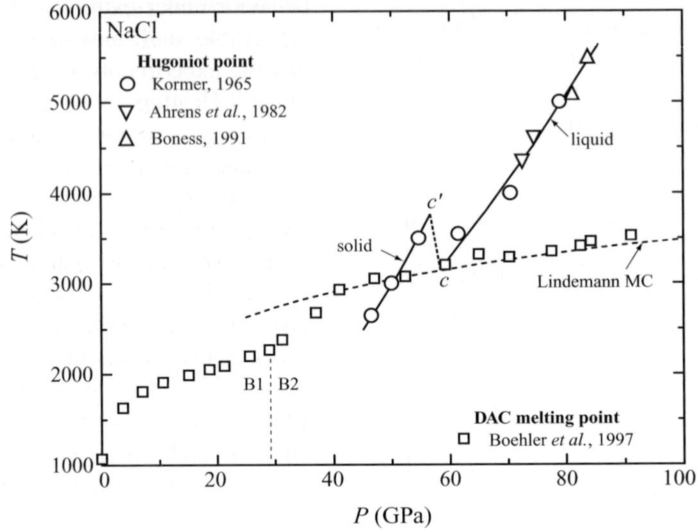

Fig. 5. Shock-state temperature and DAC melting point measurements for NaCl.

abcd where bc coincides with the equilibrium melting curve. The Hugoniot represented by abc′cd indicates that shocked solid is superheated above the melting curve. As P in successive Hugoniot states increases, T reaches a maximum ($T_{c'}$) and then drops to T_c upon melting near P_c. The maximum superheating on the Hugoniot with non-equilibrium melting, is $\Theta_H^+ = T_{c'}/T_c - 1$. The reason for the preconception of equilibrium melting on Hugoniot is two-fold: equilibrium thermodynamics was assumed previously for convenience, and more importantly, the pioneering shock temperature measurements on NaCl (Kormer, 1965) demonstrate sharp $T - P$ slope changes between 50 and 65 GPa (Fig. 5), thus appeared to lend experimental support to the equilibrium interpretation. As shown in the following analysis of static and dynamic melting experiments, superheating-melting upon shock loading is a dominant feature in shock melting experiments of non-porous materials.

The techniques employed to detect shock melting are sound-speed and temperature measurements at the shock states. Upon melting, the sound-speed drops from longitudinal to bulk sound-speed due to the loss of rigidity. Similarly, the shock temperature in successive Hugoniot states drops as the solid melts due to latent heat (e.g., Fig. 6). Other techniques such as transient electron diffraction are important diagnostics to detect melting from the loss of long-range order within the shocked crystal (Williamson et al., 1984).

Shock melting experiments have been conducted on alkali halides (NaCl, CsBr and KBr), silicates (fused

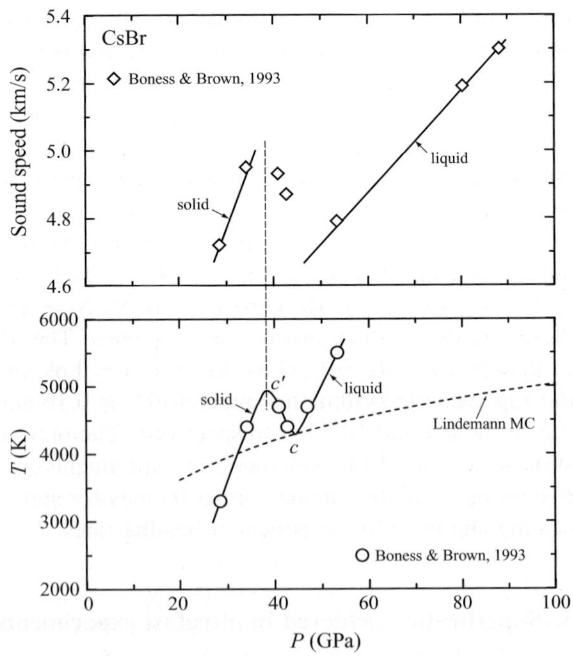

Fig. 6. Shock-state sound-speed and temperature measurements for CsBr. The thin dashed line denotes the onset of melting.

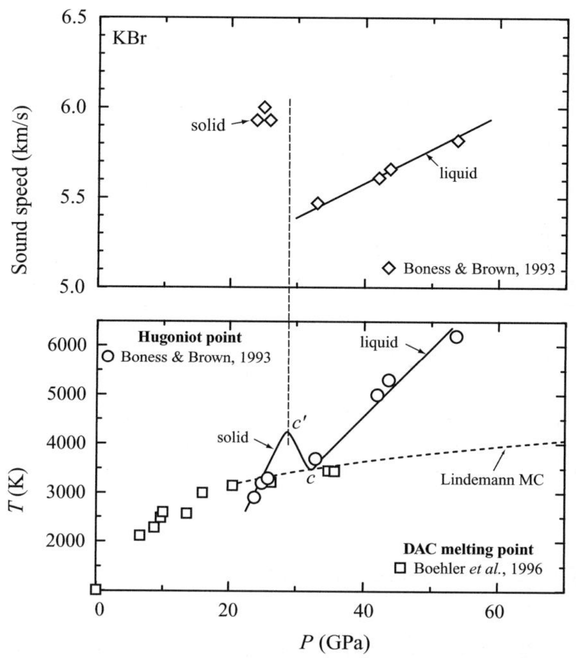

Fig. 7. Shock-state sound-speed and temperature, and DAC melting point measurements for KBr. The thin dashed line denotes the onset of melting.

quartz, quartz and forsterite), and transition metals (Fe, V, Mo, Ta). Alkali halides are advantageous for studying melting behavior due to their optical transparency and relatively low melting temperatures upon shock wave loading. The $P - T$ conditions on Hugoniots are accessible in DAC apparatus. Simultaneous measurements of shock-state sound-speed and temperature for CsBr and KBr (Boness, 1991; Boness and Brown, 1993) are shown in Figs. 6 and 7. The simultaneous drop in sound-speed and temperature at ~38 GPa (CsBr) and ~29 GPa (KBr) signals melting. The amount of superheating is estimated to be 0.16 and 0.20 for CsBr and KBr, respectively. If we assume the energy barrier β of the high-pressure B2 phases of CsBr and KBr to be the same as ambient values, the superheating systematics predict superheating (Θ_{ns}^{+}) of 0.20 and 0.18 respectively with $Q = 10^{12}$ K/s (Table 3). (P_c, T_c) is assumed to locate on the equilibrium melting curve, and it coincides with the DAC results (Boehler et al., 1996) for KBr. We also construct a Lindemann melting curve (discussed in next section) integrated from (P_c, T_c). The static data agree almost exactly with the Lindemann curve. For NaCl, we interpret the shock wave data (Ahrens et al., 1982; Kormer, 1965) as non-equilibrium melting (Fig. 5), and the estimated superheating agrees with the superheating systematics. The constructed Lindemann curve also agrees exactly with the DAC results. Thus, by considering superheating, we in principle can obtain one point, (P_c, T_c), on the melting curve.

Similar features have been observed in shock-state sound-speed and/or temperature measurements on other materials. These measurements include sound-speed and temperature measurement on fused quartz, quartz (Lyzenga et al., 1983; McQueen, 1992) (Figs. 8a and 9), Fe (Brown and McQueen, 1982, 1986; Williams et al., 1987; Yoo et al., 1993) (Fig. 10) and V (Dai et al., 2001) (Fig. 11), temperature measurement on forsterite (Holland and Ahrens, 1997) (Fig. 8b),

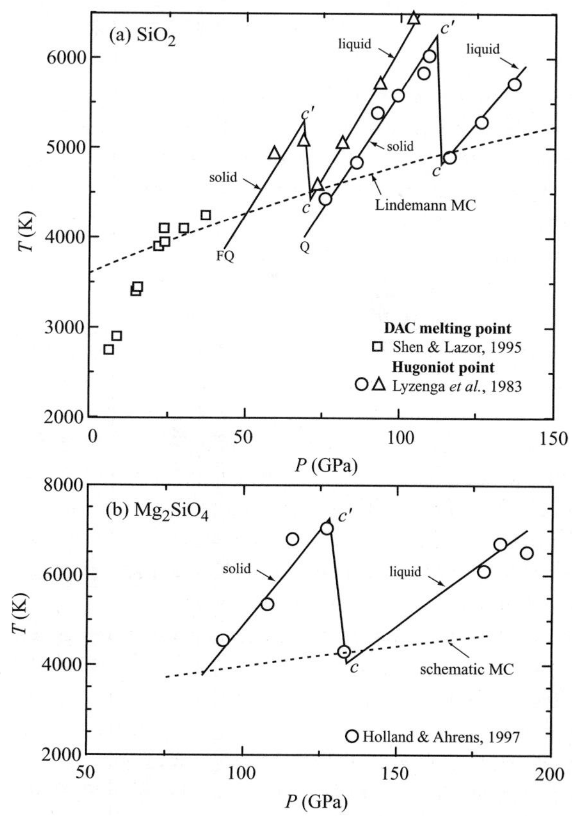

Fig. 8. Shock-state temperature measurements for fused quartz and quartz (a) and forsterite (b). The DAC measurement on silica melting curve is also shown in (a). The Lindemann MC is constructed from (P_c, T_c) on the principal Hugoniot of fused quartz (FQ), and that from quartz Hugoniot (Q) is about 150 K lower.

Table 3
The maximum superheating achieved in shock wave loading, and the deduced melting parameters based on the Lindemann law and $R \ln 2$ rule

Starting phase	Ending phase	Superheating		T_c (K)	P_c (GPa)	ρ_0 (g/cm^3)	s[a]	C_0[a] (km/s)	γ_0[a]	q[b]	dT_m/dP[c] (K/GPa)	$\Delta V/V$[c] (%)	$\Delta S(R)$[c]
		Θ^+_H	Θ^+_{ns}										
NaCl (B1)	B2	0.20	0.18	3125	57.8	2.232	1.22	3.62	0.77	1.0	10.88	0.85	0.704
KBr (B1)	B2	0.20	0.18	3500	33.0	3.071	1.20	3.14	0.75	1.0	21.92	1.02	0.708
CsBr (B2)	B2	0.16	0.20	4260	43.0	4.440	1.25	2.25	0.81	1.0	22.09	0.91	0.705
SiO_2 (fused)	Rutile	0.17	–	4500	70.0	4.310	1.08[d]	12.25[d]	1.35[e]	2.6[e]	10.87	1.58	0.744
SiO_2 (quartz)	Rutile	0.30	–	4800	113.0	4.310	1.24[d]	9.66[d]	1.35[e]	2.6[e]	9.33	1.47	0.730
Fe (b.c.c.)	Unknown	0.25	0.23	5800	270.0	8.280[f]	1.47[f]	4.64[f]	1.70[g]	1.0[g]	8.53	1.19	0.732
V (b.c.c.)	Unknown	0.24	0.26	6150	220.0	6.110	1.23[h]	5.06[h]	2.00[i]	1.0	14.98	1.83	0.761

[a] Obtained from finite strain, linear U_S–u_p relation (Jeanloz and Grover, 1987) unless stated otherwise.
[b] Assumed to be 1 unless stated otherwise.
[c] Evaluated at (P_c, T_c).
[d] Metastable Hugoniot centered on the high-pressure phase.
[e] Luo et al. (2002b).
[f] Brown and McQueen (1982).
[g] Brown (2001).
[h] Gathers et al. (1983).
[i] Dai et al. (2001).

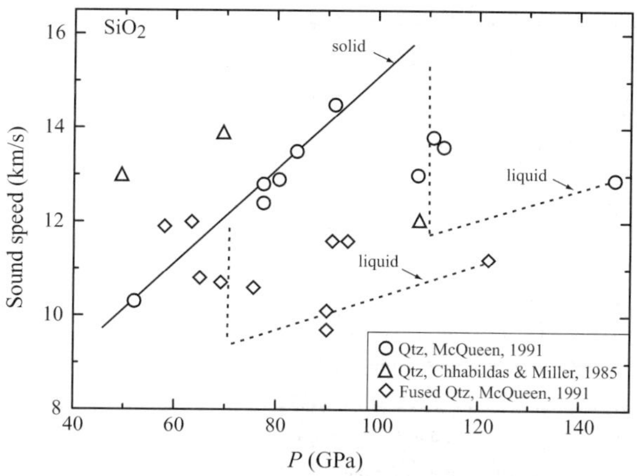

Fig. 9. Shock-state sound-speed measurements for fused quartz and quartz. The vertical dotted lines at ~70 (on the principal Hugoniot of fused quartz) and 110 GPa (on the principal Hugoniot of quartz) indicate melting of stishovite (McQueen, 1992).

and sound-speed measurement on Mo (Hixson et al., 1989) (Fig. 12) and Ta (Shaner et al., 1984) (Fig. 13). The superheating achieved on these Hugoniots (except for forsterite) is in reasonable agreement with the predictions of the superheating systematics (Table 3). The superheating achieved on the principal Hugoniot of forsterite is estimated to be $\Theta_H^+ = 0.63$ which corresponds to $\beta \sim 20$, an energy barrier exceptionally higher than those for any other material tested so far. Although we argued that β does not vary significantly at high pressures, we cannot rule out the possibility of $\beta \sim 20$ for forsterite (or new phases from decomposition) under shock compression. Theoretical efforts (e.g., MD simulations) in resolving variations of β (essentially γ_{sl}) with pressure, are of immediate necessity. Other causes could also well account for such a large superheating, such as underestimated emissivity (Luo et al., 2004b).

Intense laser irradiation experiments have been conducted on Al (Williamson et al., 1984), Pb(1 1 1) (Herman and Elsayed-Ali, 1992), Bi(1 0 0 0) (Murphy et al., 1993) and GaAs (Fabricius et al., 1986), and the superheating achieved compares favorably to the present predictions of the superheating systematics.

The shock melting experiments demonstrate a common feature of superheating with sharp drops in achieved shock temperatures, a behavior deviating from equilibrium melting. For complete melting of the solid (in the shock front) near P_c, the drop in T relative to T_m due to the latent heat, can be approximated to the first order as $\Delta S_m/C_p \sim 1/3$. Here ΔS_m is the heat of fusion per mole atoms and assumed to be $\sim R$, and C_p is $\sim 3R$. This temperature excess is comparable to the predictions of the superheating systematics, and the superheating achieved along Hugoniot assuming that (P_c, T_c) lies along the equilibrium melting curve. The amount of maximum superheating achieved is sufficient not only to initiate the melting at certain Q, but also to compensate latent heat for complete melting, although these two processes are different physically. Thus, such an interpretation of shock melting experiments is reasonable, and an important point on the high-pressure melting curve, (P_c, T_c) can be resolved.

4. Melting parameters induced from shock wave experiments

Given (P_c, T_c) from shock melting experiments, next we deduce other important thermodynamic parameters of melting, such as the melting curve, changes in entropy and volume upon melting at high pressures. In the absence of first-principles theories, the Lindemann law and the $R \ln 2$ rule (systematics of the entropy of melting) are adopted.

4.1. The R ln 2 rule for entropy of melting

First we briefly examine the validity of the $R \ln 2$ rule. Previously, efforts have been made to establish the systematics of the entropy of melting (Oriani, 1951; Stishov, 1975; Tallon and Robinson, 1982; Poirier, 1991). The entropy of melting per mole atoms, ΔS_m, can be written as

$$\Delta S_m = \Delta S_D + \Delta S_V \quad (12)$$

where ΔS_V is the entropy of fusion associated with volume and vibrational changes, and ΔS_D the entropy of disorder (Oriani, 1951). Later, a $R \ln 2$ rule was proposed (Stishov, 1975; Tallon and Robinson, 1982) for simple atomic and some binary and ternary substances, as

$$\Delta S_m = R \ln 2 + \alpha K \Delta V_m \quad (13)$$

where R is the gas constant, α the thermal expansivity of volume, K the bulk modulus, and ΔV_m the volume

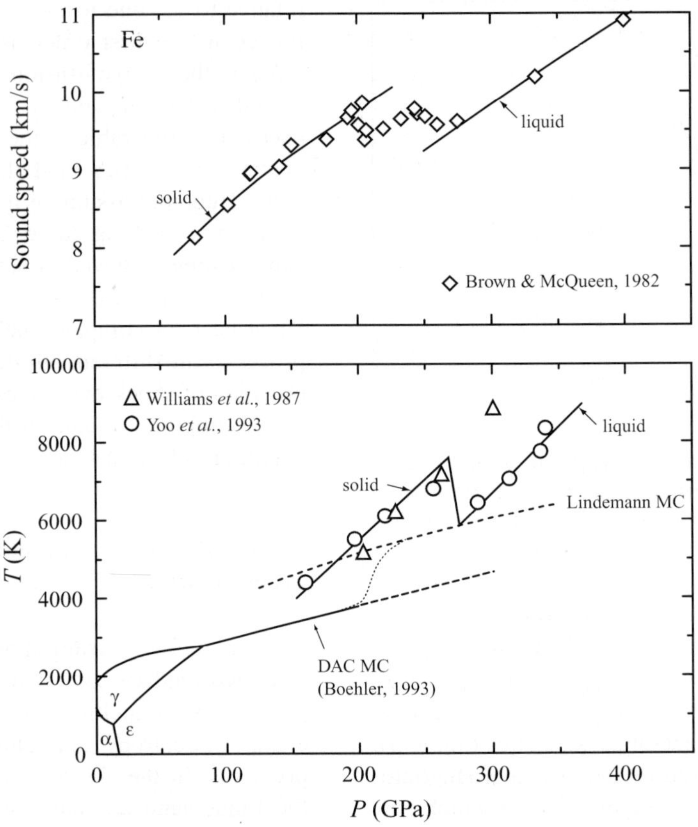

Fig. 10. Shock-state sound-speed and temperature, and DAC melting point measurements for Fe. The dotted curve is the speculated melting curve connecting ϵ-iron and a possible unknown solid phase.

change upon melting. Comparing Eqs. (12) and (13) yields that ΔS_D is essentially $R \ln 2$, although Oriani (1951) used a different approach to quantify disordering upon melting. The $R \ln 2$ rule was justified for atomic systems with a simple cell model (Stishov, 1975) and by calculating the topological entropy (Rivier and Duffy, 1982).

The $R \ln 2$ rule has been checked against limited data and the universality is not well established. We take the entropies for elements (Table 4) (Oriani, 1951) and plot ΔS_m versus ΔS_V. As shown in Fig. 14, the $R \ln 2$ rule applies as expected. Note that there are exceptions, e.g., Ga (not plotted). The behavior of silicates is of particular interest. For the five silicates investigated (Table 5) (Poirier, 1991), the data points of forsterite, fayalite, pyrope and enstatite cluster around the $R \ln 2$ line, except quartz for which ΔS_D is essentially 0 (Fig. 14). An appreciable amount of data for alkali halides was collected in Table 6 (Jackson, 1977). Entropies for KX (X = F, Cl, Br and I) agree

Table 4
The $R \ln 2$ rule: entropy of melting for metals (values of entropies are from Oriani (1951) and normalized by R)

	Ag	Al	Au	Cu	Pb	Li	Na	K	Cs	Fe	Cd	Mg	Sn
$\Delta S_m(R)$	1.102	1.378	1.142	1.152	1.026	0.770	0.855	0.855	0.830	0.991	1.237	1.177	1.715
$\Delta S_V(R)$	0.407	0.528	0.654	0.367	0.397	0.050	0.100	0.099	0.091	0.100	0.513	0.538	0.774
$\Delta S_D(R)$	0.695	0.850	0.488	0.785	0.629	0.720	0.755	0.756	0.739	0.891	0.724	0.639	0.941

Table 5
The $R \ln 2$ rule: entropy of melting per mole atoms for silicates (n is the number of atoms per formula)

Silicate	n (cm^3/g)	ΔV_m [a] (cm^3/g)	$\Delta V_m/V$ [a] (cm^3/g)	α [b] ($\times 10^{-6} K^{-1}$)	K [c] (GPa)	$\Delta S_m(R)$	$\Delta S_V(R)$	$\Delta S_D(R)$
Quartz (SiO_2)	3	2.50	0.105	24.3	37.8	0.221	0.092	0.129
Forsterite (Mg_2SiO_4)	7	3.80	0.081	30.6	129.5	1.203	0.259	0.944
Fayalite (Fe_2SiO_4)	7	3.70	0.076	26.1	134.0	1.047	0.222	0.825
Pyrope ($Mg_3Al_2Si_3O_{12}$)	20	18.3	0.158	19.9	172.8	0.975	0.378	0.597
Enstatite ($MgSiO_3$)	5	5.40	0.160	24.1	107.8	0.989	0.337	0.652

[a] Poirier (1991).
[b] Fei (1995).
[c] Bass (1995).

with the $R \ln 2$ rule, while ΔS_D is 0 or negative for LiX. The values of ΔS_D for NaX lie between the $R \ln 2$ and 0 line (Fig. 14). It is not surprising if we consider the possible large differences in disordering upon melting. Instead of a strict $R \ln 2$ rule, we propose that

$$\Delta S_m = xR \ln 2 + \alpha K \Delta V_m \quad (14)$$

where $0 \leq x \leq 1$, although by eliminating large uncertainties in calculating ΔS_V (e.g., NaX), data points may cluster near the $x = 1$ and 0 lines. The structure-dependence of x, should be investigated in a fundamental way and is an important issue for future investigation.

For the materials investigated in shock melting experiments (essentially, melting of high-pressure phases of Fe, V, KBr, CsBr, NaCl and silica), we assume that the $R \ln 2$ rule applies ($x = 1$) as there is no apparent contradiction.

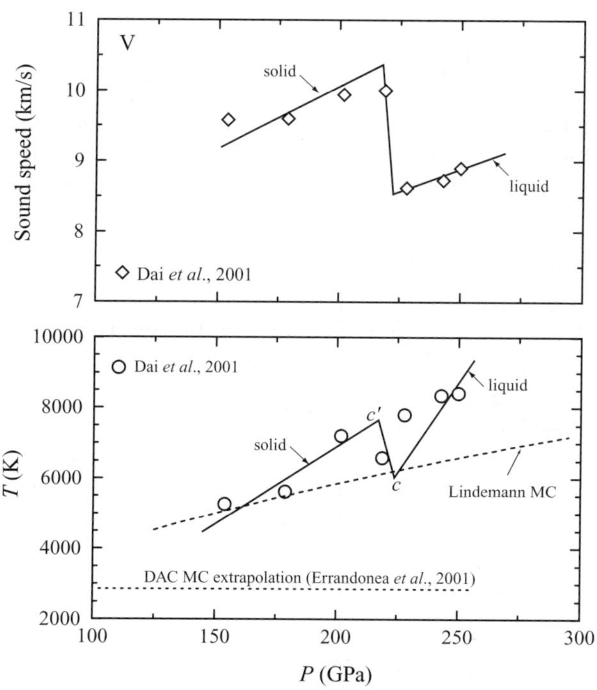

Fig. 11. Shock-state sound-speed and temperature measurements for V. DAC melting data of V (Errandonea et al., 2001) is extrapolated to higher pressures.

4.2. High-pressure melting parameters based on the Lindemann law and $R \ln 2$ rule

Now we deduce the parameters of high-pressure melting which are implied by the previously discussed shock temperature, and sound-speed data for elements, alkali halides and minerals.

Eq. (13) can be written alternatively as

$$\Delta S_m = R \ln 2 + \gamma C_V \frac{\Delta V_m}{V} \quad (15)$$

where γ is the Grüneisen parameter and C_V the specific heat per mole atom. Both Eqs. (13) and (15) will be used for evaluating ΔS_m with either γC_V or αK. ΔS_m is related to ΔV_m via the Clausius–Clapeyron equation

$$\Delta V_m = \frac{dT_m}{dP} \Delta S_m \quad (16)$$

The slope dT_m/dP along melting curve can be calculated using the Lindemann law (Stacey and Irvine, 1977)

$$\frac{d \ln T_m}{dP} = \frac{2\gamma}{K} \quad (17)$$

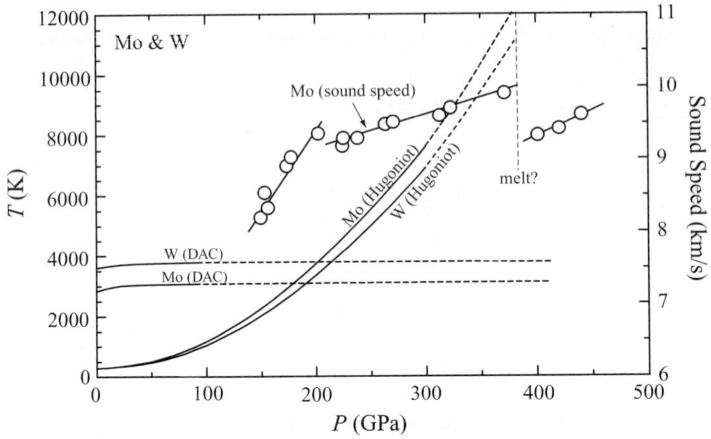

Fig. 12. Shock-state sound-speed measurements for Mo (Hixson et al., 1989) and calculated $P-T$ Hugoniots for Mo and W (Hixson and Fritz, 1992). DAC melting data of Mo and W (Errandonea et al., 2001) are extrapolated to higher pressures.

Since in most of the materials of interest (Table 3), melting of a high-pressure phase occurs, we require the Grüneisen parameter and equation of state (EOS) of the high-pressure phase. We thus employ in most cases the bulk modulus that is derived from the principal Hugoniot centered at the zero-pressure density of the metastable high-pressure phase, and assume a Murnaghan EOS (Poirier, 1991). The shock wave EOS, can be described by a linear shock velocity (U_S)–particle velocity (u_p) relation

$$U_S = C_0 + s u_p \tag{18}$$

where C_0 and s are the parameters. The bulk modulus is $K_0 = \rho_0 C_0^2$, and its pressure derivative is $K_0' = 4s - 1$. The subscript 0 denotes ambient condition. The Grüneisen parameter, γ, is assumed to be volume dependent only

$$\gamma(V) = \gamma_0 \left(\frac{V}{V_0}\right)^q \tag{19}$$

where $\gamma_0 = (162s^2 - 360s + 215)/18s$ (Jeanloz and Grover, 1987). Given a reference state (P_c, T_c) resolved from shock melting experiments (Table 3),

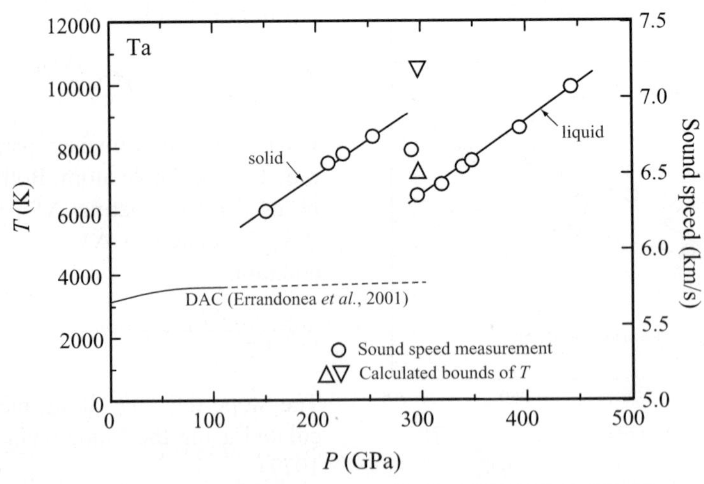

Fig. 13. Shock-state sound-speed measurements for Ta, and the upper and lower bound of calculated shock temperature (Shaner et al., 1984) near the melting pressure.

Table 6
The $R \ln 2$ rule: entropy of melting for alkali halides (heat capacity and entropy refer to per mole atoms)

	γ_0[a] (cm^3/g)	ΔV_m[a] (cm^3/g)	$\Delta V_m/V$ (cm^3/g)	$C_V(R)$[b]	$\Delta S_m(R)$[a]	$\Delta S_V(R)$	$\Delta S_D(R)$
LiF	1.63	3.47	0.356	2.52	1.451	1.458	−0.007
LiCl	1.81	5.63	0.277	2.89	1.356	1.449	−0.093
LiBr	1.94	6.22	0.248	2.94	1.290	1.417	−0.127
LiI	2.19	7.22	0.222	3.01	1.187	1.465	−0.278
NaF	1.51	4.75	0.320	2.82	1.592	1.361	0.231
NaCl	1.62	7.01	0.259	3.04	1.567	1.277	0.290
NaBr	1.65	7.67	0.239	3.09	1.539	1.220	0.319
NaI	1.71	8.82	0.216	3.14	1.519	1.161	0.358
KF	1.52	5.04	0.219	2.95	1.501	0.981	0.520
KCl	1.49	7.02	0.187	3.09	1.529	0.860	0.669
KBr	1.50	7.72	0.178	3.15	1.524	0.843	0.681
KI	1.49	8.86	0.167	3.18	1.514	0.789	0.725

[a] Jackson (1977).
[b] JANAF (Malcolm, 1998).

the melting curve can be calculated by integrating Eq. (17). The Lindemann melting curves are plotted in Figs. 5–11. ΔS_m and ΔV_m at certain pressure, e.g., P_c can be readily calculated as shown in Table 3. Note that the volume change ($\Delta V_m/V$) upon melting at high-pressure, varies between 0.9 and 1.8% for the materials investigated, thus ΔS_m is close to $R \ln 2$ due to the small changes in volume. The Lindemann melting curve for Fe suggests that Fe melts at ~6277 K and 330 GPa.

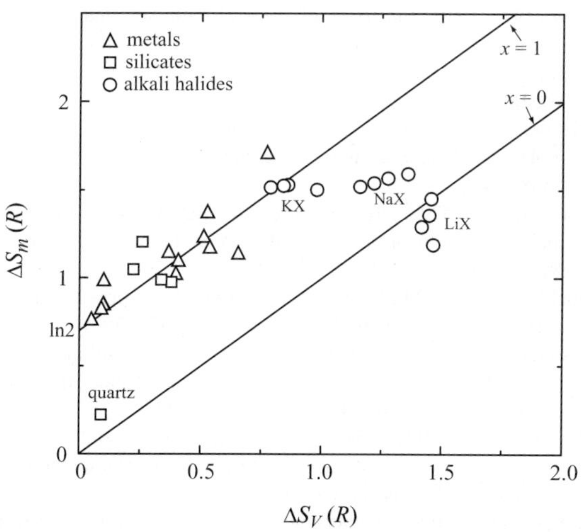

Fig. 14. The $R \ln 2$ rule of entropy of melting: ΔS_m vs. ΔS_V for metals, silicates and alkali halides. Also refer to Tables 4–6. Solid lines have slopes of 1 and intercepts of $R \ln 2$ ($x = 1$) and 0 ($x = 0$), respectively. X denotes F, Cl, Br and I.

The Lindemann melting curves constructed via the above procedures demonstrate self-consistency and good agreement with the DAC experiments using laser-heating techniques. The two Lindemann curves for stishovite obtained from fused and crystal quartz agree closely (within 150 K). The Lindemann curve for fused quartz also agrees with the DAC data (Shen and Lazor, 1995) at 25–40 GPa. There is excellent agreement between DAC experiments (Boehler et al., 1996, 1997) and Lindemann curves from shock wave results for KBr and NaCl.

The above comparison illustrates the encouraging convergence between the static and dynamic data for silica and alkali halides. But significant discrepancies exist for transition metals. Shock melting pressures for Fe, V, Mo, W and Ta are significantly higher than those achieved in DAC melting experiments, thus a comparison requires large extrapolation of the DAC melting data. If we extrapolate the DAC melting curves (Errandonea et al., 2001) for Fe, V, Mo, W and Ta to 200–400 GPa where shock melting occurs, shock temperature measurement and calculation would indicate $\Theta_H^+ \sim 0.7$–2.0 (Figs. 9–13). These large values of superheating are not consistent with the superheating systematics, even when the uncertainties in shock temperature and β variations at high pressures are taken into account. The discrepancies could be reconciled by possible solid–solid phase transitions at high pressures which may steepen the melting curves at higher pressures, and bring the convergence of static and dynamic experimental results. In particular, Fe could undergo a solid–solid phase transition at ∼200 GPa (Brown, 2001) and melt at ∼270 GPa upon shock wave loading.

5. Conclusions

Superheating systematics $\beta = (A_0 - b \log Q)\theta_c(\theta_c - 1)^2$ based on undercooling experiments incorporate heating rates and predict the tightest upper bound among a host of methods reported to date. For $Q \sim 10^{12}$ K/s, the maximum superheating predicted for crystalline materials, varies between 0.05 and $0.43T_m$. We examined the shock superheating-melting data on silicates, alkali halides and metals against the superheating systematics, and validate the method of resolving the equilibrium melting point (P_c, T_c) on Hugoniots. The agreement between shock-induced superheating and the superheating systematics indicates that the assumption of homogeneous nucleation of melt upon shock wave loading of single crystals is viable.

We parameterize the $R \ln 2$ rule for disordering entropy of melting as $\Delta S_D = xR \ln 2$. $x = 0$ for quartz and lithium halides at ambient pressure. x is taken to be 1 for the high-pressure phases investigated. With (P_c, T_c) resolved from shock wave experiments, the melting curves at high pressures, entropy and volume changes upon melting were obtained for stishovite, KBr, CsBr, NaCl (B2 phases), Fe and V based on the Lindemann law and $R \ln 2$ rule. The volume change ($\Delta V_m/V$) upon melting at P_c, varies between 0.9 and 1.8% for the materials investigated, thus ΔS_m is close to $R \ln 2$ due to the small changes in volume. For alkali halides and silica, the values for (P_c, T_c) and corresponding Lindemann melting curves, are in excellent agreement with the DAC results. For transition metals, significant discrepancies exist between shock wave results and the extrapolation of DAC data. Simple extrapolations of the DAC melting curves to the pressure range of shock melting indicate superheating of $\Theta_H^+ = 0.7$–2.0 for Fe, V, Mo, W and Ta. These are far beyond the predictions of superheating systematics, even if we consider the uncertainties in shock temperature measurement (or estimation) and pressure-induced changes in β. Possible high-pressure solid–solid phase change—thus a change in slope, dT_m/dP—could reconcile these discrepancies, although the Hugoniots do not suggest such phase changes except for Fe and Mo. This work suggests that Fe melts at ∼270 GPa upon shock loading, and the equilibrium T_m is ∼6300 K at 330 GPa, and that the first drop in shock-state sound-speed measurements could be due to a solid–solid phase change at ∼200 GPa as previously suggested (Brown, 2001).

Acknowledgements

This work has been supported by NSF Grant No. EAR-0207934. We have benefited from the discussion with R. Boehler. Constructive comments by T. Duffy and two reviewers helped to improve the manuscript. Contribution No. 8927, Division of Geological and Planetary Sciences, California Institute of Technology.

References

Ahrens, T.J., Lyzenga, G.A., Mitchell, A.C., 1982. Temperature induced by shock waves in minerals. In: Akimoto, S., Manghnani, M.H. (Eds.), High-Pressure Research in Geophysics, Adv. Earth Planet. Sci., vol. 12, pp. 579–594.

Ahrens, T.J., Holland, K.G., Chen, G.Q., 2002. Phase diagram of iron, revised-core temperatures. Geophys. Res. Lett. 29 (7) (art. no. 1150).

Allen, M.P., Tildesley, D.J., 1987. Computer Simulation of Liquids. Clarendon Press, London.

Bass, J.D., 1995. Elasticity of minerals, glasses and melts. In: Ahrens, T.J. (Ed.), Mineral Physics and Crystallography: A Handbook of Physical Constants, American Geophysical Union, Washington, DC, pp. 45–63.

Belonoshko, A.B., Ahuja, R., Johansson, B., 2000. Quasi-ab initio molecular dynamics study of Fe melting. Phys. Rev. Lett. 84, 3638–3641.

Boehler, R., 1993. Temperatures in the Earth's core from melting-point measurements of iron at high static pressures. Nature 363, 534–536.

Boehler, R., Ross, M., Boercker, D.B., 1996. High-pressure melting curve of alkali halides. Phys. Rev. B 53, 556–563.

Boehler, R., Ross, M., Boercker, D.B., 1997. Melting of LiF and NaCl to 1 mbar: systematics of ionic solids at extreme conditions. Phys. Rev. Lett. 78, 4589–4592.

Boness, D.A., 1991. Shock wave experiments and electronic band-structure calculations of materials at high temperature and pressure. Ph.D. Thesis. University of Washington.

Boness, D.A., 1999. Metastability in shocked iron: controversy with regard to sound velocity and temperature measurements. In: Furnish, M.D., Chhabildas, L.C., Hixson, R.S. (Eds.), Shock Compression of Condensed Matter. Am. Inst. Phys., pp. 77–80.

Boness, D.A., Brown, J.M., 1993. Bulk superheating of solid KBr and CsBr with shock-waves. Phys. Rev. Lett. 71, 2931–2934.

Brown, J.M., 2001. The equation of state of iron to 450 GPa: another high pressure solid phase? Geophys. Res. Lett. 28, 4339–4342.

Brown, J.M., McQueen, R.G., 1982. The equation of state for iron and the Earth's core. In: Akimoto, S., Manghnani, M.H. (Eds.), High-Pressure Research in Geophysics, Adv. Earth Planet. Sci., vol. 12, pp. 611–623.

Brown, J.M., McQueen, R.G., 1986. Phase transitions, Grüneisen parameter, elasticity for shocked iron between 77 and 400 GPa. J. Geophys. Res. 91, 7485–7494.

Christian, J.W., 1965. The Theory of Transformations in Metals and Alloys. Pergamon Press, Oxford.

Dai, C., Jin, X., Zhou, X., Liu, J., Hu, J., 2001. Sound velocity variations and melting of vanadium under shock compression. J. Phys. D: Appl. Phys. 34, 3064–3070.

Davidchack, R.L., Laird, B.B., 2000. Direct calculation of the hard-sphere crystal/melt interfacial energy. Phys. Rev. Lett. 85, 4751–4754.

Errandonea, D., Schwager, B., Ditz, R., Gessmann, C., Boehler, R., Ross, M., 2001. Systematics of transition-metal melting. Phys. Rev. B 63, 132104.

Fabricius, N., Hermes, P., von der Linde, D., Pospieszczyk, A., Stritzker, B., 1986. Observation of superheating during laser melting. Solid State Comm. 58, 239–242.

Fecht, H.J., Johnson, W.L., 1988. Entropy and enthalpy catastrophe as a stability limit for crystalline material. Nature 334, 50–51.

Fei, Y., 1995. Thermal expansion. In: Ahrens, T.J. (Ed.), Mineral Physics and Crystallography: A Handbook of Physical Constants, American Geophysical Union, Washington, DC, pp. 29–44.

Gathers, G.R., Mitchell, A.C., Holmes, N.C., 1983. Hugoniot measurements in vanadium using the LLNL two-state light-gas gun. In: Asay, J.R., Graham, R.A., Straub, G.K. (Eds.), Shock Compression of Condensed Matter. Elsevier, Amsterdam, Netherlands, pp. 89–90.

Herman, J.W., Elsayed-Ali, H.E., 1992. Superheating of Pb(1 1 1). Phys. Rev. Lett. 69, 1228–1231.

Hixson, R.S., Boness, D.A., Shaner, J.W., Moriaty, J.A., 1989. Acoustic velocities and phase transitions in molybdenum under strong shock compression. Phys. Rev. Lett. 62, 637–640.

Hixson, R.S., Fritz, J.N., 1992. Shock compression of tungsten and molybdenum. J. Appl. Phys. 71, 1721–1728.

Holland, K.G., Ahrens, T.J., 1997. Melting of $(Mg,Fe)_2SiO_4$ at the core-mantle boundary of the Earth. Science 275, 1625–1653.

Hoover, W.G., Ree, F.H., 1968. Melting transition and communal entropy for hard spheres. J. Chem. Phys. 49, 3609–3617.

Hughes, D.S., McQueen, R.G., 1958. Density of basic rocks at very high pressures. Trans. Am. Geophys. Union 39 (5), 959–965.

Jackson, I., 1977. Melting of some alkali-earth and transition-metal fluorides and alkali fluoroberyllates at elevated pressures: a search for melting systematics. Phys. Earth Planet. Int. 14, 143–164.

Jeanloz, R., Grover, R., 1987. Birch Murnaghan and U_S–u_p equations of state. In: Schmidt, S.C., Holmes, N.C. (Eds.), Shock Compression of Condensed Matter. Elsevier, Amsterdam, Netherlands, pp. 69–72.

Kelton, K.F., 1991. Crystal nucleation in liquids and glasses. Solid State Phys. 45, 75–177.

Kormer, S.B., 1965. Optical study of the characteristics of shock-compressed dielectrics. Sov. Phys. Usp. 21, 689–700.

Lu, K., Li, Y., 1998. Homogeneous nucleation catastrophe as kinetic stability limit for superheated crystal. Phys. Rev. Lett. 80, 4474–4477.

Luo, S.-N., Cagin, T., Strachan, A., Goddard II, W.A., Ahrens, T.J., 2002a. Molecular dynamics modeling of stishovite. Earth Planet. Sci. Lett. 202, 147–157.

Luo, S.-N., Mosenfelder, J.L., Asimow, P.D., Ahrens, T.J., 2002b. Direct shock wave loading of stishovite to 235 GPa: implications for perovskite stability relative to an oxide assemblage at lower mantle conditions. Geophys. Res. Lett. 29 (14) (art. no. 1691).

Luo, S.-N., Strachan, A., Swift, D., 2004a. Non-equilibrium melting and crystallization of a model Lennard-Jones system. J. Chem. Phys., submitted.

Luo, S.-N., Akins, J.A., Ahrens, T.J., Asimow, P.D., 2004b. Shock compressed $MgSiO_3$ glass, enstatite, olivine and quartz: Optical emission, temperatures and melting. J. Geophys. Res., submitted.

Luo, S.-N., Ahrens, T.J., 2003. Superheating systematics of crystalline solids. Appl. Phys. Lett. 82 (12), 1836–1838.

Luo, S.-N., Ahrens, T.J., Cagin, T., Strachan, A., Goddard III, W.A., Swift, D., 2003. Maximum superheating and undercooling: Systematics, molecular dynamics simulations and dynamic experiments. Phys. Rev. B, 68 (art. no. 134206).

Lyzenga, G.A., Ahrens, T.J., Mitchell, A.C., 1983. Shock temperatures of SiO_2 and their geophysical implications. J. Geophys. Res. 88, 2431–2444.

Malcolm Jr., W.C., 1998. NIST-JANAF Thermochemical Tables. Am. Chem. Soc./Am. Inst. Phys. Nat. Inst. Stand. Tech.

McQueen, R.G., 1992. The velocity of sound behind strong shocks in SiO_2. In: Schmidt, S.C., Dick, R.D., Tasker, D.G. (Eds.), Shock Compression of Condensed Matter. Elsevier, Amsterdam, Netherlands, pp. 75–78.

Motorin, V.I., Musher, S.L., 1984. Kinetics of the volume melting: nucleation and superheating of metals. J. Chem. Phys. 81, 465–469.

Murphy, E.A., Elsayed-Ali, H.E., Herman, J.W., 1993. Superheating of Bi(1 0 0 0). Phys. Rev. B 48, 4921–4924.

Oriani, R.A., 1951. The entropies of melting of metals. J. Chem. Phys. 19, 93–97.

Poirier, J.P., 1991. Introduction to the Physics of the Earth's Interior. Cambridge University Press, Cambridge, 102 pp.

Porter, D.A., Easterling, K.E., 1981. Phase Transformations in Metals and Alloys. Van Nostrand Reinhold, Wokingham, 192 pp.

Rethfeld, B., Sokolowski-Tinten, K., von der Linde, D., Anisimow, S.I., 2002. Ultrafast thermal melting of laser-excited solids by homogeneous nucleation. Phys. Rev. B 65, 092103.

Rivier, N., Duffy, D.M., 1982. On the topological entropy of atomic liquids and the latent heat of fusion. J. Phys. C 15, 2867–2874.

Shaner, J.W., Brown, J.M., McQueen, R.G., 1984. Melting of metals above 100 GPa. Mater. Res. Soc. Symp. Proc. 22, 137–141.

Shao, Y., Spaepen, F., 1996. Undercooling of bulk liquid silicon in an oxide flux. J. Appl. Phys. 79 (6), 2981–2985.

Shen, G., Lazor, P., 1995. Measurement of melting temperatures of some minerals under lower mantle pressures. J. Geophys. Res. 100, 17699–17713.

Stacey, F.D., Irvine, R.D., 1977. Theory of melting: thermodynamic basis of Lindemann law. Aust. J. Phys. 30, 631–640.

Stishov, S.M., 1975. The thermodynamics of melting of simple substances. Sov. Phys. Usp. 117, 625–643.

Strachan, A., Çağın, T., Gülseren, O., Mukherjee, S., Cohen, R.E., Goddard III, W.A., 2001. First-principles force-field for metallic tantalum, preprint.

Tallon, J.L., Robinson, W.H., 1982. A model-free elasticity theory of melting. Phys. Lett. A 87, 365–368.

Tallon, J.L., 1989. A hierarchy of catastrophes as a succession of stability limits for the crystalline state. Nature 342, 658–660.

Taylor, R.D., Pasternak, M.P., Jeanloz, R., 1991. Hysteresis in the high-pressure transformation of b.c.c.-iron to h.c.p.-iron. J. Appl. Phys. 69, 6126–6128.

Turnbull, D., Fisher, J.C., 1949. Rate of nucleation in condensed systems. J. Chem. Phys. 17, 71–73.

Turnbull, D., 1952. Kinetics of solidification of supercooled liquid mercury droplets. J. Chem. Phys. 20, 411–424.

Uhlmann, D.R., 1980. On the internal nucleation of melting. J. Non-Cryst. Solids 41, 347–357.

Walton, A.G., 1969. Nucleation in liquids and solutions. In: Zettlemoyer, A.C. (Ed.), Nucleation. Marcel Dekker, New York, pp. 225–307.

Williams, Q., Jeanloz, R., Bass, J., Svendsen, B., Ahrens, T.J., 1987. The melting curve of iron to 250 gigapascals: a constraint on the temperature at Earth's center. Science 236, 181–182.

Williamson, S., Mourou, G., Li, J.C.M., 1984. Time-resolved laser-induced phase transition is aluminum. Phys. Rev. Lett. 52, 2364–2367.

Yoo, C.S., Holmes, N.C., Ross, M., Webb, D.J., Pike, C., 1993. Shock temperatures and melting of iron at Earth core conditions. Phys. Rev. Lett. 70, 3931–3934.

Available online at www.sciencedirect.com
SCIENCE @ DIRECT®
ELSEVIER
Physics of the Earth and Planetary Interiors 143–144 (2004) 387–395
PHYSICS OF THE EARTH AND PLANETARY INTERIORS
www.elsevier.com/locate/pepi

Hydrous phase stability and partial melt chemistry in H_2O-saturated KLB-1 peridotite up to the uppermost lower mantle conditions

Tatsuhiko Kawamoto [a,b,*]

[a] *Bayerisches Geoinstitut, Universität Bayreuth, D-95440 Bayreuth, Germany*
[b] *Institute for Geothermal Sciences, Graduate School of Science, Kyoto University, Beppu 874-0903, Japan*

Received 18 December 2002; received in revised form 22 June 2003; accepted 24 June 2003

Abstract

Stability of hydrous phases and partial melt chemistry in 13.6 wt.% H_2O bearing mantle peridotite are reported in a pressure range up to 24 GPa and a temperature range from 900 to 1400 °C. Various hydrous phases like phase E, phase D, superhydrous B, hydrous β and γ phases of olivine stoichiometry are observed between 14 and 24 GPa. The phase boundary between the β phase and γ phase of olivine under hydrous conditions is found to be at higher pressure than under the dry conditions. The H_2O-undersaturated solidus temperature of a model mantle is estimated up to 11 GPa based on (1) the reported solubility of H in nominally anhydrous minerals and (2) linearity of H_2O-undersaturated solidus between the dry and wet solidus [Science 276 (1997) 240] with respect to H_2O abundance at a constant pressure.

Under H_2O-saturated conditions at pressures greater than 10 GPa, the $(Mg + Fe)/Si$ atomic ratios of partial melts decrease with increasing temperature: from 2.5 at 1100 °C to 1.6 at 1300 °C at 14 GPa, from 3.3 at 1100 °C to 1.8 at 1300 °C at 17 GPa, and from 2.8 at 1200 °C to 1.7 at 1400 °C at 20 GPa. This is in contrast to the reported constant value of dry partial melts (1.1 between 10 and 18 GPa [J. Geophys. Res. 99 (1994) 17729]). A partial melt coexisting with Mg-perovskite at 24 GPa and 1400 °C has $(Mg + Fe)/Si$ and Ca/Al atomic ratios of 2.2 and 11, respectively. Ultramafic hydrous melts similar to those observed experimentally under uppermost lower mantle conditions may have contributed to chemical differentiation between the upper and lower mantle.

Keywords: Mantle; H_2O; Melt; Magma; Hydrous phase; High pressure

1. Introduction

The existence of liquid H_2O is one of the most unique features characterizing the planet earth. H_2O has a significant effect on the properties of earth-forming materials, and especially on the generation and the differentiation processes of magmas. All terrestrial magmas have variable amounts of H_2O before eruption, because hydrogen is present over the whole mantle and is preferentially partitioned into silicate melts under pressure. The production and subsequent eruption of magmas is an effective degassing mechanism in planetary systems. Therefore, it is important to know the possible distribution of H_2O or OH in the earth's mantle in order to understand the differentiation processes of the planet (Thompson, 1992; Okuchi, 1997). There have been many exper-

* Tel.: +81-977-22-0713; fax: +81-977-22-0965.
E-mail address: kawamoto@bep.vgs.kyoto-u.ac.jp (T. Kawamoto).

0031-9201/$ – see front matter © 2004 Elsevier B.V. All rights reserved.
doi:10.1016/j.pepi.2003.06.003

imental works to determine the stability of hydrous phases in the $MgO–SiO_2–H_2O$ system (see comprehensive reviews (Mysen et al., 1998; Frost, 1999; Ohtani et al., 2001)). In addition to experiments in the $MgO–SiO_2–H_2O$ system, Asahara and Ohtani (2001) and Litasov and Ohtani (2002) conducted a series of experiments in the $CaO–MgO–Al_2O_3–SiO_2–H_2O$ system, and Kawamoto et al. (1995, 1996) and Kawamoto and Holloway (1997) conducted a series of experiments in a more complex system using a model mantle peridotite. The present paper reports the stability of hydrous phases and the chemistry of partial melts in an H_2O-saturated model mantle peridotite at conditions up to the uppermost part of the lower mantle.

2. Experimental

A series of experiments were carried out in a pressure range from 14 to 24 GPa and a temperature range from 900 to 1400 °C by the use of Kawai-type multi-anvil high-pressure apparatus at Bayerisches Geoinstitut mainly following Ito et al.'s (1984) technique. As a starting material, I used a mixture of Mg free KLB-1 peridotite gel and brucite (13.6 wt.% H_2O, 0.89 Mg/(Mg + Fe), 1.46 (Mg + Fe)/Si atomic ratio, Table 2). The powders were loaded into welded Ag, Au, or Pt capsules of 0.15 mm wall thickness. The assembly consisted of an MgO octahedron, $LaCrO_3$ cylindrical furnace and a W3%Re/W25%Re thermocouple covered by Cu coils. The pressure calibration was undertaken using phase transitions in the Mg_2SiO_4 system at 1600 °C (Katsura and Ito, 1989; Ito and Takahashi, 1989). The pressure of 24 GPa reported in the present study is determined less precisely than lower pressures. It represents the pressure where Mg-perovskite is found. Recovered quenched samples were embedded in epoxy and polished. In order to identify phases of olivine and enstatite stoichiometry, unpolarized Raman spectra of the polished samples were measured with a Raman microscope consisting of 532 nm YAG laser, holographic transmission gratings, and CCD detector for 200–4500 cm^{-1} spectral range (Kaiser HoloLab 5000 system) at the Institute for Geothermal Sciences of Kyoto University at Beppu. Quenched phases were also observed with an electron microprobe analyzer and their chemical compositions were determined with wavelength dispersive system (JEOL 8900 superprobe).

3. Phase relations

Almost all experimental products have variable amounts of a portion showing dendritic texture (Fig. 1)

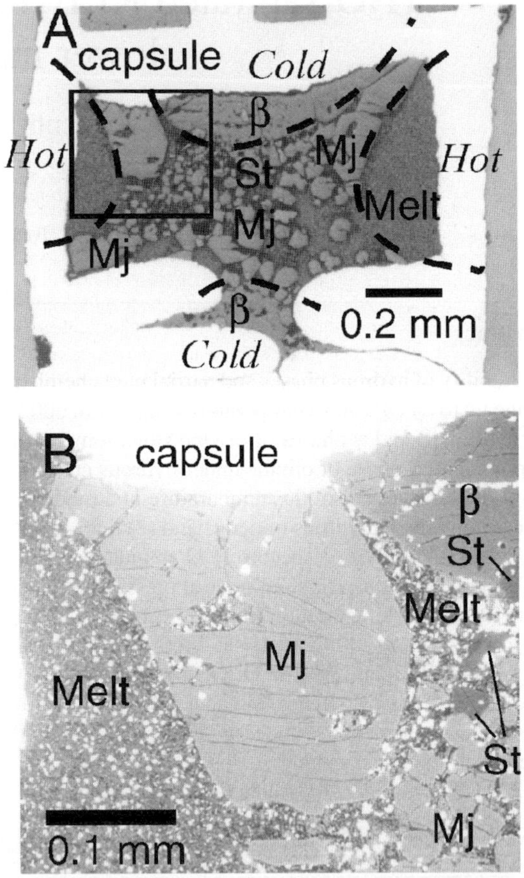

Fig. 1. (A) Reflected light photomicrograph of a polished experimental charge of run #413 (20 GPa, 1300 °C). A chemically heterogeneous distribution is observed along a possible thermal gradient. Portions showing dendritic texture (Melt) are observed along capsule wall, which is close to the heating furnace and is placed in the center of the experimental charge. Majoritic garnet (Mj) crystallizes next to melt pools. β phase of olivine (β) crystallizes preferentially at capsule roof and base. Stishovite (St) crystals are scattered with Mj and also in β. (B) Backscattered electron image of the part indicated by the solid rectangle in (A). Capsule material is Pt. Mj (bright), β (gray), and St (dark) are each contacted with melt. Small white dots are artificial contaminants of silver during the polishing process.

Table 1
Summary of experimental conditions and observed phase assemblages

Run number	P (GPa)	T (°C)	Capsule	t (h)	Melt	Melting (%)	Fe-loss (%)	Crystals
423	14	1000	Pt	6	Small M			En, PhE, Gt, Cac (trace), Mgs (trace)
405	14	1100	Pt	6	M2.5	11	52	PhE, Gt, En
409	14	1200	Ag	7	M1.8	37	2	En, β, Gt
419	14	1300	Ag	7	M1.6	44	18	En, β
430	14.5	900	Ag	4	n.d.			St, PhE, En, Mgs (trace)
428	17	1000	Ag	7	Small M			PhD, ShyB, Gt, St, Mgs (trace)
408	17	1100	Pt	9	M3.3	12	29	β, Gt, St (trace in β)[a], PhE (trace)
406	17	1200	Pt	8.7	M2	30	38	Gt, β, St (trace in β)[a]
418	17	1300	Pt	7	M1.8	38	30	Gt, β, En (trace)
434	20	1000	Ag	5	n.d.			PhD, Ca-Pv, ShyB, Mw, Mgs (trace)
429	20	1100	Ag	4.5	Small M			PhD, ShyB
421	20	1200	Pt	5	M2.8	23	20	St, Gt, γ
413	20	1300	Pt	6	M2.2	40	12	β, Mj, St
415	20	1400	Pt	6	M1.7	55	24	Mj, St (trace in Mj)[a]
403	24	1300	Ag	10	Small M			Mg-Pv, Ca-Pv, ShyB, St (trace in Mg-Pv)[a]
414	24	1400	Ag	4	n.d.			Mg-Pv (Mgs, PhD, ShyB at 1200 °C)[b]
432	24	1400	Au	8	M2.2	16	30	Mg-Pv, St (trace in Mg-Pv)[a]

M: melt; small M: too small amount of melt to be analyzed; M2.5: melt has 2.5 (Mg + Fe)/Si; n.d.: not detected; En: enstatite; PhE: phase E; Cac: calcium carbonate; Mgs: magnesite; Gt: garnet; β(γ) : β(γ) phase of olivine; St: stishovite; PhD: phase D; Mw: magnesium wüstite; Ca-Pv: calcium perovskite; Mj: majoritic garnet; ShyB: superhydrous phase B; Mg-Pv: magnesium perovskite.
[a] Found only inside of another phase.
[b] Observed at the cold end, around 1200 °C.

except for three experiments (900 °C at 14.5 GPa, 1000 °C at 20 GPa, 1400 °C at 24 GPa, Table 1). Chemically heterogeneous distributions are observed along the thermal gradient in almost all experimental charges (Fig. 1). It is difficult to distinguish whether the dendritic portions are quench products of partial melts or H_2O fluids dissolving silicate components at high temperature and pressure conditions. In a single experimental product, two types of dendritic texture are not observed to coexist. This may indicate that there is a continuous change from silicate rich H_2O fluids at lower temperature to H_2O rich partial melts at higher temperature. If this is the case in the present pressure range, the H_2O-saturated solidus temperature may represent a temperature where the concentration of silicate components dissolved into aqueous fluids increase drastically (Kawamoto et al., 2003). In an H_2O–peridotite system, the pressure and temperature conditions for the second critical point, beyond which those two fluids cannot be distinguished, remains to be investigated in the future. In the present study, the chemical compositions of these sample regions represent quenched partial melts. In order to obtain the partial melt compositions, we determined major element compositions of the dendritic portions over a 30 μm × 40 μm scanning area. The analytical procedures are the same as in previous work with the same starting material at lower pressures (Kawamoto and Holloway, 1997). The chemical compositions of these melts (Table 2) will be discussed later.

A trace amount of carbonate is observed in a few experimental charges suggesting the presence of carbon in the starting material (Table 1). This can be due to remnant carbon during the gel preparation and/or the absorption of atmospheric CO_2 by $Mg(OH)_2$. I used Ag, Au, and Pt tubes for sample containers, and these materials absorb some FeO from the sample during the experimental durations. The degree of Fe-loss was estimated by mass balance calculations. In these calculations, the iron contents always exhibit the poorest results. Therefore, first I calculated mass balance without iron abundance and then re-calculated possible iron abundance for the bulk sample using the obtained mass fraction. The calculated iron abundances for the bulk compositions are always less than those of the starting material, and these discrepancies give percentages of possible iron loss (Table 1). Some trace phases are found only in a single phase and may not

Table 2
Chemical compositions of starting material, and crystalline phases and partial melts of the experimental products

Run number		SiO_2	TiO_2	Al_2O_3	FeO[a]	MgO	CaO	Na_2O	Cr_2O_3	Total	Mg/(Mg + Fe)	(Mg + Fe)/Si
	KLB-1-14	44.69	0.17	3.37	8.59	38.96	3.96	0.26		100.00	0.89	1.46
405	PhE	37.68	0.14	2.15	3.73	44.46				88.15	0.96	1.84
	Gt	44.49	0.17	16.78	6.57	23.88	5.00			96.89	0.87	0.92
	En	57.68		0.16	2.03	38.61				98.48	0.97	1.03
	Melt	12.86	0.26	2.18	4.12	19.09	13.18	0.18		51.86	0.89	2.53
	Stdev[b]	1.68	0.07	1.70	0.33	2.17	0.71	0.01		2.28	0.01	0.49
409	Gt	45.15		16.04	7.33	25.47	3.79		0.89	98.67	0.86	0.98
	En	58.35		0.15	2.68	37.69				98.87	0.96	1.00
	β	41.09		0.38	8.57	46.07			0.25	96.36	0.91	1.85
	Melt	24.65	0.24	4.01	9.10	24.50	7.05	0.24		69.79	0.83	1.80
	Stdev	1.64	0.03	0.66	0.68	0.77	1.91	0.09		1.95	0.01	0.09
419	En	58.19		0.14	2.81	37.84				98.98	0.96	1.01
	β	41.23		0.30	8.40	47.20				97.13	0.91	1.88
	Melt	27.79	0.23	4.64	6.27	25.31	6.53	0.14		70.91	0.88	1.56
	Stdev	2.76	0.04	1.42	0.43	1.79	1.91	0.06		3.47	0.01	0.17
408	PhE	36.66	0.16	1.43	6.58	43.28				88.12	0.92	1.91
	β	41.85		0.21	5.77	47.93				95.76	0.94	1.82
	Gt	49.09		12.13	5.61	29.10	2.82			98.75	0.90	0.98
	St	96.69		0.25	0.26	0.19				97.39	0.45	0.01
	Melt	11.88	0.38	0.49	5.48	22.95	11.79	0.30		53.27	0.88	3.26
	Stdev	0.42	0.03	0.04	0.38	1.13	1.44	0.04		3.10	0.00	0.10
406	β	42.71		0.25	4.29	49.22			0.17	96.64	0.95	1.80
	Gt	51.84		10.02	4.05	30.99	1.92		0.63	99.45	0.93	0.96
	St	96.04		0.36	0.08	0.01				96.49	0.14	0.00
	Melt	23.12	0.31	0.58	5.42	28.56	7.25	0.31		65.55	0.90	2.04
	Stdev	0.41	0.03	0.03	0.12	0.54	0.61	0.04		0.74	0.00	0.03
418	β	41.93		0.19	4.81	49.58				96.51	0.95	1.86
	En	58.90		0.08	1.54	38.77				99.29	0.98	1.00
	Gt	53.74		6.99	4.37	32.68	1.47		0.20	99.45	0.93	0.97
	Melt	26.52	0.24	0.44	5.21	28.89	5.67	0.16		67.14	0.91	1.79
	Stdev	0.92	0.03	0.02	0.29	0.75	0.86	0.03		1.27	0.00	0.04
421	St	95.64		0.48	0.09	0.02				96.23	0.28	0.00
	Gt	49.09		13.79	4.73	28.03	2.83		0.33	98.81	0.91	0.93
	γ	41.25		0.11	8.55	48.38				98.28	0.91	1.92
	Melt	17.46	0.41	0.90	5.29	30.19	8.45	0.25		62.96	0.91	2.83
	Stdev	0.63	0.03	0.06	0.23	1.21	0.63	0.03		1.71	0.00	0.12
413	β	42.15		0.24	4.66	51.10				98.14	0.95	1.90
	Mj	52.40		6.82	4.90	33.19	1.59			98.90	0.92	1.02
	St	96.80		0.34	0.14	0.01				97.29	0.07	0.00
	Melt	24.97	0.24	0.65	8.08	31.56	5.92	0.28		71.70	0.87	2.16
	Stdev	0.60	0.03	0.04	0.24	0.67	0.44	0.06		0.80	0.00	0.08
415	St	94.77		0.47	0.15	0.00				95.40	0.02	0.00
	Mj	52.55		7.66	3.91	33.41	1.19		0.53	99.25	0.94	1.01
	Melt	31.15	0.19	0.74	5.99	31.33	4.71	0.14		74.25	0.90	1.66
	Stdev	0.86	0.03	0.08	0.85	1.31	0.78	0.03		1.38	0.01	0.09
432	St	97.21		1.06	0.08	0.01				98.35	0.12	0.00
	Mg-Pv	54.54	0.13	4.65	5.24	35.92	0.13		0.29	100.90	0.92	1.06
	Melt	17.87	0.26	1.04	4.91	23.28	12.59	0.15		60.10	0.89	2.20
	Stdev	2.17	0.07	0.16	0.80	0.80	2.20	0.03		3.12	0.02	0.29

[a] Total iron as FeO.
[b] Standard deviation (1σ) for melt data.

be equilibrated with the other phases (Table 1). In the mass balance calculations for four run products (#408, #418, #415, #432), the mass balance for the least abundant phase yielded a negative percentage. To avoid this unreasonable result, the mass balance was recalculated without including this phase. The observed Fe-loss results in higher Fe_2O_3/FeO ratios in the experimental products than the starting material treated at the Ni–NiO buffer. The Mg/(Mg + total Fe) ratios of coexisting phases systematically change and decrease in the following order: En > β > Gt, Mj > melt (Table 2). This suggests that Fe^{3+}/total Fe of those phases should increase in the following order: En < β < Gt, Mj, if $Mg/(Mg+Fe^{2+})$ can be equally partitioned among those phases.

4. Stability of hydrous phases

Various hydrous phases are observed in the present experimental conditions (Tables 1 and 2 and Fig. 2). Phase E (Kanzaki, 1991) is stable between 9 (Kawamoto et al., 1996) and 17 GPa and has 1.5–2 wt.% Al_2O_3. Phase D (Yang et al., 1997), which is also called as phase F (Kanzaki, 1991) and phase G (Ohtani et al., 1997a), is observed between 17 and 24 GPa and has 4–8 wt.% Al_2O_3. Superhydrous B is found also between 17 and 24 GPa and has 0.5–1.5 wt.% Al_2O_3. Hydrous β and γ phases of olivine stoichiometry are observed between 14 and 20 GPa. The phase boundary between the β phase and γ phase of olivine in hydrous conditions is found to be at higher pressure than in the dry conditions (Fig. 2). This is consistent with previous experimental results in 1, 2, and 3 wt.% H_2O bearing Mg_2SiO_4 system (Inoue et al., 2001).

There are two different results at 24 GPa and 1400 °C: one without hydrous phase (#432) and the other with phase D and superhydrous B (#414). The latter (#414) has phase D and superhydrous B at the coldest part of the capsule, which can be at around 1200 °C. In addition to this discrepancy, at 1300 °C and 24 GPa only superhydrous B is found (#403). In Fig. 2, dry assemblage at 24 GPa and 1400 °C (#432), superhydrous B at 24 GPa and 1300 °C (#403) and superhydrous B and phase D at 1200 °C (the coldest position of #414) are plotted in order to have an internal consistency among these experimental results.

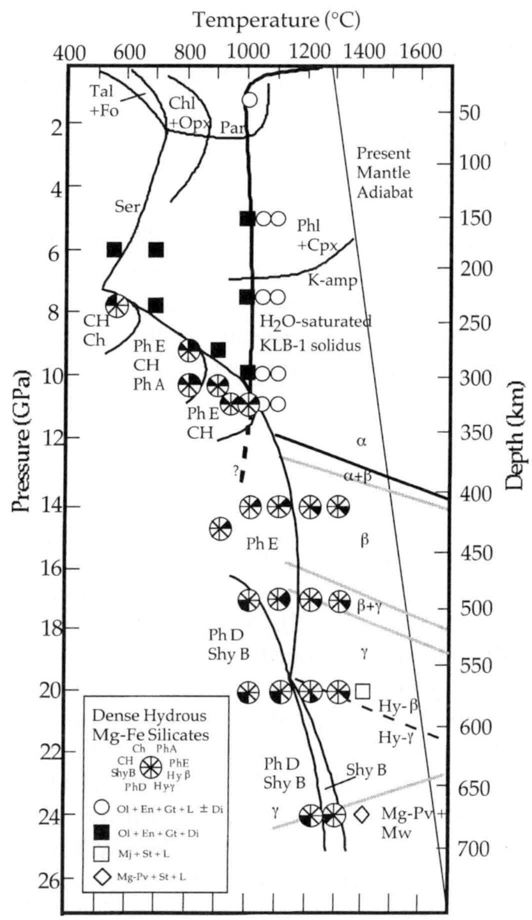

Fig. 2. Pressure–temperature projection of the stability of dense hydrous Mg–Fe silicates in H_2O-saturated KLB-1 peridotite. Data at a pressure range lower than 11 GPa are from Kawamoto and Holloway (1997) and Kawamoto et al. (1996). Open circles represent the phase assemblage of liquid (L), olivine (Ol), garnet (Gt), Ca-poor pyroxenes (En), with or without Ca-rich pyroxene (Di). Closed squares represent sub-solidus assemblages: olivine, garnet, and two pyroxenes at 5–10 GPa. Experiments at 1 GPa are after Hirose (1997); the stabilities of pargasite (Par), chlorite (Chl), serpentine (Ser), talc (Tal) and phlogopite (Phl)–K-amphibole (K-amp) are after Schmidt and Poli (1998) and Sudo and Tatsumi (1990); phase boundaries among dry α, α + β, β, and β + γ-$(Mg_{0.9}Fe_{0.1})_2SiO_4$ and γ-$(Mg_{0.9}Fe_{0.1})_2SiO_4$ and Mg-perovskite (Mg-Pv) + magnesium wüstite (Mw) in dry conditions are after Katsura and Ito (1989) and Ito and Takahashi (1989), respectively. Hy-β and Hy-γ are hydrous β and γ phase of olivine stoichiometry, respectively. The phase boundary of Hy-β and Hy-γ is at higher pressure than in the dry conditions. Ch, CH, and PhA are chondrodite and, clinohumite, and phase A, respectively. The other abbreviations are shown in Table 1.

Superhydrous B is thermally more stable than phase D (Fig. 2). This is consistent with previous experimental works (Frost, 1999) and Schreinemakers' bundle analysis (Komabayashi et al., 2002). At 24 GPa, a melt was obtained at 1300 °C (#403) but not at 1400 °C (#414). There could be a fossil of a melt somewhere inside of the capsule (#414), which was not observed on the surface. At 24 GPa and 1400 °C, in the other experiment (#432) Mg-perovskite is observed together with stishovite and a quenched partial melt showing dendritic texture. In these conditions, ferropericlase was not observed.

5. Solidus temperature

The H_2O-saturated solidus temperature of mantle peridotite meets the stability field of dense hydrous phases at 1000 °C and 11 GPa (Kawamoto and Holloway, 1997; Fig. 2). At pressures greater than 11 GPa, the H_2O-saturated solidus temperature does not have geological importance, because a realistic H_2O abundance in mantle peridotite (within a few weight percentage) can be entirely accommodated by dense hydrous phases. The 13.6 wt.% H_2O of the present starting material is greater than the possible water-carrying capacity in the observed hydrous crystalline phases. Therefore, there are always free aqueous fluids in the experimental charges at high-pressure and temperature conditions. These fluids dissolve silicate components and their effects on the phase relations and melt chemistry remains uncertain in the present study.

An H_2O-undersaturated solidus temperature under these conditions can be determined only after hydrous phase relations are determined more precisely in future work. In contrast, at pressures lower than 11 GPa, the melting temperature of an H_2O-undersaturated mantle peridotite can be calculated (Fig. 3), based on (1) the H_2O-saturated solidus (Kawamoto and Holloway, 1997), (2) pressure dependence of hydrogen solubility in nominally anhydrous minerals, and (3) an assumption of linearity between the H_2O-undersaturated solidus temperature and the H_2O abundance at a given pressure. An H_2O-undersaturated solidus temperature of mantle peridotite should lie between the H_2O-saturated solidus and the dry one (Takahashi et al., 1993; Zhang and Herzberg, 1994). The H_2O-undersaturated solidus is assumed to have a linear relation with respect to H_2O concentration. The amount of H_2O necessary to saturate mantle peridotite depends on the H solubility in nominally anhydrous minerals at a given pressure. Although K_2O also stabilizes phlogopite and K-richterite in a hydrous peridotite system and may control the melting temperature (Trønnes et al., 1988; Sudo and Tatsumi, 1990), its quantitative role remains uncertain. Infra-red spectroscopy demonstrated that olivine and pyroxene crystals of natural mantle xenoliths have up to several hundreds ppm H_2O (Smyth et al., 1991; Bell and Rossman, 1992). Experimental

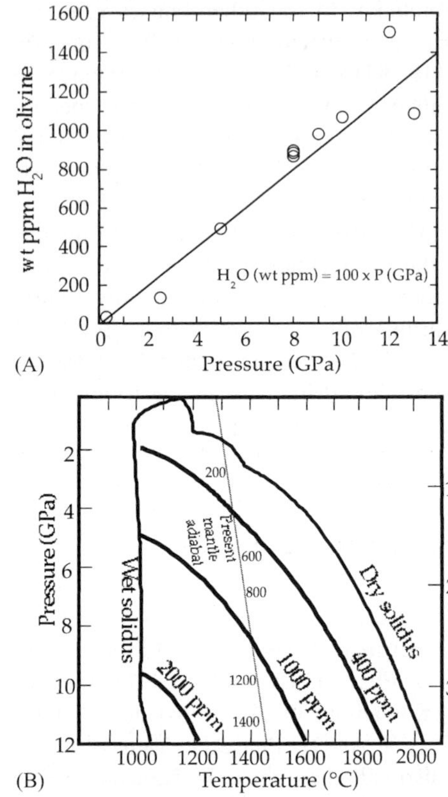

Fig. 3. (A) H_2O solubilities in olivine as a function of pressure. Data are from Bai and Kohlstedt (1992) and Kohlstedt et al. (1996). H_2O solubilities in olivine can be roughly written as C_{H_2O} (ppm) $= 100 \times P$ (GPa). (B) Calculated melting temperature of H_2O-undersaturated mantle peridotite (H_2O abundance in ppm). The numbers along the present mantle adiabat indicate ppm H_2O concentrations of mantle peridotites that can start melting at that pressure. The H_2O-saturated (wet) solidus is from Kawamoto and Holloway (1997).

studies demonstrated that H_2O solubility in Fo_{91} olivine increases from 34 ppm H_2O at 0.3 GPa to more than 1000 ppm at 13 GPa (Bai and Kohlstedt, 1992; Kohlstedt et al., 1996). The H_2O concentration to saturate a model mantle peridotite can be expressed as C_{H_2O} (ppm) = 220 (±50) × P (GPa) for garnet peridotite and C_{H_2O} (ppm) = 290 (±80) × P (GPa) for spinel peridotite with assumption of the following: (1) H_2O solubilities in olivine can be roughly written as C_{H_2O} (ppm) = 100 × P (GPa) (Fig. 3A; Bai and Kohlstedt, 1992; Kohlstedt et al., 1996), (2) partition coefficients of H between enstatite and olivine, diopside and olivine, and garnet and olivine are 3.75 (±1.25), 7.5 (±2.5) and 1, respectively (Hirth and Kohlstedt, 1996) and there is no H in spinel (Bell and Rossman, 1992), (3) modal abundances of 55 wt.% olivine, 20% enstatite, 10% diopside, and 15% garnet for garnet peridotite and 55% olivine, 20% diopside, 22% enstatite, and 3% spinel for spinel peridotite. By the use of the present estimation, the H_2O-undersaturated solidus–pressure diagram can be drawn as a function of H_2O concentration in a model mantle peridotite (Fig. 3B).

Melt inclusion studies suggest that mid-oceanic ridge basalts, Hawaiian hot spots basalts, and Archean komatiites can be formed through partial melting of 100–200, 250–450, and 600–5000 ppm H_2O bearing mantle peridotite, respectively (Michael, 1988; Hirose and Kawamoto, 1995; McDonough and Danyushevsky, 1995; Kamenetsky, 1996; Shimizu et al., 2001). If these hydrous compositions follow the present mantle adiabat under mid-oceanic ridges, the partial melting should start at depths equivalent to 2.5 GPa (75 km) beneath mid-oceanic ridges, 3.5 GPa (105 km) under Hawaiian hot spots, and 5.5 GPa (165 km) and greater pressures to produce komatiites, respectively (Fig. 3B).

6. Partial melt chemistry

(Mg + Fe)/Si atomic ratios of partial melts in the H_2O-saturated KLB-1 peridotite increase from 1 at 5 GPa to 2 at 11 GPa (Kawamoto and Holloway, 1997). At 7.5 GPa, (Mg+Fe)/Si ratios of partial melts increase with increasing temperature (Fig. 4). In contrast, at pressures greater than 10 GPa, (Mg + Fe)/Si ratios of partial melts decrease with increasing temperature (Fig. 4). (Mg + Fe)/Si ratios of partial melts decrease from 2.5 at 1100 °C to 1.6 at 1300 °C at 14 GPa, and from 3.3 at 1100 °C to 1.8 at 1300 °C at 17 GPa. At 20 GPa, (Mg + Fe)/Si ratios decrease from 2.8 at 1200 °C to 1.7 at 1400 °C. Because similar behavior was also observed in the $MgO–SiO_2–H_2O$ system at 12 and 15.5 GPa (Inoue, 1994), various ratios of Fe^{3+}/total Fe may have little effect on the (Mg + Fe)/Si variations as a function of pressure. Since the peridotite itself has an (Mg+Fe)/Si ratio of 1.46, the partial melts are more ferromagnesian than the bulk system at 10 GPa. This change is surprising when compared to the much smaller change seen in dry melts of KLB-1 peridotite: (Mg + Fe)/Si ratio of 1 at 5 GPa, 1.1 at 10 GPa, and 1.1 at 18 GPa (Fig. 4; Herzberg and Zhang, 1996). At a given temperature, (Mg + Fe)/Si ratios of partial melts increase with increasing pressure. It is likely that the structure of wet silicate melts changes drastically as a function of pressure, therefore, spectroscopic observation of wet melt compositions should be conducted to explain this conspicuous feature.

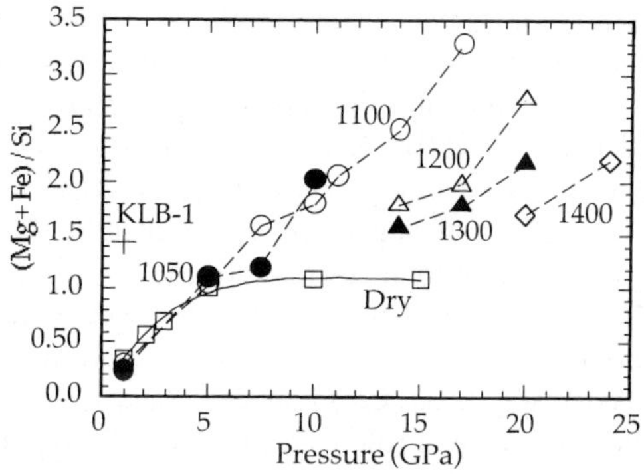

Fig. 4. Molar (Mg + Fe)/Si in the partial melts of H_2O-saturated KLB-1 peridotite as functions of pressure and temperature. Data at a pressures lower than 11 GPa are from Kawamoto and Holloway (1997) and Kawamoto et al. (1996). The composition of KLB-1 peridotite is represented by the cross. Numerals represent the experimental temperatures in °C. The data for dry partial melts of KLB-1 near to the solidus temperature (open square) are from Hirose and Kushiro (1993) for 1–3 GPa, and Herzberg and Zhang (1996) for pressures greater than 5 GPa.

A partial melt coexisting with Mg-perovskite at 24 GPa and 1400 °C has (Mg + Fe)/Si atomic ratio of 2.2. Ultramafic hydrous melts similar to those observed experimentally under uppermost lower mantle conditions may have contributed to chemical differentiation between the upper and lower mantle. Similar implications of H_2O-rich fluids with high Mg/Si ratios were also discussed by previous workers (Stalder et al., 2001; Mibe et al., 2002). The difference in Ca/Al ratios between a chondritic model mantle (0.75) and suggested upper mantle (0.9) is also another important open question (e. g., Palme and Nickel, 1985). Partial melts in the wet peridotite system are also characterized by high Ca/Al ratio (4.5 at 10 GPa and 1100 °C, 11 at 24 GPa and 1400 °C) and have a potential role to resolve this discrepancy.

Effects of H_2O on the partition coefficients of siderophile elements between metallic iron-rich melt and silicate melt remain to be investigated. Righter and Drake (1999) conducted an experimental study at 1 GPa and 1300 °C, and concluded that H_2O has little effect on the partition coefficients of siderophile elements between metallic iron-rich melt and silicate melt. However, recent high-pressure experiments suggest that H_2O can have more effects on melt structure at higher pressure (Stalder et al., 2001; Mibe et al., 2002). Therefore, siderophile elements may be more strongly partitioned into silicate melts in the wet system than the dry system at pressures greater than 10 GPa. This is because H_2O depolymerizes silicate melts and siderophile elements are more strongly partitioned into a less polymerized melt (Hillgren et al., 1996), and also because H_2O lowers the activities of divalent cations in silicate melts and most siderophile elements are divalent in silicate melts in the presence of metallic iron (Hillgren et al., 1996). The partition coefficients of siderophile elements between ultramafic hydrous residual melt and silicate lower mantle minerals will therefore be greater than in the dry system (Ohtani and Yurimoto, 1996; Ohtani et al., 1997b). Effects of H_2O on the partitioning of elements between silicates and iron melts remains to be determined over a wide pressure range to provide further constraints on the chemical evolution of the earth's mantle.

Acknowledgements

I thank B. Poe, N. Bolfan-Casanova, Y. Xu, D. Rubie, D. Krauße, and H. Schulze for their help to conduct experiments at Bayreuth. I am grateful to D. Rubie and Y. Tatsumi for their efforts for financial support and encouragement during this study. T. Komabayashi kindly let me know of his work. Careful reviews by P. Ulmer, T. Duffy, and an anonymous reviewer significantly improved the manuscript. This work was supported by Visiting Scientists' Program of BGI, the Mako Post-Bank Foundation, Kyoto University, the Nissan Science Foundation, and the Kurata Memorial Hitachi Science and Technology Foundation.

References

Asahara, Y., Ohtani, E., 2001. Melting relations of the hydrous primitive mantle in the CMAS–H_2O system at high pressures and temperatures, and implications for generation of komatiites. Phys. Earth Planet. Int. 125, 31–44.

Bai, Q., Kohlstedt, D.L., 1992. Substantial hydrogen solubility in olivine and implications for water storage in the mantle. Nature 357, 672–674.

Bell, D.R., Rossman, G.R., 1992. Water in earth's mantle: the role of nominally anhydrous minerals. Science 255, 1391–1396.

Frost, D.J., 1999. The stability of dense hydrous magnesium silicates in earth's transition zone and lower mantle. In: Fei, Y., Bertka, C., Mysen, B.O. (Eds.), Mantle Petrology: Field Observations and High-pressure Experimentation. A Tribute to Francis R. (Joe) Boyd. Special Publication. Geochemical Society, pp. 283–296.

Herzberg, C., Zhang, J., 1996. Melting experiments on anhydrous peridotite KLB-1: compositions of magmas in the upper mantle and transition zone. J. Geophys. Res. 101, 8271–8295.

Hillgren, V.J., Drake, M.J., Rubie, D.C., 1996. High pressure and high temperature metal–silicate partitioning of siderophile elements: the importance of silicate liquid composition. Geochim. Cosmochim. Acta 60, 2257–2263.

Hirose, K., 1997. Melting experiments on lherzolite KLB-1 under hydrous conditions and generation of high-magnesian andesitic melts. Geology 25, 42–44.

Hirose, K., Kawamoto, T., 1995. Hydrous partial melting of lherzolite at 1 GPa and genesis of basaltic magmas. Earth Planet. Sci. Lett. 133, 463–473.

Hirose, K., Kushiro, I., 1993. Partial melting of dry peridotites at high pressures: determination of compositions of melts segregated from peridotite using aggregates of diamond. Earth Planet. Sci. Lett. 114, 477–489.

Hirth, G., Kohlstedt, D.L., 1996. Water in oceanic upper mantle: rheology, melt extraction and evolution of the lithosphere. Earth Planet. Sci. Lett. 144, 93–108.

Inoue, T., 1994. Effect of water on melting phase relations and melt composition in the system Mg_2SiO_4–$MgSiO_3$–H_2O up to 15 GPa. Phys. Earth Planet. Int. 85, 237–263.

Inoue, T., Higo, Y., Ueda, T., Tanimoto, Y., Irifune, T., 2001. The effect of water on the high-pressure phase boundaries in the system Mg_2SiO_4–Fe_2SiO_4. In: Ito, E. (Ed.), Conference Abstract of Transport of Materials in the Dynamic Earth, Kurayoshi, Japan, pp. 128–129.

Ito, E., Takahashi, E., 1989. Postspinel transformations in the system Mg_2SiO_4–Fe_2SiO_4 and some geophysical implications. J. Geophys. Res. 94, 10637–10646.
Ito, E., Takahashi, E., Matsui, Y., 1984. The mineralogy and chemistry of the lower mantle: an implication of the ultrahigh-pressure phase relations in the system MgO–FeO–SiO_2. Earth Planet. Sci. Lett. 67, 238–248.
Kamenetsky, V., 1996. Methodology for the study of melt inclusions in Cr-spinel, and implications for parental melts of MORB from FAMOUS area. Earth Planet. Sci. Lett. 142, 479–486.
Kanzaki, M., 1991. Stability of hydrous magnesium silicates in the mantle transition zone. Phys. Earth Planet. Int. 66, 307–312.
Katsura, T., Ito, E., 1989. The system Mg_2SiO_4–Fe_2SiO_4 at high pressures and temperatures: precise determination of stabilities of olivine, modified spinel, and spinel. J. Geophys. Res. 94, 15663–15670.
Kawamoto, T., Leinenweber, K., Hervig, R.L., Holloway, J.R., 1995. Stability of hydrous minerals in H_2O-saturated KLB-1 peridotite up to 15 GPa. In: Farley, K. (Ed.), Volatiles in the Earth and Solar System. American Institute of Physics, pp. 229–239.
Kawamoto, T., Hervig, R.L., Holloway, J.R., 1996. Experimental evidence for a hydrous transition zone of the early earth's mantle. Earth Planet. Sci. Lett. 142, 587–592.
Kawamoto, T., Holloway, J.R., 1997. Melting temperature and partial melt chemistry of H_2O-saturated mantle peridotite to 11 gigapascal. Science 276, 240–243.
Kawamoto, T., Matsukage, K., Mibe, K., Isshiki, M., Nishimura, K., Ishimatsu, N., Ono, S., 2003. H_2O-rich magmas or silicate-rich H_2O fluids? A perspective from high pressure and temperature experiments. European Geophysical Society, American Geophysics Union, European Union of Geoscience, Joint Assembly, EAE03-A-01670.
Kohlstedt, D.L., Keppler, H., Rubie, D.C., 1996. Solubility of water in α, β and γ phases of $(Mg, Fe)_2SiO_4$. Contrib. Miner. Petrol. 123, 345–357.
Komabayashi, T., Omori, S., Maruyama, S., 2002. Petrogenetic grid in the system MgO–SiO_2–H_2O up to 30 GPa, 1600 °C: implications for water transportation by slab peridotite and for deep-focus seismicity. Am. Geophys. Union Trans. 83, F1378.
Litasov, K., Ohtani, E., 2002. Phase relations and melt compositions in CMAS–pyrolite–H_2O system up to 25 GPa. Phys. Earth Planet. Int. 134, 105–127.
McDonough, W.F., Danyushevsky, L.V., 1995. Water and sulfur contents of melt inclusions from Archean komatiites. Am. Geophys. Union Trans. 76, 266.
Mibe, K., Fujii, T., Yasuda, A., 2002. Composition of aqueous fluid coexisting with mantle minerals at high pressure and its bearing on the differentiation of the earth's mantle. Geochim. Cosmochim. Acta 66, 2273–2285.
Michael, P.J., 1988. The concentration, behavior and storage of H_2O in the suboceanic upper mantle. Geochim. Cosmochim. Acta 52, 567–577.
Mysen, B.O., Ulmer, P., Konzett, J., Schmidt, M.W., 1998. The upper mantle near convergent plate boundaries. In: Hemley, R.J. (Ed.), Ultrahigh-pressure Mineralogy: Physics and Chemistry of the Earth's Deep Interior. Reviews in Mineralogy, pp. 97–138.
Ohtani, E., Mizobata, H., Kudoh, Y., Nagase, T., Arashi, H., Yurimoto, H., Miyagi, I., 1997a. A new hydrous silicate a water reservoir in the upperpart of the lower mantle. Geophys. Res. Lett. 24, 1047–1050.
Ohtani, E., Toma, M., Litasov, K., Kubo, T., Suzuki, A., 2001. Stability of dense hydrous magnesium silicate phases and water storage capacity in the transition zone and lower mantle. Phys. Earth Planet. Int. 124, 105–117.
Ohtani, E., Yurimoto, H., 1996. Element partitioning between metallic liquid, magnesiowustite, and silicate liquid at 20 GPa and 2500 °C: a secondary ion mass spectrometric study. Geophys. Res. Lett. 23, 1993–1996.
Ohtani, E., Yurimoto, H., Seto, S., 1997b. The element partitioning between metallic liquid, silicate liquid, and lower mantle minerals: implications for the core formation of the earth. Phys. Earth Planet. Int. 100, 97–114.
Okuchi, T., 1997. Hydrogen partitioning into molten iron at high pressure: implications for earth's core. Science 278, 1781–1784.
Palme, H., Nickel, K., 1985. Ca/Al ratio and the compositions of the earth's upper mantle. Geochim. Cosmochim. Acta 49, 2123–2132.
Righter, K., Drake, M.J., 1999. Effect of water on metal–silicate partitioning of siderophile elements: a high pressure and temperature terrestrial magma ocean and core formation. Earth Planet. Sci. Lett. 171, 383–399.
Schmidt, M., Poli, S., 1998. Experimentally based water budgets for dehydrating slabs and consequences for arc magma generations. Earth Planet. Sci. Lett. 163, 361–379.
Shimizu, K., Komiya, T., Hirose, K., Shimizu, N., Maruyama, S., 2001. Cr-spinel, an excellent micro-container for retaining primitive melts—implications for a hydrous plume origin for komatiites. Earth Planet. Sci. Lett. 189, 177–188.
Smyth, J.R., Bell, D.R., Rossman, G.R., 1991. Incorporation of hydroxyl in upper-mantle clinopyroxenes. Nature 351, 732–734.
Stalder, R., Ulmer, P., Thompson, A.B., Günther, D., 2001. High pressure fluids in the system MgO–SiO_2–H_2O under upper mantle conditions. Contrib. Miner. Petrol. 140, 607–618.
Sudo, A., Tatsumi, Y., 1990. Phlogopite and K-amphibole in the upper mantle: implication for magma generation in subduction zones. Geophys. Res. Lett. 17, 29–32.
Takahashi, E., Shimazaki, T., Tsuzaki, Y., Yoshida, H., 1993. Melting study of a peridotite KLB-1 to 6.5 GPa, and the origin of basaltic magmas. Philos. Trans. R. Soc. London A 342, 105–120.
Thompson, A.B., 1992. Water in the earth's upper mantle. Nature 358, 295–302.
Trønnes, R.G., Takahashi, E., Scarfe, C.M., 1988. Stability of K-richterite and phlogopite to 14 GPa. Trans. Am. Geophys. Union 69, 1510–1511.
Yang, H., Prewitt, C.T., Frost, D.J., 1997. Crystal structure of the dense hydrous magnesium silicate, phase D. Am. Miner. 82, 651–654.
Zhang, J., Herzberg, C., 1994. Melting experiments on anhydrous peridotite KLB-1 from 5.0 to 22.5 GPa. J. Geophys. Res. 99, 17729–17742.

Available online at www.sciencedirect.com
SCIENCE DIRECT®

ELSEVIER

Physics of the Earth and Planetary Interiors 143–144 (2004) 397–406

PHYSICS OF THE EARTH AND PLANETARY INTERIORS

www.elsevier.com/locate/pepi

Melting experiments of mantle materials under lower mantle conditions with implications for magma ocean differentiation

E. Ito*, A. Kubo, T. Katsura, M.J. Walter

Institute for Study of the Earth's Interior, Okayama University, Misasa, Tottori-ken 682-0193, Japan

Received 1 February 2003; received in revised form 22 August 2003; accepted 4 September 2003

Abstract

Liquidus phase relations and major element partitioning have been determined for fertile peridotite and CI chondritic mantle material to 33 and 35 GPa, respectively. In peridotite, the first liquidus phase changes from ferropericlase to Mg-perovskite at 31 GPa, and at 33 GPa liquidus Mg-perovskite is successively followed down temperature by ferropericlase and Ca-perovskite within a small temperature range. In CI chondritic mantle, Mg-perovskite is the liquidus phase followed down temperature by Ca-perovskite at pressures higher than 28 GPa. Ferropericlase is absent at the liquidus. Differentiation by crystal fractionation of Mg-perovskite, ferropericlase, and Ca-perovskite in a deep magma ocean has been examined for a CI chondritic and two peridotitic bulk silicate earth models. Mass balance indicates that subtraction of about 40% Mg-perovskite and 2% Ca-perovskite from a CI chondritic bulk silicate earth yields a residual melt with some characteristics of model fertile upper mantle composition. A crystal layer composed of Mg- and Ca-perovskites would pile up to a depth ca. 1400 km, and may be characterized as an enriched and possibly heat-producing reservoir by the high capability of Ca-perovskite to accommodate large cations such as La and alkaline elements. The observed effect of pressure on element partitioning indicates that better mass balance solutions may be obtained for higher pressure liquidus phase compositions. For peridotitic bulk silicate earth models, fractionation would be quite limited, up to 10 wt.% of Mg-perovskite in addition to trace amount of Ca-Pv.

Keywords: Sintered diamond; Lower mantle conditions; Melting relations; Peridotite; CI chondritic mantle; Magma ocean differentiation

1. Introduction

It is generally accepted that the earth was covered by a deep magma ocean during early accretion. The sources of heat required for such extensive melting were supplied by the released gravitational potential energy supplied by the kinetic energy from large impacts, including a possible giant moon-forming-impact (Wetherill, 1990; Melosh, 1990), as well as potential energy release associated with segregation of the central metallic core from uniform primordial material (Birch, 1965; Flasar and Birch, 1973). Cooling of the silicate magma ocean could have resulted in crystal fractionation that would have produced large-scale chemical stratification in the mantle.

Melting experiments on model mantle compositions at high pressures corresponding to a deep magma ocean provide important information for clarifying the geochemical effects of crystal fractionation. Melting relations and element partitioning among coexisting liquidus phases and melt in peridotitic materials have been studied extensively (Takahashi,

* Corresponding author. Tel.: +81-858433753; fax: +81-858433450.
E-mail address: eiito@misasa.okayama-u.ac.jp (E. Ito).

0031-9201/$ – see front matter © 2004 Elsevier B.V. All rights reserved.
doi:10.1016/j.pepi.2003.09.016

1986; Ito and Takahashi, 1987; Zhang and Hertzberg, 1994; McFarlane et al., 1994; Trønnes and Frost, 2002). Due to limitations in the mechanical strength of carbide anvils, experimental pressures achieved in a Kawai-type of multi-anvil apparatus have been limited to ca. 26 GPa, corresponding to the uppermost part of the lower mantle. These experiments have shown that the liquidus phase of fertile peridotite changes from olivine to majorite garnet (Mj) at 15 GPa, and at a pressure of about 23 GPa, the liquidus phase changes to ferropericlase (Fp) followed by Mg-perovskite (Mg-Pv) within a small crystallization interval. In this regard it is possible that Ito and Takahashi (1987) misjudged the crystallization sequence of Mg-Pv and Fp in their charge due to a relatively uniform temperature distribution caused by the diamond (originally graphite) sample capsule and $LaCrO_3$ heater.

Separation of Mj (Kato et al., 1988), Mg-Pv (Ito and Takahashi, 1987; Kato et al., 1988), or both Mg-Pv and Fp (McFarlane et al., 1994) from a terrestrial magma ocean has been examined in terms of the fractionations imparted to refractory and moderately refractory lithophile element ratios in the residual melt, under the assumption that the relative abundances of these elements in the bulk mantle are equal to those in CI-chondrite. However, crystal fractionation models have generally been unsuccessful in reproducing the element ratios observed in upper mantle peridotite or estimates of primitive mantle (i.e., the mantle before crust extraction). This implies either crystal fractionation in a magma ocean was impeded by slow crystal growth of liquidus phases and/or vigorous convection in the magma ocean (e.g., Solomatov, 2000); or that mantle differentiation caused by early crystal fractionation was erased by long term solid-state mantle convection.

However, these early models are based on a limited set of experimental data at pressures confined to the deep upper mantle or the uppermost lower mantle. Pressure can have a considerable effect on liquidus phase relations, and experiments at higher pressures may provide new information regarding the plausibility of fractionation in a deep magma ocean. In this context we have investigated melting of peridotite and a CI chondritic mantle material (CMM) based on CI carbonaceous chondrites (Hart and Zindler, 1986; Ohtani et al., 1986) under lower mantle conditions, at pressures up to 35 GPa.

2. Experimental

Natural peridotite starting material was a finely powdered split of spinel xenolith KR4003 (Xue et al., 1990; Walter, 1998) from West Kettle River, Canada, which has a bulk composition typical of fertile upper mantle peridotite. A CI chondritic mantle material was synthesized based on the model composition of Ohtani et al. (1986). Small amounts of oxides, TiO_2, MnO, NiO, and Na_2O in chondritic relative proportion were added to the original composition in the CFMAS system. A rather large amount of K_2O, 0.14%, was also added in order to accurately measure its partitioning among coexisting phases. Chemical compositions of these starting materials are given in Tables 2 and 3.

High-pressure experiments were performed in a cubic multi-anvil, Kawai-type system (Kawai and Endo, 1970) with the typical second-stage carbide anvils replaced by sintered diamond anvils. Sintered diamond cubes with 14 mm edge length and 1.5 or 2.0 mm truncated corners were employed together with an octahedral pressure medium of magnesia mixed with 5% Cr_2O_3. A cross-section of the heating assembly for 2.0 mm truncation is schematically shown in Fig. 1. Powdered starting material was loaded directly into a cylindrical Re heater, which was placed centrally in the octahedron with a $LaCrO_3$ sleeve for thermal insulation. Sample pressure versus press load in an oil chamber type press was calibrated by detecting phase transitions of Bi III–V (7.7 GPa), semiconductor to metal in ZnS (15 GPa) and GaP (22 GPa), and Zr ω–β (33 GPa) at room temperature and those of coesite-stishovite in SiO_2 (10.4 GPa) and ilmenite-perovskite in $MgSiO_3$ (23 GPa) at 1600 °C. The phase transitions at room temperature and at 1600 °C plotted virtually on a single calibration curve (Ito et al., 1998) when using a relatively thick $LaCrO_3$ sleeve with inner and outer diameters 0.6 and 2.2 mm (Ito et al., 1998). Accuracy of pressure determination based on the calibration curve was estimated to be within ± 1 up to 35 GPa. Sample temperature was monitored by a thin $W_3Rh_{97}/W_{25}Rh_{75}$ thermocouple up to 2200 °C, and the temperature/power relation was extrapolated to determine higher temperature. After the sample was kept at the prescribed pressure and at 2500 ± 100 °C for 2 or 3 min, it was quenched by shutting off the power to the heater. Polished sections

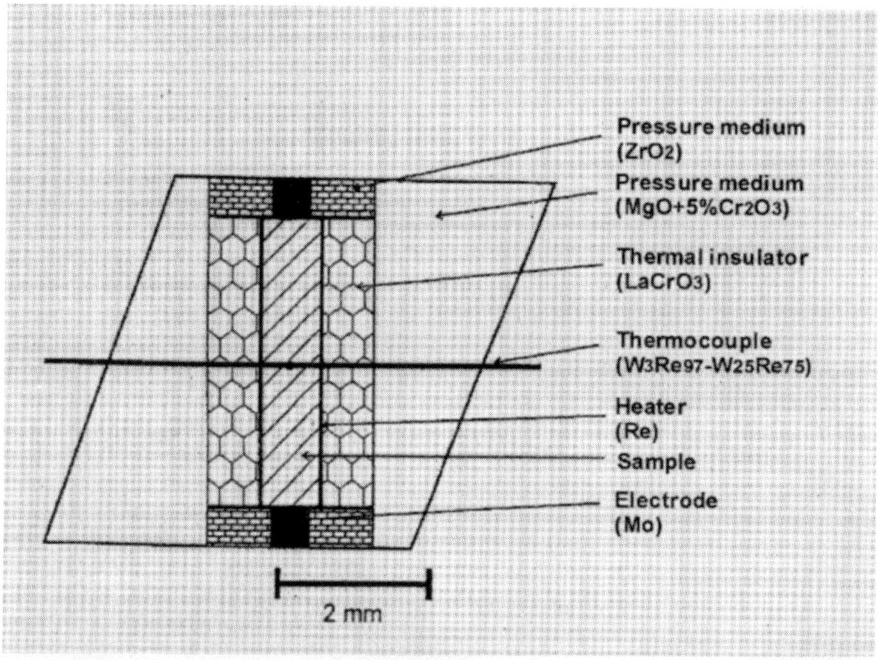

Fig. 1. A schematic drawing of the sample assembly for melting experiments and using 2.0 mm truncations. All the parts for 1.5 mm truncation are slightly scaled-down. The thermocouple junction is put on the outer surface of the Re heater.

of the recovered charges were examined by electron microscopy and then analyzed using a JEOL, Super Probe 8800 electron micro probe analyzer operated at an accelerating voltage of 15 keV and a beam current of 10 nA.

3. Phase relations

Results of melting experiments are summarized in Table 1. A cross-section of the quenched charge is shown in Fig. 2. Occurrence of melting was confirmed

Table 1
List of melting experiments

Run number	Starting material	P (GPa)	T (°C)[a]	Duration (min)	Phase assemblage[b] (in crystallization sequence)
Pe26	Peridotite	26	2500	3.0	L → Fp → Mg-Pv
Pe29	Peridotite	29	2500	2.5	L → Fp → Mg-Pv
Pe31	Peridotite	31	2500	2.5	L → Mg-Pv → Fp → Ca-Pv
Pe32	Peridotite	32	2500	3.0	L → Mg-Pv → Fp → Ca-Pv
Pe33	Peridotite	33	2500	3.0	L → Mg-Pv → Fp → Ca-Pv
CI22	CI mantle	22	2500	3.0	L → Mj → Fp → Ca-Pv → β(γ)
CI28	CI mantle	28	2500	2.5	L → Mg-Pv → Ca-Pv
CI35	CI mantle	35	2500	2.5	L → Mg-Pv → Ca-Pv

[a] Temperature of 2500 °C was estimated by extrapolating the relation between temperature and power supply constructed to 2200 °C. The uncertainty is expected within ±150 °C.

[b] Abbreviations: L, melt; Mg-Pv, Mg-perovskite; Fp, ferropericlase; Ca-Pv, Ca-perovskite; Mj, majorite garnet; β(γ), modified spinel (or spinel). In Pe31 run at 31 GPa, both Mg-Pv and Fp coexisted on the liquidus.

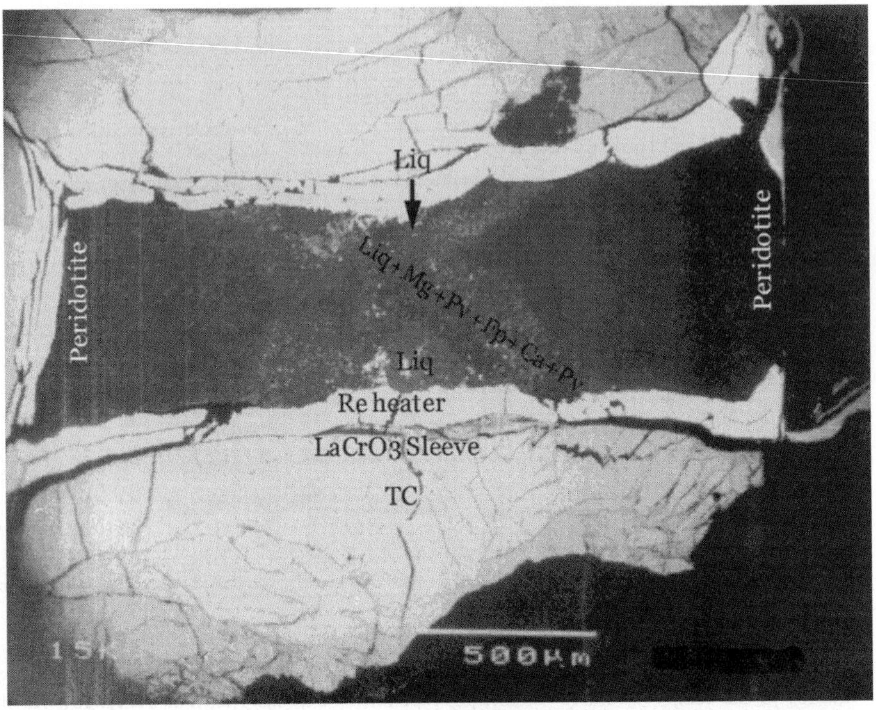

Fig. 2. A whole view (back-scattered electron image) of the sectioned charge (run Pe32). Variation in phases present from melt in the central portion to the starting peridotite at both end portions is due to the steep temperature gradient through the charge. Abbreviations: Liq, melt; Mg-Pv, Mg-perovskite; Fp, ferropericlase; Ca-Pv, Ca-perovskite; TC, relic of the thermocouple.

by the presence of dendritic texture in the central portions of the charge quenched from ca. 2500 °C. In both the end portions about 1 mm away from the center, however, the starting material showed no apparent reaction, suggesting that the temperature there was less than 1000 °C. In this case, we estimate that the temperature gradient through the sample was 1–2 °C/μm (see also Ito and Takahashi, 1989). Most of the melt concentrated in the central, high-temperature region of the charge, which would have been caused by the dissolution of solid phase components at high temperature and precipitation of those components at lower temperature due to the saturation gradient (Lesher and Walker, 1988) in the steep temperature gradient. This makes location of the liquidus very clear, whereas the solidus is defined more ambiguously in terms of grain sizes; larger grains in the super-solidus region and the presence of psuedomorphs of the starting minerals in the sub-solidus region. Identification of the super-solidus phases was carried out based on morphology and chemical composition. The temperature gradient in the charge makes it possible to establish the crystallization sequence of different phases with decreasing temperature from the liquidus (see Table 1). Melting relations of both peridotite and the CI chondritic mantle material were determined from characteristic features of the run products as follows.

3.1. Peridotite

Melting experiments on fertile peridotite were made in a pressure range of 26–33 GPa. A back-scattered electron image of the quenched run product from 33 GPa is shown in Fig. 3(a). Ferropericlase (Fp) is the liquidus phase up to about 30 GPa, which is successively followed at lower temperature by magnesian perovskite (Mg-Pv) and calcium perovskite (Ca-Pv). Both Fp and Mg-Pv coexist along the liquidus in the charge quenched at 31 GPa, indicating multiple saturation at the liquidus of these phases. At 33 GPa, however, the front of Fp grains moves back from the liquidus to the slightly lower temperature region and

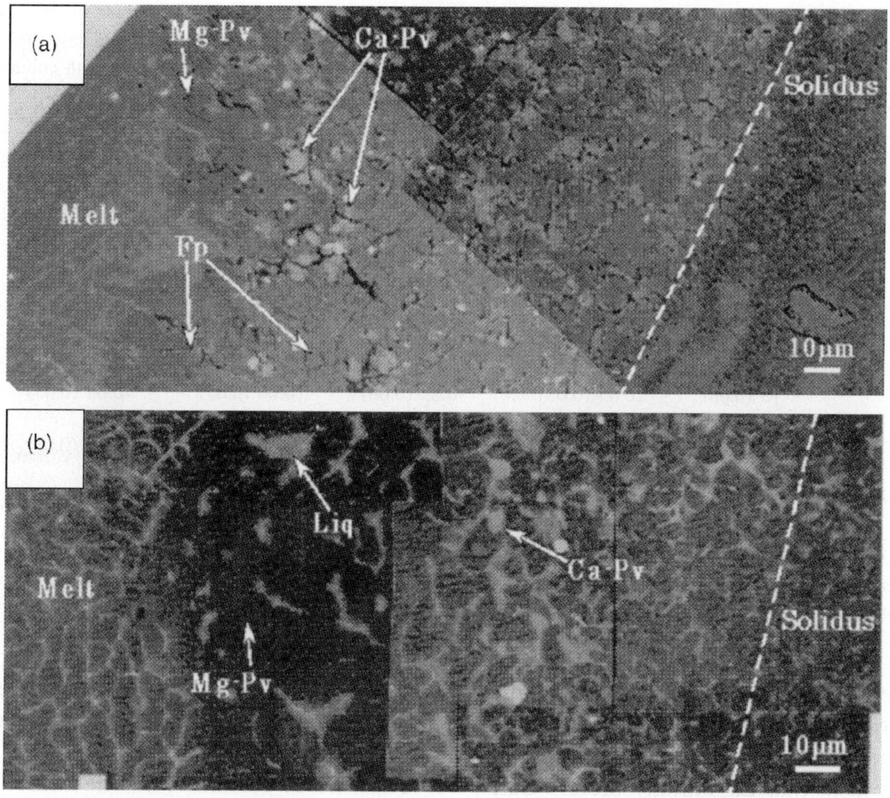

Fig. 3. Back-scattered electron images of the regions including both the solidus and liquidus: (a) peridotite quenched at ca. 33 GPa and 2500 °C (run Pe33); (b) CI chondritic mantle material quenched at ca. 35 GPa and 2500 °C (run CI35). Abbreviations: Liq, melt; Mg-Pv, Mg-perovskite; Fp, ferropericlase; Ca-Pv, Ca-perovskite.

Mg-Pv becomes the liquidus phase (Fig. 3(a)). Ca-Pv crystallizes at a somewhat lower temperature than Fp and Mg-Pv at pressures up to ca. 29 GPa. However, the crystallization temperatures of Fp and Ca-Pv become closer with increasing pressure, perhaps within only a few degrees at 33 GPa. The change in liquidus phase relations in peridotite indicates that the composition of the initial melt formed in peridotitic mantle deeper than 900 km should be more basic (or more deficient in silica) than the peridotitic composition.

The experimental charge was usually somewhat contaminated by the $LaCrO_3$ sleeve. Chemical compositions of melt, Mg-Pv, Fp, and Ca-Pv coexisting near the liquidus at 33 GPa are shown in Table 2, together with those of Mg-Pv and Fp close to the solidus. The central "melt pools" in the run products were relatively homogeneous with an average composition very close to the initial peridotite. However, the FeO content is lower by more than 1%. It is remarkable that both Mg-Pv and Fp on the liquidus have conspicuously higher Mg/Fe ratio than those located close to the solidus (Table 2). The increase in Fe/Mg ratio of the crystalline phases from the liquidus to the solidus was also observed by McFarlane et al. (1994) in the melting of KLB-1 peridotite at lower pressure (24.5 GPa). The feature would be caused by rapid reaction of the super-solidus Mg-Pv and Fp with the initial melt whose composition was controlled by the saturation gradient dictated by the steep temperature gradient, i.e., enrichment of MgO in high-temperature portion and that of FeO in lower temperature portion.

Element partition coefficients between the solid phases and melt coexisting near the liquidus are summarized in Table 3. Partition coefficients between Mg-Pv, Fp and melt for Si, Mg, Fe, and Ni do not show prominent pressure dependence when compared

Table 2
Chemical compositions of starting material KR4003 and those of melt (Liq), Mg-perovskite (Mg-Pv), ferropericlase (Fp), and Ca-perovskite (Ca-Pv) coexisting at 33 GPa (2nd to 5th columns), and of Mg-Pv and Fp located near the solidus (7th and 8th columns)

	KR4003[a]	Liq ($n = 16$)	Mg-Pv ($n = 8$)	Fp ($n = 8$)	Ca-Pv ($n = 7$)	Mg-Pv[b] ($n = 6$)	Fp[b] ($n = 4$)
SiO_2	44.9	44.86 (0.75)	56.37 (0.76)	0.21 (0.03)	41.95 (1.30)	53.45 (0.92)	0.63 (0.47)
TiO_2	0.16	0.17 (0.03)	0.14 (0.04)	–	0.19 (0.03)	0.17 (0.10)	0.01 (0.01)
ReO_2	–	1.52 (0.30)	0.32 (0.10)	0.31 (0.11)	1.22 (0.77)	0.16 (0.11)	0.3 (0.13)
Cr_2O_3	0.41	1.26 (0.20)	0.64 (0.09)	1.11 (0.16)	0.63 (0.12)	0.34 (0.05)	1.2 (0.29)
Al_2O_3	4.26	4.13 (0.19)	3.77 (0.17)	1.66 (0.10)	4.34 (0.30)	5.37 (0.78)	2.25 (0.12)
La_2O_3	–	1.05 (0.59)	0.02 (0.02)	–	11.84 (0.97)	–	0.01 (0.01)
FeO[c]	8.02	6.26 (0.54)	2.65 (0.14)	10.73 (1.27)	2.21 (0.33)	7.25 (0.34)	29.78 (3.90)
MgO	37.3	37.96 (0.65)	36.92 (0.38)	86.64 (2.03)	4.10 (0.67)	35.08 (0.59)	65.43 (4.90)
CaO	3.45	2.88 (0.21)	1.36 (0.23)	0.09 (0.01)	33.84 (1.11)	0.56 (0.36)	0.19 (0.02)
NiO	0.24	0.18 (0.05)	0.05 (0.03)	0.89 (0.06)	0.06 (0.05)	0.08 (0.05)	1.41 (0.12)
CoO	–	–	–	0.06 (0.04)	0.02 (0.03)	–	0.16 (0.06)
Na_2O	0.22	0.07 (0.02)	0.01 (0.02)	0.18 (0.06)	0.16 (0.04)	0.03 (0.03)	1.13 (0.26)
K_2O	0.09	0.02 (0.02)	0.01 (0.01)	–	0.32 (0.20)	–	0.01 (0.01)
Total	99.05	100.36	102.26	101.88	100.88	102.54	102.54

[a] After Xue et al. (1990).
[b] Compositions of Mg-Pv and Fp located near the solidus.
[c] All Fe was assumed to be ferrous.

with those determined at lower pressures (McFarlane et al., 1994; Trønnes and Frost, 2002). However, Al, Cr, and Ti tend to be "incompatible" in Mg-Pv at 33 GPa. It is evident that La strongly prefers Ca-Pv rather than melt, presumably occupying of the large 12-coordinated 'Ca-site' in the structure. The high concentrations of alkaline elements (K and Na) in Ca-Pv may be coupled with the substitution of La for Ca. Therefore, it is inferred that Ca-Pv has a high capability to accommodate large univalent and trivalent cations, especially when both are combined.

Table 3
Partition coefficients between Mg-perovskite and melt ($D^{Mg\text{-}Pv/Liq}$), ferropericlase and melt ($D^{Fp/Liq}$), and Ca-perovskite and melt ($D^{Ca\text{-}Pv/Liq}$) coexisting close to the liquidus at 33 GPa

	$D^{Mg\text{-}Pv/Liq}$	$D^{Fp/Liq}$	$D^{Ca\text{-}Pv/Liq}$
Si	1.26 (0.02)	0.005 (0.001)	0.94 (0.03)
Ti	0.84 (0.20)	0.006	1.11 (0.25)
Re	0.08 (0.03)	–	0.8 (0.5)
Cr	0.51 (0.10)	0.88 (0.19)	0.50 (0.08)
Al	0.91 (0.04)	0.40 (0.03)	1.05 (0.09)
La	–	–	11.3 (3.2)
Fe	0.42 (0.04)	1.71 (0.25)	0.35 (0.02)
Mg	0.97 (0.01)	2.28 (0.06)	0.11 (0.02)
Ca	0.47 (0.08)	0.031 (0.005)	11.74 (2.03)
Ni	0.3 (0.2)	4.83 (1.41)	0.33 (0.30)
Co	–	–	–
Na	0.2 (0.3)	–	2.3 (0.9)
K	0.6 (1.0)	–	21.0 (27)

3.2. The CI chondritic mantle material

Melting experiments on CMM were carried out at 22, 28, and 35 GPa (Table 1). The quenched charges also showed a concentration of melt in the central, high-temperature region and a decrease in Mg/(Mg + Fe) ratio of liquidus phases from the liquidus to the solidus. At 22 GPa, the first liquidus phase is Mj which is successively followed by small amounts of Fp, Ca-Pv and wadsleyite or ringwoodite. This observation is consistent with previous studies (Ohtani et al., 1986; Ohtani and Sawamoto, 1987) up to 25 GPa. At 28 GPa, however, Mj and Fp completely disappear, and the liquidus phase is Mg-Pv followed down temperature by Ca-Pv. Melting relations at 35 GPa are the same as those at 28 GPa. A back-scattered electron image of the charge quenched from 35 GPa is shown in Fig. 3(b), in which Mg-Pv and Ca-Pv appearing above the solidus are indicated.

Chemical compositions of the central melt, super-solidus Mg-Pv and Ca-Pv at 35 GPa are presented in Table 4. The melt has a similar composition to the starting CMM composition, as in the cases of melting of peridotite. There are many small patches

Table 4
Chemical compositions model CI mantle composition and those of melt (Liq), Mg-perovskite (Mg-Pv), and Ca-perovskite (Ca-Pv) coexisting at 35 GPa

	CI mantle[a]	Liq ($n = 3$)	Mg-Pv ($n = 10$)	Ca-Pv ($n = 5$)
SiO_2	49.83	49.91 (0.61)	57.13 (0.78)	36.89 (1.89)
TiO_2	0.16	0.14 (0.01)	0.17 (0.04)	0.06 (0.04)
ReO_2	–	0.20 (0.15)	0.12 (0.13)	0.24 (0.24)
Cr_2O_3	–	1.89 (0.08)	1.20 (0.10)	0.96 (0.13)
Al_2O_3	3.71	5.54 (0.25)	4.92 (0.40)	6.15 (0.52)
La_2O_3	–	0.91 (0.16)	0.12 (0.14)	23.55 (4.29)
FeO[b]	7.02	4.91 (0.41)	2.31 (0.41)	1.79 (0.52)
MgO	35.46	35.05 (1.20)	36.98 (0.61)	5.08 (1.34)
CaO	2.80	2.32 (0.35)	1.63 (0.40)	23.60 (2.39)
MnO	0.11	0.13 (0.07)	0.12 (0.04)	0.09 (0.05)
NiO	0.20	0.06 (0.01)	0.03 (0.04)	0.02 (0.03)
Na_2O	0.28	0.19 (0.01)	0.04 (0.02)	0.49 (0.12)
K_2O	0.14	0.07 (0.06)	0.04 (0.03)	1.27 (0.40)
Total	99.71	101.32	104.81	100.19

[a] Compounded flowing (Ohtani et al., 1986) and adding small amounts of minor components.
[b] All Fe was assumed to be ferrous.

of interstitial melt between the liquidus and solidus, which are compositionally different from the central melt due to growth of the surrounding crystals on quenching. Composition of the liquidus Mg-Pv is similar to that of the liquidus Mg-Pv in peridotite at 33 GPa. This charge was also contaminated by $LaCrO_3$, with the same marked preference of La, Na, and K into Ca-Pv over melt observed for peridotite melting.

4. Fractionation of the mantle in a magna ocean

New results on melting of peridotite and CMM obtained in the present study lead us to reexamine the geochemical effects of crystal fractionation in a magma ocean. Here, we use a simple mass balance approach to assess whether observed liquidus phases can be subtracted in reasonable proportions from the bulk silicate earth (BSE) to derive a residual that is consistent with fertile peridotite as sampled in the uppermost mantle (FPER). Least-squared regressions are made with respect to five components according to the mass balance equation:

$$\mathrm{BSE} = a_1\mathrm{FPER} + a_2\mathrm{Mg\text{-}Pv} + a_3\mathrm{Fp} + a_4\mathrm{Ca\text{-}Pv} \quad (1)$$

where BSE and FPER denote bulk silicate earth and fertile peridotite, respectively, and a_1, a_2, a_3, and a_4 are fitting parameters. Results of mass balance calculations are summarized in Table 3.

We have chosen three model BSE compositions, including a model based on a bulk earth with CI chondrite composition (Hart and Zindler, 1986), and two primitive mantle compositions that were constructed on the basis of correlations among refractory element ratios in selected sets primitive peridotite samples from the upper mantle and chondritic meteorites; i.e. 'pyrolite' (McDonough and Sun, 1995) and PRIMA (Allègre et al., 1995). These primitive mantle compositions have near-chondritic refractory lithophile element ratios with the exception of Mg/Si. The non-chondritic Mg/Si ratio in these models is considered to be a primary feature of BSE, due either to volatilization of these elements in cosmochemical processes (McDonough and Sun, 1995) or partitioning of Si into the core (Ringwood, 1966; Allègre et al., 1995). We adopt the least depleted ultramafic xenolith model by Jagoutz et al. (1979) as the standard fertile peridotite of the "upper" mantle (FPER), which is non-chondritic in refractory element ratios such as Ca/Al, Al/Ti as well as Mg/Si. Such a composition, if representative of a large reservoir in the mantle, could represent the residual liquid of crystal fractionation from a deep magma ocean. Admittedly there is circularity in using compositions for BSE and FEPR that are both constructed on the basis of upper

Table 5
Major element mass balance regression for differentiation of the Mg-perovskite (Mg-Pv), Ferropericlase (Fp), and Ca-perovskite (Ca-Pv) from bulk silicate earth

	BSE[a]			FPER[a]	Crystals[a]			Residual melts[b]		
	CMM	Pyrolite	PRIMA		Mg-Pv	Fp	Ca-Pv	CMM	Pyrolite	PRIMA
SiO_2	50.16	45.52	46.73	45.73	55.77	0.21	48.54	46.34	45.51	45.7
Al_2O_3	3.61	4.5	4.14	4.02	3.73	1.67	5.02	3.46	4.5	4.19
FeO	7.23	8.14	7.58	7.92	2.62	10.8	2.55	10.61	8.17	8.16
MgO	36.14	38.24	38.27	38.78	36.53	87.23	4.74	37.1	38.41	38.54
CaO	2.86	3.59	3.27	3.54	1.34	0.09	39.15	2.48	3.41	3.42
$(MgO/SiO_2)_n$[c]	0.72	1.17	1.14	1.18	0.91	572.58	0.14	1.11	1.17	1.17
$(CaO/Al_2O_3)_n$[c]	0.79	1.01	1	1.11	0.45	0.07	9.85	0.91	0.96	1.03
$(Al_2O_3/MgO)_n$[c]	0.1	1.18	1.09	1.04	1.02	0.19	10.6	0.94	1.17	1.09
Mg#	89.91	89.33	90	89.73	96.13	93.51	76.81	86.18	89.34	89.39

Mass balance regression was carried out with respect to SiO_2, Al_2O_3, FeO, MgO, and CaO, following the formula: BSE $= a_1$FPER $+ a_2$Mg-Pv $+ a_3$Fp $+ a_4$Ca-Pv. The combinations of a_i's for the best fit are (0.574, 0.400, –, 0.023), (0.989, 0, 0, 0.005), and (0.894, 0.102, 0, 0.002) with associated residuals ($\sqrt{\sum R^2}$) of 2.0, 0.61, and 0.33 for CMM, 'pyrolite', and PRIMA, respectively. Residual melts are derived by subtracting the crystals from BSEs in the "best fit" proportions.

[a] See text for data sources and comments.

[b] Abbreviations are corresponding to those of BSE.

[c] Ratio normalized to CI chondrite except CMM.

mantle peridotite. We are forced to use such compositions in the absence of any independent constraints on the composition of BSE, and view the results only as an exercise in identifying possible solutions.

For the fractionating phases in Eq. (1), the liquidus Mg-Pv, Fp, and Ca-Pv in peridotite at 33 GPa were adopted, because these phases coexisted with peridotitic melt. Mass balance was made only on SiO_2, Al_2O_3, FeO, MgO, and CaO, and all compositions were re-normalized to these five compositions.

First we consider the results using CI chondritic BSE. Mass balance results in an unrealistic solution, with negative coefficients for both Fp and Ca-Pv. However, phase relations for CI chondritic composition show that Fp is not a liquidus phase. If Fp is eliminated from the mass balance, then a solution is generated in which BSE is balanced by 58% FPER, 40% Mg-Pv, and 2% Ca-Pv (Table 5). At face value such a solution would indicate that the present upper mantle could be a residual liquid produced by over 40% fractionation of Mg-Pv and Ca-Pv. This would correspond to a crystal mass extending from the bottom of the mantle to form a layer attaining a depth ca. 1400 km. Such a perovskite layer could be permanently stabilized in the lower mantle due to intrinsically high density and rheologic stability. Such a perovskite layer, however, would be expected to become complexly disturbed over geologic time by interaction with subducted slabs and by the formation of upwelling plumes. These structural features of the deep mantle may be depicted as envisioned by Kellog et al. (1999) in their Fig. 1. The perovskite layer may also be characterized by the high capability of Ca-Pv for accommodating rare earth elements, and perhaps heat-producing potassium, which in some geochemical models are required to be held in a hidden reservoir, most often taken in the deep mantle (e.g., Hofmann, 1997).

In spite of these interesting features marked difficulties exist in that the mass balance solution has high residuals, especially for both FeO and CaO, the residual melt would have chondrite-normalized CaO/Al_2O_3 and Al_2O_3/MgO ratios of less than unity and would have a low Mg# of about 86. These features are not consistent with the composition of primitive upper mantle. Nevertheless, it is worth pointing out that the partition coefficients at 33 GPa (Table 3) yield many element ratios that are closer to the upper mantle composition due to perovskite fractionation starting from CI mantle when compared to results using liquidus phase compositions from the lower pressure studies (McFarlane et al., 1994; Trønnes and

Frost, 2002). This indicates that the effect of pressure on partitioning may work in favor of a model of perovskite fractionation such that at even higher pressures a successful solution might be derived.

As expected from the closer similarity of model peridotite BSE compositions and the FEPR composition, mass balance solutions indicate much less fractionation than the CI case. Indeed, only a very minor amount of Ca-Pv fractionation is needed to create a mass balance between 'pyrolite' and FPER. In the case of PRIMA, about 10% fractionation of Mg-Pv with very minor Ca-Pv fractionation is needed for mass balance. Thus, if BSE has a composition similar to either 'pyrolite' or PRIMA, and if residual upper mantle has a composition close to FEPR, then only a minor amount of crystal fractionation could have occurred and a massive chemically distinct region in the lower mantle would not have formed; the perovskite layer would be limited to a several hundred kilometers above the core.

There are several caveats in this mass balance analysis. First, as mentioned above, the compositions chosen for BSE and fertile upper mantle peridotite are model dependent. Second, we have assumed implicitly that phase compositions are constant and representative of crystals in a deep magma ocean. Third, we have assumed static equilibrium crystallization, whereas vigorous convection in the magma ocean could cause dynamic fractionation which is not an equilibrium process. In order to simulate crystal fractionation in a deep magma ocean more accurately we have to acquire information on melting of mantle materials at still higher pressures and over a range of compositions so that phase relations and element partitioning can be modeled as a function of pressure, temperature and composition.

5. Concluding remarks

(1) Melting relations of peridotite and a CI chondrite mantle material have been studied under lower mantle conditions, up to 35 GPa, using a Kawai-type apparatus equipped with sintered diamond anvils. In peridotite, ferropericlase (Fp) is the liquidus phase up to 30 GPa, which is successively followed down temperature by magnesium perovskite (Mg-Pv) and Ca-perovskite (Ca-Pv). Both Fp and Mg-Pv coexist on the liquidus at 31 GPa, indicating multiple saturation of these phases. At 33 GPa, liquidus Mg-Pv is followed by Fp and Ca-Pv within a small crystallization temperature. In contrast, in the CI mantle material, Fp is absent on the liquidus at pressures higher than 28 GPa, and the liquidus phase is Mg-Pv, which is followed down temperature by Ca-Pv.

(2) Differentiation of Mg-Pv, Fp, and Ca-Pv in magma ocean deeper than 900 km has been examined for CI chondritic mantle and two peridotitic bulk silicate earth models, 'pyrolite' (McDonough and Sun, 1995) and PRIMA (Allègre et al., 1995) in terms of mass balance regression with respect to SiO_2, Al_2O_3, FeO, MgO, and CaO.

Preferential subtraction of about 40% Mg-Pv and 2% Ca-Pv from the CI chondritic mantle yields a residual melt with some characteristics of model upper mantle peridotite, although certain ratios are not well matched. This crystal fraction would result in a crystal pile extending from the bottom to ca. 1400 km depth in the mantle. The perovskite layer may be characterized as an enriched reservoir due to high capability of Ca-Pv to accommodate large cations such as La. The observed effect of pressure on liquidus phase compositions suggests that at higher pressures more successful mass balance between chondritic BSE and fertile upper mantle may be obtained. In the peridotite bulk silicate earth models, on the other hand, separation of very limited amounts of Ca-Pv and up to 10% Mg-Pv (in PRIMA) are suggested.

Acknowledgements

We are grateful to A. Yoneda for his discussion and comments. We also thank C. Oka and Y. Shimizu for their technical assistance. Careful review and comments by D. Frost and E. Takahashi are much appreciated. This work was financially supported by the grant from Ministry of Education, Culture, Sports, Science and Technology, Japan (#13852005).

References

Allègre, C.J., Poirier, J.-P., Humler, E., Hofmann, A.W., 1995. The chemical composition of the Earth. Earth Planet. Sci. Lett. 134, 515–526.

Birch, F., 1965. Energetics of core formation. J. Geophys. Res. 70, 6217–6221.

Flasar, F.M., Birch, F., 1973. Energetics of core formation: a correction. J. Geophys. Res. 78, 6101–6103.

Hart, S.R., Zindler, A., 1986. In search of a bulk-earth composition. Chem. Geol. 57, 247–267.

Hofmann, A., 1997. W. Mantle geochemistry: the message from ocean volcanism. Nature 385, 219–229.

Ito, E., Kubo, A., Katsura, T., Akaogi, M., Fujita, T., 1998. High-pressure transformation of pyrope ($Mg_3Al_2Si_3O_{12}$) in a sintered diamond cubic anvil assembly. Geophys. Res. Lett. 25, 821–824.

Ito, E., Takahashi, E., 1987. Melting of peridotite at uppermost lower-mantle conditions. Nature 328, 514–517.

Ito, E., Takahashi, E., 1989. Postspinel transformations in the system Mg_2SiO_4–Fe_2SiO_4 and some geophysical implications. J. Geophys. Res. 94, 10.

Jagoutz, E., Palume, H., Badenhausen, H., Blum, K., Cendales, M., Dreibus, G., Spettel, B., Lorenz, V., Wänke, H., 1979. The abundances of major and trace elements in the earth's mantle as derived from primitive ultramafic modules. Lunar Planet. Sci. X, 2031–2050.

Kato, T., Ringwood, A.E., Irifune, T., 1988. Experimental determination of element partitioning between silicate perovskites, garnets and liquids: constraints on early differentiation of the mantle. Earth Planet. Sci. Lett. 89, 123–145.

Kawai, N., Endo, S., 1970. The generation of ultrahigh hydrostatic pressure by a split sphere apparatus. Rev. Sci. Instrum. 4, 425–428.

Kellog, L.H., Hagger, B.H., van der Hilst, R.D., 1999. Compositional stratification in the deep mantle. Science 283, 1881–1884.

Lesher, C.E., Walker, D., 1988. Cumulate maturation and melt migration in a temperature gradient. J. Geophys. Res. 93, 10295–10311.

McDonough, W.F., Sun, S.-S., 1995. The composition of the Earth. Chem. Geol. 120, 223–253.

McFarlane, E.A., Drake, M.J., Rubie, D.C., 1994. Elemental partitioning between Mg-perovskite, magnesiowüstite, and silicate melt at conditions of the Earth's mantle. Geochim. Cosmochim. Acta 58, 5161–5172.

Melosh, H.J., 1990, Giant impacts and the thermal state of the early Earth. In: Newsom, H.E., Jones, J.H. (Eds.), Origin of the Earth. Oxford University Press, Oxford, pp. 69–83.

Ohtani, E., Kato, T., Sawamoto, H., 1986. Melting of a model chondritic mantle to 20 GPa. Nature 322, 352–353.

Ohtani, E., Sawamoto, H., 1987. Melting experiment on a model chondritic mantle composition at 25 GPa. Geophys. Res. Lett. 14, 733–736.

Ringwood, A.E., 1966. Chemical composition of the earth. In: Hurley, P.M. (Ed.), Advance in Earth Science. MIT Press, Cambridge, MA, pp. 287–356.

Solomatov, V.S., 2000. Fluid dynamics of a terrestrial magma ocean. In: Canup, R.M., Righter, K. (Eds.), Origin of the Earth and Moon. University of Arizona Press, pp. 323–338.

Takahashi, E., 1986. Melting of a dry peridotite KLB-1 up to 14 GPa: implications on the origin of the peridotitic upper mantle. J. Geophys. Res. 91, 9367–9382.

Trønnes, R.G., Frost, D.J., 2002. Peridotite melting and mineral-melt partitioning of major and minor elements at 22–24.5 GPa. Earth Planet. Sci. Lett. 197, 117–131.

Walter, M.J., 1998. Melting of garnet peridotite and the origin of komatiite and depleted lithosphere. J. Petrol. 398, 29–60.

Wetherill, G.W., 1990. Formation of the Earth. Ann. Rev. Earth Planet. Sci. 18, 205–256.

Xue, X., Baadsgaard, H., Irving, Scarfe, C.M., 1990. Geochemical and isotopic characteristics of lithospheric mantle beneath West Kettle River, British Columbia: evidence from ultramafic xenoliths. J. Geophys. Res. 95, 15879–15891.

Zhang, J., Hertzberg, C., 1994. Melting experiments on anhydrous peridotite KLB-1 from 5.0 to 22.5 GPa. J. Geophys. Res. 99, 17729–17742.

Available online at www.sciencedirect.com
SCIENCE DIRECT

ELSEVIER

Physics of the Earth and Planetary Interiors 143–144 (2004) 407–419

PHYSICS OF THE EARTH AND PLANETARY INTERIORS

www.elsevier.com/locate/pepi

Trace element partitioning between majoritic garnet and silicate melt at 25 GPa

Alexandre Corgne*, Bernard J. Wood

Department of Earth Sciences, University of Bristol, Wills Memorial Building, Bristol BS8 1RJ, UK

Received 3 December 2002; received in revised form 15 May 2003; accepted 11 August 2003

Abstract

We have determined the partitioning of 39 trace elements between majorite-rich garnet and silicate melt at 25 GPa and ~2300 °C. Concentrations in crystalline and quenched melt phases were determined by both SIMS and laser ablation-ICPMS techniques with, in general, very good agreement in partition coefficients. As anticipated from size and charge considerations, D_{Lu} (where $D_{Lu} = [Lu]_{Majorite}/[Lu]_{Melt} = 0.77$) is much greater than D_{La} (=0.02) and K, U and Th are strongly incompatible. Partition coefficients for isovalent cations entering either the X- or Y-site exhibit a near-parabolic dependence on the radius of the incorporated cations as predicted from the lattice strain model. This underlines the important contribution made by the crystal structure to the control of trace element partitioning. Our data indicate that anhydrous melting at high pressures in the presence of residual garnet could lead to the generation of Al-depleted komatiites. The partitioning data have also been used to assess the potential effects of majoritic garnet fractionation into the deep mantle during a putative magma ocean event in early earth history. Crystallisation of a maximum of 14% majoritic garnet would have slightly depleted the primitive upper mantle in Si and Al without disturbing the chondritic abundance patterns of other refractory elements. The resultant reservoir would have a high Lu/Hf ratio, but a chondritic Sm/Nd ratio and would not be enriched in heat producing elements.

Keywords: Majorite garnet; Partitioning; Trace elements; Crystal chemistry; Partial melting; Differentiation

1. Introduction

The primitive upper mantle (PUM, the earth's mantle before crust extraction) appears to have strong chemical affinities to the chondritic meteorites, an inference drawn from the observation that refractory lithophile elements such as the rare earths (REE) have approximately CI chondritic ratios one to another in PUM (e.g. Jagoutz et al., 1979; Palme and Nickel, 1985; McDonough and Sun, 1995; Allègre et al., 1995). The implications of this observation are that the earth accreted largely from chondrite-like materials and that differentiation of core and lower mantle had no effect on refractory lithophile element ratios in the upper mantle. The upper mantle does, of course, exhibit depletions in siderophile elements such as Ni and the Pt group which are generally accepted to have entered the core (McDonough and Sun, 1995; Allègre et al., 1995). In addition, when compared to the REE, PUM is depleted by about 28% in Si relative to CI chondrites (McDonough and Sun, 1995) and may also be depleted in Al (Palme and Nickel, 1985). Since

* Corresponding author. Present address: Geophysical Laboratory, Carnegie Institution of Washington, 5251 Broad Branch Rd. N.W., Washington, DC 20015, USA. Tel.: +1-202-478-8930; fax: +1-202-478-8901.
E-mail address: a.corgne@gl.ciw.edu (A. Corgne).

0031-9201/$ – see front matter © 2004 Published by Elsevier B.V.
doi:10.1016/j.pepi.2003.08.012

Si is only very weakly siderophile at low pressure (Kilburn and Wood, 1997) its observed depletion in PUM has led to considerable debate. The three principal hypotheses which have been advanced to explain the observation are: (1) the Si/REE ratio of the bulk earth is not chondritic due to the volatilisation of Si in the solar nebula prior to accretion (Ringwood, 1979). This hypothesis can only be tested by elimination of one or both of the other two possibilities. (2) The 'missing' Si is in the core (Allègre et al., 1995). This suggestion has been experimentally investigated by Gessmann et al. (2001), who show that, at high pressures (>25 GPa) Si becomes sufficiently siderophile to enter liquid Fe at the concentrations necessary to explain the apparent deficit in the upper mantle. The pressure–temperature conditions required for Si to become siderophile are, however, extremely restrictive and it is not yet clear that they are consistent with the near-chondritic ratios of refractory lithophile elements such as Nb/Ta in PUM (Wade and Wood, 2001). (3) The Si/REE ratio of the earth is chondritic, but the mantle is stratified into PUM and an Si-enriched deep mantle (Anderson, 1983). This hypothesis is the most easily tested and has already been the subject of a number of experimental studies (e.g. Ohtani et al., 1986; Ringwood et al., 1987; Herzberg et al., 1988; Kato et al., 1988a,b; Drake et al., 1993; Corgne and Wood, 2002). The aim of our study was to place better constraints on the extent to which Si-, Al-rich majorite garnet could have accumulated in the deep mantle without disturbing the chondritic pattern of element ratios in the overlying upper mantle.

It is probable that the earth underwent extensive melting during its early history due to the impact energy of planetesimal bombardment and the gravitational energy of differentiation (e.g. Wetherill, 1990). Under some scenarios, melting was long-lived and represented by a deep vigorously convecting magma ocean (Ohtani, 1985; Herzberg and O'Hara, 1985). As the earth cooled the magma ocean would have crystallised, leading, in principle, to a fractionally-crystallised mantle, although mantle convection may have significantly reduced such stratification (Tonks and Melosh, 1993). This process provides a mechanism for enriching the lower mantle in Si and Al if majoritic garnet crystallised at the base, depleting the residual liquid, now the PUM, in Si and Al. The possible consequences of magma ocean fractional crystallisation have been the subject of considerable debate (e.g. Kato et al., 1988a,b; Agee and Walker, 1988; Drake et al., 1993) because the lack of accurate crystal–liquid partition coefficients under deep mantle conditions has limited the ability to model appropriate earth differentiation scenarios.

In this study, partition coefficients between (Mg, Fe)SiO_3 majoritic garnet and coexisting silicate melt were precisely determined for a wide range of compatible and incompatible trace elements at 25 GPa and 2300 °C. The application of the 'lattice strain' model (Brice, 1975; Blundy and Wood, 1994) to these new data provides insights into the influence of crystal chemistry on trace element partitioning. The partitioning data enable us to assess the maximum possible fractionation of majoritic garnet during the putative magma ocean event. The possible petrogenesis of Al-depleted komatiites by anhydrous melting at high pressures in the presence of residual majoritic garnet has also been modelled using the partitioning data.

2. Materials and methods

The starting material was an Fe-, Al-rich peridotite composition prepared by mixing analytical grade oxides and carbonates (Table 1). This composition rather than a 'normal' peridotite composition was chosen to enlarge the stability field of majoritic garnet at high pressures. After grinding under acetone in an agate mortar, the mixture was slowly decarbonated in air at ~800 °C and reground before adding trace elements at 100–1000 ppm level using 1000 ppm atomic absorption standard solutions. The doped powder was then denitrated at 800 °C for 2 h and subsequently reduced in a controlled atmosphere furnace at 1000 °C for several hours at an f_{O_2} about 2 log units below the fayalite–magnetite–quartz oxygen buffer.

The experiment was performed at 25 GPa (equivalent to ~700 km depth) using a Kawai-type multianvil apparatus in a 1000 t press at the Bayerisches Geoinstitut, University of Bayreuth, Germany. The pressure scale was calibrated against the forsterite–wadsleyite and wadsleyite–ringwoodite transformations in Mg_2SiO_4 and the ilmenite–perovskite transition in $MgSiO_3$ with reference to the Jamieson et al. (1982) equation of state for gold. The composition of coexisting garnet and perovskite in the enstatite–pyrope

Table 1
Major and minor element contents of the run products and corresponding partition coefficients

Oxides (wt.%)	Peridotite	Opx seeds[a] ($n = 10$)[b]	Maj ($n = 15$)	Mw ($n = 6$)	Melt ($n = 14$)	$D^{Maj/Melt}$	$D^{Mw/Melt}$	Cations[c] (12O)	Maj
SiO_2	44.90	55.04 (50)	52.06 (30)	0.21 (6)	42.08 (184)	1.24 (5)	0.005 (1)	Si	3.61
TiO_2	0.13	0.08 (4)	0.02 (1)	0.02 (1)	0.18 (5)	0.14 (8)	0.13 (7)	Ti	0.00
Al_2O_3	5.21	3.39 (26)	8.36 (39)	1.09 (3)	2.60 (28)	3.21 (37)	0.42 (5)	Al	0.68
Cr_2O_3	0.32	0.37 (17)	0.30 (8)	0.44 (10)	0.22 (6)	1.35 (50)	1.97 (69)	Cr	0.02
FeO	12.10	6.33 (13)	6.76 (29)	33.56 (73)	20.65 (217)	0.33 (4)	1.63 (17)	Fe	0.39
MgO	33.44	33.11 (32)	30.08 (26)	61.10 (46)	27.82 (75)	1.08 (3)	2.20 (6)	Mg	3.11
CaO	3.40	0.56 (7)	2.57 (12)	0.05 (1)	3.80 (13)	0.68 (4)	0.012 (4)	Ca	0.19
MnO	–	0.26 (25)	0.18 (4)	0.19 (7)	0.31 (23)	0.59 (45)	0.63 (52)	Mn	0.01
NiO	–	0.13 (13)	0.04 (1)	0.69 (17)	0.12 (6)	0.32 (20)	5.56 (305)	Ni	0.00
Na_2O	0.50	0.03 (3)	0.32 (3)	0.43 (4)	0.83 (15)	0.39 (8)	0.52 (10)	Na	0.04
Total	100.00	99.30	100.69	97.78	98.61			Total	8.06
Mg#	0.831	0.903	0.888	0.764	0.706			$Si_{[Y]}$	0.61
								$Mg_{[X]}$	2.37
								$Mg_{[Y]}$	0.74
								[Y]	2.05

Number in parentheses represents two standard errors in terms of least units cited. Maj, majoritic garnet; Mw, magnesiowüstite; Mg# is Mg/(Mg + Fe), where Fe is total iron.

[a] Orthopyroxene seeds added to the peridotite composition (∼10 wt.%).

[b] Number of electron microprobe analyses.

[c] Number of cations normalised to 12 oxygens per formula unit. $Si_{[Y]}$, the number of Si atoms in the Y-site, is calculated assuming 3 Si atoms in the Z-site and the remaining Si atoms in the Y-site. $Mg_{[X]}$ and $Mg_{[Y]}$ are calculated assuming Ca, Fe and Na only enter the X-site and Al, Cr only enter the Y-site. This leads to individual $D_{[Y]}(Si)$ of 0.21 and $D_{[X]}(Mg)$ of 0.83.

system (Fei and Bertka, 1999) provided an additional calibration point at 26 GPa. The sample was contained in an unsealed rhenium capsule (∼1.4 mm long) surrounded by an MgO sleeve, a cylindrical straight $LaCrO_3$ heater and an outer ZrO_2 sleeve. The sample octahedron (MgO with 5% Cr_2O_3) with 10 mm edge was compressed using 32 mm WC anvils (Widia grade THM-U) with 4 mm truncations. Temperature was monitored using a $W_3Re/W_{25}Re$ thermocouple axially inserted above the capsule. The H_2O content of the sample was minimised by preliminary heating of the entire assembly for ∼12 h at 220 °C in a vacuum oven. The experiment was pressurised at room temperature in 4 h, then raised at ∼100°/min to 2300 ± 50 °C and held there for 30 min. The sample was quenched rapidly by switching off the power to the $LaCrO_3$ furnace and then gradually decompressed over approximately 17 h. In an attempt to increase product crystal size we added ∼10 wt.% of orthopyroxene seeds (Table 1) and these appear to facilitate the growth of unzoned crystals up to 50 μm in size (Fig. 1).

Major element contents of the experimental charges were measured using the 3-spectrometer CAMECA Camebax electron microprobe in the Department of Earth Sciences, University of Bristol. Analytical conditions were 15 kV accelerating voltage and 10 nA beam current with a counting time of 30 s on peak and 10 s on background. Calibration standards were andradite (Ca and Fe), synthetic $SrTiO_3$ (Ti), St. John's Island olivine (Mg and Si), synthetic spinel (Al), Ni metal (Ni), synthetic MnO (Mn) and synthetic Cr_2O_3 (Cr). Minerals were analysed with a focused beam whereas quenched melt areas were analysed with a 10 μm defocused beam to average the compositions of fine grained quench crystals that compose the melt.

Trace element concentrations were determined by secondary ion mass spectrometry (SIMS) at the University of Edinburgh and by laser ablation-ICPMS (LA-ICPMS) at the Department of Earth Sciences, University of Bristol. SIMS analyses were carried out using a Cameca IMS-4f. The O^- primary beam with ∼10 nA beam current was accelerated to ∼10 kV

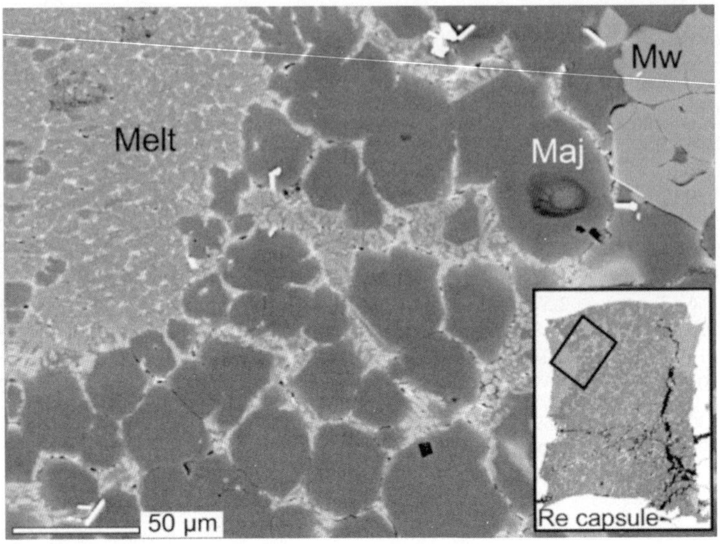

Fig. 1. Backscattered electron image of the run products: Maj, majoritic garnet; Mw, magnesiowüstite. Note the size of garnet crystals compare to SIMS pit.

before striking the gold-coated sample. To reduce transmission of molecular ion species an offset of 75 V was applied to the secondary ion accelerating voltage of 4500 V. The energy window was set to ±20 eV. With these conditions, a beam size of ~20 μm was obtained for both melt and crystal analyses. Measured isotopes were ratioed to ^{30}Si, after calibration on NIST 610 standard. Mass 130.5 was used to monitor background—all analyses reported here have zero background counts. Counting times were adjusted so as to obtain at least 10^3 total counts per isotope and were typically (for analyses consisting of 10 cycles) around 5 s per cycle for elements present at the ppm level. LA-ICPMS analyses were performed using a Nd:YAG 266 nm UV laser (Merchantek) and a VG Elemental PQ3 + S-Option quadrupole ICPMS. Depending on crystal size, the laser beam diameter was 10–20 μm. Background gas blanks were measured on all masses for 15 s before laser ablation, total laser firing time was 30 s. Si determined by electron microprobe was used as the internal standard. The reference NIST 610 and NIST 612 glasses were used as primary calibration and secondary check standards, respectively. The precision of both analytical techniques is better than 15% for all measured elements.

3. Results

Liquidus phases consist of about 60% crystals of majoritic garnet, 10% magnesiowüstite and 30% quenched melt located at the top of the capsule and trapped between the crystals (Fig. 1). (Mg, Fe)SiO_3 perovskite, which decomposed under the electron beam, and magnesiowüstite were also found at the bottom end of the capsule. Although garnet crystals are large enough (up to 50 μm) for trace element analysis the magnesiowüstite crystals are only large enough (<15 μm) for electron microprobe analysis. Major element contents of experimental products and corresponding partition coefficients (D_i, with D_i is the concentration of element i in the crystal/concentration of element i in the melt) are presented in Table 1. Trace element contents of majoritic garnet and melt and corresponding partition coefficients are presented in Table 2. Since unzoned crystals were produced, we believe that a run duration of 30 min at these pressure and temperature conditions is sufficient to approach chemical equilibrium. Also, it is important to mention that no assembly contamination occurred, as evidenced by reasonable Zr and La contents of the run products compared to initial doping (100 ppm each).

Table 2
Trace element contents (in ppm) of orthopyroxene seeds, majoritic garnet and quenched melt, and majoritic garnet–melt partition coefficients[a]

	Opx seeds	Majorite		Melt		$D^{Maj/Melt}$		
	SIMS ($n = 5$)[c]	SIMS ($n = 21$)	LA-ICPMS ($n = 10$)	SIMS ($n = 25$)	LA-ICPMS ($n = 6$)	SIMS	LA-ICPMS	Preferred[b]
Li	0.7	102 (9)	193 (47)	1570 (77)	2856 (605)	0.065 (6)	0.068 (22)	0.066 (11)
B	0.4	16 (0.3)	n.d.	124 (5)	268 (38)	0.13 (1)	–	0.13 (1)
P	16	1908 (36)	1078 (119)	1982 (17)	1296 (33)	0.96 (2)	0.83 (9)	0.90 (5)
K	6	51 (8)	n.m.	3374 (149)	n.m.	0.015 (3)	–	0.015 (3)
Sc	15	86 (1)	69 (13)	90 (0.2)	64 (6)	0.96 (1)	1.07 (23)	1.01 (11)
Ti	449	143 (6)	184 (26)	1198 (31)	1331 (76)	0.12 (1)	0.14 (2)	0.13 (1)
Cr	2876	1970 (15)	2203 (197)	1551 (11)	1310 (36)	1.27 (1)	1.68 (16)	1.48 (8)
Mn	885	719 (2)	1009 (112)	1296 (27)	1518 (35)	0.55 (1)	0.66 (8)	0.61 (4)
Ni	945	869 (5)	n.m.	2363 (65)	n.m.	0.37 (1)	–	0.37 (1)
Co	56	45 (0.3)	72 (25)	109 (3)	131 (6)	0.42 (1)	0.55 (19)	0.48 (10)
Zn	n.m.	n.m.	63 (25)	n.m.	48 (3)	–	1.29 (52)	1.29 (52)
Ga	3	155 (2)	121 (9)	245 (4)	192 (6)	0.63 (1)	0.63 (5)	0.63 (3)
Ge	9	59 (0.4)	55 (7)	103 (2)	88 (3)	0.57 (1)	0.62 (8)	0.60 (4)
Rb	3	10 (2)	19 (9)	696 (36)	1580 (150)	0.014 (3)	0.012 (6)	0.013 (3)
Sr	0.1	10 (1)	13 (2)	342 (14)	438 (25)	0.028 (3)	0.030 (6)	0.029 (3)
Y	0.3	47 (1)	49 (4)	111 (1)	109 (3)	0.43 (1)	0.45 (4)	0.44 (2)
Zr	0.4	21 (1)	22 (2)	162 (3)	173 (8)	0.13 (1)	0.13 (1)	0.13 (1)
Nb	n.d.	4.3 (7)	6.0 (18)	172 (6)	265 (20)	0.025 (4)	0.023 (7)	0.024 (4)
Cd	n.m.	n.m.	9.9 (22)	n.m.	29 (1)	–	0.34 (7)	0.34 (7)
Sn	0.6	11 (1)	12.5 (22)	111 (3)	97 (5)	0.10 (1)	0.13 (2)	0.12 (1)
Cs	0.03	1.7 (5)	6.1 (41)	181 (17)	402 (31)	0.009 (3)	0.015 (10)	0.012 (5)
Ba	0.1	32 (11)	38 (18)	2520 (138)	2792 (178)	0.013 (4)	0.013 (6)	0.013 (4)
La	0.01	4.8 (11)	3.7 (13)	187 (6)	246 (16)	0.025 (6)	0.015 (6)	0.020 (4)
Ce	0.02	3.5 (7)	5.1 (17)	144 (7)	206 (13)	0.024 (5)	0.025 (8)	0.025 (5)
Pr	0.01	4.0 (9)	4.5 (7)	139 (6)	169 (12)	0.029 (6)	0.026 (5)	0.028 (4)
Nd	n.d.	5.7 (9)	6.2 (12)	140 (5)	162 (10)	0.041 (7)	0.038 (8)	0.040 (5)
Sm	0.02	9.7 (11)	11 (1)	123 (3)	142 (10)	0.080 (9)	0.078 (11)	0.079 (7)
Eu	n.d.	14 (0.5)	15 (0.9)	118 (3)	128 (5)	0.12 (1)	0.12 (1)	0.12 (1)
Gd	n.d.	36 (2)	37 (4)	287 (5)	251 (12)	0.13 (1)	0.15 (2)	0.14 (1)
Ho	0.03	43 (2)	45 (4)	112 (1)	120 (3)	0.39 (2)	0.37 (4)	0.38 (2)
Er	0.03	42 (3)	45 (4)	97 (0.4)	92 (2)	0.43 (3)	0.49 (4)	0.46 (2)
Yb	0.09	63 (3)	63 (5)	92 (1)	94 (2)	0.68 (3)	0.67 (5)	0.68 (3)
Lu	0.3	70 (4)	69 (5)	89 (1)	93 (1)	0.79 (4)	0.75 (6)	0.77 (4)
Hf	0.03	19 (2)	21 (2)	116 (2)	138 (6)	0.17 (1)	0.15 (2)	0.16 (1)
Ta	0.02	4.6 (8)	4.2 (12)	115 (5)	251 (18)	0.040 (7)	0.017 (5)	0.028 (4)
Pb	n.d.	6.6 (15)	1.4 (3)	49 (2)	11 (0.5)	0.13 (3)	0.13 (3)	0.13 (2)
Bi	3	n.m.	0.51 (23)	n.m.	13 (0.6)	–	0.039 (18)	0.039 (18)
Th	n.d.	2.6 (9)	4.7 (15)	128 (5)	344 (23)	0.020 (7)	0.014 (5)	0.017 (4)
U	n.d.	2.8 (8)	8.7 (33)	134 (5)	431 (22)	0.021 (6)	0.020 (8)	0.020 (5)

Number in parentheses represents two standard errors in terms of least units cited. n.d.: not detected; n.m.: not measured.

[a] Note that the La and Zr contents of the run products show no evidence of contamination from the $LaCrO_3$ heater and ZrO_2 insulator sleeve. Since no D is lower than 0.01, it is likely that our data suffer from ~1% contamination of the garnet crystal analyses by interstitial melt. Elements such as La, U, Th, Ba, Rb, K and Cs probably have lower D than measured. Also, note that the trace element contents of orthopyroxene seeds analysed by SIMS are very low and thus cannot be related to the trace element contents of majoritic garnet.

[b] Preferred D values used in the text and figures are the average of the SIMS and LA-ICPMS values.

[c] Number of SIMS/LA-ICPMS analysis.

In general, agreement between trace element concentrations measured by SIMS and LA-ICPMS is good, notable exceptions being for U, Th and Pb (Table 2). Since we used the same primary standard (NIST 610 glass) for both sets of measurements, differences in absolute concentrations must be due to matrix effects which depend on the physics of the ablation process. Based on the initial doping levels used (200 ppm U, 200 ppm Th and 100 ppm Pb) it seems that the LA-ICPMS values are closer to the true concentrations for U and Th, assuming no loss of these elements. Although the SIMS value for Pb appears more reasonable, the concentrations from both techniques are too low, implying some loss of this element during the experiment. Despite the differences in absolute measured concentrations, however, it is important to note that *D* values are virtually identical for the two analytical methods, even for elements such as U, Th and Pb (Table 2). Thus, matrix effects appear to cancel out when silicate crystal concentrations are ratioed to those in the quenched melt. *D* values used hereafter in the text and figures are the average of the SIMS and LA-ICPMS values (preferred *D*, Table 2). It should be noted that there is very good agreement between the preferred *D* values obtained from SIMS and LA-ICPMS and the *D* values obtained from the electron microprobe for major and minor elements. Most elements are incompatible ($D < 1$) in majoritic garnet with the exception of Si, Al, Cr, Sc, Zn and Mg. Rare earth elements (REE) display a wide range of partition coefficients, from slightly incompatible for Lu to highly incompatible for La.

Our results for major and minor elements are broadly consistent with previous data (Ito and Takahashi, 1987; Kato et al., 1988a; Ohtani et al., 1989; Yurimoto and Ohtani, 1992; Drake et al., 1993; Herzberg and Zhang, 1996; Suzuki and Akaogi, 1998; Suzuki et al., 2000; Trønnes and Frost, 2002), the main differences being the slightly higher *D* values for Mg and Si. As for trace elements, when compared with work carried out a decade ago (e.g., Kato et al., 1988a; Ohtani et al., 1989; Yurimoto and Ohtani, 1992), there is qualitative agreement on the compatible or incompatible nature of an element but large quantitative differences in partition coefficients. Significant differences include measurements for Zn which we found to be compatible in contrast to the observations of Yurimoto and Ohtani (1992). We also found some incompatible elements notably Li, B, K, Mn, Zr, La, Sm, Gd, Yb and Hf to be more incompatible than in earlier studies. The reasons for the differences in results are three-fold. Firstly, our experimental duration of 30 min is longer than that employed by previous workers, allowing more time for equilibrium to be achieved. Secondly, measurements of minor or trace element contents made using the electron microprobe usually have large uncertainties or require a relatively high level of doping (Kato et al., 1988a; Ohtani et al., 1989). Thirdly, small amounts of melt contamination of the crystal analysis is a frequent problem when measuring partitioning behaviour of highly incompatible elements. Yurimoto and Ohtani (1992) recognised the possibility that contamination generated some analytical artefacts in their data. We successfully grew quite large garnet crystals, thereby eliminating the contamination problem for, we believe all elements with *D* values greater than 0.02. Elements with partition coefficients of 0.02 or lower, however, notably K, Rb, Cs, Ba, La, U and Th should, because of potential melt contamination, be considered as upper bounds.

The magnesiowüstite–melt partition coefficients of major and minor elements are in excellent agreement with previous studies (McFarlane et al., 1994; Ohtani et al., 1991, 1998; Trønnes and Frost, 2002). We confirm that Ni, Mg, Cr and Fe are compatible elements, whereas Mn, Na and Ti are incompatible and Si and Ca highly incompatible elements (Table 1).

4. Crystal chemistry considerations

Garnets along the majorite ($Mg_4Si_4O_{12}$)–pyrope ($Mg_3Al_2Si_3O_{12}$) join are tetragonal in the compositional range up to ~20 mol% pyrope, and cubic at higher pyrope content (e.g. Heinemann et al., 1997). With 3.61 Si atoms per formula unit (Table 1), the majoritic garnet studied here appears too pyrope rich to have the tetragonal structure. We will therefore assume that the majoritic garnet has cubic symmetry with the general structural formula $X_3Y_2Z_3O_{12}$, where the X-, Y- and Z-positions are dodecahedral, octahedral and tetragonal cation sites, respectively. In majorite-rich garnet, Mg and Si are present in both X- and Y-sites, and Y- and Z-sites, respectively. Estimated contents of Mg and Si in each site are presented in Table 1 together with individual partition coefficients.

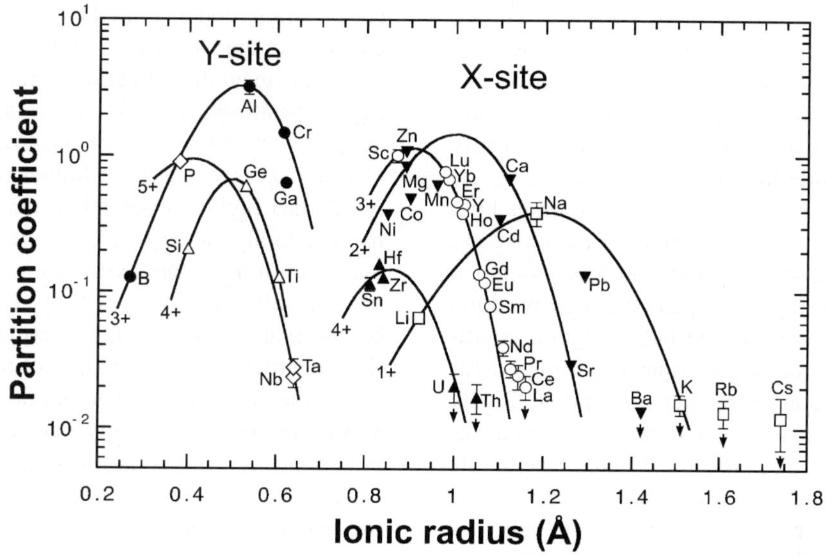

Fig. 2. 'Lattice strain' model (Brice, 1975; Blundy and Wood, 1994) fits to experimental data on majoritic garnet–melt partitioning. Ionic radii are taken from Shannon (1976). Regression parameters E, D_0 and r_0 are listed in Table 3. Individual partition coefficients for Si and Mg (Table 1) and bulk partition coefficients (Tables 1 and 2) for other elements are reported here. Bulk partition coefficients for U, Th, La, Ba, K, Rb and Cs should be considered as upper bounds; see text for discussion.

To understand the influence of crystal chemistry on trace element partitioning, we applied the 'lattice strain' model (Brice, 1975; Blundy and Wood, 1994) to our data. In this model, the partitioning of an ion i (D_i) onto a given crystal site depends on the site Young's modulus (E), and the partition coefficient (D_0) of an ion of the same charge and of radius r_0 which enters the lattice without strain:

$$D_i = D_0 \exp\left\{\frac{-4\pi E N_A}{RT} \times \left[\frac{r_0}{2}(r_i - r_0)^2 + \frac{1}{3}(r_i - r_0)^3\right]\right\} \quad (1)$$

In Eq. (1), r_i is the ionic radius of i, N_A the Avogadro's number, R the gas constant and T the temperature in K. According to the model, a near-parabolic dependence of $\log D_i$ on ionic radius should be observed. Fig. 2 shows that partition coefficients for a suite of isovalent cations entering either the X- or Y-site exhibit the predicted near-parabolic dependence on the radius of the incorporated cations. This underlines the important contribution made by the crystal structure to the control of trace element partitioning. Furthermore, as previously found for clinopyroxene and low pressure garnet (Hill et al., 2000; Van Westrenen et al., 2000a), Young's modulus, E, increases and optimum radius r_0 decreases with increasing charge of the cation incorporated at the X-site of majorite-rich garnet (Table 3).

In drawing-up Fig. 2 we considered that each ion enters only one site, with the exception of Mg and Si for which individual partition coefficients for the X- and Y-sites are reported. We allocated all ions with radius greater than that of Mg to the large X-site and all ions close to Al in size to the Y-site. This

Table 3
Regression parameters of the best fits of the 'lattice strain' model for majoritic garnet

Site	Charge	E (GPa)	r_0 (Å)	D_0	R^2
Y-site	3+	840	0.519	3.27	1
	4+	1504	0.501	0.67	1
	5+	708	0.411	0.94	1
X-site	1+	130 (17)	1.197 (18)	0.39 (1)	0.997
	2+[a]	261	0.987	1.62	1
	3+	521 (47)	0.908 (3)	1.14 (4)	0.993
	4+	522 (327)	0.856 (35)	0.15 (3)	0.945

Number in parentheses represents one standard deviation in terms of least units cited.

[a] Excluding transition elements and Ba.

leaves a few ions, notably Ni, Sc, Ga, Cr, Ti, Sn, Hf, and Zr which may, like Mg, enter both X- and Y-sites. Comparison of measured partition coefficients for these elements with the parabolas based on other elements suggests that Cr, Ga and Ti predominantly enter the Y-site while Sn, Hf and Zr enter the X-site (Fig. 2). Similarly, P and B may, like Si, enter both Y- and Z-sites. If the elements mentioned above are indeed present in two sites, individual partition coefficients for each site rather than the bulk partition coefficient should be used, the bulk partition coefficient being the sum of the individual partition coefficients. For this reason, some elements in Fig. 2 may have partition coefficients for a given site overestimated. For example, if 80% of P enters the Z-site and 20% the Y-site, the partition coefficient for the Y-site would be a factor of 5 lower than the bulk partition coefficient reported in Fig. 2. This new value would lead to different regression parameters of the best fit of the 'lattice strain' model for the 5+ cations in the Y-site. The fit parameters (Table 3) should therefore be viewed with caution in cases where the assignment of the ions to the site of interest is uncertain.

D values for some divalent elements (Ni, Co, Mn, Cd, and Pb) do not fit the trend predicted from the elastic strain model applied to Ca, Mg and Sr. This indicates that the partitioning of these elements depends on factors in addition to lattice strain. The discrepancy cannot be attributed to the incorporation of these elements in several sites since Ni, Co, Cd and Mn have observed *D* lower and not higher than predicted. Pb, which has higher observed *D* than predicted, is almost certainly too large to enter the Y-site. It is known that transition element partitioning is sensitive to crystal field effects (Burns, 1970). This has been used by Moriyama et al. (1992) to explain similar anomalies in garnet–melt partitioning of transition elements. Anomalies for Pb are also common (e.g. Blundy and Wood, 1994) and probably due to the presence of a lone pair of electrons on this ion.

Partition coefficients of trivalent cations entering the X-site are plotted against the ionic radius in Fig. 3. Data from Van Westrenen et al. (1999, 2000b) on pyrope–grossular–almandine garnet are shown for comparison. Majorite-rich garnet has most affinities with pyrope-rich garnet, $E(3+)$ and $r_0(3+)$ values for the X-site of majorite being slightly lower than for $Py_{72}Gros_{19}Alm_9$ and $Py_{82}Gros_{18}$ garnets (Fig. 3). $D_0(3+)$ of the X-site of majorite is, however, about one order of magnitude lower than that of $Py_{72}Gros_{19}Alm_9$ garnet which has the same Mg# = 0.89.

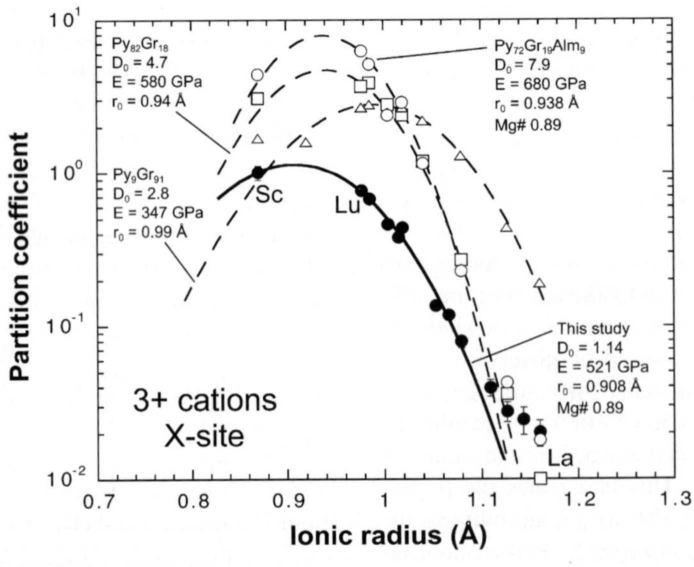

Fig. 3. 'Lattice strain' model (Brice, 1975; Blundy and Wood, 1994) fits to experimental data on partitioning of trivalent cations between the X-site of garnet and melt. Data from Van Westrenen et al. (1999, 2000b) are shown for comparison.

5. Implications for generation of komatiites and early mantle differentiation

Melting experiments on peridotitic and chondritic compositions have shown that majoritic garnet is the liquidus phase at depths of the transition zone (Ohtani et al., 1986; Ito and Takahashi, 1987; Zhang and Herzberg, 1994; Agee et al., 1995; Trønnes, 2000; Trønnes and Frost, 2002). At 23–24 GPa in a 'normal' peridotite composition (Trønnes and Frost, 2002) and at 25–26 GPa in a chondritic composition (Agee et al., 1995), magnesiowüstite becomes the first liquidus phase. In the Fe- and Al-rich peridotite composition used in this study, majoritic garnet remains the liquidus phase at 25 GPa, followed closely by magnesiowüstite. In early earth history, when the temperature of the mantle was higher than at present, partial melting of peridotite or chondrite-like material at high pressures almost certainly occurred. Melting at high temperatures has, for example, been used to explain the genesis of Archean komatiites (e.g. Green, 1975; Ohtani et al., 1989) and has been proposed to lead to a deep magma ocean (Ohtani, 1985; Herzberg and O'Hara, 1985). In the forthcoming discussion, our majoritic garnet–melt partitioning results are used to test two hypotheses: (1) the generation of Archean Al-depleted komatiites by peridotite partial melting at high pressures in the majorite stability field, and (2) the generation of Si depletion in the upper mantle through majorite fractionation in a magma ocean.

Komatiites in mid-Archean (~3.5 Ga) belts are depleted in Al and heavy REE (HREE), a signature that is attributed to garnet fractionation at some stage during their formation (e.g. Green, 1975; Sun and Nesbitt, 1978). They have subchondritic Al_2O_3/TiO_2 ratios and superchondritic Gd/Yb ratios, whereas late-Archean (~2.7 Ga) komatiites have roughly chondritic ratios of these elements. Two mechanisms have been proposed to explain the origin of the oldest komatiites: either garnet was a residual phase during the melting that generated Al- and HREE-depleted ultramafic magmas (e.g. Green, 1975; Sun and Nesbitt, 1978; Arndt, 1986; Ohtani et al., 1989), or the garnet subtraction was produced by majorite fractionation during the crystallisation of a deep magma ocean in the early history of the earth (e.g. Nisbet and Walker, 1982; Gruau and Jahn, 1983; Ohtani et al., 1989). Recently, Blichert-Toft and Arndt (1999) found that mid- and late-Archean komatiites have positive initial ε_{Hf}, indicating that the source of Al-depleted komatiites must have had high Lu/Hf. This implies, assuming $D_{Lu} > D_{Hf}$ for garnet, that the source was enriched in garnet rather than depleted as in earlier models. These authors concluded that low Al_2O_3/TiO_2 and low Lu/Hf can only be features of the komatiitic magmas rather than their mantle source and result from presence of garnet in the residue of partial melting. Since the majoritic garnet–melt partition coefficient for Lu had never been measured directly however, the effect of residual majoritic garnet on the Lu/Hf ratio has remained uncertain.

Using the equation for batch melting and the partition coefficients obtained in this study, we have modelled the effect of leaving majoritic garnet in the residue on the chondrite-normalised Al/Ti and Gd/Yb ratios in the melt. The results of the calculation, which assumes that majoritic garnet is the only major host phase for REE, are shown in Fig. 4. Our modelling suggests that anhydrous melting of peridotite in the presence of residual majoritic garnet at high pressure could contribute to the chemical signatures of Al-depleted komatiites. This is in agreement with the conclusions of Ohtani et al. (1989) and Yurimoto and Ohtani (1992) based on majoritic garnet–melt partitioning at 16 GPa and 1900 °C. Furthermore, such melts, in the presence of residual majorite would have much lower Lu/Hf than the source region. Using the data of Van Westrenen et al. (2000b) obtained at 3 GPa, it seems unlikely that Al-depleted komatiites were produced by partial melting of an anhydrous mantle source at low pressures (Fig. 4). Depending on whether or not komatiites have formed in dry parts of the mantle (e.g. Arndt et al., 1998; Parman et al., 1997), two plausible mechanisms for the production of Al-depleted komatiitic magmas can be envisaged: (1) partial melting of an anhydrous source in the deep upper mantle in the presence of residual majorite (this study; Ohtani et al., 1989; Yurimoto and Ohtani, 1992); (2) partial melting of an hydrous mantle source in the presence of residual majorite or garnet in the 5.5–16 GPa pressure range (after Inoue et al., 2000). It should be mentioned that $CaSiO_3$ perovskite, which is stable in the mantle above 20 GPa, is also a major host phase for REE (e.g. Corgne and Wood, 2002). Its effect on the melting path is negligible in this

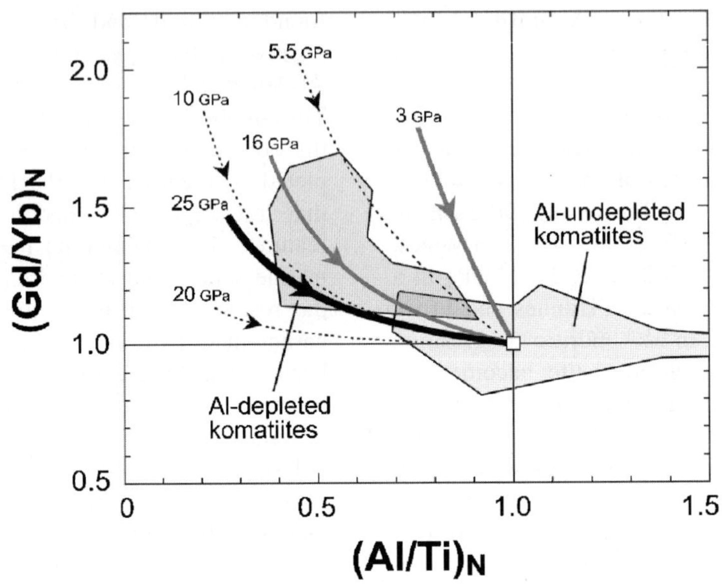

Fig. 4. Evolution of chondrite-normalised Gd/Yb and Al/Ti ratios in melt during partial melting in the presence of residual majorite (or garnet). Partition coefficients for Al, Ti, Gd and Yb in anhydrous compositions at 25, 16 and 3 GPa are taken from this study, Ohtani et al. (1989) and Van Westrenen et al. (2000b), respectively. Data for hydrous compositions at 5.5, 10 and 20 GPa (Inoue et al., 2000) are also shown. Light and dark grey areas correspond to the compositional ranges of Al-depleted and Al-undepleted komatiites after Jahn et al. (1982) and Arndt (1986).

pressure range, however, except at very small degrees of melting since it disappears within the first few percents of partial melting of peridotite.

Global melting and subsequent fractionation and gravitational settling of majoritic garnet in a magma ocean has been proposed as a mechanism for forming a primitive upper mantle with high Mg/Si and Ca/Al ratios (e.g. Ohtani et al., 1986; Herzberg et al., 1988). The difficulty with this hypothesis is that it is required to take place without significantly changing the ratios of refractory lithophile elements one to another in the residual melt (PUM). Using our new partition coefficient data we are able to examine the majorite fractionation hypothesis in detail and to place bounds on the volume of any 'hidden' layer in the deep mantle. The variations of some refractory lithophile element ratios of the residual melt as a function of majoritic garnet fractionation are shown in Fig. 5. Due to low D values, Sm and Nd are not fractionated one from another significantly during crystallisation of majorite, which means that the Sm/Nd ratio remains in the PUM range. In contrast, the Pr/Sc, Gd/Yb, Lu/Hf, and Lu/La ratios are driven outside the PUM range after less than 25% of majoritic garnet fractionation. The strongest constraint on majorite fractionation is given by the Pr/Sc ratio, for which the allowable extent of majoritic garnet fractionation is less than 14%. In other words, less than 14% fractionation of majoritic garnet would not be detected in the composition of PUM. We conclude, therefore, that given chondritic ratios of refractory lithophile elements in PUM, no more than 14% majoritic garnet fractionation may have survived from an early magma ocean. This means that the depletion, relative to carbonaceous chondrites, of PUM in Si cannot be due to majorite garnet fractionation since 60% crystallisation would be required.

Other geochemical anomalies which might conceivably be attributed to majorite fractionation are the subchondritic Nb/Ta ratio of upper mantle and crust (Rudnick et al., 2000) and the 'missing' low Lu/Hf reservoir suggested as complementary to depleted mantle and crust (Blichert-Toft and Albarède, 1997; Vervoort et al., 1999). Because Nb is highly incompatible in majoritic garnet, a majoritic 'layer' cannot balance the Nb content of the upper mantle plus crust. Similarly, since majorite prefers Lu to Hf,

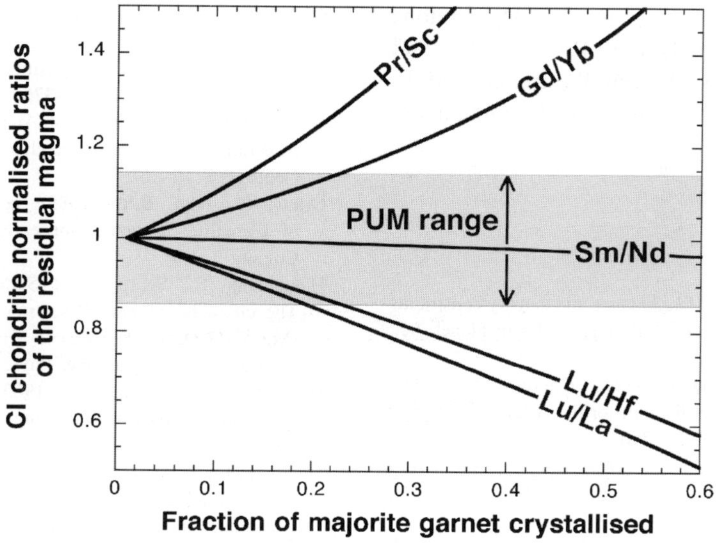

Fig. 5. Variation of refractory lithophile element ratios in the residual melt as a function of majoritic garnet fractionation during a magma ocean stage. Grey box shows range of ratios in the primitive upper mantle assuming the element concentrations are known within ±10% (e.g. McDonough and Sun, 1995).

the putative majoritic garnet layer can clearly not be a low Lu/Hf reservoir. Positive initial ε_{Hf} has been measured for Archean komatiites (Blichert-Toft and Arndt, 1999), Archean Amitsoq gneisses (Vervoort and Blichert-Toft, 1999), and the Archean Isua supracrustal belt (Blichert-Toft et al., 1999). Such results imply that the mantle source had high Lu/Hf ratio, which could be consistent with a source enriched in majoritic garnet. The same samples have positive initial ε_{Nd}, however, implying a high Sm/Nd in the source region. Such a signature could not, as shown in Fig. 5, be produced by majorite enrichment.

6. Conclusions

Majorite-rich garnet exhibits similar patterns of crystal–melt partitioning to pyrope-rich garnet, the major difference being that partition coefficients for all REE are less than 1 in majoritic garnet, as are those for Ti, Zr, Hf, Nb, Ta, U and Th. Partition coefficients for isovalent cations entering either the X- or Y-site exhibit the near-parabolic dependence on the radius of the incorporated cations predicted from the lattice strain model. This underlines the important contribution made by the crystal structure to the control of trace element partitioning. As previously suggested by Ohtani et al. (1989) and Yurimoto and Ohtani (1992), we find that anhydrous partial melting at high pressures in the presence of residual majoritic garnet may contribute to the chemical signature of early Archean Al-depleted komatiites. We also find that fractionation of 14% of majoritic garnet in a magma ocean during early earth history could have led, without changing the chondritic ratios of refractory lithophile elements in PUM, to a layer with a high Lu/Hf ratio but a chondritic Sm/Nd ratio. Such a reservoir would not, however, be large enough to explain the observed depletion in Si of PUM, nor would it constitute a region enriched in heat producing elements, or a region enriched in Nb relative to Ta.

Acknowledgements

AC acknowledges the support of a postgraduate scholarship from the University of Bristol. We acknowledge support of the NERC for purchase of the LA-ICPMS and for use of the SIMS facility at the University of Edinburgh. Experiments in Bayreuth were performed under the EU 'IHP-Access to Research Infrastructures' Programme (Contract No.

HPRI-1999-CT-00004 to D.C. Rubie). Discussions with B.A. Paterson, R.W. Hinton, R.G. Trønnes and M.J. Walter, reviews by W. van Westrenen and E. Takahashi, and editorial handling by E. Ohtani are greatly appreciated.

References

Agee, C.B., Walker, D., 1988. Mass balance and density constraints on early differentiation of chondritic mantle. Earth Planet. Sci. Lett. 90, 144–156.

Agee, C.B., Li, J., Shannon, M.C., Circone, S., 1995. Pressure–temperature phase diagram for the Allende meteorite. J. Geophys. Res. 100, 17725–17740.

Allègre, C., Poirier, J.P., Humler, E., Hofmann, A., 1995. The chemical composition of the earth. Earth Planet. Sci. Lett. 134, 515–526.

Anderson, D.L., 1983. Chemical composition of the mantle. J. Geophys. Res. 88, 1341–1352.

Arndt, N.T., 1986. Komatiites: a dirty window to the Archean mantle. Terra Cognita 6, 59–66.

Arndt, N.T., Ginibre, C., Chauvel, C., Albarède, F., Cheadle, M., Herzberg, C., Jenner, G., Lahaye, Y., 1998. Were komatiites wet? Geology 26, 739–742.

Blichert-Toft, J., Albarède, F., 1997. The Lu–Hf isotope geochemistry of chondrites and the evolution of the mantle crust system. Earth Planet. Sci. Lett. 148, 243–258.

Blichert-Toft, J., Arndt, N.T., 1999. Hf isotope compositions of komatiites. Earth Planet. Sci. Lett. 171, 439–451.

Blichert-Toft, J., Albarède, F., Rosing, M., Frei, R., Bridgwater, D., 1999. The Nd and Hf isotopic evolution of the mantle through the Archean. Results from the Isua supracrustals, West Greenland, and from the Birimian terranes of West Africa. Geochim. Cosmochim. Acta 63, 3901–3914.

Blundy, J.D., Wood, B.J., 1994. Prediction of crystal–melt partitioning coefficients from elastic moduli. Nature 372, 452–454.

Brice, J.C., 1975. Some thermodynamic aspects of the growth of strained crystals. J. Cryst. Growth 28, 249–253.

Burns, R.G., 1970. Mineralogical Applications of Crystal Field Theory. Cambridge University Press, Cambridge, 224 pp.

Corgne, A., Wood, B.J., 2002. $CaSiO_3$ and $CaTiO_3$ perovskite–melt partitioning of trace elements: implications for gross mantle differentiation. Geophys. Res. Lett. 29, 1903. 10.1029/2001GL014398.

Drake, M.J., McFarlane, E.A., Gasparik, T., Rubie, D.C., 1993. Mg-perovskite/silicate melt and majorite garnet/silicate melt partition coefficients in the system $CaO–MgO–SiO_2$ at high temperatures and pressures. J. Geophys. Res. 98, 5427–5431.

Fei, Y., Bertka, C.M., 1999. Phase transitions in the earth's mantle and mantle mineralogy. In: Fei, Y., Bertka, C.M., Mysen, M.O. (Eds.), Mantle Petrology: Field Observations and High-pressure Experimentation: A Tribute to Francis R. (Joe) Boyd. Geochemical Society Special Publication No. 6, pp. 189–207.

Gessmann, C.K., Wood, B.J., Rubie, D.C., Kilburn, M.R., 2001. Solubility of silicon in liquid metal at high pressure: implications for the composition of the earth's core. Earth Planet. Sci. Lett. 184, 367–376.

Green, D.H., 1975. Genesis of Archean peridotitic magmas and constraints on Archean geothermal gradients and tectonics. Geology 3, 15–18.

Gruau, G., Jahn, B.M., 1983. Significance of REE typology of komatiite: argument for early mantle stratification. Terra Cognita 3, 130.

Heinemann, S., Sharp, T.G., Seifert, F., Rubie, D.C., 1997. The cubic–tetragonal phase transition in the system majorite ($Mg_4Si_4O_{12}$), pyrope ($Mg_3Al_2Si_3O_{12}$), and garnet symmetry in the earth's transition zone. Phys. Chem. Miner. 24, 206–221.

Herzberg, C.T., O'Hara, M.J., 1985. Origin of mantle peridotite and komatiite by partial melting. Geophys. Res. Lett. 12, 541–544.

Herzberg, C., Zhang, J., 1996. Melting experiments on anhydrous peridotite KLB-1: compositions of magmas in the upper mantle and transition zone. J. Geophys. Res. 101, 8271–8295.

Herzberg, C., Feigenson, M., Skuba, C., Ohtani, E., 1988. Majorite fractionation recorded in the geochemistry of peridotites from South Africa. Nature 332, 823–826.

Hill, E., Wood, B.J., Blundy, J.D., 2000. The effect of Ca-Tschermarks component on trace element partitioning between clinopyroxene and silicate melt. Lithos 53, 203–215.

Inoue, T., Rapp, R.P., Zhang, J., Gasparik, T., Weidner, D.J., Irifune, T., 2000. Garnet fractionation in a hydrous magma ocean and the origin of Al-depleted komatiites: melting experiments of hydrous pyrolite with REEs at high pressure. Earth Planet. Sci. Lett. 177, 81–87.

Ito, E., Takahashi, E., 1987. Melting of peridotite at uppermost lower-mantle conditions. Nature 328, 514–517.

Jagoutz, E., Palme, H., Baddenhausen, H., Blum, K., Cendales, M., Dreibus, G., Spettel, B., Lorenz, V., Wänke, H., 1979. The abundances of major, minor and trace elements in the earth's mantle as derived from primitive ultramafic nodules. Proc. Lunar Planet. Sci. Conf. 10, 2031–2050.

Jahn, B.M., Gruau, G., Glikson, A.Y., 1982. Komatiite of Onverwacht Group, South Africa: REE geochemistry, Sm/Nd age and mantle evolution. Contrib. Miner. Petrol. 80, 25–40.

Jamieson, J.C., Fritz, J.N., Manghnani, M.H., 1982. Pressure measurement at high temperature in X-ray diffraction studies: gold as a primary standard. In: Akimoto, S., Manghnani, M.H. (Eds.), High-pressure Research in Geophysics. CAP, Tokyo, pp. 27–48.

Kato, T., Ringwood, A.E., Irifune, T., 1988a. Experimental determination of element partitioning between silicate perovskites, garnets and liquids: constraints on early differentiation of the mantle. Earth Planet. Sci. Lett. 89, 123–145.

Kato, T., Ringwood, A.E., Irifune, T., 1988b. Constraints on element partition coefficients between $MgSiO_3$ perovskite and liquid determined by direct measurements. Earth Planet. Sci. Lett. 90, 65–68.

Kilburn, M.R., Wood, B.J., 1997. Metal–silicate partitioning and the incompatibility of S and Si during core formation. Earth Planet. Sci. Lett. 152, 139–148.

McDonough, W.F., Sun, S.S., 1995. The composition of the earth. Chem. Geol. 120, 223–253.

McFarlane, E.A., Drake, M.J., Rubie, D.C., 1994. Element partitioning between Mg-perovskite, magnesiowüstite and silicate melt at conditions of the earth's mantle. Geochim. Cosmochim. Acta 58, 5161–5172.

Moriyama, J., Kawabe, I., Fujino, K., Ohtani, E., 1992. Experimental study of element partitioning between majorite, olivine, merwinite, diopside and silicate melts at 16 GPa and 2000 °C. Geochem. J. 26, 357–382.

Nisbet, E.G., Walker, D., 1982. Komatiites and the structure of the Archean mantle. Earth Planet. Sci. Lett. 60, 103–113.

Ohtani, E., 1985. The primordial terrestrial magma ocean and its implications for stratification of the mantle. Phys. Earth Planet. Int. 78, 70–80.

Ohtani, E., Kato, T., Sawamoto, H., 1986. Melting of a model chondritic mantle to 20 GPa. Nature 322, 352–353.

Ohtani, E., Kawabe, I., Moriyama, J., Nagata, Y., 1989. Partitioning of elements between majorite garnet and melt and implications for petrogenesis of komatiite. Contrib. Miner. Petrol. 103, 263–269.

Ohtani, E., Kato, T., Ito, E., 1991. Transition metal partitioning between lower mantle and core materials at 27 GPa. Geophys. Res. Lett. 18, 85–88.

Ohtani, E., Moriwaki, K., Kato, T., Onuma, K., 1998. Melting and crystal–liquid partitioning in the system Mg_2SiO_4–Fe_2SiO_4 to 25 GPa. Phys. Earth Planet. Int. 107, 75–82.

Palme, H., Nickel, K., 1985. Ca/Al ratio and the composition of the earth's upper mantle. Geochim. Cosmochim. Acta 49, 2123–2132.

Parman, S.W., Dann, J.C., Grove, T.L., deWit, M.J., 1997. Emplacement conditions of komatiite magmas from the 3.49 Ga Komati Formation, Barberton Greenstone Belt, South Africa. Earth Planet. Sci. Lett. 150, 303–323.

Ringwood, A.E., 1979. Origin of the Earth and Moon. Springer-Verlag, New York, 295 pp.

Ringwood, A.E., Kato, T., Irifune, T., 1987. Minor element partition relationships among high pressure minerals and implication for gross mantle differentiation. EOS 68, 1548.

Rudnick, R.L., Barth, M., Horn, I., McDonough, W.F., 2000. Rutile-bearing refractory eclogites: missing link between continents and depleted mantle. Science 287, 278–281.

Shannon, R.D., 1976. Revised effective ionic radii and systematic studies of interatomic distances in halides and chalcogenides. Acta Crystallogr. A 32, 751–767.

Sun, S.S., Nesbitt, R.W., 1978. Petrogenesis of Archean ultrabasic and basic volcanics: evidence from rare earth elements. Contrib. Miner. Petrol. 65, 301–325.

Suzuki, T., Akaogi, M., 1998. Element partitioning between some mantle minerals and the coexisting silicate melt under high pressure. In: Manghnani, M.H., Yagi, T. (Eds.), Properties of Earth and Planetary Materials at High Pressure and Temperature. Geophysics Monographic Series, vol. 101. American Geophysics Union, pp. 261–270.

Suzuki, T., Akaogi, M., Nakamura, E., 2000. Partitioning of major elements between garnet-structured minerals and silicate melt at pressure of 3–15 GPa. Phys. Earth Planet. Int. 120, 79–92.

Tonks, W.B., Melosh, H.J., 1993. Magma ocean formation due to giant impacts. J. Geophys. Res. 98, 5319–5333.

Trønnes, R.G., 2000. Melting relations and major element partitioning in an oxidized bulk earth model composition at 15–26 GPa. Lithos 53, 233–245.

Trønnes, R.G., Frost, D.J., 2002. Peridotite melting and mineral–melt partitioning of major and minor elements at 21–24 GPa. Earth Planet. Sci. Lett. 197, 117–131.

Van Westrenen, W., Blundy, J.D., Wood, B.J., 1999. Crystal-chemical controls on trace element partitioning between garnet and anhydrous silicate melt. Am. Miner. 84, 838–847.

Van Westrenen, W., Allan, N.L., Blundy, J.D., Purton, J.A., Wood, B.J., 2000a. Atomistic simulation of trace element incorporation into garnets—comparison with experimental garnet–melt partitioning data. Geochim. Cosmochim. Acta 64, 1629–1639.

Van Westrenen, W., Blundy, J.D., Wood, B.J., 2000b. Effect of Fe^{2+} on garnet–melt trace element partitioning: experiments in FCMAS and quantification of crystal-chemical controls in natural systems. Lithos 53, 191–203.

Vervoort, J.D., Blichert-Toft, J., 1999. Evolution of the depleted mantle: Hf isotope evidence from juvenile rocks through time. Geochim. Cosmochim. Acta 63, 533–556.

Vervoort, J.D., Patchett, P.J., Blichert-Toft, J., Albarède, F., 1999. Relationships between Lu–Hf and Sm–Nd isotopic systems in the global sedimentary system. Earth Planet. Sci. Lett. 168, 79–99.

Wade, J., Wood, B.J., 2001. The earth's missing niobium may be in the core. Nature 409, 75–78.

Wetherill, G.W., 1990. Formation of the earth. Annu. Rev. Earth Planet. Sci. 18, 205–256.

Yurimoto, H., Ohtani, E., 1992. Element partitioning between majorite and liquid: a secondary ion mass spectrometric study. Geophys. Res. Lett. 19, 17–20.

Zhang, J., Herzberg, C.T., 1994. Melting experiments on anhydrous KLB-1 from 5.0 to 22.5 GPa. J. Geophys. Res. 99, 17729–17742.

Available online at www.sciencedirect.com
SCIENCE DIRECT
ELSEVIER
Physics of the Earth and Planetary Interiors 143–144 (2004) 421–432
PHYSICS OF THE EARTH AND PLANETARY INTERIORS
www.elsevier.com/locate/pepi

Phase relations of a carbonaceous chondrite at lower mantle conditions

Yuki Asahara*, Tomoaki Kubo, Tadashi Kondo

Institute of Mineralogy, Petrology and Economic Geology, Tohoku University, Sendai 980-8578, Japan

Received 2 December 2002; received in revised form 1 August 2003; accepted 28 October 2003

Abstract

The phase relations of Allende meteorite, which belongs to the CV carbonaceous chondrites, were investigated at 22–25 GPa and 1600–2300 °C. With increasing pressure, the liquidus phase changes from garnet to Mg-perovskite at about 24.5 GPa. Magnesiowüstite crystallizes as the second phase throughout the investigated pressure range, and the solidus curve coincides with the appearance of ringwoodite below 23 GPa, Mg-perovskite at 23–24 GPa and Ca-perovskite above 24 GPa. Immiscible silicate and metallic liquid phases are present in the entire melting range, and a single metallic liquid persists below the solidus down to at least 1600 °C. The $Kd(Fe/Mg)^{crystal/silicateliq.}$ is well below unity for magnesiowüstite (0.7–0.8), garnet (0.35–0.39) and Mg-perovskite (0.34). Ni and S are strongly partitioned into the metallic liquid, and the Ni and Fe partition coefficients between metallic liquid and coexisting minerals are relatively constant with variations in pressure or temperature. An important factor controlling $D^{metalliq./crystal}$ of Fe and Ni is the oxygen fugacity. Our results provide further constraints on models for early Earth differentiation and core formation.

Keywords: Phase relation; Carbonaceous chondrite; High pressure; Magma ocean

1. Introduction

The refractory element ratios of Earth's mantle are close to carbonaceous chondrite values. One of the ideas on Earth's bulk composition is that the Earth is made of carbonaceous chondrite-like material (e.g., Ringwood, 1979; Wänke et al., 1984; Allègre et al., 2001), but the redox state of Earth, i.e., the ratio of metallic Fe to total Fe (Fe^0/Fe = 0.85–0.89 (McDonough and Sun, 1995; Allègre et al., 2001), is closer to that of enstatite chondrite (Fe^0/Fe = 0.8) which is much more reduced than carbonaceous chondrite (Fe^0/Fe = 0–0.35) (e.g., Van Schmus and Wood, 1967; Jarosewich, 1990). The variation of bulk compositions, oxidation state and other characteristics among chondrite groups have been regarded as representing localized heterogeneity of the solar nebular. The redox variation in a chondrite group may be due to the secondary oxidizing or reducing process after condensation (e.g., Rubin et al., 1988; Brearley and Jones, 1997). Both the homogeneous accretion (e.g., Ringwood, 1979) and the heterogeneous accretion models (e.g., Wänke et al., 1984) assume that a reduced component of the Earth-forming materials is highly reduced material free from volatile elements but containing all other elements in C1 abundance ratios. The chondritic materials might have been reduced by metamorphism in planetesimals (e.g., Rubin

* Corresponding author. Tel.: +81-22-217-6662; fax: +81-22-217-6675.
E-mail address: asahara@ganko.tohoku.ac.jp (Y. Asahara).

0031-9201/$ – see front matter © 2004 Elsevier B.V. All rights reserved.
doi:10.1016/j.pepi.2003.10.011

et al., 1988), shock-induced devolatilization (e.g., Tyburczy et al., 2001) or secondary reheating by some supersolar wind at an early stage (Allègre et al., 2001). The reduction of FeO could also occur in the terrestrial magma ocean, with elemental carbon as a reducing agent (e.g., Kuramoto and Matsui, 1996; Nagahara and Ozawa, 2001). Therefore, a thorough evaluation of early differentiation processes in planets requires knowledge of the phase relations of appropriate model compositions under variable Fe/FeO ratios at high pressures.

If the bulk Earth has a carbonaceous chondritic composition from the beginning, the FeO content of silicate represents the redox state of the growing Earth. The Allende meteorite (CV chondrite) composition may be appropriate as a high oxygen fugacity model of the proto-Earth, whereas peridotite–metal mixtures may represent lower oxygen fugacity models. Agee et al. (1995) conducted melting experiments on Allende meteorite at 1–27 GPa, although their pressure values are likely to be overestimated by about 2 GPa in the pressure range above 20 GPa. The purpose of the present investigation of the melting and subsolidus phase relations of Allende meteorite at high pressures is to derive further constraints on the interactions of metal and silicate phases, which is important for considering the differentiation process in the deep interior of the early Earth.

Table 1
Composition of the Allende meteorite in weight percent after Jarosewich (1990)

SiO_2	34.23
TiO_2	0.15
Al_2O_3	3.27
Cr_2O_3	0.52
FeO	27.15
MgO	24.62
MnO	0.18
CaO	2.61
Na_2O	0.45
K_2O	0.03
P_2O_5	0.23
H_2O	<0.10
Fe	0.17
Ni	0.36
Co	0.01
FeS	4.03
C	0.29
Other	1.68
Total	99.98

2. Experiments

We used natural Allende meteorite as the starting material acquired from R.A. Langheinrich Meteorites, New York, USA (http://www.nyrockman.com). The composition of the Allende meteorite from Jarosewich (1990) is shown in Table 1. We crushed and grounded the sample material to fine powder using an agate mortar and pestle. Before experiments, the sample powder packed in a carbon capsule was dried in vacuum oven at 130 °C for 30 min to remove the absorbed water, and other components of the furnace assembly were fired for 6 h at 850 °C.

The phase relations of the Allende meteorite were investigated in the pressure range from 22 to 25 GPa, at 1600–2300 °C. In each experiment, the sample was first compressed to the desired pressure, then the temperature was raised at a rate of 100 °C/min to the desired value. The experiments were performed in a Kawai-type multianvil apparatus, using WC-anvils with 2 mm long truncation edges. Semi-sintered zirconia was used as the pressure medium, and the samples were contained in graphite capsules. Cylindrical heaters of $LaCrO_3$ and Re were used for subsolidus and melting experiments, respectively. The cross-sections through the furnace assemblies are shown in Fig. 1. We used Re instead of $LaCrO_3$ as a heating material for melting experiments, because the $LaCrO_3$ sometimes contaminates silicate liquid probably due to breakage of the carbon capsule associated with the graphite–diamond transformation. Temperatures of experiments using $LaCrO_3$ heaters were measured with a W3%Re–W25%Re thermocouple. The uncertainty in temperature is about ±30 °C in the experiments using $LaCrO_3$ (Fig. 1A). The experiments with Re heaters were conducted without thermocouples, and the relative temperatures were estimated from the electric power to the furnace, and an uncertainty of about ±60–100 °C. We determined the relation between the generated temperature and input power of the furnace assembly using cylindrical Re heaters (Fig. 1B) based on the melting points of Fe, 2000 °C at 23 GPa (Shen et al., 1998), and Mg-perovskite, 2500 °C at 23 GPa (Ito and Katsura, 1992), and was found to be nearly linear.

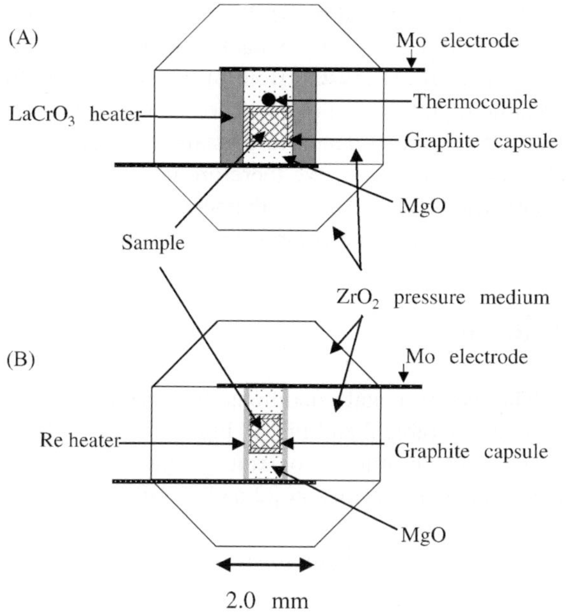

Fig. 1. Furnace assembly used in the subsolidus experiments (A) and that used in the melting experiments (B).

The uncertainty in temperature in the experiments using Re heaters is about ±60 °C at around 2000 °C. It is presumed that the uncertainty becomes large at higher temperatures (about ±100 °C at 2300 °C) because the reaction between liquid Fe–S alloy in the sample and the Re heater causes a change in the temperature–power relation of this furnace (see Table 4). We did not use thermocouples for the experiments using Re heaters because a thermocouple often made the heater unstable in our preliminary experiments.

Pressure calibrations were made at high temperature (1750 °C) using phase relations determined by in situ X-ray diffraction measurements: the ilmenite–perovskite phase transition in $MgSiO_3$ (Hirose et al., 2001) and the Al_2O_3 content of Mg-perovskite in equilibrium with garnet (Hirose, 2002). Both of these calibrations are based on the equation of state of gold (Jamieson et al., 1982), and are therefore compatible with the pressure scale used by most other multianvil laboratories. The uncertainty in pressure in the present experiments is about ±0.3 GPa, which is based on the reproducibility of the semiconductor–metal phase transition of GaP (22 GPa) at room temperature.

Microfocused X-ray diffractometry and electron microprobe analysis (EPMA) were used for phase identification and compositional analysis of run products. Electron microprobe analysis was performed with a JEOL JXA-8800M using wavelength-dispersive techniques under operating conditions of 20 kV and 10 nA specimen current. A counting time of 20 s for peaks and 10 s for background were used for all elements

Table 2
Experimental conditions and results

Run no.	*P* (GPa)	*T* (°C)	Time (min)	Heater	Phase assembly
A29	22	2300	1	Re	Silicate liq., metal liq.
A06	22	2000	0.5	Re	Silicate liq., Gt, metal liq.
A23	22	1900	10	Re	Silicate liq., Gt, Mw, metal liq.
A05	22	1800	60	$LaCrO_3$	Gt, Mw, Rw, metal liq.
A09	22	1600	180	$LaCrO_3$	Gt, Mw, Rw, metal liq.
A22	23	2000	2	Re	Silicate liq., Gt, Mw, metal liq.
A31	23	1800	180	$LaCrO_3$	Gt, Mw, Rw, Mg-Pv, metal liq.
A21	24	2000	1.5	Re	Silicate liq., Gt, Mw, metal liq.
A30	24	1900	5	Re	Gt, Mw, Mg-Pv, metal liq.
A26	24	1800	360	$LaCrO_3$	Gt, Mw, Mg-Pv, metal liq.
A33	24	1600	360	$LaCrO_3$	Mg-Pv, Mw, Ca-Pv, metal liq.
A28	25	2300	1.5	Re	Silicate liq., metal liq.
A16	25	2000	2.5	Re	Silicate liq., Mg-Pv, Mw, metal liq.
A32	25	1900	180	$LaCrO_3$	Mg-Pv, Mw, Ca-Pv, metal liq.
A15	25	1770	360	$LaCrO_3$	Mg-Pv, Mw, Ca-Pv, metal liq.
A12	25	1600	360	$LaCrO_3$	Mg-Pv, Mw, Ca-Pv, metal liq.

Silicate liq., silicate liquid; metal liq., metallic liquid; Gt, garnet; Rw, ringwoodite; Mw, magnesiowüstite; Mg-Pv, Mg-perovskite; Ca-Pv, Ca-perovskite.

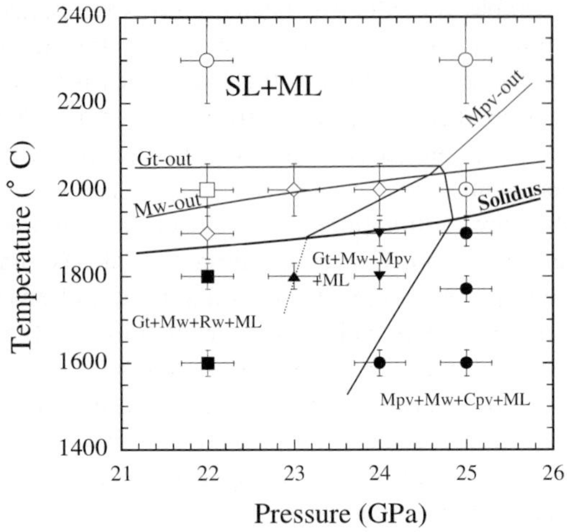

Fig. 2. *P–T* phase diagram of Allende meteorite at 22–25 GPa: (○) silicate liquid (SL) + metallic liquid (ML); (□) SL + garnet (Gt); (⊙) SL + Mg-perovskite (MPv) + magnesiowüstite (Mw) + ML; (◇) SL+Gt+Mw+ML; (■) ringwoodite (Rw)+Gt+Mw+ML; (▼) Gt + Mw + MPv + ML; (●) MPv + Mw + Ca-perovskite (CPv)+ML; (▲) Gt+Mw+MPv+Rw+ML; 'Gt-out' indicates the high-temperature stability limit of garnet.

except O in the metal phase. We used a counting time of 40 s for peaks and 20 s for background for O in metal using an LDE1H crystal. Both of the silicate and the metallic liquids were not quenched to a homogeneous phase, but formed dendritic crystals with several uncharacterized phases. We therefore used a defocused beam from 10 to 30 μm in diameter for EPMA analysis to determine the compositions of liquid phases.

3. Results

The experimental conditions and results are summarized in Table 2 and Fig. 2. Fig. 3 shows backscattered electron images of run products of melting experiments performed at 22 and 25 GPa.

3.1. Achievement of equilibrium

Heating durations varied from 0.5 to 10 min in melting experiments, and from 60 to 360 min at subsolidus conditions for silicates. Subsolidus experiments required long heating durations to achieve equilibrium

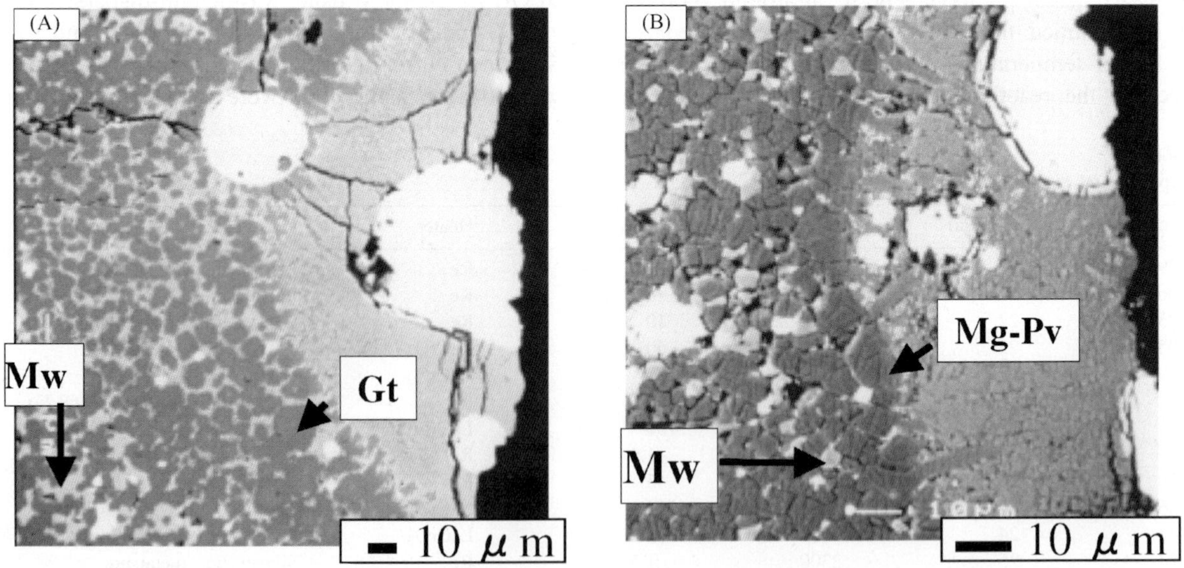

Fig. 3. Backscattered electron images of the products of melting experiments: high-temperature side in contact with the diamond capsule wall next to the Re heater is on the right and consists of quenched metal liquid (bright spheres) and silicate liquid. The crystallization sequence along the temperature gradient is observed. (A) Run product of A23 at 22 GPa and 1900 °C. Garnet (dark gray plain grains surrounded by quench liquid) is the liquidus phase and magnesiowüstite (light gray grains) is the second phase to crystallize. (B) Run product of A16 at 25 GPa and 2000 °C. Mg-perovskite (dark gray cracked grains) is the liquidus phase and magnesiowüstite crystallizes slightly away from liquidus interface.

and to obtain a sufficiently large grain size for EPMA analysis. Melting experiments did not require such long times to achieve equilibrium, and, furthermore, long heating durations caused the Fe–S liquid alloy to infiltrate the carbon capsule and react with the Re tube heater at temperatures above 2000 °C (e.g., Thibault and Walter, 1995). In the run products of melting experiments, chemical zoning of the coexisting minerals was not observed, the silicate liquid was homogeneous and the metallic liquid, which was initially dispersed in the sample as fine grains, accumulated into several rounded pools. Therefore, short heating durations, from 0.5 to 2 min, were considered to be enough to achieve equilibrium.

3.2. Phase relations

The liquidus phase changes from garnet to Mg-perovskite at about 24.5 GPa but magnesiowüstite remains as the second crystallizing phase throughout the 22–25 GPa range. In case of peridotite, the liquidus phase changes from garnet to magnesiowüstite at 23 GPa (Trønnes and Frost, 2002), which is replaced by Mg-perovskite at 33 GPa (Ito et al., 2001). This difference may be due to a change in the melting temperature of coexisting minerals with increasing FeO content in the system. The solidus curve coincides with the appearance of ringwoodite below 23 GPa, Mg-perovskite at 23–24 GPa and Ca-perovskite above 24 GPa. Immiscible silicate and metal liquid phases are present in the entire melting range, and a single metal liquid persists below the solidus down to at least 1600 °C.

The compositions of coexisting silicate liquid and minerals are listed in Table 3. We could not measure the Ca-perovskite composition because the grain size of this phase was too small for accurate compositional analysis (<5 μm). The cation sums of Mg-perovskite, ranging from 8.08 to 8.18, indicate the presence of a significant amount of ferric iron, i.e., Fe^{3+}/Fe_{total}

Table 3
Chemical compositions of silicate liquid and coexisting minerals in weight percent

Run no.	*P* (GPa)	*T* (°C)	Phase	SiO_2	TiO_2	Al_2O_3	Cr_2O_3	FeO	MnO	MgO	CaO	Na_2O	P_2O_5	SO_3	NiO	Total
A29	22	2300	SL	42.83	0.132	4.71	0.596	16.12	0.223	31.32	3.228	0.293	0.323	0.071	0.24	99.98
A06	22	2000	SL	38.85	0.173	3.82	0.576	22.76	0.193	28.48	2.692	0.142	0.243	0.162	0.344	98.53
			Gt	51.24	0.052	8.124	0.7612	8.718	0.133	28.67	2.6262	0.094	0.275	n.d.	0.147	100.61
A23	22	1900	SL	38.35	0.252	2.31	0.384	28.96	0.212	24.44	3.7823	0.356	0.344	0.204	0.395	99.78
			Gt	51.27	0.053	9.28	0.728	9.54	0.132	27.23	2.9024	0.1214	0.263	n.d.	0.134	101.46
			Mw	0.33	n.d.	0.705	0.777	48.25	0.192	49.03	0.082	0.251	n.d.	n.d.	1.709	101.31
A22	23	2000	SL	39.95	0.153	4.12	0.557	21.98	0.182	29.14	3.0516	0.253	0.272	0.162	0.397	100.12
			Gt	49.24	0.031	9.424	0.628	8.315	0.125	28.34	2.4047	0.143	0.233	n.d.	n.d.	98.79
			Mw	0.55	n.d.	1.51	1.027	34.47	0.144	63.32	0.073	0.164	n.d.	n.d.	1.611	102.72
A21	24	2000	SL	40.24	0.152	4.31	0.496	18.95	0.203	30.63	3.0710	0.222	0.253	0.122	0.297	99.03
			Gt	50.06	n.d.	10.52	0.638	6.41	0.082	29.43	1.847	0.132	0.237	n.d.	0.066	99.43
			Mw	0.263	n.d.	1.61	0.838	27.83	0.143	69.16	0.031	0.142	n.d.	n.d.	1.406	101.36
A30	24	1900	Gt	50.63	0.081	7.22	0.8111	12.56	0.1811	23.74	4.677	0.396	0.5114	n.d.	0.230	100.87
			Mg-Pv	53.49	0.282	3.18	0.4521	12.49	0.168	29.83	1.677	0.113	0.041	n.d.	0.112	101.44
			Mw	0.41	0.062	0.255	0.5911	65.119	0.282	32.019	0.091	0.542	n.d.	n.d.	1.883	101.33
A28	25	2300	SL	39.837	0.184	4.2314	0.556	20.111	0.204	29.46	3.112	0.204	0.233	0.092	0.36	98.47
A16	25	2000	SL	36.25	0.161	3.63	0.443	29.34	0.231	22.111	3.7624	0.323	0.5720	0.203	0.4512	97.85
			Mg-Pv	50.87	0.184	4.86	0.5017	13.25	0.152	28.910	1.0435	0.061	n.d.	n.d.	0.142	99.73
			Mw	0.31	0.0555	1.31	0.8911	48.418	0.225	46.318	0.1311	0.527	n.d.	n.d.	1.9311	100.15
A32	25	1900	Mg-Pv	49.013	0.184	4.79	0.5112	14.314	0.144	29.010	0.8032	0.083	0.042	n.d.	0.102	98.77
			Mw	0.163	0.022	0.893	1.455	62.35	0.364	32.713	0.072	1.119	n.d.	n.d.	1.66	100.91
A15	25	1770	Mg-Pv	49.65	0.176	4.962	0.469	13.87	0.194	29.510	0.9874	0.051	0.011	n.d.	0.074	100.19
			Mw	0.32	0.041	0.51	0.988	65.722	0.343	30.014	n.d.	1.015	n.d.	n.d.	1.8812	100.78
A12	25	1600	Mg-Pv	50.421	0.208	4.318	0.4510	12.627	0.156	29.126	0.444	0.062	n.d.	n.d.	0.106	99.07
			Mw	0.21	n.d.	0.421	0.985	66.78	0.313	30.19	n.d.	1.022	n.d.	n.d.	1.468	101.34

The italicized numbers are uncertainties on the last digits based on standard deviations of the analyses.

Table 4
Chemical composition of metallic liquid in weight percent and oxygen fugacity

Run no.	P (GPa)	T (°C)	Metal liq.								ΔIW[a]	ΔIW[b]
			Fe	Ni	Co	Re	P	S	O	Total		
A29	22	2300	60.66 ± 0.81	9.99 ± 1.14	0.42 ± 0.02	5.52 ± 0.48	n.d.	21.34 ± 1.34	0.78 ± 1.27	98.71	−0.97[c]	−1.29
A06	22	2000	56.56 ± 4.14	13.62 ± 6.04	0.42 ± 0.05	1.27 ± 0.82	0.06 ± 0.04	25.49 ± 2.94	0.95 ± 0.64	99.19	−0.54[c]	−0.86
A23	22	1900	57.97 ± 2.12	11.99 ± 2.12	0.40 ± 0.04	0.13 ± 0.1	0.01 ± 0.01	26.98 ± 3.4	3.25 ± 2.46	100.65	−0.23	−0.59
A22	23	2000	54.34 ± 1.59	10.58 ± 2.23	0.34 ± 0.03	5.30 ± 2.92	n.d.	26.98 ± 2.5	3.32 ± 1.76	100.86	−0.56	−0.82
A30	24	1900	53.75 ± 1.91	11.05 ± 5.03	0.45 ± 0.02	n.d.	n.d.	30.04 ± 1.58	0.86 ± 0.74	96.28	0.13	−0.19[d]
A28	25	2300	61.98 ± 0.97	9.41 ± 0.67	0.41 ± 0.02	7.34 ± 2.53	n.d.	17.59 ± 1.28	2.10 ± 1.42	98.95	−0.74	−1.10
A16	25	2000	58.18 ± 0.81	11.22 ± 1.14	0.38 ± 0.02	2.39 ± 0.48	n.d.	22.65 ± 1.34	5.73 ± 1.27	100.59	−0.20	−0.54
A32	25	1900	56.85 ± 1.14	13.20 ± 2.44	0.46 ± 0.02	n.d.	0.08 ± 0.04	27.20 ± 1.87	0.95 ± 0.37	98.75	0.05	−0.27[d]
A15	25	1770	53.25 ± 2.59	14.32 ± 3.25	0.49 ± 0.02	n.d.	n.d.	29.49 ± 3.43	1.95 ± 0.82	99.61	0.20	−0.12[d]
A12	25	1600	55.82 ± 0.4	12.16 ± 2.2	0.50 ± 0.02	0.09 ± 0.01	n.d.	29.38 ± 2.3	1.43 ± 0.5	99.40	0.16	−0.16[d]

Values given in average in about 10 measurements ± error (standard deviation in the measurements); ΔIW, oxygen fugacity relative to iron–wüstite buffer calculated by $\Delta IW = 2\log(X_{FeO}/X_{Fe})$, where X_{FeO} is the mole fraction of FeO in magnesiowüstite (footnote 'a' except those values indicated by footnote 'c') or in silicate liquid (footnote 'b' except those values indicated by footnote 'd'), and X_{Fe} is the mole fraction of Fe in metallic liquid.

[a] Ideality of FeO in magnesiowüstite is assumed.

[b] Ideality of FeO in silicate liquid is assumed.

[c] ΔIW calculated by $\Delta IW = 2\log[\gamma_{FeO}^{\text{silicate liq.}} X_{FeO}^{\text{silicate liq.}} / X_{Fe}^{\text{metallic liq.}}]$, where $\gamma_{FeO}^{\text{silicate liq.}} = 1.44$.

[d] ΔIW calculated by $\Delta IW = 2\log[\gamma_{FeO}^{Mw} X_{FeO}^{Mw} / X_{Fe}^{\text{metallic liq.}}]$, where $\gamma_{FeO}^{MW} = 0.69$.

Table 5
Partition coefficients $Kd^{metalliq./silicateliq.}$ and $D^{Xs/silicateliq.}$ at 2000 °C

Run no.	P (GPa)	$Kd^{metalliq./silicateliq.}$ (average ± error)		Phase	$D^{Xs/silicateliq.}$ (average ± error)						$Kd(Fe/Mg)^{Xs/silicateliq.}$
		Ni/Fe	S/Fe		Si	Al	Mg	Ca	Na	Ni	
A06	22	15.67 ± 7.33	121.60 ± 22.16	Gt	1.20 ± 0.03	1.94 ± 0.57	0.92 ± 0.03	0.89 ± 0.23	0.60 ± 0.27	0.37 ± 0.19	0.38 ± 0.08
A22	23	10.93 ± 3	132.50 ± 2.85	Gt	1.17 ± 0.02	2.14 ± 0.56	0.92 ± 0.02	0.75 ± 0.15	0.54 ± 0.13	n.d.	0.38 ± 0.08
				Mw	0.01 ± 0.01	0.37 ± 0.02	2.26 ± 0.03	0.02 ± 0.01	0.67 ± 0.2	4.33 ± 0.73	0.71 ± 0.03
A21	24	[a]	[a]	Gt	1.10 ± 0.02	2.15 ± 0.08	0.85 ± 0.01	0.53 ± 0.01	0.50 ± 0.09	n.d.	0.35 ± 0.01
				Mw	0.01 ± 0.001	0.38 ± 0.02	2.25 ± 0.03	0.04 ± 0.02	1.74 ± 0.25	4.64 ± 1.08	0.64 ± 0.02
A16	25	12.45 ± 3.56	109.13 ± 18.58	Mg-Pv	1.25 ± 0.02	1.20 ± 0.17	1.17 ± 0.07	0.23 ± 0.08	0.16 ± 0.04	0.28 ± 0.09	0.34 ± 0.02
				Mw	0.01 ± 0.002	0.38 ± 0.03	2.25 ± 0.13	0.04 ± 0.03	1.74 ± 0.31	4.64 ± 1.24	0.78 ± 0.06

Error is calculated from the standard deviation in concentrations.
[a] We could not obtain the chemical composition of metal in run A21 because large metal grains were lost during polishing.

ratios of 30–50%, and/or an oxygen-deficient substitution component involving Al (e.g., Navrotsky, 1999; Trønnes, 2000; Frost and Langenhorst, 2002). Similar perovskite cation sums were observed for experiments, using Re capsules on the more oxidized bulk composition, of Trønnes (2000). This possibly indicates that the $Fe^{3+}AlO_3$ component of perovskite is not sensitive to the oxygen fugacity.

3.3. Oxygen fugacity

Table 4 summarizes compositions of the metallic liquid and the oxygen fugacity of the experiments. For simplicity, we assumed that the activity of Fe in the Fe–S alloy is the same as the mole fraction of Fe. Ideality of activities of FeO and Fe was also assumed for magnesiowüstite and for silicate and metallic liquids. Then, we calculated ΔIW (log fO_2 relative to the iron–wüstite buffer) from the concentration of Fe in Fe–S alloy and FeO in silicate liquid or FeO in magnesiowüstite, employing the relation $\Delta IW = 2\log[X_{Fe}]^{metal}/[X_{FeO}]^{silicateliq.orMw}$ where X is the mole fraction (see also Hillgren et al., 1994; Geßmann and Rubie, 1998; Li and Agee, 2001). The oxygen fugacity obtained from the composition of magnesiowüstite was higher than that calculated using the composition of silicate liquid by about 0.3 log units. This discrepancy in the calculated oxygen fugacities may be due to a difference in activity coefficients of FeO in silicate liquid and magnesiowüstite. The ratio of the activity coefficient in magnesiowüstite and silicate liquid, $\gamma_{FeO}^{silicate\ liq.}/\gamma_{FeO}^{Mw}$, is calculated as 1.44 ± 0.09 using the results of runs A23, A22 and A16. Assuming that the ratio does not change with pressure, temperature, and compositions, we calculated two ΔIW values for ideality of silicate liquid and magnesiowüstite in Table 4. The difference between the oxygen fugacities obtained from ideal activity and those from 1 bar activity coefficients in magnesiowüstite (Srečec et al., 1987) does not exceed a half log unit. Activity coefficients for both magnesiowüstite and silicate liquid at high pressure and temperature are needed to correct the discrepancy.

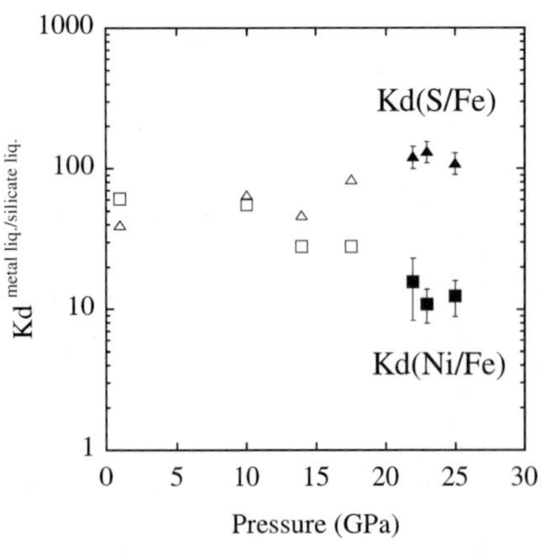

Fig. 4. Partition coefficients, Kd(S/Fe) and Kd(Ni/Fe), between metallic liquid and silicate liquid calculated from the concentrations in mol%: (△, □) partition coefficients for S and Ni, respectively, reported by Agee et al. (1995); (▲, ■) partition coefficients for S and Ni, respectively, from this study.

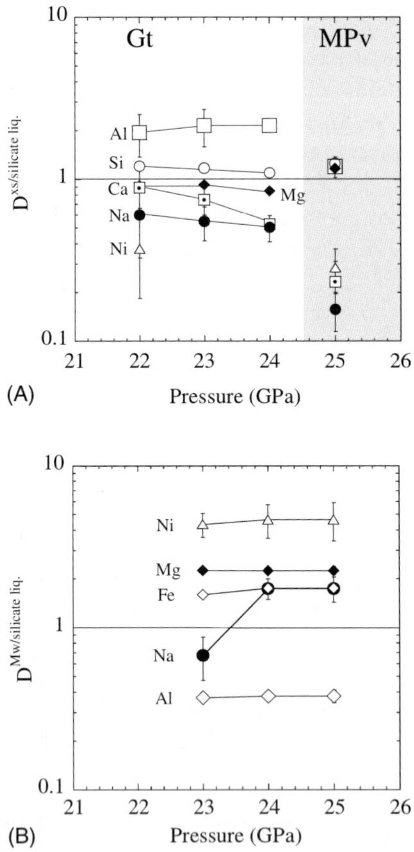

Fig. 5. Partition coefficients between coexisting minerals and silicate liquid at 2000 °C calculated from the concentrations in mol%. (A) Partition coefficient of (○) Si, (□) Al, (◆) Mg, (⊡) Ca, (●) Na and (△) Ni between liquidus phase and silicate liquid. The liquidus phases are garnet at 22–24 GPa and Mg-perovskite at 25 GPa. (B) Partition coefficients for Al, Mg, Na and Ni between magnesiowüstite and silicate liquid.

3.4. Partition coefficients

We summarize partition coefficients Kd (S/Fe)$^{metalliq./silicateliq.}$ [$Kd(S/Fe)^{A/B}=([X_S]^A/[X_{Fe}]^A)/([X_S]^B/[X_{Fe}]^B)$, where [$X$] is the mole fraction], Kd(Ni/Fe)$^{metalliq./silicateliq.}$, $D^{Xs/silicateliq.}$ ($D_i^{A/B} = [X_i]^A/[X_i]^B$) and Kd(Fe/Mg)$^{Xs/silicateliq.}$ at 2000 °C in Table 5. The FeO content of magnesiowüstite observed near the liquidus is about 48 wt.%. These values are much lower than those reported by Agee (1990): FeO content = 73 wt.% and Kd(Fe/Mg)$^{Mw/silicate\ liq.}$ = 3.6. Our Kd(Fe/Mg) values of 0.7–0.8 between magnesiowüstite and silicate liquid in the 22–25 GPa pressure range overlap with those of Trønnes (2000), for a metal-free, FeO-rich bulk Earth model composition in the same pressure range. The extremely high Kd values reported by Agee (1990) may be related to instantaneous and local non-equilibrium melting caused by reheater failure.

Fig. 4 shows partition coefficients, Kd(S/Fe) and Kd(Ni/Fe), between metallic liquid and silicate liquid from this study and that of Agee et al. (1995) as a function of pressure. The partition coefficient for S slightly increases, whereas that of Ni decreases with increasing pressure. Our results are generally in good agreement with the previous study of Agee et al. (1995).

Fig. 5 shows major element partitioning coefficients between coexisting minerals and silicate liquid at 2000 °C. The $D_{Ca}^{Gt/silicate\ liq.}$ decreases with increasing pressure. The $D^{liquidusphase/silicateliq.}$ for Al, Ca and Na decreases with change of liquidus phase from garnet to Mg-perovskite, whereas the $D_{Na}^{Mw/silicate\ liq.}$ increases with pressure and it is greater than unity at 24–25 GPa. $D^{Mg\text{-}Pv/silicateliq.}$ for Si, Mg and Al have similar values, around 1.2, at 25 GPa.

The partition coefficients, Kd(Ni/Fe), between metallic liquid and coexisting minerals are summarized in Fig. 6 and Table 6. The Kd(Ni/Fe)$^{metalliq./crystal}$ values do not show a distinct temperature dependence and the Kd(Ni/Fe)$^{metalliq./Mw}$ and the Kd(Ni/Fe)$^{metalliq./Mpv}$ increase with increasing pressure. Oxygen fugacity may be the most important

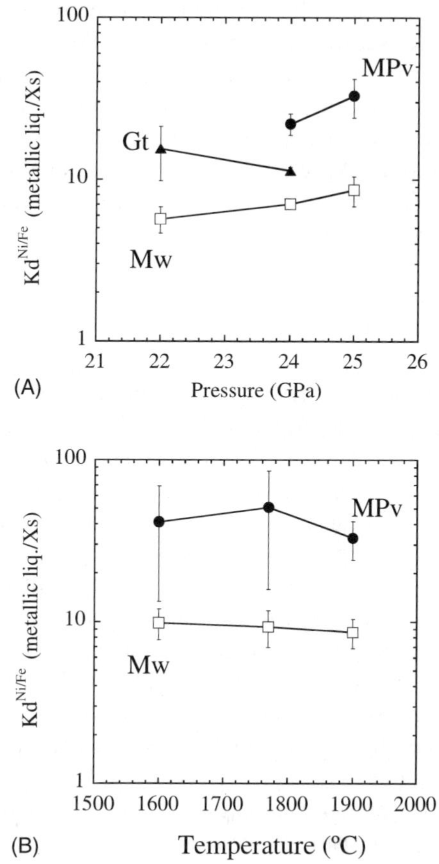

Fig. 6. Partition coefficient, Kd(Ni/Fe), between metallic liquid and coexisting minerals calculated from the concentrations in mol%. (A) Effect of changing pressure at 1900 °C: (□) Kd(Ni/Fe)$^{metalliq./Mw}$; (●) Kd(Ni/Fe)$^{metalliq./Mpv}$; (▲) Kd(Ni/Fe)$^{metalliq./Gt}$. (B) Effect of changing temperature at 25 GPa.

Table 6
Partition coefficients between metallic liquid and coexisting crystals

Run no.	P (GPa)	T (°C)	Kd$^{metalliq./Gt}$ (Ni/Fe)	Kd$^{metalliq./Mpv}$ (Ni/Fe)	Kd$^{metalliq./Mw}$ (Ni/Fe)
A23	22	1900	15.4 ± 5.6		5.7 ± 1.1
A30	24	1900	11.3 ± 0.6	22.0 ± 3.3	7.0 ± 0.3
A32	25	1900		32.8 ± 8.8	8.6 ± 1.8
A15	25	1770		51 ± 35	9.3 ± 2.4
A12	25	1600		41 ± 28	9.9 ± 2.1

Error is calculated from the standard deviation in concentrations.

factor controlling the partitioning of these elements, and in the subsolidus experiments, ΔIW was nearly constant, 0.05–0.2. The carbon (diamond) capsules used in our experiments suggest that the oxygen fugacity of the charges was buffered by the C–CO_2 reaction (e.g., Agee et al., 1995), though the reaction was not confirmed by an observation of carbonate phases in run products.

4. Discussion

Our observed phase relations differ from those of Agee (1990) and Agee et al. (1995) in the pressure range above 22 GPa. Whereas Agee et al. reported garnet and magnesiowüstite as liquidus phases near 26 GPa, we observed Mg-perovskite as the liquidus phase at 25 GPa. A possible reason for this inconsistency is that Agee et al. (1995) overestimated pressures because they did not calibrate pressures above 20 GPa. The phase assembly of their experiment at 26 GPa corresponds to our result at 24 GPa. The pressure generation efficiency in multianvil experiments decreases strongly especially in the highest pressure range generated by a given pressure cell configuration. Therefore, pressure calibration covering the entire experimental pressure range is needed.

The overestimation of pressure by Agee et al. (1995) is estimated to be about 2 GPa. A similar overestimation has been made not only by Agee et al. (1995) but also by many previous high pressure experimental studies especially over 24 GPa where there are no good pressure fixed points at high temperature. It is worth noting that this systematic pressure error may influence the conditions of the terrestrial magma ocean estimated from siderophile element partition coefficients between metal and silicate using the multi-variable linear regression method (Righter et al., 1997; Li and Agee, 2001; Walter, 2001). Precise experimental pressures at high temperature are needed to verify the depth of the magma ocean which has estimated to be at a pressure of 30–60 GPa. Our partition coefficient values are more reliable with respect to pressure than the previous results.

Fig. 7 shows a *P–T* phase diagram that compiles results of Agee et al. (1995) at 1–20 GPa and results of this study at 22–25 GPa. We did not use the results by Agee et al. (1995) for the phase relation above 20 GPa due to the pressure ambiguity. The important point of this study is that melting relations are different between peridotite and the Allende meteorite above 20 GPa. In the Allende system, Mg-perovskite replaces garnet as the liquidus phase above about 25 GPa and magnesiowüstite remains the second phase to crystallize below the liquidus in the entire 22–25 GPa range. For peridotite compositions, however, magnesiowüstite is the liquidus phase at 23–33 GPa and is replaced by Mg-perovskite at about 33 GPa (Ito et al., 2001; Trønnes and Frost, 2002). The pressure at which Mg-perovskite appears as a liquidus phase in the Allende composition is 8 GPa lower than in the peridotite system. Ca-perovskite, on the other hand, is absent below 24 GPa in the Allende meteorite subsolidus phase assemblage above 1600 °C, whereas it appears above 20 GPa in the peridotite system (Zhang and Herzberg, 1994).

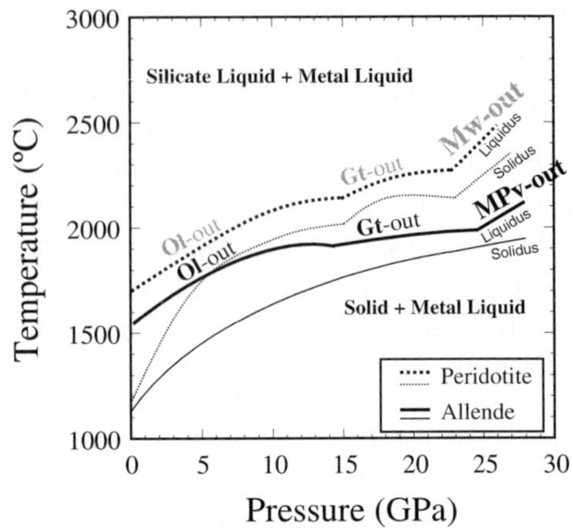

Fig. 7. *P–T* phase diagram of the Allende meteorite and peridotite. Liquidus (thick solid line) and solidus (thin solid line) of the Allende meteorite from 0 to 20 GPa are from Agee et al. (1995), and those above 20 GPa are from this study. Liquidus (thick dashed curve) and solidus (thin dashed curve) of peridotite at 5–20 GPa are from Zhang and Herzberg (1994), and those above 20 GPa are from Trønnes and Frost (2002).

Our result suggests that chemical differentiation processes could change with the FeO content of the magma ocean. If the Earth is made of carbonaceous chondrite-like material, the FeO content of the silicate may represent the redox state (Fe^0/Fe) of the growing planet. For the reduced (FeO-poor) magma ocean similar to the present Earth ($Fe^0/Fe = 0.85$–0.89), the fractionation of magnesiowüstite as the liquidus phase at 23–33 GPa would increase the Si/Mg ratio of the residual melt. In contrast, in an oxidized (FeO-rich) magma ocean with an oxidation state comparable to that of the Allende meteorite ($Fe^0/Fe = 0.01$), the liquidus phase changes from majoritic garnet ($Si/Mg = 1.2$) to Mg-perovskite ($Si/Mg = 1.1$) at 24.5 GPa and their fractionation would slightly increase the Mg/Si ratio of the melt. Mg-perovskite fractionation will produce only small changes in the Al/Si and Mg/Si ratios, but larger increases in the Ca/Si ratio of the silicate melt. The FeO content of the Allende silicate liquid is about 20–30 wt.%. This is 3–4 times higher than in the Earth's upper mantle, and reduction of FeO to metallic Fe reacting with elemental carbon as a reducing agent (e.g., Nagahara and Ozawa, 2001) is required in the terrestrial magma ocean. Fractionation of magnesiowüstite would not contribute to the necessary reduction of the FeO content, as proposed by Agee (1990), because $Kd(Fe/Mg)^{Mw/silicateliq.}$ is lower than unity.

Acknowledgements

The authors greatly acknowledge E. Ohtani for valuable comments. We also thank Y. Ito for the EPMA analysis. We are grateful to R. Trønnes and an anonymous reviewer for their helpful comments. This work was partially supported by the Grant-in-Aid for JSPS Fellows to Y. Asahara, and the Grant-in-Aid of the Scientific Research of Priority Area (B) (no. 12126201) and the Scientific Researches (no. 14102009) of Ministry of Education, Culture, Sport, Science and Technology of Japanese Governments to E. Ohtani.

References

Agee, C., 1990. A new look at differentiation of the Earth from melting experiments on the Allende meteorite. Nature 346, 834–837.

Agee, C.B., Li, J., Shannon, M.C., Circone, S., 1995. Pressure–temperature phase diagram for the Allende meteorite. J. Geophys. Res. 100, 17725–17740.

Allègre, C., Manhès, G., Lewin, É., 2001. Chemical composition of the Earth and the volatility control on the planetary genetics. Earth Planet. Sci. Lett. 185, 49–69.

Brearley, A.J., Jones, R.H., 1997. Chondrite meteorites. In: Papike, J.J. (Ed.), Reviews in Mineralogy, vol. 36. Mineralogical Society of America, Washington, DC, Chapter 3, pp. 1–398.

Frost, D.J., Langenhorst, F., 2002. The effect of Al_2O_3 on Fe–Mg partitioning between magnesiowüstite and magnesium silicate perovskite. Earth Planet. Sci. Lett. 199, 227–241.

Geßmann, C.K., Rubie, D.C., 1998. The effect of temperature on the partitioning of nickel, cobalt, manganese, chromium, and vanadium at 9 GPa and constraints on formation of the Earth's core. Geochim. Cosmochim. Acta 62, 867–882.

Hillgren, V.J., Drake, M.J., Rubie, D.C., 1994. High-pressure and high-temperature experiments on core-mantle segregation in the accreting earth. Science 264, 1442–1445.

Hirose, K., 2002. Phase transitions in pyrolitic mantle around 670 km depth: implications for upwelling of plumes from the lower mantle. J. Geophys. Res. 107, B4, 10.1029/2001, JB00597.

Hirose, K., Komabayashi, T., Murakami, M., 2001. In situ measurements of majorite–akimotoite–perovskite phase transition boundaries in $MgSiO_3$. Geophys. Res. Lett. 28, 4351–4354.

Ito, E., Katsura, T., 1992. Melting of ferromagnesium silicates under the lower mantle conditions. In: Syono, Y., Manghnani, M.H. (Eds.), High-Pressure Research: Application to Earth and Planetary Science. TERRAPUB, Tokyo, pp. 315–322.

Ito, E., Kubo, A., Katsura, T., Walter, M.J., 2001. Melting of peridotite under lower mantle conditions and its implication to differentiation of deep magma ocean. In: Abstract of Transport of Materials in the Dynamic Earth at Kurayoshi, Japan, 2–5 October. p. 150. Institute for Study of the Earth's Interior, Misasa.

Jamieson, J.C., Fritz, J.N., Manghanani, M.H., 1982. Pressure measurement at high temperature in X-ray diffraction studies: gold as a primary standard. In: Akimoto, S., Manghanani, M.H. (Eds.), High-Pressure Research in Geophysics. CAPJ, Tokyo, pp. 27–48.

Jarosewich, E., 1990. Chemical analyses of meteorites: a compilation of stony and iron meteorite analyses. Meteoritics 25, 323–337.

Kuramoto, K., Matsui, T., 1996. Partitioning of H and C between the mantle and core during the core formation in the Earth: its implications for the atmospheric evolution and redox state of early mantle. J. Geophys. Res. 101, 14909–14932.

Li, J., Agee, C.B., 2001. The effect of pressure, temperature, oxygen fugacity and composition on partitioning of nickel and cobalt between liquid Fe–Ni–S alloy and liquid silicate: implications for the Earth's core formation. Geochim. Cosmochim. Acta 65, 1821–1832.

McDonough, W.F., Sun, S.S., 1995. The composition of the Earth. Chem. Geol. 120, 223–253.

Nagahara, H., Ozawa, K., 2001. The time scale of accretion and core formation of the Earth inferred from Hf-W isotope systematics. Proc. Jpn. Acad. Ser. B 77, 167–171.

Navrotsky, A., 1999. A lesson from ceramics. Science 284, 1788–1789.

Righter, K., Drake, M.J., Yaxley, G., 1997. Prediction of siderophile element metal-silicate partition coefficients to 20 GPa and 2800 °C: the effects of pressure, temperature, oxygen fugacity, and silicate and metallic melt compositions. Phys. Earth. Planet. Interiors 100, 115–134.

Ringwood, A.E., 1979. Origin of the Earth and Moon. Springer-Verlag, New York, 295 pp.

Rubin, A.E., Fegley, B., Brett, R., 1988. Oxidation state in chondrites. In: Kerridge, J.F., Matthews, M.S. (Eds.), Meteorites and the Early Solar System. The University of Arizona Press, Tucson, pp. 488–511.

Shen, G., Mao, H., Hemley, R., Duffy, T., Rivers, M.L., 1998. Melting and crystal structure of iron at high pressures and temperatures. Geophys. Res. Lett. 25, 373–376.

Srečec, I., Ender, A., Woermann, E., Gans, W., Jacobsson, E., Eriksson, G., Rosén, E., 1987. Activity–composition relations of the magnesiowüstite solid solution series in equilibrium with metallic iron in the temperature range 1050–1400 K. Phys. Chem. Miner. 14, 492–498.

Thibault, Y., Walter, M.J., 1995. The influence of pressure and temperature on the metal-silicate partition coefficients of nickel and cobalt in a model C1 chondrite and implications for metal segregation in a deep magma ocean. Geochim. Cosmochim. Acta 59, 991–1002.

Trønnes, R.G., 2000. Melting relations and major element partitioning in an oxidized bulk earth model composition at 15–26 GPa. Lithos 53, 233–245.

Trønnes, R.G., Frost, D.J., 2002. Peridotite melting and mineral-melt partitioning of major and minor elements at 22–24.5 GPa. Earth. Planet. Sci. Lett. 197, 117–131.

Tyburczy, J.A., Xu, X., Ahrens, T.J., Epstein, S., 2001. Shock-induced devolatilization and isotopic fractionation of H and C from Murchison meteorite: some implications for planetary accretion. Earth Planet. Sci. Lett. 192, 23–30.

Van Schmus, W.R., Wood, J.A., 1967. A chemical-petrologic classification for the chondritic meteorites. Geochim. Cosmochim. Acta 31, 747–765.

Walter, M.J., 2001. Core formation in a reduced magma ocean: new constraints from W, P, Ni and Co. In: Abstract of Transport of materials in the Dynamic Earth at Kurayoshi, Japan, 2–5 October. p. 152. Institute for Study of the Earth's Interior, Misasa.

Wänke, H., Dreibus, G., Jagoutz, E., 1984. Mantle chemistry and accretion history of the Earth. In: Kröner, A., Hanson, G.N., Goodwin, A.M. (Eds.), Archaean Geochemistry. Springer-Verlag, Berlin, pp. 2–24.

Zhang, J., Herzberg, C., 1994. Melting experiments on anhydrous peridotite KLB-1 from 5.0 to 22.5 GPa. J. Geophys. Res. 99, 17729–17742.

Available online at www.sciencedirect.com

SCIENCE DIRECT®

ELSEVIER

Physics of the Earth and Planetary Interiors 143–144 (2004) 433–443

PHYSICS OF THE EARTH AND PLANETARY INTERIORS

www.elsevier.com/locate/pepi

The behaviour of sulphur in metal–silicate core segregation experiments under reducing conditions

Julien Siebert [a,b], Valérie Malavergne [a,*], François Guyot [b,c], Rossana Combes [a], Isabelle Martinez [d]

[a] Laboratoire des Géomatériaux, Université de Marne-La-Vallée, Cité Descartes, Champs-sur-Marne, 77454 Marne-La-Vallée Cedex, France
[b] Institut de Physique du Globe de Paris, 4 Place Jussieu, 75 252 Paris Cedex 05, France
[c] Laboratoire de Minéralogie-Cristallographie, Case 115, 4 Place Jussieu, 75 252 Paris Cedex 05, France
[d] Laboratoire de Géochimie des Isotopes Stables, IPGP-Université Paris 7, 2 Place Jussieu, Tour 54-64, 75 252 Paris Cedex 05, France

Received 17 December 2002; received in revised form 18 July 2003; accepted 29 July 2003

Abstract

Reactions between molten alloys and solid silicates were investigated under reducing conditions, down to four log units below the IW buffer, at 20–25 GPa and 2000 °C in the Mg–Si–Fe–O–S system. The iron-rich alloys formed in the Fe–Si–S ternary system were single phased at high-pressure and temperature, suggesting closure of the liquid immiscibility gap at these conditions. At the metal/silicate interface, the reduction of Fe^{2+} was coupled to oxidation of Si in metal, a reaction also observed in similar sulphur-free systems. Silicon and sulphur were observed to coexist simultaneously in the metal phase in contact with silicates and magnesium sulphide, MgS, was observed in the experiments. This latter phase would not segregate to the planetary cores, thus providing a reservoir of sulphur in the silicate and surficial planetary envelopes.

Keywords: Core segregation; Metal–silicate interactions; High-pressure and high-temperature experiments; Reducing conditions; Magnesium sulphide

1. Introduction

Metal/silicate interactions have been studied at pressures and temperatures corresponding to the transition zone and to the top of the lower mantle of the Earth by numerous authors (e.g., Ringwood and Hibberson, 1991; Ito et al., 1995; Othani et al., 1997; Li and Agee, 1996, 2001a,b; Righter et al., 1997; O'Neill et al., 1998; Gessmann et al., 2001; Wade and Wood, 2001). These conditions are considered pertinent in most models of Earth differentiation which imply a contribution of metal/silicate interactions at relatively moderate pressure and high-temperature conditions (e.g., Karato and Murthy, 1997). Moreover based on metal–silicate partition coefficients for siderophile elements obtained at high-pressure and high-temperature, several authors argue for an equilibrium between liquid metal and liquid silicate at the basis of a magma ocean at such pressure, temperature and oxygen fugacity conditions (e.g., Murthy, 1991; Li and Agee, 1996, 2001a; Othani et al., 1997; Righter et al., 1997; Walter et al., 2000; Drake, 2000). Understanding incorporation of silicon and sulphur in metal under these conditions is therefore important.

* Corresponding author. Tel.: +33-1-49-32-90-36; fax: +33-1-49-32-91-37.
E-mail address: malaverg@univ-mlv.fru.fr (V. Malavergne).

0031-9201/$ – see front matter
doi:10.1016/j.pepi.2003.07.022

Some geochemical mass balance calculations imply core models with significant Si amounts (MacDonald and Knopoff, 1958; Allègre et al., 1995, 2001; Javoy, 1995). It should be noticed that the missing Si could be also sequestred in a lower mantle chemically different from the upper mantle (Anderson, 1988; Javoy, 1995) or could be due to Earth accreting materials with a super-chondritic Mg/Si ratio. In this study, we investigate the first possibility, that is the presence of significant amounts of Si in metal which is not immediately consistent with the average oxygen fugacity of the Earth as deduced from the bulk Fe^{2+} content in the upper mantle. Reductive incorporation of Si in metal could explain the coexistence of a silicon-rich core with a Fe^{2+}-rich mantle according to:

$$SiO_{2\,\text{in silicates}} + 2Fe_{\text{in metal}} \leftrightarrow 2FeO_{\text{in silicates}} + Si_{\text{in metal}} \qquad (1)$$

Several authors have studied process (1) for pressures ranging between ambient values to 27 GPa and at high temperatures (Ito et al., 1995; Guyot et al., 1997; Othani et al., 1997; O'Neill et al., 1998; Gessmann et al., 1999, 2001; Li and Agee, 2001a,b; Malavergne et al., 2004). Only few have investigated the role of sulphur on process (1) (Kilburn and Wood, 1997; Li and Agee, 2001a,b). Sulphur is a major element present in metal phases of all metal-bearing chondrite groups. It has been proposed as an important element in the core of the Earth (Murthy and Hall, 1970; Walker et al., 1993; Li and Agee, 1996, 2001a,b; Othani et al., 1997). However, liquid immiscibility in the Fe–Si–S system at ambient pressure (Raghavan, 1988; Poirier, 1994) and possibly at higher pressure, as well as difficulties for incorporating simultaneously Si and S in metal (Kilburn and Wood, 1997), have led to discussions about its abundance in the Earth's core. Moreover, geochemical constraints based on elemental volatilities suggest a maximum abundance of 1–2 wt.% of S in the core (Dreibus and Palme, 1996; Allègre et al., 2001). Keeping in mind current uncertainties on core formation models, as well as the large diversity of possible scenarios, it is still of interest, however, for planetary differentiation studies, to determine how sulphur behaves in the presence of Si-bearing metal, i.e., how it may interact with process (1).

The purpose of this study has thus been to carry out (Fe–Si–S) alloy/silicate core segregation experiments. The main goal was to determine the reactivity of simultaneously Si and S rich metals (possibly biphasic) with silicates in order to compare with results obtained on Fe–S–silicate systems under more usual less reducing conditions (Walker et al., 1993; Othani et al., 1997; Li and Agee, 1996, 2001b).

2. Experimental procedure

The starting materials used were FeS (Goodfellow™, 99.9% pure), Si (Goodfellow™, 99.9% pure), natural enstatite $(Mg_{0.88}Fe_{0.12})SiO_3$ from San Carlos (Arizona, USA), hereafter labelled SC enstatite, containing 1 wt.% of Ca and 4.8 wt.% of Al, synthetic enstatite ($MgSiO_3$). Synthetic enstatite was synthesised according to procedure given by Thiéblaut et al. (1999). Fine-grained powders of FeS, Fe and Si, typically 15–20 μm, were intimately mixed and used as the metallic component. The dried materials were grounded together in appropriate proportions in an agate mortar in pure alcohol. Compositions and proportions of starting materials used in the different runs are given in Table 1.

High-pressure and high-temperature experiments were performed using a 1200 t multianvil apparatus at the Bayerisches Geoinstitut, Germany. The pressure assembly consisted of Cr-doped MgO octahedron with an edge length of 14 or 10 mm in the pressure range of 10–25 GPa. The octahedron was compressed using 32 mm carbide anvils with truncation edge length of 8 or 4 mm, with 14 and 10 mm octahedra, respectively. The samples were contained either in polycrystalline MgO or in single crystal MgO capsules. The capsule was surrounded by cylindrical resistance heater of $LaCrO_3$. The description of the cell assembly is shown in Fig. 1. Using single crystals of MgO has proven useful for reducing leakage of the low viscosity material contained in the samples at high pressures and temperatures as well as for minimising reactions between samples and capsules. Temperatures were measured with a W3%Re–W25%Re thermocouple inserted coaxially with thermocouple junction in contact with one hand of the sample capsule. Pressure calibrations for these assemblies are given by Rubie (1999). The *P–T* uncertainties are estimated to be ±1 GPa and ±100 K, respectively. In each experiment, the sample was first compressed to the desired

Table 1
Experimental runs and starting material compositions

Runs	S2852[a]	S2853[b]	S2861[b]	BT14[c]
Capsule	MgO single crystal	MgO single crystal	MgO single crystal	MgO polycrystal
Pressure, P (GPa)	20	20	25	14
Temperature, T (°C)	2000	2000	2000	1700
Duration (min)	10	10	9	45
Starting metal composition (at.%)				
Fe	35.96	24.82	24.82	83.85
S	35.96	24.82	24.82	16.15
Si	28.08	50.36	50.36	0
Starting metal composition (wt.%)				
Fe	50.80	38.45	38.45	90
S	29.20	22.10	22.10	10
Si	20.00	39.45	39.45	0

[a] Starting mixes are as follows: 4 wt.% Si metal, 16 wt.% FeS, 80 wt.% SC enstatite $(Mg_{0.88}Fe_{0.12})SiO_3$.
[b] Starting mixes are as follows: 7.5 wt.% Si metal, 11.5 wt.% FeS, 81 wt.% SC enstatite $(Mg_{0.88}Fe_{0.12})SiO_3$.
[c] Starting mixes are as follows: 33 wt.% $Fe_{0.9}S_{0.1}$, 67 wt.% synthetic enstatite $(MgSiO_3)$.

pressure and then heated to the desired temperature at a rate of 200 °C/min. The samples were quenched by switching off the electrical power. The quench rate was approximately 500 °C/s. After quenching, samples were decompressed over 12 h. The entire sample plus pressure assemblies were then mounted in epoxy, sectioned through their centers, and their surfaces were polished and eventually carbon coated for scanning electron microscopy (SEM) analyses.

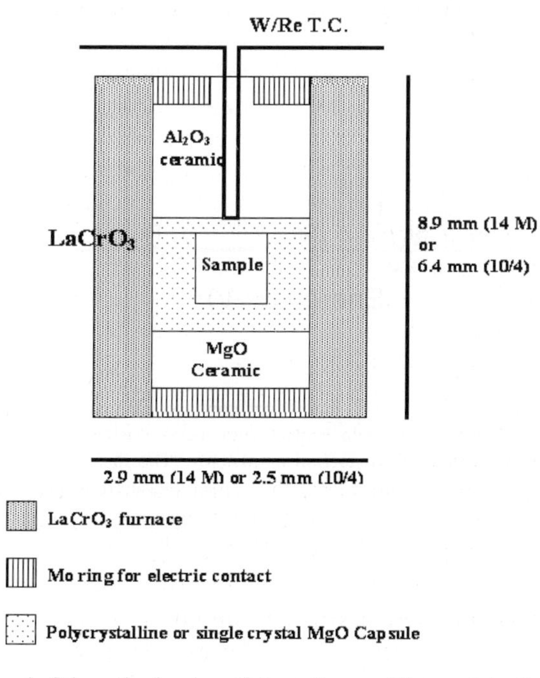

Fig. 1. Schematic drawing of the cell assemblies used in these experiments.

Samples were studied and analysed by optical microscopy, by SEM (Leo™ Steroscan 440) equipped with a Princeton™ Gamma-Tech PGT™ Spirit energy-dispersive X-ray analyser (EDX) able to analyse oxygen quantitatively, and Raman spectroscopy (XY Dilor™). Operating conditions of the SEM were an accelerating voltage of 20 kV; counting times were of about 100 s with a beam current of 0.5 nA for each EDX analysis and of 1 s per point with a beam current of 2 nA for quantitative chemical maps. Standards used were pure Fe metal and FeS for Fe and S, wollastonite or quartz and FeSi for Si, MgO for Mg, wollastonite for Ca and Al_2O_3 for Al. Raman spectra were recorded between 200 and 1400 cm^{-1} with acquisition times between 120 and 600 s and with an incident laser power between 10 and 100 mW. The argon laser beam tuned at 514 nm, was focused through microscope objectives (×100) down to approximately 1 μm.

3. Experimental results

In the three samples initially containing distinct grains of FeS and Si (samples S2852, S2853, S2861), these two phases, well separated prior to the runs, have

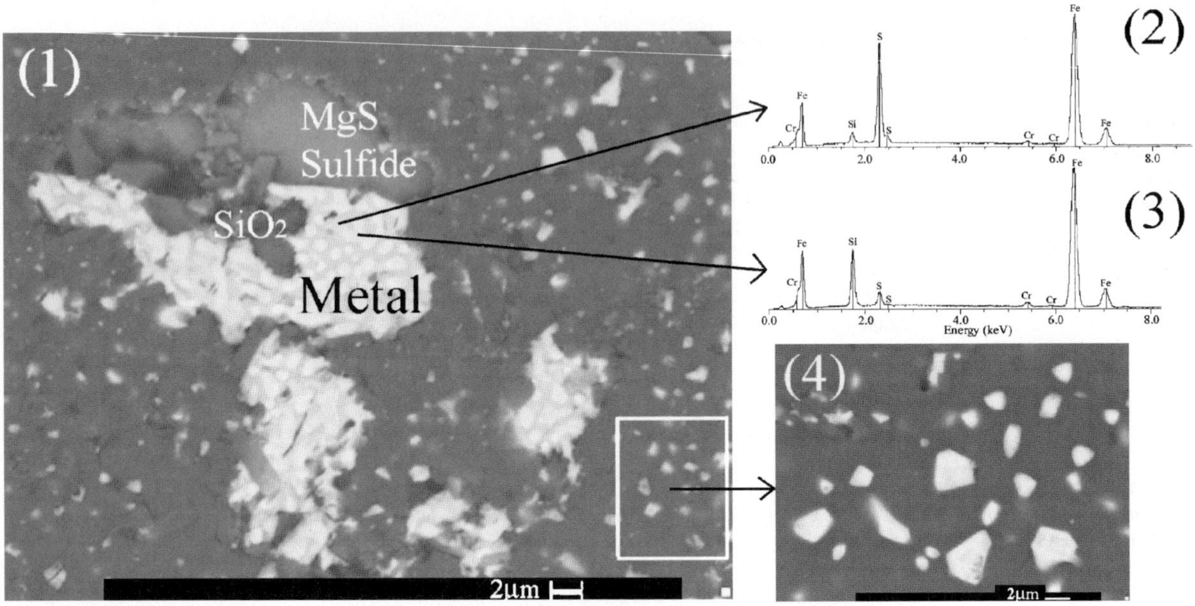

(a)

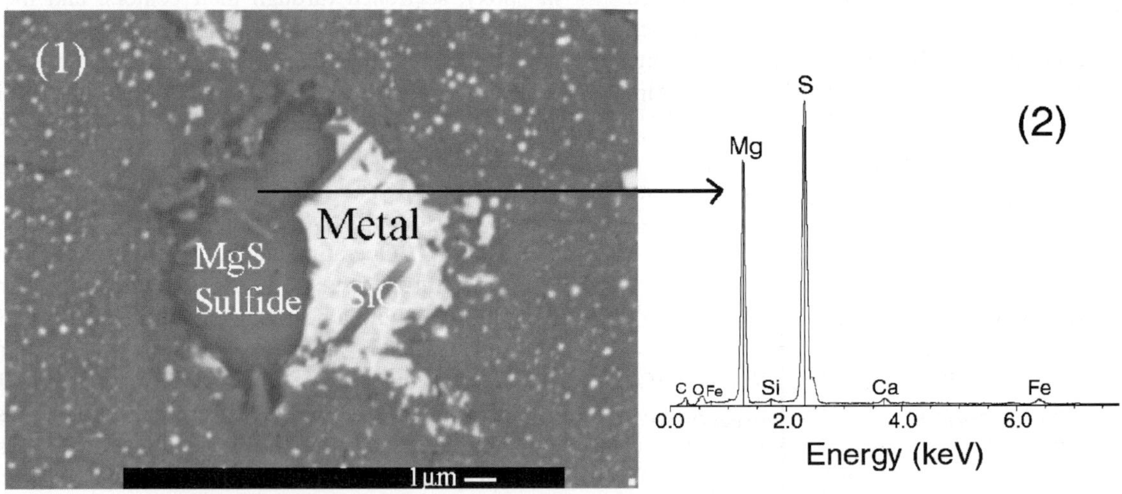

(b)

Fig. 2. (a) Backscattered electron image of sample S2861 (25 GPa) showing Fe–S rich, Si-poor phase (light grey and EDX spectrum) and Fe–Si rich, S poor phase (white and EDX spectrum). SiO_2 and MgS-sulphide are observed in contact with metal. An enlargement of the region enclosed by the white frame shows a detail of small metal particles formed by reduction of majorite. The metallic particles also contain two different contrasts corresponding to Fe–Si rich, S poor phase (white) and Fe–S rich, Si-poor phase (light grey). (b) Backscattered electron image of sample S2853 (20 GPa) showing similar phases as in 25 GPa samples. The magnesium sulphide phase has a stoichiometry close to MgS (EDX spectrum) and contains little amounts of Ca and Fe. Rods of SiO_2 stishovite, biphasic quenched metal phase and tiny biphasic metal particles are visible.

Table 2
Analyses of metal and silicates phases in sulphur bearing samples

Runs	S2852	S2853	S2861
Capsule	MgO single crystal	MgO single crystal	MgO single crystal
P (GPa)	20	20	25
T (°C)	2000	2000	2000
Duration (min)	10	10	9
ΔIW	−2.8	−3.6	−3.8
Metal liquid composition (wt.%)[a]			
Fe	66.3	63.5	61.6
S	29.1	17.0	17.7
Si	4.6	19.5	20.7
Metal liquid composition (wt.%)[b]			
Number of analyses	13	14	14
Fe	65.3 (1.0)	63.0 (3.6)	62.2 (1.0)
S	30.5 (2.3)	17.5 (3.9)	18.4 (1.2)
Si	4.2 (1.2)	19.5 (5.6)	19.4 (1.6)
Silicate composition in oxide (wt.%)			
Number of analyses	10	10	10
SiO_2	56.0 (1.2)	58.1 (0.9)	58.2 (0.6)
MgO	37.4 (0.9)	35.9 (0.7)	36.0 (0.6)
Al_2O_3	4.3 (0.2)	4.7 (0.1)	4.9 (0.1)
CaO	0.9 (0.05)	1.0 (0.05)	1.1 (0.05)
FeO	1.4 (0.1)	0.5 (0.03)	0.4 (0.03)

Numbers in parentheses are the standard deviations. Silicate was analysed for Fe, Si, Mg, S, Ca and Al. No significant amount of S was detected (less than 0.1 wt.%). ΔIW: oxygen fugacity relative to the iron-wüstite buffer based on the Fe–FeO equilibrium; for details see text.

[a] Metal composition reintegrated on the basis of the chemical maps (see text).

[b] Metal composition estimated by using a window size just smaller than the size of the metal blobs. Metal was analysed for Fe, Si, S. The presence of few amounts of Cr in a few metal blobs (<1 wt.%) is not shown. No significant oxygen was detected in the metal. However, the presence of small amounts of oxygen (<1 wt.%) in the metal cannot be ruled out.

coalesced into single large metallic blobs (Fig. 2). SEM images reveal that this metal contains two phases: an Fe–S rich, Si-poor phase and an Fe–Si rich, S poor phase (Fig. 2). The small dimensions of these two intergrown phases (typically less than 1 μm) do not allow single phase microanalysis with the EDX system of the SEM, even using a focused beam. In order to obtain analyses of the bulk metal phase, we used a window of size just smaller than the size of the metal particles (Table 2). The large standard deviation reflects some scatter in the analyses due to the non-negligible size of the two intergrown phases constituting the metal blobs, with regard to the window size, typically between 10 and 100 μm^2. In an alternative method, representative areas in each sample were selected and chemically mapped (around 7500 analyses for the metal phase in each map) by quantitative chemical mapping. Average element concentrations in the metal were then obtained (Table 2) by recombining the analyses using the SPIRIT-image analysis software.

The silicate material is mostly majorite, as revealed by its composition (Table 2) and Raman spectra (Fig. 3). It contains tiny metallic droplets, well visible in Fig. 2, which make the quantitative determination of the iron content in majorite difficult. However, some rare areas in majorite are free of metal droplets at the scale of the SEM observations, which allows obtaining the iron content in this phase. With these values, it is possible to estimate the effective oxygen fugacities that have prevailed during the experiments, relative to the iron-wüstite (IW) equilibrium according to the following equation:

$$\log fO_{2\,\text{experiment}} = \log fO_{2\,\text{IW}} + 2\log\left(\frac{a_{\text{FeO}}}{a_{\text{Fe}}}\right) \quad (2)$$

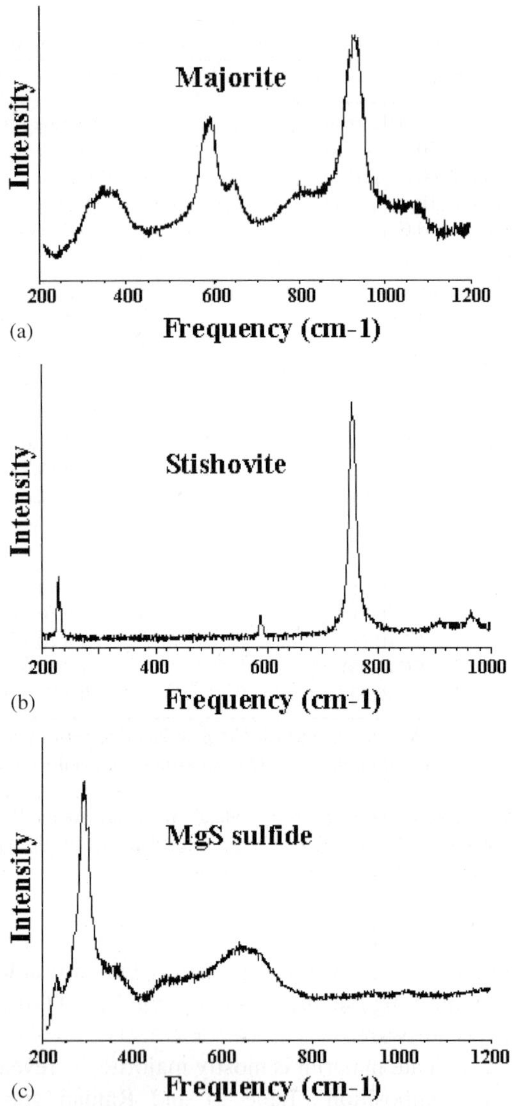

Fig. 3. (a) Raman spectrum of silicate in sample S2861. This spectrum is in good agreement with the spectrum of majorite obtained by Rauch et al. (1996). (b) Raman spectrum of the SiO_2 phase in sample 2861. This spectrum is in good agreement with the spectrum of stishovite obtained by Gillet et al. (1990). (c) Raman spectrum of the MgS-sulphide phase in sample 2861. Notice that MgS niningerite has no first order Raman spectrum.

where a_{FeO} and a_{Fe} are respectively the activities of FeO in silicate and Fe in liquid metal. In the absence of magnesiowüstite in the samples, due to lack of reactions between the MgO single crystalline capsule and the samples, fO_2 must be estimated by using the majorite-metal equilibrium. Unfortunately, the poor knowledge of activity-composition relations in majorites does not allow to easily retrieve oxygen fugacities. It is however possible to calculate the equilibrium a_{FeO} in magnesiowüstite by combining our analytical data in majorite with measurements from previous works about Fe partitioning between magnesiowüstite and majorite at relevant conditions (Tronnes, 2000; Wood, 2000; Frost, 2003). It appears that the most recent determinations (Frost, 2003) show that the partition coefficient between magnesiowüstite and majorite is close to unity at low total Fe contents. This behaviour has been observed with other silicates as well (Frost et al., 2001; Frost, 2003). At the low FeO contents of the present experiments, the activities of FeO can thus be considered to be close in magnesiowüstite and majorite. Taking into account non-ideality in magnesiowüstite (Fei et al., 1991) would lead to a maximum correction of 0.6 log units in oxygen fugacity, in agreement with previous estimates (Gessmann et al., 1999). Since this difference is not large and keeping in mind the large uncertainties which remain about activity-composition relations at high-pressure, results using a partition coefficient of 1 between majorite

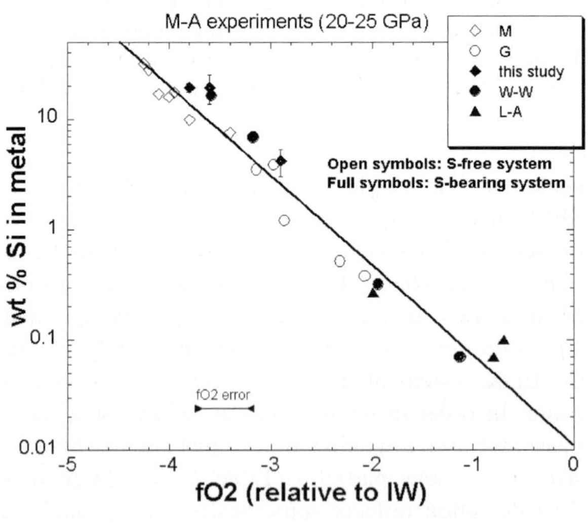

Fig. 4. wt.% Si in the metallic phase as a function of $\log fO_2$ for this study and previous studies at similar conditions. Oxygen fugacities are given relative to the IW buffer using FeO content in metal and liquid silicates (W–W, L–A) or solids (M, G, this study). Open symbols: S-free experiments. Data from M: Malavergne et al., 2004; G: Gessmann et al., 2001. Filled symbols: S-bearing experiments. Data from this study, W–W: Wade and Wood, 2001; L–A: Li and Agee, 2001b.

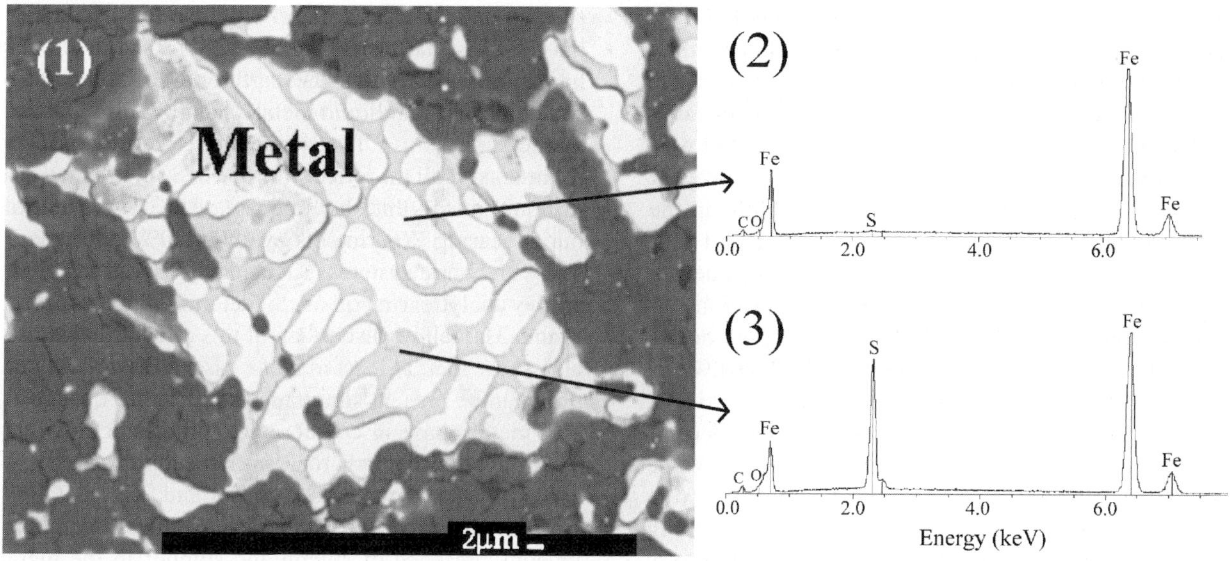

Fig. 5. Backscattered electron image of sample BT14 showing the quenched liquid metal containing two phases: Fe–S poor liquid (white and EDX spectrum) and Fe–S rich liquid (light grey and EDX spectrum). Neither stishovite nor magnesium sulphide have been evidenced in this sample.

and magnesiowüstite and an ideal model in magnesiowüstite are reported in Fig. 4, consistently with the approach of Gessmann et al. (1999).

A SiO_2 phase identified as stishovite by Raman spectroscopy (Fig. 3) is observed at the contact between the metal and majorite phases. At the contact with the metal phase, a phase of MgS composition containing minor amounts of Fe and Ca, is also observed (Fig. 2). The Raman spectrum of this phase is shown in Fig. 3; the assignment of the observed modes to a structure is not straightforward. Cubic monosulphides (space group *Fm*3*m*) have no first order Raman spectra. This magnesium sulphide phase is thus likely a high-pressure polymorph of MgS, the structure of which awaits to be determined.

Finally, a sample consisting of a starting assemblage of Fe, FeS and synthetic enstatite was transformed at 14 GPa and 1700 °C (sample BT14, Table 1). Even if the run conditions are quite different, this sample provided a useful reference for the reactivity of a similar system in the absence of reduced silicon. In this sample, the initially well separated Fe and FeS phases have coalesced into single metallic blobs, which contain two phases in the quenched samples (Fig. 5). These two phases are a Fe-rich, S poor phase and a Fe-sulphide phase close to Fe_2S in stoichiometry. Neither MgS nor stishovite have been evidenced in this sample. This Fe–S–silicates was previously studied in literature (Walker et al., 1993; Othani et al., 1997; Li and Agee, 1996, 2001a,b; Holzheid and Grove, 2002). No formation of MgS is reported from these studies.

4. Discussion

4.1. Fe–Si–S ternary system

In both Si-containing and Si-free experiments, the observed metal blobs are well segregated into one single phase after reaction at high-pressure and high-temperature, although in the starting material two different metal phases were present in each sample: Fe and FeS in the Si-free system and FeS and Si in the Si-bearing system. These metal particles are constituted by two distinct intergrown phases after quench and pressure release. These two intergrown phases are always in close association in the whole samples. This is strong evidence that the liquid metal was a single phase at high-pressure and temperature. This result has long been established in the Fe–S system; it was not obvious in the Fe–S–Si system due to the existence of a liquid immiscibility gap at

1 bar (e.g., Raghavan, 1988 for Fe–S and Fe–S–Si phase diagrams). Our observations thus suggest that, at 2000 °C and 20–25 GPa, the immiscibility gap is closed in the Fe–S–Si system, or at least considerably reduced, since our metal composition would fall well within the gap existing at 1 bar. Moreover, the sizes of the two intergrown phases within metal blobs in the four samples investigated (typically less than 1 μm) are consistent with textures developing during quench. This result is consistent with recent observations by Sanloup and Fei (2004) who observed the presence of a single metallic liquid (Fe–18 wt.% S–8.5 wt.% Si) liquid at 20 GPa and −1900 °C. It is also consistent with the report of a metal phase with 16.4 wt.% of Si and 3.5 wt.% of S at 2300 °C and 25 GPa by Wade and Wood (2001). The argument of immiscibility in the Fe–S–Si system thus cannot be used for excluding simultaneous presence of Si and S in the Earth's core.

In the three samples investigated in the Fe–S–Si metal system, some silicon was oxidised to stishovite. Abundant metal droplets are observed within the silicate phase showing that Fe^{2+} in majorite was reduced (e.g., Fig. 2). These two processes are coupled by oxygen balance as:

$$2FeSiO_{3\,\mathrm{in\,silicate}} + Si_{\mathrm{metal}} \leftrightarrow 3SiO_2 + 2Fe_{\mathrm{metal}} \quad (3)$$

The Si content in metal (Table 2) is not very different from that measured in a sulphur-free system at the same conditions (Fig. 4, Table 3). Table 3 shows that the presence of sulphur in metal does not drastically affect the final composition for similar reduction states of the starting materials. Therefore, under the conditions investigated, sulphur does not appear to have a strong effect on process (1) or (3). Liu and Fleet (2001) have shown that, at 1 bar and 1000–1400 °C, sulphur has a strong effect on the partition coefficient of Si between solid and liquid metal, changing from 0.45 in S-free systems to 3 for S contents relevant to our experiments. If all this effect was due to S–Si interactions in liquid metal, this should decrease the Si content of our S-bearing experiments by about 10% in comparison with S-free systems. Such a difference is too close to the analytical error on Si determination in metal (Table 3) to allow discussion of this effect. The results also show that, at least in some cases, silicon and sulphur can be present simultaneously in large quantities in a metal phase in contact with silicates. This result might appear contradictory with reports from experiments at lower pressures (2.5 GPa and −2000 °C) by Kilburn and Wood (1997) that would exclude simultaneous presence of silicon and sulphur in the metal phase. Insufficient reaction times which would lead to underestimated silicon or sulphur extraction in our experiments are not likely for two reasons: first, no compositional gradients are observed in the silicate matrix; second, small metal droplets formed by reduction of majorite contain Si and S which would be impossible if the equilibrium composition was fully depleted in one of the two elements. The observed coexistence of Si and S in metal could be due to the absence of an S-bearing silicate melt at 2000 °C in our experiments. In the absence of a melt, the activity of sulphide ions in the non-metal phase would be presumably larger, thus explaining larger S contents in metal. However, experiments by Wade and Wood (2001) at 20 GPa and 2300 °C seem to indicate Fe–Si–S metals even in the

Table 3
Comparison of Si contents (wt.%) in metal in sulphur-free and sulphur bearing metal/silicate experiments under highly reducing conditions

	S2852	S2853	S2861	2860[a]	2561[b]	2416[c]
P (GPa)	20	20	25	20	20	25
T (°C)	2000	2000	2000	2000	2200	2000
Duration (min)	10	10	9	1	7	60
Si/O[d] (wt.%)	0.59	0.79	0.79	0.65	0.72	0.68
Si (wt.%)[e]	4.2 (1.2)	19.5 (5.6)	19.4 (1.6)	7.60 (0.07)	17.60 (0.12)	16.00 (0.09)

[a] Starting mixes are as follows: 85 wt.% SC enstatite, 15 wt.% $Fe_{0.83}Si_{0.17}$ (Malavergne et al., 2004).
[b] Starting mixes are as follows: 85 wt.% SC enstatite, 15 wt.% FeSi (Malavergne et al., 2004).
[c] Starting mixes are as follows: 80 wt.% SC enstatite, 20 wt.% FeSi (Malavergne et al., 2004).
[d] Si/O in starting material. This ratio is an indicator of the reduction state of the sample (e.g. Gessmann et al., 1999).
[e] Uncertainties on Si determination in metal are given in parentheses. The large uncertainty on the determinations of the present study is due to the biphasic nature of the quenched metal phase.

presence of a melt. Alternatively, a pressure effect on chemical equilibria could explain differences between samples at 2.5 GPa (Kilburn and Wood, 1997) and at 20–25 GPa (Wade and Wood, 2001, and this study). Further experiments at intermediate pressures are needed to elucidate this issue. In any case, the simultaneous presence of Si and S in an Fe-alloy equilibrated with silicates cannot be ruled out under some plausible primitive Earth differentiation conditions.

4.2. MgS formation in reducing segregation

In the three samples investigated in the Fe–S–Si metal system, a magnesium sulphide (MgS) was formed at the rim of metal particles, whereas it was systematically absent in Si-free experiments. Cubic monosulphide series with the general formula (Mg,Mn,Ca,Fe)S are common phases in enstatite chondrites and in some achondrites (see Keil, 1989, for a review and extensive references). The formation of these sulphides requires exceptionally low oxygen and high sulphur fugacities (e.g., Larimer and Bartholomay, 1979). In our experiments, the formation of MgS through:

$$\mathrm{FeS_{in\ metal}} + \mathrm{MgSiO_{3\ in\ silicate}} \leftrightarrow \mathrm{MgS} + \mathrm{FeSiO_{3\ in\ silicate}} \quad (4)$$

is possible only under highly reducing (below IW-3) conditions at which it can be coupled to reaction (3). The sum reaction is then written as:

$$\mathrm{2FeS} + \mathrm{2MgSiO_3} + \mathrm{Si_{in\ metal}} \leftrightarrow \mathrm{2MgS} + \mathrm{3SiO_2} + \mathrm{2Fe_{in\ metal}} \quad (5)$$

Our results mean that reaction (5) is proceeding toward formation of MgS at 20–25 GPa and 2000–2200 °C. It is worth emphasising that, at 1 bar and 300 K, thermodynamic data ($\Delta_r G^\circ \sim -104$ kJ, Chase, 1998) suggest that MgS formation would occur as long as activity of Si in the metal phase is large enough. Carbon has been proposed as a reducing agent for MgS formation (Fleet and Mac Rae, 1987), although it might be insufficiently reducing (Larimer and Ganaparthy, 1987). Even at low pressure, Si in metal could be an alternative reducing agent efficient for magnesium and calcium sulphidation.

Magnesium sulphide, MgS, is a low density material (2.7 at 1 bar and 300 K). Using results from diffraction experiments on MgS at high-pressure by Peiris et al. (1994) and a third-order Birch–Murnaghan equation of state yields a density of 3.2 at 20–25 GPa and room temperature. Even if the MgS phase observed in our experiments is indeed a denser high-pressure polymorph, as suggested by the Raman spectra, MgS as a separate phase will remain a low density phase in comparison with metal. Therefore, this sulphur bearing phase would not segregate to the core with metal phases. If the core segregation in the Earth or in planetesimals implied Si-bearing liquid metal percolating through sub-solidus silicates, Mg–Ca sulphides would constitute a major form of sulphur sequestred in the silicate part of the planet. This would be an additional mechanism to fractionate Mg/Si in the early Earth. MgS is an highly refractory phase, however quite unstable in the presence of water. Further mixing of MgS buoyantly segregated to the shallow mantle with more oxidised hydrated components (Wänke and Dreibus, 1988), could then destabilise this phase according to:

$$\mathrm{MgS} + \mathrm{H_2O} \leftrightarrow \mathrm{MgO} + \mathrm{H_2S} \quad (6)$$

which could have delivered significant quantities of H_2S to the surface of the primitive Earth. It should be noticed, however, that numerous metal/silicate segregation models of the Earth, but not of planetesimals, imply molten instead of sub-solidus silicates. Recent experiments about the melting behaviour of enstatite chondrites (McCoy et al., 1999; Fogel, 1998; Dickinson et al., 1998) indicate that at temperatures exceeding 1300 °C at 1 bar, sulphides dissolve into an S-rich silicate melt. The consequences for sulphur behaviour could then be very different since no separate (Mg,Ca)S phase would form. We note however that Ca,Mg,Mn monosulphides have been reported to form by fractional crystallisation from S-bearing silicate melts under highly reducing conditions (McCoy et al., 1999). (Mg,Ca)S sulphides might thus be important phases even in differentiation processes involving partially or totally molten silicates.

5. Conclusion

We have performed core segregation experiments at 20–25 GPa and 2000 °C, using assemblages of sub-solidus silicates and (Fe–Si–S) metal alloys. Our observations suggest that, at these pressures and temperatures, the immiscibility gap in the ternary Fe–S–Si system is closed. The argument of immiscibility in the Fe–S–Si system thus cannot be used for excluding simultaneous presence of Si and S in planetary cores. Moreover, we show that appropriate P–T–fO_2 conditions exist under which silicon and sulphur can be present simultaneously in a metal phase in contact with silicates. This is a different conclusion than in previous studies and could be due to the absence of S-bearing silicate melts in our experiments. The simultaneous presence of Si and S in an Fe-alloy equilibrated with silicates thus cannot be ruled out under some plausible Earth or planetesimals differentiation conditions. Finally, magnesium sulphide, MgS, a low density material was formed in the experiments. If the core segregation in the Earth or in planetesimals implied Si-bearing liquid metal percolating through sub-solidus silicates, Mg–Ca sulphides would then have constituted a reservoir of sulphur in the silicate part of the planet, which upon hydration could have delivered significant quantities of H_2S to the early planetary surfaces.

Acknowledgements

High-pressure and high-temperature experiments were performed at Bayerisches Geoinstitut, Germany, under the EU "IHP-Access to Research Infrastructures" Programme (contract no. HPRI-1999-CT-00004 to D.C. Rubie). We thank Dan Frost and Stephan Borensztajn for respective assistance with multianvil experiments and SEM observations. The French INSU-CNRS Programs PNP, Planétologie, and Terre Profonde are thanked for financial support. We wish to thank C.B. Agee and K. Righter for their constructive and insightful reviews.

References

Allègre, C., Poirier, J.P., Humler, E., Hofman, A., 1995. The chemical composition of the earth. Earth Planet. Sci. Lett. 134, 515–526.

Allègre, C., Manhès, G., Lewin, E., 2001. Chemical composition of the earth and the volatility control on planetary genetics. Earth Planet. Sci. Lett. 185, 49–69.

Anderson, D.L., 1988. Composition of the earth. Science 243, 367–370.

Chase, M.W. Jr., 1998. NIST-JANAF thermochemical tables, 4th ed. J. Phys. Chem. Ref. Data, Monograph 9, 1–1951.

Dickinson, T.L., McCoy, T.J., Lofgren, G.E., 1998. Melting of indarch (EH4) powders: phase relations and elemental exchange. In: Proceedings of the 29th Lunar and Planetary Science Conference.

Drake, M.J., 2000. Accretion end primary differentiation of the earth: a personal journey. Geochim. Cosmochim. Acta 64, 2363–2370.

Dreibus, G., Palme, H., 1996. Cosmochemical constraints on the sulfur content in the earth's core. Geochim. Cosmochim. Acta 60, 1125–1130.

Fei, Y., Mao, H.-K., Mysen, B., 1991. Experimental determination of element partitioning and calculation of phase relation in the MgO–FeO–SiO_2 system at high pressure and high temperature. J. Geophys. Res. 96, 2157–2169.

Fleet, M.E., Mac Rae, N.D., 1987. Sulfidation of Mg-rich olivine and the stability of niningerite in enstatite chondrites. Geochim. Cosmochim. Acta 51, 1511–1521.

Fogel, R.A., 1998. High sulfur-low iron silicate melts: low oxygen fugacity phenomena of importance to aubrite formation. In: Proceedings of the 61st Annual Meteoritical Society Meeting.

Frost, D.J., Langhenhorst, F., Van Aken, P.A., 2001. Fe–Mg partitioning between ringwoodite and magnesiowüstite and the effect of pressure, temperature, and oxygen fugacity. Phys. Chem. Miner. 26, 455–470.

Frost, D.J., 2003. Fe^{2+}–Mg partitioning between garnet, magnesiowüstite, and (Mg, Fe)SiO_4 phases of the transition zone. Am. Min. 88, 387–397.

Gessmann, C.K., Rubie, D.C., McCammon, C.A., 1999. Oxygen fugacity dependence of Ni, Co, Mn, Cr, V and Si partitioning between liquid metal and magnesiowustite at 9–18 GPa and 2200 °C. Geochim. Cosmochim. Acta 63, 1853–1863.

Gessmann, C.K., Wood, B.J., Rubie, D.C., Kilburn, M.R., 2001. Solubility of silicon in liquid metal at high pressure: implications for the composition of the earth's core. Earth Planet. Sci. Lett. 184, 367–376.

Gillet, P., Le Cleach, A., Madon, M., 1990. High temperature Raman spectroscopy of SiO_2 and GeO_2 polymorphs. Anharmonicity and thermodynamic properties at high temperatures. J. Geophys. Res. 81, 21635–21655.

Guyot, F., Zhang, J., Martinez, I., Matas, J., Ricard, Y., Javoy, M., 1997. P–V–T measurements of iron silicide (ε-FeSi). Implications for silicate–metal interactions in the early earth. Eur. J. Mineral. 9, 277–285.

Holzheid, A., Grove, T.L., 2002. Sulfur saturation limits in silicate melts and their implications for core formation scenarios for terrestrial planets. Am. Min. 87, 227–237.

Ito, E., Morooka, K., Ujike, O., Katsura, T., 1995. Reaction between molten iron and silicate melts at high pressure: implications for the chemical evolution of the earth's core. J. Geophys. Res. 100, 5901–5910.

Javoy, M., 1995. The integral enstatite chondrite model of the earth. Geophys. Res. Lett. 22, 2219–2222.

Karato, S., Murthy, V.R., 1997. Core formation and chemical equilibrium in the earth. I. Physical considerations. Earth Planet. Sci. Lett. 100, 61–79.

Keil, K., 1989. Enstatite meteorites and their parents bodies. Meteoritics 24, 195–208.

Kilburn, M.R., Wood, B.J., 1997. Metal–silicate partitioning and the incompatibility of S and Si during core formation. Earth Planet. Sci. Lett. 152, 139–148.

Larimer, J.W., Bartholomay, M., 1979. The role of carbon and oxygen in cosmic gases: some applications to the chemistry and mineralogy of enstatite chondrites. Geochim. Cosmochim. Acta 43, 1455–1466.

Larimer, J.W., Ganaparthy, R., 1987. The trace element chemistry of CaS in enstatite chondrites and some implications regarding its origin. Earth Planet. Sci. Lett. 84, 123–134.

Li, J., Agee, C.B., 1996. Geochemistry of mantle-core differentiation at high pressure. Nature 381, 686–689.

Li, J., Agee, C.B., 2001a. Element partitioning constraints on the light element composition of the earth's core. Geophys. Res. Lett. 28, 81–84.

Li, J., Agee, C.B., 2001b. The effect of pressure, temperature, oxygen fugacity and composition of nickel and cobalt between liquid Fe–Ni–S alloy and liquid silicate: implications for the earth's core formation. Geochim. Cosmochim. Acta 65, 1821–1832.

Liu, M., Fleet, M.E., 2001. Partitioning of siderophile elements (W, Mo, As, Ag, Ge, Ga, and Sn) and Si in the Fe–S system and their fractionation in iron meteorites. Geochim. Cosmochim. Acta 65, 671–682.

MacDonald, G.J.F., Knopoff, L., 1958. On the chemical composition of the outer core. J. Geophys. Res., Astron. Soc. 1, 284–297.

Malavergne, V., Siebert, J., Guyot, F., Gautron, L., Combes, R., Hammouda, T., Borenstajn, S., Frost, D.J., Martinez, I., 2004. Si in the core? New high pressures and high temperature experimental data. Geochim. Cosmochim. Acta, submitted for publication.

McCoy, T.J., Dickinson, T.L., Lofgren, G.E., 1999. Partial melting of the indarch (EH4) meteorite: a textural view of melting and melt migration. Meteorit. Planet. Sci. 34, 735–746.

Murthy, V.R., Hall, H.T., 1970. The origin and chemical composition of the earth's core. Phys. Earth Planet. Inter. 2, 276–282.

Murthy, V.R., 1991. Early differentiation of the earth and the problem of mantle siderophile elements: a new approach. Science 253, 303–306.

O'Neill, H.St.C., Canil, D., Rubie, D.C., 1998. Oxide-metal equilibria to 2500 °C and 25 GPa: implications for core formation and the light component in the earth's core. J. Geophys. Res. 103, 12239–12260.

Othani, E., Yurimoto, H., Seto, S., 1997. Element partitioning between metallic liquid, silicate liquid, and lower-mantle minerals: implications for core formation of the earth. Phys. Earth Planet. Inter. 100, 97–114.

Peiris, M.S., Campbell, A.J., Heinz, D.L., 1994. Compression of MgS to 54 GPa. J. Phys. Chem. Solids 1994, 419–431.

Poirier, J.P., 1994. Light elements in the earth's outer core: a critical review. Phys. Earth Planet. Inter. 85, 319–337.

Raghavan, V., 1988. Phase Diagrams of Ternary Iron Alloys. Part 2. Ternary Systems Containing Iron and Sulphur. The Indian Institute of Metals, Calcutta.

Rauch, M., Keppler, H., Hafner, W., Poe, B., Wokan, A., 1996. A pressure-induced phase transition in $MgSiO_3$-rich garnet revealed by Raman spectroscopy. Am. Min. 81, 1289–1292.

Righter, K., Drake, M.J., Yaxley, G., 1997. Prediction of siderophile element metal–silicate partition coefficient to 20 GPa and 2800 °C: the effect of pressure, temperature, fO_2 and silicate and metallic melt composition. Phys. Earth Planet. Inter. 100, 115–134.

Ringwood, A.E., Hibberson, W., 1991. Solubilities of mantle oxides in molten iron at high pressures and temperatures: implications for the composition and formation of earth's core. Earth Planet. Sci. Lett. 102, 235–251.

Rubie, D.C., 1999. Characterising the sample environment in multianvil high-pressure experiments. Phase Trans. 68, 431–451.

Sanloup, C., Fei, Y., 2004. Closure of the Fe–S–Si in liquid miscibility gap at high pressure and its implications for planetary core formation. In: Proceedings of the 35th Lunar and Planetary Science Conference. Phys. Earth planet Inter., submitted for publication.

Thiéblaut, L., Téqui, C., Richet, P., 1999. High-temperature heat capacity of grossular ($Ca_3Al_2Si_3O_{12}$), enstatite ($MgSiO_3$), and titanite ($CaTiSiO_3$). Am. Min. 84, 848–855.

Tronnes, R.G., 2000. Melting relations and major element partitioning in an oxidized bulk earth model at 15–26 GPa. Lithos 53, 233–245.

Wade, J., Wood, B.J., 2001. The earth's 'missing' niobium may be in the core. Nature 409, 75–78.

Walker, D., Norby, L., Jones, J.H., 1993. Superheating effects on metal/silicate partitioning of siderophile elements. Science 262, 1858–1861.

Walter, M.J., Newsom, H., Ertel, W., Holzeid, A., 2000. Siderophile elements in the earth and moon: metal–silicate partitioning and implications for core formation. In: Canup, R.M., Righter, K. (Eds.), Origin of the Earth and Moon. University of Arizona, Tucson.

Wänke, H., Dreibus, G., 1988. Chemical composition and accretion history of terrestrial planets. Phil. Trans. Roy. Soc. Lond. A 325, 545–557.

Wood, B.J., 2000. Phase transformations and partitioning relations in peridotite under lower mantle conditions. Earth Planet. Sci. Lett. 174, 341–354.

6. Iron and Planetary Cores

Available online at www.sciencedirect.com
SCIENCE DIRECT®

ELSEVIER

Physics of the Earth and Planetary Interiors 143–144 (2004) 445–453

PHYSICS OF THE EARTH AND PLANETARY INTERIORS

www.elsevier.com/locate/pepi

Non-collinear magnetism in iron at high pressures

R.E. Cohen*, S. Mukherjee

Geophysical Laboratory, Carnegie Institution of Washington, 5251 Broad Branch Rd., N.W., Washington, DC 20015, USA

Received 21 January 2003; received in revised form 5 June 2003; accepted 17 August 2003

Abstract

Using a first-principles based, magnetic tight-binding total energy model, the magnetization energy and moments are computed for various ordered spin configurations in the high pressure polymorphs of iron (fcc, or γ-Fe, and hcp, or ϵ-Fe), as well as ferromagnetic bcc iron (α-Fe). For hcp an antiferromagnetic spin configuration is more stable than non-magnetic ϵ iron up to about 50 GPa. Accounting for magnetism yields better agreement with the experimental equation of state, in contrast to the non-magnetic equation of state, which is in poor agreement with experiment below 50 GPa. We also studied non-collinear magnetism in γ- and ϵ-Fe. In ϵ-Fe the non-collinear effects are quite small energetically; in its stability field the collinear afmII structure is more stable than all non-collinear structures we explored.

PACS: 71.55.Ak; 64.30.+t

Keywords: Electronic structure; Iron; Fe; High-pressure; Magnetism; Tight-binding

1. Introduction

Magnetism is known to be important in the phase stability, structure and elastic properties of iron. For example, α-Fe, the ground state at ambient conditions, would be mechanically unstable if it were not magnetic. Even above the Curie temperature, T_c, there are local magnetic moments in α-Fe. Face-centered cubic iron (fcc or γ-Fe) has incommensurate magnetic correlations which change rapidly with volume, and give rise to the anti-Invar effect (large thermal expansivity) (Mryasov et al., 1992). The magnetic behavior of ϵ-Fe is important for high pressure materials research, and for interpreting high pressure experiments aimed at understanding the Earth's inner core. Hcp iron is not quenchable to zero pressure, so magnetic studies must be made in situ at high pressures; it was long thought to be non-magnetic, due to several Mössbauer experiments that showed no hyperfine splitting in ϵ-Fe, even down to helium temperatures. However, self-consistent first-principles computations show a magnetic ground state for ϵ-Fe which is stable up to about 50 GPa (Steinle-Neumann et al., 1999). Recently the hyperfine fields were computed for antiferromagnetic hcp-Fe using the LAPW method, and it was found that the predicted hyperfine fields are too small to be observed in Mössbauer due to cancellation of the valence and semi-core contributions to the magnetic field at the nucleus (Steinle-Neumann et al., 2004). Therefore the theoretical prediction of magnetism is not in contradiction to experiments. Here non-collinear magnetism in ϵ-Fe is explored using a magnetic tight-binding model fitted to first-principles calculations.

* Corresponding author. Tel.: +1-202-537-3951; fax: +1-202-537-3951.
E-mail address: cohen@gl.ciw.edu (R.E. Cohen).

0031-9201/$ – see front matter
doi:10.1016/j.pepi.2004.02.002

When magnetic moments are collinear, electrons can be considered to be "spin-up" or "spin-down" in a global sense, i.e. there is a global magnetic quantization direction. This means that one can solve for the spin-up and spin-down electrons separately, and then combine the results to compute the total charge density. When moments are oblique to each other, the spin state is said to be "non-collinear." In that case the problem does not factorize, and one must diagonalize a Hamiltonian of twice the order of the collinear case. A system can be non-collinear either in response to chemical or thermal disorder, or in order to minimize frustration. Both hcp and fcc lattices are frustrated with respect to antiferromagnetism, in that one cannot order these lattices antiferromagnetically so that all neighbors have opposite spins. This is known to lead to spin-waves and non-collinear magnetism in fcc-Fe. Also, a common example of non-collinear spins occurs on heating bcc-Fe above the Curie point, where the spins disorder dynamically. A comprehensive review of the theory of non-collinear magnetism is given in Sandratskii (1998).

2. Method

To study non-collinear magnetism in Fe, a first-principles-based non-magnetic tight-binding model (Cohen et al., 1994, 1997; Wasserman et al., 1996) is combined with a model for magnetism (Pickett, 1996). The model described here is improved relative to an earlier version (Mukherjee and Cohen, 2001) which neglected the magnetic terms that arise from the non-orthogonal basis, the fact that the s- and p-states are also spin polarized, and the output moment directions as well as magnitudes are obtained self-consistently, and both can be constrained by self-consistently determined staggered magnetic fields.

The total energy in the model is given by

$$E = E_{\mathrm{bs}} + \frac{1}{4}\sum_{jLL'} |m_{jL}| I_{jLjL'} |m_{jL'}| + \sum_{j} \vec{b}_j \cdot \vec{m}_j \tag{1}$$

where $I_{jLjL'}$ is the Stoner parameter that represents the exchange interaction on state L from state L' on atom j, $\vec{b}_j$ is an applied magnetic field on atom j, $|m_{jL}|$ is the magnitude of the magnetic moment of state L on atom j. The last term in Eq. (1) is to subtract the interaction energy with the applied field, so that the energy for given moments in the absence of a field can be obtained. The moments are obtained from the spin density matrix $\rho_{\sigma\sigma'}$

$$\rho_{jL\sigma\sigma'} = \frac{1}{2} \sum_{n,n',j,j',L'} C^{*}_{jLn\sigma} S_{jLn,j'L'n'} C_{j'L'n'\sigma'} \tag{2}$$

where S is the overlap matrix, C is the eigenvector (orbital coefficients), and n labels the different orbitals for each L. The spin density matrix can be decomposed into Cartesian moments in terms of the Pauli spinor matrices, σ_i, or into spherical coordinates by

$$|m_{jL}| = ((\rho_{jL\uparrow\uparrow} - \rho_{jL\downarrow\downarrow})^2 + 4\rho_{jL\uparrow\downarrow}\rho^{*}_{jL\downarrow\uparrow})^{1/2} \tag{3}$$

$$\theta_{jL} = \cos^{-1}\left(\frac{\rho_{jL\uparrow\uparrow} - \rho_{jL\downarrow\downarrow}}{|m_{jL}|}\right) \tag{3}$$

$$\phi_{jL} = \Im(\ln \rho_{jL\uparrow\downarrow}) \tag{4}$$

where $\theta : 0 \le \theta \le \pi$ is the polar angle and ϕ is the azimuthal angle. The moments in Cartesian coordinates are

$$\vec{m}_{jL} = (\cos(\phi_{jL})\sin(\theta_{jL}), \sin(\phi_{jL})\sin(\theta_{jL}), \cos(\theta_{jL})) \tag{6}$$

E_{bs} is the band structure energy for the magnetic system

$$E_{\mathrm{bs}} = \sum_{i,k} \epsilon_{i,k} w_k f(\epsilon_{i,k} - E_{\mathrm{F}}) \tag{7}$$

where $\epsilon_{i,k}$ is the band eigenvalue i at k-point k, w_k is the k-point weight, f is the Fermi function, and E_{F} is the Fermi level, which is found so that the total number of electrons is correct,

$$N = \sum \epsilon_{i,k} w_k f(\epsilon_{i,k} - E_{\mathrm{F}}) \tag{8}$$

The order of the Hamiltonian (H) of the system is double the order n of the Hamiltonian of the non-magnetic system. The non-magnetic part is reproduced in the two diagonal order-n blocks, and the off-diagonal blocks are for non-collinear matrix elements, so that the Hamiltonian has the structure

$$\begin{pmatrix} \uparrow\uparrow & \uparrow\downarrow \\ \downarrow\uparrow & \downarrow\downarrow \end{pmatrix} \tag{9}$$

The overlap S is also doubled, with the off-diagonal blocks zero. There are 18 orbitals, s↑, 3p↑, 5d↑, s↓, 3p↓, and 5d↓ orbitals for each atom j.

The Kohn–Sham Hamiltonian is derived from the total energy equation (1) from the variational principle, and is given by

$$H_{jLj'L'} = H^0_{jLj'L'} \cdot \sigma_0 - \frac{1}{4}\sum_{L''} s_{jLj'L'}(m_{jL''}I_{jLjL''} + m_{j'L''}I_{j'Lj'L''}) \cdot \sigma_i - \frac{1}{2}\sum_{j'L'} s_{jLj'L'}(b_j + b_{j'}) \cdot \sigma_i \quad (10)$$

where H^0 is the non-magnetic Hamiltonian and σ_0 is the $2n \times 2n$ identity matrix, and σ_i are the Pauli spinor matrices mapped onto the $2n \times 2n$ space. The elements of the overlap matrix S and non-magnetic Hamiltonian H^0 are unchanged from the non-magnetic non-orthogonal tight-binding model (Cohen et al., 1994, 1997; Wasserman et al., 1996) except for being repeated in the diagonal $\uparrow\uparrow$ and $\downarrow\downarrow$ blocks. H_0 was accurately fit to eigenvalues and total energies from an extensive set of non-magnetic linearized augmented plane wave (LAPW) results within the generalized gradient approximation (GGA) (Perdew et al., 1996) in the bcc, fcc, and hcp structures as functions of pressure and strain, and has been extensively tested (Cohen et al., 1994, 1997; Wasserman et al., 1996).

The code runs in three modes: (1) the moment magnitudes and directions and total energy can be determined self-consistently, (2) the moment directions are constrained, and the moment magnitude and total energy are determined self-consistently, (3) the moment directions and magnitudes are constrained and the total energy is determined. In modes (2) and (3) the moments are constrained by applying staggered fields $\vec{b}_j$ on each site j. The fields are found using a penalty function and a self-consistent procedure. The procedure used to find the constraining fields is similar to, but different from, the procedure outlined in Ujfalussy et al. (1999). A penalty function is also used when necessary to prevent unphysical charge transfer between the atoms.

The model differs from the conventional Stoner model (Stoner, 1938), in which a ferromagnetic instability is predicted by the inequality $IN(0) > 1$, where $N(0)$ is the non-magnetic density of states at the Fermi level, and the extended Stoner model (Krasko, 1987), in which N is replaced by the effective density of states $\tilde{N}(M) = M/\delta\epsilon$, where $\delta\epsilon$ is the exchange splitting, in that our model includes the changes in band structure with spin-polarization due to different hybridization depending on magnetic state, accounting thereby to the actual magnetic structure.

3. Results and discussion

3.1. bcc

The model is in good agreement with previous self-consistent calculations for α-Fe. The lowest energy state for α-Fe for the model at ambient and at higher pressures is ferromagnetic, in agreement with the first principles LAPW self-consistent calculations (Stixrude et al., 1994). As pressure is increased the magnetization energy decreases smoothly (Fig. 1). The value of I needed for quantitative agreement is almost constant (around 1 eV) for larger atomic volumes or lower pressures but increases for higher pressures. This is consistent with the increase in the exchange interaction at higher pressures noted previously (Asada and Terakura, 1992). Even though

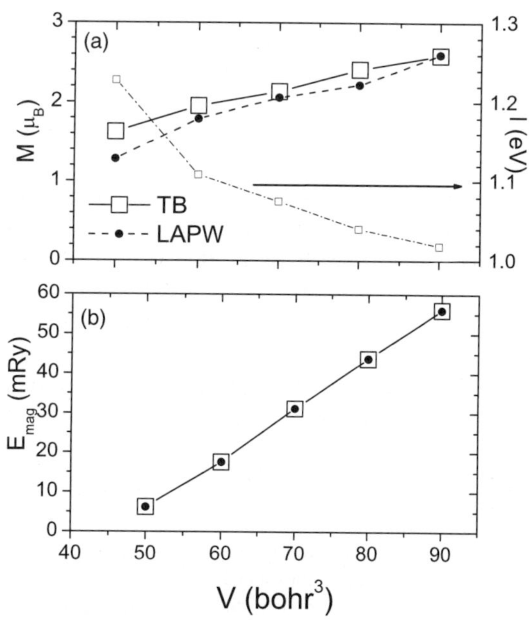

Fig. 1. Magnetization (a) and magnetization energy (b) for bcc iron as functions of volume. The self-consistent LAPW results are from Stixrude et al. (1994). I is chosen to best fit the LAPW magnetization energies and the best-fit I is also shown in (a).

the exchange interaction increases with pressure, the non-magnetic density of states near the Fermi energy decreases, giving rise to a net reduction in the magnetic moment and magnetization energy, and finally the loss of magnetism at around atomic volume of 40 bohr^3 (1 bohr (=atomic unit = au) = 0.529177 Å).

The Stoner parameter I was adjusted at each volume to give agreement with self-consistent full-potential linearized augmented plane wave magnetization energies (Steinle-Neumann et al., 1999) for bcc. I is expected to increase as the electronic density increases with decreasing atomic volume, similar to what has been seen before for γ-Fe (Krasko, 1987) and α-Fe (Fig. 1).

3.2. *fcc*

γ-Fe shows a great richness in non-collinear magnetic structures as volume is varied, and here it is not explored in detail; much study has been done, as

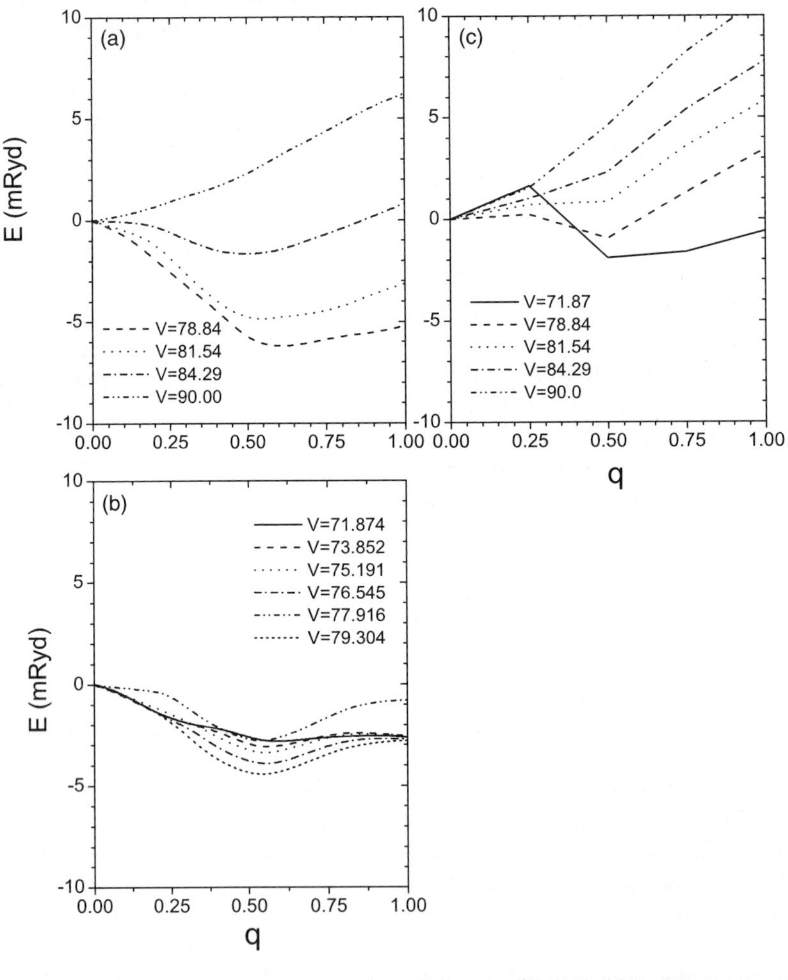

Fig. 2. Variation of magnetization energy vs. wavevector for γ-Fe for (0, 0, q), with q in units of $2\pi/a$, where a is the lattice constant. Wavevectors $q = 0$ and 1 correspond to ferromagnetic and antiferromagnetic structures, respectively, and $q = 0.5$ is a non-collinear spin state with relative angle between the neighboring spins of $\pi/2$. (a) LMTO results from Mryasov et al. (1992). (b) LAPW results from Sjöstedt and Nordström (2002). (c) Results from this study using the best-fit value of I (1.066, 1.052, 1.047, 1.041, and 1.03 eV for the volumes shown). The tight-binding results show the correct behavior of ferromagnetic at large volumes and antiferromagnetic at small volumes with a non-collinear transition region. The results of Sjöstedt and Nordström (2002) seem more different than our results and those of Mryasov et al. (1992) than can be explained by the atomic moment approximation. Note that the line dashes and volumes correspond for (a) and (c), but not for (b).

reviewed in Sandratskii (1998). The energies and moments for fcc-Fe with spiral spins along (0 0 1) and $\theta = \pi/2$ were determined using eight-atom supercells, allowing calculations for $(0\,0\,q)$ with $q = 0$, $\pi/4$, $\pi/2$, $3\pi/4$, and π. A k-point mesh of $12 \times 12 \times 4$ was used, giving 72 k-points in the irreducible wedge for tetragonal symmetry. For γ-Fe, the complex behavior of magnetic moments and the magnetization energy with increasing pressures and varying spiral spin density wave states are qualitatively reproduced by the model (Figs. 2 and 3) compared with self-consistent calculations (Mryasov et al., 1992; Uhl et al., 1994; Sjöstedt and Nordström, 2002). Quantitatively the results are sensitive to the value of I. In any case there is some variation in self-consistent results for γ-Fe, due to extreme sensitivity of the magnetic structure to basis set, k-point sampling, etc., due to the small energy scale. Furthermore, the magnetic ground state in γ-Fe is sensitive to the atomic moment approximation (Sjöstedt and Nordström, 2002) in the tight-binding model; that is the moment is really a field that varies with position in the crystal, and is not constant on each atom.

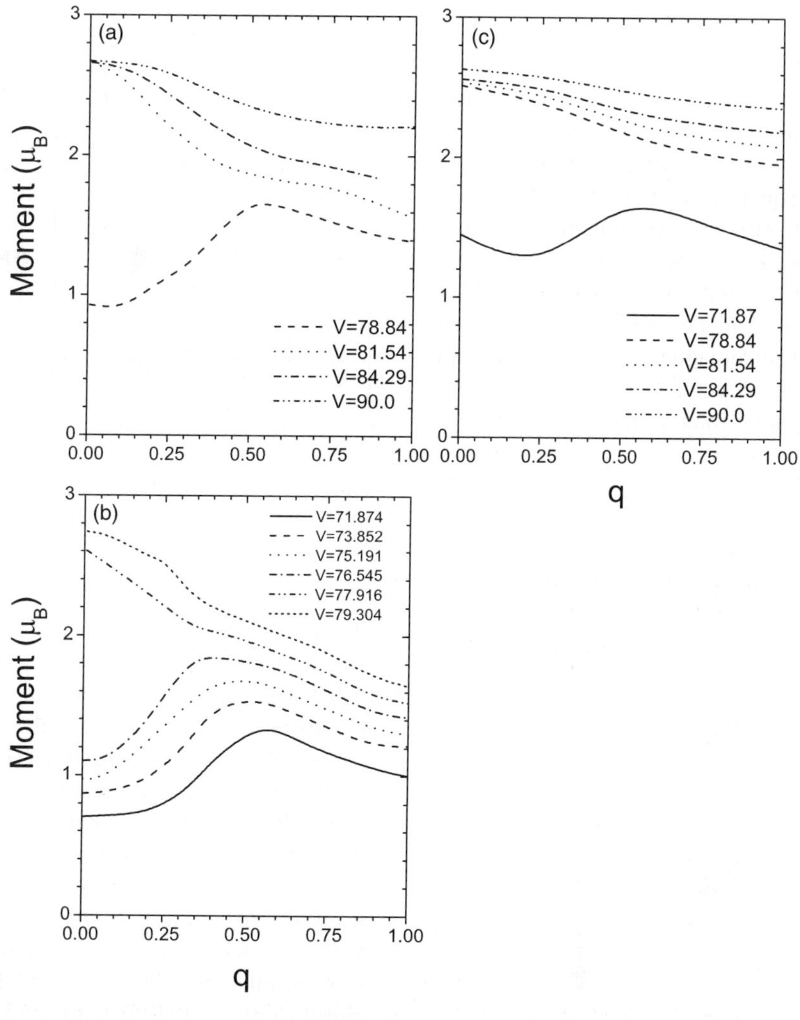

Fig. 3. Variation of moments versus wavevector for γ-Fe for $(0, 0, q)$. (a) LMTO results from Mryasov et al. (1992). (b) LAPW results from Sjöstedt and Nordström (2002). (c) Results from this study using the best-fit value of I (1.017, 1.006, 1.003, 1.000, and 0.993 eV for the volumes shown). Note that the line dashes and volumes correspond for (a) and (c), but not for (b). A high-spin low-spin transition is evident in all cases for ferromagnetic and low-q spin-waves with increasing pressure.

Within the atomic moment approximation our results are reasonably consistent with the self-consistent results. More detailed comparisons with the comprehensive non-collinear LAPW calculations of Sjöstedt and Nordström (2002) including k-point convergence tests are called for, but have not yet been done. In Sjöstedt and Nordström (2002) 4000 k-points were used in the full Brillouin zone, compared to our 4608, which seems comparable. It is more difficult to compare with Mryasov et al. (1992) since a real space multiple scattering approach and a muffin-tin potential approximation was used.

3.3. hcp

Previous collinear first-principles calculations show stability of an antiferromagnetic state, afmII, in hcp-iron (Steinle-Neumann et al., 1999). Moreover, the computed equation of state of afmII greatly improves agreement with the experimental equation of state. An hcp lattice is frustrated for antiferromagnetism; it is not possible to have perfect antiferromagnetic order on it. In the afmII structure each atom has eight antiferromagnetically oriented and four ferromagnetically oriented neighbors, maximizing the antiferromagnetic interactions. In the case of a nearest-neighbor (n.n.) Heisenberg model, which has energy $E = J_1\sum_{n.n.} \vec{m}_i \cdot \vec{m}_j$, the energy is independent of the angle between the moments of one antiferromagnetic pair and another (see Fig. 4). If non-nearest neighbor interactions are important, or if the Heisenberg model does not completely describe the energetics, the energy might be further lowered if the antiferromagnetic pairs are oblique or perpendicular to each other (i.e. $\alpha \neq 0$).

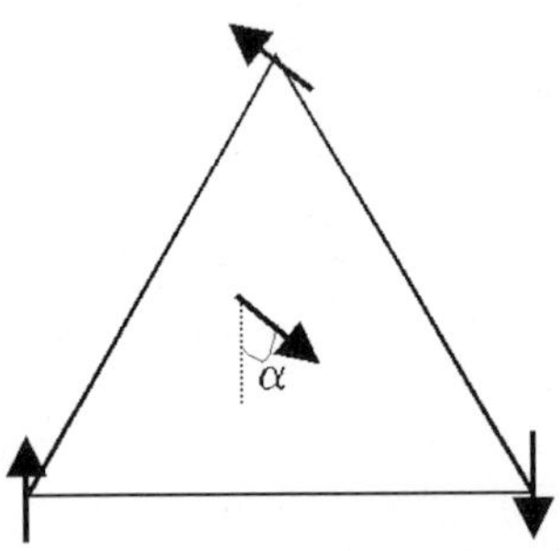

Fig. 4. The motif for an hcp lattice with antiferromagnetic interactions. One pair of antiferromagnetic iron atoms is at an angle α to another pair. If this motif is used as shown, a collinear structure with symmetry Pmmn is represented by $\alpha = 0$.

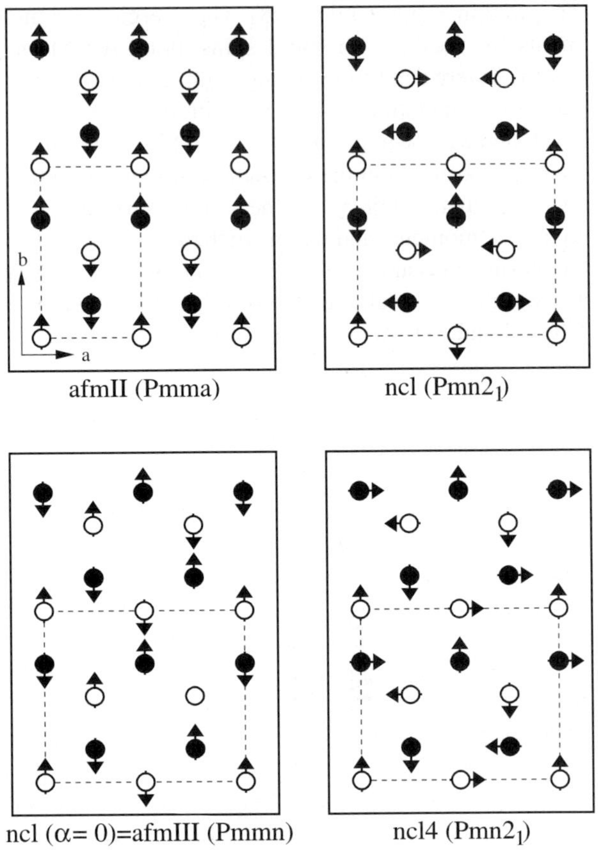

Fig. 5. Some of the different magnetic structures considered are shown here. Various rotation angles were also explored. The orthogonal cell shows the outline of the primitive cell used.

The energy does depend on which pairs are chosen to be antiferromagnetic; this gives rise to different possible collinear structures (Fig. 5). The collinear afmII structure has four atoms per unit cell in space group Pmma (Steinle-Neumann et al., 1999). In order to tile the lattice with the pattern shown in Fig. 4, cells with eight atoms are obtained. A value of 1.6 was used for c/a for all of the hcp-based calculations here, which is close to the ground state value for the volumes studied here. Tests showed that c/a does not vary significantly (<0.005) with magnetic state. A dense $12 \times 12 \times 12$ k-point mesh, giving 216 k-points in the irreducible wedge of the Brillouin zone for the orthorhombic eight-atom cells, was used. Fig. 6 shows

Table 1
Magnetization energy (mRy per atom) and moment M per atom for different atomic volumes V (bohr^3) for hcp-Fe

V (bohr^3)	I (eV)	Magnetization energy (mRy)				M (μ_B)			
		afmII		ncl	ncl4	afmII		ncl	ncl4
		TB	LAPW	TB	TB	TB	LAPW	TB	TB
90	1.030	18.6	20.2	20.1	19.1	2.3	2.2	2.3	2.4
80	1.050	7.9	9.2	8.9	8.4	1.9	1.8	2.0	1.9
75	1.066	4.1	5.5	4.5	4.3	1.5	1.4	1.7	1.6
70	1.070	1.2	2.6	0.9	1.1	1.1	1.1	1.1	1.1
65	1.085	0.1	0.4	0.0	0.0	0.6	0.4	0.1	0.4
60	1.10	0.0	0.0	0.0	0.0	0.1	0.0	0.0	0.0

The value of Stoner parameter I (eV) for the different volumes is given in the first column. The LAPW values are from Steinle-Neumann et al. (1999).

the energy versus angle α for different volumes for the ncl structure; $\alpha = 0$ corresponds to the Pmmn structure. Energies are shown relative to the afmII structure. In all cases the energy decreases when α is varied from zero, but for volumes less than or equal to about 70 bohr^3 (about 2 GPa), the afmII structure is more stable than the non-collinear structure. Table 1 shows the results of the tight-binding model for hcp Fe with ferromagnetic, afmII, and the non-collinear structure with $\alpha = 90°$ (ncl).

The ground state non-collinear total energies of ϵ-Fe were fit to a Vinet equation of state (Vinet et al., 1987) as a function of volume. Table 2 shows the results both using $K'_0 = 4$ and allowing K'_0 to vary. $K'_0 = 4$ is the value for a second-order finite strain expansion (Birch, 1978), although it has no special meaning in the Vinet equation. One expects $K'_0 > 4$ if something interesting is happening during compression, such as changes in crystal or magnetic structure with pressure. In any case, K'_0, that is the pressure derivative of the bulk modulus at zero pressure, is difficult to accurately constrain from equation of state data, and rather is a parameter that along with K_0 and V_0 describes the whole equation of state. This is especially true for hcp-Fe, which is not stable at zero pressure, so that V_0, K_0, and K'_0 are fictive. The parameters presented here are also derived from the whole equation of state, but the data contains only systematic, not random errors. For non-magnetic Fe, $K'_0 = 4$ is not a bad approximation, but when magnetism is included $K'_0 > 5$. The results show that the tight-binding model is in reasonable agreement with the self-consistent LAPW computations of Steinle-Neumann et al. (1999) (Table 2) (Fig. 7).

The above parameters are based on fitting a large range of volumes, whereas the magnetic region is restricted to volume above 60 bohr^3 (approximately 50 GPa). When the fit is restricted to this region the bulk modulus further softens (Table 2). This shows that although the Vinet equation is excellent when structural or magnetic transitions are not involved (Cohen et al., 2000), it is not perfect when there are changes in magnetic moment with compression. In any case, since the experimental zero pressure parameters were also derived from fitting the whole

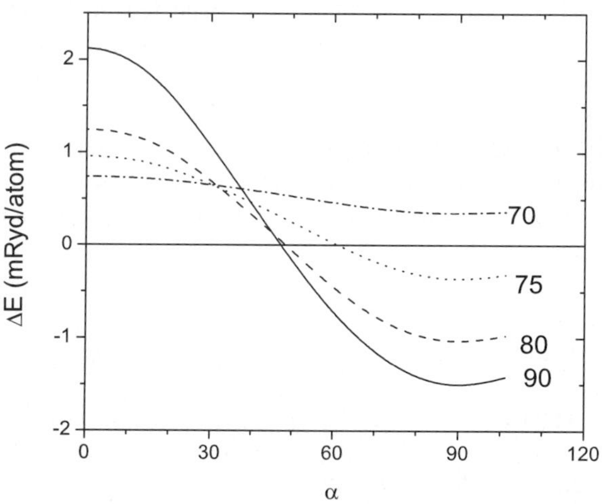

Fig. 6. Energy vs. α for a planar non-collinear structure (Fig. 4). The collinear Pmmn structure is represented by $\alpha = 0$. The labels for each curve are the volume per atom in bohr^3. Energies are relative to afmII. These structures are those considered by Diep (1992).

Table 2
Comparison of experimental and theoretical values of equilibrium volume (V_0) and bulk modulus (K_0) for ϵ-Fe

Fe (GGA)	V_0 (bohr^3)	K_0 (GPa)	K'_0
Experimental (Mao et al., 1990)	75.4	165	5.3
Non-magnetic (GGA) (Steinle-Neumann et al., 1999)	69.0	292	4.4
Non-magnetic	68.8	297	4.6
Collinear (afmII) (Steinle-Neumann et al., 1999)	71.2	209	5.2
Collinear (afmII), $V = 50$–90 au	70.4	213	6.1
Collinear (afmII), $V = 60$–90 au	70.4	210	5.9
ncl, $V = 50$–90 au	70.5	203	6.3
ncl, $V = 60$–90 au	70.9	193	5.3
ncl4, $V = 50$–90 au	70.4	210	6.2

The Vinet equation of state is used.

equation of state, comparison of the large range fit is appropriate, especially since hcp-Fe is not stable at low pressures. Fitting the lower pressure data alone decreases the apparent zero pressure bulk modulus, so a detailed comparison with experiment should be made between the actual data and computations rather than the fitting parameters. Non-coplanar magnetic structures may also have somewhat softer equations of state.

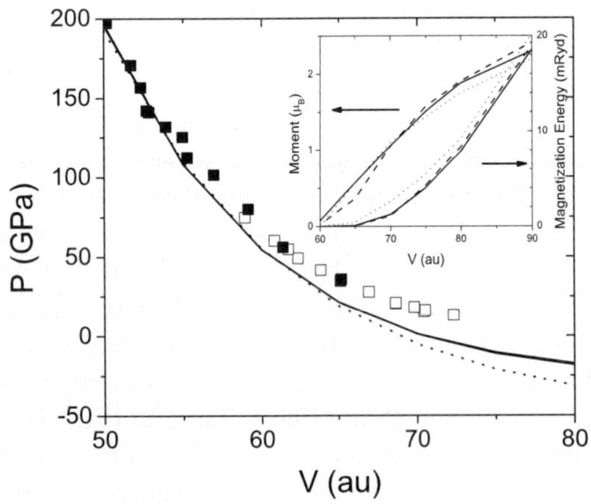

Fig. 7. Equation of state of ϵ-Fe. Symbols are experiments from Jephcoat et al. (1986) (open) and Mao et al. (1990) (solid). The non-collinear, afmII, and non-magnetic equations of state are shown as solid, dashed, and dotted lines. The afmII and ncl equations of state fit over the entire range are identical on this scale. The remaining deviation from the experimental points is primarily a shift due to errors in the GGA. The inset shows the magnetization energy per atom and moments in the non-collinear (solid) and afmII (dashed) magnetic structures versus volume. The dotted lines in the inset are the LAPW results.

In spite of theoretical evidence for magnetism in ϵ-Fe, and the great improvement in the equation of state when magnetism is included, the experimental situation is unclear. Mössbauer spectroscopy shows no evidence of magnetism in hcp Fe (Taylor et al., 1982, 1991). X-ray absorption spectroscopy also has been interpreted to show lack of magnetism (Rueff et al., 1999), but due to the absence of an absolute calibration, and sensitivity of the spectrum to changes in the density of states, the experiments show only that magnetic moments are lower in ϵ-Fe than α-Fe, a result consistent with theory. On the other hand, there is some independent evidence of magnetism from Raman spectroscopy, which shows two peaks (Merkel et al., 2000), instead of the one expected in the hcp structure. The frequencies and splitting of these peaks are predicted well from first-principles calculations for magnetic ordered afmII ϵ-Fe. Calculations shows that the absence of observed splitting in Mössbauer is due to cancellation of the valence and semi-core contributions to the hyperfine fields (Steinle-Neumann et al., 2004).

Acknowledgements

Thanks to B. Fultz, A. Goncharov, R. Hemley, A. Jephcoat, I. Mazin, H. Olijnyk, W. Pickett, G. Steinle-Neumann, L. Stixrude, and V. Struzhkin for helpful discussions. This work was supported by DOE ASCI/ASAP subcontract B341492 to Caltech DOE W-7405-ENG-48 and the National Science Foundation EAR-998002. S. Pella and D. Shook contributed to computations for bcc-Fe. Computations were performed on the Cray SV1 at the Geophysical Laboratory, supported by NSF grant EAR-9975753 and by the W.M. Keck Foundation.

References

Asada, T., Terakura, K., 1992. Cohesive properties of iron obtained by use of the generalized gradient approximation. Phys. Rev. B 46, 13599–13602.

Birch, F., 1978. Finite strain isotherm and velocities for single-crystal and polycrystalline NaCl at high pressures and 300 K. J. Geophys. Res. 95, 1257–1268.

Cohen, R.E., Mehl, M., Papaconstantopoulos, D., 1994. Tight-binding total energy method for transition and noble metals. Phys. Rev. B 50, 14694–14697.

Cohen, R.E., Stixrude, L., Wasserman, E., 1997. Tight-binding computations of elastic anisotropy of Fe, Xe, and Si under compression. Phys. Rev. B 56, 8575–8589.

Cohen, R.E., Gulseren, O., Hemley, R.J., 2000. Accuracy of equation-of-state formulations. Am. Miner. 85, 338–344.

Diep, H.T., 1992. First-order transition in the hexagonal-close-packed lattice with vector spins. Phys. Rev. B 45, 2863–2867.

Jephcoat, A.P., Mao, H.-K., Bell, P.M., 1986. Static compression of iron to 78 GPa with rare-gas solids as pressure-transmitting media. J. Geophys. Res. 91, 4677–4684.

Krasko, G.L., 1987. Metamagnetic behavior of fcc iron. Phys. Rev. B 36, 8565.

Mao, H.-K., Wu, Y., Chen, L., Shu, J., Jephcoat, A.P., 1990. Static compression of iron to 300 GPa and $Fe_{0.8}Ni_{0.2}$ alloy to 260 GPa: implications for composition of the core. J. Geophys. Res. 95, 21737–21742.

Merkel, S., Goncharov, A., Mao, H., Gillet, P., Hemley, R., 2000. Raman spectroscopy of iron to 152 gigapascals: implications for Earth's inner core. Science 288, 1626–1629.

Mryasov, O., Gubanov, V., Liechtenstein, A., 1992. Spiral-spin-density-wave states in fcc iron: linear-muffin-tin-orbitals band-structure approach. Phys. Rev. B 45, 12330–12336.

Mukherjee, S., Cohen, R.E., 2001. Tight binding based non-collinear spin model and magnetic correlations in Iron. J. Comput.-Aided Mater. Des. 8, 107–115.

Perdew, J., Burke, K., Ernzerhof, M., 1996. Generalized gradient approximation made simple. Phys. Rev. Lett. 77, 3865–3868; correction: Phys. Rev. Lett. 78 (1997) 1396.

Pickett, W., 1996. Non-collinear magnetic states: from density functional theory to model Hamiltonians. J. Korean Phys. Soc. 29, S70.

Rueff, J.P., Krisch, M., Cai, Y.Q., Kaprolat, A., Hanfland, M., Lorenzen, M., Masciovecchio, C., Verbeni, R., Sette, F., 1999. Magnetic and structural alpha-epsilon phase transition in Fe monitored by X-ray emission spectroscopy. Phys. Rev. B 60, 14510–14512.

Sandratskii, M.L., 1998. Noncollinear magnetism in itinerant-electron systems: theory and applications. Adv. Phys. 47, 91–160.

Sjöstedt, E., Nordström, L., 2002. Noncollinear full-potential studies of ϵ-Fe. Phys. Rev. B 66, 14447.

Steinle-Neumann, G., Stixrude, L., Cohen, R.E., 1999. First-principles elastic constants for the hcp transition metals Fe, Co, and Re at high pressure. Phys. Rev. B 60, 791–799.

Steinle-Neumann, G., Stixrude, L., Cohen, R.E., 2004. Magnetism in dense hexagonal iron. Proc. Natl. Acad. Sci. 101, 33–36

Stixrude, L., Cohen, R.E., Singh, D., 1994. Iron at high pressure: linearized augmented plane wave computations in the generalized gradient approximation. Phys. Rev. B 50, 6442–6445.

Stoner, E.C., 1938. Collective electron ferromagnetism. II. Energy and specific heat. Proc. R. Soc. London, Ser. A 169, 339–371.

Taylor, R., Cort, G., WIllis, J., 1982. Internal magnetic fields in hcp-iron. J. Appl. Phys. 53, 8199–8201.

Taylor, R., Pasternak, M., Jeanloz, R., 1991. Hysteresis in the high pressure transformation of bcc- to hcp-iron. J. Appl. Phys. 69, 6126–6128.

Uhl, M., Sandratskii, M.L., Kübler, J., 1994. Spin fluctuations in ϵ-Fe and in Fe_3 Pt Invar from local-density-functional calculation. Phys. Rev. B 50, 291–301.

Ujfalussy, B., Wang, X.-D., Nicholson, D.M.C., Shelton, W.A., Stocks, G.M., 1999. Constrained density functional theory for first principles spin dynamics. J. Appl. Phys. 85, 4824–4826.

Vinet, P., Ferrante, J., Rose, J., Smith, J., 1987. Compressibility of solids. J. Geophys. Res. 92, 9319–9325.

Wasserman, E., Stixrude, L., Cohen, R.E., 1996. Thermal properties of iron at high pressures and temperatures. Phys. Rev. B 53, 8296–8309.

Available online at www.sciencedirect.com

SCIENCE @ DIRECT®

Physics of the Earth and Planetary Interiors 143–144 (2004) 455–467

PHYSICS OF THE EARTH AND PLANETARY INTERIORS

www.elsevier.com/locate/pepi

In situ X-ray diffraction studies of iron to Earth-core conditions

Yanzhang Ma[a,*], Maddury Somayazulu[b], Guoyin Shen[c], Ho-kwang Mao[b], Jinfu Shu[b], Russell J. Hemley[b]

[a] *Department of Mechanical Engineering, Texas Tech University, Box 41021, Lubbock, TX 79409, USA*
[b] *Geophysical Laboratory, Carnegie Institution of Washington, 5251 Broad Branch Rd. NW, Washington, DC 20015, USA*
[c] *Consortium for Advanced Radiation Sources, University of Chicago, Chicago, IL 60637, USA*

Received 30 January 2003; received in revised form 29 May 2003; accepted 24 June 2003

Abstract

The high-pressure, high-temperature behavior of iron has been investigated to 161 GPa and 3000 K by in situ synchrotron X-ray diffraction with double-side laser-heated diamond anvil cells. We found that only α-, γ- and ε-Fe can be clearly verified as the stable solid phase in the *P–T* range studied. Only ε-Fe is observed from deep lower mantle (∼1500 km) to outer core conditions. Within the *P–T* range examined, we did not observe a significant change with pressure or temperature on the *c/a* ratio of ε-Fe. The melting curve of iron has been determined to 105 GPa. A Lindeman law fit gives a melting point of iron at the inner core boundary of 5800 (±200) K, which provides an upper bound on the temperature at that depth. We also examine numerous experimental factors that may complicate the analysis of high *P–T* diffraction data, and discuss the effects of sample stress on the X-ray diffraction results.

Keywords: Iron; Phase diagram; High pressure; High temperature; X-ray diffraction

1. Introduction

The Earth's core is mainly composed of iron and an additional light element component. Thus the structural properties and phase diagram of iron at core conditions are critical for understanding this most inaccessible region of our planet. From seismology, the inner core is known to be solid whereas the outer core is liquid. At the inner core boundary ICB, this iron-rich material crystallizes and releases latent heat and gravitational energy that at least in part drives the dynamo (Anderson, 1990). The melting temperature of iron at ultrahigh (i.e., megabar) pressures thus gives a bound on the temperature regime of the core. In addition, there is evidence for unusual elastic anisotropy of the inner core (e.g., Song and Helmberger, 1993), a property that is controlled by the crystal structure of the material. Hence, investigation of the high-pressure melting curve, phase relations, and lattice parameters of the stable phases of iron at deep Earth conditions is of fundamental geophysical importance.

Iron has been reported to have at least six phases based on X-ray diffraction at high pressure and temperature, i.e., α, body-centered-cubic (b.c.c.); γ, face-centered-cubic (f.c.c.); δ, b.c.c.; ε, hexagonal closed-packed (h.c.p.); double-h.c.p. phase (d.h.c.p.); and an orthorhombic phase. The α-, γ-, and δ-phases at lower pressures are well established and broadly accepted. The ε-phase, first observed at a high pressure and room temperature (Takahashi and Bassett,

* Corresponding author. Tel.: +1-806-742-0971; fax: +1-806-742-3540.
E-mail address: y.ma@coe.ttu.edu (Y. Ma).

0031-9201/$ – see front matter © 2004 Elsevier B.V. All rights reserved.
doi:10.1016/j.pepi.2003.06.005

1964), has been proved a dominant phase in a wide *P–T* range approaching Earth's core conditions. Other high-pressure phases of iron have been observed at temperatures near its high-pressure melting line. A phase (called β) was proposed based on reported temperature–laser power relations and high *P–T* X-ray diffraction (Saxena et al., 1995; Dubrovinsky et al., 1998, 2000). Other experiments revealed a double-hexagonal close-packed structure (d.h.c.p. or ε′) that was believed to be identical to the β-phase and identified as metastable (Yoo et al., 1995). An orthorhombic phase was also reported at 30–100 GPa and temperatures to 2370 K (Andrault et al., 1997). These results, however, could not be reproduced in other experiments (e.g., Shen et al., 1998), where it was found that the ε-h.c.p. phase persists to the melting points of iron over a range of pressures. This is consistent with the proposal that the identified phases were either metastable (Yoo et al., 1995) or incorrectly identified (Yoo et al., 1997; see also Hemley and Mao, 2001a). In view of the continuing discussion of these issues and the importance for understanding the core (e.g., Boehler, 2000), further clarification of the high-pressure solid phases of iron and their stability fields, as well as discussion of the relevant experimental details, is needed.

The high-pressure melting curve of iron is also controversial (see Hemley and Mao, 2001a). Although it has been extensively investigated by many available experimental methods, including Joule heating (Mao et al., 1987; Boehler, 1986), laser heating with visual observation (Williams et al., 1987; Boehler, 1993), laser heating with synchrotron X-ray diffraction (Shen et al., 1998), and shock compression (Brown and McQueen, 1986; Yoo et al., 1993), most of the results are inconsistent with each other, particularly at higher pressures where direct measurements become difficult. The differences in melting temperatures among different determinations can exceed 1000 K when pressures approach the megabar range. Thus, as in the case of the solid-phase studies, it is necessary to carefully re-examine earlier results and further investigate the solid–liquid equilibrium line to pressures of the Earth's core.

The continued development of synchrotron radiation and diamond cell laser heating techniques has resulted in an ideal method for the study of materials under the extreme *P–T* conditions of the Earth's core (Mao et al., 1997; Shen and Heinz, 1998; Hemley et al., 2002). However, differences in techniques, distinctions that are often subtle, can have major effects on the experimental results. With the increasing numbers of experimental studies, however, the evaluation of experimental factors becomes particularly important. For example, we find that some earlier disputes on the behavior of iron at high pressures and temperatures can be attributed to the experimental accuracy of specific methods and the degree to which key experimental factors can be manipulated and controlled. This attention to experimental detail is crucial for testing the broad range of increasingly accurate theoretical predictions of the behavior of iron (and iron alloys) at Earth's core conditions (e.g., Alfè et al., 1999, 2000; Steinle-Neumann et al., 2001).

In this paper, we present the results of experimental studies of the high *P–T* phase diagram and structural properties of iron to 161 GPa and 3000 K obtained using recent improvements in diamond cell laser heating/synchrotron X-ray diffraction techniques. Preliminary reports of some of this data have been mentioned in earlier review articles (Shen and Heinz, 1998; Hemley and Mao, 2001a). Here we present additional experimental details that are critically important for assessing the accuracy of the results, as well as new data and analyses of the high *P–T* behavior of iron to compare with recent theoretical predictions.

2. Experimental

2.1. Sample loading

We employed symmetric diamond anvil cells with flat diamonds for lower pressures experiments and beveled diamonds at higher pressures (over 60 GPa). Flat diamond culet sizes ranged from 300 to 400 μm, and the beveled diamond culets from 80 to 110 μm. Rhenium and stainless steel gaskets were used in these experiments; gaskets were pre-indented to 20 μm or less before drilling holes for the sample chambers. The diameter of the sample chambers were larger than 120 μm with flat diamonds, and were the same sizes as the diamond culet for beveled diamonds. NaCl, MgO, and Al_2O_3 served as pressure media and heating insulators in our experiments; these materials were pressed to form transparent flakes, and sandwiched the sample on either side within the sample chamber. Experiments were carried out with all three pressure media to verify their possible chemical reactions with the sample. The sample was pure iron powder with grain sizes smaller than 1 μm, compressed into a ~5 μm thick and 30–40 μm wide flake before loading. During sample loading, the iron was located at the center, and care was taken to avoid bridging the gasket by the sample, i.e., no portion of the sample was in contact with the edges of the gasket during the course of the experiment. We also removed all sample debris on the gasket within the diamond flat surface before compression. In the runs with beveled diamonds, we loaded the samples such that no gasket remained on top of the diamond culet after increasing the pressure (Fig. 1). After loading the sample, we evacuated and filled the cell with argon repeatedly over a period of several hours, and then kept it in an oven at a temperature of 90 °C in an environment of flowing Ar gas for 72 h to eliminate water and oxygen. In our experiments below 60 GPa, we also used Ar gas as a pressure medium, and 5 μm-sized ruby chips as spacers to keep the sample from contacting the diamond before filling with liquid argon. Other aspects of these runs are similar to those described above. The iron equation of state was used as a pressure scale (Mao et al., 1990).

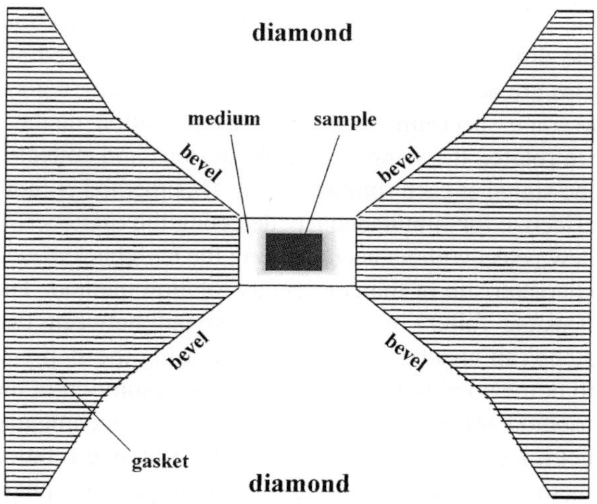

Fig. 1. Cross-section of a sample loaded in a diamond anvil cell with beveled diamonds. Note that the size of the gasket hole is the same as that of the diamond culet.

2.2. Laser heating

We conducted our experiments both at GSECARS, Sector 13 of the Advanced Photon Source (APS), Argonne National Laboratory, and beamline X17B of the National Synchrotron Light Source (NSLS), Brookhaven National Laboratory. In the experiments, we used a double-sided laser heating system with a dual imaging setup, which allows us to visually observe the sample during the heating process (Fig. 2). We measured the temperature by the thermal radiation method in these systems. In the system at APS, two Nd:YLF single-mode lasers are combined together, with a total output of 105 W, which has been proven sufficient to generate stable temperatures higher than 3000 K. One laser was operated in TEM_{00} mode and the other in the donut mode to create a homogeneous flat top power distribution (with radial gradient less than 3%) at the laser spot center. A detailed description of this system and an explanation of the experimental method has been published previously (Shen et al., 2001). At NSLS we used a multimode YAG laser with output power in the order of 50 W.

2.3. Synchrotron X-ray diffraction

We carried out our in situ high P–T X-ray diffraction measurements of iron using energy-dispersive synchrotron radiation methods. The experiments above 60 GPa were carried out at the APS. With the insertion of the linear arrays of north–south permanent magnets with alternating polarity in a straight section, the APS provides a brilliant beam with very low divergence (283 × 40 μrad). This has yielded significant improvements in the quality of in situ high P–T diffraction measurements at megabar pressures and thousands of degrees Kelvin with diamond cells, where the sample is thin and the gasket hole becomes much smaller than originally drilled. The low divergence of the X-ray beam helps to eliminate the detectable scattering from the gasket and other contamination on it. We reduced the size of the X-ray beam using two perpendicular slits with paired tungsten carbide cubes having sharp edges, and eliminated the tails of the beam with another pairs of slits made out of tungsten plates. With this configuration, we obtained X-ray beam sizes of less than 8 μm × 10 μm at the sample position. No X-ray focusing mirrors were used in experiments at

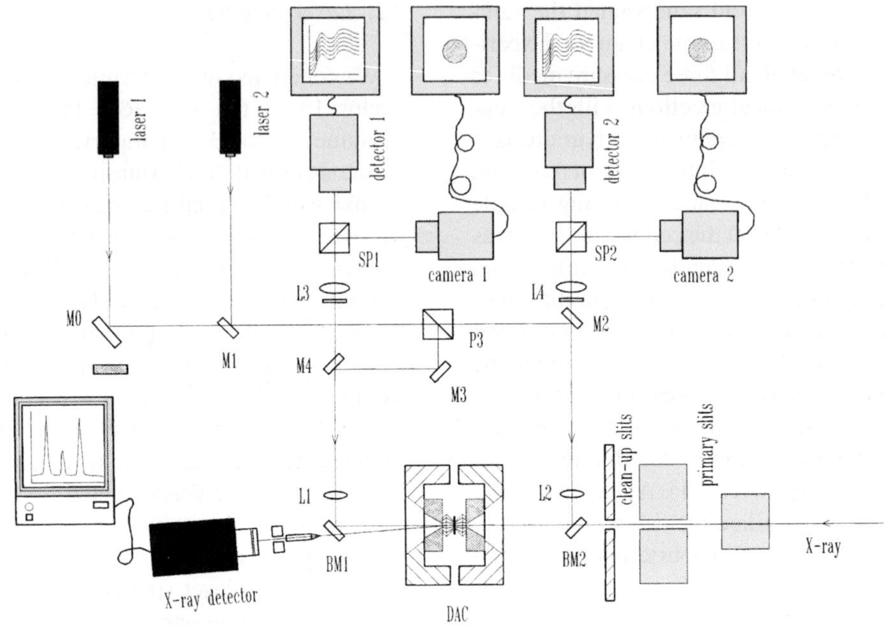

Fig. 2. Schematic diagram of synchrotron X-ray diffraction system with double-sided laser heating apparatus.

APS. Several experiments below 60 GPa were carried out at NSLS to verify the phase boundaries. In these runs, we used a Kirkpatrick-Baez mirror to enhance the beam intensity; the X-ray beam size at the sample position was 25 μm × 30 μm.

With energy-dispersive synchrotron X-ray diffraction, a well-resolved diffraction pattern of a strongly scattering sample such as iron can be collected within several seconds. This allows real-time capture of the signal that can reveal ongoing processes taking place in the sample at a given pressure and temperature. As conventionally applied, however, it does not permit any signal to be collected away from the point of detection; hence, information away from the detecting point is not available. This effect becomes more critical at high temperatures where recrystallization and crystal coarsening from an originally fine-grained powder sample can occur. To overcome such an effect in phase identification experiments, we developed a two-dimensional X-ray diffraction method for double-sided laser heating of diamond anvil cells (Ma et al., 2001), where the diamond cell is mounted on a high-precision rotation stage and the sample center coincides with its rotation center. We estimated the off-center error caused by the alignment to be less than 2–3 μm. Thus, it is impossible for the sample center position to drift out of the flat part of the heating spot, which is about 25 μm in diameter and fully covers the X-ray beam.

We used the rotating diamond cell method (both continuous and step-wise rotation) in our experiments. In the continuous-rotation method, we kept the sample rotating at a preset speed during data collection; the diffracted signals detected within all covered angles are added to make one pattern producing a good average spectrum. In the step-wise-rotation method, we rotated the sample by discrete steps and collected the signal at each rotation angle. This method is more effective for simultaneously detecting the existence of new peaks without accumulating the high background characteristic of the continuous-rotation method. The possible failure to detect any signal between two steps can also be avoided by decreasing the step size to correspond to the system's spatial resolution in the rotation angle, which is defined by the tip opening and collimation system of the detector. The step-rotation method was used to identify the high-pressure phase of iron to 161 GPa. We selected several rotation steps to detect possible peaks at angles as low as 0.5°. We could not resolve any change in the diffraction pattern at steps below 5°. At steps higher than 5°, we observed intensity changes between steps. Therefore, we chose 5° steps in most of our experiments.

The application of the cell-rotation methods is facilitated by high-resolution CCD cameras in the laser heating system, where we use the X-ray-induced visible fluorescence from the pressure medium in the cell to locate the X-ray position, then move the sample on the rotation center to this position and couple the lasers from both sides to it. Such an alignment allows the X-ray beam and the lasers from two sides of the sample to be coincident within ~2–3 μm. Another possible effect of the cell rotation is that the laser coupling position may drift during heating because of the differences in diamond thickness in different directions (i.e., within a machining error). This effect was proved minor in our experiments, and can be further eliminated by aligning the laser beam perpendicular to the diamond surfaces.

3. Results and Discussion

3.1. Artifacts and experimental errors

High-pressure/high-temperature phase identification using synchrotron X-ray and laser-heated diamond anvil cell techniques cell involves many advanced experimental methods, and error analysis becomes a most important issue. Disputes about on the high *P–T* phases of iron are commonly related to the estimation of certain experimental errors, which are hard to clarify. We feel it is highly valuable and beneficial to evaluate some common errors involved in the experiments that have long been ignored.

Fig. 3 shows X-ray diffraction patterns acquired in some of the experiments carried out to deliberately introduce systematic errors. Most of the diffraction peaks appear to show a broad shoulder. These patterns are unstable during heating, and the shoulders can diminish after some fine alignment. Such a phenomenon indicates that the shoulders, sometimes considered as diffraction peaks, are directly related to the effects of heating under pressure. The pressure gradient caused by stress in the sample chamber of a diamond anvil cell during compression has been studied since the earliest application of the device (e.g., Sung et al., 1977). A typical radial pressure gradient for NaCl, a soft pres-

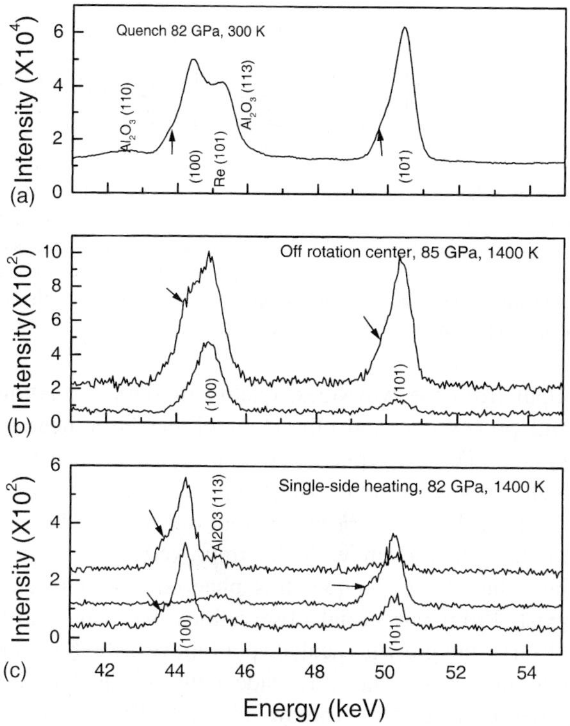

Fig. 3. Examination of errors associated with X-ray diffraction patterns of iron. (a) Temperature-quenched sample heated to 1400 K at 82 GPa through 360° rotation; (b) in situ pattern at 85 GPa and 1400 K at different rotation angles with heating spot away form the diffracting position; (c) in situ pattern at 82 GPa and 1400 K with one-sided heating. Arrows point to the shoulders in the patterns.

sure medium often used in laser heating, has been measured to be ~5 GPa/50 μm at 35–40 GPa (Meade and Jeanloz, 1988), whereas iron itself and other pressure media can give much larger gradients. Recent experimental results show that the stress difference between the axial and radial directions can cause as high as a 20% difference in the *d*-spacings of iron at 50 GPa (Mao et al., 1998). Even argon, the conventional pressure medium, has a 10% radial–axial pressure difference at 60 GPa (Mao et al., in preparation; Hemley and Mao, 2001b). At ambient temperature, the effect of such a stress can only broaden the diffraction peaks, while at high temperature, the effects of stress can be much more complicated.

Another experimental issue of concern with laser-heated diamond cell techniques is the temperature gradient across the sample, which can be affected by several factors, including the difference in power

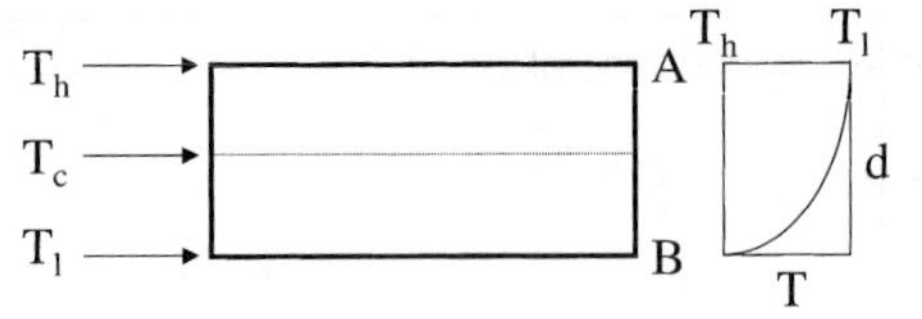

Fig. 4. Schematic diagram of the temperature-distribution effect on *d*-spacings. Left: temperature distribution of a heated spot along the axial direction illustrating the temperature difference between the two sides: T_h, highest temperature (A side); T_l, lowest temperature (B side); T_c, stress relaxation critical point. Right: *d*-spacing changes with temperature.

input from the two sides, quality of sample loading and processing, and alignment of the laser heating system. The early one-sided heating experiment typically produced large axial temperature gradient as a result of the power drop-off from one side of the cell to the other. Even with the improved double-sided laser heating technique, this phenomenon may still exist. From our visual observations, we can clearly observe the glowing heated spot (above 1200 K) from one side, while the other side with an equal power input remains dark (below 1000 K). The measured temperature difference between the two sides can be as high as 1000 K.

Experimental results also show that even though the stress is released inside the heating spot at temperatures higher than 1200 K (unpublished data), pressure gradients still exist outside and on the border of the heating spot (see also Shim et al., 2000). With the potential buildup of stress and the manner in which stress can be released with increasing temperature, we recognize that the assumed experimental condition may also cause the diffraction peaks to split. In our model we assume a laser-heated sample as shown in Fig. 4, with side A at higher temperature, T_h, and side B at lower temperature, T_l. The temperature distribution between A and B continuously decreases from T_h to T_l. Our experimental data suggests the existence of a T_c, the temperature at a critical point of stress relaxation at high pressure. When $T_l < T_c < T_h$, a two-layer sample is created: one layer with temperatures higher than T_c and another with a temperature lower than T_c. In the high-temperature layer, the stress is released, while in the low-temperature layer the stress still exists. Because of this stress release, the *d*-spacings accordingly adopt the same value in every direction, and the diffracted peak could be shifted by as much as 20%. This effect has also been extensively discussed by Kavner and Duffy (2001). The X-ray beam passes through both layers, and the detector collects the scattered signal from both the stressed and unstressed layers. Hence we can almost certainly expect a splitting of the diffraction peaks. We demonstrated this effect by heating an iron sample from one side to 1400 K (thermal emission was recorded from one side whereas the other side remained dark). We clearly observed splitting of the peaks (i.e., shoulders appeared on the iron (1 0 0) and (1 0 1) peaks), similar to patterns reported in single-sided laser heating experiments (Fig. 3c).

Moreover, even for a well-controlled axial temperature distribution (i.e., well-aligned double-sided laser heating), the radial temperature distribution may also complicate the measurements. An ideal radial temperature distribution has a flattened peak with a sharply decreasing half-Gaussian-shaped edge (Mao et al., 1997). The flat top can range from as small as several microns to several tens of microns. The temperature distribution within this spot is far from homogeneous, in the case of an X-ray measurement, the probe beam is comparable with, or even bigger than, the heating spot. We must also take into account other experimental factors, such as the mismatch of the laser heating and X-ray diffracted spots, and cell translation during and after heating, because of the typical symmetric thermal expansion of the cell and its holder. To test these effects, we intentionally mismatched the X-ray beam with the laser heating spot by 10–20 μm at high pressures; all the clearly resolved peaks produce a shoulder when the collecting time was long enough (i.e., producing diffraction patterns similar to those depicted in Fig. 3b and c). This phenomenon disappeared when the laser heating spot, X-ray beam, and sample were well aligned.

Another important factor that can adversely affect the experimental result is X-ray diffraction from the gasket. As can be inferred from Fig. 5, we can see that the source X-ray distribution also has a Gaussian-shaped space intensity distribution. Also, we need to further consider the method used to generate a highly collimated X-ray beam. For example, scattering from the slits used to define the X-ray beam size and the "wings" (extra intensity away from the beam center) generated from focusing mirrors can affect the diffraction pattern. The signals diffracted by these unexpected sources are usually negligible.

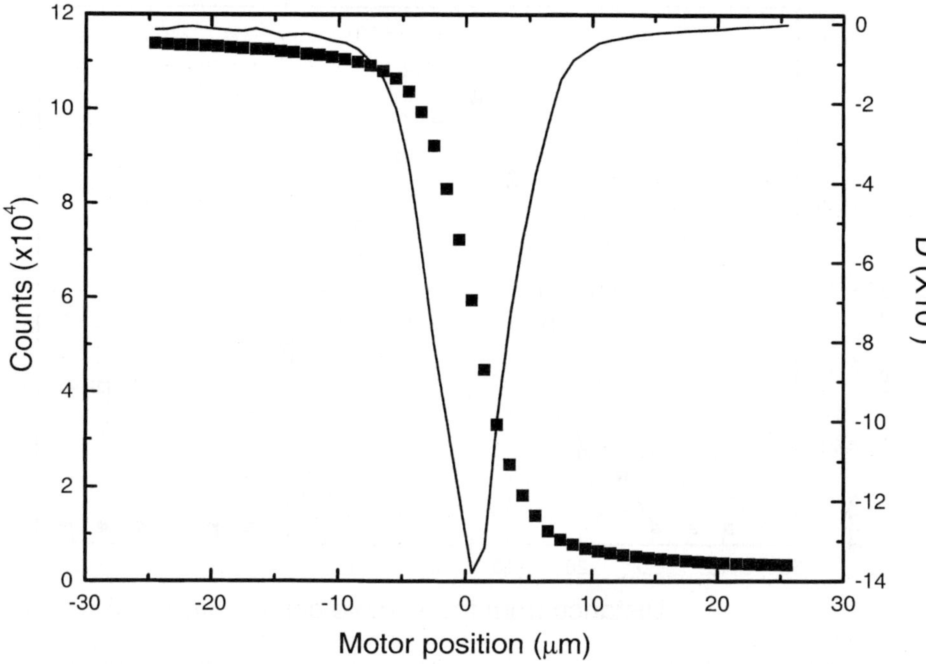

Fig. 5. Spatial distribution of the X-ray beam as shown by the sharp edge in a step scan. Left axis (solid square marks): detected counts with scan motor position; right axis (solid curve): differential of the intensity (counts) with motor position.

However, under the special circumstances associated with laser-heated diamond anvil cell experiments, these effects must be carefully considered. In most cases, the sample thickness is about one-third that of the gasket. We can reasonably expect that 10% of the peak X-ray intensity can easily generate detectable diffraction from the gasket. When a strong scatterer such as rhenium is adopted as a gasket material, the X-ray intensity required to generate a signal from the gasket can be low. The scatter due to the wings of the X-ray beam previously mentioned can also be intense; their effect on the diffraction data should be carefully checked.

We have tested these effects in our experiments. We used a synchrotron X-ray source with a beam size of 8 μm × 10 μm obtained with two pairs of slits. The rhenium gasket hole was 80 μm in diameter, and a 30 μm-sized iron sample was loaded in the center of the sample chamber. When the X-ray and laser were well aligned at the center of the sample, we clearly obtain diffraction patterns that are characteristic of the pure h.c.p. phase of iron at high pressures and high temperatures. If the X-ray beam shifts by 10 μm (the magnitude of this shift can be easily caused by the thermal expansion of parts of the cell or its mounts), the diffraction from the gasket appears as shoulders on the iron peaks (Fig. 3a). Some of these shoulders have intensities comparable to those of the iron peaks. These rhenium peaks disappear after the sample is re-centered with the X-ray beam. This phenomenon is reproducible; we conclude that the area affected by scattering from the gasket can be 20–30 μm larger in each direction than the beam size. This result strongly suggests that experiments with gasket hole diameters less than the sum of the X-ray beam size plus 2×30 μm should be carefully checked for any scattering from the gasket. Further, the mechanical mounting of the diamond cell can shift by ~10 μm on heating. Frequent checking of the position of the sample relative to X-ray and laser beams (e.g., by X-ray transmission scans) is therefore required to ensure accurate high *P–T* measurements. We conclude that a narrow, well-defined X-ray beam and a homogeneous temperature distribution are critical for high *P–T* phase X-ray diffraction for obtaining quality data.

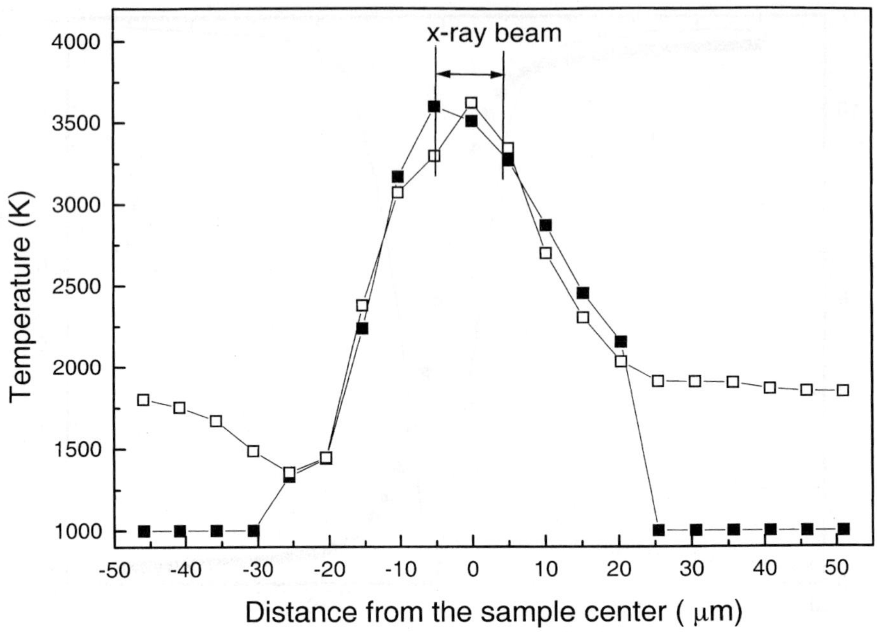

Fig. 6. Temperature distribution compared with the X-ray beam size for the laser heating at 105 GPa and 3500 K on the sample.

3.2. *P–T stability of ε-Fe*

As discussed by Shen et al. (2001), the laser heating method adopted at GSECARS has a highly improved temperature distribution relative to previous systems. The temperature difference within the dimension of the X-ray beam and between the two sides of the sample can be as low as 30 K. Fig. 6 shows a temperature distribution at 105 GPa and 3500 K. The temperature gradient within the X-ray range is about ±100 K at this pressure, which can be considered as the maximum error in our experiments. An X-ray diffraction image of iron at 161 GPa and 2410 K at rotation angles from 0 to 90° at 5° steps with data collected for 30 s at each step is presented in Fig. 7. The dark regions represent corresponding photon counts at certain energies and rotation angles. The dark lines (mostly diffraction lines) are straight and smooth, indicating homogenous

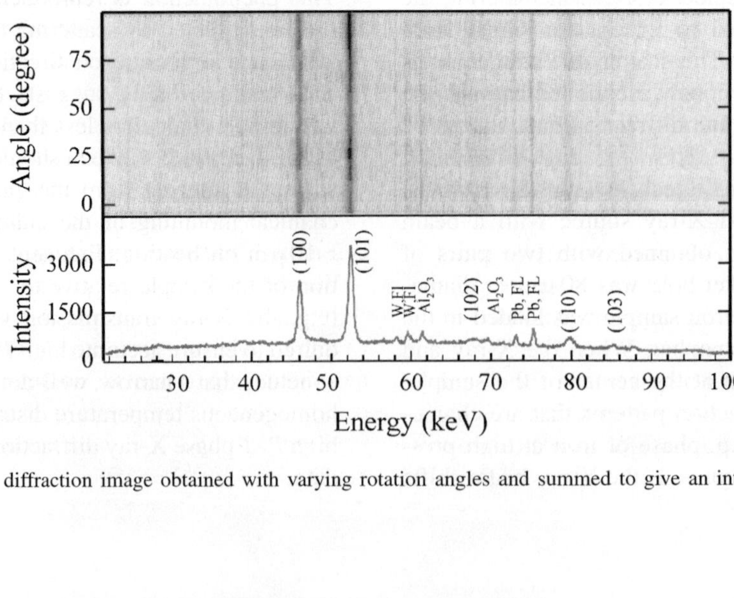

Fig. 7. X-ray diffraction image obtained with varying rotation angles and summed to give an integrated pattern.

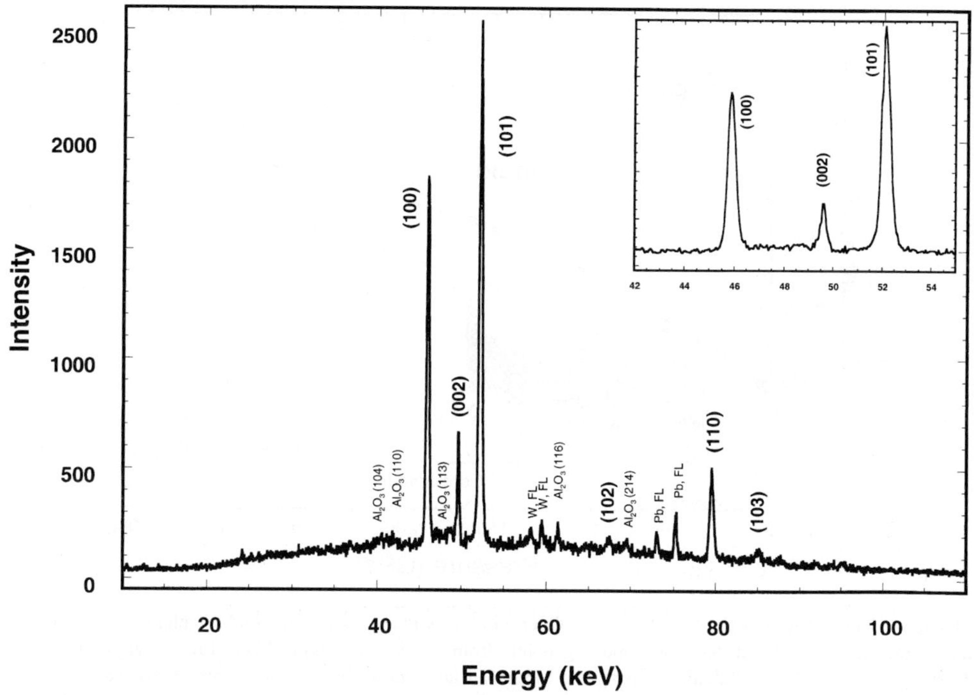

Fig. 8. X-ray diffraction pattern of iron at 161 GPa and 2450 K. Inset shows intense peaks of iron on a smaller scale. FL denotes fluorescence peaks.

P–T conditions in the sample and no coarse crystal formation. Fig. 8 is an integrated diffraction pattern of iron at 161 GPa and 2500 K. From this pattern, we can clearly observe five peaks from the h.c.p. phase, with refined parameters $a = 2.238(1)$ Å, $c = 3.594(4)$ Å, $c/a = 1.605(7)$, and $V = 15.595(18)$ Å^3. At this *P–T* point, the *c*/*a* ratio is very close to that at lower pressure (1.604) (Takahashi and Bassett, 1964). The intense peaks, (1 0 0) and (1 0 1), have linewidths (FHWM) of 378 and 384 eV, respectively, which are almost at the resolution limit of the detector (300 eV). This pattern demonstrates the existence of only the h.c.p. phase of iron under these *P–T* conditions. Furthermore, from our experiments, we can only observe three solid phases in our *P–T* range, b.c.c. at lower pressure, f.c.c. at higher temperature, and finally the h.c.p. phase. At pressures higher than 60 GPa (pressure at the ε-γ-liquid triple point), only two phases can be identified, the solid ε-h.c.p. phase and the liquid. At this point there is no compelling experimental evidence for any other solid phase of iron stable between 1500 km depth and the inner core boundary (ICB).

We clearly observed a change in the relative intensities of the iron diffraction peaks in a limited temperature interval (marked in the phase diagram (Fig. 9)). At different times, the intensity of the most intense (1 0 0) and (1 0 1) peaks vary, and the (0 0 2) peak will appear and disappear. The peak positions also drift systematically under these conditions, indicating a small pressure change. No additional features (i.e., shoulders) or relative intensity changes can be observed when the patterns were collected after a sufficiently long time had elapsed for the system to reach equilibrium and the X-ray and laser beams were well aligned at the center of the sample. We believe that this phenomenon is caused by the relaxation of the sample inside the cell at high pressures and temperatures.

We selected different pressure media (Ar, NaCl, Al_2O_3, and MgO) to examine possible reactions between sample and medium. In runs with the above-described experimental configurations, we could clearly identify the expected phases of the pressure medium and the iron sample from the diffraction pattern. We did not observe other phases of iron.

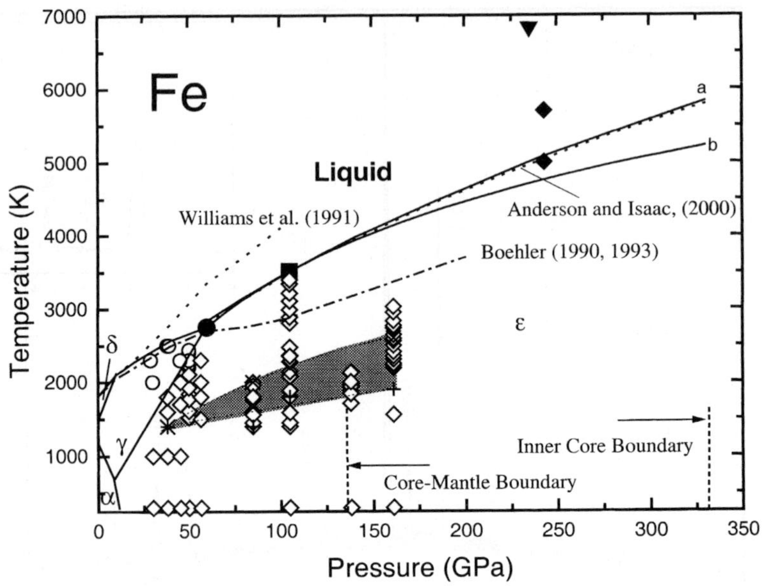

Fig. 9. Phase diagram of iron determined from this study. Open circles: γ-phase; open diamonds: ε-phase; filled circle: ε-γ-liquid triple point; filled square: melting point; filled diamonds: melting points from shock compression data; filled triangle: melting point from Yoo et al. (1995). Curve (a): melting curves calculated using the Lindeman law; curve (b): melting curve calculated using the Kraut–Kennedy law. The shaded area shows the *P–T* region where diffraction peaks change relative intensities (the cross marks the beginning temperature of these changes, and asterisk gives the ending point of such changes). Thermal pressure corrections are not accounted for this phase diagram. Williams et al., 1991; Anderson and Isaak (2000); Boehler (1990, 1993).

Discounting the peaks due to the pressure media, the diffraction patterns collected with different media are entirely consistent with each other. Thus, we found no evidence for the pressure medium reacting with the iron sample at the high *P–T* conditions studied.

3.3. *c/a* ratio of ε-Fe

The *c/a* ratio of ε-Fe has significant implications for the Earth's inner core. Recent first-principles calculations predict that the axial ratio of ε-Fe increases substantially with temperatures, reaching 1.7 at a temperature of 5700 K (Steinle-Neumann et al., 2001). Fig. 10 shows the measured temperature dependence of *c/a* at 161 GPa, together with pressure dependence at 300 K up to that pressure. No systematic variation in *c/a* with temperature can be observed. The values for *c/a* fall between 1.596 and 1.608 Å (a range of 0.011), which is comparable to our experimental error. Though the calculations were carried out at higher pressures, the theoretical results suggest that temperature dependence is larger at lower pressures (i.e., approaching those of the present study). Furthermore, the pressure effect on *c/a* also is equally weak, consistent with previous experimental results, at pressures sufficiently above the α–ε transition (Jephcoat et al., 1986; Mao et al., 1990). Additional high *P–T* data are required to directly address the question of the *c/a* ratio and the associated elastic anisotropy of iron at inner core conditions. Nevertheless, the current comparison suggests the need for improved theoretical treatment of the elastic anisotropy of iron at high pressures and temperatures.

3.4. Melting line

When the temperature was raised to 3510 K at 105 GPa, we observed the loss of all diffraction peaks from iron, while the diffraction from the diamond remained persisted during continuous rotation of diamond anvil cell. This temperature reflects the lower bound on the melting point of iron (e.g., Shen et al., 1998), and is about 700 K higher than that obtained by Boehler (1993), and appears to be lower than the

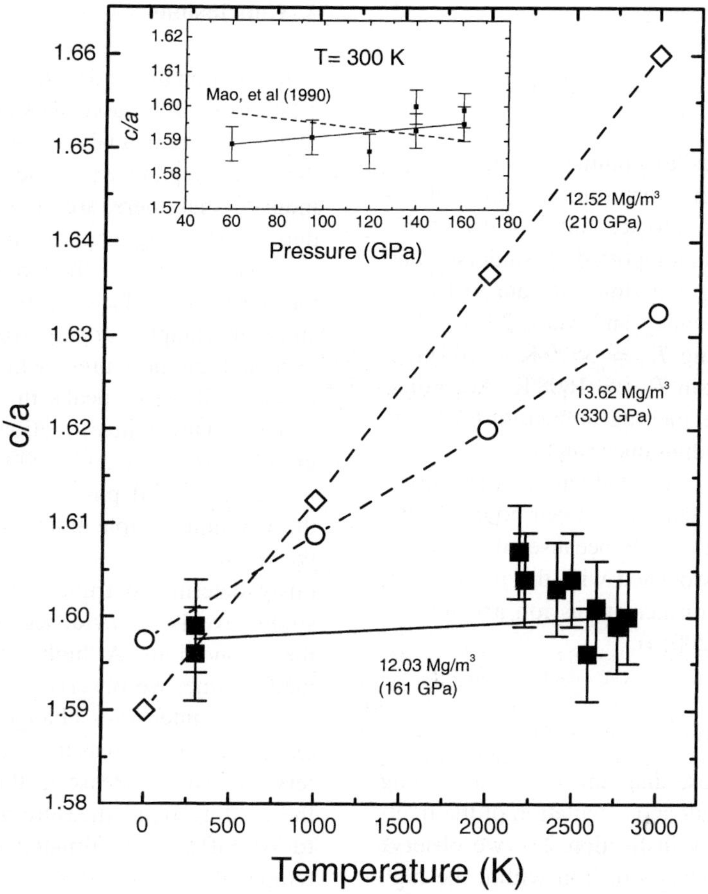

Fig. 10. Variation in the *c*/*a* ratio of ε-Fe with temperature, at 161 GPa, (solid line) and comparison to theory (dashed lines; Steinle-Neumann et al., 2001). The inset shows the pressure dependence of c/a at 300 K (solid line) and earlier results (dashed line; Mao et al., 1990).

results of shock wave experiments. We now consider the application of various melting laws to the data. A simple linear extrapolation from the triple point at 60 GPa and 2750 K (Shen et al., 1998) to higher pressure yields a melting temperature above 7200 K at 328 GPa (ICB), which of course neglects the expected curvature in the melting line.

According to the Kraut–Kennedy law (Kraut and Kennedy, 1966),

$$T_m = T_{m0}\left(1 + C\frac{\Delta V}{V_0}\right), \quad (1)$$

where T_m is the melting temperature at pressure P, T_{m0} the melting temperature at ambient pressure, $\Delta V/V_0$ the compression at pressure P, and C a constant. Using the equation of state determined by Mao et al. (1990) to calculate the volume compression, along with our data above the ε-γ-liquid triple point, we determined the constant coefficients T_{m0} and C in Eq. (1) to be 551 K and 21.2, respectively. The extrapolated melting temperature of iron at 330 GPa is therefore 5220 K. Assuming that the uncertainty of our temperature measurement is ±100 K, the error in melting temperature at the ICB is 320 K. In this calculation, thermal pressure and thermal expansion effects are not included.

Finally, consider the application of the Lindeman melting law, which has been more widely used. This melting relation is conventionally written as (Anderson and Isaak, 2000)

$$\frac{dT_m}{dP} = 2\left(\frac{T_m}{K_T}\right)\left(\gamma - \frac{1}{3}\right).$$

The equation can be recast as

$$\frac{T_m}{T_{m0}} = \left(\frac{V}{V_0}\right)^{2/3} \exp\left\{\frac{2\gamma_0}{q}\left[1-\left(\frac{V}{V_0}\right)^q\right]\right\},$$

where V/V_0 corresponds to volume compression at high P and T, i.e., thermal pressure and thermal expansion have been taken into account. Our use of this formulation is close to that reported by Anderson and Isaak (2000), who used experimental data to fix T_m at low pressure (see Hemley and Mao, 2001a). Assuming $q = 1$, and using $T_m = 3510$ K at 105 GPa, and $\gamma_0 = 1.92$, we obtain $T_{m0} = 1608$ K. According to this formulation, the melting temperature of iron at ICB is 5830 K. Including uncertainties in the measured temperatures, the calculated temperature ranges from 5600 and 6000 K. This is an upper bound on the actual temperature of the ICB because of the melting point depression associated with the presence of other components; for further discussion and review, see Hemley and Mao (2001a).

3.5. Phase diagram

Fig. 9 shows the phase diagram of iron according to our experimental result and calculation of the melting line. As pointed out in Section 2.4, we observe only the α-, γ-, and ε-phases of iron within the high P–T range studied. At lower pressures (<60 GPa), our experimental results are consistent with those of Shen et al. (1998); thus no adjustment in the lower pressure phase boundaries, including the triple point, is required. The calculated melting curves from both the Kraut–Kennedy and Lindeman laws based on our experimental data are shown in the phase diagram with the curve calculated from thermochemical data by Anderson and Isaak (2000) for comparison. The γ-ε-liquid triple point from our calculation using the Lindeman law is 2835 K, which is also within the experimental error range when compared with the result of Shen et al. (1998). The calculated melting curve also falls within the range determined by Brown and McQueen (1986) based on shock wave experiments. This analysis also assumes that no new high P–T phases exist; there is as yet no experimental evidence for such phases, though this has been the subject of some discussion in the literature (see Hemley and Mao, 2001a).

4. Conclusions

Resolution of the debates about the high P–T phase relations of iron based on static pressure experiments requires careful analyses of experimental details in these challenging experiments. Our systematic tests indicate that there are many factors that can produce misleading experimental results, and care is required to evaluate their reliability. We believe that the most crucial factor is the existence of significant stress in samples arising from the high strength of iron at high pressures, which can produce considerable splitting of peaks in the measured diffraction patterns. Direct investigation of the iron phase diagram to 161 GPa and 3000 K reveals that the ε-Fe is the only solid phase of iron at pressures beyond the ε-γ-liquid triple point. Direct measurements on the α- and γ-phases were consistent with the previously determined triple point. Careful experimental study reveals no evidence of other solid phases at these conditions. At high temperatures but below the melting line, we observe a region where the diffraction peak intensities change. We ascribe this to the existence of a pressure-dependent temperature interval of stress release in the material. We extended the melting point measurement by X-ray diffraction to 105 GPa, and estimated the upper bound on the temperature at the inner core boundary to be between 5600 and 6000 K. Finally, measurements of the temperature dependence of the c/a ratio at high pressure indicate the need to examine theoretical predictions of the origin of the elastic anisotropy of the inner core.

Acknowledgements

We would like to thank the staff of GSECARS at the Advanced Photon Source for their help with setting up the apparatus, and J. Hu and Q. Guo at beamline X17B of the National Synchrotron Light Source for useful discussions and experimental assistance. Thanks are also due to T.S. Duffy, G. Fiquet, and an anonymous reviewer for constructive comments. We are particularly grateful to S. Gramsch for careful reading of the manuscript. We thank NSF, NASA, DOE-BES, DOE-NNSA, and the W.M. Keck Foundation, for support.

References

Alfè, D., Gillan, M.J., Price, G.D., 1999. The melting curve of iron at the pressures of the Earth's core from ab initio calculations. Nature 401, 462–464.

Alfè, D., Kresse, G., Gillan, M.J., 2000. Structure and dynamics of liquid iron under Earth's core conditions. Phys. Rev. B 61, 132–142.

Anderson, O.L., 1990. The high pressure triple points and their effect on the heat flow from the Earth's core. J. Geophys. Res. 95, 21697–21707.

Anderson, O.L., Isaak, D.G., 2000. Calculated melting curves for phases of iron. Am. Miner. 85, 376–385.

Andrault, D., Fiquet, G., Kunz, M., Visocekas, F., Haeusermann, D., 1997. The orthorhombic structure of iron: an in situ study at high-temperature and high-pressure. Science 278, 831–834.

Boehler, R., 1986. The phase diagram of iron to 430 kbar. Geophys. Res. Lett. 13, 1153–1156.

Boehler, R., 1993. Temperature in the Earth's core from melting point measurement of iron at high static pressure. Nature 363, 534–536.

Boehler, R., 2000. High-pressure experiments and the phase diagram of lower mantle and core materials. Rev. Geophys. 38, 221–245.

Brown, J.M., McQueen, R.G., 1986. Phase transitions, Gruneisen parameter, and elasticity for shocked iron between 77 and 400 GPa. J. Geophys. Res. 91, 7485.

Dubrovinsky, L., Saxena, S.K., Lazor, P., Weber, H.-P., 1998. Structure of β-iron at high temperature and pressure. Science 281, 11a.

Dubrovinsky, L.S., Saxena, S.K., Tutti, F., Rekki, S., Le Bihan, T., 2000. In situ X-ray study of thermal expansion and phase transition of iron at multimegabar pressure. Phys. Rev. Lett. 84, 1720–1723.

Hemley, R.J., Mao, H.K., 2001a. In situ studies of iron under pressure: new windows on the Earth's core. Int. Geol. Rev. 43, 1–26.

Hemley, R.J., Mao, H.K., 2001b. Progress in cryocrystals to megabar pressures. J. Low Temp. Phys. 122, 331–344.

Hemley, R.J., Chiarotti, G., Bernasconi, M., Ulivi, L. (Eds.), 2002. Proceedings of the International School of Physics "Enrico Fermi" on High-Pressure Phenomena, Course CXLVII. IOS Press, Amsterdam.

Jephcoat, A.P., Mao, H.K., Bell, P.M., 1986. The static compression of iron to 78 GPa with rare gas solids as pressure-transmitting media. J. Geophys. Res. 91, 4677–4684.

Kavner, A., Duffy, T.S., 2001. Pressure–volume–temperature paths in the laser-heated diamond anvil cell. J. Appl. Phys. 89, 1907–1914.

Kraut, E.A., Kennedy, G.E., 1966. Melting law for high pressure. Phys. Rev. 151, 668–675.

Ma, Y., Somayazulu, M., Mao, H.K., Hemley, R.J., Gramsch, S.A., Shen, G., Somayazulu, M., 2001. Two-dimensional energy dispersive X-ray diffraction at high pressures and temperatures. Rev. Sci. Instrum. 72, 1302–1305.

Mao, H.K., Bell, P.M., Hadidiacos, C., 1987. Experimental phase relations of iron to 360 kbar and 1400 °C, determined in an internally heated diamond anvil apparatus. In: Manghnani, M.H., Syono, Y. (Eds.), High Pressure Research in Mineral Physics. Terra Scientific/AGU, Washington, DC, pp. 135–138.

Mao, H.K., Wu, Y., Chen, L.C., Shu, J.F., Jephcoat, J.P., 1990. Static compression of iron to 300 GPa and $Fe_{0.8}Ni_{0.2}$ alloy to 260 GPa: implications for composition of the core. J. Geophys. Res. 95, 21737.

Mao, H.K., Shen, G., Hemley, R.J., Duffy, T.S., 1997. X-ray diffraction with a double hot plate laser-heated diamond cell. In: Manghnani, M.H., Syono, Y. (Eds.), High-Pressure Research in Mineral Physics: Application to Earth and Planetary Sciences. Terra Scientific/AGU, Washington, DC, pp. 27–34.

Mao, H.K., Shu, J., Shen, G., Hemley, R.J., Li, B., Singh, A.K., 1998. Elasticity and rheology of iron above 220 GPa and the nature of the Earth's inner core. Nature 396, 741–743.

Meade, C., Jeanloz, R., 1988. Yield strength of the B1 and B2 phases of NaCl. J. Geophys. Res. 93, 3270–3274.

Saxena, S.K., Dubrovinsky, L.S., Haggkvist, P., Cerenius, Y., Shen, G., Mao, H.K., 1995. Synchrotron X-ray study of iron at high pressure and temperature. Science 269, 1703–1704.

Shen, G., Heinz, D.L., 1998. High-pressure melting of deep mantle and core materials. In: Hemley, R.J. (Ed.), Ultrahigh-Pressure Mineralogy, Reviews in Mineralogy. Mineralogical Society of America, Washington, DC, pp. 369–396.

Shen, G., Mao, H.K., Hemley, R.J., Duffy, T.S., Rivers, M.L., 1998. Melting and crystal structure of iron at high pressures and temperatures. Geophys. Res. Lett. 25, 373–376.

Shen, G., Rivers, M., Wang, Y., Sutton, S., 2001. Laser heated diamond cell system at the Advanced Photon Source for in situ X-ray measurements at high pressure and temperature. Rev. Sci. Instrum. 72, 1273–1282.

Shim, S.H., Duffy, T.S., Shen, G., 2000. The equation of state of $CaSiO_3$ perovskite to 108 GPa at 300 K. Phys. Earth Planet. Interiors 120, 327–338.

Song, X.D., Helmberger, D.V., 1993. Anisotropy of Earth's inner core. Geophys. Res. Lett. 20, 2591–2594.

Steinle-Neumann G., Stixrude, G.L., Cohen, R.E., Gülseren, O., 2001. Elasticity of iron at the temperature of the Earth's inner core. Nature 413, 57–60.

Sung, C.M., Goetze, C., Mao, H.K., 1977. Pressure distribution in the diamond anvil press and the shear strength of fayalite. Rev. Sci. Instrum. 48, 1386–1391.

Takahashi, T., Bassett, W.A., 1964. High pressure polymorph of iron. Science 145, 483–486.

Williams, Q., Jeanloz, R., Bass, J., Svendsen, B., Ahrens, T.J., 1987. The melting curve of iron to 250 GPa: a constraint on temperature at Earth's center. Science 236, 181–183.

Yoo, C.S., Holmes, N.C., Ross, M., 1993. Shock temperatures and melting of iron at Earth core conditions. Phys. Rev. Lett. 70, 3931–3934.

Yoo, C.S., Akella, J., Campbell, A.J., Mao, H.K., Hemley, R.J., 1995. Phase diagram of iron by in situ X-ray diffraction: implication for the Earth's core. Science 270, 1473–1475.

Yoo, C.S., Akella, J., Campbell, A.J., Mao, H.K., Hemley, R.J., 1997. Detecting phases of iron. Science 275, 94–96.

Available online at www.sciencedirect.com
SCIENCE DIRECT®
ELSEVIER
Physics of the Earth and Planetary Interiors 143–144 (2004) 469–479
PHYSICS OF THE EARTH AND PLANETARY INTERIORS
www.elsevier.com/locate/pepi

Phase relationships and equations of state for FeS at high pressures and temperatures and implications for the internal structure of Mars

Satoru Urakawa[a,*], Keiko Someya[a], Hidenori Terasaki[b,1], Tomoo Katsura[c], Syo Yokoshi[c], Ken-ichi Funakoshi[d], Wataru Utsumi[e], Yoshinori Katayama[e], Yu-ichiro Sueda[f], Tetsuo Irifune[f]

[a] *Department of Earth Sciences, Okayama University, Okayama 700-8530, Japan*
[b] *Geoscience Institute, University of Tsukuba, Tsukuba 305-8571, Japan*
[c] *Institute for Study of the Earth's Interior, Okayama University, Misasa 682-0193, Japan*
[d] *Japan Synchrotron Radiation Research Institute, Mikaduki 679-5198, Japan*
[e] *Japan Atomic Energy Research Institute, Mikaduki 679-5198, Japan*
[f] *Geodynamic Research Center, Ehime University, Matsuyama 790-8577, Japan*

Received 12 February 2003; received in revised form 24 November 2003; accepted 1 December 2003

Abstract

In situ X-ray diffraction experiments on FeS up to 22 GPa and 1600 K were carried out using large volume multianvil apparatus, combined with synchrotron radiation at SPring-8. We investigated phase stability relationships of FeS and determined the straight phase boundaries between FeS III (monoclinic phase) and FeS IV (hexagonal phase) to be T (K) $= 20P$ (GPa) $+ 170$ and between FeS IV and FeS V (NiAs-type phase) to be T (K) $= 39.6P$ (GPa) $+ 450$. We also found anomalous behavior in the c/a ratio, thermal expansion, and isothermal compression of FeS V as well as FeS IV, in the pressure range 4–12 GPa. These anomalies in FeS can be attributed to the spin-pairing transition of Fe, and divides FeS IV and FeS V into the high-spin low-pressure phase (LPP) and the possibly low-spin high-pressure phase (HPP). In order to investigate the internal structure of Mars, we evaluated the equations of state for FeS IV (HPP) and FeS V (HPP). A least square fit to the experimental data yielded $K_{0T} = 62.5 \pm 0.9$ GPa at $T = 600$ K and $(\mathrm{d}K_0/\mathrm{d}T)_P = -0.0208 \pm 0.0028$ GPa/K for FeS IV (HPP), and $K_{0T} = 54.3 \pm 1.0$ GPa at $T = 1000$ K and $(\mathrm{d}K_0/\mathrm{d}T)_P = -0.0117 \pm 0.0015$ GPa/K for FeS V (HPP) with fixed $K' = 4$. Thermal expansion coefficients were $\alpha = 7.16 \times 10^{-5} + 6.08 \times 10^{-8}T$ for FeS IV (HPP) and $\alpha = 10.42 \times 10^{-5}$ for FeS V (HPP), respectively. Using these equations of state, we examined the internal structure of Mars that has a model mantle composition [Meteoritics 20 (1985) 367] and Fe–FeS core. Our models show that an Mg-silicate perovskite-rich lower mantle is stable only with the Fe-rich core having less than 20 wt.% sulfur. The polar moment of inertia factor C derived from Mars Pathfinder data [Science 278 (1997) 1749] is consistent with any compositions between Fe and FeS for the Martian core, but it excludes the presence of a crust thicker than 100 km.

Keywords: Iron sulfide; Spin transition; Martian core; Bulk modulus; Thermal expansion

* Corresponding author. Tel.: +81-86-251-7885; fax: +81-86-251-7895.
E-mail address: urakawa@cc.okayama-u.ac.jp (S. Urakawa).
[1] Present address: Bayerisches Geoinstitut, University of Bayreuth, D-95440 Bayreuth, Germany.

0031-9201/$ – see front matter © 2004 Elsevier B.V. All rights reserved.
doi:10.1016/j.pepi.2003.12.015

1. Introduction

Iron sulfide (FeS) has been thought to be one of the major constituents of the cores of terrestrial planets such as Mars. Phase stability relations and density of FeS at high pressure and high temperature are fundamental to evaluate the internal structure of planets having a sulfur-bearing core, especially Mars (e.g., Fei et al., 1995; Kavner et al., 2001). However, uncertainties in Martian interior models are still large due to the ambiguity of phase stability fields and inaccuracy in the density of FeS. Thus, we require accurate phase relationships and equations of state for FeS at pressures and temperatures corresponding to the deep interior of Mars and other planets.

FeS is known to undergo several structural and electro-magnetic phase transformations up to 25 GPa and between 300 and 1000 K (e.g., Taylor and Mao, 1970; King and Prewitt, 1982; Fei et al., 1995; Kusaba et al., 1997, 1998a,b, 2001; Kobayashi et al., 1997; Rueff and Kao, 1999; Takele and Hearne, 1999; Urakawa et al., 2002). Troilite (FeS I), a stable stoichiometric FeS phase at ambient condition, has a superstructure of the NiAs-type structure with a hexagonal unit cell ($\sqrt{3}a$, $2c$), where a and c are the fundamental unit cell parameters of the NiAs-type structure. Room temperature compression of FeS causes the successive phase transition of FeS I into FeS II (MnP-type phase) at 3.4 GPa and then into FeS III (monoclinic phase) at 6.7 GPa (King and Prewitt, 1982; Fei et al., 1995; Kusaba et al., 1997, 1998a,b). Mössbauer spectroscopy and X-ray emission spectroscopy revealed that FeS III adopts a low-spin state configuration of Fe, and the structural phase transition of FeS II to FeS III involves the electronic transition from a high-spin state to a low-spin state (Kobayashi et al., 1997; Rueff and Kao, 1999; Takele and Hearne, 1999). Above room temperature, FeS transforms to FeS IV and then to FeS V with increasing temperature in the pressure range 0–20 GPa (e.g., Urakawa et al., 2002). FeS IV has a hexagonal NiAs-type superstructure with a ($2a$, c) unit cell, whereas FeS V shows the simple NiAs-type unit cell (Fei et al., 1995). Fei et al. (1995) and Kusaba et al. (1998b) found anomalous behaviors during compression of FeS IV, in which the isothermal compression curve is inflected and the c-axis is abruptly shortened around 7 GPa. Both authors speculated that these anomalies can be attributed to the electronic transition of FeS IV without any structural change. It is, therefore, worth elucidating whether FeS V shows similar high-pressure behavior as FeS IV, because the high-temperature phase is dominant in planetary interiors. We must take the possibility of the electronic transition into consideration in order to evaluate the equation of state for FeS relevant to the interior of Mars. In this study, we examine the phase relationships of FeS and the equations of state for FeS IV and V using in situ X-ray diffraction observation and propose models of the internal structure of Mars.

2. Experiments

Experiments were performed at SPring-8, using large volume high-pressure apparatus: SMAPII at BL14B1 and SPEED-1500 at BL04B1 (Utsumi et al., 1998). SMAPII is a 180 t DIA-type cubic press, which was used for experiments up to 10 GPa, and is equipped with six WC anvils with a 3 mm anvil truncation and a boron-epoxy resin pressure transmitting cube with a 6 mm edge length. SPEED-1500, a KAWAI-type 6–8 double stage apparatus driven by 1500 t press, was used for experiments at higher pressures than 10 GPa. A set of WC cubic anvils with a 3 mm truncated edge length and MgO octahedral pressure transmitting medium with a 9 mm edge were used. High temperature was generated by resistance heating using a graphite tube heater below 10 GPa and a titanium carbide–diamond sheet heater above 10 GPa. Sample temperature was measured by a W97%Re3%–W75%Re25% thermocouple, and the effect of pressure on emf was ignored.

Starting material was polycrystalline FeS (troilite), which was synthesized from pure Fe and S sealed in an evacuated glass tube at 903 K for 2 weeks (Morimoto, 1976; King and Prewitt, 1982). The unit cell parameters of the troilite were $a = 5.969(1)$ Å and $c = 11.760(6)$ Å. The chemical formula of the troilite was determined to be $Fe_{0.999}S$ from the d_{114} value (Yund and Hall, 1969). The powdered FeS (5–10 μm grain size) was mixed with powdered NaCl in a 1:1 volume ratio, and then packed in an hBN sample container. NaCl was used as the pressure standard, and also acted as a diluent to prevent grain growth of FeS at high temperature and to reduce the deviatric stress on the sample.

X-ray diffraction was conducted by an energy dispersive method using white X-rays up to 150 keV with a Ge solid state detector, in a transmitting geometry. The Bragg angle was fixed at 4.5° or 5° to cover the main diffraction peaks of the FeS polymorphs and NaCl. The typical exposure time was 5 min, providing the data of sufficient quality for phase identification and unit cell parameter evaluation. X-ray diffraction profiles of the powdered mixture of FeS and NaCl were acquired at the position close to the hot junction of the thermocouple. Sample pressures were evaluated from the unit cell volume of NaCl using Brown's equation of state (Brown, 1999).

First, the sample was pressurized by applying oil pressure to a target value. Next, the sample was heated to the highest desired temperature at a given oil pressure, in order to minimize the effects of deviatoric stress on the observed unit cell parameters. Then the sample temperature was decreased to room temperature at a constant oil pressure, with X-ray diffraction patterns of the sample being acquired at an interval of 100 K. This procedure was repeated to the maximum press load, and X-ray diffraction patterns of FeS were collected at about 150 pressure and temperature conditions up to 22 GPa and 1600 K.

3. In situ X-ray diffraction observations and stability of FeS polymorphs

We observed four stable phases, FeS I (troilite), FeS III (monoclinic phase), FeS IV (hexagonal phase), and FeS V (NiAs-type phase) by in situ X-ray diffraction experiments up to 22 GPa and 1600 K. We, however, could not find FeS II, which was reported to be stable between FeS I and FeS III at room temperature (e.g., King and Prewitt, 1982; Fei et al., 1995; Marshall et al., 2000). FeS II was

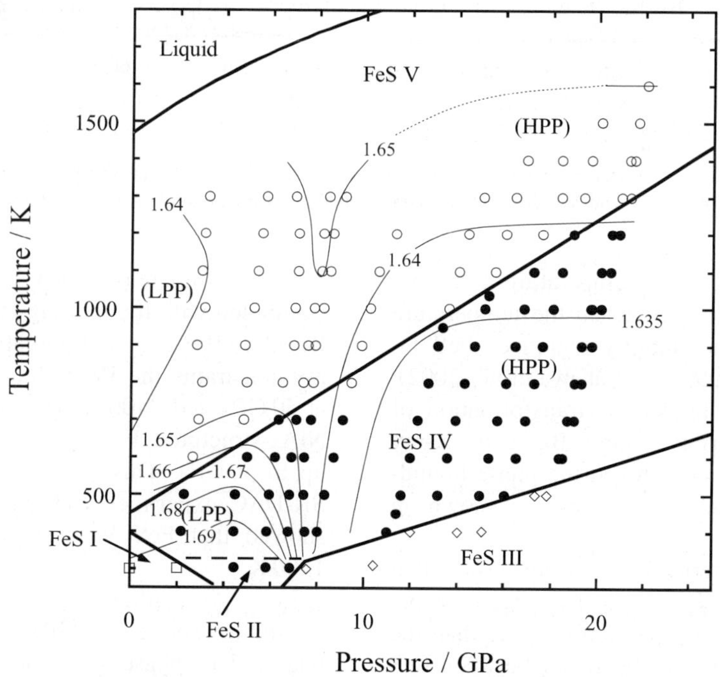

Fig. 1. Summary of the phase identifications obtained by in situ X-ray observations and the stability fields of FeS polymorphs with the iso-c/a ratio curves (thin lines) of the fundamental NiAs unit cell. Open squares are FeS I (troilite), open diamonds are FeS III (monoclinic phase), solid circles are FeS IV (hexagonal phase), and open circles are FeS V (NiAs-type phase). FeS II was not found in this study (see text). Iso-c/a ratio curves are consistent with those of Kusaba et al. (1998b) who reported iso-c/a ratio curves up to 10 GPa and 700 K. The curve with $c/a = 1.64$ at low pressure is constrained using data of Kusaba et al.

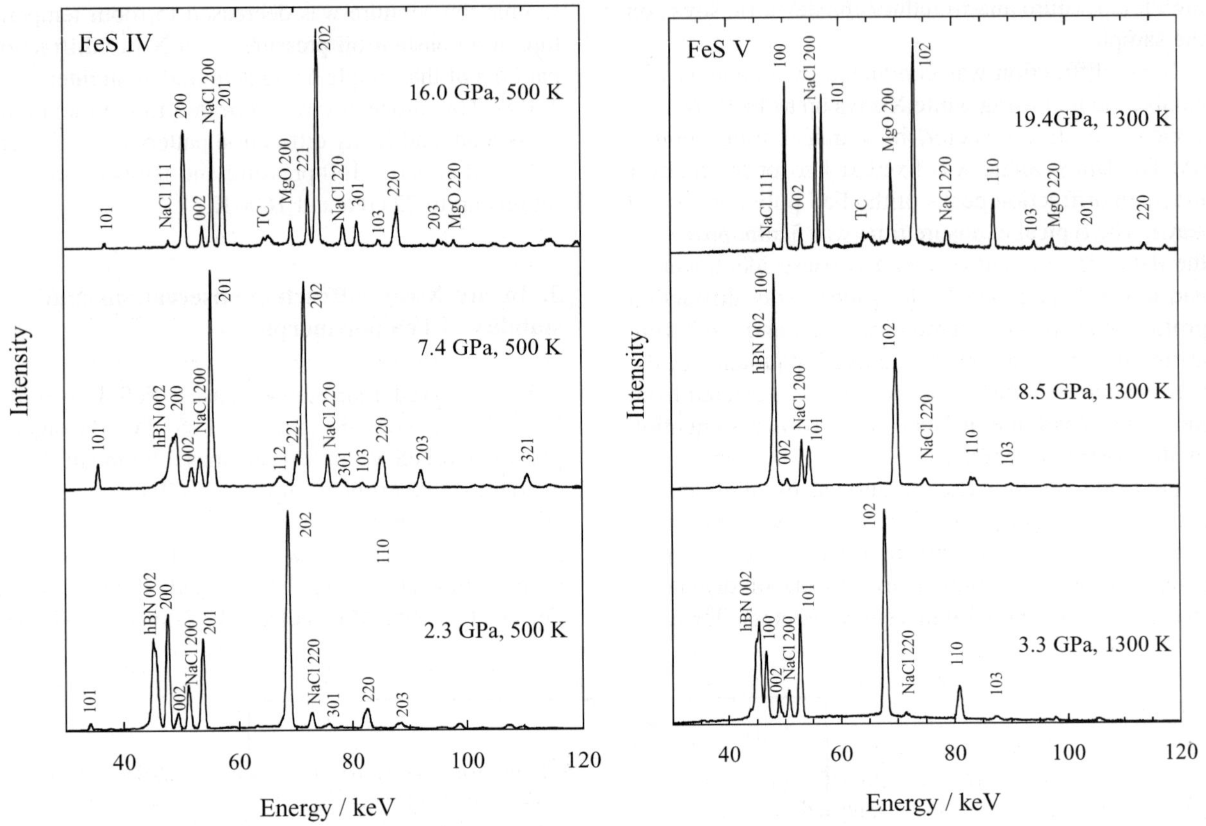

Fig. 2. Examples of X-ray diffraction spectra of FeS IV (hexagonal phase) and FeS V (NiAs-type phase) at high temperatures and high pressures. Crystal structures of FeS IV and V are not affected by anomalous compression behavior between 4 and 12 GPa.

always detected during room temperature compression, but it was not observed when the temperature was decreased from the stability region of FeS IV (Kusaba et al., 1997, 1998a,b; Urakawa et al., 2002). Kinetic barriers might hinder the transformation of FeS IV into FeS II during cooling. Based on X-ray diffraction profiles, we determined the phase boundaries between the four FeS phases, as shown in Fig. 1.

X-ray diffraction patterns (Fig. 2) clearly show that FeS V, with a simple NiAs-type structure, is stable up to at least 22 GPa at temperatures higher than the stability field of FeS IV. The boundary between FeS IV and FeS V is defined as a straight line, $T\,(\mathrm{K}) = 39.6P\,(\mathrm{GPa}) + 450$. The straight phase boundary is more consistent with that reported by Kusaba et al. (1998a) rather than the curved boundary predicted by Fei et al. (1995), but has a shallower slope than the previous estimate (47.2 K/GPa, Kusaba et al., 1998a). Combined with the melting curve of FeS reported by Boehler (1992), an extrapolation of the IV–V boundary constrains the IV–V liquid triple junction to lie at 50 GPa and 2400 K. This suggests that the simple NiAs-structured FeS V can exist as a stable phase up to 50 GPa. This is consistent with the results of SR-DAC experiments (Kavner et al., 2001), which showed that FeS V is stable up to 35 GPa above 1500 K.

Low-spin configurated FeS III (monoclinic phase) is stable above 7.5 GPa as a low temperature phase. The phase boundary between FeS III and IV is determined to be $T\,(\mathrm{K}) = 20P\,(\mathrm{GPa}) + 170$ (Fig. 1). The slope of the III–IV boundary is consistent with that reported by Fei et al. (1995), but the absolute temperature is lower by about 50 K.

4. Isostructural phase transition of FeS IV and FeS V

Anomalous high-pressure behavior of FeS IV was reported by Fei et al. (1995) and Kusaba et al. (1998b). In both these studies, it was found that the isothermal compression curve of FeS IV cannot be defined by a single equation of state and FeS IV exists as two phases, a low-pressure phase (LPP) and a high-pressure phase (HPP). They also reported an abrupt change in the *c*/*a* ratio of FeS IV, corresponding to an inflection of the compression curve. An anomaly was also found in the thermal expansivity of FeS IV by Kusaba et al. (1998b, 2001), who reported that thermal expansivity has a minimum at 5.7 GPa. In both studies, these phenomena were attributed to an electronic transition of FeS IV without structural change, in which the spin configuration of Fe changes from the high-spin state to the low-spin state. Although this interpretation has not been confined yet, the electronic state of FeS III, stable phase at room temperature above 7 GPa, strongly supports this hypothesis.

We found similar anomalies not only in FeS IV but also in FeS V. Isothermal compression curves for both FeS IV and FeS V can be divided into three parts separated by two inflections (Fig. 3): a high-spin LPP region, a transient region, and possibly a low-spin HPP region. High-pressure phases of both FeS IV and FeS V are clearly denser than the low-pressure phases. Inflection points seem to move toward higher pressure with increasing temperature. Hereafter, we denote the low-pressure and the high-pressure phases of FeS IV and V as FeS IV (LPP), FeS V (LPP), FeS IV (HPP), and FeS V (HPP), respectively.

The observed changes in the *c*/*a* ratio of FeS IV along the isothermal compression curve at 500 K are consistent with those reported by Fei et al. (1995) as shown in Fig. 4(B). The change in the *c*/*a* ratio of FeS V at 1200 K is small and unclear (Fig. 4(A)), but iso-*c*/*a* ratio curves in Fig. 1 show that FeS IV (LPP) and FeS V (LPP) have larger interlayer distances than FeS IV (HPP) and FeS V (HPP). At pressures above 10 GPa, the *c*/*a* ratios of HPP are close to ideal closed-packed values and increase slightly with temperature from 1.63 at 500 K to 1.64 at 1300 K. This means that the linear thermal expansivity along the *c*-axis is larger than that along the *a*-axis. On the

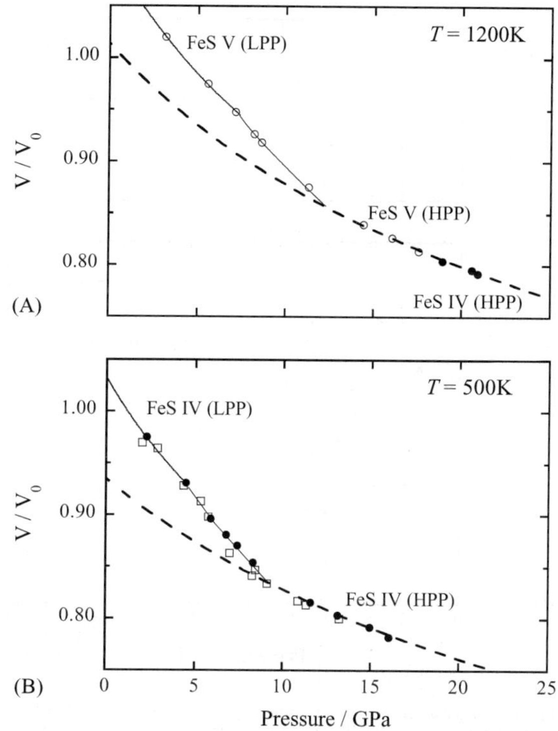

Fig. 3. Isothermal compression curves of FeS at 1200 K (A) and 500 K (B). Solid circles are FeS IV and open circles are FeS V. Open squares in (B) are data for FeS IV at 473 K from Kusaba et al. (1998b). Dashed lines are compression curves for the high-pressure phases calculated by using the parameters listed in Table 1. Solid lines have been drawn as guides for the eye. Compression data for each phase can be divided into three parts; the low-pressure phase (LPP) region, a transient region, and the high-pressure phase (HPP) region.

other hand, the iso-*c*/*a* curves are complex in the LPP region and the transient region below 10 GPa.

We calculated the isobaric thermal expansivities of FeS IV and V by fitting a linear equation to the volume data at constant pressure (Fig. 5). Combined with the data from Kusaba et al. (1998b, 2001), a steep depression of thermal expansivity of FeS IV was observed in the pressure range 4–9 GPa. FeS V also shows an anomalous pressure dependence of thermal expansivity below 9 GPa, which increases with pressure between 4 and 9 GPa by a factor of 2. On the other hand, above 9 GPa, the thermal expansivities of both FeS IV (HPP) and FeS V (HPP) have similar values and decrease monotonously with pressure.

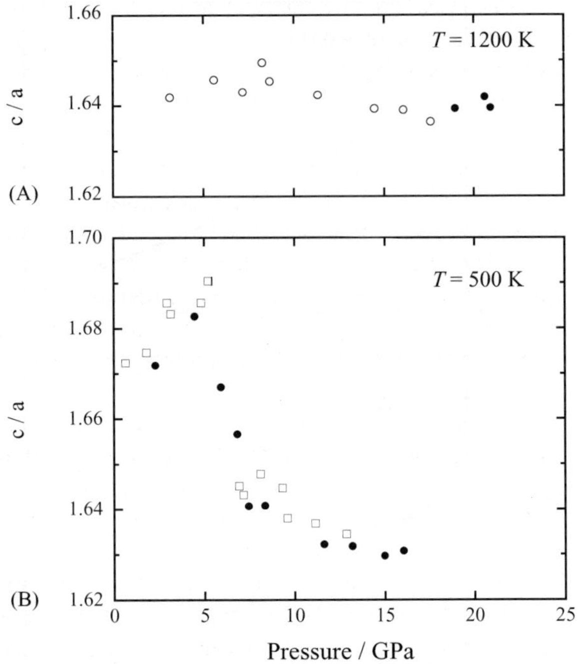

Fig. 4. The *c/a* ratios for the fundamental NiAs unit cell of FeS plotted along the isothermal compression curves at 1200 K (A) and 500 K (B). Closed circles are FeS IV and open circles are FeS V. Open squares at 500 K are data from Fei et al. (1995), which are consistent with our results.

This high-pressure behavior indicates that both FeS IV and FeS V undergo pressure induced isostructural phase transitions. While the compression curves and the *c/a* ratios indicate that this phase transition proceeds in the pressure range 5–9 GPa for FeS IV and 7–12 GPa for FeS V, thermal expansivities show that the transient region of FeS V is identical to that of FeS IV and lies between 4 and 9 GPa. All anomalous behavior correlates with the interlayer distance in the fundamental NiAs unit cell, and occurs in the transient region between the LPP stability field and the HPP stability field. If this isostructural transition of FeS is due to the electronic transition of Fe from the high-spin state to the low-spin state, the two electronic states might be mixed in the transient region and the proportions of the electronic state might change with both pressure and temperature. In this case, anomalous thermoelastic properties of FeS IV and FeS V can be explained by the interaction between Fe atoms along the *c*-axis of the fundamental NiAs unit cell.

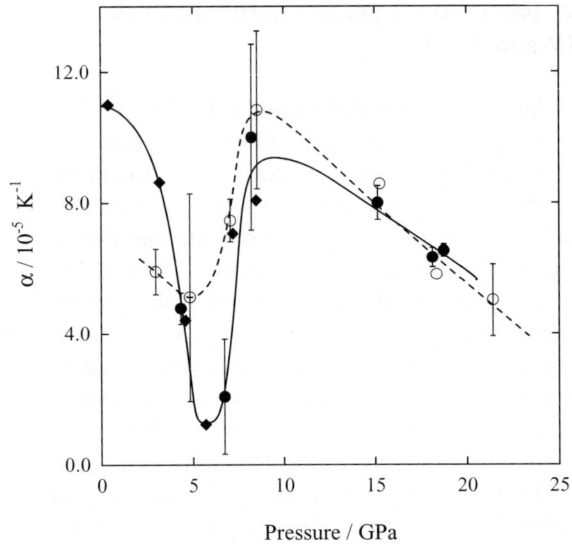

Fig. 5. Pressure dependence of the volume thermal expansivity of FeS. Solid circles are FeS IV, open circles are FeS V, and solid diamonds are data for FeS IV from Kusaba et al. (1998b, 2001). For the calculation of thermal expansivity, we used the data obtained at a constant pressure (±0.2 GPa).

Anomalies in FeS IV and FeS V affect not only thermoelastic properties but also phase relationships in the system including FeS. Inflections of the eutectic melting curves in the Fe–FeS binary system and the Fe–FeS–FeO ternary system in the pressure range 5–10 GPa (Usselman, 1975; Fei et al., 1997; Urakawa et al., 1987) must correlate with the isostructural phase transition from FeS V (LPP) to FeS V (HPP).

5. Equations of state for FeS IV (HPP) and FeS V (HPP)

As shown above, thermoelastic properties of FeS IV and FeS V change over the pressure range 4–12 GPa. For applications to the deep interior of the terrestrial planets, density data for FeS IV (HPP) and FeS V (HPP) are necessary. We evaluate high-temperature equations of state for FeS IV (HPP) and FeS V (HPP) using the volume data determined above 11.7 GPa using SPEED-1500 at BL04B1.

We fitted the following Birch–Murnaghan equation of state to the volume data:

Table 1
Parameters of equations of state for FeS IV (HPP) and FeS V (HPP)

	FeS IV	FeS V
T_0 (K)	600	1000
$K_{0T}(T_0)$ (GPa)	62.5 ± 0.9	54.3 ± 1.0
dK_0/dP[a]	4.0	4.0
dK_0/dT (GPa/K)	−0.0208 ± 0.0028	−0.0117 ± 0.0015
$V_0(T_0)$[b] (Å^3)	228.02 ± 0.41	59.92 ± 0.15
α_0 ($\times 10^{-5}$ K^{-1})	7.16 ± 0.51	10.42 ± 0.49
α_1 ($\times 10^{-8}$)	6.08 ± 1.31	–

[a] This parameter is fixed to be 4.0 in least square fitting.
[b] Unit cell volume for FeS IV ($Z = 8$) and for FeS V ($Z = 2$).

$$P(V,T) = \frac{3}{2}K_{0T}(T)\left\{\left[\frac{V}{V_0(T)}\right]^{-7/3} - \left[\frac{V}{V_0(T)}\right]^{-5/3}\right\} \times \left\{1-\frac{3}{4}\left[4-\left(\frac{\partial K_0}{\partial P}\right)_T\right]\left[\left[\frac{V}{V_0(T)}\right]^{-2/3}-1\right]\right\} \quad (1)$$

$$K_{0T}(T) = K_{0T}(T_0) + \left(\frac{\partial K_0}{\partial T}\right)_P (T - T_0) \quad (2)$$

$$V_0(T) = V_0(T_0)\exp\left(\int_{T_0}^{T}\alpha\,dT\right) \quad (3)$$

Thus, bulk modulus, its temperature derivative, thermal expansivity, and the unit cell volume at 0.1 MPa and T_0 for both FeS IV (HPP) and FeS V (HPP) were determined by a non-linear least squares fit with a fixed pressure derivative of the bulk modulus $dK_0/dP = 4$. We did not use volume data from near the phase transition pressure in this procedure. We assumed the temperature dependency of thermal expansivity to be $\alpha = \alpha_0 + \alpha_1 T$, and calculated for both temperature-dependent and -independent cases. Parameter data sets, that are consistent with the experimentally determined phase boundary between FeS IV and FeS V, are $K_{0T} = 62.5$ GPa at $T = 600$ K, $(dK_0/dT)_P = -0.0208$ GPa/K, and $\alpha = 7.16 \times 10^{-5} + 6.08 \times 10^{-8}T$ for FeS IV (HPP) and $K_{0T} = 54.3$ GPa at $T = 1000$ K, $(dK_0/dT)_P = -0.0117$ GPa/K, and $\alpha = 10.42 \times 10^{-5}$ for FeS V (HPP) (Table 1). Equations of state at 500 and 1200 K are shown in Fig. 3 together with the experimental data.

Bulk modulus and thermal expansivity of FeS IV (HPP) were also determined in previous studies by Fei et al. (1995) and Kusaba et al. (1998b). The former study gave $K_{0T} = 54 \pm 6$ GPa at $T = 800$ K, which is consistent with our value of 58.3 GPa at 800 K, whereas the latter reported a significant lower value of $K_{0T} = 49 \pm 2$ GPa at $T = 573$ K. Kusaba et al. (1998b) used volume data close to the phase transition pressure in their fit, so that their K_{0T} is relatively low due to the large volume change near the phase boundary, as shown in Fig. 3. Our estimation of thermal expansivity for FeS IV (HPP) is slightly larger than that of Fei et al. (1995).

The present study firstly reports the thermoelastic properties of FeS V (HPP). The bulk modulus and thermal expansivity of FeS V (HPP) are close to those of FeS IV (HPP), because the phase transition between FeS IV and V is at a second order (Kusaba et al., 2001). The equation of state for FeS V (HPP) can be compared with the density data reported by Kavner et al. (2001), which were determined by a laser-heated SR-DAC technique. The reported density of FeS V (HPP) of 5.96 Mg/m^3 at 17 GPa and 1600 K is clearly higher than the present result of 5.75 Mg/m^3 by 4%, corresponding to a temperature difference of ~500 K at constant pressure. We suggest that Kavner et al. (2001) may have overestimated the density of FeS V (HPP) due to a large temperature error associated with laser heating.

6. Implications for the structure of the Martian interior

Studies of the internal structure of Mars have been based on cosmochemical models combined with high-pressure experiments (Ohtani and Kamaya, 1992; Kamaya et al., 1993; Fei et al., 1995; Bertka and Fei, 1997, 1998a,b; Kavner et al., 2001; Sanloup et al., 2002), assuming an Fe-rich mantle and an Fe–FeS core. Estimated density profiles of the Martian mantle are consistent with each other and there is a consensus on the Martian mantle structure, although there has been a debate concerning whether or not Mars has an Mg-silicate perovskite-bearing lower mantle. The presence of an Mg-perovskite layer would influence the thermal history of Mars, because this layer controls the mode of mantle convection and acts as a stable thermal boundary layer at the base of the mantle. A recent value of the dimensionless polar moment of inertia factor $C = 0.3662 \pm 0.0017$, evaluated from Mars Pathfinder data (Folkner et al., 1997), constrains the bulk composition of Mars. Bertka and Fei (1998a,b) showed that models of the interior of Mars that are consistent with C determined from the Mars Pathfinder data are not consistent with a chondritic bulk Fe content or Fe/Si ratio. On the other hand, Zharkov and Gudkova (2000) argued that the addition of H to the Fe–FeS core and the reduction of the Fe number of the mantle make a chondritic bulk composition possible. This controversy arises mainly from uncertainties in the density profiles of the Martian core. Previous work on the density of FeS at high pressures (Fei et al., 1995; Kavner et al., 2001) was semi-quantitative and there were large uncertainties in the density estimation at Martian core conditions. The present study of the equations of state of FeS IV (HPP) and FeS V (HPP) provides an opportunity to examine the Martian interior models with an Fe-rich mantle and an Fe–FeS core with improved accuracy.

In order to evaluate the internal structure of Mars, we adopt the Bertka and Fei (1998a) density model for the Martian mantle. In this model, a mantle composition, which was derived from the chemical composition of the SNC meteorites (Dreibus and Wänke, 1985), was assumed and the density profiles were calculated along the temperature distribution in the Martian interior determined by Longhi et al. (1992). Our results show that FeS V (HPP) is a stable phase under the Martian core conditions proposed by Longhi et al. (1992). The Fe–FeS binary system, however, may have a eutectic melting relation at the pressures of the Martian core (e.g., Fei et al., 2000). Thus, the density of liquid Fe–FeS must be taken in account, but is not available in the required pressure range. Therefore, we used the density of a mixture of γ-Fe and FeS V (HPP) instead of the liquid density for intermediate compositions in the Fe–FeS system. This assumption should be appropriate because the density difference between liquid and solid in Fe–FeS system is expected to be very small at high pressures due to the gentle Clapeyron slope of the melting curve (e.g., Urakawa et al., 1987). We used the equation of state for γ-Fe by Besson and Nicol (1990), which is consistent with the volume data reported by Boehler et al. (1989, 1990) and Funamori et al. (1996).

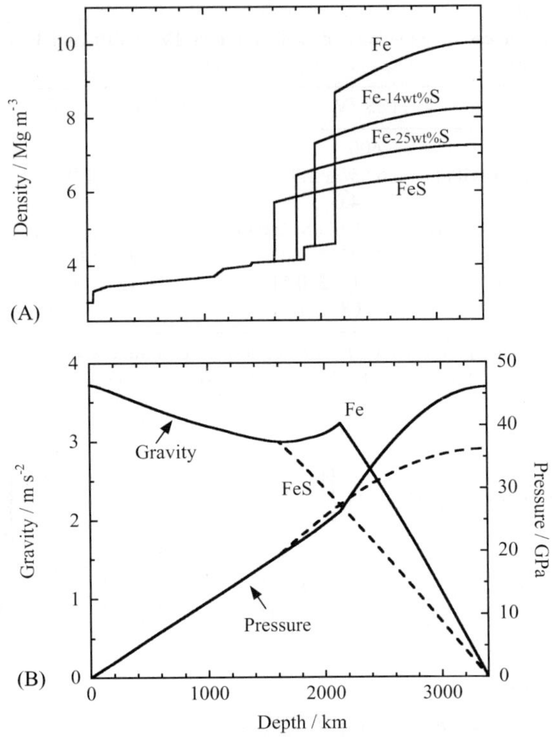

Fig. 6. Internal structure models for the Mars. (A) Density distribution profiles for four different core compositions with a 50 km thick crust. Mantle density is from Bertka and Fei (1998a). (B) Gravity and pressure in the Marian interior corresponding to Fe (solid line) and FeS (dashed line) core models in (A). Pressure in the Martian core largely depends on its composition.

Using these data, we calculated internal models of Mars that satisfy its mass and radius for given core compositions (from Fe to FeS), crustal density (3.0 Mg/m^3) and crustal thicknesses (from 0 to 150 km). We did not constrain the moment of inertia factor and bulk Fe/Si ratio in the calculations. Fig. 6 shows density, gravity and pressure profiles of the Martian interior with 50 km thick crust. For a given core composition, there are clear trade-offs between the core radius and pressure in the core. Pressure at the center of Mars changes by about 10 GPa with increasing S content, from 36.3 GPa for a FeS core to 46.1 GPa for a pure Fe core, which is consistent with previous estimates by Schubert and Spohn (1990). The results of the present calculations are summarized in Table 2, including bulk Fe content, Fe/Si ratio, and moment of inertia factor.

Table 2
Calculated bulk Fe content, bulk Fe/Si weight ratio, core mass fraction (M_{core}), core radius (R_{core}), pressures at CMB (P_{CMB}) and the center of Mars (P_C), polar moment inertia factor (C) for Mars models with given core composition and crustal thickness (h_{crust})

Core composition	h_{crust} (km)	Bulk Fe (wt.%)	Fe/Si	M_{core} (wt.%)	R_{core} (km)	P_{CMB} (GPa)	P_C (GPa)	C
Fe	0	23.7	1.29	11.4	1239	26.4	45.4	0.3656
Fe–14 wt.% S	0	24.0	1.35	14.0	1410	24.0	41.2	0.3665
Fe–25 wt.% S	0	24.3	1.41	17.0	1568	21.9	38.4	0.3674
FeS[a]	0	24.4	1.49	21.1	1750	19.7	36.0	0.3688
Fe	50	24.2	1.32	11.9	1259	26.3	46.1	0.3638
Fe–14 wt.% S	50	24.5	1.39	14.7	1433	23.9	41.7	0.3647
Fe–25 wt.% S	50	24.8	1.46	17.9	1593	21.7	38.8	0.3657
FeS	50	24.9	1.54	22.1	1779	19.4	36.3	0.3672
Fe	100	24.7	1.36	12.6	1280	26.3	46.8	0.3618
Fe–14 wt.% S	100	25.1	1.43	15.6	1459	23.7	42.2	0.3628
Fe–25 wt.% S	100	25.4	1.51	18.8	1619	21.5	39.2	0.3639
FeS	100	25.5	1.60	23.3	1810	19.2	36.6	0.3655
Fe	150	25.4	1.41	13.3	1304	26.2	47.6	0.3597
Fe–14 wt.% S	150	25.8	1.49	16.5	1486	23.5	42.7	0.3608
Fe–25 wt.% S	150	26.0	1.56	19.8	1648	21.3	39.7	0.3620
FeS	150	26.1	1.67	24.6	1843	18.8	36.9	0.3637

[a] FeS includes 36.5 wt.% sulfur.

Spherically symmetrical models of the Martian interior are constrained only by the mean moment of inertia factor I. The polar moment of inertia factor C is one of three principal moments of inertia, and is slightly different from I. Here we use the polar moment of inertia C as the constraint for the present model instead of I, because there is certain uncertainty in the estimated I value (e.g., Zharkov and Gudkova, 2000; Schubert et al., 2001).

The polar moment of inertia factor C, calculated as a function of given crustal thickness and core composition, is plotted against depth of the core–mantle boundary (CMB) in Fig. 7. The polar moment of inertia factor C depends strongly on both crustal thickness and core composition, whereas the depth of CMB is less sensitive to crust thickness and depends mainly on the core composition. The post-spinel transition occurs at around 1850 km depth in the Martian mantle (Fig. 6), so that an Mg-silicate perovskite-bearing lower mantle appears only when the Fe-rich core contains less than 20 wt.% sulfur, although Kavner et al. (2001) concluded that the lower mantle exists for a core composition close to FeS. The polar moment of inertia factor $C = 0.3662 \pm 0.0017$ derived from Mars Pathfinder data (Folkner et al., 1997) is consistent with any compositions between Fe and FeS for the Martian core, but it excludes a crust thicker than 100 km. A thick crust is also unlikely from a petrological point of view if the Martian crust has a basaltic composition, because this would transform into dense eclogite at a pressure of ~1 GPa (i.e., 90 km in depth). In fact, topography and gravity data, measured by Mars Global Surveyor mission, provided an estimation of average crustal thickness of 50 km (Zuber et al., 2000). In this case, the Martian core must contain more than 10 wt.% sulfur (Fig. 7).

A notable difference between our study and that of Bertka and Fei (1998a,b) or Kavner et al. (2001) is that in the density estimation for the FeS phase. Bertka and Fei used FeS IV (HPP), instead of FeS V (HPP), in their models (Bertka and Fei, 1998a,b), while Kavner et al. (2001) clearly overestimated the density of FeS V (HPP) as discussed above. Our improved equation of state for FeS V (HPP) gives a lower density for FeS at Martian core conditions than previous studies, corresponding to a 100 km difference in the depth of CMB for a FeS core (Fig. 7). This leads to a larger bulk Fe/Si ratio at the same sulfur content. The Fe/Si ratios for the bulk Martian composition calculated in the present models (Table 2), however, are still smaller than the chondritic value of 1.73, with the maximum value being 1.60 for a FeS core and a 100 km thick crust, which are within the acceptable limits constrained by the observed C value.

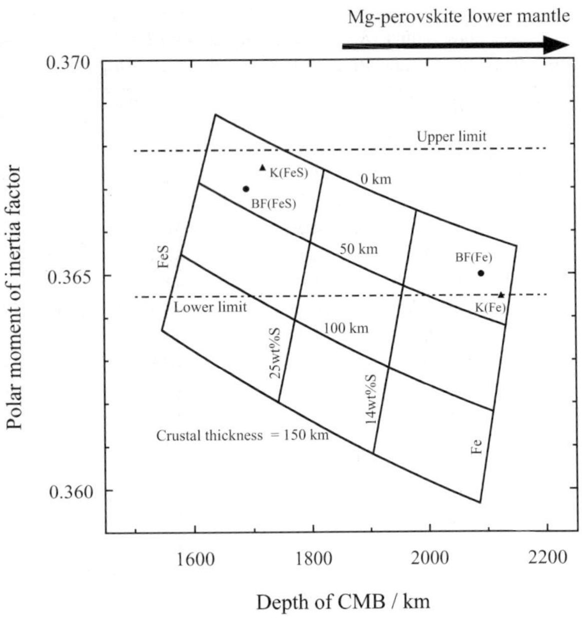

Fig. 7. Depths of the core–mantle boundary of Mars plotted for several models with different core compositions and crustal thicknesses. An Mg-perovskite lower mantle is stable if the Martian core includes less than 20 wt.% sulfur. Allowable limits of the polar moment of inertia factor *C* shown as dot-dashed lines (Folkner et al., 1997) do not constraint the chemical composition of the Martian core, but exclude a crust thicker than 100 km. Previous models based on a crustal thickness of 50 km are also shown as solid circles (Bertka and Fei, 1998a) and solid triangles (Kavner et al., 2001).

Acknowledgements

The authors would like to appreciate K. Kusaba for the preparation of troilite sample. They are also grateful to T. Okada for his help in synchrotron radiation experiments. Comments of an anonymous reviewer, P.S. Balog and D. Rubie helped to improve the manuscript. This study has been performed under approval of the advisory committee of SPring-8.

References

Bertka, C.M., Fei, Y., 1997. Mineralogy of the Martian interior up to core–mantle boundary pressures. J. Geophys. Res. 102, 5251–5264.

Bertka, C.M., Fei, Y., 1998a. Density profiles of an SNC model Martian interior and the moment-of-inertia factor of Mars. Earth Planet. Sci. Lett. 157, 79–88.

Bertka, C.M., Fei, Y., 1998b. Implications of Mars Pathfinder data for the accretion history of the terrestrial planets. Science 281, 1838–1840.

Besson, J.M., Nicol, K., 1990. An equation of state of γ-Fe and some insight about magnetoelastic effects on measurements of the α–β–γ triple point and other transitions. J. Geophys. Res. 95, 21717–21720.

Boehler, R., 1992. Melting of the Fe–FeO and Fe–FeS systems at high pressure: constraints on core temperature. Earth Planet. Sci. Lett. 111, 217–227.

Boehler, R., Besson, J.M., Nicol, M., Nielsen, M., Itie, J.P., Weil, G., Johnson, S., Grey, F., 1989. X-ray diffraction of γ-Fe at high temperatures and pressures. J. Appl. Phys. 65, 1795–1797.

Boehler, R., von Bargen, N., Chopelas, A., 1990. Melting, thermal expansion, and phase transitions of iron at high pressures. J. Geophys. Res. 95, 21731–21736.

Brown, M.J., 1999. The NaCl pressure standard. J. Appl. Phys. 86, 5801–5808.

Dreibus, G., Wänke, H., 1985. Mars: a volatile-rich planet. Meteoritics 20, 367–382.

Fei, Y., Prewitt, C.T., Mao, H.-K., Bertka, C., 1995. Structure and density of FeS at high pressure and high temperature and the internal structure of Mars. Science 268, 1892–1894.

Fei, Y., Bertka, C.M., Finger, L.W., 1997. High-pressure iron sulfur compound, Fe_3S_2, and melting relations in the system Fe–FeS at high pressure. Science 275, 1621–1623.

Fei, Y., Bertka, C., Prewitt, C.T., 2000. Structure type and bulk modulus of Fe_3S, a new iron–sulfur compound. Am. Mineral. 85, 1830–1833.

Funamori, N., Yagi, T., Uchida, T., 1996. High pressure and high temperature in situ X-ray diffraction study of iron to above 30 GPa using MA8-type apparatus. Geophys. Res. Lett. 23, 953–956.

Folkner, W.M., Yoder, C.F., Yuan, D.N., Standish, E.M., Preston, R.A., 1997. Interior structure and seasonal mass redistribution of Mars from radio tracking of Mars Pathfinder. Science 278, 1749–1752.

Kamaya, N., Ohtani, E., Kato, T., Onuma, K., 1993. High pressure phase transitions in a homogeneous model Martian mantle. In: Takahashi, E., Jeanloz, R., Rubie, D. (Eds.), Evolution of the Earth and Planets. American Geophysical Union, Washington, DC, pp. 19–26.

Kavner, A., Duffy, T.S., Shen, G., 2001. Phase stability and density of FeS at high pressures and high temperatures: implications for the interior structure of Mars. Earth Planet. Sci. Lett. 185, 25–33.

King, H.E., Prewitt, C.T., 1982. High-pressure and high-temperature polymorphism of iron sulfide (FeS). Acta Cryst. B 38, 1877–1887.

Kusaba, K., Syono, Y., Kikegawa, T., Shimomura, O., 1997. Structure of FeS under high pressure. J. Phys. Chem. Solids 58, 241–246.

Kusaba, K., Syono, Y., Kikegawa, T., Shimomura, O., 1998a. Structures and phase equilibria of FeS under high pressure and temperature. In: Manghnani, M.H., Yagi, T. (Eds.), Properties of Earth and Planetary Materials at High Pressure and Temperature. American Geophysical Union, Washington, DC, pp. 297–305.

Kusaba, K., Syono, Y., Kikegawa, T., Shimomura, O., 1998b. High pressure and temperature behavior of FeS. J. Phys. Chem. Solids 59, 945–950.

Kusaba, K., Utsumi, W., Yamakata, M., Shimomura, O., Syono, Y., 2001. Second-order phase transition of FeS under high pressure and temperature. J. Phys. Chem. Solids 61, 1483–1487.

Kobayashi, H., Sato, M., Kamimura, T., Sakai, M., Onodera, H., Kuroda, N., Yamaguchi, Y., 1997. The effect of pressure on the electronic states of FeS and Fe_7S_8 studied by Mössbauer spectroscopy. J. Phys.: Condens. Matter 9, 515–527.

Longhi, J., Knittle, E., Holloway, J.R., Wänke, H., 1992. The bulk composition, mineralogy and internal structure of Mars. In: Kieffer, H.H., Jakosky, B.M., Synder, C.W., Matthews, M.S. (Eds.), Mars, University of Arizona Press, Tucson, pp. 184–208.

Marshall, W.G., Nelmes, R.J., Loverday, J.S., Klotz, S., Besson, J.M., Hamel, G., Parise, J.B., 2000. High-pressure neutron-diffraction study of FeS. Phys. Rev. B 61, 11201–11204.

Morimoto, N., 1976. Nonstoichiometry and structure-disorder of pyrrhotite ($Fe_{1-x}S$). J. Miner. Soc. Jpn. 12, 326–338 (in Japanese).

Ohtani, E., Kamaya, N., 1992. The geochemical model of Mars: an estimation from the high pressure experiments. Geophys. Res. Lett. 19, 2239–2242.

Rueff, J.-P., Kao, C.-C., 1999. Pressure-induced high-spin to low-spin transition in FeS evidenced by X-ray emission spectroscopy. Phys. Rev. Lett. 82, 3284–3287.

Sanloup, C., Guyot, F., Gillet, P., Fei, Y., 2002. Physical properties of liquid Fe alloys at high pressure and their bearing on the nature of metallic planetary cores. J. Geophys. Res. 107. 10.1029/2001JB000808.

Schubert, G., Spohn, T., 1990. Thermal history of Mars and the sulfur content of its core. J. Geophys. Res. 95, 14095–14104.

Schubert, G., Turcotte, D.L., Olson, P., 2001. Mantle Convection in the Earth and Planets. Cambridge University Press, Cambridge, 940 pp.

Takele, S., Hearne, G.R., 1999. Electrical transport, magnetism, and spin-state configurations of high-pressure phases of FeS. Phys. Rev. B 60, 4401–4403.

Taylor, L.A., Mao, H.-K., 1970. A high-pressure polymorph of troilite FeS. Science 170, 850–851.

Urakawa, S., Kato, M., Kumazawa, M., 1987. Experimental study on the phase relations in the system Fe–Ni–O–S up to 15 GPa. In: Manghnani, M.H., Syono, Y. (Eds.), High-pressure Research in Mineral Physic, American Geophysical Union, Washington, DC, pp. 95–111.

Urakawa, S., Hasegawa, M., Yamakawa, J., Funakoshi, K., Utsumi, W., 2002. High-pressure phase relationships for FeS. High Pressure Res. 22, 491–494.

Usselman, T.M., 1975. Experimental approach to the state of the core. Part I. The liquidus relations of the Fe-rich portion of Fe–Ni–S system from 30 to 100 kb. Am. J. Sci. 275, 278–290.

Utsumi, W., Funakoshi, K., Urakawa, S., Yamakata, M., Tsuji, K., Konishi, H., Shimomura, O., 1998. SPring-8 beamlines for high pressure science with multi-anvil apparatus. Rev. High Pressure Sci. Technol. 7, 1484–1486.

Yund, R.A., Hall, H.T., 1969. Hexagonal and monoclinic pyrrhotites. Econ. Geol. 64, 420–423.

Zharkov, V.N., Gudkova, T.V., 2000. Interior structure models, Fe/Si ratio and parameters of figure for Mars. Phys. Earth Planet. Inter. 117, 407–420.

Zuber, M.T., Solomon, S.C., Phillips, R.J., Smith, D.E., Tyler, G.L., Aharonson, O., Balmino, G., Banerdt, W.B., Head, J.W., Johnson, C.L., Lemoine, F.G., MaGovern, P.J., Neumann, G.A., Rowlands, D.D., Zhong, S., 2000. Internal structure and early thermal evolution of Mars from Mars Global Surveyor topography and gravity. Science 287, 1788–1793.

ELSEVIER

Available online at www.sciencedirect.com

SCIENCE DIRECT

Physics of the Earth and Planetary Interiors 143–144 (2004) 481–495

PHYSICS OF THE EARTH AND PLANETARY INTERIORS

www.elsevier.com/locate/pepi

The structure of amorphous iron at high pressures to 67 GPa measured in a diamond anvil cell

Guoyin Shen [a,*], Mark L. Rivers [a], Stephen R. Sutton [a], Nagayoshi Sata [a], Vitali B. Prakapenka [a], James Oxley [b], Kenneth S. Suslick [b]

[a] *Consortium for Advanced Radiation Sources, University of Chicago, Chicago, IL 60637, USA*

[b] *School of Chemical Sciences, University of Illinois at Urbana-Champaign, Urbana, IL 61801, USA*

Received 18 December 2002; received in revised form 24 April 2003; accepted 2 May 2003

Abstract

We report the results of room temperature compression studies of amorphous iron up to 67 GPa. The experiments were performed with a modified diamond anvil cell (DAC) that allows for measurements of X-ray scattering to a maximum momentum of 86.6 nm^{-1} using a monochromatic beam at a wavelength of 0.3311 Å. It is shown that accurate structural determination can be made for amorphous materials in a DAC to ultra high pressures. The dense random-packing model is favored to interpret our observations for amorphous iron. The resultant structure factor and the pair distribution function show that the dominant peak does not significantly change in shape and in intensity with the increase of pressure, except for peak positions. This observation is consistent with essentially zero pressure dependence of the coordination number and ratios of distances derived from the obtained pair distribution functions at high pressures. It is thus proposed that the structural contraction with the increase of pressure for amorphous iron is isotropic, which allows us to estimate volume changes as a function of pressure from the positions of the first peak in the pair distribution function. The compression behavior is found to be similar to those of b.c.c.-Fe and h.c.p.-Fe.

Keywords: Amorphous structure; Iron; High pressure; Diamond anvil cell

1. Introduction

Iron is believed to be the major constituent of the Earth's core (Birch, 1964) of which more than 96% by volume is in a liquid state. It is therefore of great geophysical interest to study the behavior of liquid iron at high pressures and high temperatures. Extensive investigations have been made by shock-wave (Ahrens and Johnson, 1995; Brown et al., 2000), static compression (Sanloup et al., 2000) and theoretical simulations (Alfe et al., 2000, 2002; Anderson and Ahrens, 1994). Information on the liquid structure provides a basis for investigating numerous macroscopic physical properties such as viscosity and self-diffusion (atomic transport) (Alfe et al., 2000; Brazhkin and Lyapin, 2000; Zhang and Guo, 2000; Zhang et al., 2000), electrical resistivity (electron transport) (Ashcroft and Lekner, 1965; March, 1990), compressibility (Ascarelli, 1968; Hasegawa and Watabe, 1972; Trigger et al., 1994) and thermal expansion (Hasegawa and Watabe, 1974). Such properties of the outer core are key parameters needed for modeling

* Corresponding author. Tel.: +1-630-252-0429; fax: +1-630-252-0436.
E-mail address: shen@cars.uchicago.edu (G. Shen).

0031-9201/$ – see front matter
doi:10.1016/j.pepi.2003.05.004

the dynamo, geomagnetism, composition and thermal state.

The structure of liquid iron has been extensively studied by theory (Alfe et al., 2000, 2002; Hausleitner and Hafner, 1989; Stixrude and Brown, 1998). Direct experimental study of the structure of liquid iron is limited to ambient pressure (Waseda and Suzuki, 1970) and to a pressure of 5 GPa with a large volume press (Sanloup et al., 2000) by X-ray scattering. Two major difficulties exist in liquid scattering experiments at high pressures. One is technical, arising from the extreme pressure–temperature (*P–T*) conditions required. The other difficulty with liquid scattering is the large background signal generated by the pressure vessel. Recently major progress has been made in the generation and control of extreme *P–T* conditions both in large volume presses (Kubo et al., 2004) and in the diamond anvil cell (DAC) (Dubrovinsky et al., 2000; Mao and Hemley, 1996; Shen et al., 2001). It is now possible to extend the measurements to extreme conditions approaching those of the outer core. To overcome the second difficulty of large background, spatial collimation is generally introduced to reduce the scattering from the pressure vessel. For example, with the energy dispersive technique, X-ray scattering is measured at a fixed angle (Katayama et al., 2000; Kruger and Meade, 1997; Meade et al., 1992; Tsuji et al., 1989). This method involves data collection at several 2θ angles and intensity normalization to the X-ray source spectrum. Recently Eggert et al. (2002) used monochromatic X-ray beam and reported structure factors of liquids (argon and water) in a DAC. Multichannel collimators have been developed with large volume presses (Mezouar et al., 2002; Yaoita et al., 1997), which allows structural studies of non-crystalline materials with the angle dispersive technique. The use of a monochromatic beam greatly increases the quality of structure factor determination for amorphous materials. In this paper, we present the structure determinations of amorphous iron with a monochromatic X-ray beam.

The structure of amorphous metals (metallic glasses) is very close to that of liquid metals (Waseda, 1980). There is a strong geophysical motivation to study the behavior of amorphous iron at high pressure. In contrast to the extensive experimental high pressure studies of crystalline iron (Boehler, 1993; Brown et al., 2000; Dubrovinsky et al., 2000; Hemley and Mao, 2001; Jephcoat et al., 1986; Mao et al., 1990; Shen et al., 1998), no high pressure work has been reported for amorphous iron, presumably because of the lack of availability of samples and difficulties involved in X-ray scattering measurements. Amorphous iron of high purity has been successfully synthesized sonochemically (Suslick et al., 1991). Neutron scattering has been used to study the structure of the amorphous iron at ambient pressure (Bellissent et al., 1993). The structure of thin films of amorphous iron have been investigated using transmission electron diffraction (Ichikawa, 1973; Leung and Wright, 1974). In this article we report the results of a room temperature compression study on amorphous iron up to 67 GPa using a monochromatic synchrotron beam and a modified DAC suitable for studies of amorphous materials at very high pressures. Accurate structure determinations are presented and the effect of pressure on the structure is shown.

2. Experiment

The amorphous iron sample was prepared sonochemically (Suslick et al., 1991). Bulk elemental analysis determined the iron powder to be >96 wt.% iron with a trace amount of carbon (<3 wt.%) and oxygen (<1 wt.%). The sample can be easily oxidized. Therefore, the sample loading was performed in a glove box filled with inert gas (argon), which provides an atmosphere of <1 ppm O_2.

A boron nitride seat (purchased from Linatec, Ukraine) was employed for high X-ray transmission over the full aperture of the DAC. The DAC used in this study is a symmetrical cell (Mao et al., 1998) with a full access opening angle of 60°. In the case of loading axis parallel to the incident X-ray beam in the present study, the maximum 2θ is 30°. With a monochromatic beam at a wavelength of 0.3311 Å, the maximum momentum transfer ($Q_{max} = 4\pi \sin\theta_{max}/\lambda$) is ~98 nm^{-1}. In the present study, the value of Q_{max} is found to be 86.6 nm^{-1}.

The sample configuration is shown in Fig. 1. Type-IA diamond was used for the anvils, with a culet size of 300 μm in diameter. A stainless steel gasket with initial thickness of 250 μm was used, and pre-indented to a thickness of ~30 μm. A hole of 300 μm in diameter was drilled and then filled

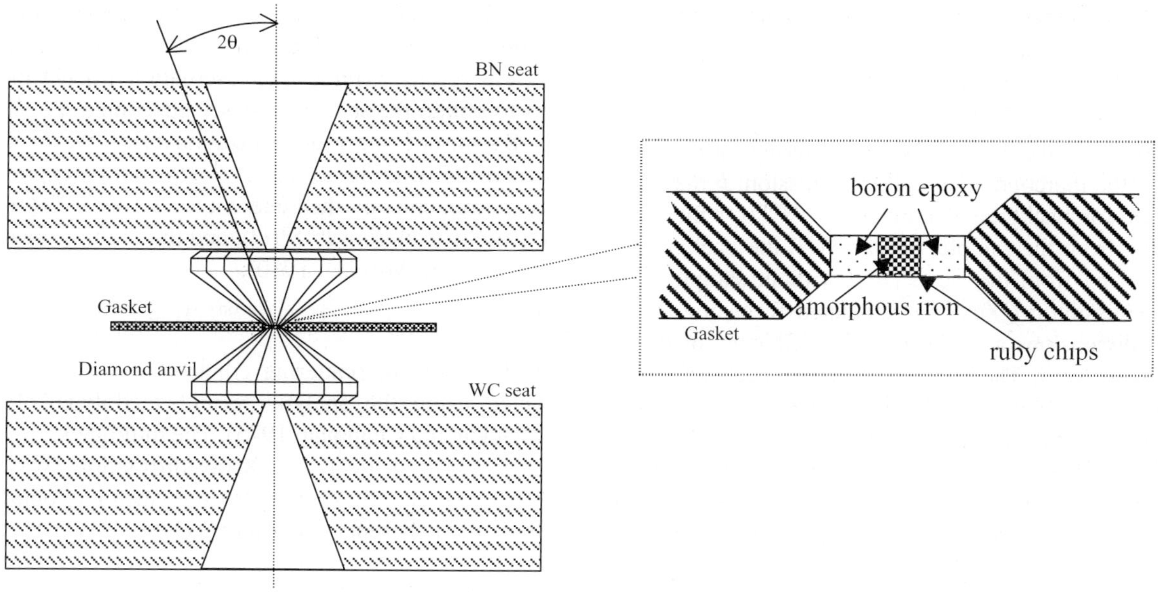

Fig. 1. Schematic of the diamond anvil cell used in this study. One seat is made of boron nitride for a large X-ray access opening. Enlarged figure shows the amorphous-boron insert gasket configuration, which helps in increasing the sample thickness and in performing proper background subtraction.

with amorphous-boron epoxy (epoxy to boron ratio of 1:2 by weight). The filled boron epoxy was then compressed by diamond anvils to a few GPa, and a 100 μm hole was drilled in the pressed boron epoxy. Finally the amorphous sample was loaded in the hole. A few ruby chips were also loaded for pressure measurement. The whole assemblage was sealed in a glove box by increasing pressures to a few GPa. The boron insert gasket helps to collect weak X-ray scattering from amorphous materials free of contamination from gasket diffraction. It also helps to make reliable background subtraction using the empty cell, without worries about gasket hole deformation.

X-ray diffraction experiments were performed at 13 BM-D at the advanced photon source. We used a 1 m long focusing mirror in the vertical direction and 0.1 m long mirror in the horizontal direction to focus the beam at the sample position to a size of 10 μm (horizontal) × 27 (vertical) μm at full width half maximum. A Bruker 2K-CCD detector was used to collect X-ray scattering. Typical exposure time was 10 min. Multiple exposures (2–3 times) were made for better statistics at high pressures. The sample-detector distance and the detector tilt angles were calibrated using powder diffraction from a CeO_2 standard that was placed at the sample position with the help of a microscope. The 2D images were angularly integrated by using the program FIT2D (Hammersley et al., 1996), and all geometric and polarization corrections were made during the integration. In cases where a few diffraction spots arose from the ruby chips, masks were used in the integration.

After reaching to a maximum pressure of 67 GPa, pressure was released and the sample was removed from the gasket hole. The same gasket was then put back in the cell, and the empty cell reference was measured with geometry identical to that at high pressures with help of a kinematic base.

3. Data analysis

3.1. Data reduction to atomic units

The description of the atomic distribution in non-crystalline materials usually employs the concepts of the structure factor and the pair distribution function in atomic units. The measured X-ray scattering intensity is in arbitrary units and needs to be converted into atomic units to describe the structure.

The observed X-ray scattering intensity is the combination of sample and background contributions:

$$I^{obs}(Q) = a(Q)I^{samp}(Q) + bI^{back}(Q) \quad (1)$$

where $Q = 4\pi \sin(\theta)/\lambda$ is the scattering momentum, $a(Q)$ the diamond anvil cell transmission, b the scale factor for background correction, $I^{samp}(Q)$ the scattering from the sample, and $I^{back}(Q)$ the background scattering from the same DAC without sample. The transmission $a(Q)$ can be calculated according to the cell dimensions and absorption coefficients of materials involved. The factor b is determined as follows (Fig. 2). At values of Q larger than $86.6\,nm^{-1}$, the sample signal is blocked by the cell body. The intensity in this region should be zero, a criterion used for determining the factor b. The subtracted pattern is also checked for reasonableness in the small-Q region. When an oscillation was observed in the small-Q region due to slight mislocation of the beam stop, a linear fit was applied to remove the oscillation. Eq. (1) allows us to obtain the scattering from the sample: $I^{samp}(Q)$.

By introducing a normalization factor N, the total scattering from the sample $I^{samp}(Q)$ can be expressed in atomic units by the coherent scattering $I^{coh}(Q)$, incoherent scattering $I^{incoh}(Q)$, and the multiple scattering $I^{mul}(Q)$ (Waseda, 1980):

$$NI^{samp}(Q) = I^{coh}(Q) + I^{incoh}(Q) + I^{mul}(Q) \quad (2)$$

The incoherent scattering contribution can be computed using the analytic formulas (Balyuzi, 1975; Thakkar and Chapman, 1975). The multiple scattering is generally neglected in X-ray diffraction (Waseda, 1980). By definition, the structure factor is obtained from the coherent scattering: $S(Q) \equiv I^{coh}(Q)/f^2(Q)$,

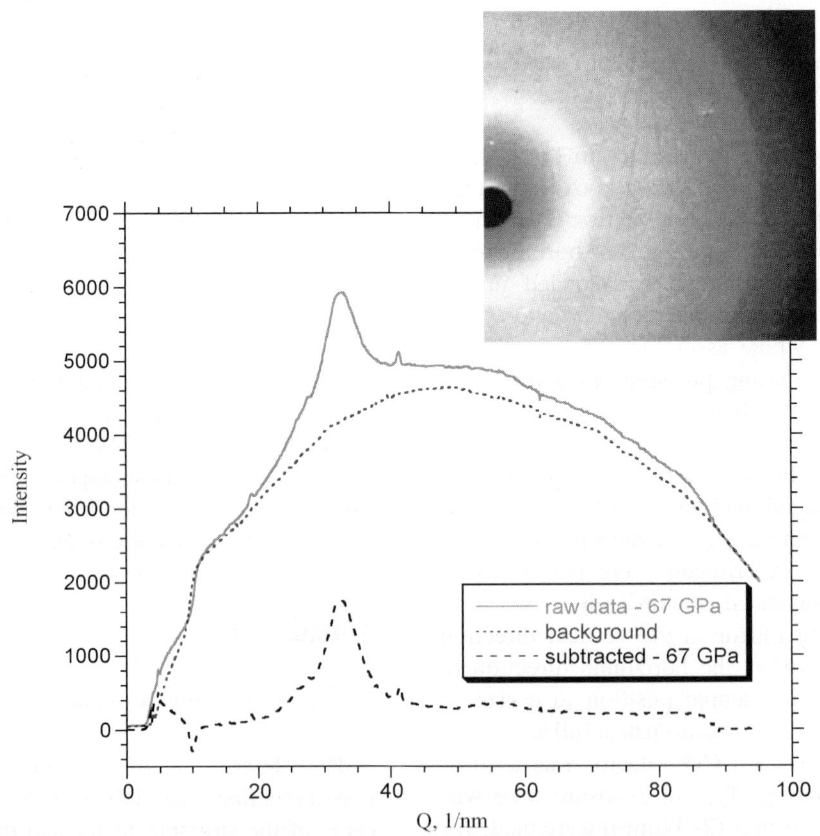

Fig. 2. Illustration of the background subtraction. An integrated pattern of X-ray scattering for amorphous iron at 67 GPa is plotted together with the empty cell background and the background subtracted pattern, with the factor b in Eq. (1) being 0.715. The insert is the background subtracted image at 67 GPa.

where $f(Q)$ is the atomic scattering factor (International Tables of Crystallography).

We determine the normalization factor N by using Krogh–Moe–Norman method (Krogh-Moe, 1956; Norman, 1957)

$$N = \frac{\int_0^{Q_{max}} Q^2 \left\{[f^2(Q) + I^{incoh}(Q)] \frac{\exp(-\gamma Q^2)}{f^2(Q)}\right\} \times dQ - 2\pi^2 n_0}{\int_0^{Q_{max}} Q^2 \left[I^{samp}(Q) \frac{\exp(-\gamma Q^2)}{f^2(Q)}\right] dQ} \quad (3)$$

where n_0 is the average atomic number density of the sample. The damping factor $\exp(-\gamma Q^2)$ is introduced to reduce errors in the high-Q region due to the factor Q^2 in Eq. (3) (Waseda, 1980). γ of 0.001 was used in the present study. By definition, the structure factor is then calculated by

$$S(Q) = \frac{1}{f^2(Q)} \left\{ N \left[\frac{I^{obs}(Q) - bI^{back}(Q)}{a(Q)} \right] - I^{incoh}(Q) \right\} \quad (4)$$

The Fourier transformation of $Q[S(Q) - 1]$ is the distribution function in real space:

$$F(r) \equiv 4\pi r n(r) - 4\pi r n_0 = \frac{2}{\pi} \int_0^{Q_{max}} Q[S(Q) - 1] \sin(rQ)\, dQ \quad (5)$$

where $n(r)$ is the function of atomic number density. From $F(r)$, the pair distribution function $g(r)$ is obtained:

$$g(r) = 1 + \frac{F(r)}{4\pi n_0 r} \quad (6)$$

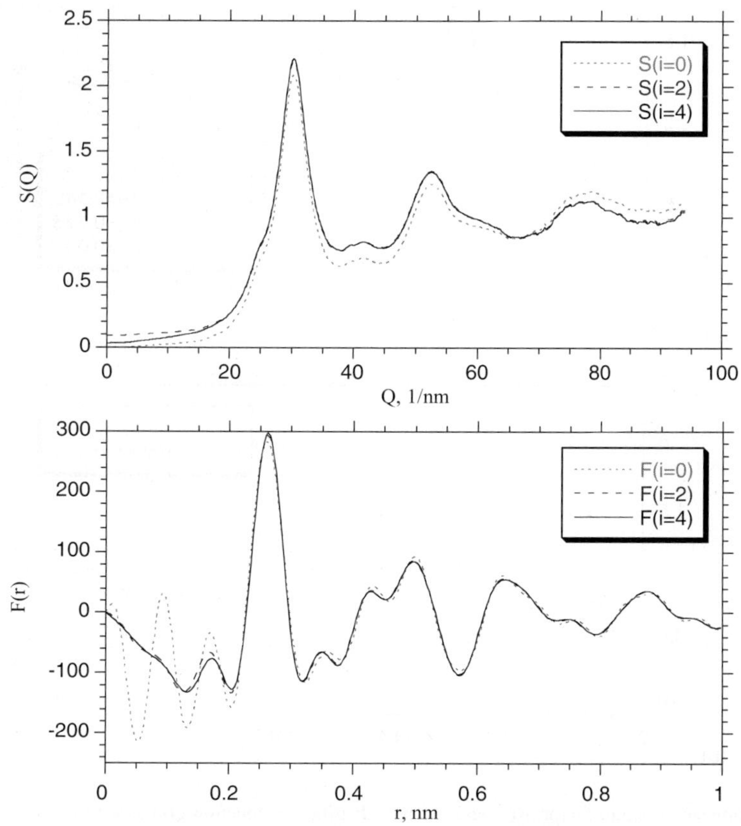

Fig. 3. Iteration evolutions of the structure factor $S_{(i)}(Q)$ and the distribution function $F_{(i)}(r)$ for amorphous iron at ambient pressure. $F_{(0)}(r)$ shows oscillations in the small-r region due to errors in $S(Q)$. Note that these oscillations quickly converge after a few iterations.

3.2. Optimization procedure

Since the atoms do not approach each other within the atomic core diameter, $n(r)$ should be zero in this region. Therefore, from Eq. (5) we have

$$F(r) = -4\pi r n_0 \quad (\text{small } r) \tag{7}$$

From this Kaplow et al. (1965) and Eggert et al. (2002) proposed a refinement procedure for $S(Q)$ and $g(r)$. The refinement procedure is based on the observation that errors in $F(r)$ due to uncertainties in $S(Q)$ are dominant in the small-r region. The relation (Eq. (7)) can be used as an iterative feedback to determine a corrected value for $S(Q)$. Following their approaches, we established an iterative procedure for analyzing the amorphous iron data. According to Eqs. (5) and (7), we have

$$F_{(i)}(r) = \frac{2}{\pi}\int_0^{Q_{\max}} Q[S_{(i)}(Q) - 1]\sin(rQ)\,\mathrm{d}Q \tag{8}$$

$$\Delta F_{(i)}(r) = F_{(i)}(r) + 4\pi r n_0 \quad (r < r_{\min}) \tag{9}$$

where $r_{\min}$ is a value close to the atomic radius, i denotes iteration number. An improved $S_{(i+1)}(Q)$ can then be obtained by applying the Fourier transformation of $\Delta F_{(i)}(r)$:

$$S_{(i+1)}(Q) = S_{(i)}(Q) - \frac{1}{Q}\int_0^{r_{\min}} \Delta F_{(i)}(r)\sin(rQ)\,\mathrm{d}r \tag{10}$$

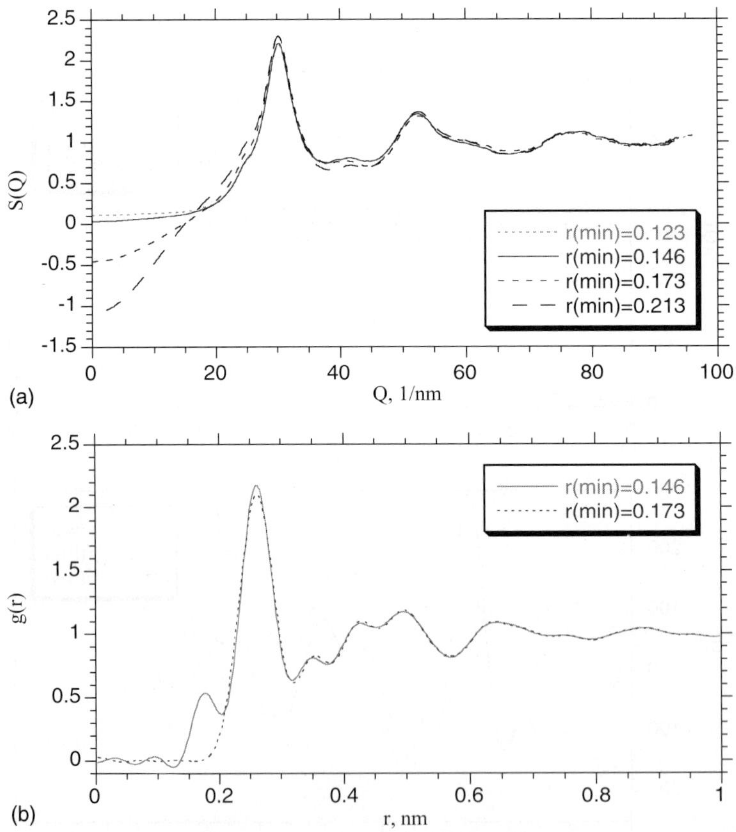

Fig. 4. Behavior of the optimized structure factor $S(Q)$ and the pair distribution function $g(r)$ as a function of $r_{\min}$ (Eqs. (9) and (10)) for amorphous iron at ambient pressure. The small peak below the dominant peak in $g(r)$ may be removed by using values of $r_{\min}$ larger than 0.176 nm. However, the resultant $S(Q)$ becomes unphysical. Reasonable $S(Q)$ is obtained with $r_{\min}$ of 0.146 nm, a number close to atomic core radius. Numbers in the figure are in units of nm.

This process is found to be effective and convergence is rapid as shown in Fig. 3.

The above process requires the value of atomic density n_0. In this study, n_0 at high pressures are from a separate work using the X-ray absorption method. This method has been successfully used for determinations of densities of liquids in a DAC (Shen et al., 2002a). Eggert et al. (2002) estimated n_0 by minimizing a chi-squared misfit: $\chi^2 \equiv \int_0^{r_{\min}} [\Delta F_{(i)}(r)]^2\,\mathrm{d}r$, for high pressure fluids (argon and water), where i is the iteration number. This could be significant because it provides a method for measuring equations of state for amorphous materials from determinations of the structure factor. However, we found that the values of χ^2 is extremely sensitive to $r_{\min}$ in Eq. (10). The determined n_0 in this way could be unphysical without strong constraints on $r_{\min}$.

3.3. Effects of $r_{\min}$ and $Q_{\max}$

While spurious oscillations in the distribution function $F(r)$ at small values of r can be effectively removed by applying the above optimization procedure (Fig. 3), it is necessary to perform reliability checks for the corrected $S(Q)$. We make use of two criteria: (1) in the small-Q region, $\lim_{Q\to 0} S(Q) = S(0) = n_0 k_B T\beta$, where k_B is the Boltzmann constant, β the isothermal compressibility and T the absolute temperature; (2) the corrected $S(Q)$ must be compared with the original data within experimental errors.

For example, as shown in Fig. 4, a peak at around 0.16 nm below the first peak in $g(r)$ can be removed by having $r_{\min}$ larger than 0.176 nm. However, the resultant $S(Q)$ with $r_{\min}$ larger than 0.176 nm becomes negative at small-Q, in conflict with the criterion (1) mentioned above. The optimum $S(Q)$ is then found at

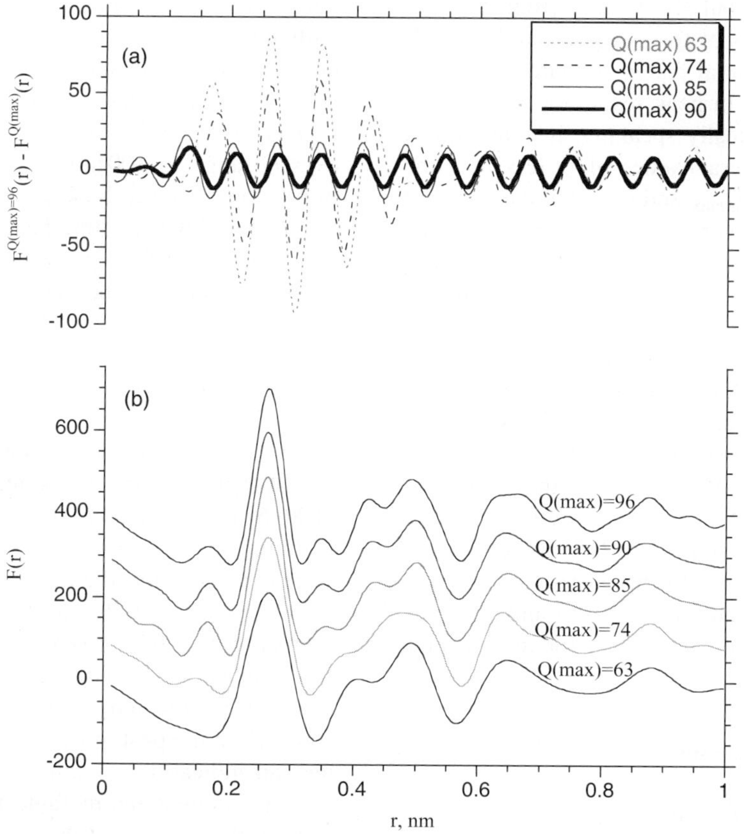

Fig. 5. Truncation effect of $Q_{\max}$ on the distribution function $F(r)$. The differences in $F(r)$ are shown in (a). Note that the large oscillations are in the vicinity of the first dominant peak. Numbers in the figure are in units of nm^{-1}.

r_{min} 0.146 nm, a number close to atomic core radius, leaving a small peak in $g(r)$ at around 0.16 nm which might arise from the truncation effect in Fourier transformation as discussed below.

The truncation of the experimental data at Q_{max} is a limiting factor to derive accurate information in real space. This is especially true for high pressure studies because of the limited scattering angle accessible with high pressure instruments. The truncation could lead to ghost peaks in $F(r)$, and their positions are functions of Q_{max}. For example, based on Waseda (1980), the ghost maxima are at distances about $\pm(5/4)(2\pi/Q_{max}), \pm(9/4)(2\pi/Q_{max}), \ldots$ from the main peak position. As shown in Fig. 5b, the distribution functions in real space are plotted from the data with different values of Q_{max}. With increasing values of Q_{max}, two sub-peaks beside the dominant peak shift close to the main peak, and their positions are close to the ghost maxima. Thus, these two sub-peaks are interpreted as not real but ghost peaks from the truncation effect. At values of Q_{max} larger than 85 nm^{-1}, we found that the general pattern remains the same, with a slight increase in peak resolution as value of Q_{max} increases. Practically speaking, it is helpful to terminate $S(Q)$ at several values of Q less than the experimental limit of Q_{max} and to check the resultant effects on $F(r)$. However, it is still difficult to state the accuracy of the obtained distribution functions. Large Q-range coverage will clearly help on this matter.

The differences in $F(r)$ with various Q_{max} values, the relative to that with Q_{max} of 96 nm^{-1}, are plotted in Fig. 5a. The resultant error results in oscillations with an approximate period of $2\pi/Q_{max}$. It is noted that the greatest oscillation occurs in the vicinity of the first peak, where it holds the most critical information (Fig. 5a). Again it shows that large Q coverage is important in experiments. In the present study, Q_{max} value of 96 nm^{-1} is used for the data at ambient pressure and Q_{max} value of 86.6 nm^{-1} for data at high pressures.

4. Results and discussions

4.1. Data at ambient pressure

The prepared sample is very sensitive to oxygen, to a level of only a few ppm (Suslick et al., 1991). For high pressure measurements, the sample was well sealed by two diamond anvils and free from oxygen contamination. Interestingly, in our experiments after compression to a maximum pressure of 67 GPa, the sample was removed from the DAC in air and the X-ray scattering pattern was found to be consistent with those for the sample in the cell and also with the neutron scattering data (Bellissent et al., 1993). This indicates that the recovered sample is still amorphous iron, providing X-ray scattering data at ambient pressure. The oxidation resistance could be due to high compression under ultra high pressures for this initially porous sample. After compression in the DAC, the compact sample has two surfaces as shiny as diamond anvils, which may retard the sample oxidation. This is the first time that such sample has been preserved in air, and it is achieved with high pressure compression.

The recorded data were analyzed with the procedures mentioned above. Since the data were collected with the sample outside the DAC, $a(Q)$ and b in Eq. (1) are both unity. The sample's self-absorption effect is neglected in this study because the sample thickness is only on the order of 10 μm. Change in absorption over Q-range to 100 nm^{-1} is less than 0.1% for an iron sample with a thickness of 10 μm at 37.44 keV. Fig. 6 shows the structure factor and the pair distribution function of the amorphous iron at ambient pressure, together with the data of liquid iron at 1833 K (Waseda and Suzuki, 1970) for comparison.

4.2. High pressure data

The structure factor and the pair distribution function at high pressures are summarized in Fig. 7. Q_{max} of 86.6 nm^{-1} is used to derive $g(r)$ for all pressures, including that at ambient pressure. It is found that with increasing pressure the peak height of $S(Q)$ (Fig. 7a) remains essentially the same, although the peak position shifts to a higher Q value. Similar features were found in $g(r)$ in real space. The peak height in $g(r)$ does not significantly change with increase of pressure, while the peak position shifts to a smaller r values, an indication of compression. Peak positions in $g(r)$ are summarized in Table 1.

With the concept of the nearest coordination number (CN), we may describe the picture of the nearest neighbor atoms in a quantitative way. There is no

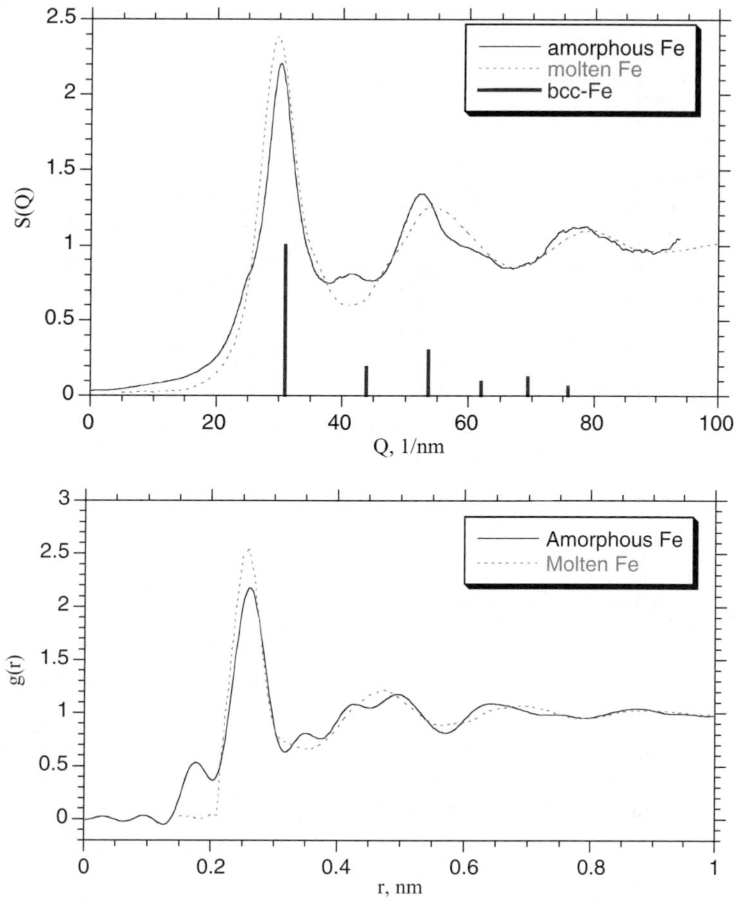

Fig. 6. The structure factor $S(Q)$ and the pair distribution function $g(r)$ of amorphous iron at ambient pressure. Data for liquid iron (Waseda and Suzuki, 1970) and crystalline b.c.c.-Fe are shown for comparison.

unique method of estimating the coordination number. In this study, it is assumed that the coordination shell is symmetric about a radius that defines the maximum in the $4\pi r^2 g(r)$ curve, the so-called radial distribution function. Thus the coordination number is determined from the integration:

$$CN = 2n_0 \int_{r_0}^{r_{max}} 4\pi r^2 g(r)\, dr \qquad (11)$$

where r_0 is the left-hand edge of the first peak and r_{max} the peak value of r in the $4\pi r^2 g(r)$ curve. The calculated coordination numbers as a function of pressure are plotted in Fig. 8. No clear coordination number change is found with increase of pressure within experimental uncertainties, a result consistent with the observation that the peak height and the peak width in $g(r)$ do not change significantly with pressure.

4.3. The structure of amorphous iron

As shown in Fig. 6, the general structural feature of the amorphous state is similar to that of the liquid state except for a few weak peaks. In the liquid state, the amplitude of the atomic vibration is large, leading to a large uncertainty in the position of the lattice point, thus resulting in an averaged smooth distribution in both $S(Q)$ and $g(r)$ (Fig. 6). On the other hand, the small amplitude of vibration of atoms in the amorphous state contributes to the construction of a more fixed atomic arrangement, causing some small peaks in $S(Q)$ and $g(r)$ (Fig. 6).

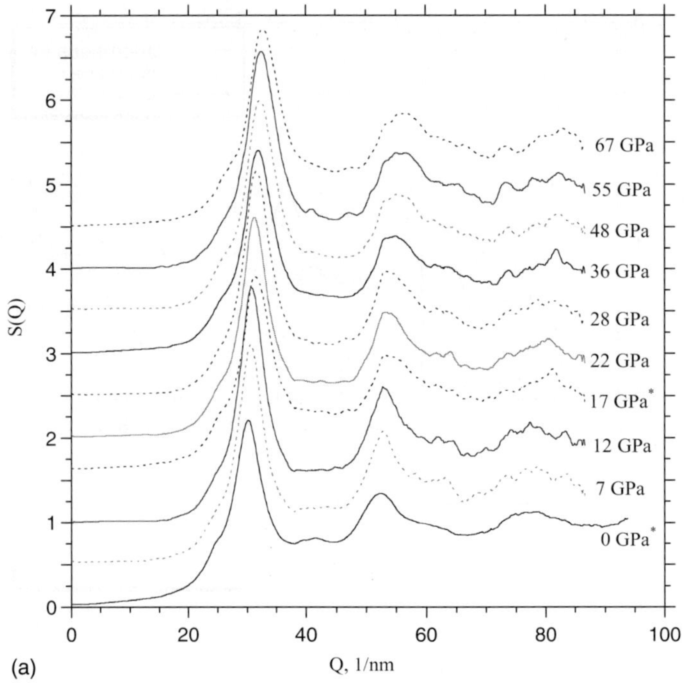

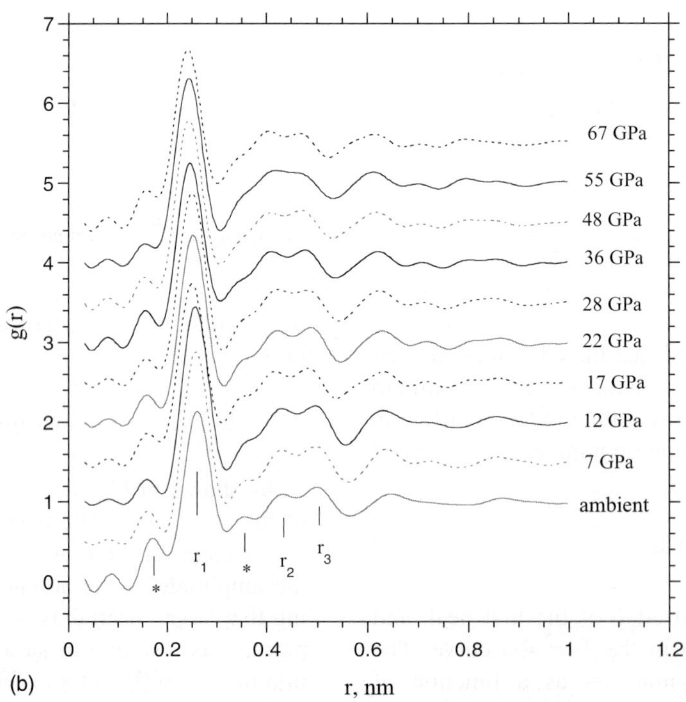

Fig. 7. (a) The structure factor $S(Q)$ at high pressures. (b) The pair distribution function $g(r)$ at high pressures. An offset of 0.5 in both $S(Q)$ and $g(r)$ is applied between pressures for clarity. Peaks marked with (∗) may arise from the truncation effect (see text). No clear change in shape or intensity is observed for the dominant first peak in $S(Q)$ and $g(r)$, while the peak positions vary with pressure.

Table 1
Peak positions and calculated ratios of distances in $g(r)$ at various pressures

P (GPa)	CN atoms	r_1 (nm)	r_2 (nm)	r_3 (nm)	r_2/r_1	r_3/r_1
0[a]	9.3(4)	0.2644(13)	0.428(3)	0.500(4)	1.63	1.89
		0.256[b]	0.434[b]	0.496[b]	1.69[b]	1.93[b]
	10.6[c]	0.258[c]	0.48[c]		1.67[d,e]	1.96[d,e]
7	10.1(5)	0.2615(15)	0.430(3)	0.500(4)	1.64	1.91
12	11.0(5)	0.2616(15)	0.429(5)	0.499(5)	1.64	1.91
17[a]	9.9(5)	0.2547(15)	0.417(3)	0.485(3)	1.64	1.90
22	10.6(5)	0.2570(15)	0.421(3)	0.489(3)	1.64	1.90
28	10.9(5)	0.2549(15)	0.419(3)	0.483(3)	1.64	1.89
36	10.0(5)	0.2507(15)	0.408(3)	0.480(3)	1.63	1.91
48	10.5(5)	0.2489(15)	0.404(3)	0.474(3)	1.62	1.90
55	11.1(5)	0.2492(15)	0.414(5)	0.476(4)	1.66	1.91
67	10.5(5)	0.2462(15)	0.404(3)	0.469(3)	1.64	1.90

[a] Data collected on decreasing pressure.
[b] Bellissent et al. (1993).
[c] Waseda and Suzuki (1970).
[d] Leung and Wright (1974).
[e] Ichikawa (1973).

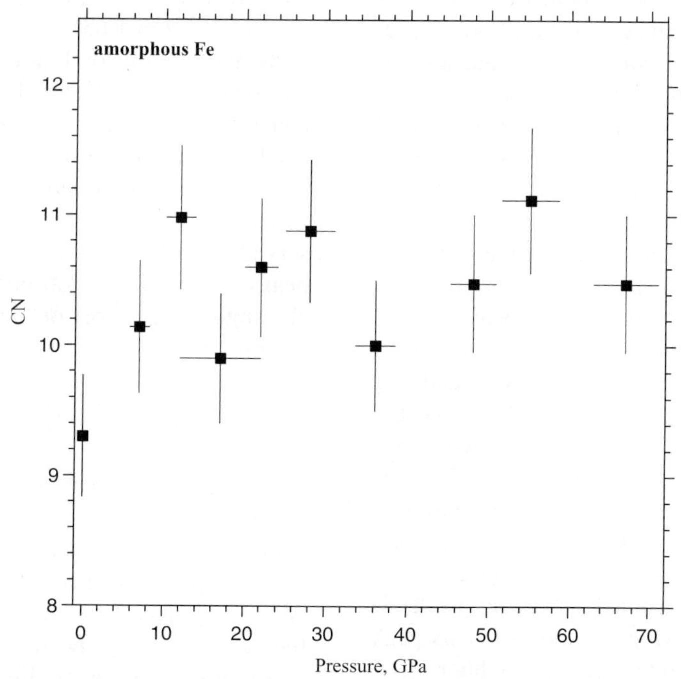

Fig. 8. The nearest neighbor coordination (CN) number as a function of pressure, determined according to Eq. (11). No pressure dependence is observed within experimental uncertainties, with an average number at 10.5(5). Errors in CN are estimated to be 5%. Errors in pressures are from multiple ruby fluorescence measurements in the region of 30 μm around the sample.

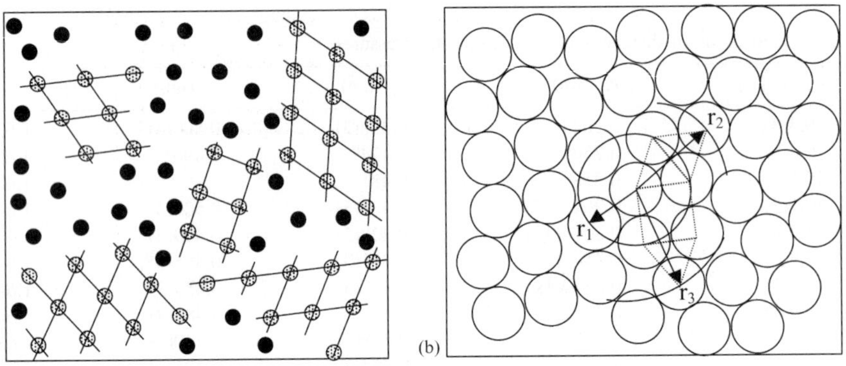

Fig. 9. Models for the amorphous structure: (a) a structure consisting of microcrystalline fragments randomly orientated and distributed; (b) a dense random-packing (DRP) model. The DRP model is favored from this study. r_1, r_2, and r_3 shown in the figure are assigned to peaks in $g(r)$ shown in Fig. 7b.

There are two models in the literature for amorphous structures, the microcrystalline disorder model and the topological disorder model (Waseda, 1980) (Fig. 9). Studies on thin films of amorphous iron showed that the later type describes the amorphous structure well (Ichikawa, 1973; Leung and Wright, 1974). Ichikawa (1973) suggested a dense random-packing model (DRP), in which the structure is built by packing of essentially irregular assemblages of basic polyhedra, such as tetrahedra and octahedra, without vacancy and long-range order. From this study, our data also support the concept of a truly amorphous phase devoid of crystalline order based on the following findings. (1) No noticeable change in shape is observed in $S(Q)$ or $g(r)$ with increase of pressure (Fig. 7). The ratios of the distances in Table 1 are essentially constant as a function of pressure. These observations disfavor the microcrystalline disorder model because crystalline iron undergoes a phase transition at around 12 GPa (b.c.c. to h.c.p. phase transition), and the phase change would affect the pattern shape and the radial distances in $g(r)$ according to the calculations of Ichikawa (1973). (2) As it can be seen in Table 1, the ratios of the distances between our data at ambient pressure and those from thin films result are in good agreement, suggesting an analogous structure.

The dominant peak (denoted as the r_1 peak) is found at around 0.25 nm (Figs. 6 and 7b). This peak can clearly be attributed to the nearest neighbors. The calculated coordination numbers (Fig. 8) of about 10.5(5) indicate that the amorphous state is densely packed.

According to the DRP model, the structure is constructed by serially and densely piling up regular or slightly distorted polyhedra (tetrahedra and octahedra) around a seed cluster (Ichikawa, 1973). The ratios of the distances of r_2/r_1 and r_3/r_1 for the amorphous iron in this study are found to be in the vicinity of 1.64 and 1.91, respectively (Table 1). The ratio of r_2/r_1 is very close to the c/a ratio in closed packed hexagonal structure ($c/a = 1.63$). The peak of r_2 is thus assigned to the basic tetrahedral unit. The close comparison to the ideal value indicates that the tetrahedra are regular by average. As shown in Fig. 9b, by combining octahedra and tetrahedra, the distance r_3 is $\sim 1.90 r_1$. This combination produces a ratio close to the measured value of r_3/r_1. Therefore, the basic polyhedra observed consist of tetrahedra and octahedra. The broad peaks in $g(r)$, always observed in amorphous materials, suggest that most of these polyhedra are slightly deformed.

4.4. High pressure compression

One feature of our high pressure data is that the dominant peak, both in $S(Q)$ and $g(r)$, does not significantly change with increase of pressure, except that the peak position shifts to a higher Q value and to a smaller r value, respectively. This observation is further confirmed by the derived coordination number and ratios of distances (Fig. 8, Table 1). Essentially no pressure dependence is observed for these numbers within experimental uncertainty. This implies that the contraction of the amorphous iron is isotropic under

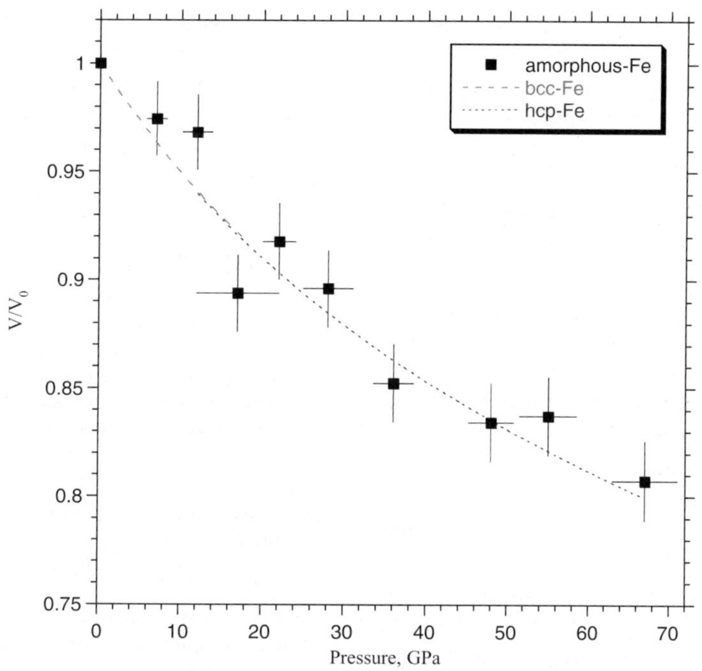

Fig. 10. High pressure volume compression of amorphous iron measured by the peak shift of the first peak in $g(r)$ (Fig. 7b). The validity of using the peak shifts to represent the volume changes is based on constant packing density as a function of pressure (Eq. (12)). The overall compression is comparable to the behaviors of b.c.c.-Fe and h.c.p.-Fe (Jephcoat et al., 1986; Mao et al., 1990).

pressure. By adopting the concept of the packing density (Ashcroft and Lekner, 1965)

$$\eta \equiv n_0 \frac{4}{3}\pi \left(\frac{1}{2}\sigma\right)^3 = \frac{1}{6} n_0 \pi \sigma^3 \qquad (12)$$

where σ is the hard-sphere diameter, isotropic contraction means a constant packing density η with pressure. Indeed, by applying the hard-sphere model, coupled with the Percus–Yevick equation (Ashcroft and Lekner, 1965), to fit the first peak in $S(Q)$, we obtained η of 0.43(1) for all pressures up to 67 GPa.

From Eq. (12), isotropic contraction makes it possible to estimate densities of amorphous iron at high pressures according to the peak positions of r_1. Although the absolute values of r_1 might be different from the hard-sphere diameter σ in Eq. (12), the change of r_1 with pressure represents the decrease of σ. Fig. 10 shows the plot of $V/V_0[= r_1^3/(r_1^3)_{\mathrm{ambient}}]$ as a function of pressure compared with the compression behaviors of b.c.c.-Fe (Jephcoat et al., 1986) and h.c.p.-Fe (Mao et al., 1990). Error bars in volume are from uncertainties in the peak fitting with a Gaussian function. Pressure error bars are from the gradients within 30 μm around the sample center. The overall compression is comparable to the behaviors of b.c.c.-Fe and h.c.p.-Fe. The data appear to be too scattered to fit with an equation of state. The scattering of data could be due to the stress effect arising from the pressure gradient, because no pressure medium was used in the present study to avoid unwanted scattering from other materials. The broad peak in $g(r)$ also limits the precision in determining the peak position.

It should be noted that the coordination number has been found to increase with pressure for many materials (Brown and Barnett, 1972; Shen et al., 2002b; Urakawa et al., 1999), especially for those with open structures (Funamori and Tsuji, 2002). In that case, caution should be taken in using Eq. (12) for estimating densities at high pressures (Brown and Barnett, 1972).

5. Conclusions

We have measured the structure of an amorphous iron sample of high purity at room temperature and at pressures up to 67 GPa. A modified diamond anvil cell is used, allowing measurements of the X-ray scattering to a maximum momentum Q_{max} of 86.6 nm^{-1} with a monochromatic beam at a wavelength of 0.3311 Å. It is demonstrated that the weak scattering from the amorphous sample can be measured with the DAC up to ultra high pressures. By following the approaches of Eggert et al. (2002) and Kaplow et al. (1965), an optimization procedure is used to obtain accurate structural data of amorphous materials at high pressures. It is shown that spurious oscillations in $g(r)$ at the small-r region can be effectively corrected by applying this optimization procedure. It is also noted that truncation of the experimental data at Q_{max} is a limiting factor. Large Q coverage in experiments is desirable. Insufficient Q-range can cause inaccurate data in the pair distribution function $g(r)$, especially in the critical region around the first peak (Fig. 5).

The structure of amorphous iron is similar to that of liquid iron. A few peaks in $g(r)$ for the amorphous state are attributed to the construction of a more fixed atomic arrangement than in liquid. It is found that the dense random-packing model (Ichikawa, 1973) is favored to interpret our observations. From both the structure factor $S(Q)$ and the pair distribution function $g(r)$, the dominant peak does not significantly change in shape or intensity with increase of pressure, except that the peak position shifts to a higher Q value and to a smaller r value, respectively. This observation is consistent with the essentially zero pressure dependence of the coordination number (Fig. 8) and ratios of distances listed in Table 1, resulting in a conclusion of structurally isotropic contraction with pressure for amorphous iron. The isotropic contraction allows us to estimate the volume changes as a function of pressure from the positions of the first peak in the pair distribution function $g(r)$. The compression behavior at room temperature is found to be similar to those of b.c.c.-Fe and h.c.p.-Fe (Jephcoat et al., 1986; Mao et al., 1990) (Fig. 10).

Acknowledgements

Thanks are due to Jennifer Jackson, Jason Nicholas, and Stas Sinogeikin for the help during sample loading, Peter Eng in X-ray scattering experiments. Comments by Jennifer Jackson, Satoru Urakawa, and Nobumasa Funamori improved the manuscript. This work is supported by NSF-EAR 0001149 and 0229987. The GSECARS sector is supported by the National Science Foundation (Earth Sciences Instrumentation and Facilities Program) and Department of Energy-Basic Energy Sciences (Geoscience Program).

References

Ahrens, T.J., Johnson, M.L., 1995. Shock wave data for minerals. In: Ahrens, T.J. (Ed.), A Handbook of Physical Constants: Mineral Physics and Crystallography. American Geophysical Union, Washington, DC, pp. 143–184.

Alfe, D., Kresse, G., Gillan, M.J., 2000. Structure and dynamics of liquid iron at the pressures of the Earth's core conditions. Phys. Rev. B 61, 132–142.

Alfe, D., Price, G.D., Gillan, M.J., 2002. Iron under Earth's core conditions: Liquid-state thermodynamics and high-pressure melting curve from ab initio calculations. Phys. Rev. B 65, 165118.

Anderson, W.W., Ahrens, T.J., 1994. An equation of state for liquid iron and implications for the Earth's core. Rev. Geophys. 99, 4273–4284.

Ascarelli, P., 1968. Velocity of sound and compressibility in liquid metals. Phys. Rev. 173, 271–274.

Ashcroft, N.W., Lekner, J., 1965. Structure and resistivity of liquid metals. Phys. Rev. 145, 83–90.

Balyuzi, H.H.M., 1975. Analytic approximations to incoherent scattered X-ray intensities. Acta Cryst. A 31, 600–602.

Bellissent, R., Galli, G., Grinstaff, M.W., Migliardo, P., Suslick, K.S., 1993. Neutron diffraction on amorphous iron powder. Phys. Rev. B 48, 15797–15800.

Birch, F., 1964. Density and composition of the mantle and core. J. Geophys. Res. 69, 4377–4388.

Boehler, R., 1993. Temperatures in the Earth's core from melting-point measurements of iron at high static pressures. Nature 363, 534–536.

Brazhkin, V.V., Lyapin, A.G., 2000. Universal viscosity growth in metallic melts at megabar pressures: the vitreous state of the Earth's inner core. Phys. Uspekhi 43, 493–508.

Brown, J.M., Fritz, J.N., Hixson, R.S., 2000. Hugoniot data for iron. J. Appl. Phys. 88, 5496–5498.

Brown, K.H., Barnett, J.D., 1972. X-ray diffraction studies on liquids at very high pressures along the melting curve. II. Sodium. J. Chem. Phys. 57, 2016–2021.

Dubrovinsky, L.S., Saxena, S.K., Tutti, F., Rekhi, S., Lebehan, T., 2000. In situ X-ray study of thermal expansion and phase transition of iron at multimegabar pressure. Phys. Rev. Lett. 84, 1720–1723.

Eggert, J.H., Weck, G., Loubeyre, P., Mezouar, M., 2002. Quantitative structure factor and density measurements of high-pressure fluids in diamond anvil cells by X-ray diffraction: argon and water. Phys. Rev. B 65, 174105.

Funamori, N., Tsuji, K., 2002. Pressure-induced structural change of liquid silicon. Phys. Rev. Lett. 88, 255508.

Hammersley, A.P., Sevsson, S.O., Hanfland, M., Fitch, A.N., Häusermann, D., 1996. Two-dimensional detector software: from real detector to idealised image or two-theta scan. High Pressure Res. 14, 235–248.

Hasegawa, M., Watabe, M., 1972. Theory of compressibility of simple liquid metals. J. Phys. Soc. Jpn. 32, 14–28.

Hasegawa, M., Watabe, M., 1974. Theory of thermodynamic properties of liquid metals. J. Phys. Soc. Jpn. 36, 1510–1515.

Hausleitner, C., Hafner, J., 1989. A theoretical study of the melting curve of iron to very high pressure. J. Phys.: Condensed Matter 1, 5243–5352.

Hemley, R.J., Mao, H.K., 2001. In situ studies of iron under pressure: new windows on the Earth's core. Int. Geol. Rev. 43, 1–30.

Ichikawa, T., 1973. Electron diffraction study of the local atomic arrangement in amorphous iron and nickel films. Phys. Stat. Solidi A 19, 707–716.

Jephcoat, A.P., Mao, H.K., Bell, P.M., 1986. Static compression of iron to 78 GPa with rare gas solids as pressure-transmitting media. J. Geophys. Res. 91, 4677–4684.

Kaplow, R., Strong, S.L., Averbach, B.L., 1965. Radial density functions for liquid mercury and lead. Phys. Rev. 138, A1336–A1345.

Katayama, Y., et al., 2000. A first order liquid–liquid phase transition in phosphorus. Nature 403, 170–173.

Krogh-Moe, J., 1956. A method for converting experimental X-ray intensities to an absolute scale. Acta Cryst. 9, 951–953.

Kruger, M.B., Meade, C., 1997. High-pressure structural study of GeI_4. Phys. Rev. B 55, 1–3.

Kubo, A., et al., 2003. Phase equilibrium study of iron using sintered diamond (SD) anvils: absence of beta phase. PEPI, this volume.

Leung, P.K., Wright, J.G., 1974. Structural investigations of amorphous transition element films. II. Chromium, iron, manganese and nickel. Phil. Mag. 30, 995–1008.

Mao, H.K., Hemley, R.J., 1996. Experimental studies of Earth's deep interior: accuracy and versatility of diamond-anvil cells. Phil. Trans. Roy. Soc. Lond. A 354, 1315–1332.

Mao, H.K., Shen, G., Hemley, R.J., Duffy, T.S., 1998. X-ray diffraction with a double hot plate laser heated diamond cell. In: Manghnani, M.H., Yagi, T. (Eds.), Properties of Earth and Planetary Materials. AGU, Washington, DC, pp. 27–34.

Mao, H.K., Wu, Y., Chen, L.C., Shu, J.F., Jephcoat, A.P., 1990. Static compression of iron to 300 GPa and $Fe_{0.8}Ni_{0.2}$ alloy to 260 GPa: implications for composition of the core. J. Geophys. Res. 95, 21737–21742.

March, N.H., 1990. Liquid Metals: Concepts and Theory. Cambridge University Press, Cambridge, 492 pp.

Meade, C., Hemley, R.J., Mao, H.K., 1992. High pressure X-ray diffraction of SiO_2 glass. Phys. Rev. Lett. 69, 1387–1390.

Mezouar, M., et al., 2002. Multichannel collimator for structural investigation of liquids and amorphous materials at high pressures and temperatures. Rev. Sci. Instrum. 73, 3570–3574.

Norman, N., 1957. The Fourier transform method for normalizing intensities. Acta Cryst. 10, 370–373.

Sanloup, C., et al., 2000. Structural changes in liquid Fe at high pressures and high temperatures from synchrotron X-ray diffraction. Europhys. Lett. 52, 151–157.

Shen, G., Mao, H.K., Hemley, R.J., Duffy, T.S., Rivers, M.L., 1998. Melting and crystal structure of iron at high pressures. Geophys. Res. Lett. 25, 373–376.

Shen, G., Rivers, M.L., Wang, Y., Sutton, S.J., 2001. A laser heated diamond cell system at the advanced photon source for in situ X-ray measurements at high pressure and temperature. Rev. Sci. Instrum. 72, 1273–1282.

Shen, G., Sata, N., Newville, M., Rivers, M.L., Sutton, S.R., 2002a. Molar volumes of liquids measured in a diamond anvil cell. Appl. Phys. Lett. 81, 1411–1413.

Shen, G., et al., 2002b. Melting studies of indium: determination of structure and density of melts at high pressures and high temperatures. J. Phys.: Condensed Matter 14, 1–9.

Stixrude, L., Brown, J.M., 1998. The Earth's core. In: Hemley, R.J. (Ed.), Ultrahigh-pressure Mineralogy. Mineralogical Society of America, Washington, DC, pp. 261–282.

Suslick, K.S., Choet, S.B., Cichowlas, A.A., Grinstaff, M.W., 1991. Sonochemical synthesis of amorphous iron. Nature 353, 414–416.

Thakkar, J., Chapman, D.C., 1975. A new analytic approximation to atomic incoherent X-ray scattering intensities. Acta Cryst. A 31, 391–392.

Trigger, S.A., Bobrov, V.B., Vlasov, Y.P., 1994. Liquid metal compressibility theory using the OCP model as a reference system. Physica B 193, 291–294.

Tsuji, K., Yaoita, K., Imai, M., Shimomura, O., Kikegawa, T., 1989. Measurements of X-ray diffraction for liquid metals under high pressure. Rev. Sci. Instrum. 60, 2425–2428.

Urakawa, S., Igawa, N., Shimomura, O., Ohno, H., 1999. High pressure X-ray diffraction study on the structure of NaCl melt using synchrotron radiation. Am. Mineral. 84, 341–344.

Waseda, Y., 1980. The Structure of Non-crystalline Materials. McGraw-Hill, International Book Co, New York.

Waseda, Y., Suzuki, K., 1970. Atomic distribution and magnetic moment in liquid iron by neutron diffraction. Phys. Stat. Solidi 39, 669–678.

Yaoita, K., Katayama, Y., Tsuji, K., Kikegawa, T., Shimomura, O., 1997. Angle-dispersive diffraction measurement system for high-pressure experiments using a multichannel collimator. Rev. Sci. Instrum. 68, 2106–2110.

Zhang, Y.G., Guo, G.J., 2000. Molecular dynamics calculation of the bulk viscosity of liquid iron–nickel alloy and the mechanism for the bulk attenuation of seismic waves in the Earth's outer core. Phys. Earth Planet. Inter. 122, 289–298.

Zhang, Y.G., Guo, G.J., Nie, G.Z., 2000. A molecular dynamics study of the bulk and shear viscosity of liquid iron using embedded-atom potential. Phys. Chem. Mineral. 27, 164–169.

7. Technical developments

Available online at www.sciencedirect.com
SCIENCE DIRECT
ELSEVIER
Physics of the Earth and Planetary Interiors 143-144 (2004) 497–506
PHYSICS OF THE EARTH AND PLANETARY INTERIORS
www.elsevier.com/locate/pepi

A large-volume high-pressure and high-temperature apparatus for in situ X-ray observation, '*SPEED-Mk.II*' ☆

Tomoo Katsura [a,*], Ken-ichi Funakoshi [b], Atsushi Kubo [a], Norimasa Nishiyama [c], Yoshinori Tange [d], Yu-ichiro Sueda [c], Tomoaki Kubo [e], Wataru Utsumi [f]

[a] *Institute for Study of the Earth's Interior, Okayama University, Misasa, Tottori-ken 682-0193, Japan*
[b] *Japan Synchrotron Radiation Research Institute, Kouto 1-1-1, Mikazuki-cho, Sayo-gun, Hyogo 678-5198, Japan*
[c] *Geodynamic Research Center, Ehime University, Bunkyo-cho 2-5, Matsuyama 790-8577, Japan*
[d] *Department of Earth and Planetary Sciences, Tokyo Institute of Technology, 2-12-1 Ookayama, Meguro-ku 152-8511, Japan*
[e] *Institute of Mineralogy, Petrology and Economic Geology, Tohoku University, Sendai 980-8578, Japan*
[f] *Japan Atomic Energy Research Institute, Kouto, Mikazuki-cho, Sayo-gun 679-5148, Japan*

Received 20 January 2003; received in revised form 1 July 2003; accepted 17 July 2003

Abstract

SPEED-Mk.II, the second KAWAI-type high *P–T* apparatus for in situ X-ray observation that was installed at the bending magnet beam line at SPring-8, is described. The guide block system was designed so that the change of the relative dimension of the cubic compression space with press load can be minimized by repeated adjustments. The hydraulic system was designed so as to enable smooth compression and decompression. These precise controls should be advantageous for high-pressure generation, especially when sintered diamond (SD) anvils are used. An oscillation system was equipped for the first time in a large volume press, making it possible to obtain high-quality diffraction patterns even when the number of sample grains is limited. The use of the oscillation system also reduces errors in pressure determination that may be caused by insufficient averaging of diffraction angles over grains in a limited diffraction volume, because the oscillating grains should sweep through the 2θ range that is allowed by the finite widths of the optical slits.

Keywords: High-pressure and high-temperature experiment; In situ X-ray observation; KAWAI-type apparatus; Oscillation system

1. Introduction

A large-volume high-pressure and high-temperature apparatus for in situ X-ray observation of Earth materials, named *SPEED-1500*, was installed at the bending magnet beam line *BL04B1* at the third-generation synchrotron radiation facility *SPring-8* in 1997 (Utsumi et al., 1998). *SPEED-1500* is a KAWAI-type high-pressure apparatus[1] with a DIA-type guide block

☆ "New Developments in High Pressure Mineral Physics and Applications to the Earth's Interior," Special Issue of *Physics of the Earth and Planetary Interiors*, Guest Editors: D. Rubie, T.S. Duffy and E. Ohtani.

* Corresponding author. Tel.: +81-858-43-3754; fax: +81-858-43-2184.
E-mail address: tkatsura@misasa.okayama-u.ac.jp (T. Katsura).

[1] "KAWAI-type apparatus" is a general name for a two-stage multi-anvil apparatus that compresses an octahedral pressure medium with eight truncated cubes. The name was proposed by Yagi [2001] after its original developer, the late professor N. KAWAI. This type of apparatus has been referred to as "6–8 type apparatus," "MA8-type apparatus," "cubic-octahedral anvil press," "split-sphere apparatus," or "Walker-type apparatus" as well in the literature.

0031-9201/$ – see front matter © 2004 Elsevier B.V. All rights reserved.
doi:10.1016/j.pepi.2003.07.025

system (Shimomura et al., 1992) and a maximum press load of 15 MN. Since its installation, a number of important experimental results have been obtained (e.g. Irifune et al., 1998, 2000; Kubo et al., 2002; Ohtaka et al., 2001a,b; Ono et al., 2000, 2001; Hirose et al., 2001a,b; Terasaki et al., 2001; Urakawa et al., 2001, 2002; Suzuki et al., 2002; Katsura et al., 2003).

The combination of its large volume capacity and the high X-ray brilliance of BL04B1 at SPring-8 is the key to the great success of *SPEED-1500*. The KAWAI-type high-pressure apparatus, along with the diamond anvil cell, is one of the most popular high-pressure apparatuses, especially in geosciences. For routine experiments, press loads are often from 3 to 8 MN in the case of the guide blocks to compress an assembly of the second stage anvils in the (1 1 1) direction. High pressures up to 25 GPa can be generated in large volumes on the order of $\sim 10^{-8}$ to 10^{-5} m^3. Such large volumes make it possible to generate fairly homogeneous *P–T* conditions, and to conduct sophisticated experiments with a complex sample assembly. However, for conventional in situ X-ray diffraction (with energies below 100 keV), the large volume capacity of KAWAI-type apparatus also creates a serious problem, due to the absorption of X-rays by the pressure media. X-ray absorption is substantial with MgO and pyrophyllite, the often-adopted light-element materials, as pressure media and gasket, respectively. To alleviate this problem, amorphous boron bonded by epoxy (B-epoxy) is often used as an alternative pressure medium. However, B-epoxy is a very reactive material, especially with metals. Its decomposition leads to a large pressure drop during heating. Use of B-epoxy thus limits the design of high *P–T* experiments. The use of highly brilliant X-rays with energies up to 160 keV at BL04B1 of *SPEED-1500* eliminated the problem of X-ray absorption, without having to resort to the reactive B-epoxy.

Fig. 1. Photograph of the press of *SPEED-Mk.II*.

Despite the great success of *SPEED-1500*, further developments of high *P–T* in situ X-ray observation in a large-volume KAWAI-type apparatus are desirable to achieve a greater pressure range and to obtain high-quality diffraction data. *SPEED-1500* was originally designed for experiments using tungsten carbide (WC) anvils that limit the generated pressures to ca. 30 GPa. In order to extend the pressure range, experiments have also been conducted using sintered diamond (SD) anvils in *SPEED-1500* (e.g. Kubo et al., 2003). However, blow-outs frequently occur, rendering it impractical for routine experiments with expensive sintered diamond anvils (about $3000 per piece with an edge length of 14 mm). In the DIA-type guide block system (Shimomura et al., 1992) employed in *SPEED-1500*, the horizontal dimensions of the compression space generally increase relative to the vertical dimension with increasing press load because of deformation of the guide blocks. This asymmetrical compression will result in unbalanced forces on the gasket. In addition, the press load is controlled with a precision of only 20–30 kN so that the compression and decompression follow a saw-tooth-like path during automatic control. These features could be the causes for the frequent occurrence of blow-outs. A well-designed guide block system and load control are desirable to prevent frequent blow-outs.

In addition, grain growth is generally rapid at high temperatures in a KAWAI-type apparatus, possibly as a result of low differential stresses within the sample, as inferred from the relations between stress and grain sizes in dynamically crystallized olivine (Karato et al., 1980; Van der Val et al., 1993). Rapid grain growth is a problem for powder diffraction, because the incident X-rays have to be collimated to small dimensions, typically 0.1 mm × 0.05 mm, in which the number of grains available can be too small, causing the disappearance of many diffraction lines. In order to obtain diffraction patterns with a reasonable number of lines, a method to oscillate the sample is desirable.

A second KAWAI-type apparatus for in situ X-ray observation, *SPEED-Mk.II*, was designed and installed at the same beam line *BL04B1* in 2002 (Fig. 1). While its basic design follows that of *SPEED-1500*, several important improvements have been made to alleviate the above problems. In this paper, we describe this new system, and present some performance test results. Particular attention is paid to the oscillation system, because this is the first time such a system has been installed.

2. Description of *SPEED-Mk.II*

SPEED-Mk.II consists of a high-pressure vessel, a uniaxial press with a hydraulic system, a heating system, a five-axis stage, an incident slit system, and a horizontal goniometer with a solid-state detector (SSD, Fig. 2).

2.1. High-pressure vessel

Like *SPEED-1500*, the high-pressure vessel was designed on the basis of the DIA-type guide block system (cf. Osugi et al., 1964; Shimomura et al., 1992). Six first-stage anvils form a cubic space, compressing a KAWAI-type of eight second-stage cubic anvils, each with a truncated corner. For experiments with second-stage SD anvils, the first-stage anvils are made of WC with 27 mm edge lengths, and the second-stage SD anvils have an edge length of 14 mm. For experiments with second-stage WC anvils, the first-stage anvils are made of hardened steel (HS) with an edge length of 50 mm, and the second-stage WC anvils have an edge length of 26 mm. The incident and diffracted X-rays pass through gaps between the second stage anvils, and grooves are made on the side of the first-stage anvils. These grooves allow a maximum diffraction angle of 28°.

As mentioned above, the main problem of the guide block system of *SPEED-1500* is that the relative dimension of the cubic space compressed by the first-stage anvils changes with press load. The top and bottom of the guide block system supporting the first-stage anvils are weaker than the sides in *SPEED-1500*. In order to fix this problem, we design the *SPEED-Mk.II* so that the strength supporting the first-stage anvils on the top and bottom are adjustable. The top and bottom anvils are supported by separate columns that are mechanically independent of the guide blocks (Fig. 3). The guide blocks have a hole in the center to accommodate the supporting columns. The system is adjusted in the following way. If the compressed space is found to be flattened in the horizontal direction, several holes are opened in the

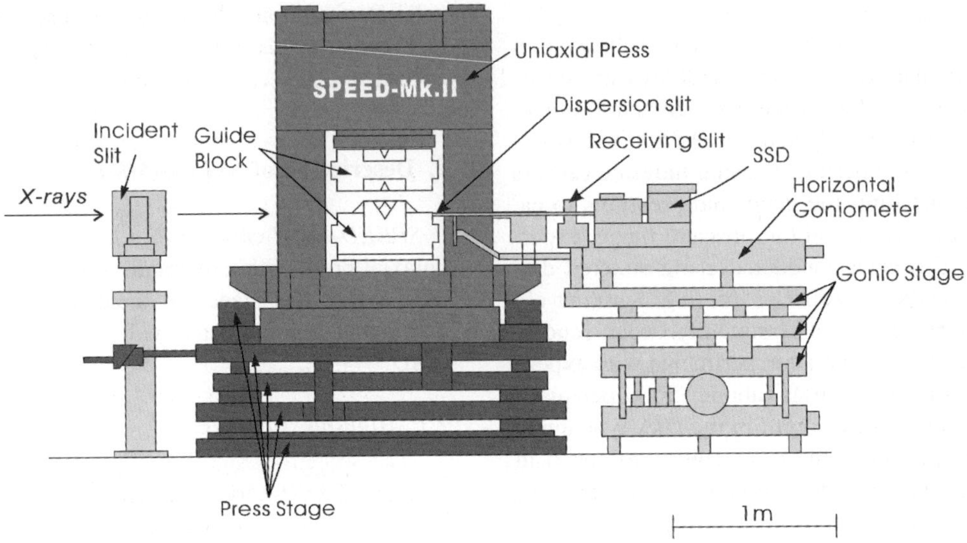

Fig. 2. Schematic drawing of *SPEED-Mk.II* in an X-ray hutch. The hydraulic and heating systems are placed outside the hutch.

supporting columns to lower its pseudo-Young's modulus. Whereas, if it is designed to be elongated in the vertical directions, the holes in the guide blocks are enlarged, and the columns are replaced with thicker

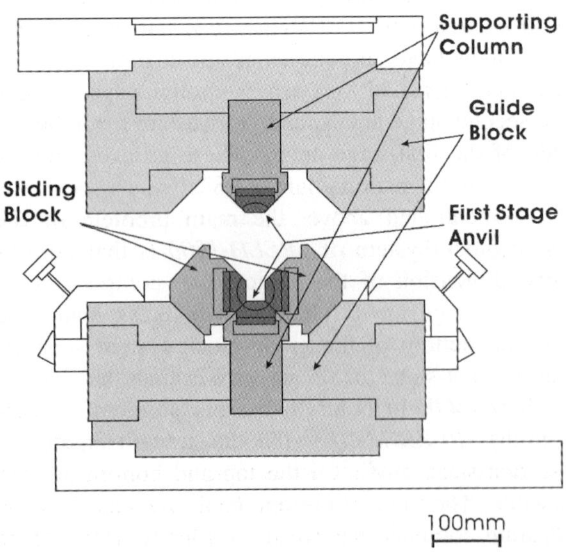

Fig. 3. Schematic drawing of the high-pressure vessel. The columns that support the upper and lower first-stage anvils (supporting columns) are separated from the main parts of the guide blocks so that the change of the relative dimension of the cubic compression space with press load can be minimized through adjustment.

ones. This procedure is repeated until the relative deformation of the compressed space is minimized.

2.2. *Uniaxial press*

The uniaxial press, with a maximum load of 15 MN, is controlled by a hydraulic system with a maximum oil pressure of 68 MPa. The hydraulic system digitally measures the oil pressure with a precision of 12.5 kPa, enabling press load control with a precision of 3 kN.

The hydraulic system is equipped with two separate inverter pumps for compression and decompression separately. The inverter pumps, controlled by a microprocessor, rotate in proportion to the difference between the set and real loads to allow smooth compression and decompression.

2.3. *Heating system*

Heating is conducted using a power supply whose voltage is proportional to signals from a function synthesizer. The AC currents generated by the power supply are sinusoidal with low distortion. This system produces minimum induction noise from the heater to the thermocouple. The function synthesizer is controlled by a computer so that an operator can control either the voltage, the power or the temperature.

2.4. Press stage

The uniaxial press with the guide blocks is fixed on a five-axis stage. The five axes, starting from the bottom, are *Y1*–κ–*X*–*Y2*–*Z*. The directions of the *X*-, *Y*- and *Z*-axes, respectively, are that of the X-ray incidence, that perpendicular to the incidence in the horizontal plane, and the vertical direction, respectively. The κ-axis is the rotation around the *Z*-axis, used for controlling sample oscillation. The *Y1*-axis allows one to set the rotation (κ) center on the incident X-ray direction. The *X*-, *Y2*- and *Z*-axes are used for positioning the samples. The resolution of the *Y1*-, *X*-, *Y2*- and *Z*-axes is 1 μm. The range and resolution of the κ-axis are from −7 to 13 and 0.001°, respectively. The shift of the sample position due to the rotation is less than 50 μm.

2.5. X-ray optics

The *SPEED-Mk.II* system has a horizontal goniometer with 2θ range from −10 to 23° with a resolution of 0.0001°. The goniometer sits on an *X*–*Y*–*Z* stage. The *X*-axis is used to adjust the rotation center to the diffraction area. A Ge solid-state detector is used for energy dispersive X-ray diffraction.

The incident X-rays are collimated by a set of horizontal and vertical incident slits. The diffracted X-rays are collimated by a scattering slit (referred to as a collimator) and horizontal and vertical receiving slits. The typical widths of the horizontal incident, scattering and receiving slits are 0.05, 0.05 and 0.2 mm, respectively. The distance between the scattering and receiving slits, that is, the length of the collimator, is 960 mm. Hence the divergence of the diffracted X-rays is 0.015°.

The diffracted X-rays are collected with a Ge SSD, and recorded by a multi-channel analyzer with 4096 channels. Each channel of the multi-channel analyzer corresponds to ~0.03–0.04 keV. Nominally, we can obtain a diffraction pattern with an energy range from 0 to ~120–160 keV. In practice, absorption of X-rays by the gasket and pressure media is high in the energy range below 30 keV. The X-ray flux from the bending magnet of SPring-8 decreases significantly above 100 keV, and the collection rate of the Ge SSD is low at energies above 120 keV. For these reasons, the X-ray energy range for practical use is from 40 to 120 keV.

The energy resolution of the present optics system is not equal to the energy of one channel of the solid-state detector. For example, the peak width at the half height is about 7 and 14 channels at 40 and 120 keV, respectively for the diffraction pattern of Si taken with a scattering and receiving slits with width of 0.05 and 0.2 mm, respectively. The peak widths for minerals are usually larger.

3. Performance test

3.1. Adjustment of the high-pressure space

The dimensions of the high-pressure space were evaluated by compressing stainless steel blocks and measuring their dimensions after decompression. The blocks are cubes with largely truncated edges. The edge length of the face was originally 61.0 mm, and reduced to 20.0 mm after the truncation.

The differences between the vertical and horizontal dimensions of the high-pressure space thus evaluated are plotted against press load in Fig. 4. In the case of *SPEED-1500*, the horizontal dimension increases relative to the vertical one with increasing press load at a rate of 13 μm/MN. In contrast, in the case of *SPEED-Mk.II* before adjustment of the guide blocks, the horizontal dimension decreases relatively at a rate of −13 μm/MN. We have adjusted the relative deformation of the guide blocks by increasing the effective cross section of the supporting columns as described above. The deformation rate was suppressed to −8 μm/MN after the first adjustment, and reached a final value of +4 μm/MN after repeated adjustment.

3.2. Precision of diffraction angle

Diffraction angles (2θ) were examined by taking diffraction patterns of Si (a = 0.54308 nm). The standard errors of the 2θ estimation from fitting each of the Si peaks are about 0.0008°. Fig. 5 shows the measured 2θ values at 6° after moving the goniometer from 4° (forward) and from 8° (backward), respectively. These values are reproducible within 0.0007°, suggesting that the diffraction angles can be set within the precision of estimation from diffraction patterns.

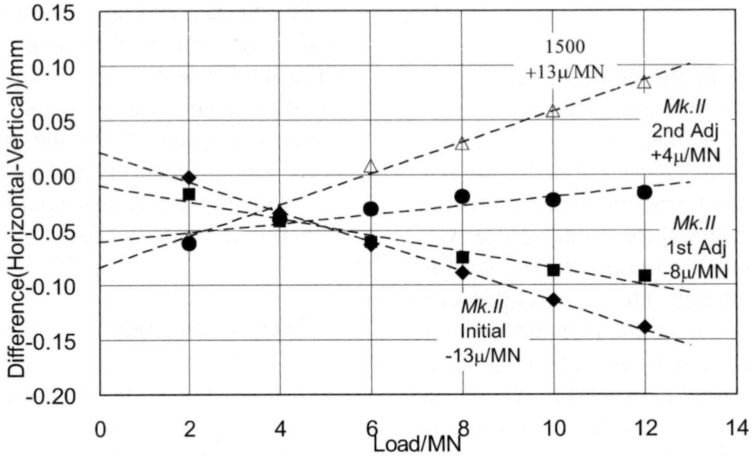

Fig. 4. The difference of the horizontal and vertical lengths of the compression space formed by the first-stage anvils. In *SPEED-1500*, this difference increases with increasing press load at a rate of +13 μm/MN: open triangles and dashed lines marked by "1500 + 13 μ/MN." In *SPEED-Mk.II*, it decreased at a rate of −13 μm/MN before adjustment: solid diamonds and dashed lines marked by "*Mk.II* initial −13 μ/MN." After the first and second adjustments, the rates were decreased to −8 and +4 μm/MN, respectively: solid squares and circles, and dashed lines marked by "*Mk.II* 1st Adj −8 μ/MN" and "*Mk.II* 2nd Adj +4 μ/MN," respectively.

3.3. High-pressure generation

We have tested pressure generation using WC and SD anvils with truncations of 3.0 and 1.5 mm, respectively, at ambient temperature. The generated pressures are measured using an MgO pressure standard with the equation of state proposed by Matsui et al. (2000), and are plotted against press load in Fig. 6.

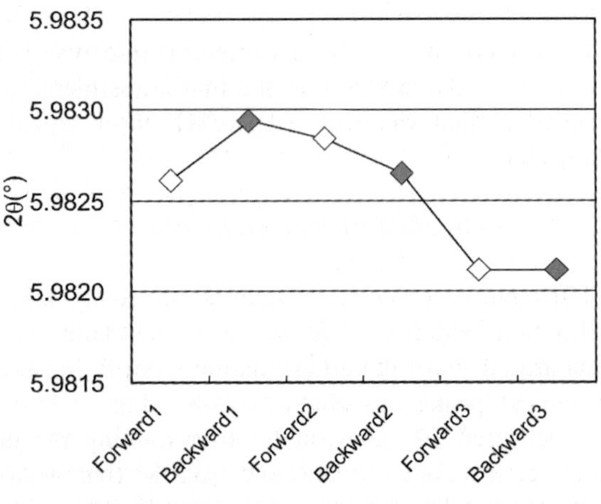

Fig. 5. Reproducibility of diffraction angles evaluated using Si diffraction at a nominal 2θ of 6°. 'Forward' denotes that the 2θ of 6° was set from a starting value of 4°, and 'Backward' denotes the 2θ of 6° was set from a the starting value of 8°.

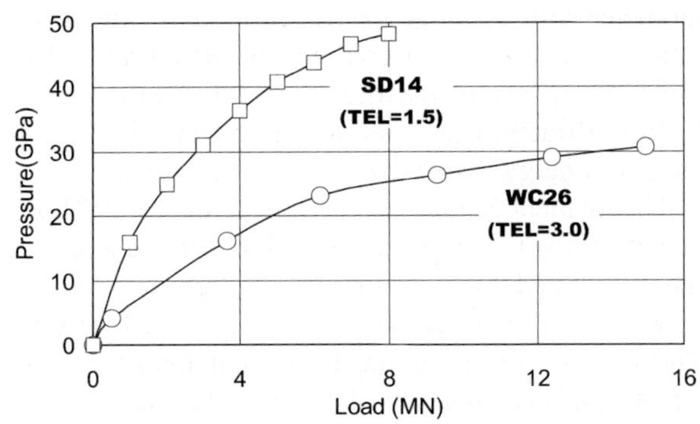

Fig. 6. Generated pressures by *SPEED-Mk.II*. The curve labeled as 'WC26' denotes pressures generated using 26 mm tungsten carbide anvils with truncated edge length of 2.5 mm, and that labeled as 'SD14' denotes pressures generated using 14 mm sintered diamond anvils with a truncated edge length of 1.5 mm.

Using WC anvils, we generated 31 GPa at a press load of 15 MN. Using SD anvils, we generated 48 GPa at a press load of 8 MN. As far as we aware, these two pressures are close to the maximum values that have been achieved so far by a KAWAI-type apparatus with WC and SD anvils[2]. The generated pressure increases

[2] In February 2004, 63 GPa was generated at a press load of 10 MN using SD anvils by E. Ito to ISEI, Okayama University.

only slightly with increasing press load at high loads of ~12–15 MN using WC anvils, but increases considerably near 8 MN using SD anvils.

3.4. Oscillation

The performance of the oscillation system was demonstrated by taking diffraction patterns of MgO with and without oscillation of the press at elevated temperatures to 2170 K with a 100 K step and pressures of about 2.5 GPa. In this experiment, nothing was mixed with MgO in order to allow maximum grain growth of MgO at high temperatures. The 2θ was about 7.017°. The range of the oscillation was from −5 to 9°. We acquired each X-ray diffraction pattern for 280 s, during which one back-and-forth oscillation was conducted at a rate of 0.1°/s.

At 1070 K, the intensity ratios of the MgO diffraction peaks were almost the same in the cases with and without oscillation, and were reasonable for the energy dispersive method in a multi-anvil press (Fig. 7a).

However, above this temperature, many peaks disappeared or weakened significantly if the oscillation system was not in operation. A diffraction pattern at 2070 K in the case without oscillation is shown in Fig. 7b. The (1 1 1), (2 0 0), (2 2 0), (2 2 2), (4 0 0), (3 3 1) and (4 2 0) peaks completely disappeared, and only the (3 1 1) and (4 2 2) peaks were recognized. Many of these missing peaks appear if the oscillation system was operated. Fig. 7c is a diffraction patterns at 2070 K with oscillation, showing that all peaks but (3 3 1) can be observed. However, the intensity ratios differed markedly from those expected assuming a randomly oriented powder. For example, the (2 0 0) peak, expected to be the strongest, was weaker than the (2 2 0) peak.

Thus, the oscillation system is quite useful when the number of grains is limited, allowing observation of many of the otherwise missing diffraction peaks. As mentioned above, the divergence of the diffracted X-rays to the X-ray optics is 0.015°, and therefore, the oscillation between −5 and 9° virtually increases the number of grain orientations by 3 orders of magnitude. It should be noted, however, that because the present oscillation system is one-dimensional with a maximum range of 20°, it is still not sufficient to obtain reliable information about peak intensities.

In addition to the recovery of otherwise missing peaks, the use of the oscillation system also increases the precision of pressure determination. Fig. 8 shows the "error" of pressures from the unit cell volumes of MgO against experimental temperature. The "errors" were obtained from standard deviations of the pressure values calculated from all the observed diffraction peaks of MgO. The errors in the case with the oscillation were smaller than 0.05 GPa, whereas those in the case without the oscillation were often larger than 0.05 GPa and sometimes exceeded 0.1 GPa.

The reason for the better precision of the pressure determination with oscillation can be explained as follows. Because the divergence of the diffracted X-rays received by the present X-ray optics system is about 0.015°, the actual diffraction angle of each grain ranges from $2\theta - 0.007°$ to $2\theta + 0.007°$ for a nominal diffraction angle of 2θ. If the number of grains is too small, the averaged diffraction angle may deviate from the nominal value, and the extent of such deviation could differ from peak to peak, resulting in large scatter in the resultant pressure values. When the sample is oscillated, the oscillating grains will continuously sweep the diffraction angle from $2\theta - 0.007°$ to $2\theta + 0.007°$. This will result in better averaged diffraction angles, and thus less scatter in the pressure values determined from different peaks. Thus the oscillation system is useful for precise determinations of pressures and equations of state of samples.

Utilizing the oscillation system, we determined the phase boundary of the B1–B2 transition in NaCl at temperatures from 1100 to 2000 K and pressures from 18 to 23 GPa (Nishiyama et al., 2003). The upper temperature bound in this experiment is close to the melting point of NaCl at 20 GPa (2000–2100 K) (Boehler et al., 1997). We could not observe any peak of B1 or B2 phases at high temperatures, unless the oscillation system was in operation.

We have also determined the thermal expansion coefficient of Mg_2SiO_4 ringwoodite at temperatures of 300–2000 K and pressures around 21 GPa (Katsura et al., submitted). In this experiment, we loaded the Mg_2SiO_4 sample and MgO pressure markers separately and without mixing with any other materials. Samples and pressure markers are often mixed with other materials to suppress grain growth. However, this is quite problematic for accurate determination of the *P–V–T* relations of minerals, because in a composite

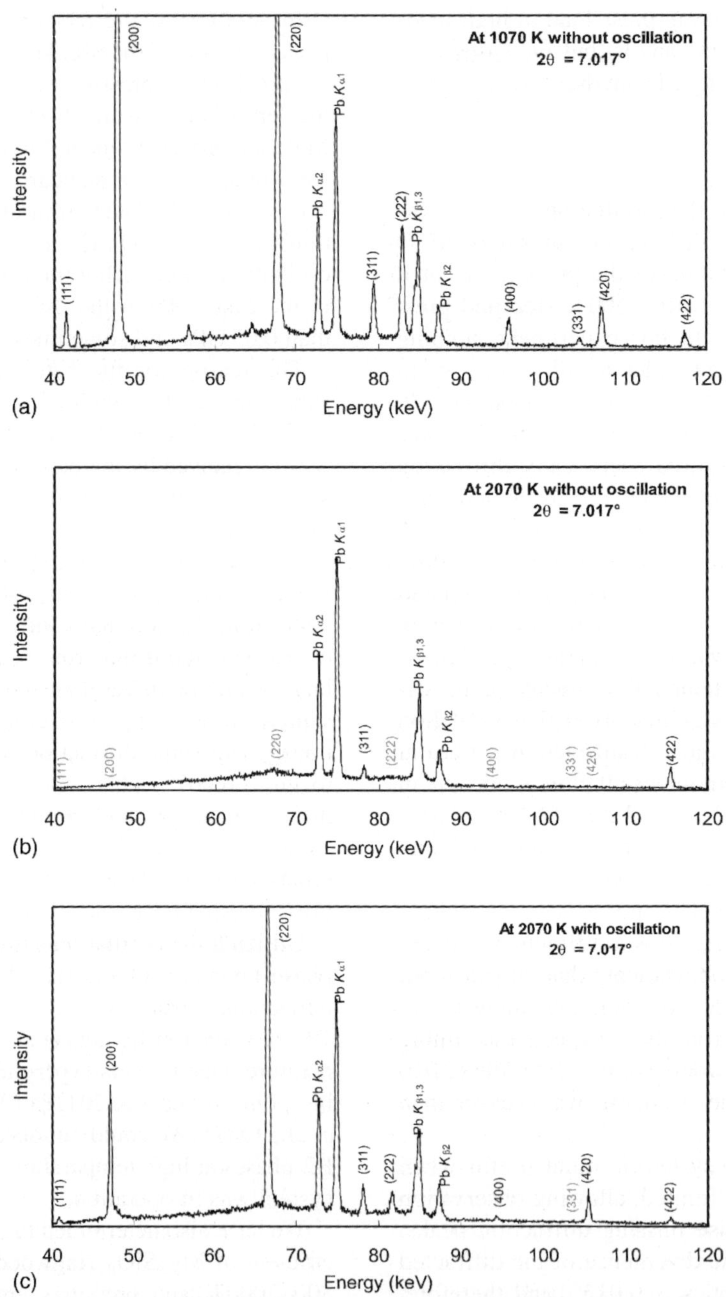

Fig. 7. Diffraction patterns of MgO with and without oscillation where $2\theta = 7.017°$: (a) at 1070 K where diffraction patterns are almost the same with and without oscillation; (b) at 2070 K without oscillation; and (c) at 2070 K with oscillation. The MgO peaks are labeled by the lattice index. The peaks by Pb fluorescence are labeled as Pb $K\alpha_1$ and so on.

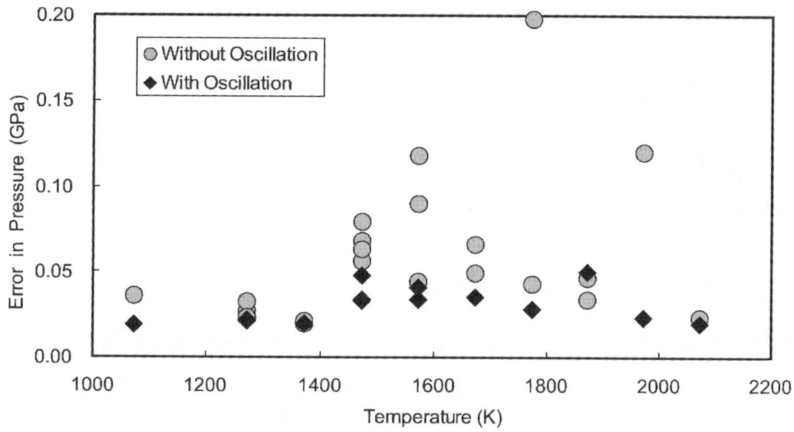

Fig. 8. Errors in pressure determination using MgO against temperature with and without oscillation.

mixed material, pressure applied on each constituent in the mixture are expected to be different from the bulk pressure (e. g. Sato et al., 1973). Use of the oscillation system then allowed us to obtain high-quality diffraction patterns without suppressing grain growth. The volumes of Mg_2SiO_4 ringwoodite and the pressures determined from the volumes of MgO in this study are, thus, free from the above problem.

4. Concluding remarks

In this paper, we have described a new KAWAI-type high *P–T* apparatus for in situ X-ray observation, *SPEED-Mk.II*. While the design of *SPEED-Mk.II* followed that of *SPEED-1500*, we have made significant improvements for the construction of *SPEED-Mk.II*. We have designed the guide block system to be adjustable to reduce uneven deformation of the high-pressure space. We also constructed a hydraulic system that can control press loads very precisely enabling smooth compression and decompression. This is particularly crucial for the development of the sintered diamond anvil technique with the KAWAI-type apparatus. We have already achieved the highest pressures attainable by a KAWAI-type high-pressure apparatus using WC and SD anvils.

SPEED-Mk.II is the first large volume press that is equipped with an oscillation system. This system enables us to obtain high-quality diffraction patterns at high temperatures, at which sample grain growth may otherwise cause disappearance of many of diffraction peaks. It also allows for more accurate pressure determination, because the oscillation of the sample grains compensates for the errors due to the finite width of the optical slits. One potential application of the oscillation system is the determination of phase boundaries at high temperatures near the melting curve where grains growth is rapid.

Acknowledgements

SPEED-Mk.II was constructed by research grants acquired by E. Ohtani and E. Takahashi from Ministry of Education and Sciences, and research fund of W. Utsumi and O. Shimomura at Japan Atomic Energy Research Institute. The following high-pressure workers are acknowledged for their discussions and suggestions: E. Ito, E. Ohtani, T. Irifune, T. Inoue, S. Urakawa, O. Ohtaka, T. Yagi and M. Akaogi. The *SPEED-Mk.II* system was produced by the following companies: press and stages; Sumitomo Heavy Industries Co., hydraulic system; Riken Seiki Co., receiving system; Kohzu Seiki Co. and incident slit system; Yonekura Seisakusho Co. The designing engineers from these companies are highly appreciated. Setup of the system was conducted using the following beam time of SPring-8: 2002A0296-ND2-np, 2002B0044-CD2-np and 2003A0088-ND2-np. We thank Y. Fei, T. Duffy, X. Xue and an anonymous referee for their comments that improved the manuscript.

References

Boehler, R., Ross, M., Boercker, D.B., 1997. Melting of LiF and NaCl to 1 Mbar: systematics of ionic solids at extreme conditions. Phys. Rev. Lett. 78, 4589–4592.

Hirose, K., Fei, Y., Ono, S., Yagi, T., Funakoshi, K., 2001a. In situ measurements of the phase transition boundary in $Mg_3Al_2Si_3O_{12}$: implications for the nature of the seismic discontinuities in the Earth's mantle. Earth Planet Sci. Lett. 184, 567–573.

Hirose, K., Komabayashi, T., Murakami, M., Funakoshi, K., 2001b. In situ measurements of the majorite-akimotoite-perovskite phase transition boundaries in $MgSiO_3$. Geophys. Res. Lett. 28, 4351–4354.

Irifune, T., Nishiyama, N., Kuroda, K., Inoue, T., Isshiki, M., Utsumi, W., Funakoshi, K., Urakawa, S., Uchida, T., Katsura, T., Ohtaka, O., 1998. The postspinel phase boundary in Mg_2SiO_4 determined by in situ X-ray diffraction. Science 279, 1698–1700.

Irifune, T., Miyashita, M., Inoue, T., Ando, J., Funakoshi, K., Utsumi, W., 2000. High-pressure phase transformation in $CaMgSi_2O_6$ and implications for origin of ultra-deep diamond inclusions. Geophys. Res. Lett. 27, 3541–3544.

Karato, S.-I., Toriumi, M., Fujii, T., 1980. Dynamic recrystallization of olivine single crystals during high-temperature creep. Geophys. Res. Lett. 9, 649–652.

Katsura, T., Yamada, H., Shinmei, T., Kubo, A., Ono, S., Kanzaki, M., Yoneda, A., Walter, M.J., Urakawa, S., Ito, E., Funakoshi, K., Utsumi, W., 2003. Post-spinel transition in Mg_2SiO_4 determined by in situ X-ray diffractometry. Phys. Earth Planet. Int. 136, 11–24.

Katsura, T., Yokoshi, S., Song, M.-S., Kawabe, K., Tsujimura, T., Kubo, A., Ito, E., Tange, Y., Tomioka, N., Saito, K., Nozawa, A., Funakoshi, K., Thermal expansion of Mg_2SiO_4 ringwoodite at high pressures. J. Geophys. Res. B, submitted.

Kubo, A., Ito, E., Katsura, T., Shinmei, T., Yamada, H., Nishikawa, O., Song, M., Funakoshi, K., 2003. In situ X-ray observation of iron using KAWAI-type apparatus equipped with sintered diamond: absence of beta phase up to 44 GPa and 2100 K, Geophys. Res. Lett. 30, 10.1029/2002GL016394.

Kubo, T., Ohtani, E., Kato, T., Urakawa, S., Suzuki, A., Kanbe, K., Funakoshi, K., Utsumi, W., Kikegawa, T., Fujino, K., 2002. Mechanisms and kinetics of the post-spinel transformation in Mg_2SiO_4, 129, 153–171.

Matsui, M., Parker, S.C., Leslie, M., 2000. The MD simulation of the equation of state of MgO: application as a pressure calibration standard at high temperature and high pressure. Am. Mineral. 85, 312–316.

Nishiyama, N., Katsura, T., Funakoshi, K., Kubo, A., Tange, Y., Sueda, Y., Kubo, T., 2003. Precise determination of the phase boundary between B1 and B2 phases in NaCl by in situ X-ray diffraction experiments. Phys. Rev. B 68, 134109.

Ohtaka, O., Fukui, H., Kunisada, T., Fujisawa, K., Funakoshi, K., Utsumi, W., Irifune, T., Kuroda, T., Kikegawa, T., 2001a. Phase relations and volume changes of hafnia under high pressure and high temperature. J. Am. Ceram. Soc. 84, 1369–1373.

Ohtaka, O., Fukui, H., Kunisada, T., Fujisawa, T., Funakoshi, K., Utsumi, W., Irifune, T., Kuroda, T., Kikegawa, T., 2001b. Phase relations and equations of state of ZrO_2 under high temperature and high pressure. Phys. Rev. B 63, 174108.

Ono, S., Ito, E., Katsura, T., Yoneda, A., Walter, M.J., Urakawa, S., Utsumi, W., Funakoshi, K., 2000. Thermoelastic properties of high pressure phase of SnO_2 determined by in situ X-ray observations up to 30 GPa and 1400 K. Phys. Chem. Mineral. 27, 618–622.

Ono, S., Katsura, T., Ito, E., Kanzaki, M., Yoneda, A., Walter, M.J., Urakawa, S., Utsumi, W., Funakoshi, K., 2001. In situ observation of ilmenito-perovskite phase transition in $MgSiO_3$ using synchrotron radiation. Geophys. Res. Lett. 28, 835–838.

Osugi, J., Shimizu, K., Inoue, K., Yasunami, K., 1964. A compact cubic anvil high pressure apparatus. Rev. Phys. Chem. Jpn. 34, 1–6.

Sato, Y., Akimoto, S., Inoue, K., 1973. Pressure intensification in the composite material. High Temp. -High Press. 5, 289–297.

Shimomura, O., Utsumi, W., Taniguchi, T., Kikegawa, T., Nagashima, T., 1992. A new high pressure and high temperature apparatus with sintered diamond anvils for synchrotron radiation use, in: Syono, Y., Manghnani, M.H., High-Pressure Research: Application to Earth and Planetary Sciences, vols. 3–11, Terra Scientific Publishing Co., Tokyo.

Suzuki, A., Ohtani, E., Funakoshi, K., Terasaki, H., Kubo, T., 2002. Viscosity of albite melt at high pressure and high temperature. Phys. Chem. Miner. 29, 159–165.

Terasaki, H., Kato, T., Urakawa, S., Funakoshi, K., Suzuki, A., Okada, T., Maeda, M., Sato, J., Sasai, S., 2001. The effect of temperature, pressure and sulfur content on viscosity of the Fe-FeS melt. Earth Planet Sci. Lett. 190, 93–101.

Urakawa, S., Terasaki, H., Funakoshi, K., Kato, T., Suzuki, A., 2001. Radiographic study on the viscosity of the Fe-FeS melts at the pressure of 5 to 7 GPa. Am. Mineral. 86, 578–582.

Urakawa, S., Hasegawa, M., Yamakata, J., Funakoshi, K., Utsumi, W., 2002. High-pressure phase relationships for FeS. High Press. Res. 22, 491–494.

Utsumi, U., Funakoshi, K., Urakawa, S., Yamakata, M., Tsuji, K., Konishi, H., Shimomura, O., 1998. SPring-8 beamlines for high pressure science with multi-anvil apparatus. Rev. High Press. Sci. Technol. 7, 1484–1486.

Van der Val, D., Chopra, P., Driry, M., Fitz Gerald, F., 1993. Relationships between dynamically recrystallized grain size and deformation conditions in experimentally deformed olivine rocks. Geophys. Res. Lett. 20, 1479–1482.

Available online at www.sciencedirect.com
SCIENCE DIRECT
ELSEVIER
Physics of the Earth and Planetary Interiors 143–144 (2004) 507–514
PHYSICS OF THE EARTH AND PLANETARY INTERIORS
www.elsevier.com/locate/pepi

A new large-volume multianvil system

D.J. Frost[a,*], B.T. Poe[a,b], R.G. Trønnes[c], C. Liebske[a], A. Duba[a,d], D.C. Rubie[a]

[a] Bayerisches Geoinstitut, Universität Bayreuth, D-95440 Bayreuth, Germany
[b] Dipartimento di Scienze della Terra, Università degli Studi "G. D'Annunzio"-Chieti, 66013 Chieti Scalo, Italy
[c] Nordic Volcanological Institute, Grensásvegur 50, IS-108 Reykjavík, Iceland
[d] American Museum of Natural History, Central Park West at 79th St., NYC, New York 10024, USA

Abstract

A scaled-up version of the 6–8 Kwai-type multianvil apparatus has been developed at the Bayerisches Geoinstitut for operation over ranges of pressure and temperature attainable in conventional systems but with much larger sample volumes. This split-cylinder multianvil system is used with a hydraulic press that can generate loads of up to 5000 t (50 MN). The six tool-steel outer-anvils define a cubic cavity of 100 mm edge-length in which eight 54 mm tungsten carbide cubic inner-anvils are compressed. Experiments are performed using Cr_2O_3-doped MgO octahedra and pyrophyllite gaskets. Pressure calibrations at room temperature and high temperature have been performed with 14/8, 18/8, 18/11, 25/17 and 25/15 OEL/TEL (octahedral edge-length/anvil truncation edge-length, in millimetre) configurations. All configurations tested reach a limiting plateau where the sample-pressure no longer increases with applied load. Calibrations with different configurations show that greater sample-pressure efficiency can be achieved by increasing the OEL/TEL ratio. With the 18/8 configuration the GaP transition is reached at a load of 2500 t whereas using the 14/8 assembly this pressure cannot be reached even at substantially higher loads. With an applied load of 2000 t the 18/8 can produce $MgSiO_3$ perovskite at 1900 °C with a sample volume of $\sim$20 mm^3, compared with $<$3 mm^3 in conventional multianvil systems at the same conditions. The large octahedron size and use of a stepped $LaCrO_3$ heater also results in significantly lower thermal gradients over the sample.

Keywords: Multianvil; Large-volume; High pressure; Pressure calibration

1. Introduction

Multianvil devices in a number of different configurations have been used for many years to generate pressures in the range 3–28 GPa while simultaneously heating to temperatures up to 3000 K. The pressure range can be extended to over 50 GPa using inner-anvils of sintered diamond (Ito, Kubo, and Funakoshi personal communication). The 6–8 type multianvil system (Kawai and Endo, 1970; Kawai et al., 1973; Ohtani et al., 1987; Walker et al., 1990) has been particularly successful due to its ability to apply stable high pressures and temperatures for significant periods of time (up to several weeks in some cases) with sample volumes large enough to be analysed using a range of conventional characterization methods. This system employs six tool-steel outer-anvils and eight tungsten carbide cubic inner-anvils to focus an applied load on an octahedral high-pressure chamber formed as a result of corner truncations on the inner-anvils. The anvils are compressed using a hydraulic press often with a 1000–2000 t (10–20 MN) capacity. By varying the corner truncation size of the inner-anvils, various sample-pressure ranges can be attained. Using

* Corresponding author. Tel.: +49-921-553737; fax: +49-921-553769.
E-mail address: dan.frost@uni-bayreuth.de (D.J. Frost).

0031-9201/$ – see front matter © 2004 Elsevier B.V. All rights reserved.
doi:10.1016/j.pepi.2004.03.003

press loads of approximately 900 t, for example, and inner-anvil truncations of 3 mm it is possible to reach sample-pressures of approximately 26 GPa. With such small inner-anvil truncations, however, the sample volume at these conditions is small, (on the order of 2 mm^3 or less), and accurate control over the temperature environment is difficult due to high thermal gradients. A number of modern analysis techniques such as NMR, calorimetry and neutron diffraction require much larger sample yields. In addition, many in situ techniques require large assembly volumes for precise measurements, for the incorporation of multiple thermocouples and in order to establish well-controlled thermal environments (e.g. Xu et al., 2004).

Here we describe a scaled-up version of the 6–8 multianvil system, which has been designed with the aim of providing significantly larger sample volumes than those of conventional systems while still operating over a similar pressure and temperature range. A particular aim of the system is to reach pressures compatible with the lower mantle and to facilitate the synthesis of (Mg, Fe)SiO_3 perovskite, the dominant phase in Earth's lower mantle, with sample volumes in excess of 10 mm^3. Extrapolation of existing calibration curves suggests that loads on the order of 3000–4000 t may be necessary to produce pressures in excess of 20 GPa with such sample volumes and for this reason the system was designed for use with a 5000 t hydraulic press.

In addition to describing the dimensions of the device and the materials employed in construction of the guide blocks and anvils, we also discuss the development and calibration of high-pressure assemblies and describe briefly some experimental applications.

2. Experimental description

In the large-volume multianvil system two "guide blocks", with dimensions that have been scaled-up in comparison to conventional systems, are compressed using a 5000 t (50 MN) hydraulic press. Each guide block consists of three tool-steel outer-anvils separated by three smaller support wedges (Fig. 1) inside a 1.2 m × 1.2 m × 0.196 m steel platen. A finite-element calculation showed that using three separate support wedges, as shown in Fig. 1, significantly reduces the tensile stresses in the surrounding outer platen. After initial testing, additional 40 mm thick tool-steel load distribution disks were included below each guide

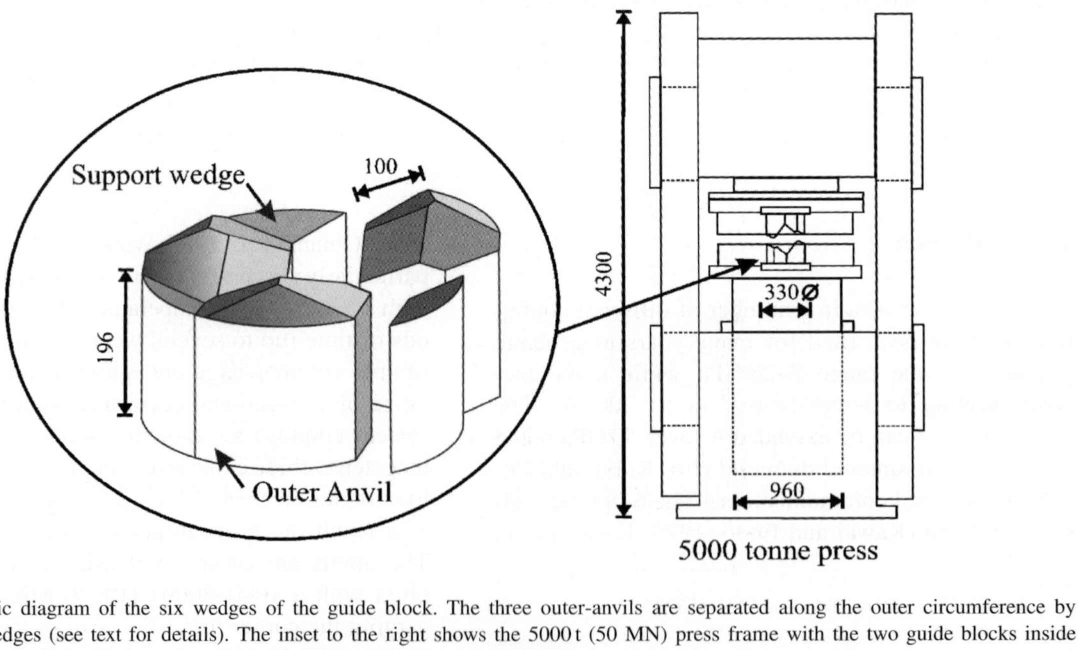

Fig. 1. Schematic diagram of the six wedges of the guide block. The three outer-anvils are separated along the outer circumference by three support wedges (see text for details). The inset to the right shows the 5000 t (50 MN) press frame with the two guide blocks inside steel platens on top of the 960 mm diameter ram. All dimensions are in millimetre.

Table 1

Assembly	Capsule size		Gaskets		Estimated	Load at P_{max} (T)
OEL/TEL (mm)	Ø (mm)	Length (mm)	Thickness (mm)	Width (mm)	P_{max} (GPa)	
25/17	5	4	3.7	6.0	13	3000
25/15	4	4	4.7	6.0	16	2500
18/11	2	3.5	3.0	5.0	19	2000
18/8	2	3.5	4.7	5.0	23	2500
14/8	1.5	2.5	2.5	5.0	21	2000

block. The three main outer-anvils define an inner cubic cavity of 100 mm edge-length in which the eight second-stage tungsten carbide inner-anvils are compressed. The large inner cubic cavity dimension enables forces up to 5000 t (50 MN) to be applied while ensuring that stresses at this interface lie well below the strength of the outer-anvils. Toshiba "F" grade tungsten carbide second-stage anvils are used, each with an edge-length of 54 mm. The dimensions of the inner-anvil truncations, octahedral assemblies and pyrophyllite gaskets employed are given in Table 1. Due to the combined weight of the eight second-stage anvils (18.5 kg) the cubes are assembled inside the lower guide block. Balsa-wood spacers are used between the cubes to provide initial alignment of the anvils. Epoxy-impregnated fibreglass laminate sheets 105 mm × 105 mm × 0.8 mm are used to electrically insulate the inner-anvils from the guide blocks. Cardboard and Teflon tape are placed on the surface of the cubes behind the gaskets to electrically insulate the thermocouple wires and to provide extra support to the gaskets.

The octahedral pressure assemblies used in the large-volume system are similar to those described elsewhere (Liebermann and Wang, 1992; Walter et al., 1995; Rubie, 1999). Some modifications, similar to those made on a piston cylinder assembly by Wade and Wood (2002), have been made, however, in order to develop an assembly that can reach temperatures >2400 °C. A diagram of this assembly is shown in Fig. 2. In order to avoid reaction or eutectic melting between various ceramic materials at high temperature, the $LaCrO_3$ furnace is completely surrounded by MgO and the thermocouple sleeve, which is normally Al_2O_3, is also made from MgO. A thin ZrO_2 sleeve for thermal insulation is inserted around the furnace, but is separated from it by an MgO sleeve. A single crystal MgO sample capsule is employed, which is particularly effective in keeping metallic liquids enclosed due to the absence of wettable grain boundaries. Ultramafic silicate liquids can also be contained for short periods of time at high pressures and temperatures because such liquids are close to being in chemical equilibrium with MgO. The resulting

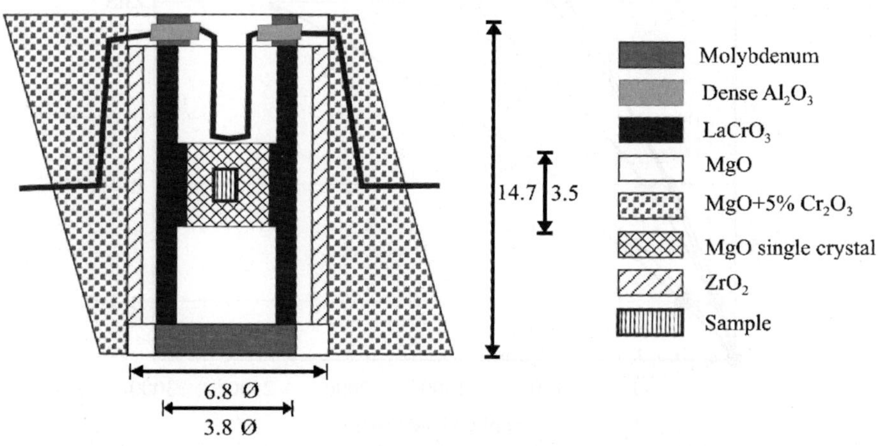

Fig. 2. Cross-section through the 18/8 assembly, modified for high temperatures. All assemblies used in the press follow a similar design.

assembly is less power-efficient than when a single thick ZrO_2 sleeve is used around the furnace, but can be heated stably to much higher temperatures (∼2800 °C).

3. Calibration

Calibrations have been performed using 25/17, 25/15, 18/11, 18/8 and 14/8 (octahedral edge-length/truncation edge-length, in millimetre) assemblies. Room temperature calibrations were performed using the Bi I–II and III–V transitions at 2.52 ± 0.05 GPa and 7.7 Gpa, respectively (Getting, 1998; Lloyd, 1971), ZnS at 15.5 ± 0.7 GPa, GaAs at 18.8 ± 0.8 GPa (Onodera and Ohtani, 1980) and GaP at 22.5 GPa (Dunn and Bundy, 1978). At high temperature the transformations of $CaGeO_3$ garnet to perovskite (Susaki et al., 1985), SiO_2 coesite to stishovite (Zhang et al., 1996), Mg_2SiO_4 forsterite to wadsleyite (Morishima et al., 1994), wadsleyite to ringwoodite (Suzuki et al., 2000) and $MgSiO_3$ majorite to perovskite (Fei et al., 1990) were bracketed in a series of quench experiments. The high temperature calibrations were performed using mixtures of both high- and low-pressure phases in the starting material.

4. Results

Room temperature calibrations for five assemblies are shown in Fig. 3. For all assemblies, the pressure reaches a plateau where the application of considerably higher loads does not produce a significant increase in the sample-pressure. This can be seen most clearly in the 18/8 calibration, where a doubling of the applied load is required to raise the pressure by 4.2 GPa from the GaAs to the GaP transition. This severe reduction in pressure efficiency defines the useful high-pressure limit of the assembly (P_{max} in Table 1). This limit is also clearly indicated by the transitions that cannot be attained using the assemblies up to 3500 t. The ZnS (15.5 GPa) transition in the 25/17, the GaAs transition in the 25/15 and the GaP transformation in the 14/8 or 18/11 are not observed with the application of loads up to 3500 t.

High-temperature pressure calibrations are shown in Fig. 4. For each assembly a higher efficiency in

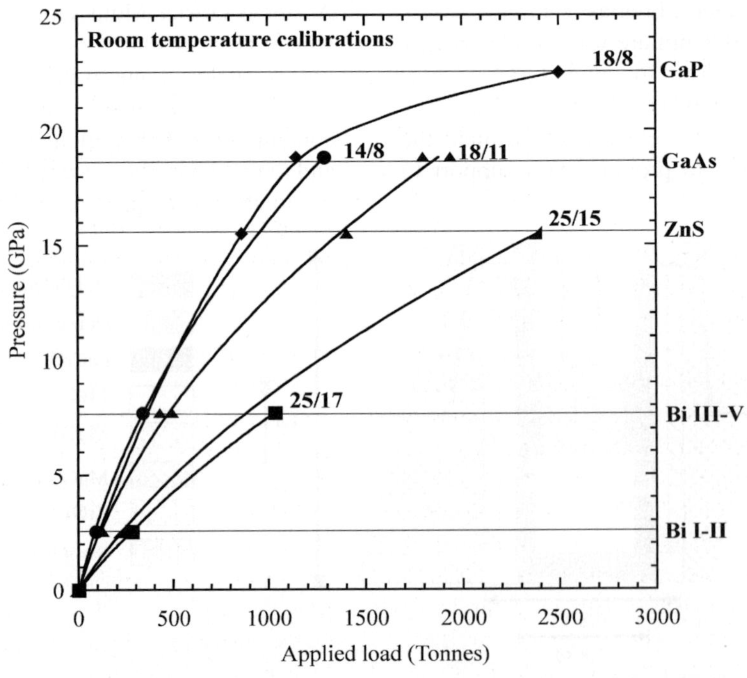

Fig. 3. Pressure calibration curves obtained at room temperature. See text for details on pressure uncertainties.

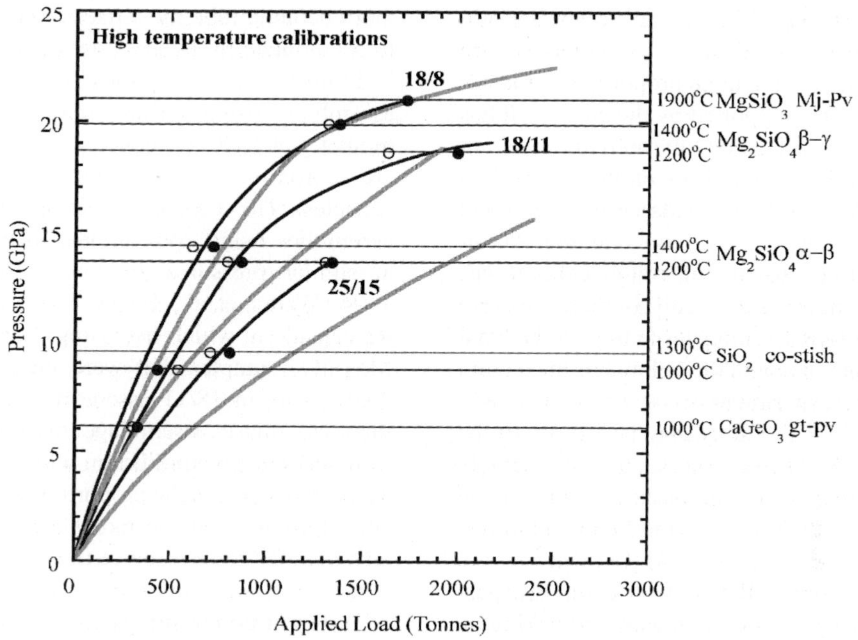

Fig. 4. High-temperature pressure calibration curves are shown (black lines). Room temperature calibrations (grey) plot below the high temperature curves of each assembly. The transformations on which the calibrations are based are $CaGeO_3$ garnet to perovskite (gt–pv), SiO_2 coesite to stishovite (co–stish), Mg_2SiO_4 forsterite to wadsleyite (α–β) and wadsleyite to ringwoodite (β–γ), and $MgSiO_3$ majorite to perovskite (Mj–Pv). The occurrence of either the high or low-pressure phase at each transformation is indicated by either filled or opened circles respectively.

generating pressure is observed at high temperature compared with room temperature but the difference between the high- and low-temperature calibration curves decreases as the size of the assembly decreases. For the 18/8, the high- and low-temperature curves are almost identical. Although there is a considerable reduction in the efficiency of pressure generation when using the 18/8 assembly above 18 GPa, samples of pure $MgSiO_3$ perovskite can be synthesized from enstatite starting materials in this assembly between 1800 and 1900 °C and with applied loads of 2000 t.

Uncertainties in the absolute pressures of room temperature and high temperature transformations, particularly at pressures above 10 GPa, are difficult to assess but are most likely at least 1 GPa. A range of determined pressures can be found in the literature for each room temperature fixed point transition above 10 GPa (see Onodera and Ohtani, 1980). This range probably arises from differences in pressure calibrant employed and due to varying states of non-hydrostatic stress. Similarly, calibrations of high temperature transformations have uncertainties associated with the bracketing of the reaction and the pressure calibrant employed. However, because we are mainly interested in the differences in pressure efficiency between the different assemblies and we have tried to use similar transformations for all assemblies, large absolute pressure uncertainties will not significantly affect our conclusions.

5. Discussion

The plateau in pressure generation that occurs for all assemblies at high loads results in an effective upper limit in the attainable sample-pressure (P_{max} in Table 1). Based on Fig. 4, for example, the 18/11 assembly is unlikely to produce sample-pressures that significantly exceed 20 GPa, no matter how large a press is employed. This plateau is most likely related to the gasketing. Each incremental increase in pressure must be associated with a decrease in volume of the assembly and a decrease in thickness of the gaskets. The plateau most likely occurs at loads where both extrusion and compressibility of the gaskets are significantly reduced and a large proportion of the additional load becomes supported by the gaskets. It has previously been recognised that pyrophyllite may not be the optimal gasket material for multianvil devices at pressures above 10 GPa (Yoneda et al., 1984) and that composite gaskets may be more suitable. An interesting feature of the room-temperature calibrations, however, is that the 18/8 assembly is more pressure efficient than the 14/8 assembly, even though the anvil truncations are the same and the volume of the 18 mm assembly after the experiment is larger. The increased efficiency of the 18/8 assembly may be a result of the larger amount of MgO from the octahedral assembly that becomes extruded into the gasket in comparison with the 14/8. In all assemblies some MgO is extruded into the gasket but the amount of MgO extrusion will increase with increasing OEL/TEL. A composite gasket is formed during compression with stiffer MgO in the higher-pressure region and softer pyrophyllite at the lower pressure side. Both 14/8 and 18/8 gaskets have a similar width after an experiment (~13 mm) but the increased amount of stronger MgO in the 18/8 gaskets probably results in a steeper pressure gradient across the gasket and therefore a smaller loss of load into the gasket region. This small increase in pressure efficiency makes a crucial difference because, in comparison with the 18/8, the 14/8 assembly cannot reach the GaP transformation at room temperature or the $MgSiO_3$ perovskite stability field at high temperature.

The survival rate of the tungsten carbide inner-anvils is very good, with anvils only very occasionally being damaged after experiments at more extreme conditions e.g. 18/8 experiments with long heating times (>5 h) at high temperatures (>2000 °C) and pressures (>20 GPa). Blow-outs, (i.e. instantaneous losses in pressure usually resulting in the fracture of tungsten carbide anvils), are rare and occur almost exclusively during compression with anvils which were most likely internally damaged during a previous experiment at the extreme conditions described above. In initial experiments using the 25/15 assembly several anvils were fractured after each experiment. This problem was alleviated, however, by replacing the alumina 4-bore thermocouple tubing with similar tubing of MgO. The relatively incompressible alumina thermocouple tube most likely caused higher stresses to accumulate in the axial direction of the assembly.

Temperature calibrations for the 18/8 and 18/11 assemblies using enstatite-diopside pyroxene thermometry at 1300 °C show that axial thermal gradients are approximately 50 °C/mm along the length of the capsules. The thermal gradients in 10/5 and 10/4 assemblies using conventional 32 mm anvil systems at similar pressures are about 100 °C/mm (Canil, 1994; Walter et al., 1995). Fig. 5 shows a multiple Re capsule in which five sample chambers contained Mg_2SiO_4 samples that were melted at 18 GPa and 2250 °C in an 18/11 assembly. In each chamber an identical phase assemblage of approximately 10% MgO crystals in equilibrium with melt was produced. The consistency in the proportion of MgO in each capsule indicates that the radial thermal gradients were less than 20°/mm, even at these relatively high temperatures. Using this type of multi-capsule, a range of different compositions can be melted in the same experiment such that relative melting temperatures can be accurately determined.

The use of larger assemblies at high pressures facilitates synthesis experiments in which large sample volumes are essential for subsequent characterization at ambient conditions (for example, using NMR—see Kohn et al., 2002). Large sample volumes are also essential for many physical property measurements, such as thermal diffusivity, electrical conductivity and elastic properties using ultrasonic measurements (e.g. Xu et al., 2004). In addition, such in-situ measurements often require the use of more than one thermocouple, which is also facilitated by using a large sample assembly. Thermal diffusivity measurements, for example, require both a large sample volume and the insertion of two thermocouples. One thermocouple on the outside surface of the sample is used to control and fluctuate the temperature in a sinusoidal manner, while a second thermocouple in the centre of the sample records the time lag in the temperature curve that results from thermal diffusion. Using Angstrom's method (Khedari et al., 1995), the thermal diffusivity can be determined from analysis of this time lag in relation to the distance between the thermocouples, which is also more accurately established in larger assemblies. Such measurements have been made on wadsleyite and ringwoodite in the large-volume press up to 20 GPa and 1100 °C (Xu et al., 2004). For ultrasonic measurements the

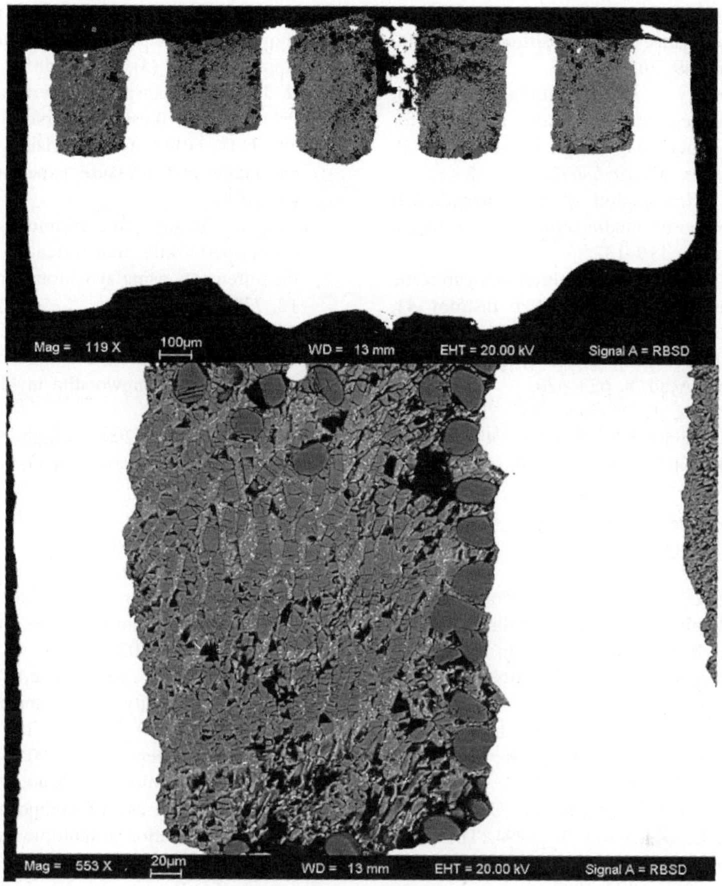

Fig. 5. A section through a 2 mm diameter multi-chamber Re sample capsule (top) containing five samples of Mg_2SiO_4 melted at 18 GPa and 2200 °C. The enlargement of the far left chamber (below) shows rounded crystals of periclase in a melt that has crystallised during quenching. Each chamber is 0.2 mm in diameter and contains an almost identical proportion of crystals and melt, which indicates very minor temperature differences between the chambers and therefore a low radial temperature gradient. The thermocouple was in contact with the bottom side of the capsule and the top section has been removed in order to determine the correct orientation for a cross-section exposing all chambers. Such capsules are used to determine relative melting temperatures of different compositions.

larger sample volume increases the precision of velocity determinations and allows better control over the sample environment such that sample deformation can be avoided. Larger volumes also benefit chemical diffusion experiments, particularly when diffusion is either rapid, such as in silicate melts, (e.g. Reid, 2003) or strongly temperature dependent.

Acknowledgements

We would like to express our thanks to G. Herrmannsdörfer, E. Ohtani, the firm of Voggenreiter (Mainleus), the firm of Paschhold (Hochstadt), M. Schmidt, H. Fischer, J. Peyronneau, Y. Xu, C. Holzapfel and J. Kung for help in the development and testing of the 5000 t multianvil system. The original manuscript was improved by the constructive comments of E. Ito and an anonymous reviewer. Funding was generously provided by the Free State of Bavaria.

References

Canil, D., 1994. Stability of clinopyroxene at pressure-temperature conditions of the transition zone. Phys. Earth Planet Int. 86, 25–34.

Dunn, K.J., Bundy, F.P., 1978. Materials and techniques for pressure calibration by resistance-jump transitions up to 500 kilobars. Rev. Sci. Instrum. 49, 365–370.

Fei, Y., Saxena, S.K., Navrotsky, A., 1990. Internally consistent thermodynamic data and equilibrium phase relations for compounds in the system $MgO–SiO_2$ at high pressure and high temperature. J. Geophys. Res. 95, 6915–6928.

Getting, I.C., 1998. New determination of the bismuth I-II equilibrium pressure: a proposed modification to the practical pressure scale. Metrologia 35, 119–132.

Kawai, N., Endo, S., 1970. The generation of ultrahigh hydrostatic pressures by a split sphere apparatus. Rev. Sci. Instrum. 41, 1178–1181.

Kawai, N., Togaya, M., Onodera, A., 1973. New device for pressure-vessels. Proc. Jpn. Acad. 8, 623–626.

Khedari, J., Benigni, P., Rogez, J., Mathieu, J.C., 1995. New apparatus for thermal diffusivity measurements of refractory solid materials by the periodic stationary method. Rev. Sci. Instrum. 66, 193–198.

Kohn, S.C., Brooker, R.A., Frost, D.J., Slesinger, A.E., Wood, B.J., 2002. Ordering of hydroxyl defects in hydrous wadsleyite (β-Mg_2SiO_4). Am. Miner. 87, 293–301.

Liebermann, R.C. and Wang, Y., 1992. Characterization of sample environment in a uniaxial split-sphere apparatus. In: Syono, Y., Manghnani, M.H. (Eds.), High-Pressure Research: Application to Earth and Planetary Sciences. AGU, Washington, DC, pp. 19–31.

Lloyd, E.C., 1971. Accurate characterization of the high pressure environment. NBS Special Publication No. 326, Washington DC, pp. 1–3.

Morishima, H., Kato, T., Suto, M., Ohtani, E., Urakawa, S., Utsumi, W., Shimomura, O., Kikegawa, T., 1994. The phase boundary between α and β-Mg_2SiO_4 determined by in situ X-ray observation. Science 265, 1202–1203.

Ohtani, E., Irifune, T., Hibberson, W.O., Ringwood, A.E., 1987. Modified split-sphere guide block for practical operation of a multiple-anvil apparatus. High Temp. High Pressures 19, 523–529.

Onodera, A., Ohtani, A., 1980. Fixed points for pressure calibration above 100 kbars related to semiconductor-metal transitions. J. Appl. Phys. 51 (5), 2581–2585.

Reid, J., 2003. Transport Properties of Silicate Liquids at High Pressure. PhD thesis, University of Bayreuth, 2003.

Rubie, D.C., 1999. Characterising the sample environment in multianvil high-pressure experiments. Phase Transitions 68, 431–451.

Susaki, J., Akaogi, M., Akimoto, S., Shimomura, O., 1985. Garnet-perovskite transformation in $CaGeO_3$: in-situ X-ray measurements using synchrotron radiation. Geophys. Res. Lett. 12, 729–732.

Suzuki, A., Ohtani, E., Morishima, H., Kubo, T., Kanbe, Y., Kondo, T., 2000. In situ determination of the phase boundary between wadsleyite and ringwoodite in Mg_2SiO_4. Geophys. Res. Lett. 27, 803–806.

Wade, J., Wood, B.J., 2002. A high-temperature (3000 K) assembly for piston cylinder experiments. Geochem. Geophys. Geosyst. 3, 1006.

Walter, M.J., Thibault, Y., Wei, K., Luth, R.W., 1995. Characterizing experimental pressure and temperature conditions in multianvil apparatus. Can. J. Phys. 73, 273–286.

Walker, D., Carpenter, M.A., Hitch, C.M., 1990. Some simplifications to multianvil devices for high pressure experiments. Am. Miner. 75, 1020–1028.

Xu, Y., Shankland, T.J., Rubie, D.C., Langenhorst, F., 2004. Thermal diffusivity measurements of olivine, wadsleyite and ringwoodite to 20 GPa and 1100 °C. Phys. Earth Planet Inter. doi: 10.1016/j.pepi.2004.03.005.

Yoneda, A., Yamamoto, S., Kato, M., Swamoto, H., Kumazawa, M., 1984. The use of composite metal gaskets to improve pressure generation in multiple anvil devices. High Temp. High Pressures 16, 637–656.

Zhang, J., Li, B., Utsumi, W., Liebermann, R.C., 1996. In situ X-ray observations of the coesite-stishovite transition: reversed phase boundary and kinetics. Phys. Chem. Miner. 23, 1–10.

Available online at www.sciencedirect.com
SCIENCE DIRECT®
ELSEVIER
Physics of the Earth and Planetary Interiors 143–144 (2004) 515–526
PHYSICS OF THE EARTH AND PLANETARY INTERIORS
www.elsevier.com/locate/pepi

A critical evaluation of pressure scales at high temperatures by in situ X-ray diffraction measurements

Yingwei Fei [a,*], Jie Li [a,1], Kei Hirose [b], William Minarik [a], James Van Orman [a,2], Chrystele Sanloup [a,3], Willem van Westrenen [1,4], Tetsuya Komabayashi [b], Ken-ichi Funakoshi [c]

[a] *Geophysical Laboratory, Carnegie Institution of Washington, 5251 Broad Branch Rd., Washington, DC 20015, USA*
[b] *Department of Earth and Planetary Sciences, Tokyo Institute of Technology, 2-12-1 Ookayama, Tokyo 152-8551, Japan*
[c] *Japan Synchrotron Radiation Research Institute (JASRI), SPring-8, Koto 1-1-1, Mikazuki, Sayo, Hyogo 679-5198, Japan*

Received 6 February 2003; received in revised form 23 August 2003; accepted 28 September 2003

Abstract

We conducted multi-anvil experiments at simultaneous high pressures and temperatures using multiple internal pressure standards including Au, Pt, MgO, W, Mo, Pd, and Ag. Extensive synchrotron X-ray diffraction data for Au, Pt, and MgO were collected at pressures up to 28 GPa and temperatures between 300 and 2173 K. We compare pressures calculated from different pressure scales and demonstrate large discrepancies in pressure determination using different pressure standards or different thermal equations of state for the same standard. The comparison allows us to quantitatively determine the differences in pressure using different pressure scales in the high *P–T* experiments. Using the MgO scale of [J. Geophys. Res. 106 (2001) 515] as a reference pressure scale, new Au and Pt scales are presented that are consistent with the MgO scale. We further examined the validity of the assumption of constant *q* value (volume dependence of the Grüneisen parameter in the Mie–Grüneisen relation) for the calculations of thermal pressures, and show that an expression of *q* as a function of temperature and pressure may be necessary to best fit the simultaneous high *P–T* data.

Keywords: Equation of state; X-ray diffraction data; High-pressure behavior; Periclase; Gold; Platinum

1. Introduction

A correct pressure scale is fundamentally important for interpreting geophysical observations using laboratory experimental data obtained at high pressure and temperature. It also allows us to make comparisons of high-pressure results produced in different laboratories using different experimental and analytical

* Corresponding author. Tel.: +1-202-478-8936; fax: +1-202-478-8901.
E-mail addresses: fei@gl.ciw.edu (Y. Fei), jackieli@uiuc.edu (J. Li), kei@geo.titech.ac.jp (K. Hirose), jav12@cwru.edu (J. Van Orman), sanloup@ccr.jussieu.fr (C. Sanloup), willem.vanwestrenen@erdw.ethz.ch (W. van Westrenen), funakosi@spring8.or.jp (K.-i. Funakoshi).
[1] Present address: Department of Geology, University of Illinois at Urbana Champaign, Urbana, IL 61801, USA.
[2] Present address: Department of Geological Sciences, Case Western Reserve University, 112 A.W. Smith Bldg., Cleveland, OH 44106, USA.
[3] Present address: Laboratoire MAGIE, Université Pierre et Marie Curie, Case 110, 75252 Paris Cedex 05, France.
[4] Present address: Institut für Mineralogie und Petrographie, ETH Zürich, Sonneggstrasse 5, ETH-Zentrum, CH 8092, Zurich, Switzerland.

0031-9201/$ – see front matter
doi:10.1016/j.pepi.2003.09.018

techniques. Metals such as Au, Pt, W, Mo, Pd, Ag, and Cu, whose equations of state are established based on shock compression experiments and thermodynamic data, are commonly used as pressure standards in high-pressure experiments. Commonly used non-metal pressure standards include MgO and NaCl. At room temperature, the ruby fluorescence pressure gauge is extensively used in diamond-anvil experiments. The ruby gauge was calibrated by simultaneously measuring the shift of ruby R_1 luminescent line and specific volume of metal standards (Cu, Mo, Pd, and Ag) as a function of pressure. The established calibration curve based on equations of state of metal standards (Mao et al., 1986) has proven to be accurate, confirmed by direct measurements of pressure by combining Brillouin scattering and X-ray diffraction techniques (Zha et al., 2000).

Accurate determination of pressure at high temperature is more difficult because of large uncertainty in calculating the thermal pressure. Gold (Au) has been extensively used as an internal pressure standard in high-pressure and high-temperature experiments. However, there are at least four different thermal equations of state of gold that predict significantly different pressures at high temperature for a measured specific volume (Jamieson et al., 1982; Heinz and Jeanloz, 1984; Anderson et al., 1989; Shim et al., 2002). The difference in calculated pressures can be as large as 2.5 GPa at 25 GPa and 2000 K (e.g., Hirose et al., 2001a,b; Ono et al., 2001; Hirose, 2002).

Another common internal pressure standard is MgO whose thermal equation of state has been studied extensively by shock and static compression experiments (e.g., Jamieson et al., 1982; Duffy and Ahrens, 1995; Utsumi et al., 1998a; Fei, 1999; Hama and Suito, 1999; Dewaele et al., 2000; Speziale et al., 2001) and by theoretical calculations (e.g., Inbar and Cohen, 1995; Karki et al., 1999; Matsui et al., 2000). At 25 GPa and 2000 K, the MgO pressure scale proposed in the early study of Jamieson et al. (1982) yields a pressure about 2 GPa below the recent scale proposed by Speziale et al. (2001). The thermal equation of state for MgO proposed by Matsui et al. (2000) is in general agreement with the results of Speziale et al. (2001) over a wide pressure and temperature range. It yields pressures about 0.5 GPa lower than those of Speziale et al. (2001) at pressure and temperature conditions near the 660 km seismic discontinuity.

Fig. 1 illustrates the discrepancy in pressure determination using different thermal equations of state of Au and MgO based on the same X-ray diffraction data. Such a large discrepancy could lead to serious geophysical consequence in understanding the nature of the 660 km seismic discontinuity in the Earth's mantle.

Decker's NaCl pressure scale (Decker, 1971) is widely used in synchrotron based multi-anvil experiments prior to 1998, when pressure generation was limited to less than 20 GPa. Brown (1999) recently revised the pressure scale and found pressure differences at high temperatures (>700 K) generally below 0.3 GPa.

The Pt pressure scales proposed by different authors (Jamieson et al., 1982; Holmes et al., 1989) are in general agreement. The differences in pressure at high temperatures are less than 0.5 GPa. However, these scales were developed based on shock wave data only, and have not been critically evaluated by comparing static high *P–T* data.

Establishment of a reliable pressure scale is severely hindered by lack of consensus within one pressure standard, regarding the thermal equation of state. The differences are largely due to reliance by some on shock wave data and by others on independently derived thermodynamic parameters. The equations of state for Au by Jamieson et al. (1982) and Anderson et al. (1989) show two end-member cases. Jamieson et al. (1982) relied on the shock wave data alone for the development of the Au scale, whereas Anderson et al. (1989) relied heavily on thermodynamic parameters consistent with high-temperature elasticity measurements at ambient pressure.

The discrepancies within each pressure standard make consistency between different pressure standards at high temperatures even more problematic. In some cases, the calculated pressures based on different standards could differ as much as 4 GPa (Fig. 1). To address these issues, we performed high *P–T* experiments using a multi-anvil apparatus installed at the SPring-8 synchrotron beamline. X-ray diffraction data for multiple internal pressure standards (Au, Pt, MgO, W, Mo, Pd, Ag) were obtained under simultaneous high *P–T* conditions up to 28 GPa and 2173 K. These data were used to evaluate pressure scales based on the existing equations of state of Au, Pt, MgO, W, Mo, Pd, and Ag. A set of equations of state is proposed to produce mutually consistent

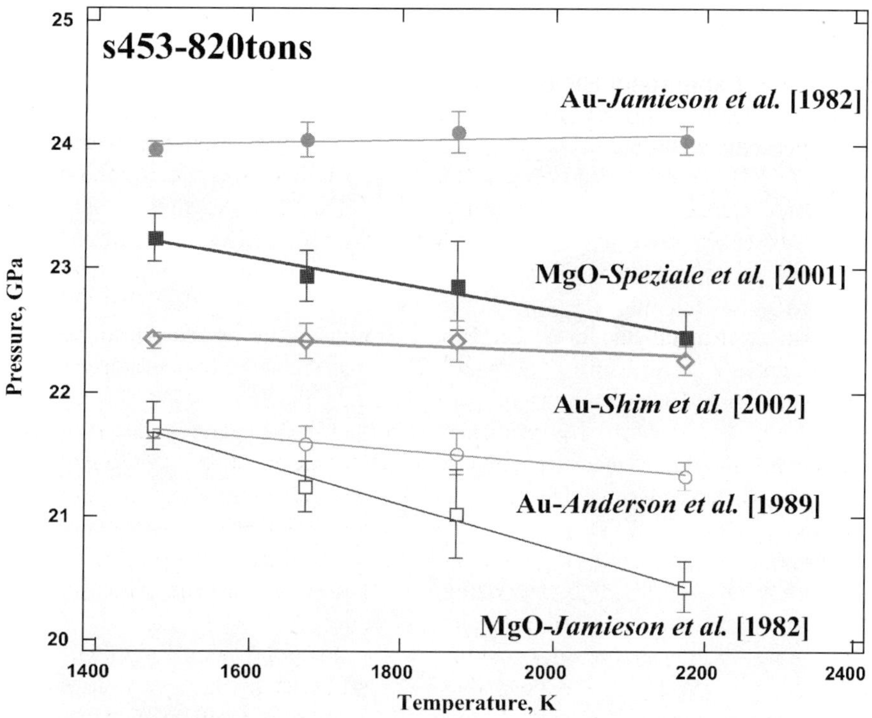

Fig. 1. Calculated pressures using MgO and Au pressure scales at high temperatures. Solid circles, open diamonds, and open circles represent pressures calculated from Au scales by Jamieson et al. (1982), Shim et al. (2002), and Anderson et al. (1989), respectively. Solid and open squares represent pressure from MgO scales by Speziale et al. (2001) and Jamieson et al. (1982), respectively.

results using different internal pressure standards at high pressures and temperatures.

2. Experimental procedures

The experiments were conducted at the BL04B1 high-pressure and high-temperature beam line in the SPring-8 synchrotron facility (Japan). A 1500 t hydraulic press with an MA8 double-stage system was used to generate pressures. The second-stage assembly consists of eight corner-truncated tungsten carbide (WC) cubes that form an octahedral cavity for the cell assembly. The synchrotron X-ray is accessed through the gap between the WC cubes. A detailed description of the facility was given by Utsumi et al. (1998b). The cell assembly used in this study is similar to the 8/3 (octahedron edge length/truncated edge length) assembly described by Bertka and Fei (1997). It contains a cylindrical Re heater with a wall thickness of 25 μm and a fitted $LaCrO_3$ sleeve outside the heater as thermal insulator. Along the X-ray path, $LaCrO_3$ was replaced with either MgO or Al_2O_3 to maximize the X-ray flux through the sample chamber. Sample temperature was measured with an axially introduced W5%Re–W26%Re (Type C) thermocouple. Reported temperatures were not corrected for the effect of pressure on emf. The selected pressure standards were placed next to the contact point of the thermocouple wires. Fig. 2 shows a back-scattered electron image of the sample chamber. The sample chamber was divided into two compartments separated by MgO powder. Each compartment was loaded with one metal standard mixed with powdered MgO. The temperature gradient in this assembly is about 30 °C/500 μm (Bertka and Fei, 1997; van Westrenen et al., 2003). We limited the total length of the sample chamber to about 250 μm to avoid a large temperature gradient.

The internal pressure standards used in this study include Au, Pt, Pd, Mo, W, Ag, and MgO. All metal

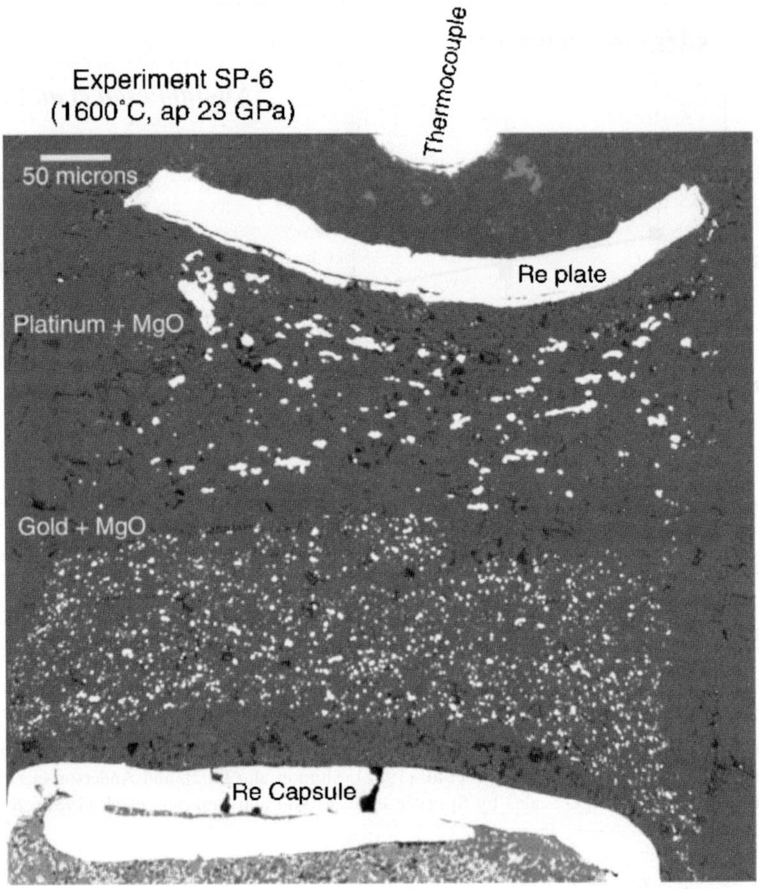

Fig. 2. A representative back-scattered electron image of the sample chamber quenched from 23 GPa and 1873 K. The sample chamber contains multiple internal pressure standards, Au, Pt, and MgO.

standards were mixed with ultra pure MgO powder to minimize re-crystallization of metal standards at high temperature. The weight ratio of MgO to metal is about 4:1. The mixtures were loaded into a MgO capsule. In each experiment, a Au–MgO mixture was loaded into one of the two sample compartments and another metal–MgO mixture occupies the other compartment. The two compartments were separated by MgO powder to avoid metal alloying. Extensive X-ray diffraction data on Au, MgO, and Pt were collected under the same *P–T* conditions.

An energy-dispersive diffraction technique was used for unit-cell volume determination, using polychromatic (white) synchrotron radiation. The diffraction data were collected with a single-element (Ge) solid-state detector, at a fixed 2θ angle. The energy-channel number relationship is determined by measuring the energies of well-determined X-ray emission lines ($K\alpha$ and $K\beta$) of Cu, Mo, Ag, Ta, Pt, Au, Pb, Ba, ^{109}Cd, ^{152}Eu, and ^{57}Co. The 2θ angle is calibrated by measuring the energies of diffraction peaks, corresponding to the known interplanar spacings $d_{h\,k\,l}$, of solids, such as gold and platinum, at ambient conditions. Typical beam size is 50 μm × 100 μm. Typical time for collecting a diffraction pattern is about 300 s.

3. Experimental results

Over 60 X-ray diffraction data points were collected at pressures up to 28 GPa and temperatures ranging from 300 to 2173 K. Because all pressure standards

Table 1
Measured unit cell parameters of MgO and Au at high pressures and temperatures

Run no.	T (K)	a_{MgO} (Å)	P_{MgO} (GPa)[a]	a_{Au} (Å)	P_{Au} (GPa)[b]	P_{Au} (GPa)[c]	P_{Au} (GPa)[d]
s456003	1273	4.1337(10)	16.14(15)	4.0152(4)	15.33(6)	15.76	15.86
s701014	1473	4.2011(10)	8.64(11)	4.0779(3)	8.45(4)	8.46	8.46
s456004	1473	4.1490(7)	15.28(10)	4.0281(9)	14.78(13)	15.22	15.32
s454003	1473	4.0991(11)	22.88(18)	3.9861(17)	21.30(30)	22.01	22.18
s453001	1473	4.0968(12)	23.25(19)	3.9838(4)	21.69(7)	22.42	22.61
s457002	1473	4.0963(3)	23.34(7)	3.9799(2)	22.36(4)	23.11	23.31
s451011	1473	4.0889(8)	24.58(13)	3.9743(4)	23.36(7)	24.14	24.33
s693017	1673	4.1235(1)	20.36(2)	4.0036(10)	19.73(16)	20.48	20.65
s453003	1673	4.1071(13)	22.94(20)	3.9920(8)	21.59(14)	22.42	22.62
s454006	1673	4.1046(11)	23.34(17)	3.9911(8)	21.74(13)	22.58	22.77
s460005	1673	4.1041(18)	23.43(25)	3.9902(12)	21.89(20)	22.73	22.93
s695007	1673	4.0972(9)	24.56(15)	3.9823(7)	23.22(13)	24.11	24.33
s456005	1873	4.1674(1)	15.45(2)	4.0513(6)	14.41(8)	14.77	14.84
s693019	1873	4.1299(4)	20.74(6)	4.0114(10)	19.85(16)	20.65	20.83
s455007	1873	4.1208(15)	22.13(22)	4.0050(3)	20.83(5)	21.68	21.87
s457010	1873	4.1190(38)	22.41(59)	4.0013(9)	21.39(15)	22.29	22.49
s696009	1873	4.1174(10)	22.66(15)	4.0044(15)	20.91(23)	21.78	21.97
s454008	1873	4.1164(7)	22.82(11)	4.0031(1)	21.11(2)	21.99	22.19
s453004	1873	4.1161(24)	22.87(36)	4.0005(10)	21.51(16)	22.42	22.62
s451013	1873	4.0995(27)	25.56(43)	3.9830(6)	24.35(10)	25.42	25.67
s454009	2023	4.1244(6)	22.60(9)	4.0101(14)	21.03(22)	21.93	22.13
s700010	2023	4.1218(10)	23.00(15)	4.0058(22)	21.68(34)	22.63	22.83
s693021	2073	4.1411(5)	20.43(7)	4.0211(10)	19.77(16)	20.57	20.74
s459002	2173	4.1380(20)	21.56(28)	4.0177(6)	20.91(9)	21.81	22.00
s453005	2173	4.1320(15)	22.46(21)	4.0148(7)	21.33(11)	22.26	22.46
s694007	2173	4.1296(26)	22.82(37)	4.0138(8)	21.47(13)	22.42	22.62

$a_0 = 4.2118(5)$ for MgO. $a_0 = 4.0786(2)$ for Au.
[a] Pressures were calculated using the MgO scale of Speziale et al. (2001).
[b] Pressures were calculated using the Au scale of Anderson et al. (1989).
[c] Pressures were calculated using the Au scale of Shim et al. (2002).
[d] Pressures were calculated using the Au scale of this study.

are of cubic symmetry, the diffraction patterns are relatively simple with well-resolved diffraction peaks. Tables 1 and 2 list experimental conditions, measured unit cell parameters of internal pressure standards, and calculated pressures. The unit cell parameters for MgO, Au, and Pt were derived from at least three observed diffraction lines. Typical uncertainty for unit cell parameter measurements is about 0.0015 Å (cf. Tables 1 and 2). The calculated pressures based on Au as the internal standard vary widely, depending on the choice of Au pressure scales. The differences are illustrated in Fig. 1 and also discussed in recent literature (e.g., Hirose et al., 2001a,b; Ono et al., 2001; Hirose, 2002; Shim et al., 2002; Matsui and Nishiyama, 2002; Fei et al., 2004). The emphasis of this study is to compare different internal standards and quantitatively determine the differences in calculated pressures using the different standards. We have obtained extensive data for Au, MgO, and Pt over a wide pressure and temperature range. These data allow us to evaluate the differences and develop equations of state that produce mutually consistent results in pressure determination at high temperatures. Limited data were also obtained for W, Mo, Pd, and Ag.

3.1. MgO–Au

For the experiments in which both MgO and Au were used as the internal standards, we calculated the pressures using the existing equations of state for MgO and Au. Fig. 1 shows the calculated pressures as temperature increased from 1473 to 2173 K in one

Table 2
Measured unit cell parameters of Au and Pt at high pressures and temperatures

Run no.	T (K)	a_{Au} (Å)	P_{Au} (GPa)[a]	P_{Au} (GPa)[b]	a_{Pt} (Å)	P_{Pt} (GPa)[c]	P_{Pt} (GPa)[d]
s424056/7	300	4.0502(16)	3.70(23)	3.70	3.9065(9)	3.58(20)	3.63
s424054/5	300	4.0309(7)	6.48(11)	6.45	3.8962(6)	5.90(14)	5.97
s424052/3	300	4.0015(16)	11.17(28)	11.04	3.8757(8)	10.85(20)	10.92
s424050/1	300	3.9842(15)	14.19(27)	14.00	3.8646(5)	13.73(13)	13.77
s424048/9	300	3.9713(12)	16.60(23)	16.33	3.8541(13)	16.59(36)	16.58
s424046/7	300	3.9571(6)	19.39(12)	19.03	3.8445(11)	19.32(31)	19.24
s424042/3	300	3.9471(9)	21.46(18)	21.02	3.8370(4)	21.54(12)	21.39
s376004	1473	4.0390(8)	13.26(11)	13.70	3.9022(5)	12.68(10)	13.37
s377002	1473	4.0122(9)	17.10(14)	17.79	3.8866(8)	16.31(19)	17.06
s381006	1473	3.9988(10)	19.19(15)	20.00	3.8743(10)	19.35(25)	20.13
s379003	1473	3.9837(5)	21.71(9)	22.62	3.8632(15)	22.25(40)	23.03
s380008	1473	3.9769(7)	22.90(13)	23.85	3.8616(11)	22.68(30)	23.46
s383004	1473	3.9539(3)	27.17(5)	28.24	3.8437(1)	27.70(3)	28.43
s375014	1673	4.0552(8)	12.55(10)	12.86	3.9104(16)	12.26(34)	13.01
s377003	1673	4.0236(10)	16.75(15)	17.45	3.8909(11)	16.67(27)	17.53
s381007	1673	4.0107(5)	18.64(7)	19.49	3.8833(10)	18.50(25)	19.39
s378006	1673	4.0005(12)	20.22(18)	21.17	3.8741(8)	20.79(21)	21.70
s379005	1673	3.9940(7)	21.25(11)	22.27	3.8722(20)	21.28(51)	22.19
s380009	1673	3.9879(6)	22.26(11)	23.34	3.8695(13)	21.98(32)	22.89
s383005	1673	3.9634(10)	26.58(18)	27.85	3.8532(9)	26.38(25)	27.28
s376012	1873	4.0684(10)	12.38(12)	12.54	3.9168(13)	12.28(26)	13.10
s376013	1873	4.0683(5)	12.38(6)	12.55	3.9156(10)	12.54(20)	13.36
s375016	1873	4.0657(9)	12.69(11)	12.89	3.9174(9)	12.16(18)	12.96
s376006/7	1873	4.0635(6)	12.94(7)	13.18	3.9116(12)	13.39(24)	14.24
s382009	1873	4.0575(6)	13.65(7)	13.99	3.9100(20)	13.73(43)	14.60
s382006/7	1873	4.0546(7)	14.00(9)	14.38	3.9098(11)	13.78(24)	14.64
s377007	1873	4.0373(9)	16.21(12)	16.84	3.8992(23)	16.13(53)	17.07
s377005	1873	4.0370(6)	16.25(8)	16.88	3.8983(6)	16.34(13)	17.28
s377006	1873	4.0368(11)	16.28(15)	16.91	3.8989(28)	16.20(66)	17.14
s377004	1873	4.0348(3)	16.54(4)	17.21	3.8998(6)	16.00(13)	16.93
s381016	1873	4.0213(1)	18.41(1)	19.26	3.8887(25)	18.58(60)	19.57
s378008	1873	4.0138(11)	19.50(16)	20.44	3.8830(10)	19.96(25)	20.98
s378010	1873	4.0138(9)	19.50(13)	20.44	3.8825(2)	20.08(5)	21.10
s378011	1873	4.0137(7)	19.48(10)	20.46	3.8828(12)	20.01(29)	21.03
s379009	1873	4.0056(14)	20.73(21)	21.77	3.8803(5)	20.62(12)	21.65
s379007	1873	4.0047(4)	20.86(7)	21.92	3.8821(18)	20.18(44)	21.20
s379011	1873	4.0042(10)	20.95(16)	22.01	3.8781(17)	21.17(42)	22.20
s379018	1873	4.0040(19)	20.98(29)	22.04	3.8770(1)	21.45(3)	22.48
s380018/9	1873	3.9989(7)	21.76(11)	22.89	3.8750(13)	21.95(32)	22.99
s380012/3	1873	3.9969(6)	22.08(10)	23.23	3.8756(7)	21.80(17)	22.84
s380010	1873	3.9960(4)	22.23(7)	23.39	3.8759(8)	21.72(21)	22.76
s383010	1873	3.9721(10)	26.23(19)	27.66	3.8600(7)	25.89(19)	26.95

$a_0 = 3.9231(5)$ for Pt. $a_0 = 4.0786(2)$ for Au.
[a] Pressures were calculated using the Au scale of Anderson et al. (1989).
[b] Pressures were calculated using the Au scale of this study.
[c] Pressures were calculated using the Pt scale of Holmes et al. (1989).
[d] Pressures were calculated using the Pt scale of this study.

experiment at constant load oil pressure. The MgO scale of Jamieson et al. (1982) predicts the lowest pressures over this temperature range, whereas the Au scale of Jamieson et al. (1982) gives the highest pressures. The pressures calculated from the MgO scale of Speziale et al. (2001) are about 1.5 GPa higher than those from the Au scale of Anderson et al. (1989). Matsui and Nishiyama (2002) reported similar discrepancy between the Au and MgO scales. The difference in the calculated pressures between the MgO scale of Speziale et al. (2001) and the revised Au scale of Shim et al. (2002) is relatively small (<0.5 GPa) at 25 GPa and 1873 K.

The MgO scale of Speziale et al. (2001) was developed by taking all available experimental data (static and shock wave) into account. The model with variable q parameter (logarithmic volume derivative of the Grüneisen parameter) can reproduce the experimental data over a wide pressure and temperature range. The proposed thermal equation of state is also in general agreement with results from molecular dynamics simulation (Matsui et al., 2000). The first primary pressure scale (extended to ultra-high pressure) at room temperature was established based on simultaneous measurements of elasticity and volume of MgO at high pressure (Zha et al., 2000). At the present time, we are lacking a practical pressure scale at high temperature that can be utilized to compare high P–T data collected by different laboratories. We are in favor of using the MgO scale of Speziale et al. (2001) as the pressure scale for consistency and inter-laboratory comparison. Of course, the MgO scale of Speziale et al. (2001) needs to be further verified by redundant equation-of-state measurements (e.g., simultaneous X-ray diffraction and acoustic measurements).

Using the MgO scale of Speziale et al. (2001) as a reference, we plotted our new compression data of Au over the temperature range of 1473–2173 K (Fig. 3). Shim et al. (2002) used the hydrostatic compression data at room temperature (Takemura, 2001) and shock wave Hugoniot data as constraints to derive the thermal equation of state for Au. Their equation of state misfits our data by about 0.5 GPa. We can fit our data by adjusting the q value. The best-fitted q values are 0.15, 0.27, 0.41, and 0.81 for the data at 1473, 1673, 1873, and 2127 K, respectively, assuming that all the other parameters are the same as those of Shim et al. (2002). The q value of 1.0 appears to fit best to the Hugoniot data (Shim et al., 2002). Our data indicate a q value of less than 1.0 with some temperature dependence. Within the uncertainties of the experimental data, a q value of 0.7 (±0.3) gave reasonable representation of both static and shock wave data.

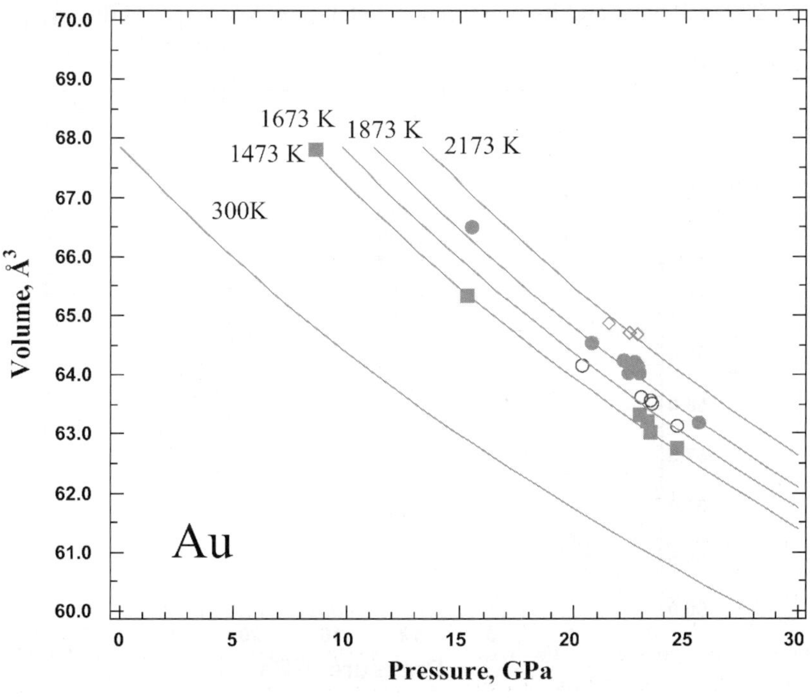

Fig. 3. Calculated isotherms for Au at 300, 1473, 1673, 1873, and 2173 K. Solid squares, open circles, solid circles, and open diamonds represents experimental data collected at 1473, 1673, 1873, and 2173 K, respectively.

3.2. Au–Pt

We designed experiments to compare pressure determination using Au and Pt as the internal standards. There are two proposed thermal equations of state for Pt (Jamieson et al., 1982; Holmes et al., 1989). Both were derived from shock wave data, but Jamieson et al. (1982) gave consistently lower pressures. For example, the difference in the calculated pressures between the two studies is about 0.5 GPa at 25 GPa and 2273 K. All discussions in this paper regarding Pt will be based on the equation proposed by Holmes et al. (1989).

Before we evaluate the Pt pressure scale at high temperature, it is important to examine the bulk modulus of Pt at room temperature. Fig. 4 shows our compression data of Pt at 300 K. These data were collected during decompression at room temperature, after the sample was annealed at 227 K and 25 GPa. A least-squares fit to the *hkl* *d*-spacings indicates no measurable deviatoric stress. The pressures were calculated using the equation of state of Au derived from the hydrostatic compression data at room temperature (Takemura, 2001). A least-squares fit to the compression data of Pt yielded a 300 K isothermal bulk modulus $K_{OT} = 290 \pm 10$ GPa and its pressure derivative $K'_{OT} = 2.7 \pm 0.9$. However, the compression range is too small to give a reliable K'_{OT} value. Other studies indicated the K'_{OT} value close to 5. Simultaneously fitting the 300 K and high-T data yielded a higher value of $K'_{OT} = 4.8$ with bulk modulus $K_{OT} = 273 \pm 3$ GPa (see below).

Because of the possible effect of deviatoric stress at low temperature, we collected X-ray diffraction data of Au and Pt only at temperatures above 1473 K. The maximum pressure of this study is about 28 GPa. The Pt scale of Holmes et al. (1989) and the Au scale of Anderson et al. (1989) predicted similar pressures at temperatures between 1473 and 1873 K. As discussed above, the Au scale of Anderson et al. (1989) gives lower pressures than the revised Au scale of Shim et al. (2002) over the *P–T* range of this study. We further refined the Au scale of Shim et al. (2002) to produce mutually consistent pressures with the MgO scale of Speziale et al. (2001). In order to establish a Pt scale that is consistent with our Au scale as well as the MgO scale of Speziale et al. (2001), we re-determined the thermal parameters for the equation of state of Pt by fitting to the compression data at 300, 1473, 1673, and 1873 K. The optimized parameters are $K_{OT} = 273$ GPa, $K'_{OT} = 4.8$, $\theta_0 = 230$ K,

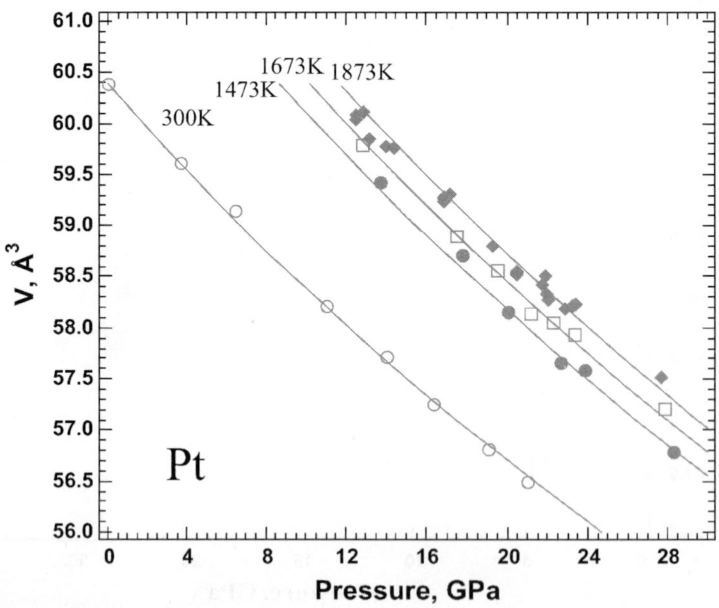

Fig. 4. Calculated isotherms for Pt at 300, 1473, 1673, and 1873 K. Open circles, solid circles, solid circles, open squares, and solid diamonds represents experimental data collected at 300, 1473, 1673, and 1873, respectively.

Table 3
Model parameters for the equations of state of MgO, Au, and Pt

Parameters	MgO[a]	Au[b]	Pt[c]
V_0 (Å^3)	74.71(1)	67.850(4)	60.38(1)
K_{OT} (GPa)	160.2(2)	167(3)	273(3)
K'_{OT}	3.99(1)	5.0(2)	4.8(3)
θ_0 (K)	773	170	230
γ_0	1.524(25)	2.97(3)	2.69(3)
q_0	1.65(40)	0.7(3)	0.5(5)
q_1	11.8(2)	0	0
3R (J/g K)	0.12664	0.12500	0.12786

[a] All parameters are from Speziale et al. (2001) ($q = q_0(V/V_0)^q_1$).
[b] All parameters except q value are from Shim et al. (2002).
[c] This study.

$\gamma_0 = 2.69$, and $q = 0.5$. Fig. 4 shows the experimental and calculated isotherms of Pt. Table 3 summarizes the thermodynamic parameters of MgO, Au, and Pt that produce mutually consistent pressures at high temperatures.

3.3. Other pressure standards, W, Mo, Pd, and Ag

In addition to the evaluation of consistency among the MgO, Au and Pt pressure standards, we also conducted several experiments using W, Mo, Pd, and Ag as the internal standards. We chose these metals as the pressure standards because they are of simple cubic structure and some of them were used for establishing the ruby pressure scale. However, they are not ideal pressure standard materials at high temperature because they are chemically reactive and easily oxidized under high oxygen fugacity conditions. We only collected a few useful data points at temperatures below 1473 K. Fig. 5 shows a comparison of the calculated pressures from the different standards. The pressures were calculated using the equations of state by Hixson and Fritz (1992) for W and Mo, and by Carter et al. (1971) for Ag and Pd. All metal standards underestimate pressures relative to the MgO scale of Speziale et al. (2001) at high temperatures. The Pd scale gives the lowest pressures.

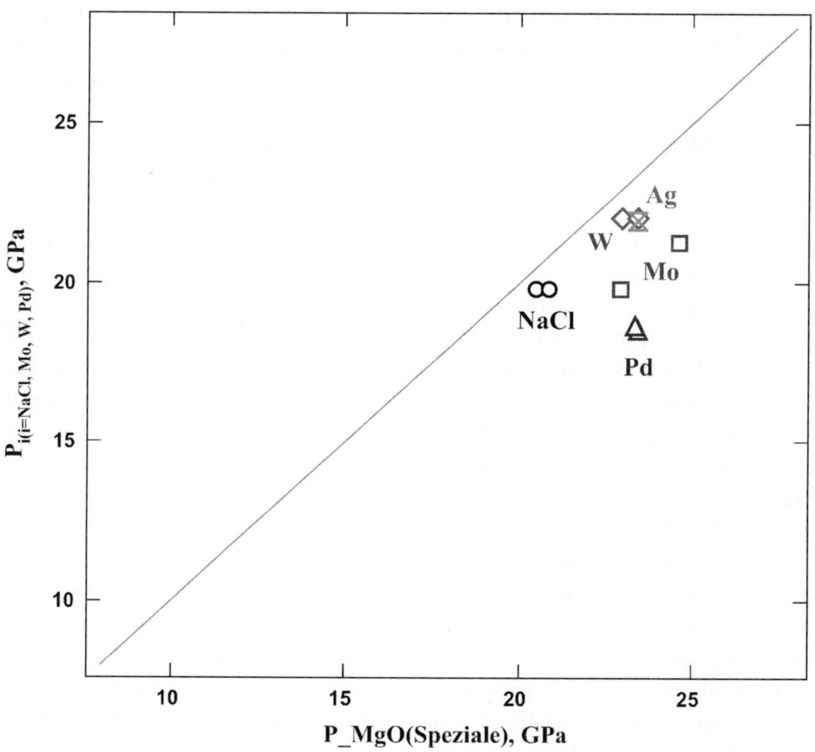

Fig. 5. Calculated pressures from the MgO scale of Speziale et al. (2001) compare to the pressures calculated from different internal pressure standards (W, open diamonds; Mo, open squares; Pd, open triangles; Ag, double triangles; and NaCl, open circles) at 1473 K.

More experimental data are needed to further constrain the thermal equations of state for W, Mo, Pd, and Ag.

4. Discussion

4.1. Comparison of internal pressure standards

The existing thermal equations of state of pressure standards such as Au, MgO, and Pt, predict a range of pressures at high temperatures for samples under the same pressure and temperature conditions. The maximum differences in the calculated pressures could be as large as 3 GPa at *P–T* conditions corresponding to the boundary between the transition zone and the lower mantle (cf. Fig. 1). The MgO scale of Jamieson et al. (1982), who adopted the thermodynamic parameters given by Carter et al. (1971), predicted the lowest pressure relative to the other pressure scales. This equation of state for MgO does not reproduce more recent static and shock wave data (e.g., Duffy and Ahrens, 1995; Fei, 1999) and should not be used for pressure calculations at high temperatures. Speziale et al. (2001) proposed an equation of state for MgO based on analysis of all available static and shock wave data. The new MgO scale predicts pressures between those of the Au scales of Anderson et al. (1989) and Jamieson et al. (1982), and in general agreement with the results from molecular dynamic simulation of the equation of state of MgO (Matsui et al., 2000). Recently, Shim et al. (2002) pointed out that Anderson et al. (1989) equation of state did not reproduce the shock wave data and Jamieson et al. (1982) used a value for the Grüneisen parameter that is too high. The revised Au scale by Shim et al. (2002) and the MgO scale of Speziale et al. (2001) predict similar pressures at high temperatures. The differences in the calculated pressures by the two pressure scales are about 0.5 GPa at 25 GPa and 2000 K. Using the MgO scale of Speziale et al. (2001) as a practical scale for consistency, we further refine the Au scale of Shim et al. (2002) using our new high *P–T* data. We can adjust the 0.5 GPa difference by either increasing the Grüneisen parameter γ_0 or decreasing the q value. Shim et al. (2002) used a q value of 1.0 to fit the shock compression data up to 580 GPa. The shock wave data below 200 GPa seem to indicate a q value of less than 1.0 that gives better fit to our static high *P–T* data. Our modified Au scale, with a q value of 0.7, gives pressures that are consistent with those calculated from the MgO scale of Speziale et al. (2001).

Pressures calculated from the Pt scale of Holmes et al. (1989) and the Au scale of Anderson et al. (1989) give similar pressures at high temperatures. Because pressures calculated from the Au scale of Anderson et al. (1989) are about 1.0 GPa lower than those of Shim et al. (2002) and about 1.5 GPa lower than those from the MgO scale of Speziale et al. (2001) at conditions corresponding to the 660 km discontinuity, the Pt scale underestimates pressures relative to the Au scale of Shim et al. (2002) and the MgO scale of Speziale et al. (2001). Our new thermal equation of state for Pt is consistent with our modified Au scale and the MgO scale of Speziale et al. (2001) as well.

4.2. Thermal equation of state

The Mie–Grüneisen relation is commonly used to calculate the thermal pressure in shock wave studies. Static compression data at room temperature are often compared to the reduced shock wave Hugoniot data. In the Hugoniot calculations, the Grüneisen parameter is a function of volume ($\gamma = \gamma_0(V/V_0)^q$), assuming a constant q value. The q value is usually assumed to be 1 in shock wave data reduction. Recent study of MgO demonstrates that variable q is required to satisfy the high *P–T* static and shock wave data (Speziale et al., 2001). First-principles theoretical calculations (Inbar and Cohen, 1995) and thermodynamic analysis (Anderson et al., 1993) also indicate that the q value is a function of pressure. Our high *P–T* data on Au are best fitted with a q value that varies as a function of temperature. Shock wave data lie along a high *P–T* trajectory. A constant q value may be sufficient to describe the high *P–T* shock wave data because of the trade-off between the pressure and temperature effects. The difficulty of simultaneously fitting the static high *P–T* and shock wave data may indicate inadequacy of the formulation of the Mie–Grüneisen relation with a constant q value. With more static high *P–T* data available, the assumption of constant q value should be further examined.

4.3. Practical pressure scale versus absolute pressure scale

Pressure determinations at high temperature rely on *P–V–T* equations of state of pressure standards such as Au, Pt, MgO, and NaCl. These thermal equations of state are derived from shock wave and thermodynamic data. With recent advances in synchrotron radiation and high-pressure and high-temperature techniques, static high *P–T* data have accumulated at an increasingly fast rate. It is essential to ensure consistency in pressure determination at high temperature if different pressure standards were used in high *P–T* experiments. We have demonstrated that calculated pressures using the existing equations of state of pressure standards could differ by as much as 3 GPa at pressure and temperature conditions corresponding to upper parts of the Earth's lower mantle. The thermal equation of state of MgO is the least controversial one from recent theoretical and experimental studies. We are in favor of using the MgO scale of Speziale et al. (2001) as a reference pressure scale for consistency and inter-laboratory comparison. We further established thermal equations of state for Au and Pt that are consistent with the MgO scale of Speziale et al. (2001).

This study provides us with a means to compare high *P–T* data obtained using different pressure standards such as MgO, Au, and Pt. The ultimate goal for pressure calibration is to establish an absolute pressure scale. Such a goal can be achieved by redundant equation-of-state measurements, i.e. simultaneous density and elasticity determination at high pressure and temperature. MgO is an ideal material for the attempt of redundant equation-of-state measurements (e.g., simultaneous X-ray diffraction and acoustic measurements). Zha et al. (2000) successfully measured the volume and elasticity simultaneously at room temperature by combining Brillouin scattering and X-ray diffraction techniques. Extending these measurements to high temperatures would provide a direct check for the MgO scale of Speziale et al. (2001).

5. Concluding remarks

The existing thermal equations of state for pressure standards such as Au, MgO, and Pt, predict a range of pressures at high temperatures for samples under the same pressure and temperature conditions. For high *P–T* experiments, especially at synchrotron facilities, there is an urgent need to establish a reliable and self-consistent pressure scale at high temperature. We evaluated different pressure scales in a series of high *P–T* experiments with multiple internal pressure standards and concluded that the use of a single practical pressure scale such as the MgO scale of Speziale et al. (2001) would be beneficial for consistency check and data comparison. We determined the relative differences among different pressure scales and established new Au and Pt scales that are consistent with the MgO scale of Speziale et al. (2001). These scales allow us to compare high *P–T* data that were collected using different internal pressure standards.

Speziale et al. (2001) and also this study showed that it is often difficult to reconcile the static high *P–T* data or thermal expansion data at ambient pressure with the shock wave Hugoniot data using the Mie–Grüneisen relation with a constant volume dependence q value. The temperature and pressure dependence of the q value may not be easily delineated from the Hugoniot data because Hugoniot are in a high *P–T* trajectory. Isothermal compression data over a wide pressure and temperature range should provide tight constraints on the q value and its possible temperature and pressure dependence. Further experimental data at simultaneous high pressure and temperature are required to determine the q value as a function of pressure and temperature.

Acknowledgements

We thank Dean Presnall, Masanori Matsui, and Sang-Heon Shim for valuable comments of the manuscript. This research was supported by NSF grants to YF (EAR9873577) and to the Consortium for Materials Properties Research in the Earth Sciences (COMPRES) and by the Carnegie Institution of Washington. The experiments were performed at SPring-8 under proposal Nos. 2000A0024-ND-np, 2000B0581-ND-np, 2001B0487-ND-np, and 2001A0290-CD-np.

References

Anderson, O.L., Isaak, D.G., Yamamoto, S., 1989. Anharmonicity and the equation of state for gold. J. Appl. Phys. 65, 1534–1543.

Anderson, O.L., Oda, H., Chopelas, A., Isaak, D.G., 1993. A thermodynamic theory of the Grüneisen ratio at extreme conditions: MgO as an example. Phys. Chem. Miner. 19, 369–380.

Bertka, C.M., Fei, Y., 1997. Mineralogy of the Martian interior up to core-mantle boundary pressures. J. Geophys. Res. 102, 5251–5264.

Brown, J.M., 1999. The NaCl pressure standard. J. Appl. Phys. 86, 5801–5808.

Carter, W.J., Marsh, S.P., Fritz, J.N., McQueen, R.G., 1971. The equation of state of selected materials for high-pressure references. In: Lloyd, E.C. (Ed.), Accurate Characterization of the High-Pressure Environment. National Bureau of Standards, Spec. Publ. 326, Washington, DC, pp. 147–158.

Decker, D.L., 1971. High-pressure equation of state for NaCl, KCl, and CsCl. J. Appl. Phys. 42, 3239–3244.

Dewaele, A., Fiquet, G., Andrault, D., Haüsermann, D., 2000. *P–V–T* equation of state of periclase from synchrotron radiation measurements. J. Geophys. Res. 105, 2869–2877.

Duffy, T.S., Ahrens, T.J., 1995. Compressional sound velocity, equation of state, and constitutive response of shock-compressed magnesium oxide. J. Geophys. Res. 100, 529–542.

Fei, Y., 1999. Effects of temperature and composition on the bulk modulus of (Mg, Fe)O. Am. Miner. 84, 272–276.

Fei, Y., Van Orman, J., Li, J., van Westrenen, W., Sanloup, C., Minarik, W., Hirose, K., Komabayashi, T., Walter, M., Funakoshi, K., 2004. Experimentally determined postspinel transformation boundary in Mg_2SiO_4 using MgO as an internal pressure standard and its geophysical implications, J. Geophys. Res. 109, B2305, doi:10.1029/2003JB002562.

Hama, J., Suito, K., 1999. Thermoelastic properties of periclase and magnesiowüstite under high pressure and high temperature. Phys. Earth Planet. Int. 114, 165–179.

Heinz, D.L., Jeanloz, R., 1984. The equation of state of the gold calibration standard. J. Appl. Phys. 55, 885–893.

Hirose, K., 2002. Phase transitions in pyrolitic mantle around 670 km depth: Implications for upwelling of plumes from the lower mantle. J. Geophys. Res. 107, 10.1029/2001JB000597.

Hirose, K., Fei, Y., Ono, S., Yagi, T., Funakoshi, K., 2001a. In situ measurements of the phase transition boundary in $Mg_3Al_2Si_3O_{12}$: implications for the nature of the seismic discontinuities in the Earth's mantle. Earth Planet. Sci. Lett. 184, 567–573.

Hirose, K., Komabayashi, T., Murakami, M., Funakoshi, K., 2001b. In situ measurements of the majorite–akimotoite–perovskite phase transition boundaries in $MgSiO_3$. Geophys. Res. Lett. 28, 4351–4354.

Hixson, R.S., Fritz, J.N., 1992. Shock compression of tungsten and molybdenum. J. Appl. Phys. 71, 1721–1728.

Holmes, N.C., Moriarty, J.A., Gathers, G.R., Nellis, W.J., 1989. The equation of state of platinum to 660 GPa (6.6 Mbar). J. Appl. Phys. 66, 2962–2967.

Inbar, I., Cohen, R.C., 1995. High pressure effects on thermal properties of MgO. Geophys. Res. Lett. 22, 1533–1536.

Jamieson, J.C., Fritz, J.N., Manghnani, M.H., 1982. Pressure measurement at high temperature in X-ray diffraction studies: gold as a primary standard. In: Akimoto, S., Manghnani, M.H. (Eds.), High Pressure Research in Geophysics. Center for Academic Publications, Tokyo, pp. 27–47.

Karki, B.B., Wentzcovitch, R.M., De Gironcoli, S., Baroni, S., 1999. High-pressure lattice dynamics and thermoelasticity of MgO. Phys. Rev. B 61, 8793–8800.

Mao, H.K., Xu, J., Bell, P.M., 1986. Calibration of the ruby pressure gauge to 800 kbar under quasihydrostatic conditions. J. Geophys. Res. 91, 4673–4676.

Matsui, M., Nishiyama, N., 2002. Comparison between the Au and MgO pressure calibration standards at high temperature. Geophys. Res. Lett. 29, 10.1029/2001GL014161.

Matsui, M., Parker, S.C., Leslie, M., 2000. The MD simulation of the equation of state of MgO: application as a pressure standard at high temperature and high pressure. Am. Miner. 85, 312–316.

Ono, S., Katsura, T., Ito, E., Kanzaki, M., Yoneda, A., Walter, M.J., Urakawa, S., Utsumi, W., Funakoshi, K., 2001. In situ observation of ilmenite–perovskite phase transition in $MgSiO_3$ using synchrotron radiation. Geophys. Res. Lett. 28, 835–838.

Shim, S., Duffy, T.S., Takemura, K., 2002. Equation of state of gold and its application to the phase boundaries near 660 km depth in the Earth's mantle. Earth Planet. Sci. Lett. 203, 729–739.

Speziale, S., Zha, C., Duffy, T.S., Hemley, R.J., Mao, H.K., 2001. Quasi-hydrostatic compression of magnesium oxide to 52 GPa: implications for the pressure–volume–temperature equation of state. J. Geophys. Res. 106, 515–528.

Takemura, K., 2001. Evaluation of the hydrostaticity of a helium-pressure medium with powder X-ray diffraction techniques. J. Appl. Phys. 89, 662–668.

Utsumi, W., Weidner, D.J., Liebermann, R.C., 1998a. Volume measurement of MgO at high pressures and high temperature. In: Manghnani, M.H., Yagi, T. (Eds.), Properties of Earth and Planetary Materials at High Pressure and Temperature. American Geophysical Union, Washington, DC, pp. 327–333.

Utsumi, W., Funakoshi, K., Urakawa, S., Yamakata, M., Tsuji, K., Konishi, H., Shimomura, O., 1998b. SPring-8 beamlines for high pressure science with multi-anvil apparatus. Rev. High Pressure Sci. Technol. 7, 1484–1486.

van Westrenen, W., Van Orman, J., Watson, H., Fei, Y., Watson, E.B., 2003. Assessment of temperature gradient in multi-anvil assemblies using spinel layer growth kinetics, Geochem. Geophys. Geosyst. 4(4), 1036, doi:10.1029/2002GC000474.

Zha, C., Mao, H.K., Hemley, R.J., 2000. Elasticity of MaO and a primary pressure scale to 55 GPa. Proc. Natl. Acad. Sci. 97, 13494–13499.

Available online at www.sciencedirect.com

SCIENCE DIRECT®

ELSEVIER

Physics of the Earth and Planetary Interiors 143–144 (2004) 527–539

PHYSICS OF THE EARTH AND PLANETARY INTERIORS

www.elsevier.com/locate/pepi

Temperature gradients and evaluation of thermoelastic properties in the synchrotron-based laser-heated diamond cell

Abby Kavner[a,*], Wendy R. Panero[b]

[a] *Earth and Space Science Department, Institute of Geophysics and Planetary Physics, UCLA, Los Angeles, CA 90095-1567, USA*

[b] *Department of Geological Sciences, University of Michigan, Ann Arbor, MI 48109, USA*

Received 30 November 2002; received in revised form 3 December 2003; accepted 6 December 2003

Abstract

Reliable determinations of thermoelastic properties inside the laser-heated diamond cell depend critically on accurate measurement of temperature, temperature gradients, and the relationship between the hotspot and the X-ray beam. Here we examine these issues and their relative importance in interpreting data from synchrotron-based studies of materials at high pressures and temperatures. When combining laser heating and X-ray diffraction, temperature gradients must be measured and compared with the size of the X-ray beam in synchrotron-based laser-heated diamond anvil cell experiments in order to avoid systematically overestimating the actual average temperature of the X-rayed volume. For a laser-heated sample with a hotspot that drops to 75% of its value at the edge of the X-ray beam, the temperature measured at the hotspot center overestimates the actual average temperature by 17%, leading to up to a 20% underestimate of thermal expansion. As temperature gradients become steeper with increasing pressure, the temperature measured at the hotspot center increasingly overestimates the average temperature of the X-rayed volume, which further distorts measurements of higher order thermoelastic parameters. This can help to explain a wide variety of anomalous experimental results, including low thermal expansion values, and a higher than expected decrease of thermal expansion with increasing pressure without necessarily calling upon constant volume conditions in the cell. To help address these issues, we present four semi-independent methods to calculate temperature gradients inside the laser-heated diamond cell by taking advantage of the Planck's law relationships between spectral intensity, integrated intensity, and temperature. Together, these methods yield a precise determination of the radial temperature gradient and provide an internal consistency check on the accuracy of the absolute temperature measurement. We show how these methods can be used to recover precise measurements of temperature gradients even in the presence of optical chromatic aberrations.

Keywords: Laser heating; Spectroradiometry; Chromatic aberration; Diamond anvil cell; Thermoelastic properties; Temperature measurement

1. Introduction

To model and predict planetary structure and dynamics, and to properly interpret results from seismic data, we require accurate and precise measurements of the behavior of minerals at the high pressures and temperatures relevant to planetary interiors. The laser-heated diamond anvil cell is a powerful experimental tool to measure material properties in situ, including pressure–volume–temperature (P–V–T) equations of state (Fiquet et al., 1998; Dewaele et al., 2000; Shim et al., 2000), phase boundaries (Boehler et al., 1990; Saxena et al., 1994; Kavner

* Corresponding author. Tel.: +1-310-206-3675; fax: +1-310-825-2779.
E-mail addresses: akavner@ucla.edu, akavner@igpp.ucla.edu, akavner@ess.ucla.edu (A. Kavner).

0031-9201/$ – see front matter © 2004 Elsevier B.V. All rights reserved.
doi:10.1016/j.pepi.2003.12.016

et al., 2001; Shim et al., 2001), melting curves (Williams and Jeanloz, 1990; Williams et al., 1991; Boehler, 1992; Lazor et al., 1993; Sweeney and Heinz, 1993; Yoo et al., 1993; Boehler and Ross, 1997; Boehler et al., 1997; Kavner and Jeanloz, 1998a,b; Shen et al., 1998; Sitaud and Thevenin, 1999) and transport properties (Andrault, 2001). Because temperature plays a tremendous role in governing the physical behavior of materials, the ability to interpret data obtained in the laser-heated diamond cell hinges upon a critical and quantitative evaluation of both precision and accuracy of the temperature measurement.

Synchrotron-based experiments designed to simultaneously measure crystal structure and volume inside the diamond cell are especially sensitive to the presence of temperature gradients which are present under most laser-heating conditions (Boehler et al., 1990; Williams et al., 1991; Boehler and Chopelas, 1991; Jeanloz and Kavner, 1996; Kavner et al., 1998; Andrault and Fiquet, 2001; Shen et al., 2001; Yagi et al., 2001; Panero and Jeanloz, 2001a). The desire for uniform temperatures in an experiment designed to measure equilibrium thermodynamic properties has generated a significant experimental effort to minimize temperature gradients across the width of the X-ray beam and thickness of the sample. These efforts, including heating with a high-powered CO_2 laser (Boehler and Chopelas, 1991) combining external resistance heating and laser heating (Li et al., 1996), heating both sides of the sample simultaneously using double-hot-plate heating (Shen et al., 2001), mixing the TEM_{01} and TEM_{00} laser modes (Shen et al., 2001), and increasing the gasket thickness (Boehler et al., 1997; Panero and Jeanloz, 2001b) have yielded mixed success.

Partially as a response to technical difficulties in measuring temperatures across intensity gradients, and partially because of some success in minimizing these gradients, the temperature measured at the center of the hotspot is often taken as a proxy of the temperature of the X-rayed sample (Boehler et al., 1990; Fiquet et al., 1996, 1998, 2000; Shen et al., 1998; Dewaele et al., 2000; Shim et al., 2000; Andrault and Fiquet, 2001; Kavner and Duffy, 2001). In all of these experiments, the central temperature is certainly an upper bound on the actual average temperature experienced by the sample volume that has a strong downward bias due both axial and radial temperature gradients (Kavner et al., 1998).

In this paper we examine the sensitivity of thermoelastic properties (measured by combined laser heating and X-ray diffraction) to the temperature gradients. To help address this issue, we propose several alternative methods to measure the radial temperature gradients present in the laser-heated diamond anvil cell.

2. Effects on thermoelastic properties

Measurements of thermoelastic properties in the laser-heated diamond cell have exposed anomalous effects, such as lower than expected thermal expansion parameter for MgO (Fiquet et al., 1996) and/or unusually strong pressure-dependence of thermal expansion for both MgO (Fiquet et al., 1996) and iron (Boehler et al., 1990). These results have been ascribed to the presence of thermal pressure inside the diamond anvil cell, and may also arise from global pressure changes during the heating process. Table 1 shows that biases in temperature measurements due to the presence of a gradient can magnify the corresponding underestimate of thermal expansion determined by lattice parameters measured by X-ray diffraction. Measurements of well-calibrated internal standards in the hotspot show that the sample is generally found to follow a P–V path intermediate to constant volume and constant pressure conditions (Kavner and Duffy, 2001). In the case of constant volume, the thermal pressure is a maximum and thermal expansion is a minimum. In the case of constant pressure, the thermal expansion is a maximum and the thermal pressure is a minimum. In either case, assuming proper use of an internal standard, neglecting the influence of temperature gradients within the X-ray spot systematically underestimates the thermal expansion (Table 1).

The ratio of the half-width at half maximum (HWHM) of the X-ray beam to the hotspot HWHM spans a wide range depending on the experimental system, but should increase as pressure increases due to compression of the sample and insulation layer coupled with the high thermal conductivity of diamond (Bodea and Jeanloz, 1989; Kavner et al., 1998; Manga and Jeanloz, 1996; Panero and Jeanloz, 2001a,b). If this effect is not taken into account, the

Table 1
Calculations of thermoelastic properties for laser-heating experiments for a range of ratios of hotspot FWHM to the width of a square X-ray beam

T at X-ray beam edge (K)	Hotspot FWHM/size of X-ray beam (~square)	T, volume average (K)	Temperature underestimate (%)	Thermal expansion underestimate[a] (%)	Thermal pressure required to explain underestimate[b] (GPa)	Constant V: Pthermal	Thermal expansion underestimate (%)
2000	No temperature gradient across X-ray spot	2000	0	0	0	8.7	0
1900	0.95	1933	3.4	4	0.3	8.3	−3.9
1800	0.90	1865	6.8	8	0.7	8.0	−7.9
1500	0.75	1660	17	20	1.7	7.0	−19.5
1000	0.5	1312	34	40	3.5	5.2	−40.5

The average temperature of the X-rayed volume is calculated assuming a central hotspot temperature of 2000 K. Thermal expansion, α (in K^{-1}), is calculated the center temperature using as a proxy for the average temperature, and apparent thermal pressure (GPa) is calculated for the case of a material with $\alpha_0 = 30 \times 10^{-6}$ and $K_{0T} = 170$ GPa in the diamond anvil cell, ignoring the temperature gradient.

[a] Assuming constant pressure.

[b] Assuming sample chamber is intermediate between constant P and constant V.

apparent thermal expansion values show an anomalously steep decrease with pressure.

This is consistent with data from pioneering measurements on Fe (Boehler et al., 1990) and MgO (Fiquet et al., 1996) where the peak temperature was taken as a proxy for the sample temperature. In the case of iron, the results were interpreted as intrinsic thermoelastic properties of Fe (i.e. a steep decrease of thermal expansion with increasing pressure). In the past, thermal pressure has been invoked explain this (Heinz, 1990; Fiquet et al., 1996, 1998; Andrault et al., 1998; Dewaele et al., 1998). Thermal pressure, the pressure increase of a sample heated under constant volume conditions, is especially dependent on an accurate evaluation of the sample temperature because overestimating the average temperature of an X-rayed volume yields results that are indistinguishable from a thermal pressure. In order to differentiate the effects of thermal pressure and thermal expansion, special care must be taken to evaluate the temperature profile with respect to the volume of the X-rayed sample.

To calculate the average temperature of the X-rayed volume, the three-dimensional temperature profile must be integrated over the dimensions of the X-rayed spot. For a cylindrically symmetric laser-heated spot centered within the X-ray beam

$$T_{\rm avg} = \frac{1}{XYt}\int_0^t \int_0^Y \int_0^X T(r, z)\,{\rm d}x\,{\rm d}y\,{\rm d}z \qquad (1)$$

where x and y describe the size and shape of the X-ray spot, t is the thickness of the sample, and $T(r, z)$ describes how the temperature profile varies across the X-ray volume, both radially (r) and axially (z). For a constant-intensity rectangular X-ray beam (X by Y), as would result from narrowing the beam using slits (Shen et al., 2001), no axial temperature gradients (double-sided laser heating in well-insulated opaque samples), and a radial Gaussian temperature profile with HWHM of σ

$$T_{\rm avg} = (T_{\rm m} - T_{\rm b})\pi\frac{\sigma^2}{XY}{\rm erf}\left(\frac{X}{\sigma}\right){\rm erf}\left(\frac{Y}{\sigma}\right) + T_{\rm b} \qquad (2)$$

$T_{\rm m}$ is the peak (central) temperature and $T_{\rm b}$ the background temperature (300 K). As shown in Fig. 1 and Table 1, the average temperature of the X-ray volume is biased by the low temperatures at the perimeter of the X-ray spot. This bias will depend on the relative size and shape of both the X-ray beam profile and the temperature profile.

In practice, several factors conspire to complicate the volume integration of the complete temperature distribution $T(r, z)$ throughout an X-rayed sample. In addition to radial gradients, axial temperature gradients are always present in semi-transparent oxide and silicate samples (Manga and Jeanloz, 1996; Panero and Jeanloz, 2001a) and in samples that are heated from one side only in the case of more strongly absorbing samples. Such axial gradients will further reduce the

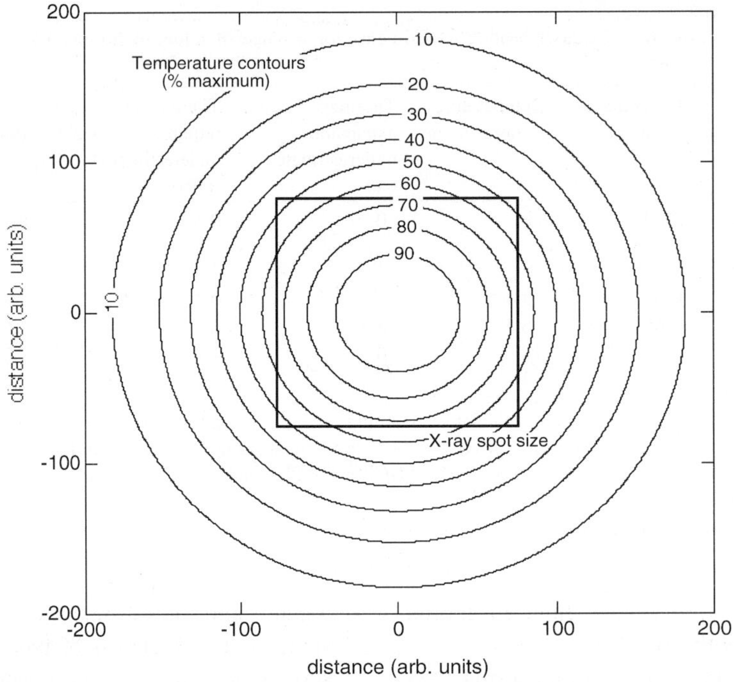

Fig. 1. Two-dimensional comparison of a Gaussian temperature profile with the size of a square X-ray beam.

average temperature of the X-rayed volume compared with the temperature measured in the center of the hotspot. In addition, the actual average temperature is very sensitive to the shape and intensity distribution of the X-ray beam and the spatial relationship between the laser beam and the X-ray beam. A square or rectangular X-ray beam will significantly lower the average temperature compared to the peak temperature because of the large volume contributed at the corners of the X-rayed sample, where temperature gradients may be significant; however, this effect is mitigated if X-ray intensity falls off towards the outer part of the X-ray beam. The exact shape and intensity profile of the X-ray beam must be accounted for in this integration.

Experimental results from double-sided heating of a platinum foil performed at GSECARS (Fig. 2) show representative temperature gradients relative to the size of the X-ray beam encountered during measurements of laser-heated Pt foil within the diamond cell. As pressure increases, the width of the hotspot decreases, as predicted by model calculations. The X-ray beam size for this experiment was 15 μm $\times$ 15 μm. Therefore, at 12 GPa, the average temperature across the X-ray volume is 96% of the peak temperature. As pressure is increased to 20 GPa, the hotspot width decreases to 37 μm, and the average temperature drops to 93% of the temperature at the hotspot temperature.

These gradients are qualitatively similar to others reported in the literature. Temperature gradients for Fe are reported to be $\sim$15 K/μm, heated in conjunction with a 20–30 μm diameter X-ray beam (Boehler et al., 1990). Assuming that the laser hotspot is perfectly centered within a rectangular-shaped X-ray beam, and that there are no axial temperature gradients, the average temperature under these conditions is 93–95% of the measured temperature at the hotspot center. Measurements on MgO were performed with an X-ray beam size of 50 μm $\times$ 50 μm with a hotspot ranging from 75 to 100 μm in diameter (Fiquet et al., 1996). Taking the diameter as the FWHM, and peak temperatures of 2000 K, the actual average temperature sampled by the X-ray beam ranges from 1585 K for a 75 μm hotspot to 1747 K for a 100 μm hotspot in these experiments, assuming a uniformly intense X-ray beam.

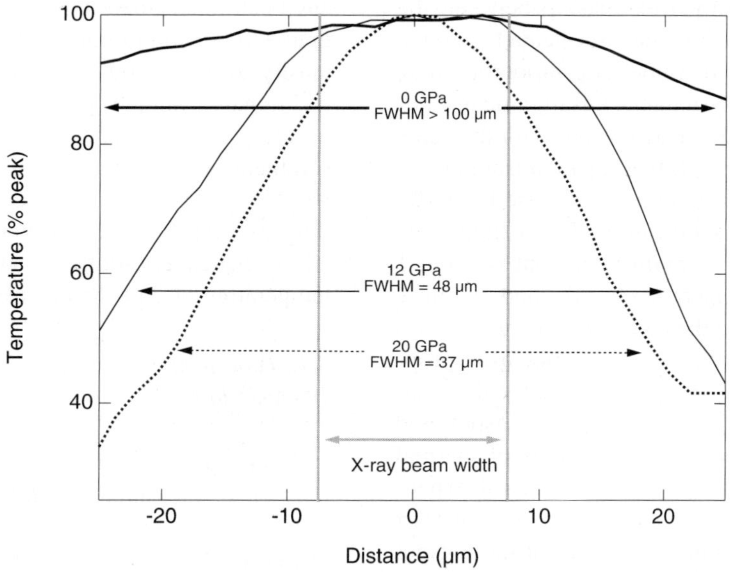

Fig. 2. Temperature measurements as a function of pressure for a platinum foil during double-sided laser heating with a Nd:YLF laser. The temperature was measured at each point of the hotspot image collected by a 100 μm entrance slit to an imaging spectrometer and CCD.

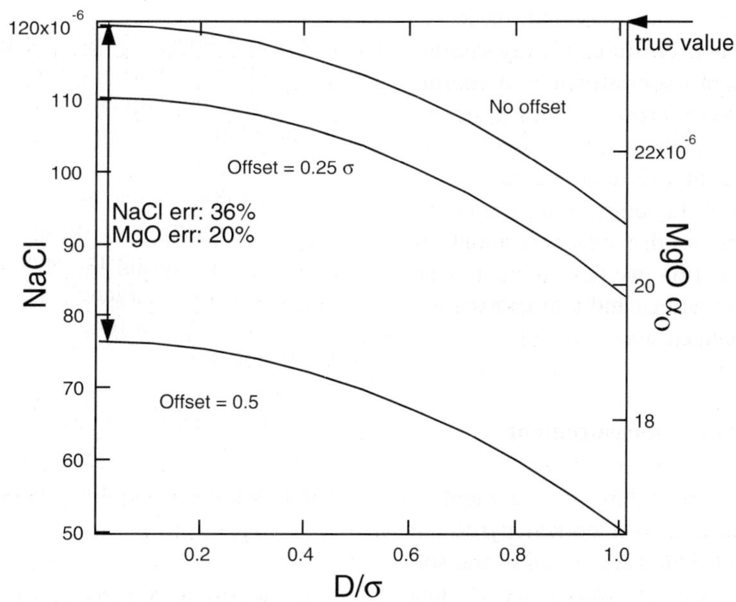

Fig. 3. Calculations of how misalignment between the X-ray and hotspot affects the thermal expansion measurement of NaCl and MgO assuming a Pt internal standard, 300 K pressure of 30 GPa, and peak temperature of 2000 K. The X-axis is a dimensionless measure of the size of the hotspot compared with the size of the X-ray beam. The X-ray beam is modeled as a square, with length D, and σ is the half-width at half maximum of the hotspot intensity. For the case of no offset, the curve shows the effect of not taking the temperature variations within the sample volume into account.

Because of the sensitive relationship between the center of the X-ray spot and the laser-heated spot, the most serious errors in thermoelastic properties may arise from small but systematic misalignments. For example, for a 30 μm-HWHM hotspot with the same width as the X-ray spot (D) a 2 μm misalignment causes the average temperature to decrease by ~4%. For a typical system magnification of 12×, and CCD pixel size of 25 μm, a 2 μm misalignment is about 1 pixel in a measured temperature profile, and therefore not readily detectable in the data. Alignment becomes even more critical for measuring thermoelastic properties of materials. Fig. 3 shows explicitly the relationship between the X-ray spot and the hotspot, and how misalignment distorts measurements of thermal expansion. For example, to measure thermal expansions within 10%, the alignment must be within $1/4\sigma$ (where σ is the half-width at half-max of the hotspot intensity profile) or at GSECARS, better than 5 μm. Inaccurate measurements of the temperature gradient contribute a greater error. In contrast to random errors in temperature measurement, failure to account for the temperature gradients and misalignments in the experiment will lead to systematic errors. If the temperature system is properly aligned and calibrated, these systematic errors will always overestimate X-ray sampled temperatures, resulting in underestimates of thermal expansion parameters. Since pressure tends to steepen the temperature gradients, these effects become increasingly more significant with pressure. Therefore, a precise measurement of the temperature gradient is required to evaluate pressure-dependent thermoelastic properties. In the next section, we describe a new approach to measure temperatures and temperature gradients inside the laser-heated diamond cell.

3. Analysis of temperature measurement

The pioneering high-temperature measurements in the diamond anvil cell relied on optical pyrometry (Ming and Bassett, 1974) while almost all of the subsequent work has used spectral radiometry to measure temperature (Jeanloz and Heinz, 1984; Heinz and Jeanloz, 1987). In this section we delve into the spectroradiometric measurement, discussing a suite of techniques based on Planck's law to use the measured intensity data from a laser-heated spot in the diamond anvil cell to calculate a central temperature and radial gradients. In addition to a direct fit of the spectral intensity to a graybody version of Planck's law, three additional techniques provide methods to evaluate the accuracy, precision and validity of a temperature measurement at high pressures within the diamond cell, and to make a precise estimate of the radial temperature gradient. For each method we present data from a hypothetical two-dimensional hotspot with a central temperature of 2500 K and a HWHM of 18 μm.

3.1. Peak temperature: full spectral fit to Planck's law

Temperature at each distance step across the hotspot is measured by fitting the calibrated and corrected spectral intensity (Schneider and Goebel, 1984) to Wien's approximation to Planck's law. Planck's law relates the spectral intensity ($I(\lambda)$), emissivity (ε), wavelength (λ) and temperature (T) of a blackbody-like thermal emitter through the relation:

$$I(\lambda, T) = 2\varepsilon(\lambda, T)\pi hc^2\lambda^{-5}\left[\exp\left(\frac{hc}{\lambda kT}\right) - 1\right]^{-1} \quad (3)$$

where c, k and h are the speed of light, Boltzmann's constant and Planck's constant, respectively.

Wien's approximation ($e^{hc/\lambda kT} \gg 1$) provides a linear relationship between spectral intensity and frequency, assuming constant emissivity (Jeanloz and Heinz, 1984; Heinz and Jeanloz, 1987). Temperature as a function of distance across the hotspot is calculated by fitting the spectral intensity to Wien's approximation of Planck's law, assuming a constant emissivity (ε) as a function of wavelength (λ) across the visible range:

$$I(\lambda, T) = 2\varepsilon\pi hc^2\lambda^{-5}\left[\exp\left(\frac{-hc}{\lambda kT}\right)\right] \quad (4)$$

where the normalized intensity is defined as

$$J \sim \ln(I\lambda^5) \quad (5)$$

and the normalized frequency is

$$\omega = -\frac{hc}{\lambda k} \quad (6)$$

The linearized form of (4) allows temperature to be determined from the slope of normalized intensity

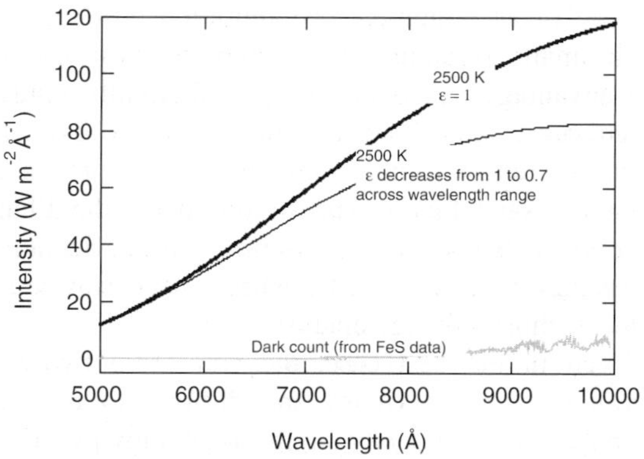

Fig. 4. Intensity as a function of wavelength for synthetic data set at 2500 K. Bold line shows spectral intensity for a blackbody. The thin black line shows calculated spectral intensity for a material whose emissivity decreases by 30% over the spectral range. The gray line shows experimental measurements obtained at GSECARS of dark noise on the CCD in an area away from an FeS hotspot.

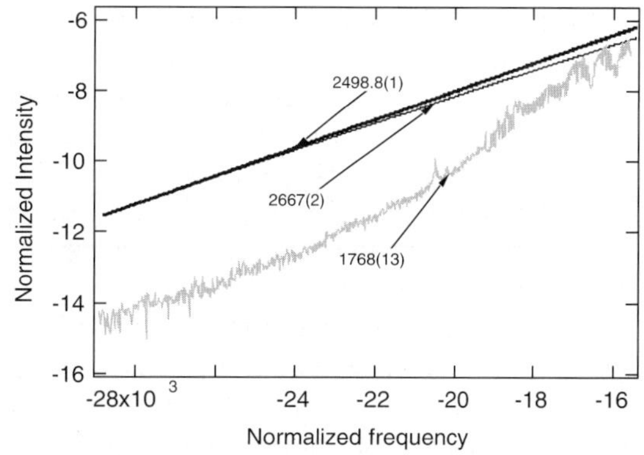

Fig. 5. Wien's plots of spectral data from the hotspot center. Normalized intensity (J) is plotted as a function of normalized frequency (see Eqs. (3) and (4) in text). Synthetic data set showing the Wien's plots are shown for the three cases in Fig. 4. Bold line: emissivity is constant; thin line: emissivity falls by 30%; gray line: Wien's plot of dark noise (once corrected for system response).

(Eq. (5)) plotted as a function of normalized frequency (Eq. (6)). The error introduced by using Wien's approximation to Planck's law is well understood, and ranges from less than 0.2% below 3500 K to less than 0.6% at 4500 K. A direct fit to Planck's law (assuming graybody behavior) is mathematically equivalent to fitting to the linearized Wien's law, because both methods fit only the same two adjustable parameters, T and ε, and the fitting errors are identical.

Linearity of the Wien plot is a necessary, but not sufficient, condition to assess the accuracy of a temperature measurement. The interdependence of the J and ω axes can hide serious errors such as those arising from chromatic aberration or diffraction effects within a linear-appearing Wien's plot. For example, a material with a 30% decrease in emissivity over the measured wavelength region (Fig. 4) will yield a temperature overestimate of 7%, or 165 K for a material emitting at 2500 K which is not resolved on the Wien plot, where the data appear linear (Fig. 5). To consider a worst-case example: even a spectrum of dark noise will generate a monotonically increasing function on a Wien plot (Fig. 5), and a straight-line fit through the data yields a temperature with an apparent precision of 0.7%. This is because the normalization technique magnifies the long wavelength intensity compared with the lower wavelength intensity. This is an inherent problem using this normalization technique, providing motivation to (1) crosscheck the raw data and (2) explore additional and complementary techniques to measure temperature from spectral intensity data.

3.2. Peak temperature: two-color pyrometry

Two-color pyrometry is a complementary method to analyze temperature data that may underscore wavelength-dependent irregularities arising from either emissivity variations or optical misalignments and miscalibrations. Two-color pyrometry is based on the same intensity normalization scheme outlined in the previous section, except instead of calculating a single temperature from the complete spectral data set, a series of temperatures are calculated from pairs of intensities separated by a fixed spectral difference, $\omega_1 - \omega_2$:

$$T = \frac{\omega_2 - \omega_1}{J(\omega_2) - J(\omega_1)} \quad (7)$$

J and ω are defined in Eqs. (5) and (6). By scanning the spectral window across the measured spectral range, temperature can be determined as a function of wavelength, depending on the width of the spectral window. The wavelength interval needs to be small enough so that the temperature measurements are relatively unaffected by possible wavelength-dependent emissivity, but large enough so that the corresponding intensity differences are significant.

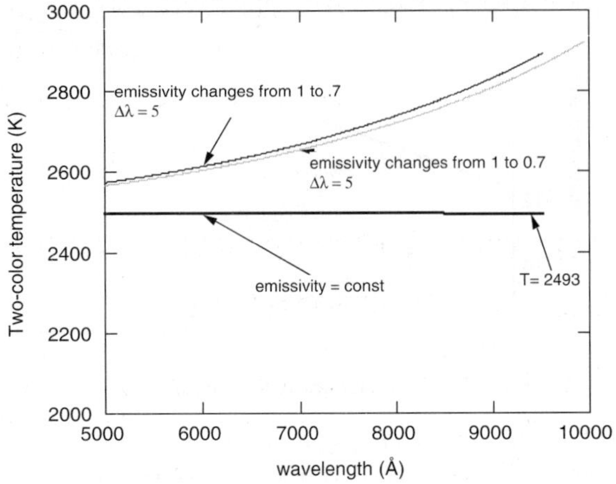

Fig. 6. Plots of two-color temperature as a function of wavelength for the spectral data shown in Fig. 4. The bold line shows the case for a graybody emitter. The thin lines show how temperature varies using a two-color pyrometer for the cases where emissivity decreases linearly across the spectral range, with the black line calculated using a window size of 500 Å, and the gray line using a window size of 50 Å.

In the ideal case, graybody radiation will yield a constant temperature across the wavelength range (Fig. 6). However, wavelength-dependent effects such as a slight miscalibration of the wavelength, optical misalignments, and wavelength-dependent emissivity will all introduce structure into a plot of temperature versus wavelength, some of which might bias the average slope on a Wien plot to higher or lower values, though not necessarily due to lower or higher temperatures. For example, the synthetic data showing a decrease in emissivity across the spectral range shows both an offset and an increasing temperature as a function of wavelength on the two-color plot (Fig. 6). For comparison, a tungsten filament decreases in emissivity only 10% over this range [http://www.pyrometry.com/relative_emissivity.html], yielding an even smaller effect than the example given in Fig. 6. Additionally, the dark noise data looks just like noise on the two-color plot. Thus, two-color pyrometry can be used to earmark meaningless data, and to help gauge the accuracy of temperatures determined by spectroradiometry by directly testing the graybody assumption. If the temperature is constant across the spectral range, the values can be accepted. Varying temperatures in the spectral range can indicate a problem—either wavelength-dependent emissivity effects, or an alignment or calibration problem with a chromatic signature. The pyrometry plot also has the advantage in its sensitivity to thermally induced electronic excitations that fluoresce visible light. For example, luminescence from stressed pressure media such as NaCl that is superimposed on the Planck function will show up as extreme thermal deviations at specific wavelengths if a narrow band wavelength filter is chosen for the analysis.

Even in the best case, the data from two-color pyrometry may be noisy and difficult to interpret. Perhaps the complexity of the two-color pyrometry plots are the reason that even though this method has been previously noted (Heinz and Jeanloz, 1987) it is rarely employed. Temperature analysis by two-color pyrometry could be misleading if it were not used in conjunction with other temperature measurement techniques. It is problematic because it magnifies noise and other errors. Unlike a direct fit to Wien's approximation to Planck's law (Fig. 5), the results from two-color pyrometry are a sensitive indicator of errors arising from miscalibration of wavelength and system response. Therefore, the temperature variation across a two-color pyrometry plot is a more realistic assessment of the precision in the temperature measurement than the goodness of fit to Planck's law. The use of varying-sized windows and spectral neighborhood averaging may allow for more interpretable results from two-color pyrometry methods.

3.3. Temperature gradients: spectral intensity

The above two methods make a temperature measurement only at a single place on the hotspot, usually the center, where intensity is brightest, the gradient at a minimum, and is least contaminated by possible spectral effects from the optical system. The assumption that the central temperature is best-determined is implicit in most of the laser-heating experiments (Andrault and Fiquet, 2001; Shen et al., 2001; Watanuki et al., 2001; Yagi et al., 2001). Temperature gradient measurements using the spectroradiometry-based analysis usually require either scanning an aperture across the hotspot, which physically selects different areas of the hotspot (Boehler, 2000), or by employing a two-dimensional detector in conjunction with an imaging spectrometer that can measure temperature gradients in one dimension (Kavner and Jeanloz, 1998b). The next two methods take advantage of the relationship between intensity and temperature described by Planck's law to use the central temperature as an anchor point to determine the radial temperature gradient. With these techniques an accurate measurement of the radial temperature gradient, relative to the central part of the hotspot, can be determined and the extent of chromatic aberrations can be quantitatively evaluated, as seen in the final section of the paper.

Assuming constant emissivity with temperature, Planck's law (Eq. (3)) yields a simple relative relationship between temperature and the intensity at a given wavelength:

$$\log\left(\frac{I_2(\lambda_0)}{I_{\max}(\lambda_0)}\right) \propto \frac{1}{T_{\max}} - \frac{1}{T_2} \tag{8}$$

$I_{\max}$ and $T_{\max}$ refer to the intensity and temperature, respectively, from the hotspot center. Since imaging spectroradiometry measures intensity as a function of wavelength and distance across the hotspot, this relationship can be used to calculate the temperature gradient, relative to the peak temperature, by using the peak temperature and intensity as an anchor point to scale the temperature with the intensity. Because of the strong dependence of intensity on temperature (a 50% change in intensity yields less than an 18% change in temperature), this technique generates an extremely precise measurement of the temperature gradient, relative to the central hotspot temperature. Using this technique, temperatures can be measured at much lower intensities than are allowable by spectroradiometry where the temperature calculation becomes extremely sensitive to both chromatic aberrations and dark noise at the hotspot edge, as shown in the last section of this paper.

Assuming counting statistics for uncertainties in the intensity measurements, the uncertainty in the temperatures measured through this technique are:

$$\frac{\delta T_2}{T_2} = \sqrt{\left(\frac{\delta T_{\mathrm{m}} T_2}{T_{\mathrm{m}}^2}\right)^2 + \left(\frac{\lambda k}{hc}\frac{T_2}{\sqrt{I_2}}\right)^2} \tag{9}$$

3.4. Temperature gradients: integrated intensity

The final method is similar to the previous method, but the sum of all of the measured intensities is used to calculate a temperature gradient instead of using only a single wavelength. Integrating the differential form of Wien's approximation to Planck's law in the spectral range of interest yields:

$$\int_{\lambda_1}^{\lambda_2} I\,\mathrm{d}\lambda = 2\varepsilon\pi c^2 h \left(\frac{kT}{hc}\right)^4 (f(\lambda_2) - f(\lambda_1)) \tag{10}$$

$$f(\lambda) = \left(\left(\frac{m}{\lambda}\right)^3 + 3\left(\frac{m}{\lambda}\right)^2 + 6\left(\frac{m}{\lambda}\right) + 6\right) \mathrm{e}^{-m/\lambda} \tag{11}$$

where $m = hc/kT$. In the limit of $\lambda_1 = 0$ and $\lambda_2 = \infty$, this reduces to the Stephan–Boltzmann equation. The results are very similar to the previous method. Because summing the intensities across the spectral range increases the signal to noise ratio of the data, better temperature measurements towards the lower-temperature edges of the hotspot can be made with this technique. This process has virtually the same effect as averaging the results from the previous method.

4. Application to synthetic one-dimensional temperature measurement

To show how these last two analysis methods remove the potency of the chromatic aberration issue, we show here (1) how small amounts of chromatic aberration and optical misalignment can create a false belief of a "flat-top" hotspot, and (2) how the scaling methods outlined above can be applied to generate an improved, accurate measurement of the temperature profile.

Intensity as a function of distance and wavelength (4000–8950 Å) was calculated according to Planck's law for a Gaussian temperature profile (HWHM = 15 μm) with peak temperature, $T_{\max} = 3000$ K. We then model the effects of incorrect background subtraction, poor focusing and CCD rotation has on the resultant temperature measurement. Codes for the synthetic calculations can be found in the supplemental data. Fig. 7a shows the original temperature profile (thick solid line), plus temperature profiles generated via a Wien's law fit (thin solid line). Corresponding temperature-dependent, wavelength-independent apparent emissivity values are shown in Fig. 7b. To

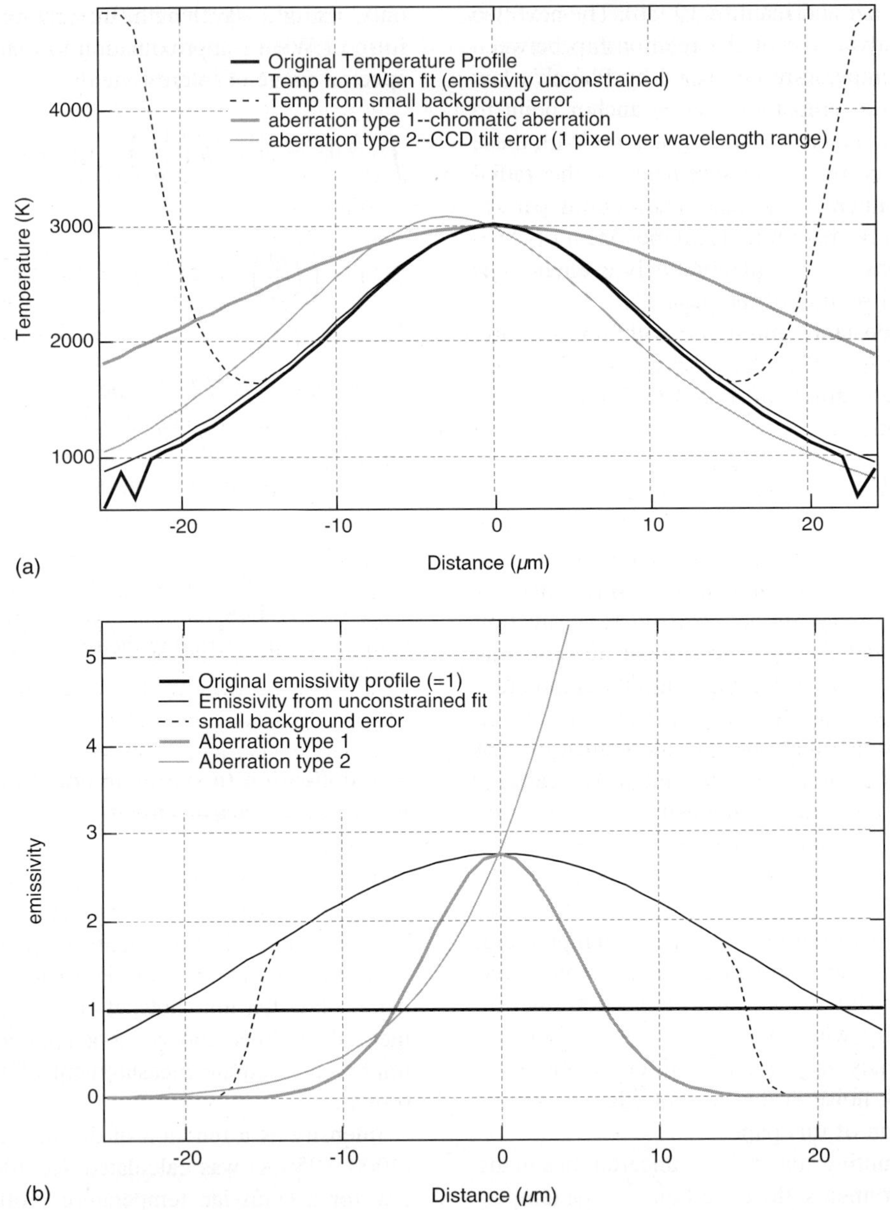

Fig. 7. Systematic errors in temperature measurements. (a) Modeled one-dimensional temperature profiles. The thick solid black line is the original temperature gradient, with a peak temperature of 3000 K and a half-width at half maximum of 15 μm. Thin black line: temperature as a function of distance recalculated using emissivity as a free parameter. Dashed line: Temperature calculated with an error in background correction (see text). Thick gray line: Temperature calculated using method 1 for a hotspot with chromatic aberration due to focusing. Thin gray line: Temperature calculated assuming a tilt error on a two-dimensional detector (see text). (b) Corresponding apparent temperature-dependent emissivities.

examine effects of small optical aberrations and misalignments on a synthetic hotspot measurement, the intensity profile at each distance was then recalculated with the wavelength-dependent Gaussian parameters adjusted to simulate measurement errors.

To simulate problems with incorrect background calibration, intensity values below a cutoff of 10 were automatically set to equal 10. Since the maximum intensities are $\sim 10^6$, this represents a 10 ppm error in background calibration. The effect on the inferred temperature is dramatic: with a temperature a non-monotonic function of distance, and showing an obviously non-physical result of temperature increasing with distance from the center of the hotspot (dashed line, Fig. 7a). Even at σ (15 μm), the temperature errors are significant, in excess of 400 K. Note the "wings" generated in the temperature profile; a feature sometimes observed in experimental measurements (e.g. Jeanloz and Kavner, 1996).

For an optical system that focuses red more tightly than blue, the effects are to give a significantly broader hotspot (thick gray line, Fig. 7a). We simulate this effect with a linear decrease in the intensity versus distance Gaussian width as a function of wavelength, with a Gaussian profile 16% wider at 4000 Å and 11% narrower at 8950 Å than when focused. This extent of chromatic aberration is in line with alternate calculations (Walter et al., this issue). The effect is to produce a nearly flat temperature profile, with apparent temperatures 1000 K greater than the actual temperature at 15 μm, alternately interpreted as a hotspot with a HWHM of 33 μm.

A slight rotation of the CCD image plane with respect to the spectral distribution has yet another effect on the inferred temperature distribution. If the CCD is tilted such that the peak intensity at 4000 Å hits the CCD just 1 pixel below where the peak intensity hits at 8950 Å, this results in the peak intensities having slightly different centers across the wavelength range. This corresponds approximately to the alignment resolution capable in several existing temperature measurement systems. The effect on the inferred temperature distribution is a fair representation of the temperature gradient, but the peak temperature provides a ~80 K overestimate, and is off center from the actual peak by about 3 μm. Referring to Fig. 3, if the laser position were "corrected" back to the center ($r = 0$), the now offset hotspot would give an underestimate of thermal expansions of 5% for MgO and 8% for NaCl.

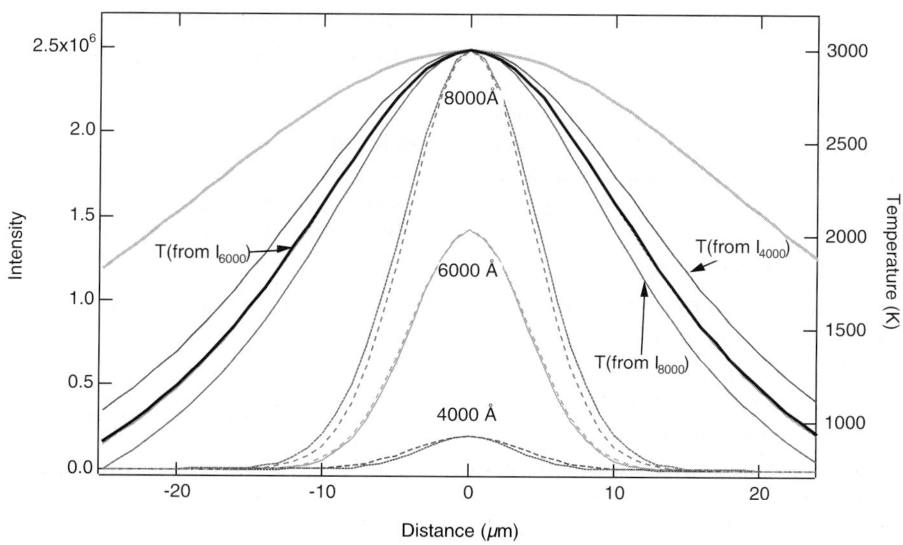

Fig. 8. Intensity profiles at 4000, 6000, and 8000 Å (dotted lines, left axis) and corresponding temperature profiles (solid lines, right axis) plotted as a function of distance. The thick black line is the original temperature profile. The thick gray line is the same as thick gray line. The thin gray lines are temperatures recalculated by scaling intensities as a function of distance, using the central temperature and intensity as a reference point (see method 3 in text). The dotted lines are the original intensity vs. distance profiles for the synthetic hotspot. The dashed lines are the intensities after applying a focusing-chromatic aberration.

As shown in all three cases, moderate aberrations and misalignments can result in extremely misleading temperature profiles. There is some information in the apparent emissivity profiles; however, even the emissivity profile of the "best" temperature measurement (thin black line) does not appear significantly qualitatively different from what might be considered the worst case: a flatter apparent hotspot than the actual temperature distribution (thick gray line).

By scaling the temperature measurement to the spectral and integrated intensity profile, as in the last two methods outlined in the previous section, these aberrations can be examined and diagnosed. Fig. 8 shows intensity profiles at 4000, 6000, and 8000 Å; both the original intensity profile, and with the above proscribed chromatic aberrations due to the red end of the wavelength having a tighter focus than the blue end. These aberrations have a small effect on the intensity distribution (Fig. 8, dashed lines); however, the intensity information combines to create strong biases in the apparent temperature determined by fitting to spectral intensity to a blackbody curve (thick gray lines in Fig. 7a and 8). Overlayed on Fig. 8 is the original temperature distribution, and temperature distributions created by scaling the intensity profiles at each of the three wavelengths to the central temperature. At a distance of 10 μm from the center part of the spot, fitting the spectral intensity data to a blackbody curve generates large errors—in this case, 545 K. This error is bypassed when the intensities, even suffering from chromatic aberrations, are scaled with the temperature and corresponding intensity at the center of the hotspot: at 4000 Å, the temperature is overestimated by only 145 K. At 8000 Å, the temperature is underestimated by 150 K. In the middle of the spectral range, the temperature profile is reproduced exactly. Therefore, a consideration of how intensity scales with temperature will significantly improve the measurement of the temperature gradient. This information will help clarify and evaluate measurements of thermoelastic properties in the synchrotron-based laser-heated diamond anvil cell.

5. Conclusion

Taken together, these four techniques, along with an examination of the raw spectral intensity data, provide a means to measure temperature, temperature gradients across a hotspot, and to realistically assess the accuracy and precision of those measurements. In the near future, the laser heating system that is currently under construction at UCLA will be employed to systematically test and evaluate these tests and corrections.

Acknowledgements

We thank Denis Andrault, Dion Heinz, Raymond Jeanloz, Mark Rivers, Guoyin Shen, and Dave Walker for comments and discussions. Portions of this work were performed at GeoSoilEnviroCARS (GSECARS), Sector 13, Advanced Photon Source at Argonne National Laboratory. GSECARS is supported by the National Science Foundation, Earth Sciences, Department of Energy, Geosciences, W.M. Keck Foundation and the United States Department of Agriculture. Use of the Advanced Photon Source was supported by the U.S. Department of Energy, Basic Energy Sciences, Office of Energy Research, under Contract No. W-31-109-Eng-38. AK acknowledges a Lamont-Doherty Earth Observatory postdoctoral fellowship and WRP acknowledges the Tuner postdoctoral fellowship from the University of Michigan for partial support of this work.

References

Andrault, D., 2001. Evaluation of (Mg, Fe) partitioning between silicate perovskite and magnesiowustite up to 120 GPa and 2300 K. J. Geophys. Res. 106, 2079–2087.

Andrault, D., Fiquet, G., 2001. Synchrotron radiation and laser heating in a diamond anvil cell. Rev. Sci. Instrum. 72, 1283–1288.

Andrault, D., Fiquet, G., Itie, J.P., Richet, P., Gillet, P., Hausermann, D., Hanfland, M., 1998. Thermal pressure in the laser-heated diamond-anvil cell: an X-ray diffraction study. Eur. J. Mineral. 10, 931–940.

Bodea, S., Jeanloz, R., 1989. Model calculations of the temperature distribution in the laser-heated diamond cell. J. Appl. Phys. 65, 4688–4692.

Boehler, R., 1992. Melting of the Fe–FeO and the Fe–FeS systems at high-pressure-constraints on core temperatures. Earth Planet. Sci. Lett. 111, 217–227.

Boehler, R., 2000. High-pressure experiments and the phase diagram of lower mantle and core materials. Rev. Geophys. 38, 221–245.

Boehler, R., Bargen, N.V., Choplas, A., 1990. Melting, thermal expansion, and phase transition of iron at high pressures. J. Geophys. Res. 95, 21731–21736.

Boehler, R., Chopelas, A., 1991. A new approach to laser-heating in high-pressure mineral physics. Geophys. Res. Lett. 18, 1147–1150.

Boehler, R., Ross, M., 1997. Melting curve of aluminum in a diamond cell to 0.8 Mbar: implications for iron. Earth Planet. Sci. Lett. 153, 223–227.

Boehler, R., Ross, M., Boerker, D.B., 1997. Melting of LiF and NaCl to 1 Mbar: systematics of ionic solids at extreme conditions. Phys. Rev. Lett. 78, 4589–4592.

Dewaele, A., Fiquet, G., Andrault, D., Hausermann, D., 2000. *P–V–T* equation of state of periclase from synchrotron radiation measurements. J. Geophys. Res. 105, 2869–2877.

Dewaele, A., Fiquet, G., Gillet, P., 1998. Temperature and pressure distribution in the laser-heated diamond-anvil cell. Rev. Sci. Instrum. 69, 2421–2426.

Fiquet, G., Andrault, D., Dewaele, A., Charpin, T., Kunz, M., Hausermann, D., 1998. *P–V–T* equation of state of $MgSiO_3$ perovskite. Phys. Earth Planet. Inter. 105, 21–31.

Fiquet, G., Andrault, D., Itie, J.P., Gillet, P., Richet, P., 1996. X-ray diffraction of periclase in a laser-heated diamond-anvil cell. Phys. Earth Planet. Inter. 95, 1–17.

Fiquet, G., Dewaele, A., Andrault, D., Kunz, M., Le Bihan, T., 2000. Thermoelastic properties and crystal structure of $MgSiO_3$ perovskite at lower mantle pressure and temperature conditions. Geophys. Res. Lett. 27, 21–24.

Heinz, D.L., 1990. Thermal pressure in the laser-heated diamond anvil cell. Geophys. Res. Lett. 17, 1161–1164.

Heinz, D.L., Jeanloz, R., 1987. Temperature measurements in the laser-heated diamond anvil cell. In: Manghnani, M., Syono, Y. (Eds.), High-pressure Research in Mineral Physics. American Geophysical Union, Washington, DC, pp. 113–127.

Jeanloz, R., Heinz, D.L., 1984. Experiments at high temperature and pressure-laser heating through the diamond cell. J. Phys. (Paris) 45, 83–92.

Jeanloz, R., Kavner, A., 1996. Melting criteria and imaging spectroradiometry in laser-heated diamond-cell experiments. Phil. Trans. R. Soc. Lond., Ser. A: Math. Phys. Eng. Sci. 354, 1279–1305.

Kavner, A., Duffy, T.S., 2001. Pressure–volume–temperature paths in the laser-heated diamond anvil cell. J. Appl. Phys. 89, 1907–1914.

Kavner, A., Duffy, T.S., Heinz, D., Shen, G., 1998. Thermoelastic behavior of platinum and MgO in the laser-heated diamond anvil cell. Trans. Am. Geophys. U. 79, S163.

Kavner, A., Duffy, T.S., Shen, G.Y., 2001. Phase stability and density of FeS at high pressures and temperatures: implications for the interior structure of Mars. Earth Planet. Sci. Lett. 185, 25–33.

Kavner, A., Jeanloz, R., 1998a. The high-pressure melting curve of Allende meteorite. Geophys. Res. Lett. 25, 4161–4164.

Kavner, A., Jeanloz, R., 1998b. High-pressure melting curve of platinum. J. Appl. Phys. 83, 7553–7559.

Lazor, P., Shen, G., Saxena, S.K., 1993. Laser-heated diamond-anvil cell experiments at high-pressure-melting curve of nickel up to 700 Kbar. Phys. Chem. Miner. 20, 86–90.

Li, X.Y., Manga, M., Nguyen, J.H., Jeanloz, R., 1996. Temperature distribution in the laser-heated diamond cell with external heating, and implications for the thermal conductivity of perovskite. Geophys. Res. Lett. 23, 3775–3778.

Manga, M., Jeanloz, R., 1996. Axial temperature gradients in dielectric samples in the laser-heated diamond cell. Geophys. Res. Lett. 23, 1845–1848.

Ming, L.C., Bassett, W.A., 1974. Laser heating in the diamond anvil press up to 2000C sustained and 3000C pulsed. Rev. Sci. Instrum. 45, 1115–1118.

Panero, W.R., Jeanloz, R., 2001a. Temperature gradients in the diamond anvil cell. J. Geophys. Res. 106, 6493–6498.

Panero, W.R., Jeanloz, R., 2001b. The effect of sample thickness and insulation layers on the temperature distribution in the laser-heated diamond cell. Rev. Sci. Instrum. 72, 1306–1308.

Saxena, S.K., Shen, G., Lazor, P., 1994. Temperatures in Earths core based on melting and phase-transformation experiments on iron. Science 264, 405–407.

Schneider, W.E., Goebel, D.G., 1984. Technical note. Laser Focus/Electro-Optics 20, 90.

Shen, G.Y., Mao, H.K., Hemley, R.J., Duffy, T.S., Rivers, M.L., 1998. Melting and crystal structure of iron at high pressures and temperatures. Geophys. Res. Lett. 25, 373–376.

Shen, G.Y., Rivers, M.L., Wang, Y.B., Sutton, S.R., 2001. Laser heated diamond cell system at the advanced photon source for in situ X-ray measurements at high pressure and temperature. Rev. Sci. Instrum. 72, 1273–1282.

Shim, S.H., Duffy, T.S., Shen, G.Y., 2000. The stability and *P–V–T* equation of state of $CaSiO_3$ perovskite in the Earth's lower mantle. J. Geophys. Res. 105, 25955–25968.

Shim, S.H., Duffy, T.S., Shen, G.Y., 2001. The post-spinel transformation in Mg_2SiO_4 and its relation to the 660-km seismic discontinuity. Nature 411, 571–574.

Sitaud, B., Thevenin, T., 1999. Melting-point measurements at high static pressures from laser heating methods: application to uranium. Int. J. Thermophys. 20, 1189–1198.

Sweeney, J.S., Heinz, D.L., 1993. Thermal-analysis in the laser-heated diamond-anvil cell. Pure Appl. Geophys. 141, 497–507.

Watanuki, T., Shimomura, O., Yagi, T., Kondo, T., Isshiki, M., 2001. Construction of laser-heated diamond anvil cell system for in situ X-ray diffraction study at SPring-8. Rev. Sci. Instrum. 72, 1289–1292.

Williams, Q., Jeanloz, R., 1990. Melting relations in the iron–sulfur system at ultra-high pressures—implications for the thermal state of the Earth. J. Geophys. Res. 95, 19299–19310.

Williams, Q., Knittle, E., Jeanloz, R., 1991. The high-pressure melting curve of iron—a technical discussion. J. Geophys. Res. 96, 2171–2184.

Yagi, T., Kondo, T., Watanuki, T., Shimomura, O., Kikegawa, T., 2001. Laser heated diamond anvil apparatus at the Photon Factory and SPring-8: problems and improvements. Rev. Sci. Instrum. 72, 1293–1297.

Yoo, C.S., Akella, J., Moriarty, J.A., 1993. High-pressure melting temperatures of uranium—laser-heating experiments and theoretical calculations. Phys. Rev. B 48, 15529–15534.

Available online at www.sciencedirect.com

SCIENCE DIRECT

PHYSICS OF THE EARTH AND PLANETARY INTERIORS

Physics of the Earth and Planetary Interiors 143–144 (2004) 541–558

www.elsevier.com/locate/pepi

The effects of chromatic dispersion on temperature measurement in the laser-heated diamond anvil cell

Michael J. Walter [a,*], Kenneth T. Koga [a,b]

[a] *Institute for Study of the Earths Interior, Okayama University, 827 Yamada, Misasa, Tottori 682-0193, Japan*

[b] *Laboratoire Sciences de la Terre, ENS-Lyon, CNRS UMR 5570, 46 Allee d'Italie 69364 Lyon Cedex 7, France*

Received 30 November 2002; received in revised form 22 July 2003; accepted 3 September 2003

Abstract

We have developed numerical simulations of spectroradiometric temperature measurement for refractive optical systems used in laser-heated diamond anvil cell (LHDAC) experiments. Chromatic dispersion in refractive optics is quantified by the wavelength-dependent radial displacement function. Results show that a combination of high peak temperature, steep temperature gradient, high spatial resolution, and large radial displacement can cause extreme errors in temperature, on the order of hundreds to thousands of degrees of Kelvin. Increasing detector bin size can dampen the magnitude of the error in peak temperature, but the average temperature produced in this way is still likely to be erroneous. Accurate peak temperature can be measured if the radial displacement is less than the radius of a region of constant temperature. Errors in accuracy can be reduced by locating the wavelength range that produces a minimum and symmetrical radial displacement. The precision of the fitted blackbody function is correlated positively with accuracy so that the optimal wavelength range can be located by minimizing the error in the blackbody fit. Modern spectroradiometric systems that utilize refractive optics are capable of accurate temperature measurement if proper steps are taken to identify and mitigate the effects of chromatic dispersion.

Keywords: Diamond anvil cell; Laser heating; Spectroradiometry; Temperature measurement; Chromatic aberration

1. Introduction

Accurate temperature measurement is critical for successful phase equilibrium experiments in the LHDAC. Temperature measurement is made typically by spectroradiometry, wherein the wavelength distribution of radiation emitted from the heated spot is measured with a spectrograph, and temperature is calculated by fitting the calibrated spectral intensity to a blackbody function (Bassett and Weathers, 1987; Heinz and Jeanloz, 1987). Because calculated temperature is sensitive to the wavelength distribution of the radiation, the optical pathway from sample to detector must faithfully preserve light intensity at all wavelengths.

Any element in the optical pathway that transmits light can potentially introduce chromatic dispersion because refractive index is a wavelength-dependent property. In practice, the elements of primary concern are the lenses used to gather and focus light into the spectrograph, as well as the diamond anvil due to its high dispersive power. An optical pathway that utilizes refracting lenses will always cause chromatic

* Corresponding author. Tel.: +81-858-43-3732; fax: +81-858-43-3450.

E-mail addresses: walter@misasa.okayama-u.ac.jp (M.J. Walter), kenneth.koga@ens-lyon.fr (K.T. Koga).

0031-9201/$ – see front matter © 2003 Elsevier B.V. All rights reserved.

doi:10.1016/j.pepi.2003.09.019

dispersion in light emitted from the sample. Reflective optics, which utilize mirrors, do not impart chromatic dispersion, although the optical pathway will still have some chromatic dispersion due to the diamond anvil (e.g. Boehler, 2000). Reflective optics can have poorer image quality that can potentially effect measured temperature (Jeanloz and Kavner, 1996). Although some LHDAC labs are using either commercially available short-focal length reflective lenses or 'home-made' longer focal length lenses (Boehler and Chopelas, 1991; Lazor et al., 1993; Sweeney and Heinz, 1998), many labs continue to use refractive optical systems (Dubrovinsky et al., 1999; Shen et al., 2001; Watanuki et al., 2001; Williams et al., 1987). This is probably because of the commercial availability of refractive lenses for high-power lasers at a variety of long focal lengths, as well as the more convenient design of using a single lens for the dual purpose of focusing the incident laser radiation and collecting and collimating the thermal emission from the sample. Refractive optical systems are also more convenient for combining LHDAC with X-ray diffraction at synchrotron facilities (Shen et al., 2001; Watanuki et al., 2001), where more severe restrictions exist on the geometry of laser focusing, the X-ray path, and light collection.

Workers have recognized that chromatic dispersion can potentially lead to large errors in calculated temperature (e.g., Boehler and Chopelas, 1991; Jeanloz and Kavner, 1996; Boehler, 2000; Shen et al., 2001), but no quantitative study has been reported that characterizes the performance of actual refractive optical systems used in LHDAC experiments. LHDAC systems equipped with refractive optical systems are used extensively for quantitative phase equilibrium and equation-of-state measurements (e.g. Knittle and Jeanloz, 1987, 1989; Fiquet et al., 1998; Shen et al., 1998; Dubrovinsky et al., 1999; Shim et al., 2001a,b), so it is important to assess the potential for uncertainty in temperature measurement in such experiments. Here, we first develop a method for quantifying the effects of chromatic dispersion on spectroradiometrically calculated temperature. Then we use numerical simulations to emulate the spectroradiometric performance in 'realistic' LHDAC experiments. Finally, using the information gained from the simulations as a guide, we present a method to optimize the accuracy of temperature measurement in actual LHDAC experiments.

2. Laser heating and imaging spectroradiometry

High temperatures are achieved in the DAC by focusing a high-powered infrared laser onto the sample, typically with a refracting objective (e.g. $f = 80$ mm). The laser-heated spot typically has a diameter of the order 50 μm. The sample is heated by absorbing the laser radiation, and it emits a continuous spectrum of thermal radiation with maximum intensities at near-infrared and optical wavelengths for temperatures in the range of 1500–5000 K. In modern LHDAC systems, temperature measurement is typically made by imaging spectroradiometry, as illustrated in Fig. 1 (see also Jeanloz and Kavner, 1996; Shen et al., 2001). Radiation from the sample (object plane) is collected and collimated by the laser focusing lens, and focused by a second long focal length refracting lens (e.g. $f = 1$ m) onto the entrance slit of an imaging spectrograph (image plane). A sample magnification of 10–20 times is typical. Within the spectrograph the divergent light is collimated by an aspheric mirror, dispersed by a diffraction grating, and re-focused by a second aspheric mirror onto a two-dimensional CCD detector (e.g. 1024 × 256 pixels), with one dimension corresponding to wavelength (e.g. 1024 pixels) and the other a spatial dimension (e.g. 256 pixels). Thus, the slit and CCD detector are at a common image plane. The spectrograph wavelength resolution is typically less than 0.5 nm. The spatial resolution at the object plane is determined by the size of the heated spot, the magnification factor, the slit-width, and the CCD pixel size as illustrated in Fig. 1b. In practice, the spatial resolution is typically on the order of 5–10 μm^2. Thus, for the example in Fig. 1, which is based on the system in our lab, the light from a 50 μm spot would be illuminated onto about 25 rows of the CCD, and a wavelength spectrum of light intensity, or spectral intensity distribution, would be acquired for each row.

A temperature is calculated for each row by fitting the spectral intensity distribution to a blackbody equation. According to Planck's law, spectral intensity is related to temperature by

$$I_\lambda = \varepsilon c_1 \lambda^{-5}\{\exp(c_2\lambda^{-1}T^{-1}) - 1\}^{-1} \quad (1)$$

where I_λ is the spectral intensity, ε the emissivity, λ the wavelength, c_1 and c_2 the physical constants, and T the temperature (K). If the emissivity of the material is independent of wavelength, then in principle the

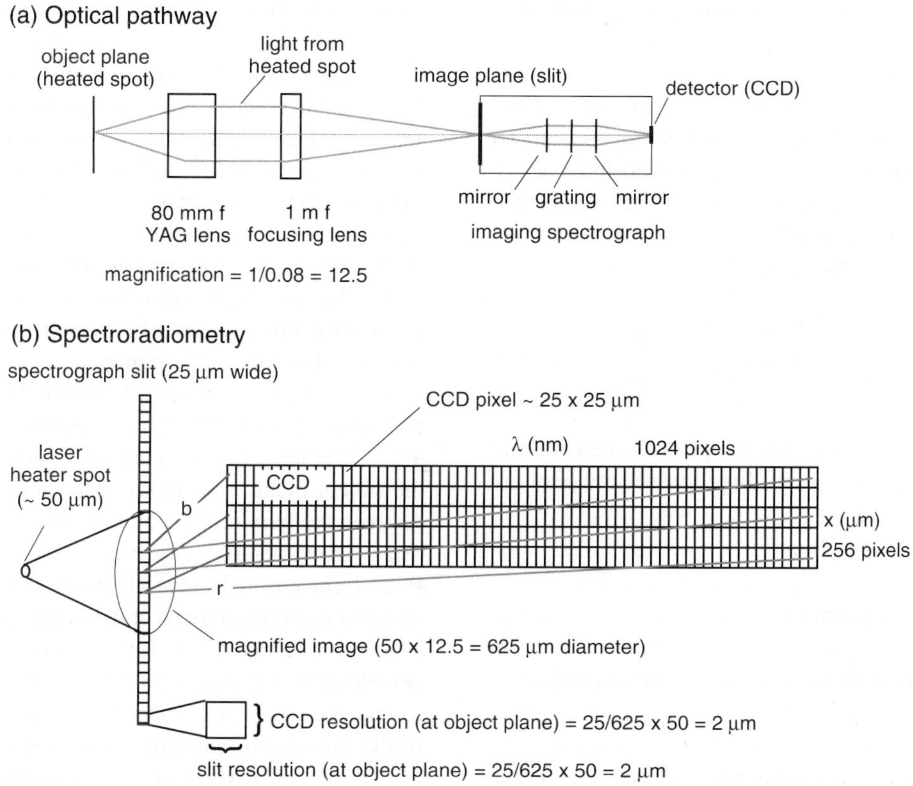

Fig. 1. Schematic illustrating the typical optical and spectroradiometry systems used for measuring temperature in the LHDAC. (a) The optical pathway of thermal radiation from the heated spot to the detector. The YAG lens collects and collimates the light and a second lens focuses the light onto the entrance slit of an imaging spectrograph. Within the spectrograph the light is diffracted and focused onto a two-dimensional CCD detector. (b) Schematic showing the principle of spectroradiometry. The image of the heated sample is magnified by the lenses and focused onto the entrance slit of the spectrograph. The slit and detector are at a common image plane (detector rows shown superimposed on the slit). The light is dispersed over the 1024 pixel dimension of the CCD. The spatial resolution at the sample, or bin size at the object plane, is determined by the image size at the slit, the CCD pixel size (~25 μm) in one dimension (256 pixels), and the slit-width (25 μm in this example) in the other dimension. In this example the bin size is 4 μm^2.

calculated temperature will be accurate and errors will depend on the accuracy and precision of the spectral intensity correction and the spectrometric measurement. There are two unknowns, ε and T, and solving for T requires non-linear curve fitting with ε as a free parameter. Temperature is more conveniently calculated using Wien's approximation to Planck's law:

$$J = \ln(\varepsilon) - \frac{\omega}{T} \tag{2}$$

using a normalized intensity, $J = \ln(\lambda^5 c_1^{-1})$, and normalized wavelength, $\omega = c_2\lambda^{-1}$ (Heinz and Jeanloz, 1987). A linear fit to a spectrum transformed to coordinates of J and ω yields both temperature (inverse slope) and emissivity (y-intercept). Wien's approximation and Planck's law give nearly identical temperatures up to about 3000 K, but diverge at higher temperatures with Wien's law giving progressively lower values (~1% at 5000 K).

The measurable quantity in spectroradiometry is the spectral response of the detector to the sample radiance (in $\mathrm{W\,sr^{-1}\,(mm^2)^{-1}\,nm^{-1}}$). However, the measured spectral intensity, I_λ, is distorted relative to the actual spectral radiance due to the wavelength dependence of detector efficiency as well as interferences in the optical pathway (e.g. absorption and multiple reflections in mirrors and beamsplitters). To eliminate these distortions, the sample spectral intensity is corrected using a standard of known spectral radiance, such as a tungsten ribbon-filament lamp calibrated relative to NIST standards. The lamp is placed at the position normally occupied by the sample so that the spectral radiance of a spot on the ribbon is acquired through the optical pathway. The spectral intensity of the unknown is then divided by that of the lamp, and the dividend is multiplied by the known, calibrated spectral radiance of the lamp:

$$(I_\lambda)_{\text{unknown-corrected}} = \frac{(I_\lambda)_{\text{unknown}}}{(I_\lambda)_{\text{standard}}} \times (R_\lambda)_{\text{standard}} \quad (3)$$

This yields a calibrated sample spectral intensity that can be fitted to a blackbody function for temperature calculation.

3. Chromatic dispersion

3.1. Chromatic dispersion in refractive optical systems

The focal length of a refractive lens depends on the refractive index of the lens material. Because refractive index is wavelength-dependent, refractive lenses chromatically disperse light, referred to as chromatic aberration. In a typical refractive optical system used with LHDAC, as shown in Fig. 1 and discussed above, thermal radiation emitted from the sample is collimated by the laser focusing lens and focused onto a spectrograph slit by a second lens. In such a system, paraxial rays have negligible chromatic dispersion, but marginal rays off the optic axis are dispersed and have a wavelength-dependent focal length as shown in Fig. 2a; the further off-axis the greater the dispersion. For a given region at the object plane, some wavelength-dependent portion of the emitted light is not in focus at a reciprocal region at the image plane (i.e. slit and detector) and is "missing" from the spectrum. For example, in Fig. 2a light of wavelength λ_g is in focus at the image plane, whereas off-axis light with shorter wavelength, λ_b, has a shorter focal length and light of longer wavelength, λ_r, has a longer focal length.

Consider out-of-focus light of a single wavelength, λ_r, emitted from a small region at the object plane. Depending on the distance from the optic axis, λ_r light comes to focus at variable distances behind the image plane (Fig. 2b). The absolute focal deviation depends on the specifications of the lens system. However, any region at the image plane, for example one pixel on a CCD detector, still receives one "lens-worth" of λ_r light in focus. That is, the out-of-focus λ_r light emitted from the region of interest at the object plane is replaced by λ_r light emitted from a different region at the object plane (Fig. 2c). The size of the region from which λ_r light will come to a common focus is defined by what we hereafter refer to as the radial displacement at the object plane.

As illustrated in Fig. 2d, radial displacement depends on the focal deviation, x, and the numerical aperture of the lens. In this example, λ_g light emitted from the object plane is in focus at the image plane. Point A gives the maximum radial displacement for λ_r light that will come to focus with λ_g light emitted from P. λ_r light emitted from all points from P to A will come to a common focus at the image plane. The maximum displacement, y, for a particular out-of-focus wavelength depends on the absolute focal deviation, x, and the angle, ϕ, where $y = x \tan\phi$. This relates to the numerical aperture of the lens (N.A. $\sim \sin\phi$), and lenses with larger numerical aperture have correspondingly greater aberrations. Thus, optical radiation over a range of wavelengths emitted along y will be in focus at the image plane at the same position as green light emitted at P. If y is at a uniform temperature, chromatic dispersion has no effect on the spectral intensity of the focused radiation because light emitted at all points has the same intensity distribution. However, if a temperature gradient exists along y, as shown in Fig. 2d, then the spectral intensity of the focused radiation will be affected according to Planck's law, as described below. The effect on intensity at a given wavelength can be quite large, as according to Stephan's law intensity increases as the 4th power of temperature. For this example, λ_g light emitted at P and at high temperature would be mixed with λ_r light emitted over a range of lower temperatures along y.

3.2. Effect of chromatic dispersion on spectroradiometrically measured temperature

Consider thermal radiation emitted from a laser-heated spot that is focused onto the image plane. For an ideal lens system, radiation at all wavelengths

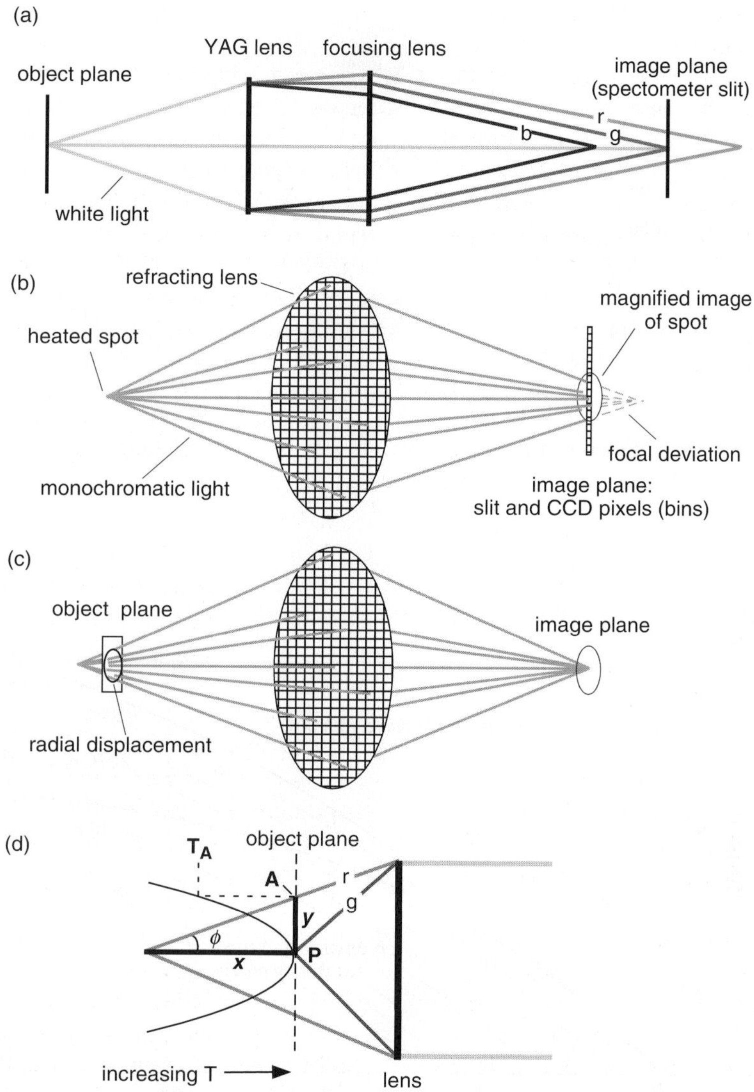

Fig. 2. (a) Schematic of a refractive optical lens system showing chromatic dispersion of non-paraxial light. In this example light of wavelength λ_g is in focus at the image plane, λ_b light has a shorter focal length, and λ_r light has a longer focal length. (b) Schematic showing the focal deviation at the image plane as a function of distance off the optic axis for monochromatic light. (c) Schematic showing the focal deviation and radial displacement at the object plane as a function of distance off the optic axis for monochromatic light. (d) Schematic showing how the radial displacement, y, for a given wavelength relates to the focal deviation, x, with respect to the object plane. In this example, P is the point of interest, and λ_g light emitted at P is in focus at the image plane. Off-axis λ_r light emitted from P is out of focus at the image plane, but is replaced by λ_r light emitted along y, with a maximum radial displacement defined by $x \tan\phi$.

collected over some spatial element at the image plane is emitted from a reciprocal region at the object plane. The spatial relation between these two 'bins' of radiation is the image magnification. As described above, in practice such a bin might correspond to an $\sim 5\,\mu m^2$ region at the object plane. Now consider a lens system with chromatic dispersion focusing light emitted from a laser-heated spot with a Gaussian temperature distribution, and with λ_b light in focus at the image plane (Fig. 3). Consider the effects of

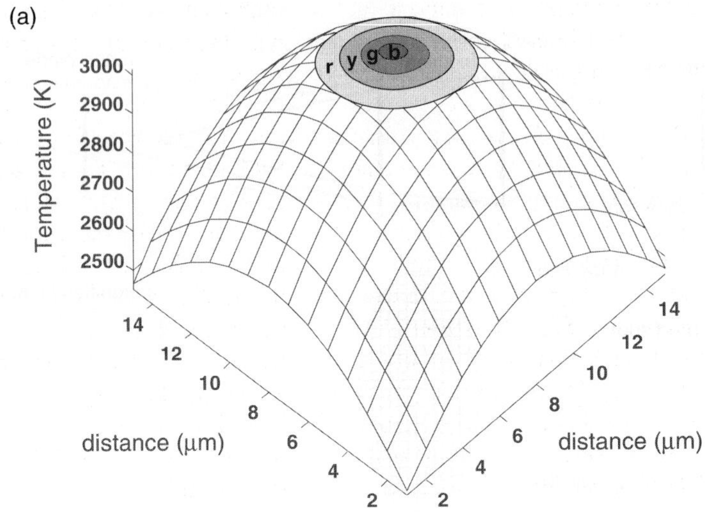

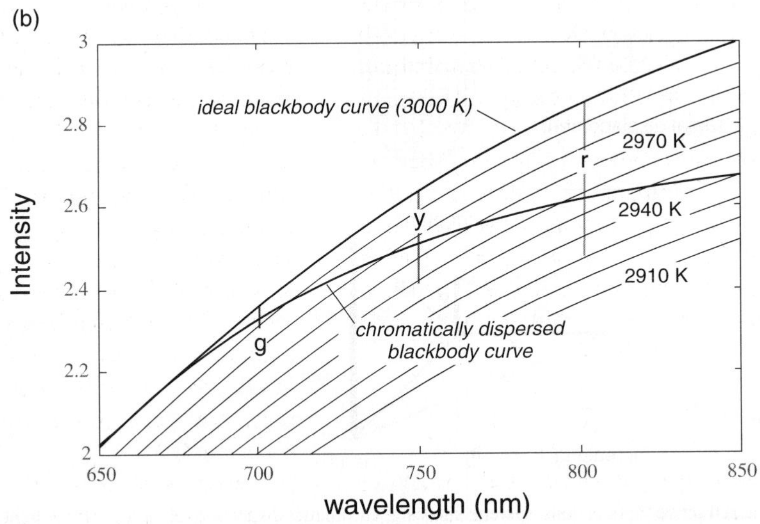

Fig. 3. (a) A schematic showing the radial displacement of light of different wavelengths emitted from the peak region of a Gaussian temperature profile. In this example, light of wavelength λ_b is in focus from the 'bin-of-interest', region b. Regions of radial displacement for light of longer wavelength, labeled as g, y, and r, that come to a common focus with light from region b are also shown. (b) A schematic showing the effect on the spectral distribution of the chromatic dispersion shown in (a). Light of wavelength λ_b is in focus and is correctly measured. The intensity of light emitted at longer wavelengths is averaged over regions defined by the radial displacement. The chromatically dispersed curve in this example has a peak in spectral intensity distribution that is shifted to shorter wavelengths relative to the ideal curve, which causes an over-calculation of temperature.

dispersion for a region, or 'bin', which corresponds to the peak temperature, labeled b in Fig. 3a. Paraxial radiation at all wavelengths comes to focus in the same detection bin at the image plane (e.g. one row of a CCD). Off-axis radiation emitted from the bin at the object plane but at wavelengths other than the focal wavelength is out of focus, and is excluded from the detection bin. This radiation is replaced by radiation emitted from lower temperature regions, at distances away from the bin-of-interest at the object plane according to the wavelength-dependent, radial displacement. In the example in Fig. 3a, out-of-focus radiation (λ_g, λ_y and λ_r) at wavelengths longer than the focal wavelength, λ_b, is replaced by lower intensity radiation emitted at lower temperatures from the flanks of the heated region. Thus, λ_b light is correctly measured, but light emitted at longer wavelengths is a mixture of light emitted over a range of temperatures and, therefore, intensities. Fig. 3b qualitatively shows the effect on the Planck distribution. Only at the focal wavelength, taken as 650 nm in this example, is intensity correctly measured at the detector. For a given out-of-focus wavelength, the intensity reflects the integrated sum of radiation emitted from a circular region (i.e. for a circular lens) with a radius dictated by the maximum displacement of the lens system for that wavelength. In the example in Fig. 3, the measured intensity distribution of the chromatically dispersed light has a peak that is shifted to shorter wavelengths relative to the actual distribution. This distortion of the spectral intensity distribution will result in a reduction in the "goodness-of-fit" to either the Planck or Wien function, and would cause an overestimation of temperature in this particular case. Notice, however, that even though the calculated temperature would be higher, the absolute intensities are lower at all out-of-focus wavelengths relative to the ideal blackbody curve.

4. Numerical methods

As long as the optical properties of the imaging and detection system are well characterized, the effect of chromatic dispersion on calculated temperature can be modeled accurately. A simple, single element refractive lens can have focal deviations on the order of several hundreds to thousands of microns over a range of optical and near-infrared wavelengths (~500–900 nm). A multiple-element refractive lens, either glued together or air-spaced (dialyte), can correct for chromatic dispersion and considerably reduce the aberration over a specific range of wavelengths (see Kingslake, 1978). A variety of achromatic (two-lens system) and apochromatic (three-lens system) lenses are commercially available for focusing high-energy infrared laser radiation while minimizing chromatic dispersion at optical and near-infrared wavelengths.

The two curves in Fig. 4a show relative focal length (normalized to 546 nm) as a function of wavelength for both an $f = 80$ mm achromat and an $f = 50$ mm apochromat, lenses which are used in our lab for focusing YAG radiation and light collection. The data points are focal deviations measured by us for the refractive lens system including the YAG lens and secondary focusing lens (see figure caption for details). The close correlation of the observed data to the vendor supplied specifications for the YAG lens shows that chromatic dispersion introduced by that lens dominates the system. From 500 to 900 nm the maximum focal deviation can be as much as 200 μm, but is much less over smaller ranges in wavelength that take advantage of the concave curve shapes.

Fig. 4a gives the deviation in focal length with wavelength, whereas calculating the effect of chromatic dispersion on spectroradiometric temperature requires determining the maximum radial displacement at the object plane for a given range of wavelengths (i.e. the magnitude of y in Fig. 2d as a function of wavelength). The radial displacement is calculated by multiplying the focal deviation for a given wavelength by the numerical aperture of the lens. Fig. 4b shows the absolute radial displacement over a 400 nm range in wavelength when 500 nm radiation is in focus for both the YAG achromats used in our lab. Note that the magnitude of the radial displacement depends critically on the chosen range of wavelength.

The diamond anvil itself also imparts chromatic dispersion. Based on the known wavelength dependence of refractive index in diamond, we have calculated that for a 2 mm thick diamond anvil coupled with the numerical aperture of our $f = 80$ mm lens, the maximum radial displacement at the object plane of light between 600 and 700 nm is less than 2 μm. The magnitude of dispersion increases at shorter wavelengths and decreases at longer wavelengths. Thus, dispersion caused by the diamond is a second-order effect relative to that of the lens.

In the examples that follow, numerical models are used to predict the effects of chromatic dispersion on calculated temperature profiles. Variables that are

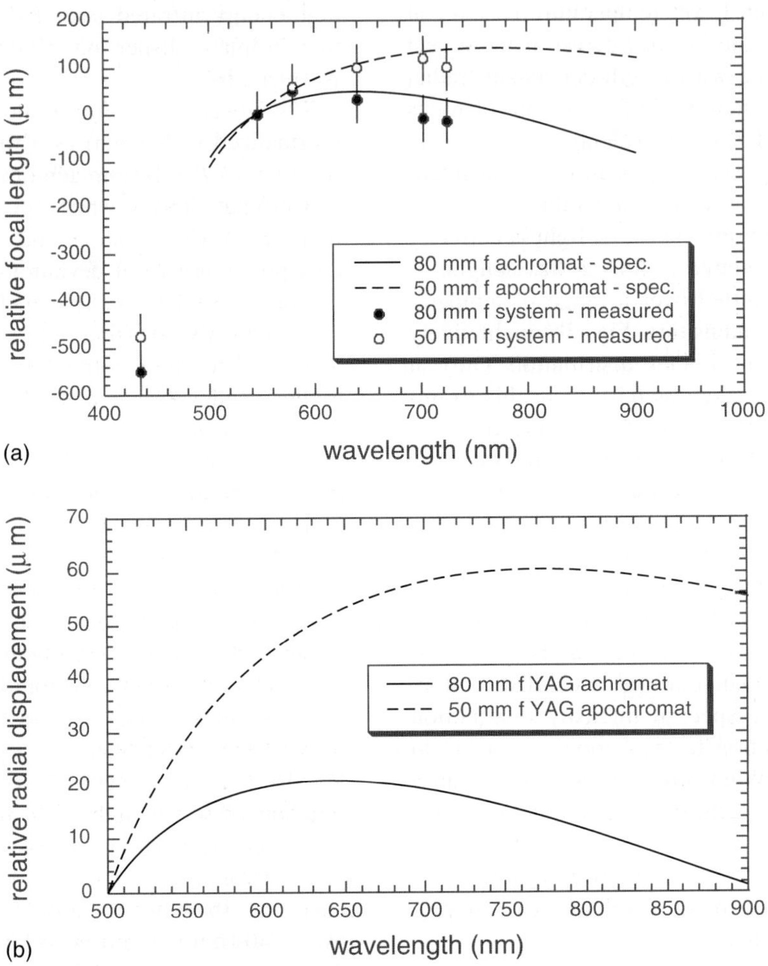

Fig. 4. (a) Focal length, normalized to 546 nm, as a function of wavelength for two YAG focusing lenses used in our lab. The curves show vendor supplied focal specifications for an $f = 50$ mm apochromat and an $f = 80$ mm achromat (data supplied by Sigma Koki, Japan). The data points are focal deviations measured in our lab for the optical system including YAG lens, $f = 1$ m achromatic secondary focusing lens, and spectrograph. A 1 mm pinhole was placed at the position normally occupied by the sample during laser heating. The pinhole was back-illuminated by either a Hg:Ar or Ne calibration lamp, so that the pinhole acted as a source. Optical elements were fixed except the YAG lens, which could be translated linearly perpendicular to the source at μm resolution. The focal length (YAG lens to pinhole) of individual calibration lines, such as the 546 nmHg line, was determined by translating the YAG lens toward and away from the pinhole and measuring the intensity of the line. The spectrometer was centered on the calibration line and acquisitions were taken for a fixed time. The position that produced maximum intensity was determined to be the focus. The method was repeated for six different calibration lines. Error bars represent 2σ uncertainty in locating the focal position. (b) Maximum radial displacement for the two YAG lenses normalized to a 500 nm focal length. Radial displacement is calculated as the focal deviation at a given wavelength times the numerical aperture of the lens, which is 0.15 for the $f = 80$ mm lens and 0.24 for the $f = 50$ mm lens.

considered include the initial temperature profile (T), focal wavelength (λ_0), wavelength range of interest (λ_r), radial displacement function (H), and spatial resolution ($s_x = x_{i+1} - x_i$, $s_y = y_{i+1} - y_i$). The following general algorithm is used to calculate the effects of chromatic dispersion:

1. A temperature profile is specified, $T = f(r, \theta)$, where r is radial coordinate and θ is azimuth. In this study we consider both Gaussian and truncated Gaussian ('flat-top') profiles, as illustrated in Fig. 5. These profiles are chosen to emulate idealized temperature profiles that could result from

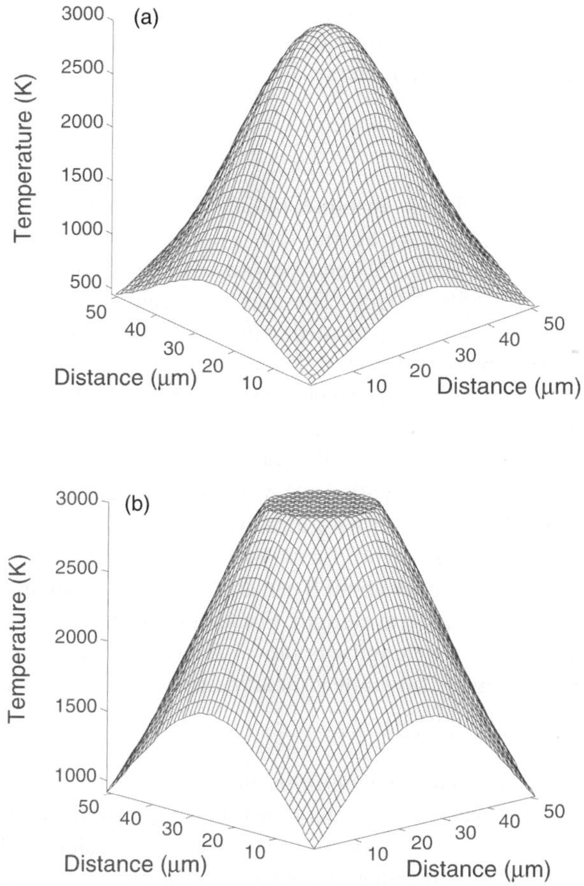

Fig. 5. Temperature profiles used in this study: (a) Gaussian profile with a peak temperature of 3000 K and an FWHM of 30 μm; (b) truncated Gaussian or 'flat-top' profile with a peak temperature of 3000 K and an FWHM of 40 μm. Spatial resolution is 1 μm².

TEM_{00} (Gaussian) and TEM_{01^*} or mixed mode laser radiation, respectively. The temperature of the Gaussian profile as a function of radius is given by

$$T(r) = (T_{\max} - T_{\min}) \exp\left(\frac{-r^2}{2\sigma^2}\right) + T_{\min} \quad (4)$$

for all θ and where $T_{\max}$ and $T_{\min}$ are the temperature at the hot-spot and edge of sample, respectively; σ the radius of the heated area at half $T_{\max}$, and at $r = 0, T = T_{\max}$. The truncated Gaussian profile is given by

$$T(r) = (T_{\max} - T_{\min}) \left(\exp\left(\frac{-r^2}{2\sigma^2}\right) - \exp\left(\frac{-{r_t}^2}{2\sigma^2}\right)\right) + T_{\max} \quad (5)$$

for $r > r_t$ where r_t is the truncation radius. At $r = 0, T = T_{\max}$, and $T(r) = T_{\max}$ for all $r = r_t$, as illustrated in Fig. 5. The temperature profiles are modeled by a two-dimensional square grid of points at a spatial resolution of 1 μm². Thus, the continuous polar coordinate is converted to a discrete Cartesian coordinate system, $x_i = r\cos\theta$ and $y_j = r\sin\theta$.

2. At each point in the temperature grid, the intensity of radiation, I, as a function of wavelength, λ, is calculated from Planck's function, P, at a wavelength interval of 1 nm over a specified wavelength range, where

$$I_{\lambda:x_i,y_i} = P(\lambda, T_{x_i,y_i}) \quad (6)$$

3. The maximum radial displacement at the object plane as a function of wavelength, $H(f, \lambda_0, \lambda_r)$, where f is numerical aperture, is determined for the specified wavelength range, as illustrated in Fig. 4b. In this example 500 nm is the focus (λ_0). At 600 nm and for the $f = 80$ mm lens, the radial displacement is zero at the optic axis and has a maximum of about 20 μm, which corresponds to marginal rays entering the edge of the lens. Chromatic dispersion can be represented as a power series of the distance off the axis (Kingslake, 1978). However, since we do not know the exact form of the series for our lens system, a linear deviation from the optic axis to the maximum radial displacement is assumed (0–20 μm in for 600 nm light). This is a conservative assumption, and the amount of radial displacement will be overestimated somewhat near the optic axis.
4. A new spectral intensity curve is calculated for each grid element at 1 nm wavelength resolution:

$$I^{\text{newspectra}}_{\lambda_k:x_i,y_j} = \left(\frac{s_x s_y}{\pi (H_{\lambda_k})^2}\right) \sum_{y_i} \sum_{x_i}^{H_\lambda > \sqrt{x_i^2 + y_i^2}} I_{\lambda_k:x_i y_j} \quad (7)$$

for $H > 0$. The intensity of a given wavelength is the integrated sum of all light at that wavelength (λ_k) as specified by the radial displacement function (H). However, because the model is calculated using a grid of square, not circular, elements (e.g. Fig. 3a), the radial distribution is approximated. If a grid element falls within a specified radius, it is included in the calculation. The effective resolution is about 1 μm^2.

5. A Planck (Eq. (1)) or Wien (Eq. (2)) function is fitted to the new spectral intensity curve, $I^{newspectra}$, for each grid point, x_i, y_j, by the least squares method and a new, chromatically dispersed temperature profile is calculated.

5. Spectroradiometric simulations

The number of possible model permutations is very large given the variables that are considered (T, λ_0, λ_r, H, and $s_{x:y}$). Here, we present a set of heuristic models that are designed to illustrate the important effects that individual variables have on calculated temperature profiles, but use parameters that are chosen to emulate conditions in actual LHDAC experiments. The results are used to estimate the potential magnitude of errors in actual experiments, and to provide the practitioner with a method for accurate temperature measurement.

Two types of temperature gradient are modeled, Gaussian and truncated Gaussian (flat-top), with peak temperatures of 3000 K as shown in Fig. 5. The wavelength range of interest is initially set at 650–850 nm, which corresponds generally to an efficient collection range for Si-based CCD detectors. The radial displacement function for our $f = 80$ mm YAG achromatic lens system is used as representative of a refractive optical system. First, we investigate the relationship between focal wavelength and focal range, and then use these results to determine the effects of varying other model parameters.

5.1. Effect of variable focal wavelength

Fig. 6 shows absolute radial displacement functions for the $f = 80$ mm lens system at focal wavelengths of 650, 750 and 850 nm over a wavelength range of 650–850 nm. The maximum radial displacement is about 15 μm, occurring when either 650 or 850 nm light is in focus. Thus, for a region at the object plane at which 650 nm light is in focus, light at all longer wavelengths is out-of-focus. 850 nm light from a region with a radius of about 15 μm would come to a common focus with 650 nm light, as would 750 nm light from a region with a radius of about 7 μm. Notice that for a 750 nm focus, the maximum radial displacement is only about 8 μm and is nearly symmetric about the focal wavelength. For an 850 nm focus, light at shorter wavelengths is out-of-focus, and the radial displacement function is nearly opposite to the case of a 650 nm focus.

5.1.1. Gaussian temperature profiles

Fig. 7 shows the effect that varying the focal wavelength between 650 and 850 nm has on calculated temperature profiles. When the focal wavelength is 650 nm, temperature is grossly overestimated at the peak of the temperature distribution, in this case by

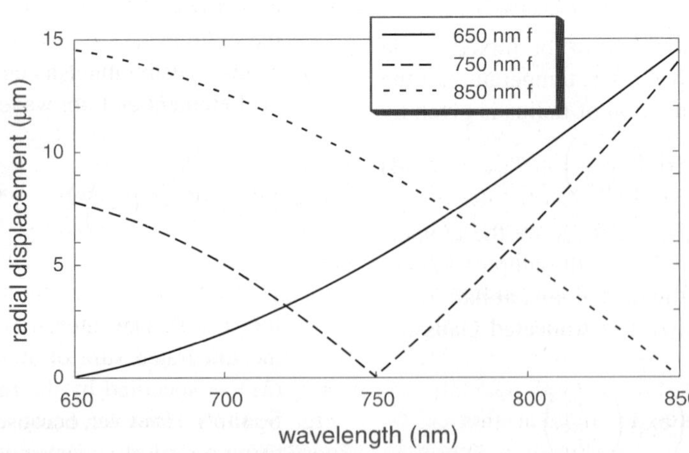

Fig. 6. Maximum radial displacement as a function of wavelength for the $f = 80$ mm YAG lens at three different focal positions: 650 nm (solid line), 750 nm (long dash line), and 850 nm (short-dashed line).

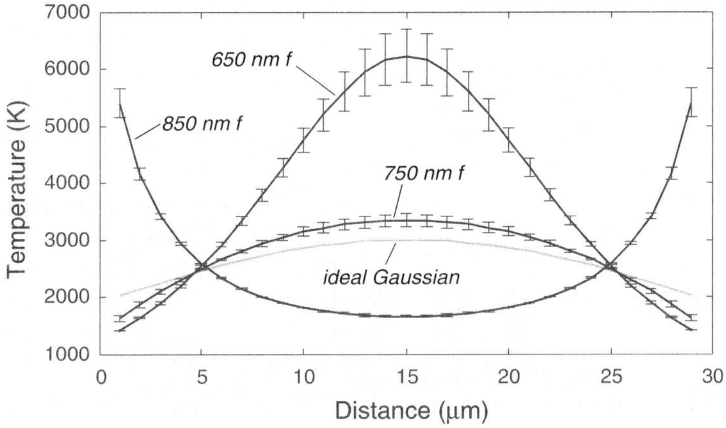

Fig. 7. Calculated temperatures (dark lines) across the peak of an ideal Gaussian temperature profile (light line) for focal lengths of 650, 750, and 850 nm, using a wavelength range of 650–850 nm. Temperatures are calculated using the Planck function. Error bars show uncertainty in the fitted function at the 95% confidence level.

about 3200 K. This occurs because out-of-focus light at wavelengths longer than the focal wavelength is replaced by lower intensity light emitted from the flanks at lower temperatures, causing an artificial shift in the peak of the spectral intensity distribution to shorter wavelengths. Notice, however, that on the flanks of the temperature profile there is a crossover, and temperature is underestimated for regions greater than about 10 μm away from the peak as the spectral intensity distribution becomes dominated by light at longer wavelengths. When the focal wavelength is 750 nm, both shorter and longer wavelengths are out-of-focus and have similar radial displacements. This results in a cancellation effect, and the calculated temperature at the peak in this particular case is only overestimated by ~270 K, and the entire temperature profile is reproduced fairly well. When the focal wavelength is 850 nm, the temperature at the peak is underestimated considerably (~1360 K), but grossly overestimated at the flanks. This occurs because out-of-focus light from the peak at shorter wavelengths is replaced by light from the flanks emitted at lower temperatures and lower intensities. This causes a shift in the peak of the spectral distribution to longer wavelengths. At some point along the flanks, the spectral intensity distribution becomes dominated by light at shorter wavelengths and temperature is overestimated. Thus, we can see from Fig. 7 that the calculated peak temperature is very sensitive to the position of the focal wavelength relative to the wavelength range, and the shape of the calculated temperature profile goes through an inversion.

Fig. 8 shows how the spectral intensity curves become distorted due to chromatic dispersion relative to ideal Wien (8a) and Planck (8b) functions for the case of a 650 nm focus (Fig. 7). Curve 1 is for the peak of the Gaussian temperature profile, and peaks 2 and 3 are 7 and 10 μm away from the peak (down temperature), respectively. At the peak temperature the spectral intensity distribution is highly distorted, resulting in a large error in both the accuracy of the calculated temperature and in the precision of the fitted Wien and Planck functions. As the calculated temperature approaches the ideal temperature (curves 2 and 3), the error in the fitted blackbody functions decrease. Curve 3 is at a position very close to where the calculated temperature crosses over from overestimation to underestimation of temperature (see Fig. 7), and the error in the fitted functions is relatively small. At the crossover point the temperature is accurate, although this is essentially a fluke in that it occurs because the spectral distribution of chromatically dispersed light emitted at both higher and lower temperatures cancels exactly.

Fig. 9 shows that accuracy and precision have a positive correlation with the radial displacement. As the radial displacement (i.e. chromatic dispersion) is minimized, both errors decrease. The general

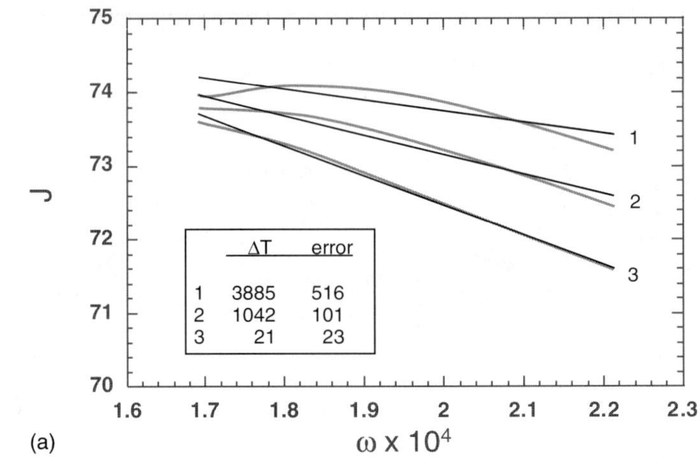

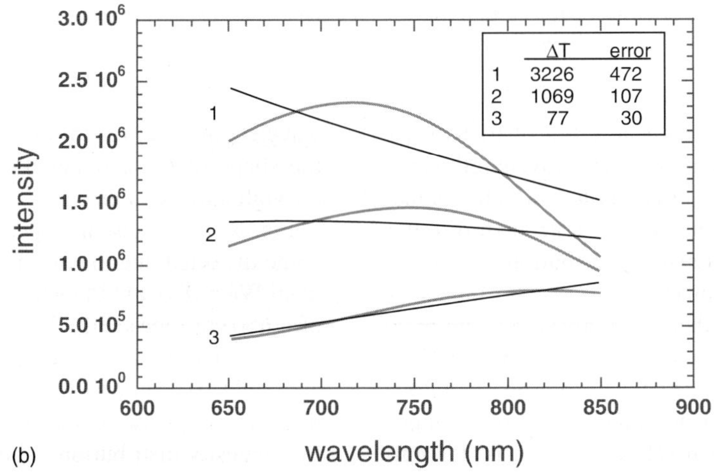

Fig. 8. Examples of fitted: (a) Wien and (b) Planck blackbody functions at three positions across the peak of the Gaussian temperature profile for the model shown in Fig. 7 with a focal position at 650 nm. Dark lines are best-fits to the calculated spectral distributions, which are shown as light lines. Line 1 is for the peak position, line 2 is 7 μm away from the peak, and line 3 is 10 μm away from the peak. The insets list the ΔT (fitted − initial) and precision (95% confidence level) of the fitted functions.

correspondence between accuracy and precision is a powerful tool for optimizing temperature measurement using refractive optical systems, as will be shown in detail below.

The models shown in Fig. 7 are for a constant temperature profile, but the maximum temperature and temperature gradient are also important parameters. Fig. 10 shows calculated peak temperature versus actual peak temperature for a Gaussian temperature distribution. We used a constant value of 30 μm for the FWHM of the distribution, meaning that as peak temperature increases so does the temperature gradient. This is justified because in real LHDAC experiments, temperature gradients tend to increase with peak temperature, as the size of the heated spot is essentially constant. The radial displacement for the $f = 80$ mm lens was used, with a focus at 650 nm and a wavelength range of 650–850 nm. Fig. 10 shows that as the peak temperature and temperature gradient increase, overestimation of the peak temperature also increases.

5.1.2. Flat-top temperature profiles

Fig. 11a and b shows calculated temperatures for a flat-top profile at focal wavelengths of 650 and 750 nm.

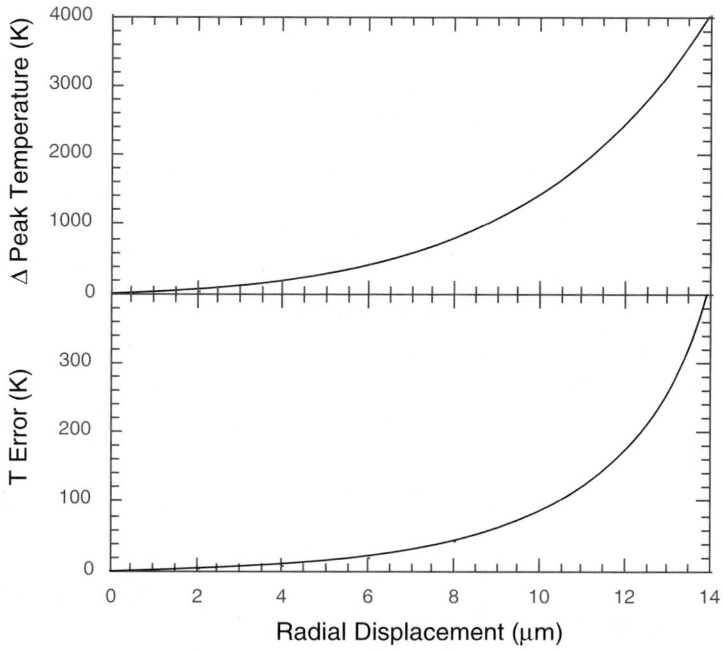

Fig. 9. Change in: (a) ΔT at the peak (fitted peak T − ideal peak T); and (b) precision in the Planck fit as a function of radial displacement. Results are calculated for the Gaussian temperature profile, and the radial displacement is linear.

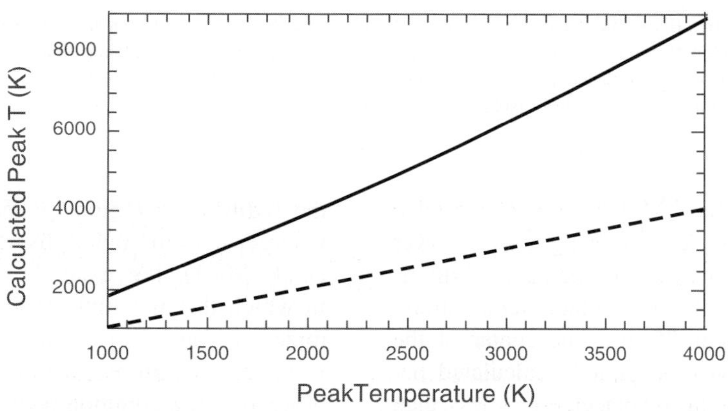

Fig. 10. Calculated peak temperature (solid line) vs. actual peak temperature for the Gaussian temperature profile. Model temperatures are calculated using the Planck function, the radial displacement for the $f = 80$ mm lens, a 650 nm focal position, and a 650–850 nm range. The FWHM of the Gaussian profile is constant at 30 μm. The dashed line is a 1:1 line.

The flat top has a considerable effect on the calculated temperature profile relative to the Gaussian case. In Fig. 11a, at a 650 nm focus the central temperature is overestimated by only about 150 K, whereas at the edges of the flat top, temperature is overestimated by about 1250 K. The bimodal 'bat-ears' distribution in maximum temperature occurs because the flat-top half-width is a large fraction of the maximum radial displacement. In the center, only a small fraction of light from the lower temperature flanks is included

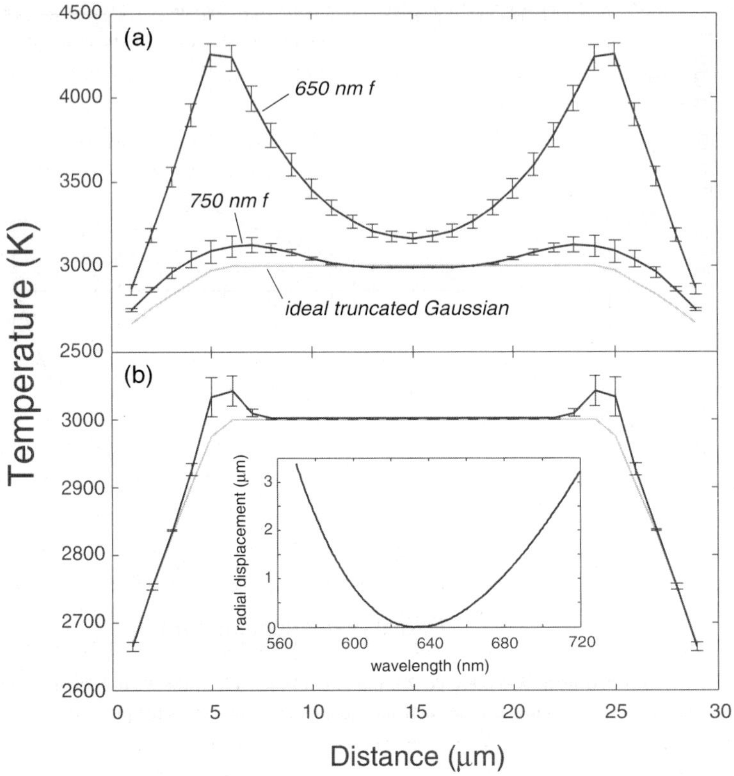

Fig. 11. Calculated temperatures (dark lines) across the peak of the ideal 'flat-top' temperature profile (light lines) for focal lengths of: (a) 650 and 750 nm, and a wavelength range of 650–850 nm. Temperatures calculated using the Planck function. Error bars show uncertainty in the fitted function at the 95% confidence level. (b) Calculated temperature profile for a 632 nm focus and a wavelength range of 570–720 nm, conditions that minimize the radial displacement (see inset).

in the spectral distribution. Moving toward the edge of the flat top, progressively more light from lower temperature regions is included in the spectral distribution, causing an increase in calculated temperature. When the focal length is 750 nm, the center of the temperature distribution is accurately calculated because the maximum radial displacement is less than the half-width of the flat top, and all light that contributes to the spectral distribution is at the same temperature. Temperature is overestimated by about 100 K at the edges of the flat-top region.

As with the Gaussian case, the precision in the fitted functions deteriorates as the error in accuracy increases, but is zero in the center of the flat top when temperature is accurately calculated (in the model case). Thus, as a general rule, 'accurate temperatures can be measured with refractive optical systems when the radius of a region of near-constant temperature is larger than the radial displacement' (see also Shen et al., 2001). For example, Fig. 11b shows a case in which the maximum radial displacement is minimized for our $f = 80$ mm lens system. We chose a focal wavelength of 632 nm because the YAG lens is designed for a common focus at 632 and 1064 nm. In order to minimize the radial displacement, a wavelength range from 570 to 720 nm was determined (see insert, Fig. 11b). The maximum radial displacement is only about 3 μm and the entire temperature profile is modeled with high accuracy. The temperature of the flat top is modeled correctly except at the edges, where it is overestimated by about 40 K. Even for the Gaussian temperature distribution this optimal condition results in an error of only about 40 K in peak temperature.

5.2. The effect of bin size

The models shown to this point were calculated at a spatial resolution of about 1 μm^2. However, in practice the size of the detection bin will vary depending on the type of detector and spatial filtering (e.g. input slit geometry). As discussed above, with an array detector the minimum size of a bin depends on CCD pixel size, image magnification, and input slit-width. Bin size can also be regulated in array detectors by spatial filters and digitally (i.e. by software). As detection bin size increases, the area on the sample that contributes to the spectral intensity distribution increases. Thus, a single temperature is calculated as the average temperature of the area. Also, as bin size increases the effect of chromatic dispersion lessens because the size of the region of radial displaced light becomes smaller relative to the size of the bin (see Jeanloz and Kavner (1996) for discussion of the effects of slit geometry).

Fig. 12 shows the relationship between bin size and calculated peak temperature with and without chromatic dispersion. The curves for the case of chromatic dispersion are calculated based on the results for a Gaussian distribution and 650 and 750 nm focal lengths (e.g. Fig. 7). Binning is one-dimensional, bin size is varied along a 1 μm wide strip across the peak of the temperature profile to emulate binning using a two-dimensional CCD detector with a vertical slit, and the bin is centered on the peak. As bin size increases, the calculated temperature for the bin decreases. Thus, the general rule is that when the bin in centered on peak temperature, larger bins produce lower calculated temperatures, and for the case of chromatic dispersion, the magnitude of the error in accuracy can be dampened considerably. Indeed, it is ironic that the better the spatial resolution of the detection system, the greater the potential magnitude of error in measured temperature. Therefore, using a spatial filter to view only a small region at peak temperature at the center of the sample does not eliminate the effects of chromatic dispersion, but enhances it, because light from the flanks remains in focus with light from the region of interest. As discussed by Jeanloz and Kavner (1996), small spatial filters can also add undesirable diffraction effects.

6. Method for accurate temperature measurement using refractive optics

Accurate temperature measurement using a refractive optical system requires minimizing the radial displacement over the wavelength range of interest. In the last section we showed that for our $f = 80$ mm lens system, temperature can be measured very accurately if the focal wavelength and focal range are optimized to produce a minimum and symmetrical displacement (Fig. 11b). During an actual LHDAC experiment, however, it is difficult to know the focal

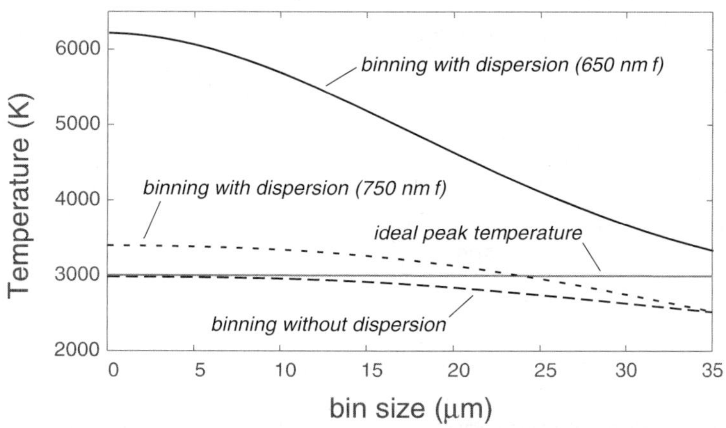

Fig. 12. Calculated Planck temperature at the peak of the Gaussian temperature profile as a function of bin size. The solid line is for the $f = 80$ mm YAG lens with a focal position at 650 nm and the short-dashed line is for a focus of 750 nm. Temperature calculated using a wavelength range of 650–850 nm. The long-dashed line is for the dispersion-free case. Binning is one-dimensional, and bin size is varied along a 1 μm wide strip across the peak of the temperature profile with the bin centered on the peak.

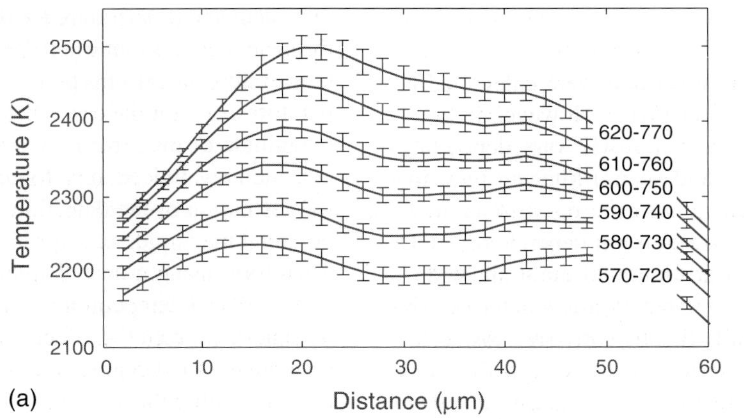

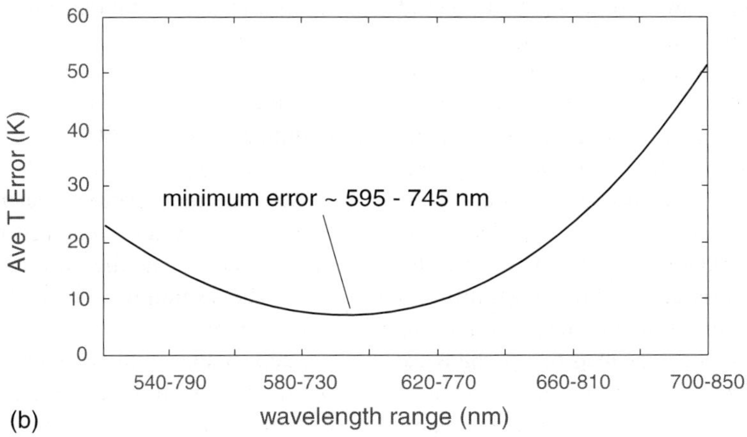

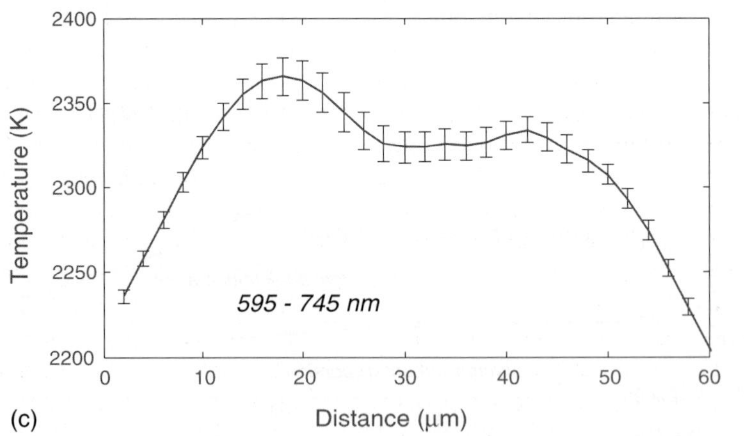

Fig. 13. Method for accurate temperature calculation when using a refractive optical system in an actual LHDAC experiment. A 10 μm thick piece of tantalum foil with a diameter of about 50 μm was pressurized to ~30 GPa in an Al_2O_3 pressure medium, using diamonds with 300 μm culets and a gasket hole about 100 μm in diameter. Spectra were obtained with a spectrometer equipped with a 1024 × 256 format Si-CCD array detector, using an input slit-width of 20 μm and an acquisition time of 500 ms. Temperatures are measured along an approximately 3 μm wide strip across the center of the heated spot, and the bin size is approximately 9 μm^2: (a) calculated Planck temperatures across the hot-spot using a 150 nm range of wavelength, and each curve represents a different wavelength range; (b) the average error in the Planck fit as a function of the of wavelength range used in the calculation; (c) the 'best-fit' temperature profile for the wavelength range of 595–745 nm.

wavelength because the focus is typically achieved visually with a white light source. Further, it is common to defocus the laser to some degree to heat a wider region. In any case, the exact focal wavelength is not known. However, the relationship between accuracy and precision in the fitted blackbody function can permit a direct determination of the optimal wavelength range for accurate temperature measurement.

Fig. 13 shows the results of temperature measurement during a laser heating experiment in the DAC. A 10 μm thick piece of tantalum foil with a diameter of ~50 μm was pressurized to about 30 GPa in an Al_2O_3 pressure medium. The sample was heated with a multimode YAG laser with an approximately 40 μm diameter spot, and held at constant peak temperature using a feedback system. Spectroradiometry was made using a spectrometer equipped with a 1024 × 256 format Si-CCD array detector with an input slit-width of 20 μm. Spectral intensity calibration was made with a W-lamp radiance standard. Temperatures were measured along an approximately 3 μm wide strip across the center of the heated spot, and the bin size was approximately 9 μm^2. Fig. 13a shows calculated temperatures across the hot-spot for a single 500 ms acquisition. Temperatures are calculated using a 150 nm range of wavelength, and each curve represents a different wavelength range. Fig. 13b shows how the average error in the fitted blackbody function varies as a function of wavelength range. The average error in fitted temperature (precision) goes through a distinct minimum, reaching a value of ±8 K for a wavelength range of 595–745 nm. Notice that this wavelength range is very close to the range determined as best for our system at a focal wavelength of 632 nm. The shift to a somewhat higher wavelength range probably means the focal wavelength was somewhat lower than 632 nm during heating. A precision of ±8 K approaches the maximum precision of ±3 K observed for our system based on replicate analysis of the W-lamp used as an unknown source of uniform temperature (i.e. no effects of chromatic dispersion). Fig. 13c shows the 'best-fit' temperature profile for the wavelength range of 595–745 nm. The bimodal temperature maxima are within about 30 K of the flat, central portion of the distribution, and this may be a mild effect of chromatic dispersion due to the sharp temperature gradient at the edges or to variations in laser mode intensity across the hot-spot.

It is important to point out that another source of error in both accuracy and precision can occur due to deviations from wavelength-independent sample emissivity (i.e. deviations from graybody emissivity). Here we assume a constant emissivity for temperature calculation, even though this is unlikely to be the actual case. If the sample emissivity varies linearly as a function of wavelength then there will be no effect on the precision of a fitted Wien function, but the fit to a Planck function will have lower precision. If the emissivity varies in a non-linear way as a function of wavelength then precision is reduced in both Wien and Planck functions. An error in accuracy will be introduced in all cases of wavelength-dependent emissivity. Unfortunately, the dependence of emissivity on wavelength, especially at high temperature, is not known for most materials. Wavelength-dependent emissivity remains as a source of uncertainty, estimated to be in the range of 100–200 K, in all spectroradiometric temperature measurements in the DAC (Heinz et al., 1991; Jeanloz and Kavner, 1996; Shen et al., 2001).

7. Summary

We have shown through numerical simulations of spectroradiometric temperature measurement that when using a refractive optics the temperature error can be extreme. The combination of a steep temperature gradient and a large chromatic dispersion can lead to errors on the order of hundreds to thousands of degrees of Kelvin. Binning can dampen the magnitude of the error in peak temperature considerably, but with considerable loss of information about the temperature gradient.

Improvements in laser technology have resulted in more uniform heating over larger regions of the sample producing relatively flat-topped temperature profiles (Panero and Jeanloz, 2001; Shen et al., 2001). Modern spectrographs equipped with two-dimensional CCD array detectors permit light collection at high spatial resolution. These factors, in combination with the availability of a selection of achromatic lenses for focusing high-powered laser radiation, provide the potential for measuring accurate temperature profiles with refractive optical systems. Based on our models, we predict that in a modern LHDAC experiment utilizing refractive optics, accurate peak temperatures can be measured when a wavelength range is chosen that minimizes the magnitude of the radial displacement relative to a region of uniform temperature (see Fig. 11). In practice, the optimal wavelength range is determined by minimizing the error in the fitted blackbody function (Fig. 13). Placing a variable aperture, or f-stop, between the collimating and secondary focusing lenses also minimizes the magnitude of chromatic dispersion. As the aperture is reduced, marginal rays are progressively eliminated and the light that reaches the detector includes a higher proportion of less dispersed, near-paraxial rays. Reducing the aperture also means a greater depth of field and less light available for detection. The amount of 'stop-down' that can be tolerated before focus is difficult to achieve or light levels are too low will depend on the optical configuration.

In modern LHDAC experiments, a potential exists for considerable error in temperature measurement. Errors on the order of several hundred degrees K in peak temperatures can easily occur if care is not taken to minimize the effects of chromatic dispersion, as illustrated in Figs. 7 and 11. This takes on greater importance as the LHDAC is now being used for more precise phase equilibrium measurements, and is especially critical when LHDAC is used in combination with synchrotron radiation to measure pressure at high temperature using the equation-of-state of an internal standard (Fiquet et al., 1998; Shen et al., 1998; Shim et al., 2001a,b).

Acknowledgements

We thank Guoyin Shen and an anonymous reviewer for comments that improved the manuscript. Support for this work provided to KK by a JSPS postdoctoral fellowship. Chamonix Briel is thanked for stimulus.

References

Bassett, W.A., Weathers, M.S., 1987. Temperature measurement in a laser-heated diamond cell. In: Manghnani, M.H., Syono, Y. (Eds.), High-pressure Research in Mineral Physics. Terra Scientific Publishing Co., Tokyo, pp. 129–133.

Boehler, R., 2000. High-pressure experiments and the phase diagram of lower mantle materials. Rev. Geophys. 38, 221–245.

Boehler, R., Chopelas, A., 1991. A new approach to laser heating in high pressure mineral physics. Geophys. Res. Lett. 18, 1147–1150.

Dubrovinsky, L.S., et al., 1999. Study of laser heated iron using third generation synchrotron X-ray radiation facility with imaging plate at high pressures. Phys. Chem. Miner. 26, 539–545.

Fiquet, G., et al., 1998. *P–V–T* equation of state of $MgSiO_3$ perovskite. Phys. Earth Planet. Inter. 105, 21–31.

Heinz, D.L., Jeanloz, R., 1987. Temperature measurement in the laser-heated diamond anvil cell. In: Manghnani, M., Syono, Y. (Eds.), High Pressure Research in Mineral Physics. American Geophysical Union, Washington, DC, pp. 113–127.

Heinz, D.L., Sweeney, J.S., Miller, P., 1991. Al laser heating system that stabilizes and controls the temperature: diamond anvil cell applications. Rev. Sci. Instrum. 62, 1568–1575.

Jeanloz, R., Kavner, A., 1996. Melting criteria and imaging spectroradiometry in laser-heated diamond-cell experiments. Phil. Trans. Roy. Soc. London A 354, 1279–1305.

Kingslake, R., 1978. Lens Design Fundamentals. Academic Press, New York, 366 pp.

Knittle, E., Jeanloz, R., 1987. Synthesis and equation of state of (Mg, Fe)SiO_3 perovskite to over 100 gigapascals. Science 235, 668–670.

Knittle, E., Jeanloz, R., 1989. Melting curve of (Mg, Fe)SiO_3 perovskite to 96 GPa: evidence for a structural transition in lower mantle melts. Geophys. Res. Lett. 16, 421–424.

Lazor, P., Shen, G., Saxena, S.K., 1993. Laser-heated diamond anvil cell experiments at high pressure: melting curve of nickel up to 700 kbar. Phys. Chem. Miner. 20, 86–90.

Panero, W.R., Jeanloz, R., 2001. Temperature gradients in the laser-heated diamond anvil cell. J. Geophys. Res. 106, 6493–6498.

Shen, G., Mao, H.K., Hemley, R.J., Duffy, T.S., Rivers, M.L., 1998. Melting and crystal structure of iron at high pressures and temperatures. Geophys. Res. Lett. 25, 373–376.

Shen, G., Rivers, M.L., Wang, Y., Sutton, S.R., 2001. Laser heated diamond cell system at the advanced photon source for in situ X-ray measurements at high pressure and temperature. Rev. Sci. Instrum. 72, 1273–1282.

Shim, S.-H., Duffy, T.S., Shen, G., 2001a. In situ determination of the post-spinel transformation in Mg_2SiO_4 using the laser-heated diamond anvil cell. Nature 411, 571–573.

Shim, S.-H., Duffy, T.S., Shen, G., 2001b. Stability and structure of $MgSiO_3$ perovskite to 2300 km depth in Earth's mantle. Science 293, 2437–2440.

Sweeney, J.S., Heinz, D.L., 1998. Laser-heating through a diamond anvil cell: melting at high pressures. In: Manghnani, M.H. (Ed.), Properties of Earth and Planetary Materials at High Pressure and Temperature. American Geophysical Union, Washington, DC, pp. 197–213.

Watanuki, T., Shimomura, O., Yagi, T., Kondo, T., Isshiki, M., 2001. Construction of laser-heated diamond anvil cell system for in situ X-ray diffraction study at Spring8. Rev. Sci. Instrum. 72, 1289–1292.

Williams, Q., Jeanloz, R., Bass, J., Svendsen, B., Ahrens, T.J., 1987. The melting curve of iron to 250 gigapascals: a constraint of the temperature at earth's center. Science 236, 235–236.

Available online at www.sciencedirect.com
SCIENCE DIRECT®
ELSEVIER
Physics of the Earth and Planetary Interiors 143–144 (2004) 559–574
PHYSICS OF THE EARTH AND PLANETARY INTERIORS
www.elsevier.com/locate/pepi

Modern techniques in measuring elasticity of Earth materials at high pressure and high temperature using ultrasonic interferometry in conjunction with synchrotron X-radiation in multi-anvil apparatus

Baosheng Li [a,*], Jennifer Kung [a], Robert C. Liebermann [a,b]

[a] *Mineral Physics Institute, Stony Brook University, Stony Brook, NY 11794-2100, USA*
[b] *Department of Geosciences, Stony Brook University, Stony Brook, NY 11794-2100, USA*

Received 13 February 2003; received in revised form 4 July 2003; accepted 25 September 2003

Abstract

In the past 6 years, state-of-the-art techniques have been developed for the studies of elastic properties of polycrystalline and single-crystal materials using simultaneous ultrasonic and X-ray diffraction techniques at high pressures and temperatures in multi-anvil apparatus. The capability of these techniques is illustrated using recent data for mantle phases such as forsterite, wadsleyite, $CaSiO_3$ perovskite, and (MgFe)O ferripericlase. Adaptation of the X-radiography technique enables direct measurements of sample length at high pressures and temperatures, enabling studies of unquenchable phases, liquids, and multi-phase aggregates. The use of dual-mode transducers produces P and S wave data at the same pressure and temperature conditions in a single acoustic experiment, thereby eliminating the uncertainties caused by the need for interpolating travel-time data to calculate the elastic wave velocities, and bulk and shear moduli at high pressure and temperature. Implementation of the transfer function method for travel-time measurements opens new opportunities for studying time-dependent processes using ultrasonic measurements, such as phase transformations and plastic deformation.

Keywords: Elasticity; Ultrasonic interferometry; Equation-of-state; X-radiography; Multi-anvil apparatus

1. Introduction

Elastic properties of Earth materials at high pressure and high temperature are crucial for understanding the composition, structure and dynamics of the Earth. Seismological investigations provide the primary source of information about the properties and processes of the Earth's interior in the form of density and velocity profiles as a function of depth. Seismic tomography on a global scale reveals lateral variations of compressional (P) and shear (S) wave velocities and densities at various depths of the Earth's interior, reflecting differences of chemical composition, mineralogy, and/or temperature. Successful interpretation of these seismic models in terms of these variables requires experimental and theoretical information on the elasticity of deep Earth materials under elevated conditions of pressure and temperature.

Ultrasonic techniques (e.g., McSkimin, 1950) have been used extensively in the last four decades to

* Corresponding author. Tel.: +1-631-632-9642; fax: +1-631-632-8410.
E-mail address: baosheng.li@sunysb.edu (B. Li).

0031-9201/$ – see front matter © 2004 Elsevier B.V. All rights reserved.
doi:10.1016/j.pepi.2003.09.020

obtain the elastic properties of Earth materials at high pressure and/or high temperature, from early experiments at $P < 1$ GPa (e.g., Gieske and Barsch, 1966; Schreiber and Anderson, 1966; Kumazawa and Anderson, 1969) to subsequent work at pressures up to 3 GPa (e.g., Jackson and Niesler, 1982; Rigden et al., 1988, 1992; Gwanmesia et al., 1990; Webb, 1989), and to recent experiments at $P > 10$ GPa (e.g., Li et al., 1998a; Knoche et al., 1998). Ultrasonic interferometry at simultaneous high pressures and temperatures has been made at gigahertz frequencies in diamond-anvil cells (e.g., Bassett et al., 2000; Jacobsen et al., 2002; Spetzler et al., 1993, 1996; Reichmann et al., 1998). Although the GHz technique has the advantage of requiring only very small specimens, the maximum pressure and temperature range achieved to date are still limited. At the same time, there has been continuous effort to develop a "transducer and sample friendly" design using large-volume, high-pressure apparatus in a number of Japanese laboratories (e.g., Yoneda, 1990; Yoneda and Morioka, 1992; Fujisawa and Ito, 1985; Fujisawa, 1998). However, these experiments shared a common limitation, i.e., the piezoelectric transducers were all subjected to high pressure/stress, causing these measurements to have low signal to noise ratios, and lead eventually to the failure of the experiment due to the failure of the piezoelectric transducer at high pressure.

Our innovations in sample cell design and the "stress-free" transducer environment have resulted in great improvements in the adaptation of ultrasonic interferometry techniques for use in multi-anvil high-pressure apparatus (Li et al., 1998a). The "sample friendly" cell assembly enables us to perform reproducible measurements on a single sample. In the few years since the successful implementation of ultrasonic interferometry in our laboratory, acoustic velocities for many mantle phases have been measured to mantle pressures using both polycrystalline and single-crystal samples (Chen et al., 1996; Flesch et al., 1998; Li et al., 1996; Liebermann and Li, 1998; Gwanmesia et al., 1998; Chen et al., 1999; Liu et al., 2000; Li and Liebermann, 2000).

Earlier ultrasonic experiments at simultaneous high-pressure and high-temperature conditions of the Earth's mantle demonstrated the feasibility of performing measurements of both P and S wave velocities to pressures above 10 GPa and simultaneous temperatures above 1300 K using large-volume, high-pressure apparatus (e.g., Knoche et al., 1998; Li et al., 1998a). However, two major difficulties in these experiments remained to be overcome: (1) the conventional method for determining pressure (i.e., a calibration curve fitted through a few points calibrated using known phase transformations) is not precise enough for the determination of the pressure and temperature dependencies of the elastic moduli and cannot readily be applied at high temperatures; (2) the estimation of the sample length (which required to convert travel times to velocities at high pressure and high temperature) using previous equation-of-state (EOS) data is not reliable, especially when the sample environment deviates from hydrostatic conditions and plastic deformation occurs at high temperatures.

In the last 5 years, we have successfully overcome these challenges by adapting the multi-anvil ultrasonic techniques for use in a DIA-type, cubic-anvil apparatus (SAM 85) installed on the superconducting wiggler beamline (X-17B1, now X-17B2) at the National Synchrotron Light Source (NSLS) of the Brookhaven National Laboratory (BNL) (e.g., Liebermann and Li, 1998). These developments have enabled P and S wave velocities and X-ray measurements to be performed simultaneously to pressures of 8 GPa and temperatures of 1600 K on both single-crystal (Chen et al., 1998; Decremps et al., 2001) and polycrystalline specimens (Li et al., 1998b, 2001; Sinelnikov et al., 1998; Kung et al., 2002). Recent implementation of X-radiography to the existing setup allows us to monitor sample lengths directly at elevated pressure and temperature (Li and Weidner, 1999; Kung et al., 2002; Li et al., 2004). In this paper, we will describe our state-of-the-art techniques for experimental studies of the elasticity of deep Earth materials at high pressure and high temperature using combined ultrasonic interferometry, X-ray diffraction, and X-radiography in large-volume apparatus.

2. P–V–V_P–V_S–T measurements using simultaneous ultrasonics and X-ray diffraction

2.1. Ultrasonic interferometry in SAM 85

The ultrasonic techniques described above have been adapted for use in SAM 85 (Weidner et al.,

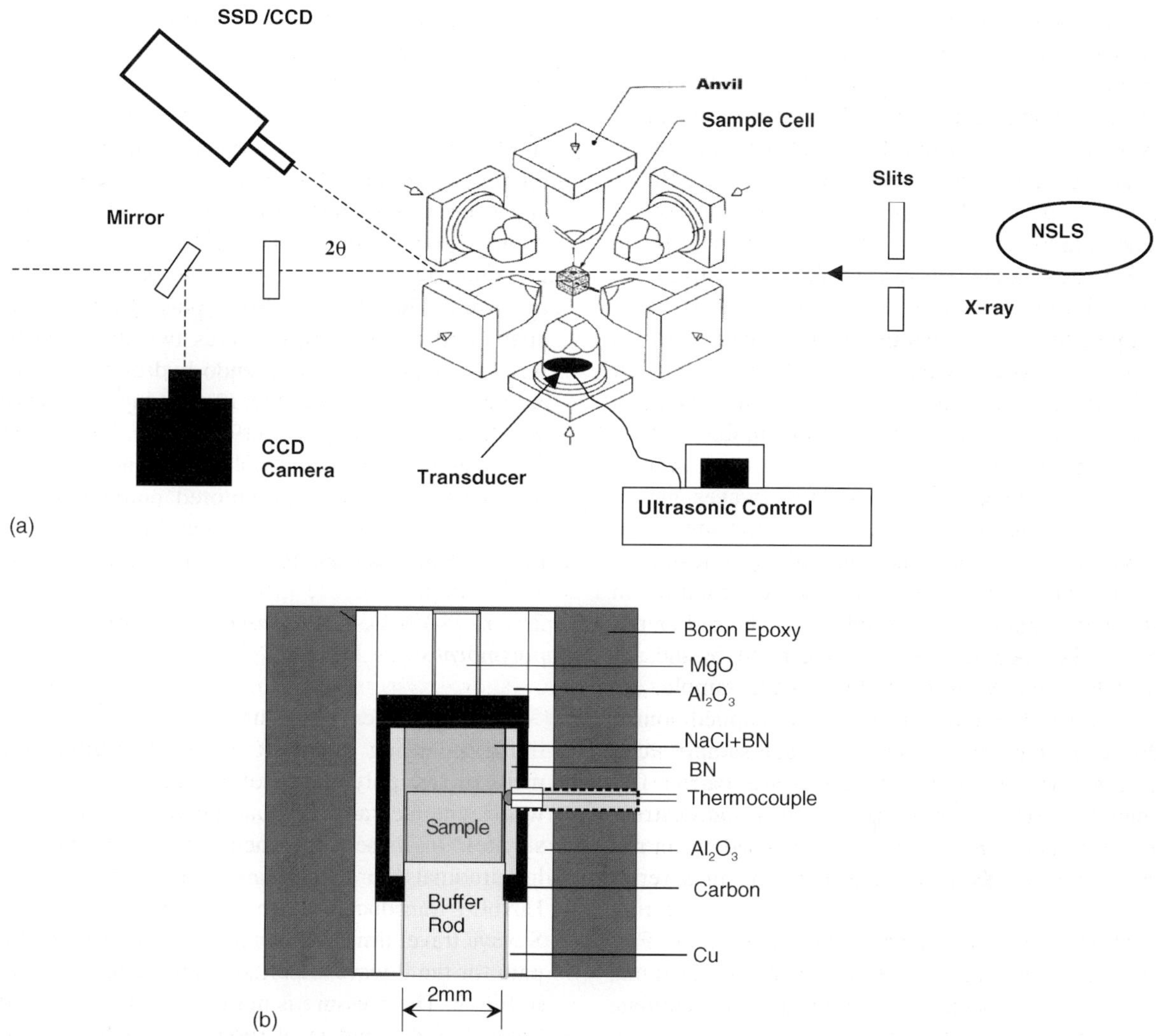

Fig. 1. (a) Schematic diagram of the experimental configuration for simultaneous ultrasonic, X-ray diffraction, and X-ray imaging measurements using the SAM 85 apparatus installed on beamline X-17B1 (B2) at the NSLS. (b) Details of the cell assembly used in the present study.

1992). In Fig. 1a, we show the key elements of the experimental setup on X-17B1 (currently reconstructed as X-17B2) at NSLS/BNL in schematic form: (a) incident X-ray beam with slits and monochromator (optional); (b) the multi-anvil press (SAM 85); (c) a solid-state detector (SSD) or a charge-coupled device (CCD) detecting system; (d) a YAG crystal and a CCD camera to record the X-radiographic images; (e) the ultrasonic system. The ultrasonic piezoelectric transducer–tungsten carbide anvil arrangement and the high-temperature cell assembly used in SAM 85 are shown in Fig. 1a and b; in this configuration, the $LiNbO_3$ transducer is mounted onto the backside of the WC anvil (Toshiba Grade F) and connected to the ultrasonic interferometer by coaxial cables (Fig. 1a). The WC anvil serves as an acoustic buffer rod to transmit the high-frequency signals (20–70 MHz) into the cell assembly. The acoustic waves pass through the bottom anvil and propagate into another buffer rod (fused silica glass, alumina or other material with suitable acoustic impedance) and then into the sample (Fig. 1b). The observed acoustic signals are a series of reflections from the interfaces perpendicular to the wave propagation path, including those from the anvil/buffer rod, buffer rod/sample, and sample/backing material (e.g., NaCl) interfaces.

The phase-comparison method of ultrasonic interferometry, using the Australian Scientific Instruments Ultrasonic Interferometer (ANUTECH), has been employed in our laboratory for a number of years (see details in Niesler and Jackson (1989) and Rigden et al. (1992)). The output from a continuous wave source is gated to produce a pair of phase-coherent, high-frequency pulses which serve as input signals to excite the transducer. The elastic waves generated by each pulse are reflected and transmitted at the buffer rod/sample interface, and the transmitted portion reverberates inside the sample, resulting in a series of 'sample' echoes following the buffer rod echo. If the applied pulses are separated by the apparent two-way travel time through the sample, the first buffer echo (B) from the second applied source pulse will superimpose with the first sample echo (S) from the first source pulse. As the carrier frequency is varied, alternate constructive and destructive interference between the superimposed buffer rod and sample signals will occur, resulting in a series of maxima and minima in the amplitude spectrum modulated by the transducer response envelope. Frequencies for pth and $(p+n)$th interference extrema, f_p and f_{p+n} respectively, can be used to estimate the apparent travel time by $t_{\rm est} = n/(f_{p+n} - f_p)$; then the p value is calculated from $p = f_p t_{\rm est}$ and the closest half or integral value is subsequently assigned to frequency f_p; all remaining extrema can be assigned sequentially. In practice, the interference minima are normally used to reduce the travel-time data because they are sharper than the maxima. For the situations in which the amplitude ratio of the first buffer rod echo and the first sample echo (B1/S1) is very different from unity, the perturbation to travel time from the transducer response envelope has to be taken into consideration, especially for interference maxima (Jackson et al., 1981; Niesler and Jackson, 1989). The typical precision for the measured travel time in this configuration is better than 0.25%.

2.2. Cell assembly for high-pressure, high-temperature acoustic experiments

As shown in Fig. 1b, the polycrystalline or single-crystal sample is centered within the cubic pressure medium of boron–epoxy mixture (4:1 wt.%) and is surrounded by a boron nitride sleeve. An amorphous carbon sleeve is used as the heating element to minimize contamination of the X-ray diffraction spectrum. The insert of NaCl plus BN (10:1 wt.%) at the end of the sample serves two important purposes: (1) providing a pseudo-hydrostatic pressure environment for the sample; (2) serving as a pressure standard using the Decker (1971) scale. X-ray diffraction spectra of both the polycrystalline sample and the NaCl medium are monitored continuously; the former provides *PVT* data to compliment the velocity measurements and the latter, the pressure standard.

2.3. P–V–V_P–V_S–T measurements on Mg_2SiO_4 polymorphs

The comprehensive nature of these experiments is illustrated using the work on polycrystalline specimens of two polymorphs of Mg_2SiO_4: forsterite and wadsleyite (see also Li et al. (1998b, 2001) on wadsleyite). In these experiments, we used pure-mode longitudinal (36° Y-cut) and transverse (41° X-cut) $LiNbO_3$ transducers to generate and measure P and S wave travel times separately. The experimental *P–T* path for the P wave experiment on forsterite is shown in Fig. 2. The pressure is normally increased to a designated peak pressure at room temperature and then the temperature is increased, with a consequent decrease of pressure due to relaxation and flow of the cell materials (see Fig. 2a in Li et al. (2001)). We also experimented with *P–T* paths that involve heating the cell assembly at moderate temperatures (<673 K) during the pressurization to relax the stress on the buffer rod and sample, as shown by the dashed lines in Fig. 2; this type of *P–T* path is very effective in improving signal-to-noise ratio when glass buffer rods are used (see discussion in Li et al. (2001)). After reaching the peak *P* and *T* conditions, the experiments were continued via a series of cooling/heating cycles at various pressures; after each cooling stage, the pressure is released slowly and then another heating stage is commenced. In the forsterite experiments, a maximum

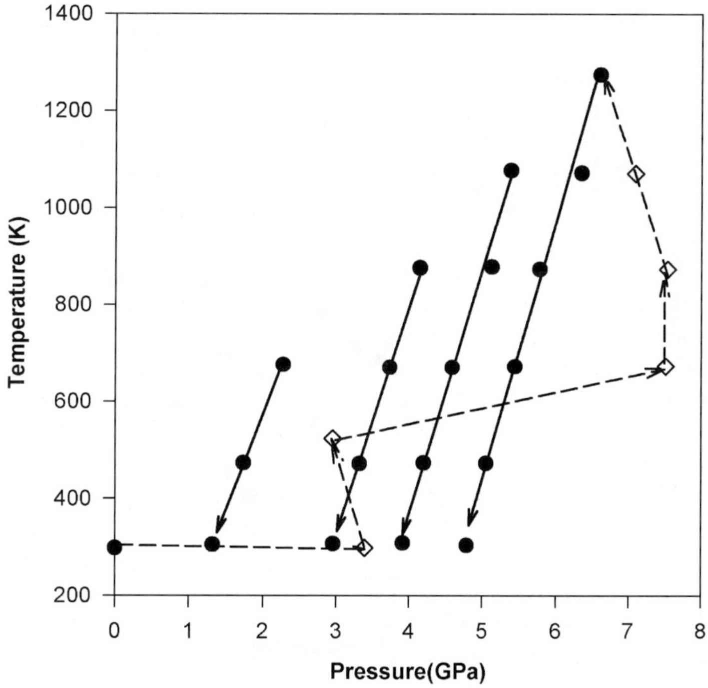

Fig. 2. Experimental *P–T* path for the P wave experiment on forsterite. Dashed lines and open squares represent the compression and heating stages before reaching the peak *P–T* conditions. Solid lines represent the sequential cooling/heating cycles as the pressure is reduced. Travel-time and X-ray diffraction data were collected at each point in *P–T* space indicated by a filled circles.

temperature of 1273 K was reached in both the P and S wave experiments. However, the sample recovered from the first experiment (P wave) had been shortened about 1.6%, presumably when the temperature was at the maximum of 1273 K. In this case, we used the recovered sample length to deduce the velocities on heating/cooling up to 1073 K during decompression, whereas the original sample length was used to reduce the velocity data prior to reaching 1273 K.

In the experiments on wadsleyite, the experimental range of P and T was limited by the thermodynamic stability of this high-pressure polymorph (which is metastable below pressures of ~13 GPa). Consequently, the peak temperatures were limited to 873 K to prevent back-transformation to the olivine phase. The wadsleyite sample recovered from these experiments was examined in detail: no changes in dimension were observed (which would have reflected permanent plastic deformation) nor were any macroscopic cracks evident; X-ray spectra indicated that the specimen was still pure wadsleyite phase.

In Fig. 3, we plot the unit cell volumes at high pressure and temperature for forsterite up to 6.7 GPa and 1273 K from the P wave experiment. These data can be analyzed independently to obtain the parameters of the *P–V–T* equation-of-state; they also provide a measurement of the length change of the polycrystalline sample under the assumption of isotropic compression and expansion (in which case $l/l_0 = (V/V_0)^{1/3}$, where l is the sample length and V the unit cell volume; subscript zero represents value at ambient conditions). If the bulk modulus and its pressure derivative are constrained to be $K_S = 125$ GPa and $K'_S = 4.5$ from our acoustic data obtained in current experiments, a fit to the data in Fig. 3 using a third-order Eulerian EOS (Birch-Murnaghan) yields the temperature derivative of bulk modulus, $(\partial K_T/\partial T)_P = -0.022(2)$ GPa/K, linear thermal expansion ($\alpha = a + bT$) coefficients, $a = 2.4 \times 10^{-5}$ K^{-1} and $b = 3.0 \times 10^{-8}$, and the unit cell volume at ambient conditions $V_0 = 289.80$ Å^3. These values are compatible with previous data at high

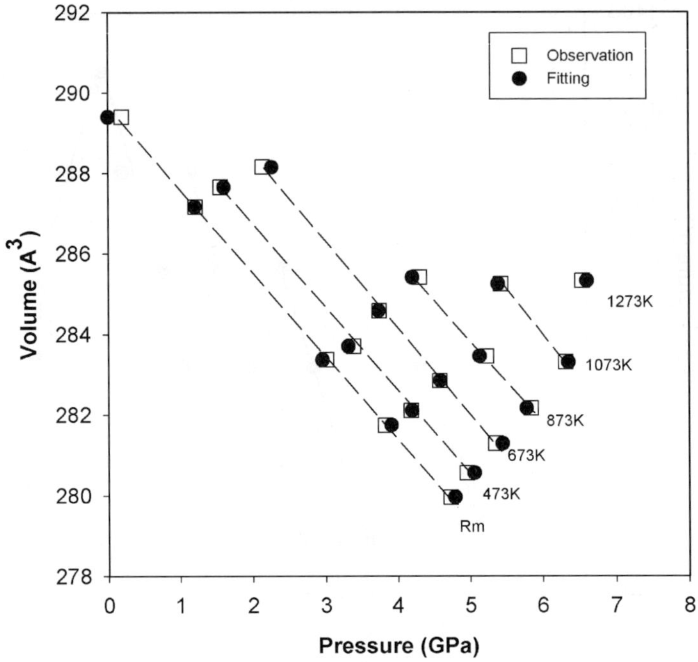

Fig. 3. Unit cell volumes of forsterite Mg_2SiO_4 as a function of pressure and temperature obtained in the P wave experiment: (□) measured values; (●) values resulting from fitting the measured data to a modified third-order Eulerian finite strain equation-of-state.

temperatures (e.g., Meng et al., 1993; Duffy and Anderson, 1989).

Travel-time data for the P and S wave velocities in forsterite are plotted in Fig. 4a and b with maximum pressures of 6.7 and 9.2 GPa, respectively. These data are converted to velocities and plotted in Fig. 5a and b using the length determinations described above. With these P–V–V_P–V_S–T data, we have obtained direct measurements of the adiabatic bulk (K_S) and shear (G) moduli and their first pressure and temperature derivatives for forsterite phase of Mg_2SiO_4 following the same procedure described for wadsleyite (Li et al., 2001). The inferred parameters, $K_S = 125(2)$ GPa, $G = 81(1)$ GPa, $(\partial K_S/\partial P)_T = 4.5(2)$, $(\partial G/\partial P)_T = 1.4(1)$, $(\partial K_S/\partial T)_P = -0.014(1)$ GPa/K and $(\partial G/\partial T)_P = -0.017(1)$ GPa/K, are compatible with previous data obtained at high pressures and/or high temperatures (e.g., Isaak et al., 1989; Duffy et al., 1995; Meng et al., 1993; Li et al., 1996) (Table 1). The combined velocity and unit cell volume data also provide the opportunity to evaluate the reliability of the Decker NaCl pressure scale (e.g., Liu, 2000), and determine

Table 1
Elastic properties of forsterite

	1	2	3	4	5	6
K_S (GPa)	125(2)	128.8(5)	128	129	129	(127.4)
$(\partial K_S/\partial P)_T$	4.5(2)	–	4.44(2)	4.1	4.2	(4.8)
$(\partial K_S/\partial T)_P$ (GPa/K)	−0.014(1)	−0.0157(4)	–	–	–	−0.021(2)
G (GPa)	81(1)	81.8(2)	80	82	81	–
$(\partial G/\partial P)_T$	1.4(1)	–	1.32(3)	1.6	1.4	–
$(\partial G/\partial T)_P$ (GPa/K)	−0.017(1)	−0.0135(2)	–	–	–	–

(1) This study: ultrasonic interferometry and X-ray diffraction, 0–9 GPa and 1273 K. (2) Isaak et al. (1989): resonance ultrasonic spectroscopy, room P and 298–1700 K. (3) Li et al. (1996): ultrasonic interferometry, room T and 0–12.5 GPa. (4) Yoneda and Morioka (1992): ultrasonic interferometry, room T and 0–6 GPa. (5) Duffy et al. (1995): Brillouin scattering, room T and 0–16 GPa. (6) Meng et al. (1993): X-ray diffraction. Results are isothermal values. 6.6–7.6 GPa and 1019–1478 K. Data in parentheses were used to derive $(\partial K_S/\partial T)_P$.

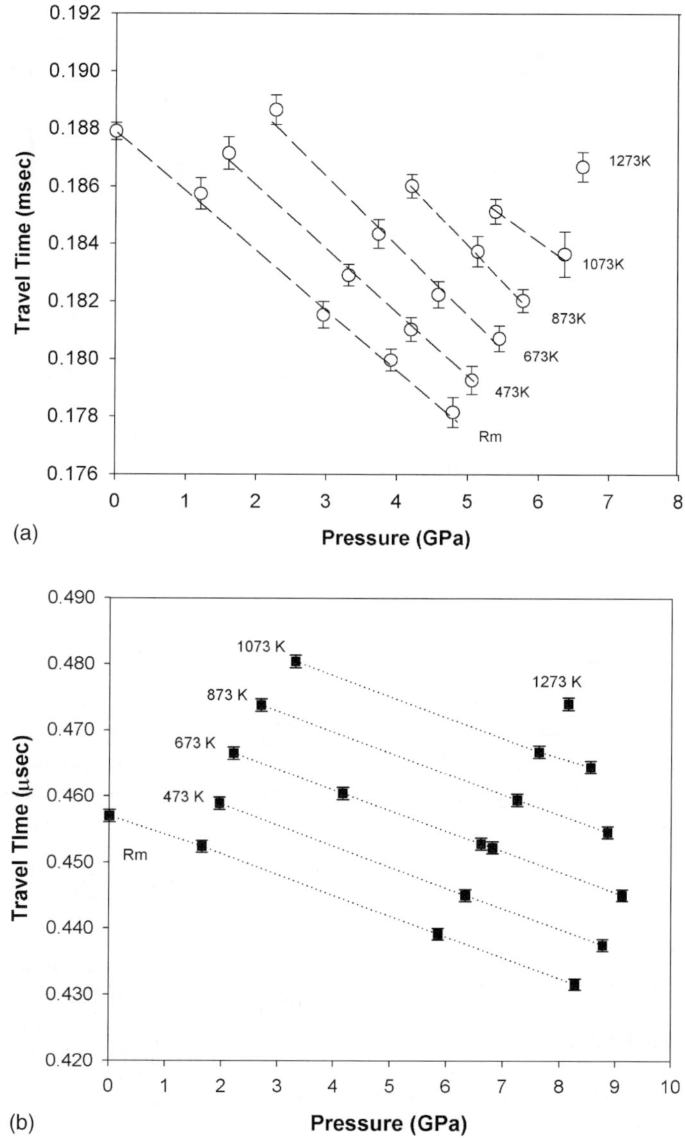

Fig. 4. (a) P wave travel times in forsterite as a function of pressure to 6.7 GPa and temperature to 1273 K. The dashed lines represent data for different isotherms. (b) S wave travel times in forsterite as a function of pressure to 9 GPa and temperature to 1273 K. The dotted lines represent data for different isotherms.

pressure derivatives without using secondary pressure standard (e.g., Kung et al., 2002).

2.4. Dual-mode transducers

In the experiments on forsterite and wadsleyite, the P and S wave travel times were measured separately using pure-mode transducers; one difficulty in this approach is that it is not simple, in practice, to ensure that the P and S travel times are measured at the same cell pressures in separate experimental runs. Recently, we have used dual-mode $LiNbO_3$ transducers (10° Y-cut) in our experiments; these transducers generate both longitudinal and transverse acoustic signals with unspecified polarization direction of the transverse signal. The size of the tungsten carbide cube

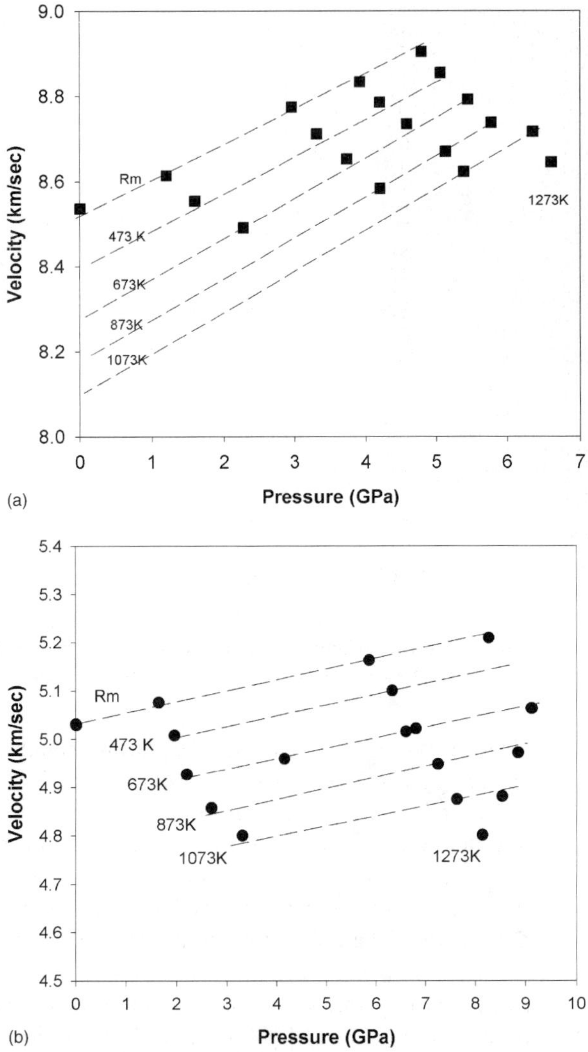

Fig. 5. (a) P wave velocities in Mg_2SiO_4 forsterite as a function of pressure to 6.7 GPa and temperature to 1273 K. The dashed lines represent data for different isotherms. (b) S wave velocity in Mg_2SiO_4 forsterite as a function of pressure to 9 GPa and temperature to 1273 K. The dashed lines represent data for different isotherms.

buffer rod (∼35 mm) provides a long delay line, so that the P and S wave echo patterns can be clearly separated in time (Fig. 6a). Travel times of a polycrystalline alumina sample (2.460 mm in length) were measured under ambient conditions using a dual-mode transducer mounted on a WC cube, a pure-mode P wave transducer mounted on a sapphire buffer rod, and a pure-mode S wave transducer mounted on a

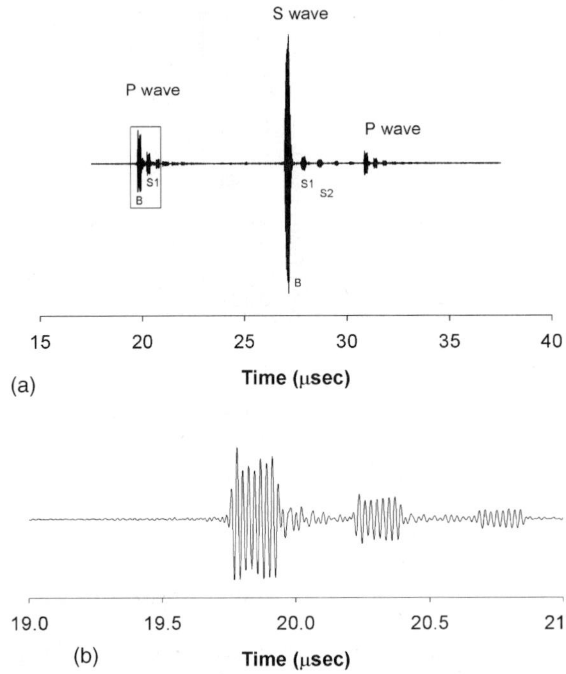

Fig. 6. (a) Example of acoustic signals at 45 MHz using dual-mode $LiNbO_3$ transducers, showing groups of both P and S wave echoes from the buffer rod (B) and the samples (S1 and S2). (b) Details of the buffer rod and sample echoes in the first P wave group.

steel buffer rod. The travel-time determinations from the pure-mode P and S wave transducers and those from dual-mode transducer are consistent within 0.1% which is within the system uncertainty. Other examples of the use of the dual-mode transducers in our laboratory may be found in Kung et al. (2002), Darling et al. (2004), and Sinelnikov et al. (2004).

3. X-radiography in DIA apparatus

3.1. X-ray imaging

X-radiography has been used in the study of viscosity of the molten materials at high pressure and high temperature using a falling sphere method (e.g., Mattischek and Sobczak, 1994). We have successfully incorporated this technique for use in our simultaneous ultrasonics and X-ray diffraction experiments (e.g., Kung et al., 2002). Briefly, the imaging system consists of a YAG crystal, a mirror, and a CCD camera

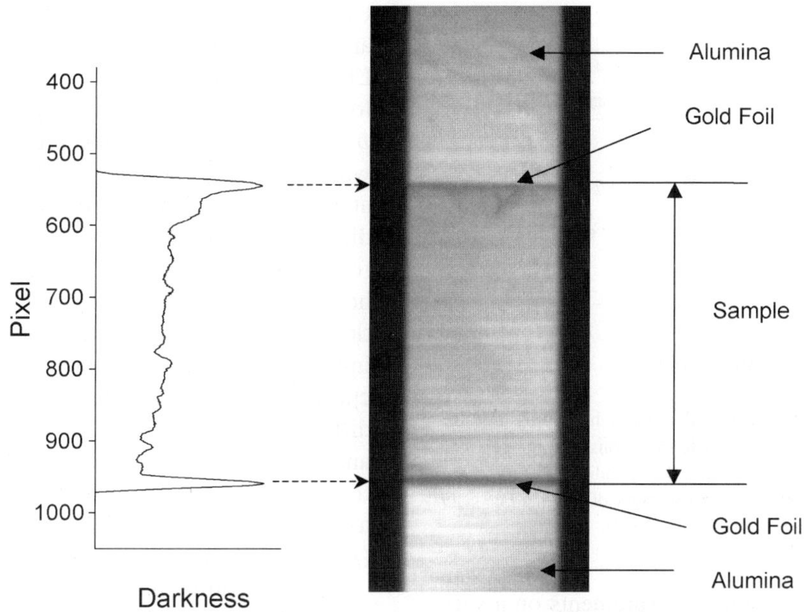

Fig. 7. Example of the X-radiographic images recorded in the high *P–T* experiments as seen between the gap in the WC anvils (on right). Image file read as gray scale (on left); the two peaks in the darkness scale correspond to the gold foils at the top and bottom of the sample.

as shown in Fig. 1a. After recording the X-ray diffraction spectra, the front slits are removed to maximize the size of the X-ray beam, which passes through the gap between two of the horizontal WC anvils and then cell assembly, and illuminates a fluorescent YAG crystal. The visible light generated by the YAG crystal is reflected by a mirror into the CCD camera where an image of the cell assembly is captured. An example of the captured image is shown in Fig. 7.

On these X-radiographic images, the brightness contrast from region to region results from the difference in the X-ray absorption coefficients of the various materials in the high-pressure cell assembly. If the absorption coefficient for the sample is sufficiently different from those of the adjacent materials (buffer rod or NaCl), the sample region can be clearly distinguished from the rest of the cell assembly. For cases in which the sample and the other parts of the cell assembly have similar absorption coefficients, metal foils with high X-ray absorption (e.g., gold) are inserted at both the top and bottom surfaces of the sample. For our ultrasonic measurements, such gold foils (0.002 mm thick) are routinely inserted at the sample/buffer rod interface to enhance the mechanical coupling between the buffer rod and sample; these also serve well in delineating the sample boundaries, as seen by the dark horizontal bands in Fig. 7.

For the calibration of sample length with respect to pixel number, one can obtain the calibration factor by comparing the pixel numbers from the last image (after the experiment has reached zero pressure) with the length of the recovered sample measured using a precise micrometer. This last image can then be used as the starting image to retrieve the relative change in pixels between neighboring *P–T* conditions by cross-correlation of the intensity profiles at sample boundaries. Typically, a change of 0.25 pixels or less can be resolved over a sample dimension of about 1000 pixels. If the sample length is directly measured from an image, it has a total uncertainty about 2–4 pixels, which gives 0.2–0.4% in precision. As seen in Fig. 8, the lengths of a polycrystalline sample of $(Mg_{0.83}Fe_{0.17})O$ at high pressures obtained from the two different methods described above (one from direct X-ray imaging measurements and the other derived from X-ray diffraction determinations of sample volumes) exhibit remarkable agreement (see also Fig. 4 of Kung et al., 2002).

Simultaneous X-ray diffraction and ultrasonic measurements coupled with X-radiography thus enable us

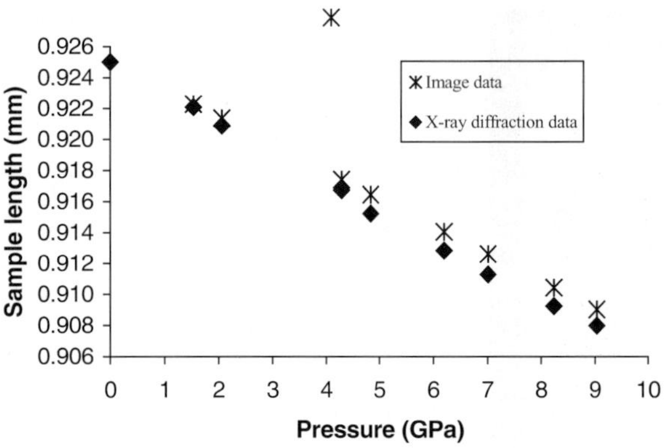

Fig. 8. Comparison of the sample lengths at high pressures derived from the X-ray diffraction determinations of unit cell volumes (◆) with those measured using the X-radiographic imaging technique (✳) for a polycrystalline specimen of $(Mg_{0.83}Fe_{0.17})O$. From experimental data of Kung et al. (2002).

to conduct precise elasticity measurements on a variety of materials and processes, including materials undergoing mode softening/phase transformations (e.g., Decremps et al., 2000, 2001) or plastic deformation (e.g., Darling et al., 2004), as well as melts or multi-phase aggregates (e.g., Mistler et al., 2002).

3.2. Application of P–V–V_P–V_S–T–L measurements to $CaSiO_3$ perovskite

The elastic properties of $CaSiO_3$ perovskite under mantle conditions are very important in understanding its possible contribution to the 520 km seismic discontinuity as well as its visibility in the velocity structure of the lower mantle (e.g., Wang et al., 1996; Hama and Suito, 1998; Karki and Crain, 1998; Mao et al., 1989; Tarrida and Richet, 1989). Since this high-pressure phase is unquenchable (i.e., cannot be recovered in its high-pressure form at ambient conditions), measurements of its properties have to be conducted after in situ synthesizing in its stability field (see phase diagram in Gasparik et al. (1994)). Another complication is that the low-pressure wollastonite phase of $CaSiO_3$ undergoes a process of conversion to an amorphous state with increasing pressure, making the estimation of the sample length for the $CaSiO_3$ perovskite phase impossible without direct measurement.

We have successfully conducted P and S wave velocity measurements on the cubic-structure phase of $CaSiO_3$ perovskite within its stability field at 13.5 GPa and 1123 K (see Li et al., 2004). As demonstrated in Li et al. (2004), the starting material, a semi-sintered low-pressure sample of $CaSiO_3$ wollastonite (triclinic, $\rho = 2.97\,g/cm^3$), become progressively amorphous with increasing pressure; an image of the sample at 9.6 GPa and 300 K and the corresponding X-ray diffraction pattern are shown in Fig. 9a and c. The formation and growth of the perovskite phase during heating was monitored by real-time X-ray diffraction; after 1.5 h at 13.5 GPa and 1123 K, the starting material was completely transformed into the cubic-structured perovskite as shown in the X-ray diffraction pattern and the sample image (Fig. 9b and d). The substantial length reduction is obvious in the images before and after the phase transformation, as a consequence of the high density of the new phase ($\rho = 4.26\,g/cm^3$).

From the direct measurements of length and travel times as described above, P and S wave velocities for $CaSiO_3$ perovskite were obtained for pressures between 8 and 11.5 GPa at room temperature (Fig. 10). When the pressure was decreased below 8 GPa, the WC anvil on which the transducer was mounted broke and no further acoustic data could be recorded. Comparing with our experimental data for the P and S wave velocities with those calculated by Karki and Crain (1998) from first-principles methods in Fig. 10, we see that the theoretical values are higher about 10%, presumably because the theoretical calculations are for 0 K and ignore the zero-point energy. Measurements at a wider range of pressure and temperature are necessary in order to provide more robust determinations of the pressure and temperature derivatives.

4. Sound velocity measurement using transfer function method

Ultrasonic interferometry has been used widely in the study of elastic properties of solid and liquid materials since the 1950s. Over the years, different methods have been developed for the measurement of the time of wave propagation, such as the pulse echo overlap (PEO), phase comparison (PC), and pulse superposition (PSP) methods (see reviews in Truell et al. (1969)). In these methods, a radio-frequency (rf) pulse is transmitted and received by a piezoelectric

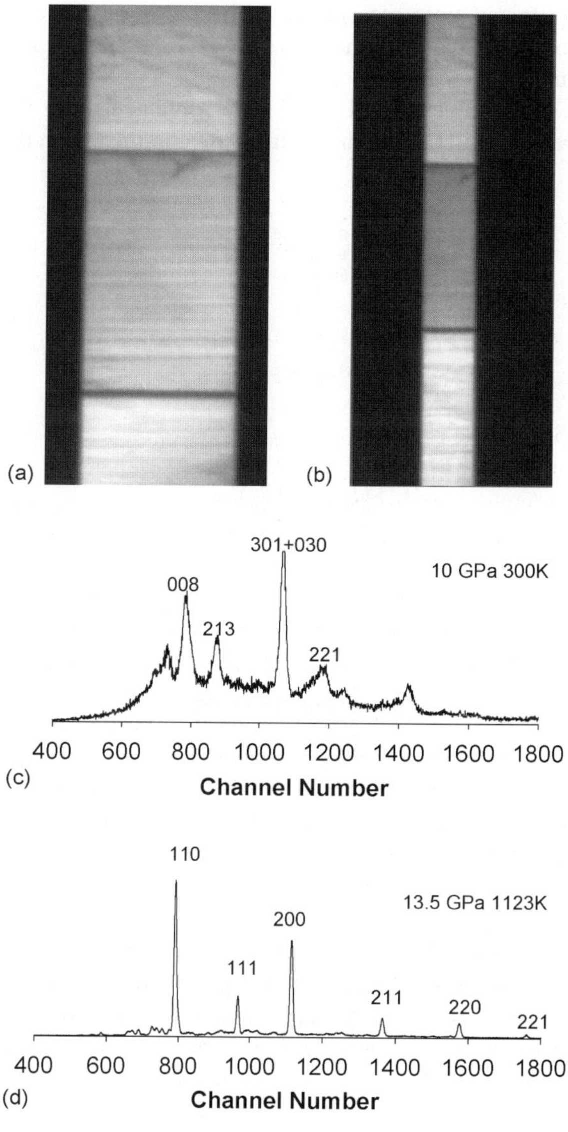

Fig. 9. X-radiographic images for $CaSiO_3$ at ~10 GPa and 300 K (a) and 13.5 GPa and 1123 K (b). The diffraction pattern taken at ~10 GPa and 300 K (c) indicates remnant diffraction peaks of the wollastonite starting material and an amorphous phase produced on compression at room temperature. The X-ray diffraction pattern taken at 13.5 GPa and 1123 K (d) indicates that most of the sample was transformed to the perovskite phase; *hkl* labels are the diffraction peaks for the cubic perovskite phase of $CaSiO_3$.

transducer attached to the sample or a buffer rod, and the propagation time is obtained by measuring the delay between two consecutive echoes as described in Section 3.

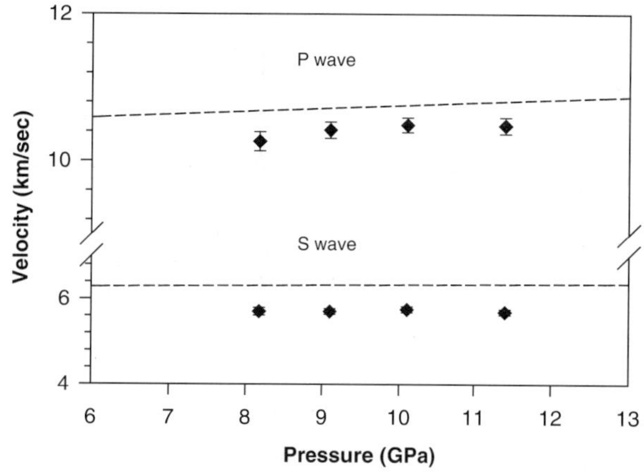

Fig. 10. Measured P and S wave velocities of $CaSiO_3$ perovskite in the pressure range of 8–12 GPa at room temperature (◆). The dashed lines are from the theoretical calculations of Karki and Crain (1998).

Recently, the development of fast analogue/digital acquisition cards has made it possible to use transfer function methods based on digital signal processing (such as cross-correlation function (CCF), Hilbert transform (HT), and continuous wavelet transform (CWT) (Moreno et al., 1999)) rather than relying on the classical pulse superposition or phase comparison methods. In this section, we describe sound velocity measurements using the transfer function method.

The transfer function depicts the response of the system as a function of frequency (e.g., piezoelectric transducers) (Huemer et al., 1999). After a wavelet $x(t)$ is sent to the buffer rod and sample, the received signal $y(t)$ is the convolution of the input signal $x(t)$ and the system response $h(t)$ along the wave propagation path, i.e., $y(t) = x(t) * h(t)$. Fast Fourier transform (FFT) of the convoluted signal yields $Y(f) = X(f)H(f)$. The transfer function of the piezoelectric transducer, buffer rod, and sample ensemble can thus be obtained, $H(f) = Y(f)/X(f)$. Subsequently, convoluting the transfer function with another input signal yields the response of the sample system to the input pulse for frequencies inside the bandwidth.

A typical received waveform for an alumina polycrystal on a tungsten carbide buffer rod is shown in Fig. 11. To simulate the PEO measurement of phase velocity, a 45 MHz sine wave toneburst (8 cycles) was convoluted with the transfer function; thus, the system response to such 45 MHz input signal is

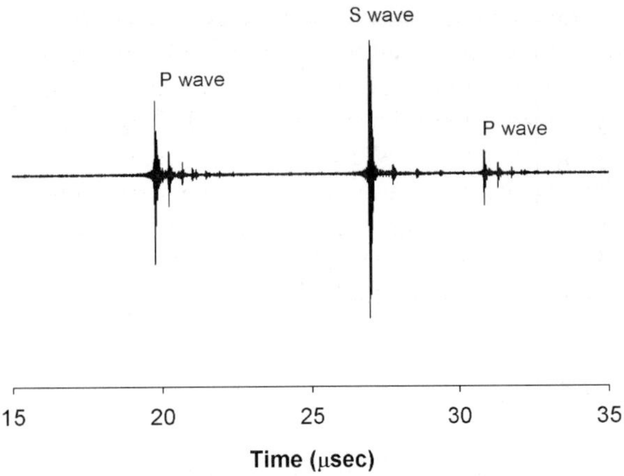

Fig. 11. Received ultrasonic signals for P and S waves using the transfer function method for a polycrystalline Al_2O_3 sample mounted on a tungsten carbide buffer rod.

reproduced (Fig. 6a). Fig. 6b shows the expanded view of the buffer rod echo and two sample echoes. The time delay is obtained by measuring the time shift of the second copy of the echo pattern relative to the original one until the buffer rod and sample echoes overlap.

As demonstrated in Li et al. (2002), the phase comparison measurements can also be reproduced using the recorded transfer function by sweeping through the frequency range and recording the interference amplitude between the buffer rod and sample (e.g., Fig. 12a). Moreover, using the amplitudes of the buffer rod (A_B), sample (A_S), and their interference (A_I), the effect of the transducer response envelope can be removed from the recorded interference spectrum using the relationship (e.g., Spetzler et al., 1993)

$$\cos\theta = \frac{A_I^2 - A_B^2 - A_S^2}{2A_B A_S} \quad \text{with}\ \theta = \theta_s + \theta_b$$

where θ is the phase difference between the buffer rod and sample echoes caused by the additional travel inside the sample (θ_s) and the effect of the bonding material (θ_b). In Fig. 12b, we show the results after the removal of the transducer response from the interference pattern in Fig. 12a. Fitting the spectrum in Fig. 12b with a cosine function yields precise determination of travel time and the phase shift of the bond. The travel times obtained from both methods are compared in Fig. 12c. Using the transfer function method, such simulations can be conducted multiple times for optimized time measurements by adjusting the data collection window and step size of the sweeping frequency.

We also compared travel times obtained using the transfer function method with those obtained from the

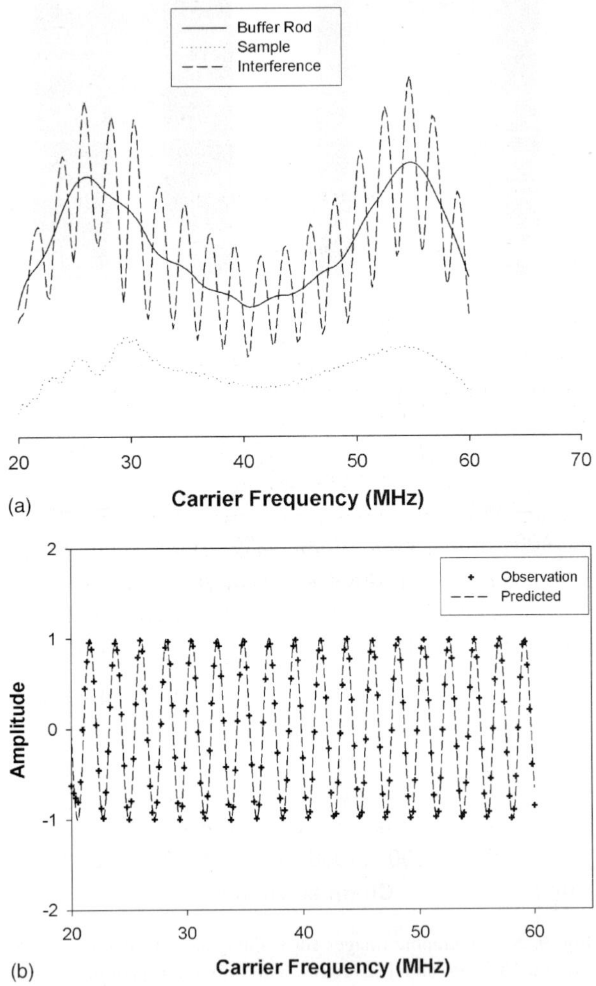

Fig. 12. (a) Amplitudes of the buffer rod and first sample echoes, and their interference, as a function of carrier frequency for the acoustic signals in Fig. 11. (b) The amplitudes of the interference of the buffer rod and sample echoes in (a) after the transducer envelope has been removed. Crosses are the recorded data and the dashed lines are from curve-fitting using a cosine function as described in the text. (c) Travel times as a function of frequency for the data in (a); these are obtained by assigning integer or half-integer values to the interference maxima. Solid line is the mean travel time over the frequency range; dashed line is the travel time obtained from the curve-fitting in (b). Dotted lines represent the standard deviation of these travel time determinations.

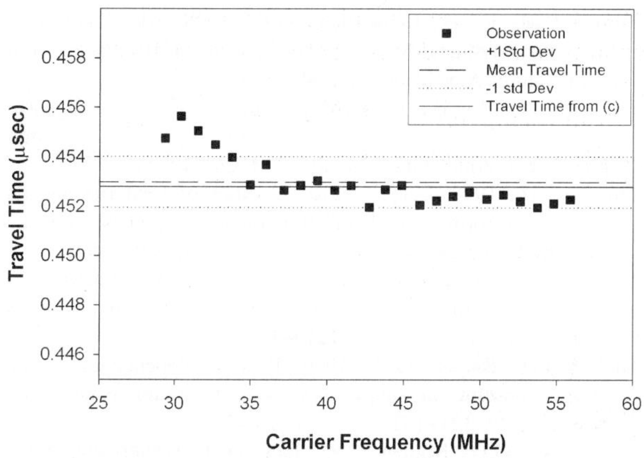

Fig. 12. (*Continued*).

phase comparison method using the ANUTECH interferometer. In Fig. 13, we plot the travel times for a polycrystalline $(Mg_{0.83}Fe_{0.17})O$ at high pressures measured using two different ultrasonic techniques (from Kung et al., 2002). The agreement is within 0.25%, comparable to the uncertainty of the individual techniques. However, compared with real-time measurements, the transfer function method reduces data collection times from minutes to seconds, which is a very valuable attribute in conducting experiments at elevated temperatures and/or pressures. The transfer function method also has the advantage of recording the system response to all frequencies inside the passband at the same time, in contrast to sweeping through frequencies in the pulse echo overlap and phase comparison methods; this is especially important when travel times are being used to study time-dependent processes, such as plastic deformation of materials (e.g., Li and Weidner, 1999) or phase transformations (Decremps et al., 2000, 2001).

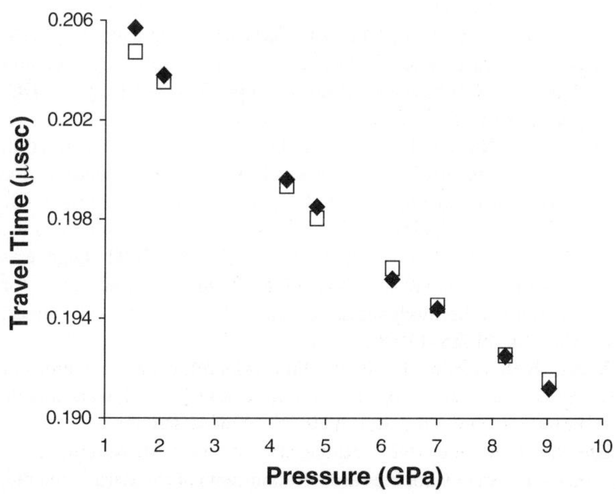

Fig. 13. Comparison of P wave travel times in a polycrystalline specimen of $(Mg_{0.83}Fe_{0.17})O$ measured using the two different ultrasonic systems: (□) from the transfer function method using the Stony Brook system; (◆) from the phase comparison method using the ANUTECH ultrasonic interferometer.

5. Conclusion

We have progressively developed state-of-the-art techniques for the studies of elastic properties of polycrystalline and single-crystal materials using simultaneous ultrasonic and X-ray diffraction techniques at high pressures and temperatures in multi-anvil apparatus, in conjunction with synchrotron X-radiation. The capability of these techniques has been illustrated using our data for mantle minerals such as the forsterite and wadsleyite phases of Mg_2SiO_4. Incorporation of X-radiography techniques enables us to directly measure sample length with a precision better than 0.5% and the power to resolve relative length changes at 2.5×10^{-4} using current setup, regardless of the sample state at high pressure and high temperature.

The implementation of the transfer function method for sound wave velocity measurements opens new opportunities for studying time-dependent processes using ultrasonic measurements, such as phase transformations and plastic deformation. Real-time signals can be reproduced off-line for travel-time measurement using the pulse echo overlap and phase comparison methods over a wide frequency range. The use of dual-mode transducers produces P and S wave data at the same pressure and temperature conditions in a single acoustic experiment, eliminating the uncertainties caused by the need for interpolating travel-time data to calculate the elastic wave velocities and bulk and shear moduli at high pressure and temperature (as was necessary in our earlier experiments without X-rays). Successful application of these techniques to the study of the behavior and physical properties of Earth materials, including unquenchable mantle phases, liquids, and multi-phase aggregates, will greatly improve our interpretation of seismic observations and our understanding about the structure, composition, and dynamics of the Earth's interior.

Acknowledgements

This research was supported by National Science Foundation under grants EAR00-03340 to BL and EAR 99-80491 to RCL. We thank two anonymous reviewers and the editor who helped to improve this manuscript. We also thank J. Chen, M.T. Vaughan, and D.J. Weidner for their support in these experiments. The in situ ultrasonic and X-ray experiments were carried out at the X-17B1 beamline of the National Synchrotron Light Source (NSLS), which is supported by the US Department of Energy, Division of Materials Sciences and Division of Chemical Sciences under Contract No. DE-AC02-76CH00016. Mineral Physics Institute Publication No. 347.

References

Bassett, W.A., Reichmann, H.J., Angel, R.J., Spetzler, H., Smyth, J.R., 2000. New diamond anvil cells for gigahertz ultrasonic interferometry and X-ray diffraction. Am. Miner. 85, 283–287.

Chen, G., Li, B., Liebermann, R.C., 1996. Selected elastic moduli of single-crystal olivines from ultrasonic experiments to mantle pressures. Science 272, 979–980.

Chen, G., Liebermann, R.C., Weidner, D.J., 1998. Elasticity of single-crystal MgO to 8 GPa and 1600 K. Science 280, 1913–1916.

Chen, G., Cook, J.A., Gwanmesia, G.D., Liebermann, R.C., 1999. Elastic wave velocities of $Mg_2Al_2Si_3O_{12}$-pyrope garnet to 10 GPa. Am. Miner. 84, 384–388.

Darling, K.L., Gwanmesia, G.D., Kung, J., Li, B., Liebermann, R.C., 2004. Ultrasonic measurements of the sound velocities in polycrystalline San Carlos olivine in multi-anvil, high-pressure apparatus. Phys. Earth Planet. Interiors 143–144, 19–31.

Decker, D.L., 1971. High-pressure equation of state for NaCl, KCl, and CsCl. J. Appl. Phys. 42, 3239.

Decremps, F., Zhang, J., Liebermann, R.C., 2000. New phase boundary and high pressure thermoelasticity of ZnO. Europhys. Lett. 51, 268–274.

Decremps, F., Zhang, J., Li, B., Liebermann, R.C., 2001. Pressure-induced softening of shear modes in ZnO. Phys. Rev. B 6322, 4105.

Duffy, T.S., Zha, C.S., Downs, R.T., Mao, H.K., Hemley, R.J., 1995. Elasticity of forsterite to 16 GPa and the composition of the upper-mantle. Nature 378, 170–173.

Duffy, T.S., Anderson, D.L., 1989. Seismic velocities in mantle minerals and the mineralogy of the upper mantle. J. Geophys. Res. 94, 1895–1912.

Flesch, L.M., Li, B., Liebermann, R.C., 1998. Sound velocities of polycrystalline $MgSiO_3$-orthopyroxene to 10 GPa at room temperature. Am. Miner. 83, 444–450.

Fujisawa, H., Ito, E., 1985. Measurements of ultrasonic wave velocities of tungsten carbide as a standard material under high pressure. J. Appl. Phys. 24 (Suppl. 1–24), 103–105.

Fujisawa, H., 1998. Elastic wave velocities of forsterite and its beta-spinel form and chemical boundary hypothesis for the 410-km discontinuity. J. Geophys. Res. 103, 9591–9608.

Gasparik, T., Wolf, K., Smith, C.M., 1994. Experimental determination of phase relations in the $CaSiO_3$ system from 8 to 15 GPa. Am. Miner. 79, 1219–1222.

Gieske, J.H., Barsch, G.R., 1966. Pressure dependence of the elastic constants of single crystalline aluminum oxide. Phys. Stat. Sol. 29, 121–131.

Gwanmesia, G.D., Rigden, S.M., Jackson, I., Liebermann, R.C., 1990. Pressure dependence of elastic wave velocity for β-Mg_2SiO_4 and the composition of the Earth's mantle. Science 250, 794–797.

Gwanmesia, G.D., Chen, G., Liebermann, R.C., 1998. Sound velocities in $MgSiO_3$-garnet to 8 GPa. Geophys. Res. Lett. 25, 4553–4556.

Hama, J., Suito, K., 1998. High-temperature equation of state of $CaSiO_3$ perovskite and its implications for the lower mantle. Phys. Earth Planet. Interiors 105, 33–46.

Huemer, M., Koppler, A., Ruppel, C.C.W., Reindl, L., Springer, A., Weigel, R., 1999. SAW based chirp Fourier transform for OFDM systems. In: Schneider, S.C., Levy, M., McAvoy, B.R. (Eds.), IEEE Ultrasonics Symposium Proceedings. Institute of Electrical and Electronic Engineers, New York, pp. 373–376.

Isaak, D., Anderson, O.L., Goto, T., Suzuki, I., 1989. Elasticity of single-crystal forsterite measured to 1700 K. J. Geophys. Res. 94, 5895–5906.

Jackson, I., Niesler, H., 1982. The elasticity of periclase to 3 GPa and some geophysical implications. In: Manghnani, M., Syono, Y. (Eds.), High Pressure Research: Application to Earth and Planetary Sciences. Terra Scientific Publishing Co. and American Geophysical Union, Tokyo and Washington, DC, pp. 93–113.

Jackson, I., Niesler, H., Weidner, D.J., 1981. Explicit correction of ultrasonically determined elastic wave velocities for transducer-bond phase shift. J. Geophys. Res. 86, 3736–3748.

Jacobsen, S.D., Spetzler, H.A., Reichmann, H.J., Smyth, J.R., Mackwell, S.J., Angel, R.J., Bassett, W.A., 2002. Gigahertz ultrasonic interferometry at high P and T: new tools for obtaining a thermodynamic equation of state. J. Phys.-Condens. Mat. 14, 11525–11530.

Karki, B.B., Crain, J., 1998. First-principles determination of elastic properties of $CaSiO_3$ perovskite at lower mantle pressures. Geophys. Res. Lett. 25, 2741–2744.

Knoche, R., Webb, S.L., Rubie, D.C., 1998. Measurements of acoustic wave velocities at P–T conditions of the Earth's mantle. In: Manghnani, M.H., Yagi, T. (Eds.), Properties of Earth and Planetary Materials at High Pressure and Temperature. American Geophysical Union, Washington, DC, pp. 119–128.

Kumazawa, M., Anderson, O.L., 1969. Elastic moduli, pressure derivatives, and temperature derivatives of single-crystal olivine and single-crystal forsterite. J. Geophys. Res. 74, 5961–5972.

Kung, J., Li, B., Weidner, D.J., Zhang, J., Liebermann, R.C., 2002. Elasticity of $(Mg_{0.83}, Fe_{0.17})O$ ferropericlase at high pressure: ultrasonic measurements in conjunction with X-radiation techniques. Earth Planet. Sci. Lett. 203, 557–566.

Li, B., Vaughan, M.T., Weidner, D.J., 2004. Direct length measurements using X-radiography for the determination of acoustic velocities at high pressure and high temperature. Rev. Sci. Instrum., submitted for publication.

Li, B., Chen, K., Kung, J., Liebermann, R.C., Weidner, D.J., 2002. Sound velocity measurement using transfer function method. J. Phys.-Condens. Mat. 14, 11337–11342.

Li, B., Liebermann, R.C., Weidner, D.J., 2001. P–V–V_P–V_S–T measurements on wadsleyite to 7 GPa and 873 K: implications for the 410 km seismic discontinuity. J. Geophys. Res. 106, 30575–30591.

Li, B., Liebermann, R.C., 2000. Sound velocities of wadsleyite beta-$(Mg_{0.88}Fe_{0.12})_2SiO_4$ to 10 GPa. Am. Miner. 85, 292–295.

Li, B., Weidner, D.J., 1999. Rheological property measurement at high pressure and high temperature using ultrasonic and X-ray diffraction techniques. In: Manghnani, M., Nellis, W.J., Nicol, M.F. (Eds.), Science and Technology of High Pressure: Proceedings of AIRAPT-17, vol. 1, p. 135.

Li, B., Chen, G., Gwanmesia, G.D., Liebermann, R.C., 1998a. Sound velocity measurements at mantle transition zone conditions of pressure and temperature using ultrasonic interferometry in a multi-anvil apparatus. In: Manghnani, M.H., Yagi, T. (Eds.), Properties of Earth and Planetary Materials at High Pressure and Temperature. American Geophysical Union, Washington, DC, pp. 41–61.

Li, B., Liebermann, R.C., Weidner, D.J., 1998b. Elastic moduli of wadsleyite to 7 GPa and 873 K. Science 281, 675–677.

Li, B., Gwanmesia, G.D., Liebermann, R.C., 1996. Sound velocities of olivine and beta polymorphs of Mg_2SiO_4 at Earth's transition zone pressures. Geophys. Res. Lett. 23, 2259–2262.

Liebermann, R.C., Li, B., 1998. Elasticity at high pressures and temperatures. In: Hemley, R. (Ed.), Ultrahigh-Pressure Mineralogy: Physics and Chemistry of the Earth's Deep Interior, Mineralogical Society of America, Washington, DC, vol. 37, pp. 459–492.

Liu, J., Chen, G., Gwanmesia, G.D., Liebermann, R.C., 2000. Elastic wave velocities of pyrope-majorite garnets ($Py_{62}Mj_{38}$ and $Py_{50}Mj_{50}$) to 9 GPa. Phys. Earth Planet. Interiors 120, 153–163.

Liu, J., 2000. Ultrasonic wave velocities of pyrope-majorite garnets at high pressure and high temperature. Ph.D. Thesis. State University of New York, Stony Brook, NY, 167 pp.

Mao, H.K., Chen, L.C., Hemley, R.J., Jephcoat, A.P., Wu, Y., Bassett, W.A., 1989. Stability and equation of state of $CaSiO_3$-perovskite to 134 GPa. J. Geophys. Res. 94, 17889–17894.

Mattischek, H.P., Sobczak, R., 1994. A new cell for measurement of viscosity under high-pressure. Meas. Sci. Technol. 5, 782–785.

McSkimin, H.J., 1950. Ultrasonic measurement techniques applicable to small solid specimens. J. Acoust. Soc. Am. 22, 413–418.

Meng, Y., Weidner, D.J., Gwanmesia, G.D., Liebermann, R.C., Vaughan, M.T., Wang, Y., Leinenweber, K., Pacalo, R.E., Yeganeh-Haeri, A., Zhao, Y., 1993. In-situ high P–T X-ray diffraction studies on three polymorphs (α, β, γ) of Mg_2SiO_4. J. Geophys. Res. 98, 22199–22207.

Mistler, G.W., Ishikawa, M., Li, B., 2002. Elastic wave velocity measurements on mantle peridotite at high pressure and temperature. In: EOS, Trans., AGU. Fall Meeting Supplement, vol. 83, p. F862.

Moreno, E., Garcia, F., Castillo, M., 1999. Measurements of ultrasonic velocity in viscoelastic materials using the wavelet transform. In: Schneider, S.C., Levy, M., McAvoy, B.R. (Eds.), 1999 IEEE Ultrasonics Symposium Proceedings. Institute of Electrical and Electronic Engineers, New York, pp. 827–830.

Niesler, H., Jackson, I., 1989. Pressure derivatives of elastic wave velocities from ultrasonic interferometric measurements on jacketed polycrystals. J. Acoust. Soc. Am. 86, 1573–1585.

Reichmann, H.J., Angel, R.J., Spetzler, H., Bassett, W.A., 1998. Ultrasonic interferometry and X-ray measurements on MgO in a new diamond anvil cell. Am. Miner. 83, 1357–1360.

Rigden, S.M., Gwanmesia, G.D., Jackson, I., Liebermann, R.C., 1992. Progress in high-pressure ultrasonic interferometry, the pressure dependence of elasticity of Mg_2SiO_4 polymorphs and constraints on the composition of the transition zone of the Earth's mantle. In: Syono, Y., Manghnani, M. (Eds.), High Pressure Research: Application to Earth and Planetary Sciences. American Geophysical Union, Washington, DC, pp. 167–182.

Rigden, S.M., Jackson, I., Niesler, H., Ringwood, A.E., Liebermann, R.C., 1988. Pressure-dependence of the elastic wave velocities for Mg_2GeO_4 spinel to 3 GPa. Geophys. Res. Lett. 15, 605–608.

Schreiber, E., Anderson, O.L., 1966. Pressure derivatives of the sound velocities of polycrystalline alumina. J. Am. Ceram. Soc. 49, 184–190.

Sinelnikov, Y.D., Chen, G., Neuville, D.R., Vaughan, M.T., Liebermann, R.C., 1998. Ultrasonic shear wave velocities of $MgSiO_3$ perovskite at 8 GPa and 800 K and lower mantle composition. Science 281, 677–679.

Sinelnikov, Y.D., Chen, G., Liebermann, R.C., 2004. Dual mode ultrasonic interferometry in multi-anvil high-pressure apparatus using single crystal olivine as the pressure standard. Int. J. High Pressure Res., in press.

Spetzler, H., Shen, A., Chen, G., Herrmannsdoerfer, G., Schulze, H., Weigel, R., 1996. Ultrasonic measurements in a diamond anvil cell. Phys. Earth Planet. Interiors 98, 93–99.

Spetzler, H.A., Chen, G., Whitehead, S., Getting, I.C., 1993. A new ultrasonic interferometer for the determination of equation of state parameters of sub-millimeter single crystals. In: Liebermann, R.C., Sondergeld, C.H. (Eds.), Experimental Techniques in Mineral and Rock Physics, vol. 141. PAGEOPH, pp. 341–377.

Tarrida, M., Richet, P., 1989. Equation of state of $CaSiO_3$ perovskite to 96 GPa. Geophys. Res. Lett. 16, 1351–1354.

Truell, R., Elbaum, C., Chick, B., 1969. Ultrasonic Methods in Solid State Physics. Academic Press, New York, 464 pp.

Wang, Y.B., Weidner, D.J., Guyot, F., 1996. Thermal equation of state of $CaSiO_3$ perovskite. J. Geophys. Res. 101, 661–672.

Webb, S.L., 1989. The elasticity of the upper mantle orthosilicates olivine and garnet to 3 GPa. Phys. Chem. Miner. 16, 684–692.

Weidner, D.J., Vaughan, M.T., Ko, J., Wang, Y., Liu, X., Yeganeh-Haeri, A., Pacalo, R.E., Zhao, Y., 1992. Characterization of stress, pressure, and temperature in SAM 85, a DIA type high pressure apparatus. In: Syono, Y., Manghnani, M. (Eds.), High Pressure Research: Application to Earth and Planetary Sciences. American Geophysical Union, Washington, DC, pp. 13–17.

Yoneda, A., Morioka, M., 1992. Pressure derivatives of elastic constants of single crystal forsterite. In: Syono, Y., Manghnani, M. (Eds.), High Pressure Research: Application to Earth and Planetary Sciences. American Geophysical Union, Washington, DC, pp. 206–214.

Yoneda, A., 1990. Pressure derivatives of elastic constants of single crystal MgO and $MgAl_2O_4$. J. Phys. Earth 38, 19–55.

Available online at www.sciencedirect.com
SCIENCE DIRECT
ELSEVIER
Physics of the Earth and Planetary Interiors 143–144 (2004) 575–586
PHYSICS OF THE EARTH AND PLANETARY INTERIORS
www.elsevier.com/locate/pepi

Sound velocity measurements on laser-heated MgO and Al_2O_3

Stanislav V. Sinogeikin[a,*], Dmitry L. Lakshtanov[a,1], Jason D. Nicholas[a,b,2], Jay D. Bass[a,b,3]

[a] *Department of Geology, University of Illinois, 1301 W Green St., Urbana, IL 61801, USA*
[b] *Department of Materials Science, University of Illinois, 1304 W Green St., Urbana, IL 61801, USA*

Received 20 November 2002; received in revised form 22 July 2003; accepted 3 September 2003

Abstract

Brillouin spectroscopy, in conjunction with CO_2 laser-heating, was employed to measure sound velocities at high temperatures on single crystals of MgO and Al_2O_3. The acoustic velocities of MgO were measured to a temperature exceeding 2500 (1 0 0) K, the highest temperature at which the Brillouin scattering measurements have been carried out. The temperature of MgO was inferred on the basis of independent sound velocity measurements using a resistance heater. The thermal background in the Brillouin spectra remained negligible to 1900 K and became problematic only at temperatures exceeding 2300 K. Both Brillouin and Raman measurements were performed on CO_2 laser-heated samples of single-crystal α-Al_2O_3 to temperatures exceeding 2000 (1 0 0) K. The temperature dependence of the Raman shift was used to estimate the sample temperature. Our high-temperature velocity measurements in different crystallographic directions are in agreement with previous high-temperature elasticity measurements by resonance techniques. This demonstrates the usefulness of Raman scattering to characterize moderate temperatures (where the intensity of thermal radiation is low). Our results show that Brillouin scattering coupled with CO_2 laser-heating is a viable means of performing sound velocity measurements at temperatures significantly higher than those readily made using resistance heating.

Keywords: MgO; Al_2O_3; High temperature; Brillouin spectroscopy; Raman spectroscopy; CO_2 laser-heating; Elastic properties

1. Introduction

The Earth's mantle is primarily composed of silicate and oxide (e.g. (Mg,Fe)O) mineral phases and has potential temperature of 1600–2000 K. Therefore, the knowledge of the elastic properties of minerals and oxide materials at high temperatures is important for interpreting the seismically observed velocity variations in the Earth's mantle and for constructing reliable mineralogical models of the Earth's interior. The high-temperature elasticity of oxide materials is also of considerable interest to materials and ceramics sciences. Recent advances in high-temperature ultrasonic and spectroscopic techniques have allowed elasticity measurements to be performed to 1825 K using rectangular parallelepiped resonance (RPR) (Goto et al., 1989; Isaak et al., 1989), and Brillouin measurements on single crystals have been performed up

* Corresponding author. Tel.: +1-217-333-6379; fax: +1-217-244-4996.
E-mail addresses: jaybass@uiuc.edu (S.V. Sinogeikin), lakshtan@uiuc.edu (D.L. Lakshtanov), jdnicholas@lbl.gov (J.D. Nicholas), jaybass@uiuc.edu (J.D. Bass).
[1] Tel.: +1-217-333-6379; fax: +1-217-244-4996.
[2] Current address: Lawrence Berkeley National Laboratory, 1 Cyclotron Rd, MS 62-203, Berkeley, CA 94720, USA. Tel.: +1-510-486-6898; fax: +1-510-486-4881.
[3] Tel.: +1-217-333-1018; fax: +1-217-244-4996.

0031-9201/$ – see front matter
doi:10.1016/j.pepi.2003.09.017

to 2100 K (Zouboulis and Grimsditch, 1991b,c). The highest temperature measurements performed by Brillouin scattering thus far are those by Vo-Thanh et al. (1996), who measured the elastic properties of silicate melts up to 2350 K. That study employed a backscattering geometry, in which one measures the product of the compressional acoustic velocity (V_P) and refractive index of the sample. To obtain V_P, the refractive index must be either measured by independent methods or extrapolated by some means (e.g. Lorentz–Lorentz equations).

Brillouin scattering is a purely optical technique in which acoustic velocities are derived from a Brillouin frequency shift produced by the interaction of incident laser light with thermally generated phonons. It has several characteristics that make it attractive for the high-temperature measurements. First, the intensity of Brillouin peaks increases with sample temperature (Sandercock, 1982). Second, the measured acoustic velocities (in platelet geometry) are independent of the sample dimensions and refractive index, both of which change with temperature. Third, highly polished and transparent oxide crystals generally have extremely low thermal emissivities, orders of magnitude lower than those of metals (e.g. Mao et al., 1998), which results in a low background from thermal emission and allows velocity measurements to be performed to temperatures exceeding 2100 K (e.g. Zouboulis and Grimsditch, 1991c; Vo-Thanh et al., 1996).

Conventional high-temperature Brillouin scattering, which employs resistance heaters to achieve the desired temperatures, has a number of limitations. One limitation is a high thermal background in high-temperature Brillouin spectra which, at temperatures below 2000 K, is usually related to the radiation from surrounding heater elements (e.g., heating wires and polycrystalline ceramic components) rather than from the sample itself. The maximum temperature of elasticity measurements is also limited by any reactions which might take place between a sample and furnace elements or a sample holder, usually resulting in a deterioration of the sample at extreme temperatures (e.g. Zouboulis and Grimsditch, 1991b,c). In addition, there are practical limitations to the construction and operation of resistance furnaces at temperatures higher than 1500 K, especially for those with sufficient optical access for accurate Brillouin measurements. Therefore it is desirable to use non-resistive or contactless heating methods in order to achieve maximum temperatures and fully exploit the benefits of high-temperature Brillouin scattering.

The most effective contactless method for heating materials is irradiation by infrared lasers. In the mid-1970s Ming and Bassett (1974) introduced laser-heating by a pulsed ruby laser and a CW YAG laser to high-pressure mineral physics. Laser-heating was successfully combined with a diamond anvil cell (DAC) allowing the synthesis of high-pressure phases and measurements of properties at simultaneous high pressures and temperatures of several thousands of Kelvins. Short-wavelength infrared lasers (e.g. YAG, $\lambda = 1064$ nm) are only able to couple with materials that are strongly absorbing or opaque in the visible and therefore cannot be used with transparent oxide materials, which generally have very low absorption at this wavelength. An alternative is to employ longer-wavelength lasers ($\lambda \sim 10\,\mu$m). These lasers were developed in early 1960s (Patel, 1968) and are routinely used for thermal processing of silicates and oxides (e.g. Paek and Weaver, 1975; Belov et al., 1978). In the early 1990s, Boehler and Chopelas (1991) and Yagi and Susaki (1992) developed experimental methods which employed a CW CO_2 laser ($\lambda = 10.6\,\mu$m) for heating optically transparent silicate and oxide materials in a DAC. The temperature in laser-heating experiments is usually measured by fitting the thermal radiation from the sample to Planck's law. However, these radiometric temperature measurements are practical only at temperatures exceeding 1200 K.

The combination of analytical techniques (e.g. synchrotron X-ray diffraction) with infrared laser-heating has became more routine for both opaque and transparent samples. However, to our knowledge there are no published reports of combining Brillouin scattering on solids with laser-heating. In this paper we present the first Brillouin scattering measurements (combined with Raman spectroscopy) on samples heated with a CO_2 infrared laser. The purpose of this study is to address the feasibility and limitations of such measurements and to explore alternative means of temperature measurements from room temperature to ~2000 K. We show that simultaneous measurements of acoustic and optical vibrational modes by Brillouin and Raman spectroscopy can be performed using CO_2 laser-heated oxide materials to temperatures above

2000 K, whereas Brillouin measurements alone can be made to temperatures exceeding 2500 (1 0 0) K.

2. Experimental setup

The CO_2 laser-heating system used with Brillouin scattering measurements is shown schematically in Fig. 1. The main components of the Brillouin system are an argon-ion laser, a goniometer for orienting the sample, and a six-pass Sandercock-type tandem Fabry-Perot interferometer. A detailed description of the Brillouin system and experimental setup is given by Sinogeikin et al. (1998).

The laser-heating was performed with a 100 W CO_2 ($\lambda = 10.6\,\mu m$) pseudo-CW laser operating in TEM_{00} mode (MPB Technologies Inc., model ST-100). The laser is equipped with a built-in power stabilization system that pulses the laser at a frequency of 5 kHz and modulates the pulse width to provide a nominal power stability of ±0.5% (after warm-up). The output laser power was controlled by applying an external voltage through a computer analog input/output board. The laser power was measured before and after Brillouin/Raman measurements with an Ophir Laserstar powermeter equipped with a thermopile-type 150 W head calibrated for a CO_2 laser. The CO_2 laser beam was steered to the sample with two silicon mirrors. In all experiments we used an unfocused laser beam with a near-Gaussian beam profile and a full width at half maximum (FWHM) of ~2 mm. The unabsorbed and reflected portions of the laser beam was stopped with air-cooled beamstops made of anodized aluminum.

The path of the CO_2 laser beam lies in the scattering plane for Brillouin and Raman scattering. The angle between the CO_2 laser and the collection direction was ~36°. All Brillouin/Raman measurements were performed in a platelet symmetric scattering geometry (e.g. Whitfield et al., 1976) with a scattering angle of 90° for MgO and 80° for Al_2O_3. In this geometry the measured acoustic velocities are independent of the sample refractive index, which can vary significantly with temperature.

The scattered light from the sample was collected through an achromatic lens ($f = 3.5$, focal distance = 100 mm), in front of which was positioned an aperture mask. This mask reduced the horizontal range of scattering angles resulting in sharper Brillouin peaks. After the collecting lens the scattered light was directed to a Fabry-Perot interferometer. Positioning a beam splitter or a flip mirror in the path of the scattered light allows us to redirect all or part of the scattered light to a Raman or thermal radiation collection system. Such a setup allows simultaneous or sequential collection of Brillouin, Raman, and thermal radiation spectra from a single point in the

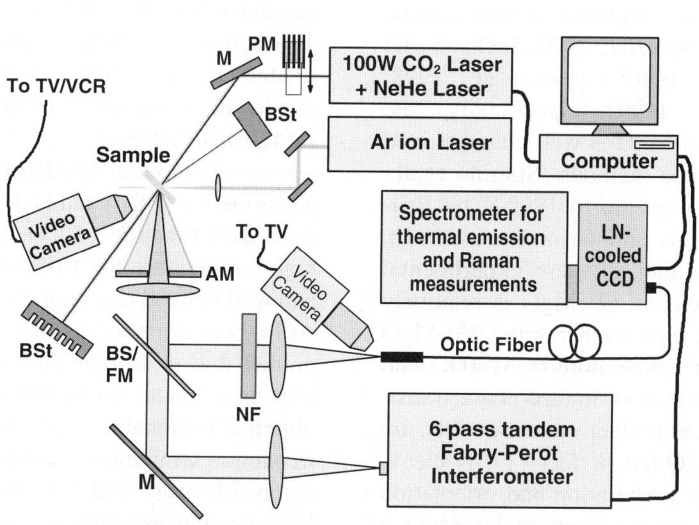

Fig. 1. Schematic diagram of the laser heating system for Brillouin measurements. Abbreviations are: AM, aperture mask; BSt, beam stop; M, mirror; BS, beam splitter; FM, flipping mirror; NF, notch filter; PM, powermeter.

sample (defined by the intersection of the focused Argon laser and the platelet sample). After the mirror/beamsplitter, the scattered light passes through a holographic super-notch filter (Kaiser Optical Systems Inc.) to attenuate the elastically scattered argon laser light in the vicinity (≈±4 nm or ±150 cm^{-1}) of the laser wavelength (514.5 nm), and is focused with an achromatic lens (focal length = 180 mm) to a 75 μm diameter silica single-fiber cable (Multimode Fiber Optics, Inc.). The signal is transmitted through the optical fiber into a SpectraPro 500i spectrometer (Acton Research) equipped with a liquid nitrogen cooled 1340 × 100 pixel back-illuminated CCD detector and interchangeable 600/1200/2400 grooves per millimeter (GPM) gratings. These gratings allow us to sample the frequency range up to ~3200 cm^{-1} with spectral resolution ranging from 2.3 to 0.43 cm^{-1} for gratings of 600–2400 GPM, respectively.

3. Laser-heated measurements

3.1. Brillouin scattering on MgO

MgO is an important lower mantle component that is well studied by many experimental and theoretical techniques. We chose it for our pilot high-temperature measurements for several reasons: (1) It is an excellent calibrant for elasticity measurements and is routinely used to assess the accuracy of new experimental techniques (e.g. Spetzler, 1970; Jackson and Niesler, 1982; Yoneda, 1990; Chen et al., 1998; Sinogeikin and Bass, 1999). Specifically, the single-crystal elasticity of MgO is well characterized up to 1800 K by resonance methods (Sumino et al., 1983; Isaak et al., 1989) and up to 1900 K by Brillouin scattering (Zouboulis and Grimsditch, 1991b; Sinogeikin et al., 2000), and all these experimental data are in mutual agreement. (2) MgO is relatively inert in air, especially at high temperatures. (3) MgO has a very high melting temperature (~3120 K) and does not undergo any high-temperature phase transitions. (4) Samples of near-perfect optical quality are readily available. (5) MgO has perfect {1 0 0} cleavage, thus making sample preparation and orientation relatively easy. All of these factors make MgO a very attractive material for exploring the feasibility of combining the Brillouin scattering and laser-heating.

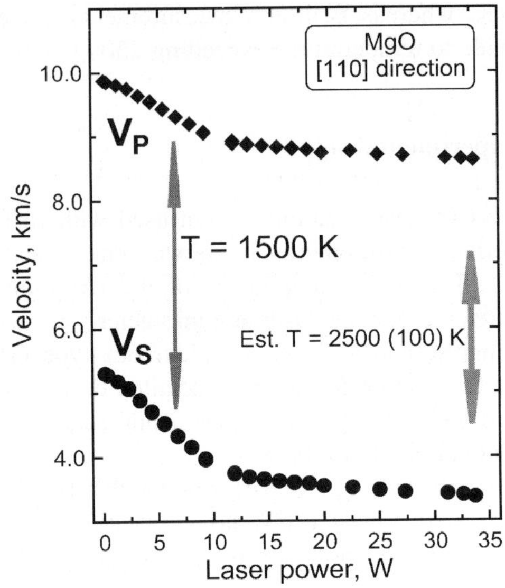

Fig. 2. Acoustic velocities in the [1 1 0] direction of MgO as a function of CO_2 laser power.

The Brillouin measurements were performed on two identical (1 0 0) slabs approximately 100 μm thick and ~1 mm × 4 mm in lateral dimensions. The slabs were polished on two parallel sides (along cleavage planes), attached to Al_2O_3 ceramic fibers with ZrO_2 cement, and mounted on a goniometer at the intersection of the argon light, CO_2 light, and the optical axis of the collection system. The crystallographic orientations of the platelets were identified by measuring the velocity distributions in (1 0 0) sample plane.

MgO is cubic and measurements of acoustic velocities in two distinct crystallographic directions are sufficient to determine the single crystal and aggregate elastic moduli. The Brillouin measurements were performed in [1 0 0] and [1 1 0] crystallographic directions as a function of CO_2 laser power (Fig. 2). At high temperature the Brillouin peaks remained sharp and well defined to a temperature of ~2400 K (Fig. 3). At higher temperatures the quality of the spectra degraded due to a high thermal background, and significantly longer collection times were necessary to obtain a reasonable signal-to-noise (S/N) ratio. The maximum temperature at which Brillouin peaks were resolvable was 2520 (1 0 0) K. Temperatures in the laser-heating experiments were obtained by comparison of the measured velocities with previous work (Sinogeikin et al., 2000). Previous measurements to

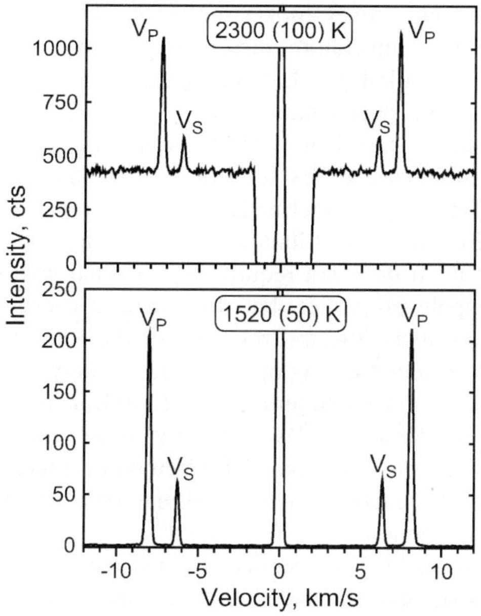

Fig. 3. Brillouin spectra of MgO in the [1 0 0] direction with CO_2 laser heating at 1520(50) and 2300 (1 0 0) K.

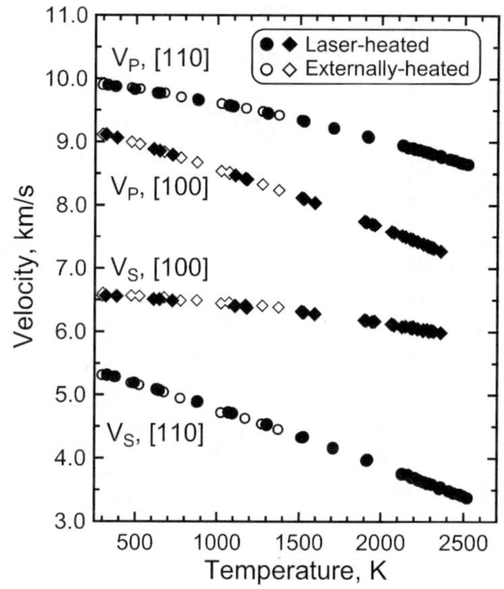

Fig. 4. Acoustic velocities in the [1 0 0] and [1 1 0] directions of MgO as a function of temperature. Solid symbols are laser-heated data, open symbols are data from Sinogeikin et al. (2000) obtained on identical samples in a resistively heated cell.

1500 K with a resistance heater were extrapolated to higher temperatures using second order polynomials. Velocities were thus used as a thermometer (Fig. 4). The temperatures calculated separately from V_P and V_S peak positions were within 80 K (Fig. 5). Moreover, for the same CO_2 laser power the temperature calculated for the [1 0 0] and [1 1 0] directions are in good agreement (Fig. 6). Using these velocity–power relations we calculated that Brillouin measurements were performed in the [1 0 0] direction to ~2300 (1 0 0) K and in the [1 1 0] direction to ~2520 (1 0 0) K (Fig. 4). The uncertainties in temperature were estimated on the basis of velocity reproducibility, the internal consistency of the power–temperature relations (e.g. Fig. 6), and agreement between temperature values calculated from V_P and V_S velocities.

The CO_2 laser power–sample temperature relations are highly non-linear, especially above ~1900 K where a large fraction of the energy is dissipated by radiation in the visible and infrared (Figs. 5 and 6). This is consistent with the fact that the thermal background in Brillouin spectra is negligible below ~1900 K, whereas above 2000 K the background is significant and increased rapidly with temperature according to the Stefan–Boltzmann law.

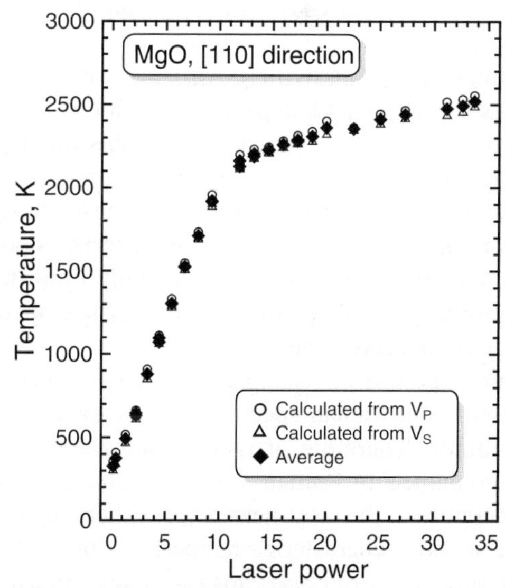

Fig. 5. Temperature of the sample as a function of CO_2 laser power calculated from V_P and V_S velocities measured in the [1 1 0] crystallographic directions.

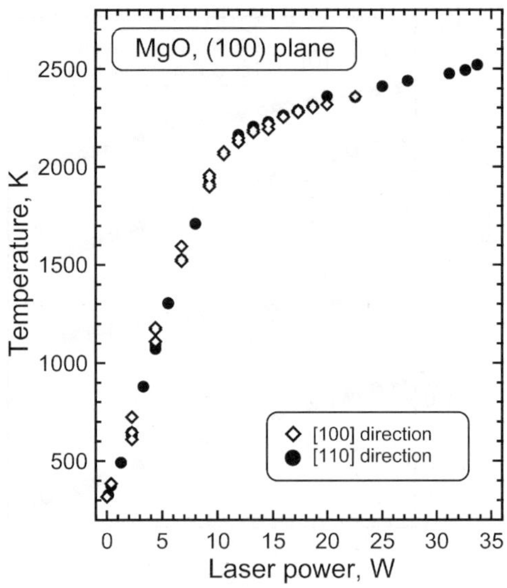

Fig. 6. Comparison of sample temperature calculated from acoustic velocities measured in the [1 0 0] and [1 1 0] direction as a function of CO_2 laser power.

3.2. Brillouin and Raman scattering on Al_2O_3

One of the drawbacks of laser-heating (especially for transparent samples) is that the temperature can only be characterized by thermal emission at $T > 1200$ K in the best of cases. This is because oxides have extremely low emissivities. Also the emissivities of oxides are wavelength dependent and not well known, which compromises the accuracy of this method. To characterize temperature in the range from ambient to 1500–2000 K, alternative ways of measuring temperature are needed. An inability to characterize moderate sample temperatures would significantly limit the applicability of CO_2 laser-heating for high-temperature elasticity measurements.

Our experimental setup allows the collection of Brillouin and Raman spectra from the same spot on the sample. Therefore, it is possible to use information contained in Raman spectra to characterize the temperature in the laser-heated spot. There are at least three ways to characterize temperature using Raman scattering: (1) from anti-Stokes–Stokes intensity ratio; (2) from the widths of Raman peaks; and (3) from the absolute positions of Raman peaks (or their temperature dependence).

The anti-Stokes–Stokes intensity ratio is widely used for temperature measurements (e.g. Herchen and Cappelli, 1991; LaPlant et al., 1996; Cui et al., 1998) with a precision of up to ± 5 K at moderate temperatures. The method is attractive and has potential because of its self-consistency: the intensity ratio is a purely thermodynamic function based on the Boltzmann distribution, and it is material independent. It does not require any external calibration of sample properties, but does require careful calibration of the spectral response of the Raman spectrometer and focusing optics. Unfortunately, its use is restricted to temperatures below 1200 K, above which the anti-Stokes–Stokes intensity ratio approaches unity and the precision of the method decreases drastically. In practice, this method may be accurate only at low temperatures. For example, Cui et al. (1998) performed high-temperature Raman measurement on diamond and concluded that the intensity ratio of Stokes to anti-Stokes lines is ill-suited for temperature measurements above $\sim$750 K.

The widths of Raman peaks increases with temperature (e.g. Herchen and Cappelli, 1991; Schmidt and Ziemann, 2000), and are potentially useful for temperature determination at moderate temperatures. Unfortunately, the increase in width is highly non-linear with temperature and requires extensive independent calibration. In addition, the quality of the Raman peaks progressively degrades by an increase of width and a decrease of S/N ratio with increasing temperature (e.g. Richet et al., 1993) thereby limiting the applicability of this technique.

From our experience, the temperature-induced Raman shift seems to be the most precise spectroscopic method of measuring moderate temperatures. Cui et al. (1998) claimed that by using the Raman shift the temperature can be measured in the range from 300 to 2000 K with an accuracy of ± 10 K. Unlike a Brillouin shift, the Raman shift is geometry (scattering angle) independent (although the relative intensity and visibility of peaks is a function of phonon direction and polarization). This gives some advantages for using the absolute Raman shift as an internal thermometer for each particular material, because it can be readily calibrated in any geometry with a variety of high-temperature devices (e.g. Zouboulis and Grimsditch, 1991a; Richet et al., 1993).

We examined the feasibility of using the temperature coefficient of the Raman line position as means of measuring sample temperature in a laser-heated spot on α-Al_2O_3 (sapphire). This material was chosen due to its extensive characterization by a variety of experimental techniques. Sapphire has a high melting temperature (2310 ± 35 K) and does not undergo any high-temperature phase transitions. Its single-crystal elasticity at high temperature was previously measured by rectangular parallelepiped resonance to 1825 K (Goto et al., 1989) and by Brillouin spectroscopy to 2100 K (Zouboulis and Grimsditch, 1991c).

Sapphire is trigonal and has seven well-defined Raman-active modes (e.g. Richet et al., 1993; Kadleíková et al., 2001), the most intense of which (417 cm^{-1}) has an FWHM of 1.3(1) cm^{-1}. The temperature coefficients of the Raman-active modes of sapphire were measured by Richet et al. (1993) to 1923 K, and were found to be linear within experimental uncertainties. However, our measurements on laser-heated α-Al_2O_3 samples revealed that the Raman shifts as a function of temperature are significantly non-linear (corresponding to a few hundreds of Kelvin difference in the inferred temperature), and thus required further determination of peak position as a function of temperature.

Preoriented single crystals of sapphire (10 mm × 10 mm × 0.5 mm, (0 0 1) orientation) were obtained from MTI Corporation. The crystals were colorless, of perfect optical quality, free of twinning, inclusions or other visible defect. The plates were cut into ~1 mm × 4 mm slabs and polished to 50–60 μm thickness.

High-temperature Raman measurements were performed in a resistively heated furnace of our own design. This compact ceramic furnace with $Pt_{30}Rh$ heating wires is a high-temperature modification of our earlier design (Sinogeikin et al., 2000), but with lower thermal gradients and more limited optical access. The details of the furnace are given by Jackson et al. (2003). To minimize temperature gradients the sample was enclosed in a 250 μm thick platinum envelope with two through holes for optical access. The temperature was measured with an R-type thermocouple.

Raman measurements were performed in near-forward scattering geometry (~15°). An argon-ion laser (λ = 514.5 nm) was used as an excitation source. Raman shifts (both Stokes and anti-Stokes) were measured to 1500 K (Fig. 7). Up to 1400 K the

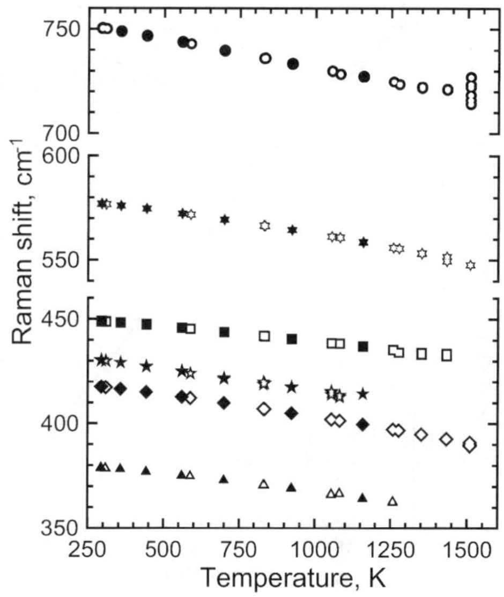

Fig. 7. Temperature dependence of the Raman-active vibrational modes of sapphire measured in an externally heated cell. Open and filled symbols are measurements performed on the same sample but in two different high-temperature runs.

positions of major Raman modes were well defined (the 645 cm^{-1} mode is not visible in this geometry). Above that temperature the thermal emission by the platinum holder, heater, and ceramic parts produces high thermal background which significantly degraded the quality of the Raman peaks, and did not allow us to precisely determine their positions.

The temperature shifts of the Raman peaks were fitted by quadratic functions. The sharp and intense peak at 417 cm^{-1} gave the most consistent positions (Fig. 8). The Raman shift versus temperature relations are given by: T (K) $= 297 - 54.31 \times \Delta\nu - 0.3409 \times \Delta\nu^2$ ($\Delta\nu$ is the shift with respect to ambient temperature). Using this function, temperatures calculated from the position of the '417 cm^{-1}' peak in the interval from 300 to 1400 K were within 10 K of temperatures measured by thermocouples.

The Brillouin measurements were performed in a symmetric platelet geometry with an 80° external scattering angle. In this geometry the angle between the CO_2 laser beam and the normal to the sample plate is only 4°. At room temperature the sample was preoriented using the velocity distribution in the sample plane, and the Chi-angle settings of the goniometer corresponding to the crystallographic axes were iden-

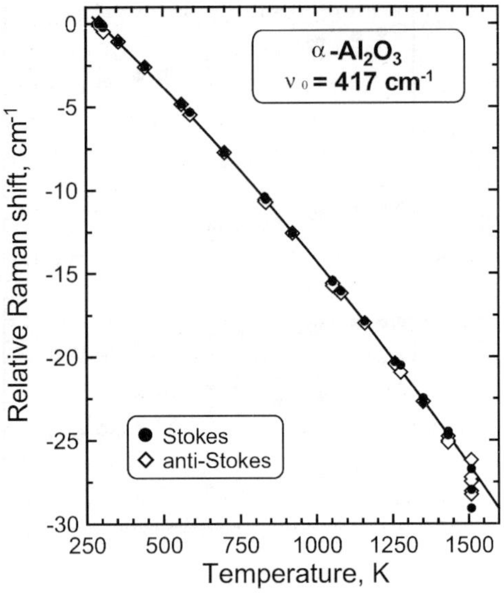

Fig. 8. Frequency shift of the most intense line ($417\,cm^{-1}$) in the Raman spectrum of sapphire as a function of temperature measured in an externally heated cell. The solid line is the quadratic least-squares best fit ($\Delta\nu = 4.95-0.0157\times T-3.4556\times10^{-6}\times T^2$). Data above 1400 K were not used in the fit. The uncertainty in temperature and peak positions are smaller than the size of the symbols.

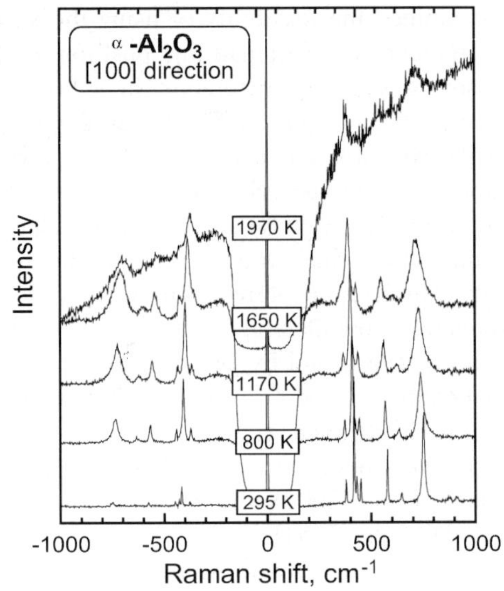

Fig. 9. Raman spectra of single-crystal sapphire at high temperatures. Temperatures are calculated from the average (Stokes–anti-Stokes) position of '$417\,cm^{-1}$' peak.

tified. Brillouin spectra were collected in the [1 0 0] and [0 1 0] crystallographic directions as a function of laser power. Raman spectra were collected in each run before and after the Brillouin measurements. Typical high-temperature Raman spectra are shown in Fig. 9.

The measured acoustic velocities in the [1 0 0] direction as a function of calculated temperature are shown in Fig. 10A as solid symbols. Fig. 10B shows the velocities measured in two distinct orientations corresponding to [0 1 0] crystallographic direction. Also shown are the velocities calculated from single-crystal elastic moduli reported in (Goto et al., 1989) (Figs. 10A and B, open squares). Our data are in agreement with those values. Our measured acoustic velocities above 1500 K are slightly slower than those of Goto et al. (1989), possibly resulting from differences in temperature measurements. It should be mentioned that similar "softening" was also observed by Zouboulis and Grimsditch (1991c) and may be a real feature reflecting either dispersion of elastic properties at high temperature, or premelting behavior (e.g. Richet et al., 1994; Fiquet et al., 1999).

4. Discussion

The limitations of high-temperature Brillouin scattering are generally caused by temperature limitations of the furnace (e.g. Sinogeikin et al., 2000), sample deterioration due to its interaction with furnace and sample holder materials (Zouboulis and Grimsditch, 1991b,c), or high thermal background. Laser-heating of samples allows one to overcome most of these problems. Samples having sufficient absorption at $\lambda = 10.6\,\mu m$ can be heated up to their melting temperature. Sample deterioration due to interaction with volatiles from the furnace elements and contact with a hot sample holder is eliminated because a laser-heated sample is attached to its holder at only one point, which is relatively cool compared to the hot spot in the sample. In addition, thermal radiation from furnace elements is eliminated. Therefore, the only thermal radiation which reaches the detector is radiation from the sample itself. Because most oxides and silicates have very low emissivities, the temperature at which the thermal radiation becomes a severe problem significantly exceeds 2000 K.

The intensity of both the Brillouin peaks and the thermal background increase with temperature but the rate of increase is quite different. The intensity of

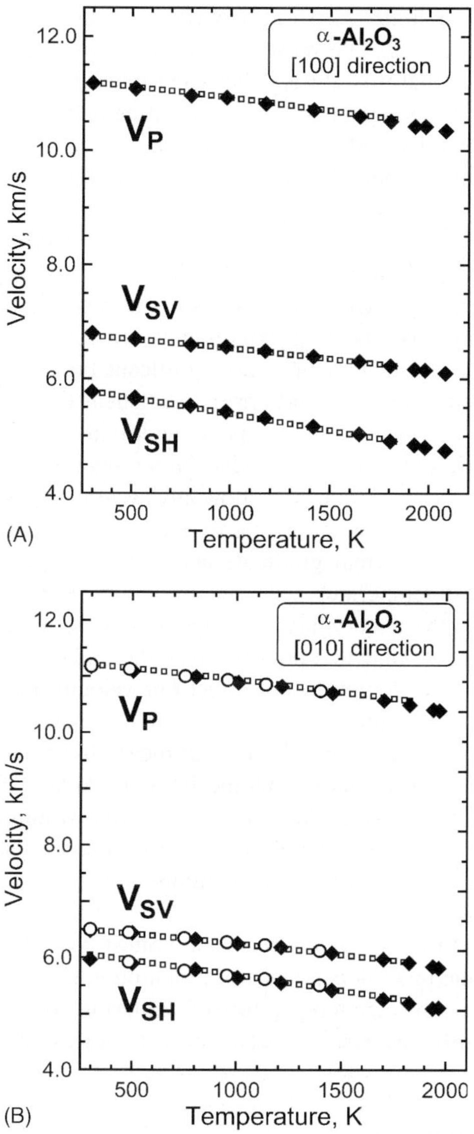

Fig. 10. Acoustic velocities in sapphire as a function of temperature in *a*-direction ([1 0 0]) and *b*-direction ([0 1 0]). Diamonds and circles are measurements from this study. The temperature was calculated from the temperature-induced frequency shift of the 417 cm^{-1} Raman peak. Open squares: data of (Goto et al., 1989) measured by RPR technique to 1825 K.

the Brillouin peaks increases linearly with temperature (Sandercock, 1982), whereas the intensity of thermal radiation is proportional to T^4. Therefore, at some temperature the Brillouin peaks will become indistinguishable from the thermal background. The maximum possible temperature depends on a number of factors. They include the emissivity of the sample at the excitation laser wavelength, the excitation laser wavelength, laser power and polarization, roughness of the sample surface, the optical quality of the bulk of the sample, and the Brillouin cross section (or photoelastic constants) of the material, among other factors. So far we have been able to achieve temperatures exceeding 2500 (1 0 0) K with an MgO sample. Note that below 1900 K the thermal background in Brillouin spectra was negligible and became significant only above 2100 K (Fig. 3). MgO and sapphire have comparably low emissivities in visible region (the thermal background in Al_2O_3 was observed only above 1900 K) and comparable Brillouin intensities, and therefore we estimate that Brillouin scattering on sapphire can be performed up to its melting temperature at ∼2300 K. We did not perform measurements to higher temperatures only because the Raman peaks were not distinguishable from the thermal background above 2000 K, even though the thermal background in Brillouin spectra remained low. We believe that with further optimization of the Brillouin system the upper temperature limit for Brillouin measurements on MgO can be increased by several hundred degrees.

Laser-heating is a localized heating method as compared to resistive heating where the entire sample is heated uniformly. Therefore it is important to consider axial (along the axis of CO_2 beam) and radial (perpendicular to the laser beam) temperature gradients. The radial thermal gradients are not a serious concern in the present experiments because we used an unfocused CO_2 laser with FWHM ∼2 mm, while the spectra were collected from an axial area of 40–60 μm.

Axial thermal gradients (normal to sample plane) are potentially a bigger problem because the sample is heated from one side only. On the other hand, the laser energy is absorbed to some degree by the entire sample volume, so long as the absorption coefficient of oxides and minerals at 10.6 μm is <1. The absorption length of the laser beam by dielectric samples is typically more than 100 μm (Fukao et al., 1968), which is comparable to or larger than the thickness of our samples. In addition, our samples are surrounded by air which has a low thermal conductivity, thereby reducing heat loss due to conduction. These arguments combined with long duration of sample heating, which results in temperature homogenizations and a more

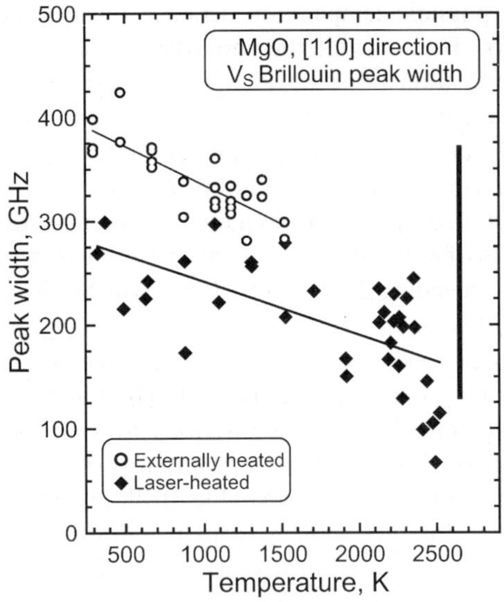

Fig. 11. Width of V_S Brillouin peaks in MgO ([1 1 0] direction) at high temperature minus the instrumental function for resistance and CO_2 laser heating. The vertical bar indicates the calculated increase in peak width corresponding to a 100 K temperature difference across the scattering volume. The difference in ambient-temperature peak width is due to the different scattering angles in externally heated and laser-heated experiments (80° and 90°, respectively), and different aperture masks.

uniform temperature distribution, suggest that the axial thermal gradients in the sample are expected to be small.

In principle it is possible to estimate the gradients based on the width of Raman or Brillouin peaks. Because the peak positions of both acoustic (Brillouin) and optical (Raman) modes change with temperature, both the thermal gradient within the sample and power fluctuations during an experiment will result in peak broadening.

The Raman peak positions at high temperature are highly reproducible (e.g., Fig. 8). At the same time the peak widths are somewhat variable even under isothermal conditions. Thus the use of the width of Raman peaks is unsuitable for estimating thermal gradients. Alternatively, Brillouin peaks (collected in an externally heated cell with practically zero static and dynamic temperature gradients) generally show a slight decrease in width with increasing temperature (Fig. 11).

Fig. 11 shows the width of the transverse V_S peak minus instrumental function in [1 1 0] direction of MgO at high temperature with resistance and CO_2 laser-heating. The reason for selecting this particular peak is that it is always sharp and intense, and shows the highest dependence of the position on temperature compared to other acoustic modes in this and other crystallographic directions ($\partial V_S/\partial T \approx 90$ (m/s)/100 K (Fig. 4)). Thus a temperature difference of 100 K in the scattering volume would increase the width of this peak by ~240 MHz. Fig. 11 shows that the rates of decrease of Brillouin peak widths in resistance heating and laser-heating experiments are approximately equal, and that there is no significant broadening of Brillouin peak beyond normal statistical fluctuations. The fluctuations in the peak width in laser-heated experiments show somewhat larger scatter, although the peak width does not show any systematic increase with increasing temperature. Thus we estimate that the axial thermal gradients across the sample at the highest temperatures do not exceed 50 K (which is within the uncertainty of temperature determination).

We also note that possible axial thermal gradients in laser-heated spot do not affect our velocity measurements as a function of temperature because the temperature is measured by optical means from the same volume in the sample as the Brillouin signal. Therefore, gradients in the scattering volume would affect both Raman and Brillouin peak position, yielding average values for both temperature and velocities. Alternatively, the thermal radiation spectra would tend to yield the temperature of the hottest part of a sample due to T^4 dependence of radiation intensity, rather than an average temperature of scattering volume, and possibly decrease the accuracy of the measurements.

5. Summary

In this study, we have shown conclusively that accurate Brillouin scattering velocity measurements on CO_2 laser-heating samples can be performed. The main obstacle to these sorts of measurements is thermal background in the Brillouin spectra, but this remained negligible to 1900 K and became significant only at temperatures exceeding ~2300 K, depending on the sample. The sound velocities in MgO were measured to temperatures exceeding 2500 (1 0 0) K, the highest temperature thus far at which Brillouin scattering measurements have been made. Velocity measurements on α-Al_2O_3 in a variety of crystallographic directions show agreement with previous high-temperature resonance measurements. Simultaneous high-temperature Raman scattering measurements are shown to be a viable method of measuring sample temperatures. Our results show that Brillouin scattering coupled with CO_2 laser-heating is a viable means of performing sound velocity measurements at extremely high temperatures, significantly beyond those readily made using resistance heating.

Acknowledgements

This research was supported by National Science Foundation, grant # EAR0002021. We thank M.H. Manghnani and an anonymous reviewer for helpful comments.

References

Belov, A.V., Bubnov, M.M., Gur'yanov, A.N., Devyatykh, G.G., Dianov, E.M., Prokhorov, A.M., Rusanov, S.Ya., Yushin, A.S., 1978. Drawing of glass-fiber waveguides using CO_2 lasers. Sov. J. Quant. Electron. 8, 1169.

Boehler, R., Chopelas, A., 1991. A new approach to laser heating in high pressure mineral physics. Geophys. Res. Lett. 18, 1147–1150.

Chen, G., Liebermann, R.C., Weidner, D.J., 1998. Elasticity of single-crystal MgO to 8 Gigapascals and 1600 K. Science 280, 1913–1916.

Cui, J.B., Amtmann, K., Ristein, J., Ley, L., 1998. Noncontact temperature measurements of diamond by Raman scattering spectroscopy. J. Appl. Phys. 83, 7929–7933.

Fiquet, G., Richet, P., Montagnac, G., 1999. High-temperature thermal expansion of lime, periclase, corundum and spinel. Phys. Chem. Miner. 27, 103–111.

Fukao, Y., Mizutani, H., Uyeda, S., 1968. Optical absorption spectra at high temperatures and radiative thermal conductivity of olivines. Phys. Earth Planet. Inter. 1, 57–62.

Goto, T., Anderson, O.L., Ohno, I., Yamamoto, S., 1989. Elastic constants of corundum up to 1825 K. J. Geophys. Res. 94, 7588–7602.

Herchen, H., Cappelli, M.A., 1991. First-order Raman spectrum of diamond at high temperatures. Phys. Rev. B 43, 11740–11744.

Isaak, D.G., Anderson, O.L., Goto, T., 1989. Measured elastic moduli of single-crystal MgO up to 1800 K. Phys. Chem. Miner. 16, 704–713.

Jackson, J.M., Palko, J.W., Andrault, D., Sinogeikin, S.V., Lakshtanov, D.L., Wang, J.Y., Bass, J.D., Zha, C.-S., 2003. Thermal expansion of natural orthoenstatite to 1473 K. Eur. J. Miner. 15, 469–473.

Jackson, I., Niesler, H., 1982. The elasticity of periclase to 3 GPa and some geophysical implications. In: Akimoto, S., Manghnani, M.H. (Eds.), High Pressure Research in Geophysics. Center for Academic Publications, Tokyo, Japan, pp. 93–113.

Kadleíková, M., Breza, J., Ueselý, M., 2001. Raman spectra of synthetic sapphire. Microelectron. J. 32, 955–958.

LaPlant, F., Laurence, G., Ben-Amotz, D., 1996. Theoretical and experimental uncertainty in temperature measurement of materials by Raman spectroscopy. Appl. Spectrosc. 50, 1034–1038.

Mao, H.K., Shen, G., Hemley, R.J., Duffy, T.S., 1998. X ray diffraction with a double hot-plate laser-heated diamond cell. In: Manghnani, M.H., Yagi, T. (Eds.), Properties of Earth and Planetary Materials at High Pressure and Temperature. AGU, Washington, DC, pp. 27–34.

Ming, L., Bassett, W.A., 1974. Laser heating in the diamond anvil press up to 2000 °C sustained and 3000 °C pulsed at pressures up to 260 kbars. Rev. Sci. Instrum. 45, 1115–1118.

Paek, U.C., Weaver, A.L., 1975. Formation of a spherical lens at optical fiber ends with a CO_2 laser. Appl. Opt. 14, 294–298.

Patel, C.K.N., 1968. High-power carbon dioxide lasers. Sci. Am. 219, 22–33.

Richet, P., Gillet, P., Pierre, A., Bouhifd, M.A., Daniel, I., Fiquet, G., 1993. Raman spectroscopy, x-ray diffraction, and phase relationship determinations with a versatile heating cell for measurements up to 3600 K (or 2700 K in air). J. Appl. Phys. 74, 5451–5456.

Richet, P., Ingrin, J., Mysen, B.O., Courtial, P., Gillet, P., 1994. Premelting effects in minerals: an experimental study. Earth Planet. Sci. Lett. 121, 589–600.

Sandercock, J.R., 1982. Trends in Brillouin scattering: studies of opaque materials, supported films and central modes. In: Cardona, M., Guntherodt, G. (Eds.), Light Scattering in Solids III: Recent Results. Springer-Verlag, New York, pp. 173–206.

Schmidt, C., Ziemann, M.A., 2000. In-situ Raman spectroscopy of quartz: a pressure sensor for hydrothermal diamond-anvil cell experiments at elevated temperatures. Am. Miner. 85, 1725–1734.

Sinogeikin, S.V., Bass, J.D., 1999. Single-crystal elasticity of MgO at high pressure. Phys. Rev. B 59, R14141–R14144.

Sinogeikin, S.V., Jackson, J.M., O'Neill, B., Palko, J.W., Bass, J.D., 2000. Compact high-temperature cell for Brillouin scattering measurements. Rev. Sci. Instrum. 71, 201–206.

Sinogeikin, S.V., Katsura, T., Bass, J.D., 1998. Sound velocities and elastic properties of Fe-bearing wadsleyite and ringwoodite. J. Geophys. Res. 103, 20819–20825.

Spetzler, H., 1970. Equation of state of polycrystalline and single-crystal MgO to 8 kbars and 800 K. J. Geophys. Res. 75, 2073–2087.

Sumino, Y., Anderson, O.L., Suzuki, I., 1983. Temperature coefficients of elastic constants of single crystal MgO between 80 and 1300 K. Phys. Chem. Miner. 9, 38–47.

Vo-Thanh, D., Polian, A., Richet, P., 1996. Elastic properties of silicate melts up to 2350 K from Brillouin scattering. Geophys. Res. Lett. 23, 423–426.

Whitfield, C.H., Brody, E.M., Bassett, W.A., 1976. Elastic moduli of NaCl by Brillouin scattering at high pressure in a diamond anvil cell. Rev. Sci. Instrum. 47, 942–947.

Yagi, T., Susaki, J.-I., 1992. A laser heating system for diamond anvil using CO_2 laser. In: Syono, Y., Manghnani, M.H. (Eds.), High-pressure Research: Application to Earth and Planetary Sciences. AGU, Washington, DC, pp. 51–54.

Yoneda, A., 1990. Pressure derivatives of elastic constants of single crystal MgO and $MgAl_2O_4$. J. Phys. Earth 38, 19–55.

Zouboulis, E.S., Grimsditch, M., 1991a. Raman scattering in diamond up to 1900 K. Phys. Rev. B 43, 12490–12493.

Zouboulis, E.S., Grimsditch, M., 1991b. Refractive index and elastic properties of MgO up to 1900 K. J. Geophys. Res. 96, 4167–4170.

Zouboulis, E.S., Grimsditch, M., 1991c. Refractive index and elastic properties of single-crystal corundum (α-Al_2O_3) up to 2100 K. J. Appl. Phys. 70, 772–777.

Available online at www.sciencedirect.com
SCIENCE DIRECT
ELSEVIER
Physics of the Earth and Planetary Interiors 143–144 (2004) 587–591
PHYSICS OF THE EARTH AND PLANETARY INTERIORS
www.elsevier.com/locate/pepi

HIP production of a diamond/SiC composite and application to high-pressure anvils

Osamu Ohtaka [a,*], Masaru Shimono [b], Naoyuki Ohnishi [a], Hiroshi Fukui [a], Hitoshi Takebe [a], Hiroshi Arima [a], Takamitsu Yamanaka [a], Takumi Kikegawa [c], Shoichi Kume [b]

[a] *Earth and Space Science, Osaka University, Toyonaka 560-0043, Japan*
[b] *Materials Chemistry, Ryukoku University, Ohtsu 520-2123, Japan*
[c] *Photon Factory, KEK, Tsukuba 305-0801, Japan*

Received 4 December 2002; received in revised form 9 April 2003; accepted 14 May 2003

Abstract

Using a hot isostatic pressing (HIP) technique, sintering of diamond/SiC composites was performed. At an HIP condition of 1450 °C and 200 MPa, a pressure much lower than that of the diamond stability field, the formation of well-sintered diamond/SiC composites was confirmed. No trace of graphite was observed. An application of the composite to high-pressure anvils was attempted. Cubes of the composites were used as the second stage anvils in a Kawai-type high-pressure apparatus. Because the diamond/SiC composites are transparent to X-rays, diffraction patterns of NaCl powder compressed in the high-pressure apparatus were successfully recorded through the anvils and thereby the generated pressures were determined using the equation of state of NaCl. The anvils were proved to be hard enough for the generation of pressures up to 30 GPa. The present anvils will be applicable not only to diffraction but also to radiographic studies under pressure, since almost the entire part of the high-pressure cell can be observed with the X-ray beam.

Keywords: Diamond composite; SiC; Hot isostatic pressing (HIP); Sintering; Kawai-type high-pressure apparatus; Synchrotron radiation

1. Introduction

The application of sintered diamond composites for use anvils for generating as high pressure has been attempted for the last few decades in order to extend the accessible pressure regime with multi-anvil apparatus (e.g., Bundy, 1975; Endo et al., 1987; Ohtani et al., 1989; Shimomura et al., 1992; Irifune et al., 1998a). Ito and Kubo (2002) showed that pressure generation up to more than 50 GPa can be achieved using eight sintered diamond cubes as the second stage anvils in a Kawai-type high-pressure apparatus (Kawai and Endo, 1970). There are two types of sintered diamond composites that are currently available. The first are conventional diamond composites such as SYNDIE from the De Beers Company made within the diamond stability field using metallic binding materials. The second is a relatively new diamond composite called ADC invented by a research group in Australia (Ringwood et al., 1989). This composite is produced at pressures slightly lower than those of the diamond stability field using a SiC binder. Because both composites are

* Corresponding author. Tel.: +81-6-6850-5797; fax: +81-6-6850-5480.
E-mail address: ohtaka@ess.sci.osaka-u.ac.jp (O. Ohtaka).

0031-9201/$ – see front matter © 2004 Elsevier B.V. All rights reserved.
doi:10.1016/j.pepi.2003.05.003

sintered in a solid state pressure-transmitting media using a large volume pressure apparatus, the size and shape of the products is limited. Furthermore, they are quite expensive for use in routine high-pressure experiments. If a method to prepare well-sintered diamond under lower pressures conditions can be developed, it would help in producing composites with large size and complex shape at a reasonable price. An HIP technique is one of the possible methods of synthesis that might meet this requirement. We have synthesized diamond/SiC composites by HIP and measured their mechanical strength (Shimono et al., 1998, 2003). The results show that both the hardness and compressive strength are higher than those of WC and the composite, therefore, seems to be well suited for use as an anvil material. It was also expected that the transparency of the composites anvils to X-rays would be high because of the low absorption of the constituent elements, C and Si. As in situ X-ray studies become more prevalent, the X-ray transparency of anvils becomes of prime importance. Here we report our recent results on HIP production of diamond/SiC composites and describe the application of this material as second stage anvils in a Kawai-type high-pressure apparatus.

2. Synthesis and characterization of composite

2.1. HIP processing

The basic steps in the HIP process for diamond/SiC composite cube production are illustrated in Fig. 1. To prepare a capsule for the synthesis of diamond composites, Si and SiC powders were mixed (70:30 in weight ratio). The mixture was molded into a cylinder with an inner cubic cavity using a cold isostatic pressing apparatus (KOBELCO PERSONAL CIP). The capsule was then semi-sintered in a vacuum at 1100 °C (Fig. 1A). Diamond powder of 12 μm average grain size was tightly packed inside the capsule so that it remained in a cubic shape. The capsule containing the diamond powder was inserted into a glass tube and sealed in a vacuum at 750 °C (Fig. 1B). Using an HIP apparatus (KOBELCO PERSONAL HIP), the assemblage was HIPed at 1450 °C and 200 MPa for 30 min according to the optimized synthesis conditions determined in our previous study (Shimono et al., 2003). Argon was the pressure-transmitting gas.

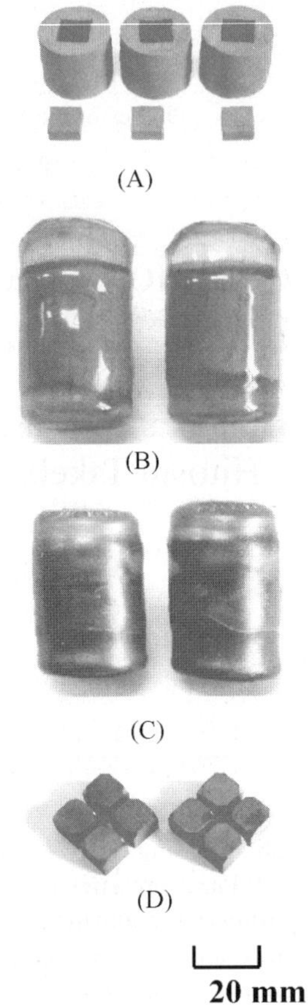

Fig. 1. An outline of the HIP process. (A) Sample capsules are made of a mixture of Si and SiC powder (70:30 weight ratio). The shape is cylindrical outside and cubic inside. (B) The capsules are glass-sealed in a vacuum. (C) The assemblage after HIP. (D) The final shape of diamond/SiC cubes with an edge length of 10 mm.

Fig. 1C shows the assemblage after HIP treatment. The product was surface ground to a cubic with an edge length of 10 mm using a lapping machine with fine diamond powders and one corner was removed to form a triangular truncation using a surface grinder. The final shape of the cube is shown in Fig. 1D.

2.2. Characterization of products

Density and constituents of the products were determined by the Archimedes' method and X-ray powder diffraction, respectively. The number of cubes produced for this study was 21 in total and the averaged density was 3.380 (0.004) g/cm^3, where the figure in parenthesis indicates the standard deviation.

Neither graphite nor Si were observed in the X-ray patterns. The area ratio of diamond to SiC was estimated at 55:45 from optical microscopy observations of the texture of a polished surface. Because this ratio gives a bulk density of 3.387 g/cm^3, which is very close to the measured density, SiC was thought to be evenly distributed through the cube. Since the starting material was diamond, SiC was most probably formed through the reaction between diamond and molten Si supplied from the capsule. In our previous work, similar composites were made from a starting material consisting of mixed diamond and Si powders (Shimono et al., 2003). The present result indicates that diamond alone can be hot pressed into a hard material if the capsule contains Si. This newly improved process yields an increase in the diamond content of composites and hence enhances the mechanical properties of the products. This process also results in only a small degree of shrinkage after HIP, and the dimensions of the product can be easily controlled.

One of the most common methods of synthesizing diamond composites is by annealing in the presence of a binder at pressures of around 5 GPa. The form of the material is, however, difficult to control during compression. Consequently, a significant amount of machining of the products is required after sintering to obtain the final article. The present study proves that HIP is more advantageous when preparing anvils for high-pressure experiments. The main points are: First, the magnitude of the required pressures for the synthesis is not so high as to require special equipment, such as the belt-type apparatus, and this makes the fabrication of the composites comparatively easy. Secondly, the shape of the sintered bodies is close to that which is required so a large amount of machining of the hard material is avoided.

3. Application to high-pressure anvils

3.1. Experimental

Compression experiments were attempted in a Kawai-type high-pressure apparatus using eight cubes with an edge length of 10 mm as the second stage anvils. In order to obtain the relationship between the press load and generated pressure, in situ X-ray diffraction experiments were performed with the MAX-III system (Funamori et al., 2000) installed at the Photon Factory in the High Energy Accelerator Research Organization. Pressure generation using anvils with a truncation edge length (TEL) of 3 and 1.5 mm was thereby examined. The pressure-transmitting medium was semi-sintered MgO and the gasket material was pyrophyllite. A powdered mixture of NaCl and Au (10:1 weight ratio) was located in a hole placed in the center of the pressure medium. The incident white X-ray beam was directed to the sample by horizontal (0.3 mm) and vertical (0.1 mm) slits. The diffracted X-ray beam, taken through one of the diamond/SiC anvils, was collimated and measured with a solid state detector (SSD) fixed on a vertically rotating goniometer. The energy range used for the analysis was approximately 30–100 keV. The 2θ angle was calibrated using the major diffraction peaks of NaCl at ambient conditions. The generated pressure was determined from the unit cell volume of NaCl and Au and their equations of state (Decker, 1971; Anderson et al., 1989). The experiments were conducted at ambient temperature.

3.2. Results and discussion

Fig. 2 shows an X-ray diffraction pattern of the mixture of NaCl and Au recorded at several loads using the anvils with TEL of 3 mm. In addition to the

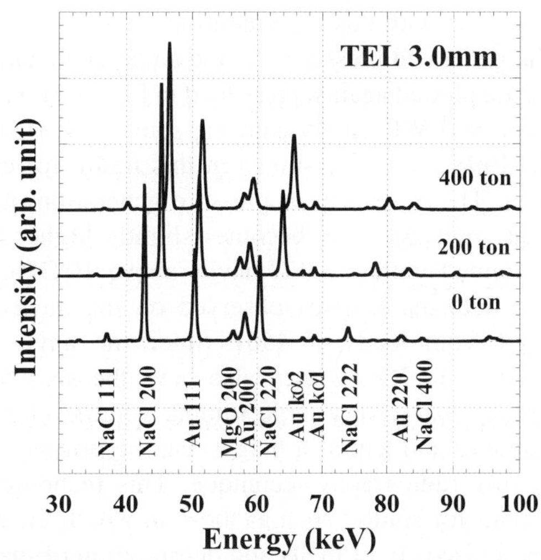

Fig. 2. Energy dispersive X-ray diffraction spectra of NaCl and Au powder at various loads. Eight cubes of diamond/SiC compact with TEL of 3 mm are used. 2θ angle is 6.017°.

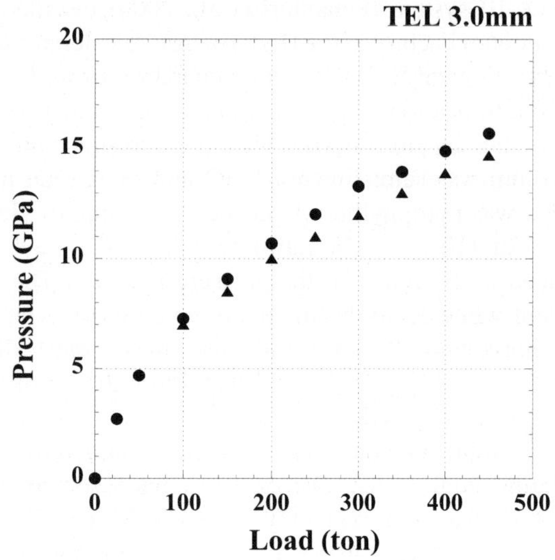

Fig. 3. Generated pressure plotted against applied load for TEL of 3 mm determined from the equation of state of NaCl (solid circles). Pressures obtained using WC anvils are also shown by solid triangles for comparison.

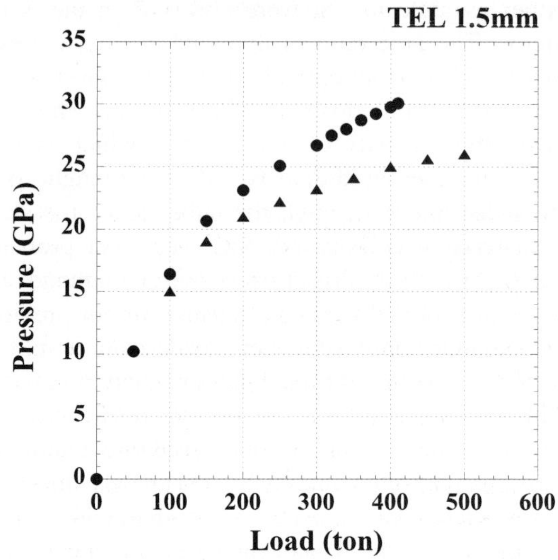

Fig. 4. Generated pressure plotted against applied load for TEL of 1.5 mm determined from the equation of state of NaCl (solid circles). Pressures obtained using WC anvils are also shown by solid triangles for comparison.

diffractions from the sample, several weak diffraction lines from the MgO pressure medium were also observed. It should be noted that these data, recorded through one of the diamond/SiC anvils, were obtained with a typical measuring time of 240 s. The strongest diffraction peak of NaCl (2 0 0) resulted from about 7500 counts with this time duration.

The generated pressures for the anvils with TEL of 3 mm are plotted against press load in Fig. 3. Those for conventional WC anvils with the same TEL (Irifune et al., 1998b) are also shown in this figure for comparison. The efficiency of pressure generation of the present anvil assembly becomes slightly higher than that attained with the WC anvils above 10 GPa. No plastic deformation was observed on the anvils recovered from loads of 450 t. Since the anvils are transparent to X-rays they also have the significant advantage that a very large sample volume (1.5 mm in diameter and 2 mm in height) can be imaged with the X-ray radiography technique. This technique is important for studies such as those in which the density and viscosity of melts and deformation of mantle materials are investigated. For second stage anvils with a 10 mm edge length, the maximum press load is limited to about 450 t by the mechanical strength of the first stage WC anvils. If the size of the diamond/SiC cubes is increased to 15 mm, which is not difficult using the present HIP technique, the limitation of press load due to the strength of the first stage WC anvils is about 1000 t. Consequently, experiments with a larger sample volume would be possible over a pressure range of 20 GPa.

The relation between the applied press load and pressure for anvils with a TEL of 1.5 mm is plotted in Fig. 4 together with that for conventional WC anvils with the same TEL (Irifune et al., 1992). At a load of 410 t, a pressure generation of 30 GPa is achieved. In comparison, the maximum pressure for conventional WC anvils is, as shown by a curve with solid triangles in this figure, less than about 27 GPa because of the limitation of the hardness of WC. No cubes were broken during loading and no plastic deformation of the recovered anvils was observed. The efficiency of pressure generation up to 30 GPa is comparable to that reported for SYNDIE anvils with the same cell assembly (Gotou et al., 2002). Since the slope of the pressure-generation curve is almost linear in the load range between 200 and 410 t, the present anvils would withstand higher pressures than 30 GPa. If the size of diamond/SiC cubes is increased to 15 mm, pressure generations far greater than 30 GPa may be possible.

4. Conclusion

Diamond/SiC composites were synthesized by the HIP of diamond powder in a Si/SiC capsule at conditions of 1450 °C and 200 MPa for 30 min. Graphite was not detected in the products although the applied pressure was well below that of the diamond stability field. Considering the melting temperature of Si (1414 °C), diamond most probably reacted with molten Si to form the diamond/SiC composite. Using these diamond/SiC composites as the second stage anvils in a Kawai-type high-pressure apparatus, in situ X-ray diffraction experiments under pressure were performed. The anvils proved to be hard enough for the generation of pressures up to 30 GPa and transparent enough as an X-ray window. Since HIP is used for the fabrication, it should be possible to obtain such composites with larger sizes and with more complex shapes.

Acknowledgements

We thank Dr. A. Hosomi of Tomei Diamond Inc. for the offer of diamond. We thank two reviewers for invaluable comments, which greatly helped to improve the present article. High-pressure in situ XRD experiments were performed with the approval of the Photon Factory Advisory Committee (Proposal No. 2002G063). This work was supported in part by a Grant-in-Aid for Scientific Research of the Japanese Society for the Promotion of Science (JSPS).

References

Anderson, O.L., Isaak, D.G., Yamamoto, S., 1989. Anharmonicity and the equation of state for gold. J. Appl. Phys. 65, 1534–1543.

Bundy, F.P., 1975. Ultrahigh pressure apparatus using cemented tungsten carbide pistons with sintered diamond tips. Rev. Sci. Instrum. 46, 1318–1324.

Decker, D.L., 1971. High-pressure equation of state for NaCl, KCl, and CsCl. J. Appl. Phys. 42, 3239–3244.

Endo, S., Akahama, Y., Utsumi, W., 1987. Pressure generation by sintered diamond multianvils. Jpn. J. Appl. Phys. 26, 1419–1420.

Funamori, F., Yagi, T., Nishiyama, N., Harada, T., Kikegawa, T., Sato, K., Kato, T., Kubo, T., Toma, M., Ozaki, S., Ohsawa, A., 2000. MAXIII. Rev. High Press. Sci. Technol. 10, 222 (in Japanese).

Gotou, H., Yagi, T., Nishiyama, N., 2002. Effect of surface roughness of sintered diamond anvil on the efficiency of pressure generation. Rev. High Press. Sci. Technol. 12, 236–241 (in Japanese).

Irifune, T., Adachi, Y., Fujino, K., Ohtani, E., Yoneda, A., Sawamoto, H., 1992. A performance test for WC anvils for multianvil apparatus and phase transformations in some aluminous minerals up to 28 GPa. In: Syono, Y., Manghnani, M.H. (Eds.), High Pressure Research in Mineral Physics: Application to Earth and Planetary Sciences. Geophys. Monogr. Ser., vol. 67. Terra Scientific Publishing, Tokyo/AGU, Washington, DC, pp. 33–36.

Irifune, T., Kuroda, K., Nishiyama, N., Inoue, T., Funamori, N., Uchida, T., Yagi, T., Utsumi, W., Miyajima, N., Fujino, K., Urakawa, S., Kikegawa, T., Shimomura, O., 1998a. X-ray diffraction measurements in a double-stage multianvil apparatus using ADC anvils. In: Syono, Y., Manghnani, M.H. (Eds.), High Pressure Research in Mineral Physics: Application to Earth and Planetary Sciences. Geophys. Monogr. Ser., vol. 67. Terra Scientific Publishing, Tokyo/AGU, Washington, DC, pp. 3–12.

Irifune, T., Nishiyama, N., Kuroda, K., Inoue, T., Isshiki, M., Utsumi, W., Funakoshi, K., Urakawa, S., Uchida, T., Katsura, T., Ohtaka, O., 1998b. The postspinel phase boundary in Mg_2SiO_4 determined by in-situ X-ray diffraction. Science 279, 1698–1700.

Ito, E., Kubo, A., 2002. Sintered diamond and research of the earth's interior. Rev. High Press. Sci. Technol. 12, 104–111 (in Japanese).

Kawai, N., Endo, S., 1970. The generation of ultrahigh pressures by a split sphere apparatus. Rev. Sci. Instrum. 41, 1178–1181.

Ohtani, E., Kagawa, N., Shimomura, O., Togaya, M., Suito, K., Onodera, A., Sawamoto, H., Yoneda, A., Tanaka, S., Utsumi, W., Ito, E., Matsumuro, A., Kikegawa, T., 1989. High-pressure generation by a multiple anvil system with sintered diamond anvils. Rev. Sci. Instrum. 60, 922–925.

Ringwood, A.E., Major, A., Willis, P., Hibberson, W.O., 1989. Diamond composite tools. In: Annual Report. Research School of Earth Sciences, ANU, pp. 38–43.

Shimomura, O., Utsumi, W., Taniguchi, T., Kikegawa, T., Nagashima, T., 1992. MAX90, a new high pressure and high temperature apparatus with sintered diamond anvil for synchrotron radiation use. In: Syono, Y., Manghnani, M.H. (Eds.), High Pressure Research in Mineral Physics: Application to Earth and Planetary Sciences. Geophys. Monogr. Ser., vol. 67. Terra Scientific Publishing, Tokyo/AGU, Washington, DC, pp. 3–12.

Shimono, M., Kazehaya, K., Kume, S., 1998. HIP sintering of Si-mixed diamond powder. J. Soc. Mater. Sci. Jpn. 47, 990–993 (in Japanese).

Shimono, M., Wada, T., Kume, S., Ohtaka, O., 2003. HIP production of diamond compact. In: Vincenzini, P. (Ed.), CIMTEC 2002, Proceedings of the 10th International Ceramics Congress Advances in Science and Technology 32, Part C, Techna Srl, pp. 525–532.

Available online at www.sciencedirect.com
SCIENCE DIRECT®

Physics of the Earth and Planetary Interiors 143–144 (2004) 593–600

PHYSICS OF THE EARTH AND PLANETARY INTERIORS

www.elsevier.com/locate/pepi

Formation of pure polycrystalline diamond by direct conversion of graphite at high pressure and high temperature

Tetsuo Irifune [a,*], Ayako Kurio [a], Shizue Sakamoto [a], Toru Inoue [a], Hitoshi Sumiya [b], Ken-ichi Funakoshi [c]

[a] *Geodynamics Research Center, Ehime University, Matsuyama 790-8577, Japan*
[b] *Itami Research Laboratories, Sumitomo Electric Industries, Itami 664-0016, Japan*
[c] *Japan Synchrotron Radiation Research Institute, Hyogo 679-5198, Japan*

Received 31 January 2003; received in revised form 21 May 2003; accepted 5 June 2003

Abstract

Polycrystalline diamond aggregates have been synthesized by direct conversion of graphite at pressures of 12–25 GPa and temperatures 1800–2500 °C using a multianvil apparatus. The synthesized diamond was optically transparent and colorless, and was of cubic symmetry as determined by micro-focus X-ray diffraction. TEM analyses revealed that the sample consists of minute crystals of typically 10–20 nm, and only a very weak and broad band at $\sim$1332 cm^{-1} was observed by Raman spectroscopy. An indentation test demonstrated that the present polycrystalline diamond possesses a Knoop hardness of up to 140 GPa, which is equivalent to or even higher than those of natural and synthetic single-crystal diamonds ($\sim$60–130 GPa) and nearly twice as high as those of synthetic polycrystalline diamonds containing binders ($\sim$50–70 GPa). The present results imply that natural polycrystalline diamonds may have been formed by a rapid transformation from metastable graphite present in cold subducted crust, when encountering in warmer regions, such as rising plumes in the mantle transition region.

Keywords: Diamond; Multianvil apparatus; High-pressure synthesis; Direct conversion; Graphite

1. Introduction

Diamond is the hardest known material with very high thermal conductivity. It is consequently valuable for industrial use, in addition to its use as gemstone. Synthesis of diamond crystals under high static pressure has been successfully made using some catalysts or solvents near the graphite–diamond phase boundary, typically at pressures 5–6 GPa and at temperatures 1500–1800 °C, since the first synthesis in the 1950s (e.g. Bundy et al., 1955; Bovenkerk et al., 1959). High-purity single crystals of diamond, even of the order of 1 cm, are now produced on a commercial basis using techniques such as the temperature gradient method (Sumiya et al., 2000).

Direct conversion of graphite to diamond without any catalysts/solvents under static pressure was conducted by flash heating (e.g. Bundy, 1963; Wakatsuki et al., 1972) or by directly passing electric current through a graphite rod (e.g. Naka et al., 1976; Higashi and Onodera, 1986) using large-volume high-pressure apparatus operated in a hydraulic press. Although the formation of diamond was observed at pressures above $\sim$10 GPa and temperatures exceeding $\sim$2500 °C, only partial conversion of graphite to diamond occurred because the heating durations in these experiments were

* Corresponding author. Tel.: +81-89-927-9645; fax: +81-89-927-8167.
E-mail address: irifune@dpc.ehime-u.ac.jp (T. Irifune).

0031-9201/$ – see front matter
doi:10.1016/j.pepi.2003.06.004

extremely limited (from microseconds to a few seconds).

More recently, the diamond anvil cell (DAC) was adopted for studying the direct conversion of graphite to diamond at room (Utsumi and Yagi, 1991) and high temperatures (Yusa et al., 1998; Yusa, 2002). In these studies, the conversion of a single crystal of graphite to hexagonal diamond (lonsdaleite) was found upon compression above 18 GPa without heating, while the formation of cubic diamond was reported at pressures above 11 GPa and at temperatures higher than ~2700 °C for the single-crystal graphite and at 17 GPa and ~2500 °C for carbon nanotubes with laser-heated DAC (LHDAC). However, the *P–T* conditions of the formation of diamond have not been well documented because of the limited number of experiments and also because only the maximum temperature of the sample was measured in the latter studies using LHDAC. Moreover, the complete transformation of graphite has not been achieved in these LHDAC experiments.

In all of these earlier studies using both large-volume high-pressure apparatus and DAC, graphite itself was used as the heating element. However, such heating may result in heterogeneous temperature distributions within the sample due to the partial transformation to diamond; when graphite partially transforms to diamond, it will not conduct electric current and/or absorb a laser beam efficiently, which leads to localized heating and also limited heating durations. We conducted high-pressure experiments on the conversion of graphite into diamond by the heating of graphite using a separate electrical heater in a multianvil apparatus, where homogeneous and longer heating is possible. The result of a reconnaissance study on the synthesis of pure polycrystalline diamonds has been reported elsewhere (Irifune et al., 2003), and here we present detailed experimental data on the synthesis conditions and the characterization of the synthesized diamonds (Table 1). Implications for the origin of natural polycrystalline diamonds are also discussed in the light of the present experimental results.

Table 1
Experimental conditions and results

Run no.	*P* (GPa)	*T* (°C)	Time (min)	Optical appearance	Phases identified by X-ray diffraction
OS822	12	2000	5	Dark gray	c-diamond + h-diamond + graphite
OS794		2000	<1	Dark gray	c-diamond + h-diamond + graphite
OS755		2300	5	Dark gray	c-diamond
OS801[a]	15	1800	<1	Dark gray	c-diamond + h-diamond + graphite
OS737[a]		2000	5	Dark gray	c-diamond + h-diamond + tr. graphite
OS799[a]		2100	1	Dark gray	c-diamond
OS818		2200	5	Dark gray	c-diamond
OS811		2300[b]	5	Transparent	c-diamond
OS795[a]		2400[b]	1	Transparent	c-diamond
OS735		2500[b]	3	Transparent	c-diamond
OS648	18	1900[b]	10	Dark gray	c-diamond + h-diamond + graphite
OS642		2000	5	Dark gray	c-diamond + h-diamond + tr. graphite
OS647		2100[b]	5	Dark gray	c-diamond
OS657		2300[b]	5	Transparent	c-diamond
OS716[a]		2400[b]	<1	Transparent	c-diamond
OS723[a]		2500	3	Transparent	c-diamond
OS790[a]		2500	1	Transparent	c-diamond
OS772	20	2100	10	Dark gray	c-diamond + h-diamond + tr. graphite
OS773		2200[b]	<1	Transparent	c-diamond
OS685	25	1800	10	Dark gray	c-diamond + h-diamond
OS693		2000	10	Transparent	c-diamond
E-956		2500	5	Transparent	c-diamond

[a] NaCl sleeve was used between Re heater and the graphite sample.
[b] Temperature was estimated based on a power–temperature relation, which may lead to an additional error of ~±50–100 °C.

2. Experimental

We used a Kawai-type multianvil apparatus (Kawai and Endo, 1970) operated in a 2000 t hydraulic ram (Orange-2000 at GRC, Ehime University), in which pressures up to 30 GPa and temperatures of 2500 °C are produced with sample volumes of a few cubic millimeters. The starting material was a pure polycrystalline graphite rod (99.9995%; Nilaco Co. Ltd.), typically 1.5 mm in diameter and 1.0 mm long, which was directly enclosed in Re or Ta tube heaters. The starting graphite consisted of small (1–2 μm) grains with random orientations but exhibited locally stacked textures. In some runs, an NaCl sleeve was also used in between the sample and the heater to achieve quasi-hydrostatic environments. Tungsten carbide anvils with truncation edge length (TEL) of 3.5 mm were used for experiments at 20 GPa and those with TEL = 2.0 mm for runs at 25 GPa.

The temperature was measured using a $W_{97}Re_3$–$W_{75}Re_{25}$ thermocouple with a precision of ±10 °C. However, no corrections of the pressure effect on e.m.f. were made, which may result in some systematic errors in the temperature measurement, probably of the order of ±50–100 °C. The pressures were calibrated on the basis of phase transitions in some reference materials at room temperature (ZnS, 15.6 GPa; GaAs, 18.7 GPa; GaP, 23 GPa) with about 10% reduction at high temperature according to our earlier studies. This procedure may yield an uncertainty of ∼±1 GPa in the pressure measurement.

Pressure was increased first over several hours, and then temperature was increased to the peak value within 20–30 min. The maximum temperature in each run was maintained for times up to 10 min, and was then quenched by turning off the electric power supply. The pressure was released slowly over about 12 h and the recovered sample was inspected by an optical microscope and a scanning electron microscope (SEM). It was further examined by a micro-focus X-ray diffractometer with a 50 μm beam and a rotating copper anode operated at 40 kV and 200 mA (M18XHF; MAC Science). Raman spectroscopic measurements were also made using 514.5 nm radiation from an argon ion laser, possessing a beam size of ∼1 μm and an output power of ∼3 mW (JOEL, JRS-System). The wavenumber was calibrated with a pure Si plate, which may have an uncertainty of about ±0.5 cm^{-1}.

A thin film of the sample for transmission electron microscope (TEM) observations was made with a focused ion beam (FIB). Textural observations of the sample were conducted using a 300 kV TEM (Hitachi, H-9000), and electron diffraction was made with a beam diameter of 400 nm. The Knoop hardness of some selected samples was measured by the indentation method using a natural diamond Knoop indenter. The indentations were made on a polished surface of the sample with a normal load of 500 g applied for 15 s at room temperature. The uncertainty of the measurement is about ±5 GPa.

3. Results

Fig. 1 shows the present experimental conditions and the results on the direct conversion of graphite to diamond. At temperatures between 1800 and 2000 °C, micro-focus X-ray diffraction demonstrated that a mixture of cubic and hexagonal diamonds (Bundy and Kasper, 1967) was formed throughout the pressure range studied, where the existence of hexagonal diamond was estimated on the basis of the presence of the (1 0 0) diffraction line (d = 2.18–2.19 Å; Bundy and Kasper, 1967). The presence of a small

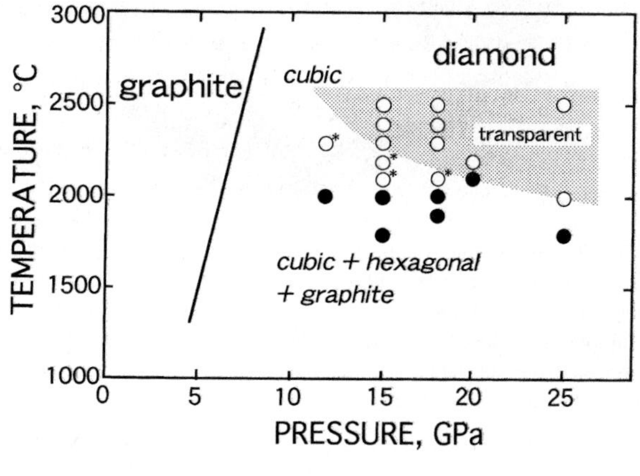

Fig. 1. Experimental conditions and the results of the present experiments. Cubic diamond was formed from graphite at temperatures above 2000 °C, rather independent of pressure, whereas the transparent and colorless diamond was synthesized at higher temperatures as shown in the shaded area. (○) Cubic diamond; (*) gray in color and opaque; (●) cubic diamond + hexagonal diamond + compressed graphite, but only cubic and hexagonal diamond were present in the product at 25 GPa and at 1800 °C (OS685).

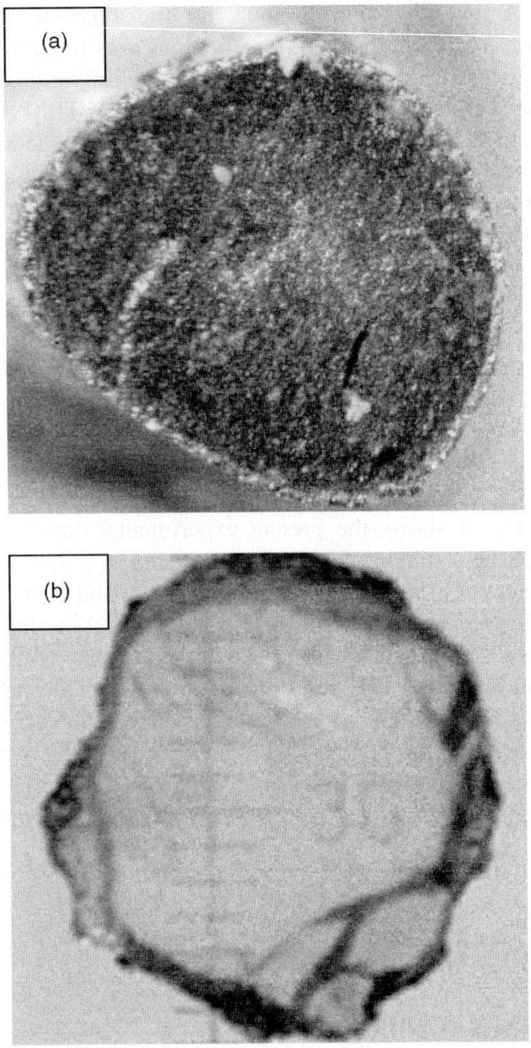

Fig. 2. Optical microscope images of (a) opaque and (b) transparent synthetic polycrystalline diamonds, which have dimensions of ~1–1.5 mm diameter and ~0.3–0.5 mm thickness. The former sample is enclosed in a Re foil heater, while this has been removed from the latter sample, which is so transparent that the scale behind it can be read through the sample.

amount of hexagonal diamond was also confirmed in some samples from electron diffraction observations (Sumiya et al., 2004). These samples exhibited a dark gray color, as shown in Fig. 2(a). A small amount of compressed graphite, with $d_{(002)} = \sim 3.15$ Å (cf. 3.36 Å for normal graphite), was also noted by X-ray diffraction analyses for the products obtained at relatively low temperatures below 2000 °C, except for that obtained at 25 GPa (OS685).

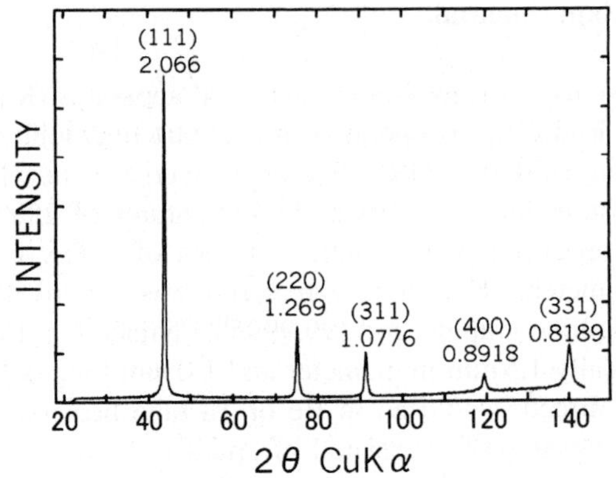

Fig. 3. An X-ray diffraction profile of a transparent sample synthesized at 15 GPa, 2400 °C (OS716). The profile corresponds to that of pure cubic diamond, and least-squares fitting of the *d*-values yielded a lattice parameter of $a = 3.579(1)$ Å (cf. 3.567 Å in JCPDS).

At temperatures higher than 2000 °C, the X-ray diffraction measurements demonstrated that only cubic diamond was present in the run products (Fig. 3). Moreover, observations under the microscope showed that thus-formed diamonds, particularly those synthesized at temperatures of 2300–2500 °C, were optically transparent and colorless (Fig. 2(b)). Such transparent diamond was formed even at lower temperatures with increasing pressure, and its formation was confirmed at about 2000 °C at 25 GPa.

The results of SEM and TEM observations for the starting material and a transparent diamond sample synthesized at 18 GPa, at 2500 °C (OS790), are shown in Fig. 4(a)–(c). The grain sizes of this sample were not resolvable in the present SEM observations (Fig. 4(b)). Actually, the diamond was found to consist of very fine granular crystals of typically 10–20 nm by TEM observations (Fig. 4(c)), which were rather randomly oriented as reported earlier based on electron diffraction measurements (Irifune et al., 2003). However, we also noted the presence of a lamellar texture with elongated crystals of up to about 200 nm in other portions of the same sample. Such a lamellar texture may be formed by rapid crystallization of diamond or represent the original stacking texture of tabular graphite crystals (Fig. 4(a)). Nevertheless, X-ray diffraction profiles of several selected parts of the

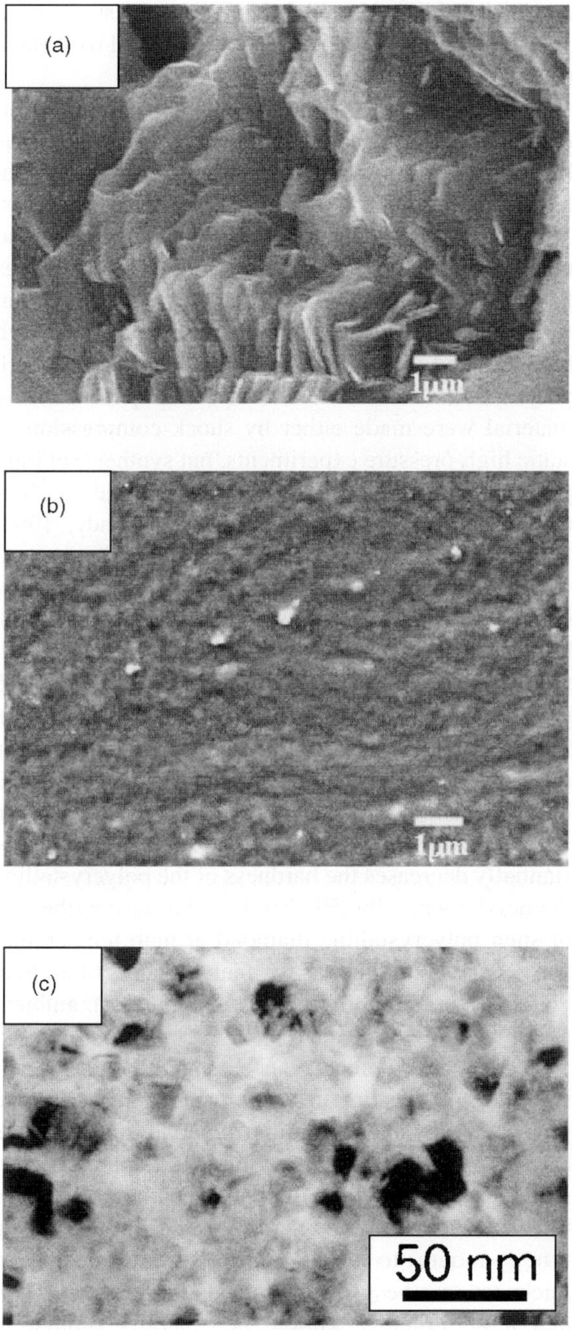

Fig. 4. SEM images of (a) starting graphite and (b) the pure polycrystalline diamond (OS790; 18 GPa, 2500 °C), and (c) a TEM image of the latter sample, which is found to consist of diamond crystals of 10–20 nm.

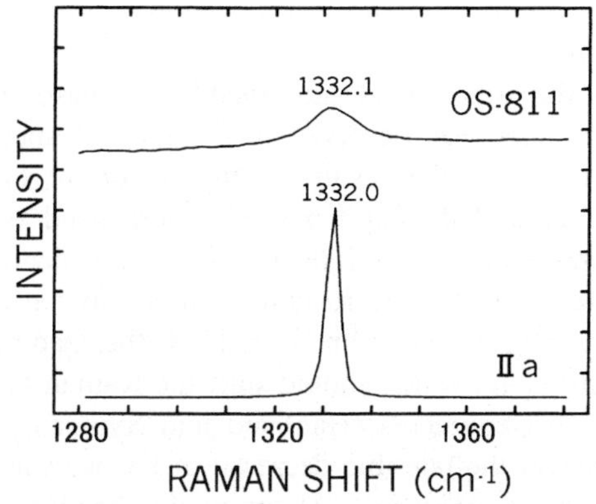

Fig. 5. A comparison of the Raman bands near 1332 cm^{-1} for polycrystalline diamond (OS811; 15 GPa, 2300 °C) and a synthetic type IIa diamond. The Raman band of the polycrystalline diamond is broader and less intense, as compared with that observed in the single-crystal sample. This band is not observed at all in other portions of the polycrystalline sample.

diamond chip using a micro-focus beam yielded ideal powder diffraction patterns, possessing the expected ratios of the (2 2 0) and (1 1 1) intensities in the range between 0.20 and 0.25 (Fig. 3). This observation indicates that no notable preferred orientation of the crystals is present within the sample as a whole. Further details on the microstructural features of the present diamond, based on TEM observations, will be published elsewhere (Sumiya et al., 2004).

Raman spectra from some arbitrarily chosen spots on the surface of a transparent sample chip, synthesized at 15 GPa and 2300 °C (OS811), showed that the intensity of the characteristic band at 1332 cm^{-1} was substantially lower than that of a synthetic type IIa single-crystal diamond (Fig. 5), and this band was even not observed in some analysis spots. This result is consistent with those of earlier studies (Namba et al., 1992; Yoshikawa et al., 1993) which demonstrated that the intensity of this band decreases with decreasing grain size of diamond, suggesting that this band may not be observable when the grain size is less than 100 nm (Namba et al., 1992). Moreover, the full width at half-maximum (FWHM) for the present sample (~10–20 cm^{-1}) was greater than that of the synthetic IIa single crystal (~2 cm^{-1}; Sumiya et al., 1997a). This also agrees well with the result of a calculation (Yoshikawa et al., 1993), which predicted that the FWHM of this band for minute diamond crystals in the range 5–20 nm may become as large as $\sim$40–10 cm^{-1}.

On the other hand, no notable decrease of the wavenumber was observed in the present specimen in contrast to the result of Namba et al. (1992), where natural diamond powders were used for the Raman spectra measurements. This may be due to the presence of some residual compressive stress in the present polycrystalline sample synthesized at very high pressure, which should shift the Raman band to higher wavenumbers (Hanfland and Syassen, 1985) and cancel the possible decrease of the wavenumber by the grain size effect. Alternatively, the shift of the Raman band toward lower wavenumbers with decreasing grain size observed for powdered diamond may be due to a local temperature increase of the sample induced by laser heating; such a temperature effect is insignificant in the present sample of pure polycrystalline diamond, because it should have very high thermal conductivity. Thus it is likely that the Raman band shift is not due to the size effect of diamond crystals.

The Knoop hardness was measured for the two transparent samples (OS790 and OS795). The hardness was also measured for the sample synthesized at lower pressure and temperature (12 GPa, 2000 °C; OS794), which also consists of polycrystalline cubic diamond but is opaque and dark gray in color. These samples were very hard to polish, and the former two transparent samples indeed had Knoop hardnesses ranging from 100 to 110 GPa for OS790 and 130 to 140 GPa for OS795 obtained at several arbitrarily selected points (cf. 60–120 GPa for natural diamonds; Brookes and Brookes, 1991), while the latter opaque sample was less hard (65–95 GPa).

4. Discussion

It is well acknowledged that diamond is the hardest material in all of known natural and synthetic materials. However, the Knoop hardness of natural single-crystal diamond was reported to vary from 60 to 110 GPa for type Ia diamond and 80 to 120 GPa for type IIa (Brookes and Brookes, 1991), depending on the measured crystallographic planes and directions. Synthetic type IIa diamond with extremely low nitrogen impurity concentrations ($\ll$ 0.1 ppm) has been reported to have very high Knoop hardness of up to 130 GPa for the $\langle 1\,0\,0 \rangle$ direction of the (1 0 0) plane (Sumiya et al., 1997b).

The occurrence of natural polycrystalline diamonds, such as carbonado, ballas, and framesite, has been reported from some diamond mines, mostly in Brazil and in Central and South Africa (e.g. Jeynes, 1978). Although their production rate is very limited ($\sim$0.1% of industrial diamond) and the quality varies from specimen to specimen, they have been important for rock drilling applications because of a very high hardness and toughness. Attempts to synthesize such polycrystalline diamond from pure graphite starting material were made either by shock compression or static high-pressure experiments, but synthesis of pure polycrystalline diamond failed mainly due to short reaction times in these experiments (Bundy, 1963; Wakatsuki et al., 1972; Naka et al., 1976; Yusa et al., 1998). Sintering of diamond powders to form massive polycrystalline diamond was also unsuccessful, because of heterogeneous stress distributions within the sample due to very hard nature of the diamond grains (e.g. Hall, 1970). Thus polycrystalline diamonds have been synthesized only by using binders consisting of either metals or inorganic materials (e.g. Katzman and Libby, 1971; Akaishi and Yamaoka, 1996). However, the presence of such materials substantially decreases the hardness of the polycrystalline diamond (generally 50–70 GPa). Moreover, the use of such polycrystalline diamond at high temperature has been very limited, as the metal binders often help diamond to transform back to graphite at ambient pressure.

Recent developments in chemical vapor deposition (CVD) techniques have made it possible to produce pure polycrystalline diamonds (e.g. Krauss et al., 2001). Fairly large films (centimeter size) made of polycrystalline diamonds can be synthesized by this method, although it generally takes a long time ($\sim$month) to make a sample of millimeter-sized thickness, as the growth rate in this method to obtain a high-quality sample is generally limited to $\sim$1 μm/h. Recent studies reported that such polycrystalline diamond can have a high Knoop hardness (e.g. Krauss et al., 2001), but the CVD diamonds are not sintered under pressure and have poor inter-grain adhesion. Thus the highest reported value of the hardness ($\sim$100 GPa) has been significantly lower than those reported for synthetic high-purity single-crystal diamonds (120–130 GPa; Sumiya et al., 1997b).

The present pure polycrystalline diamond has a hardness equivalent to or even higher than those of known natural and synthetic diamonds. Moreover, we found that such diamond persists metastably at temperatures at least up to ~1200 °C in an inert gas (Ar) atmosphere, whereas most polycrystalline diamond composites with metal binders are stable only up to 600–700 °C in a similar atmosphere. Thus the present diamond will be highly valuable in some industrial applications, if one can produce larger samples. The use of a multianvil apparatus with a larger press capacity could easily produce such samples, and other tests on the mechanical properties, such as compressional strength and transverse rupture strength, could be conducted using such samples. Moreover, further detailed studies on the pressure and temperature conditions are required for the formation of polycrystalline diamonds, and also those on the dependency of heating time, starting graphite, surrounding materials, etc., should yield samples of higher quality, and may produce a new ultrahard material for industrial and scientific instrument applications.

The present experimental results have some implications for the formation of natural polycrystalline cubic diamonds, which has been a matter of debate (e.g. Smith and Dawson, 1985; Kagi et al., 1994; Daulton and Ozima, 1996; De et al., 1998). It is likely that these diamonds were formed from crystal carbon of organic origin because of their low $\delta^{13}C$. Such carbon may have been transported into the deep mantle via the subduction of oceanic lithosphere, where it can persist to pressures far higher than the graphite–diamond equilibrium boundary because of the relatively low temperatures in the subducted slab and the slow graphite–cubic diamond reaction kinetics. If the slab is warmed to ~2000 °C at depths of the mantle transition region ($P > 13$ GPa), the metastable graphite should rapidly transform to cubic diamond according to our experiments. The excess pressure over the equilibrium phase boundary may become a driving force for the rapid nucleation and development of lamellar textures of diamond in metastable graphite, which hinders the formation of large single crystals but yields polycrystalline diamond with very high hardness and toughness. Natural polycrystalline diamonds, such as carbonados, do not contain hexagonal diamond, which is also consistent with the characteristics of the cubic polycrystalline diamonds synthesized above 2000–2300 °C, depending on pressure, in the present study. Although the grain sizes of natural polycrystalline diamonds are larger than those of the present polycrystalline diamonds by about two orders of magnitude, our experimental run durations were very limited and significant crystal growth should have occurred in the longer time scales of geological processes.

Polycrystalline diamonds formed in this way could have been brought up to the surface of the earth via kimberlite eruptions originating in rising hot plumes. The very limited regions where the occurrence of such polycrystalline diamonds have been reported may imply that they were formed in rare geodynamic settings where a cold subducted slab encountered a hot region, such as provided by a superplume, in which temperatures as high as 2000 °C can be achieved at these depths. In fact, the possible collision of a rising plume and a subducting slab is suggested on the basis of seismological studies (Geist and Richards, 1993). Thus slab–plume interaction could have played an important role in the formation of natural polycrystalline diamonds.

Acknowledgements

The authors thank M. Nakazawa and Y. Hosokawa for their contributions to some preliminary experiments, and W. Utsumi, D. Yamazaki, H. Kagi, H. Yusa, O. Fukunaga, and M. Akaishi, for valuable discussion. Comments by R.W. Luth and an anonymous reviewer greatly contributed to improve the quality of the present paper.

References

Akaishi, M., Yamaoka, S., 1996. Physical and chemical properties of the heat resistant diamond compacts from diamond–magnesium carbonate system. Mater. Sci. Eng. A 209, 54–59.

Bovenkerk, H.P., Bundy, F.P., Hall, H.T., Strong, H.M., Wentorf Jr., R.H., 1959. Preparation of diamond. Nature 184, 1094–1098.

Brookes, C.A., Brookes, E.J., 1991. Diamond I perspective: a review of mechanical properties of natural diamond. Diamond Relat. Mater. 1, 13–17.

Bundy, F.P., Hall, H.T., Strong, H.M., Wentorf Jr., R.H., 1955. Man-made diamonds. Nature 176, 51–55.

Bundy, F.P., 1963. Direct conversion of graphite to diamond in static pressure apparatus. J. Chem. Phys. 38, 631–643.

Bundy, F.P., Kasper, J.S., 1967. Hexagonal diamond—a new form of carbon. J. Chem. Phys. 46, 3437–3446.

Daulton, T.L., Ozima, M., 1996. Radiation induced diamond formation in uranium-rich carbonaceous materials. Science 271, 1260–1263.

De, S., Heaney, P.J., Hargraves, R.B., Vicenzi, E.P., Taylor, P.T., 1998. Microstructural observation of polycrystalline diamond: a contribution to the carbonado conundrum. Earth Planet. Sci. Lett. 164, 421–433.

Geist, D., Richards, M., 1993. Origin of the Columbia Plateau and Snake River plain: deflection of the Yellowstone plume. Geology 21, 789–792.

Hall, H.T., 1970. Sintered diamond, a synthetic carbonado. Science 169, 868–869.

Hanfland, M., Syassen, K., 1985. Pressure dependence of the first-order Raman mode in diamond. Phys. Rev. B 31, 6896–6899.

Higashi, K., Onodera, A., 1986. Non-catalytic synthesis of diamond from amorphous carbon at high static pressure. Physica 139/140B, 813–815.

Irifune, T., Kurio, A., Sakamoto, S., Inoue, T., Sumiya, H., 2003. Ultrahard polycrystalline diamond from graphite. Nature 421, 599–600.

Jeynes, C., 1978. Natural polycrystalline diamond. Ind. Diamond Rev. 1, 14–23.

Kagi, H., Takahashi, K., Hidaka, H., Masuda, A., 1994. Chemical properties of Central African carbonado and its genetic implication. Geochim. Cosmochim. Acta 58, 2629–2638.

Katzman, H., Libby, L.L., 1971. Sintered diamond compacts with a cobalt binder. Science 172, 1132–1134.

Kawai, N., Endo, S., 1970. The generation of ultrahigh pressure by a split sphere apparatus. Rev. Sci. Instrum. 41, 1178–1181.

Krauss, A.R., Auciello, O., Gruen, D.M., Jayatissa, A., Sumant, A., Tucek, J., Mancini, D.C., Moldovan, N., Erdemir, A., Ersoy, D., Gardos, M.N., Busmann, H.G., Meyer, E.M., Ding, M.Q., 2001. Ultrananocrystalline diamond thin films for MEMS and moving mechanical assembly devices. Diamond Relat. Mater. 10, 1952–1961.

Naka, S., Horii, K., Takeda, Y., Hanawa, T., 1976. Direct conversion of graphite to diamond under static pressure. Nature 259, 38–39.

Namba, Y., Heidarpour, E., Nakayama, M., 1992. Size effects appearing in the Raman spectra of polycrystalline diamonds. J. Appl. Phys. 72, 1748–1751.

Smith, J.V., Dawson, J.B., 1985. Carbonado: diamond aggregates from early impacts of crystal rock? Geology 13, 342–343.

Sumiya, H., Toda, N., Nishibayashi, Y., Satoh, S., 1997a. Crystalline perfection of high purity synthetic diamond crystal. J. Cryst. Growth 178, 485–494.

Sumiya, H., Toda, N., Satoh, S., 1997b. Mechanical properties of synthetic type IIa diamond crystal. Diamond Relat. Mater. 6, 1841–1846.

Sumiya, H., Toda, N., Satoh, S., 2000. High-quality large diamond crystals. New Diamond Front. Carbon Technol. 10, 233–251.

Sumiya, H., Irifune, T., Kurio, A., Sakamoto, S., 2004. Microstructural features of polycrystalline diamond synthesized directly from graphite under static high pressure. J. Mat. Sci. 39, 445–450.

Utsumi, W., Yagi, T., 1991. Light-transparent phase formed by room-temperature compression of graphite. Science 252, 1542–1544.

Wakatsuki, M., Ichinose, K., Aoki, T., 1972. Notes on compressible gasket and Bridgman-anvil-type high pressure apparatus. Jpn. J. Appl. Phys. 11, 578–590.

Yoshikawa, M., Mori, Y., Maegawa, M., Katagiri, G., Ishida, H., Ishitani, A., 1993. Raman scattering from diamond particles. Appl. Phys. Lett. 62, 3114–3116.

Yusa, H., Takemura, K., Matsui, Y., Morishima, H., Watanabe, K., Yamawaki, H., Aoki, K., 1998. Direct transformation of graphite to cubic diamond observed in a laser-heated diamond anvil cell. Appl. Phys. Lett. 72, 1843–1845.

Yusa, H., 2002. Nanocrystalline diamond directly transformed from carbon nanotubes under high pressure. Diamond Relat. Mater. 11, 87–91.

Available online at www.sciencedirect.com
SCIENCE DIRECT
ELSEVIER
Physics of the Earth and Planetary Interiors 143–144 (2004) 601–609
PHYSICS OF THE EARTH AND PLANETARY INTERIORS
www.elsevier.com/locate/pepi

Electron channelling spectroscopy of iron in majoritic garnet and silicate perovskite using a transmission electron microscope

Nobuyoshi Miyajima [a,b,*], Falko Langenhorst [b], Daniel J. Frost [b], Takehiko Yagi [a]

[a] *Institute for Solid State Physics, University of Tokyo, Kashiwa, Chiba 277-8581, Japan*
[b] *Bayerisches Geoinstitut, Universität Bayreuth, D-95440 Bayreuth, Germany*

Received 30 January 2003; received in revised form 10 June 2003; accepted 24 July 2003

Abstract

Majoritic garnet and silicate perovskite assemblages synthesised from natural pyroxene in the multi-anvil press and diamond anvil cell were studied by energy dispersive X-ray spectroscopy (EDXS) and electron energy-loss spectroscopy (EELS) in a transmission electron microscope (TEM) to quantify the composition and oxidation state of iron. In these assemblages of the (Mg, Fe)SiO_3–Al_2O_3 system, Fe^{3+} has a much stronger affinity to silicate perovskite than to majoritic garnet at the same pressure, temperature and oxygen fugacity. To determine the site occupancy of iron in majoritic garnet and silicate perovskite, we have also performed both spectroscopic techniques (EDXS and EELS) under channelling conditions. Electron channelling spectroscopy enables the octahedral sites of both phases to be distinguished from other cation sites by variations in the intensity ratios of the EDXS and EELS spectra acquired systematically at tilting conditions about the channelling-sensitive crystal plane and zone axis. In majorite almost all of ferric iron is partitioned into the octahedral site, whereas in silicate perovskite both ferrous and ferric iron substitute preferentially into the pseudo-dodecahedral site of the perovskite structure. The latter is charge balanced by aluminium in the octahedral site.

Keywords: ALCHEMI-EDXS and EELS; Transmission electron microscopy; Silicate perovskite; Majoritic garnet; Earth's lower mantle; Ferric iron

1. Introduction

The redox state of the earth's mantle is most likely buffered by reactions between the constituent minerals involving iron in both the ferric and ferrous states. The resultant oxygen fugacity imposed by such buffering reactions will be a strong function of the $Fe^{3+}/\sum Fe$ ratios of the individual minerals. To understand the redox state of the mantle therefore requires knowledge of the degree to which ferric and ferrous iron are partitioning between the various minerals.

Recently, several studies in electron energy-loss spectroscopy (EELS) have examined the valence state of Fe in silicate perovskite, which is likely to be the dominant mineral in the lower mantle. Experimental samples of (Mg, Fe)SiO_3 perovskite have been produced that have an $Fe^{3+}/\sum Fe$ ratio of approximately 20% in the presence of magnesiowüstite. Al_2O_3 bearing silicate perovskite samples, however, have been synthesised with $Fe^{3+}/\sum Fe$ ratios as high as 80% (McCammon, 1997; Lauterbach et al., 2000; Frost and Langenhorst, 2002). The stabilisation of high levels of Fe^{3+} in Al_2O_3 bearing silicate perovskite is likely

* Corresponding author. Tel.: +81-4-7136-3231; fax: +81-4-7136-3230.
E-mail address: enkichi@issp.u-tokyo.ac.jp (N. Miyajima).

0031-9201/$ – see front matter
doi:10.1016/j.pepi.2003.07.023

to have important effects on lower mantle properties, such as electrical conductivity and cation diffusion (see Navrotsky, 1999). The mechanism by which Fe^{3+} and Al^{3+} are substituted into the perovskite structure is also likely to affect the elastic properties of this phase. Mantle minerals such as silicate perovskite, majoritic garnet and silicate spinel have usually two potential cation sites on which ferric iron could exist. To clarify the behaviour of iron in mantle phases, to identify the likely substitution mechanisms and to enable the resultant solid solution to be described thermodynamically it is hence necessary to determine the site-specific affinity of ferric iron.

Here we have investigated the $Fe^{3+}/\sum Fe$ ratios of mineral assemblages of majoritic garnet and silicate perovskite synthesised under the conditions corresponding to the top of earth's lower mantle. It is now possible to qualitatively determine the $Fe^{3+}/\sum Fe$ ratios of such coexisting phases from high-pressure experiments using electron energy-loss spectroscopy, even though the mineral grain sizes are frequently less than 1 μm in size (van Aken et al., 1998; Garvie and Buseck, 1998; Gloter et al., 2000; Frost et al., 2001; Frost and Langenhorst, 2002; van Aken and Liebscher, 2002). In addition we have used the electron channelling technique in order to determine the site-specific affinity of ferric iron in majoritic garnet and the site preference of iron and aluminium in silicate perovskite. This technique, commonly termed Atom Location by CHannelling-Enhanced MIcroanalysis (ALCHEMI, Taftø and Spence, 1982; Spence and Taftø, 1983) is a combination of methods including transmission electron microscopy (TEM), energy dispersive X-ray spectroscopy (EDXS) and electron energy-loss spectroscopy.

2. Syntheses and experimental methods

2.1. Electron energy-loss spectroscopy

A natural orthopyroxene and two natural garnets with different chemical compositions were used as starting materials (Table 1). An assemblage of silicate perovskite and majorite in the (Mg, Fe)SiO_3–Al_2O_3 system was synthesised from natural orthopyroxene at 25 GPa and 2223 K using a multi-anvil apparatus (run no. H1488). To pressures of 25 GPa a 10/4-type multi-anvil assembly was used, meaning that a 10 mm edge length Cr_2O_3-doped MgO octahedron was employed with tungsten carbide cubes of 4 mm truncation edge length. In the assemblies $LaCrO_3$ heaters were employed with a W3%Re/W25%Re thermocouple. The thermocouple was inserted axially inside an alumina tube and a single 1.2 mm Re capsule was used. The high-pressure synthesis of silicate perovskite from garnet was performed at approximately 35 GPa for pyrope (run no. Prp01) and 65 GPa for almandine (run no. Alm01), using a modified Mao-Bell type diamond anvil cell (Yagi and Akimoto, 1982). Anvils with a cullet size of 0.45 and 0.35 mm, respectively, were used. In the Prp01 run the powdered pyrope sample (Pyc garnet in Miyajima et al., 1999) was inserted into the hole of the Re gasket and then directly compressed between the two anvils, whereas the powdered almandine sample (Alm01 in Table 1) was compressed with KCl as pressure medium. Heating was made using a double-sided heating system at Institute for Solid State Physics, equipped with a 100 W multimode YAG laser (LHDAC). The temperature of the sample was roughly estimated from the spectrum of the incandescent light from the heated sample. The main product phases were identified using X-ray diffraction methods. All samples were made into thin sections of less than 30 μm thickness for ion thinning and subsequent examination using ATEM. The thin sections were glued on 3 mm sized Mo grids and thinned to electron transparency by Ar ion bombardment. To reduce the amount of amorphisation of silicate perovskite, Ar ion thinning was done under liquid nitrogen cooling and at low voltages and currents (4 kV and 0.8 mA and an incidence angle of 13° in the initial stage, followed by 2.5 keV and 0.3 mA and incidence angle of 11° in the final stage) for the duration of about 30 h. The Fe-$L_{2,3}$-edge electron energy-loss near-edge structure (ELNES) of the product phases was examined in a 200 kV analytical transmission electron microscope (ATEM, Philips CM-20FEG) equipped with an energy-dispersive X-ray spectrometer (EDXS, NORAN Ge detector) and a parallel electron energy-loss spectrometer (PEELS Gatan 666). A stationary unfocused electron beam of about 500 nm spot size was used for EDXS and PEELS. To minimise electron irradiation damage during measurements, TEM thin foils were cooled to nearly liquid nitrogen temperature (ca. 100 K) in a Gatan cooling stage. EDXS spectra

Table 1
Experimental conditions and composition of the phases for ATEM analyses

Run number and conditions (*P*, *T*, *t*)	Phases	Chemical composition in atoms per formula unit							
		Mg	$\sum$Fe	Ca	Al	Si	$\sum$Cations[b]	Fe[a]	$Fe^{3+}/\sum Fe$ (%)
Experiments for two-phase assemblages consisting of majorite and perovskite									
Orthopyroxene (starting material of $(Mg_{0.84}Fe_{0.09}Ca_{0.02}Al_{0.04})(Si_{0.96}Al_{0.04}Cr_{0.01})O_3$)									
Natural	Orthopyroxene	0.837 (27)	0.094 (6)	0.019 (2)	0.081 (3)	0.958 (12)	1.997 (1)	0.090 (10)	<6 (5)
H1488 (25 GPa, 2223 K, 180 min)	Silicate perovskite[c]	3.66 (8)	0.27 (1)	0.03 (1)	0.23 (1)	3.79 (5)	8.03 (6)	0.071 (3)	48 (5)
	Garnet[c]	3.06 (10)	0.29 (2)	0.11 (1)	0.86 (15)	3.56 (7)	7.99 (3)	0.090 (5)	12 (5)
S2841 (26 GPa, 2023 K, 150 min)	Silicate perovskite[c]	3.69 (12)	0.32 (2)	0.05 (1)	0.33 (3)	3.65 (6)	8.11 (6)	0.080 (20)	40 (5)
	Garnet[c]	2.66 (5)	0.32 (1)	0.26 (1)	1.17 (7)	3.41 (3)	8.07 (2)	0.100 (4)	12 (5)
S2903 (26 GPa, 2223 K, 80 min)	Silicate perovskite[c]	3.60 (6)	0.32 (6)	0.06 (1)	0.31 (4)	3.72 (4)	8.07 (1)	0.080 (10)	29 (5)
	Garnet[c]	2.94 (4)	0.26 (9)	0.21 (2)	0.92 (8)	3.55 (4)	8.02 (1)	0.080 (20)	16 (5)
Experiments for perovskite transformed from garnet									
Pyrope (starting material of $(Mg_{2.20}Fe_{0.52}Ca_{0.33}Mn_{0.01})(Al_{1.85}Cr_{0.06})(Si_{3.01}Ti_{0.03})O_{12}$)									
Prp01 (35 GPa, 2000 K, 15 min)	Silicate perovskite[d]	3.00 (7)	0.56 (6)	–	1.14 (16)	3.16 (19)	8.06 (3)	0.15 (1)	68 (5)
Almandine (starting material of $(Fe_{2.34}Mg_{0.55}Ca_{0.09}Mn_{0.01})Al_{2.05}Si_{2.97}O_{12}$)									
Alm01 (65 GPa, 2000 K, 30 min)	Silicate $LiNbO_3$ phase[e]	1.33 (9)	2.12 (5)	–	2.10 (7)	2.38 (4)	8.04 (2)	0.61 (2)	53 (5)

Values in parentheses are experimental uncertainties by ATEM analysis. The numbers are 1 standard deviation from a few to 10 analyses in terms of least units cited.

[a] $Fe = \sum Fe/(\sum Fe + Mg + Ca)$.

[b] Including minor elements such as Mn and Ti.

[c] Coexisting with stishovite.

[d] Coexisting with $CaSiO_3$ silicate perovskite, the NAL phase and stishovite.

[e] Coexisting with $CaSiO_3$ silicate perovskite and stishovite.

were collected with 100 s live time and were quantified according to the procedure of van Cappellen and Doukhan (1994). This method corrects for X-ray absorption on the basis of the principle of electroneutrality and requires the quantification of oxygen. The k factors were calibrated, using the parameter-less correction method (van Cappellen, 1990). ELNES spectra were collected in diffraction mode with convergence and collection semiangles of $\alpha = 8$ mrad and $\beta = 2.7$ mrad, an energy dispersion of 0.1 eV channel and 20–30 s integration time per read-out. A lower extraction voltage of the filament was used to improve the energy resolution to around 0.8 eV, measured as the width of the zero-loss peak at half height. Quantification of the Fe-$L_{2,3}$-ELNES followed the procedure described by van Aken et al. (1998) and van Aken and Liebscher (2002), using an empirically calibrated universal curve. During the amorphisation of silicate perovskite we found a change in EELS signal of the Fe-$L_{2,3}$-ELNES as a function of time. In case of a mixed valence state of iron, the Fe-L_2 and Fe-L_3 edges are composed of two overlapping peaks. The intensity of the left peaks in these doublets, corresponds to the Fe^{3+}, content and increased during electron irradiation, as was also noted by Lauterbach et al. (2000). In case of such iron oxidation during EELS measurements, we have acquired and quantified time dependent series of Fe-$L_{2,3}$-ELNES spectra. The initial $Fe^{3+}/\sum Fe$ ratio of the sample was determined from a linear extrapolation of the measured $Fe^{3+}/\sum Fe$ ratios to zero irradiation.

2.2. *Atom location by channelling-enhanced microanalysis (ALCHEMI)*

ALCHEMI experiments were conducted on two separate multi-anvil samples (Table 1). The assemblages of silicate perovskite and majorite in the (Mg, Fe)SiO_3–Al_2O_3 system were synthesised from natural orthopyroxene at 26 GPa and 2023 K (no. S2841) and 2223 K (no. S2903). To generate pressures in excess of 25 GPa at 2073 K, a newly developed 7/3 type assembly was used, meaning that a 7 mm edge length Cr_2O_3-doped MgO octahedron was employed with tungsten carbide cubes of 3 mm truncation edge length. We again used $LaCrO_3$ heaters and a W3%Re/W25%Re thermocouple for temperature measurements. The sample was inserted into a single 1.0 mm Re capsule. For ALCHEMI analysis of majoritic garnet with EDXS and EELS, the axial channelling effect of the $\langle 1\,1\,1 \rangle$ zone axis was employed to determine the specific valence in the octahedral site (Otten, 1989; Lábár, 1999). The beam convergence for ALCHEMI-EDXS was kept at a value of 6 mrad. The Al/Mg + Si ratio was used as a measure of the strength of the channelling effect, which is very sensitive to the sample orientation. The orientation was finally adjusted by a small shift of the condenser aperture from the centre of the optical axis. For silicate perovskite, planar channelling experiments with systematic tilts about the $\langle 1\,1\,0 \rangle$ direction (Rossouw et al., 1988a,b) were used to clarify the site preference of aluminium and iron in the pseudo-dodecahedral site. ALCHEMI-EDXS spectra were collected in a convergent beam electron diffraction mode. This allowed to perfectly orient the crystal under study by aid of Kikuchi lines and to control for possible irradiation damage during analysis. For ALCHEMI-EELS of beam sensitive majorite and silicate perovskite we evaluated only the first spectrum of the time dependent Fe-$L_{2,3}$-ELNES spectra.

3. Results and discussion

3.1. *Fe-$L_{2,3}$-ELNES of coexisting silicate perovskite and majorite*

Changes in the valence state of iron result in chemical shifts of the white lines at the Fe-L_3 and Fe-L_2 edges. If both ferrous and ferric iron are present, this effect results in a splitting of the white lines and intensity variations, which are a quantitative measure of the ferric/ferrous ratio in the phase under study (van Aken et al., 1998; van Aken and Liebscher, 2002). The reference Fe-$L_{2,3}$-ELNES spectrum of the starting orthopyroxene reveals, for example that the iron is almost completely present as ferrous iron ($Fe^{3+}/\sum Fe = 0.06(5)$). Typical Fe-$L_{2,3}$-ELNES spectra of coexisting silicate perovskite and majorite synthesised from this orthopyroxene at 25 GPa and 2223 K are shown in Fig. 1. The Fe-$L_{2,3}$-edge spectrum of majorite reveals mainly ferrous iron but the ferrous/ferric ratio is slightly enhanced in comparison to orthopyroxene. The spectrum of silicate perovskite clearly indicates a mixed iron valence state. The exact $Fe^{3+}/\sum Fe$ ratios

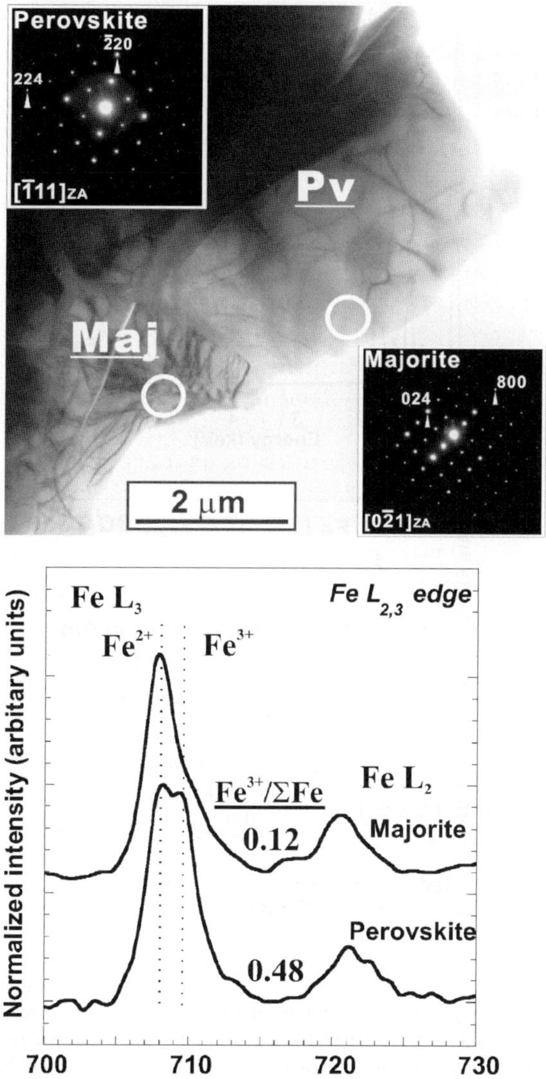

Fig. 1. Bright field TEM image (a) and typical Fe-$L_{2,3}$-edge ELNES (b) of coexisting silicate perovskite (Pv) and majorite (Maj) synthesised at 25 GPa and 2223 K for 180 min. Note insets of (a) are selected area electron diffraction (SAED) patterns of a ⟨1 1 1⟩ zone axis of silicate perovskite (upper left) and a ⟨0 2 1⟩ zone axis of majorite (lower right). Several SAED patterns of both phases are consistent with the space group of *Pbnm* and the space group of *Ia* overline(3)*d*, respectively. The open circles in the TEM image are the areas examined by EELS and by electron diffraction patterns. Spectra have been gain-normalised, background-subtracted and deconvoluted using the low-loss spectra and the $Fe^{3+}/\sum Fe$ ratio has been determined by the treatment described by van Aken et al. (1998). The each ratio includes an absolute error of 0.05.

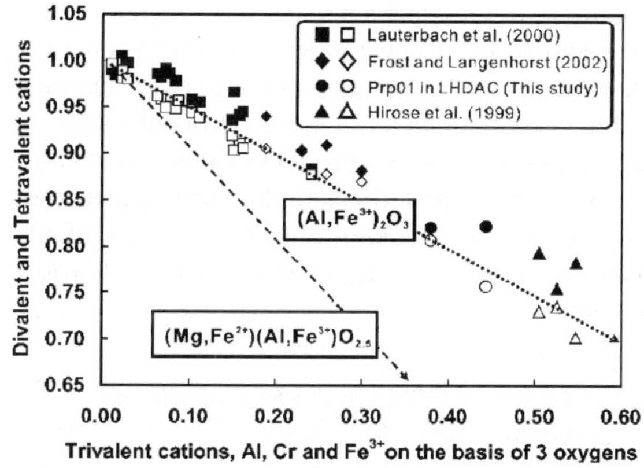

Fig. 2. The variations of cation contents in Fe-, Al-silicate perovskites with combined trivalent cations. Results are from this study including $Fe^{3+}/\sum Fe$ re-measurements of those of Hirose et al. (1999), Lauterbach et al. (2000) and Frost and Langenhorst (2002). Solid symbols denote contents of divalent cations and open symbols denote contents of tetravalent cations. Two arrows indicate likely substitution mechanisms, indicated by (Al, $Fe^{3+})_2O_3$ (dotted line) and (Mg, Fe^{2+})(Al, $Fe^{3+})O_{2.5}$ end-member components (short-dashed line).

determined from these spectra are 0.12(5) for majorite and 0.48(5) for silicate perovskite. This result indicates that oxidation took place during the synthesis of the high-pressure silicates. It additionally suggests that in the silicate perovskite–majorite assemblage of (Mg, Fe)SiO_3–Al_2O_3 system, which can be regarded as representative of the top of the lower mantle, Fe^{3+} has a distinctly stronger affinity to the silicate perovskite phase than to majorite under the same pressure, temperature and oxygen fugacity. Silicate perovskites transformed from two garnets also contain a relatively high Fe^{3+} content, compared with the precursor garnets (no. Prp01; no. Alm01). In order to elucidate the substitution mechanism of trivalent cations in Al-rich silicate perovskite, the concentrations in atoms per formula unit of divalent and tetravalent cations in this phase synthesised at different pressures (*P*), temperatures (*T*), and compositions (*C*) are plotted against the number of trivalent cations (Fig. 2). Two substitution mechanisms are indicated in this diagram by the broken curves. For the (Al^{3+}, $Fe^{3+})_2O_3$ substitution mechanism, trivalent cations charge balance each other on both pseudo-dodecahedral and octahedral sites, while for the (Mg^{2+}, Fe^{2+})(Al^{3+}, $Fe^{3+})O_{2.5}$ component trivalent cations occur on the octahedral site and are charge balanced by an oxygen vacancy.

The trend of data indicates that the latter substitution mechanism, which gives rise to a certain amount of oxygen vacancies, does not appear to be the dominant mechanism particularly at higher trivalent cation contents. In agreement with Frost and Langenhorst (2002), the $Fe^{3+}AlO_3$-coupled substitution mechanism, therefore, appears to be dominant at conditions of pressure and temperature applicable to the earth's lower mantle.

3.2. ALCHEMI-EDXS and EELS of silicate perovskite and majoritic garnet

Mantle minerals such as silicate perovskite and majorite have two potential structural sites for the location of ferric iron. The affinity of Fe^{3+} for Al_2O_3 bearing silicate perovskite indicates that it seems to be energetically favourable to have Al^{3+} on one of the cation sites charge balancing Fe^{3+} on the other. From cation site size constraints it would seem that a $FeAlO_3$ component with Al on the octahedral site is likely to be preferable to the alternative $AlFeO_3$ component. However, this needs to be confirmed by direct ALCHEMI determination of the site-specific affinity of ferric iron and aluminium in silicate perovskite and majorite.

A typical texture of coexisting majorite and silicate perovskite examined by the ALCHEMI is shown in Fig. 3. Fig. 4 shows typical EDXS and EELS spectra

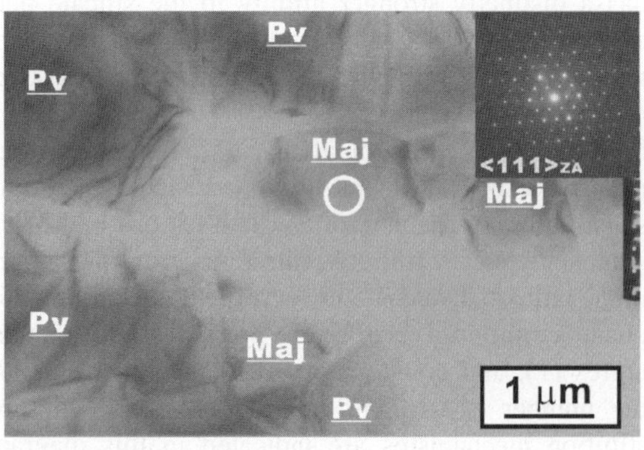

Fig. 3. Bright field TEM image of coexisting silicate perovskite (Pv) and majorite (Maj) synthesised at 26 GPa and 2023 K for 150 min. The inset (upper right) is a SAED pattern of a ⟨1 1 1⟩ zone axis of majorite in which a measured area indicated by a white cycle. The electron beam parallel to [1 1 1] direction of garnet structure axis yields the strongest axial channelling effect in EDXS and EELS.

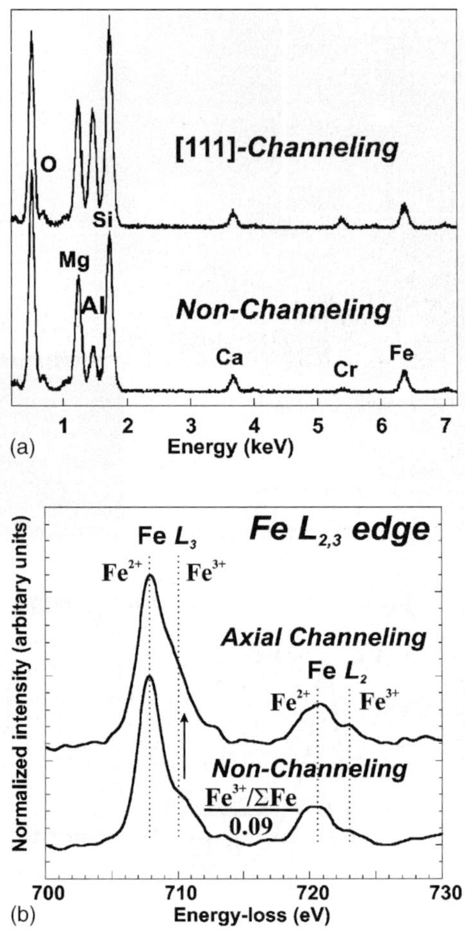

Fig. 4. Comparison of a non-channelling spectrum and an axial channelling spectrum for majoritic garnet: (a) EDXS spectra. Note the change in relative height of the Al and Cr peaks, reflecting occupants of the octahedral sites. (b) Fe-$L_{2,3}$-ELNES spectra. Note the change in relative height of the peak at ca. 710 eV, reflecting the Fe^{3+}-preference for the octahedral site.

of majorite for channelling and non-channelling conditions about the ⟨1 1 1⟩ zone axis. The increase in the intensities of Al and Cr peaks relative to those of the other elements in the channelling EDXS spectrum results from a localised standing electron wave inside the crystal that preferentially causes characteristic ionisation processes on the octahedral site. In case of the Fe-$L_{2,3}$-ELNES spectra recorded under the same diffraction conditions, one detects an increase in the peak intensity at the right shoulder of the L_3 edge (at ca. 710 eV), reflecting the higher Fe^{3+} content on the

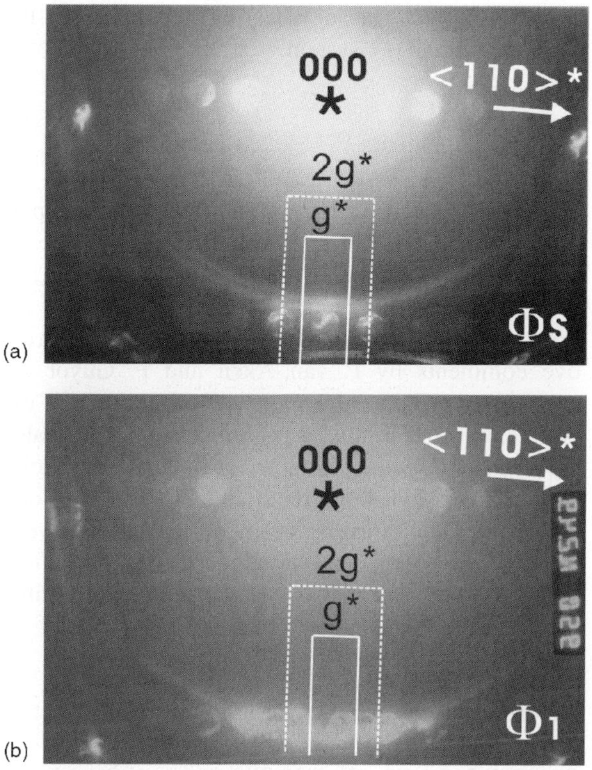

Fig. 5. Pseudo-{1 1 0} systematic-row convergent beam electron diffraction patterns of silicate perovskite (S2903), which indicate the orientations: (a) ϕ_s: $k_x/g_{110} \approx 0$ and (b) ϕ_1: $k_x/g_{110} \approx 0.5$.

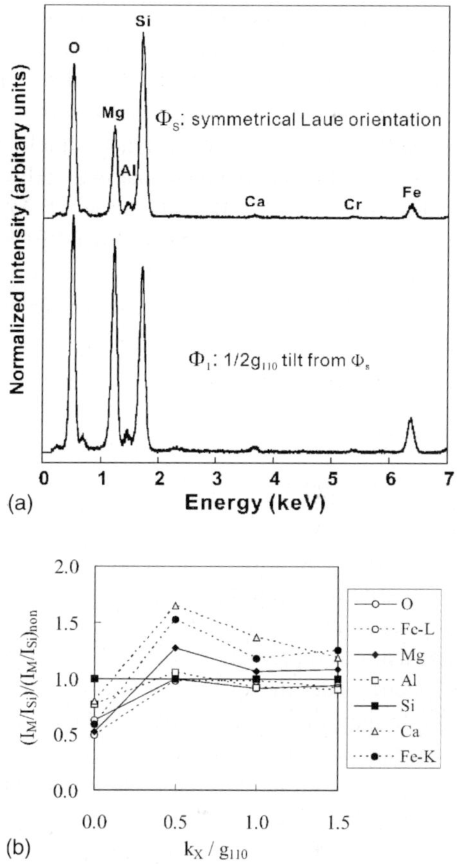

Fig. 6. (a) Comparison of energy-dispersive X-ray spectra for silicate perovskite (S2903) oriented parallel to ⟨1 1 0⟩ directions: the symmetrical Laue orientation, ϕ_s (upper) and slightly inside the Bragg angle for (1 1 0), ϕ_1 (lower), normalised to Si intensities. Note the change in relative height of the Fe and Ca peaks, reflecting occupation of pseudo-dodecahedral sites. (b) Normalised variations in the characteristic X-ray integrated intensity ratio of each element (*M*) under different diffraction conditions for silicate perovskite along the ⟨1 1 0⟩ direction.

octahedral site. This initial result suggests that the octahedral sites prefer to hold the Fe^{3+} cation, consistent with previous Mössbauer results (McCammon, 1998).

The $Fe^{3+}/\sum Fe$ ratio in silicate perovskite (S2903) is 0.29(5), on the basis of the evaluation on Fe-$L_{2,3}$-edge ELNES spectra. We performed the ALCHEMI analysis of silicate perovskite under {1 1 0} pseudo-systematic row diffraction conditions in the lattice setting of S.G. *Pbnm* (Fig. 5). Results of ALCHEMI analysis show that the EDXS spectra of silicate perovskite display remarkable changes in the peak intensity ratios at systematic tilting conditions about the ⟨1 1 0⟩ direction (Fig. 6a), consistent with the results on $CaTiO_3$ perovskite (Rossouw et al. 1988a,b). The variations in the intensities of Fe-K and Ca-K peaks relative to Si-K in the channelling EDXS spectra results from a localised standing electron wave inside the crystal that preferentially causes characteristic ionisation processes on either the octahedral site or the pseudo-dodecahedral site (Fig. 6b). The intensities of Fe-K and Ca-K peaks are strongly correlated with those of Mg-K peaks, although the intensities of Al-K peaks are weakly correlated with those of Si-K peaks. The intensities of Fe-L peaks are not strongly correlated with those of Mg-K and Fe-K peaks, because Fe-L shell excitation is less sensitive to dynamical diffraction effects, indicating greater delocalisation for this lower-energy excitation. These variations indicate that Fe appears to occupy the pseudo-dodecahedral site together with Ca. It appears, therefore, that Al prefers to occupy the octahedral site, charge balanced by Fe^{3+} in the pseudo-dodecahedral site. This initial result suggests that two different cation sites might be qualitatively distinguished with systematic planar ALCHEMI experiments along the ⟨1 1 0⟩ directions of $MgSiO_3$ silicate perovskite. Currently, ALCHEMI-EELS analysis of silicate perovskite is underway to elucidate the site-specific affinity of Fe^{3+} in the perovskite structure.

4. Conclusions

We have carried out the first ALCHEMI-EDXS and EELS experiments on mantle mineral phases such as majoritic garnet and silicate perovskite. Our preliminary results demonstrate that each octahedral site of these phases can be distinguished against the other cation sites using complementary pairs of the EDXS and EELS spectra under the specific diffraction conditions required. In majorite almost all of the ferric iron occupies the octahedral site, whereas both ferrous and ferric iron prefer to substitute into the pseudo-dodecahedral site of the perovskite structure, coupled with aluminium in the octahedral site. For simplicity, complete end-member formulae for majorite and perovskite are $^{VIII}Mg_3{}^{VI}(Mg, Si)_2{}^{IV}Si_3O_{12}$ = $^{VIII}(Mg, Fe^{2+})_3{}^{VI}(Al, Fe^{3+})_2{}^{IV}Si_3O_{12}$ and $^{VIII}Mg^{VI}SiO_3$ = $^{VIII}Fe^{3+VI}Al^{3+}O_3$, respectively. These first results are encouraging and indicate that electron-energy loss spectroscopy performed in a TEM under channelling conditions and combined with electron diffraction and imaging techniques, is a very powerful tool for the characterisation of the crystal chemistry of high-pressure mantle minerals.

For implications of the earth's lower mantle, the above-summarised results indicate that $MgSiO_3$ silicate perovskite is likely the most important carrier mineral phase for trivalent cations such as aluminium and ferric iron in the uppermost part of earth's lower mantle. By the substitution of $Fe^{3+}Al^{3+}O_3$ component into $Mg^{2+}Si^{4+}O_3$ perovskite, the possibility of stabilising a certain amount of oxygen vacancies in the silicate perovskite structure might be dismissed. In short, the incorporation mechanisms of ferric iron might restrict possible physical and chemical properties in oxygen vacancy-bearing perovskite in the $MgSiO_3$–$MgAlO_{2.5}$ system.

Acknowledgements

We are grateful for the excellent technical support by G. Herrmannsdörfer, H. Fischer, H. Schulze. J. Hopf and D. Krausse (Bayerisches Geointitut, University of Bayreuth) and M. Ichihara (Institute for Solid State Physics, University of Tokyo). Constructive comments by P. van Aken and F. Guyot are appreciated. High-temperature and high-pressure experiments using a multi-anvil press and laser heated diamond anvils were conducted at Bayerisches Geoinstitut, University of Bayreuth and Institute for Solid State Physics, University of Tokyo, respectively. We acknowledge financial support from the Visiting Scientists' Program of Bayerisches Geoinstitut, Germany.

References

Frost, D.J., Langenhorst, F., 2002. The effect of Al_2O_3 on Fe–Mg partitioning between magnesiowüstite and magnesium silicate perovskite. Earth Planet. Sci. Lett. 199, 227–241.

Frost, D.J., Langenhorst, F., van Aken, P.A., 2001. Fe–Mg partitioning between ringwoodite and magnesiowustite and the effect of pressure, temperature and oxygen fugacity. Phys. Chem. Miner. 28, 455–470.

Garvie, L.A.J., Buseck, P.R., 1998. Ferrous/ferric ratios from nanometer-sized area in minerals. Nature 396, 667–670.

Gloter, A., Guyot, F., Martinez, I., Colliex, C., 2000. Electron energy-loss spectroscopy of silicate perovskite–magnesiowüstite high-pressure assemblage. Am. Miner. 85, 1452–1458.

Hirose, K., Fei, Y., Ma, Y., Mao, H.K., 1999. The fate of subducted basaltic crust in the earth's lower mantle. Nature 397, 53–56.

Lábár, J.L., 1999. Dynamic electron scattering distinguishes dodecahedral and tetrahedral crystallographic sites in garnet structure. Appl. Phys. Lett. 75, 70–72.

Lauterbach, S., McCammon, C.A., van Aken, P., Langenhorst, F., Seifert, F., 2000. Mössbauer and ELNES spectroscopy of (Mg, Fe)(Si, Al)O_3 perovskite: a highly oxidized component of the lower mantle. Contrib. Miner. Petrol. 138, 17–26.

McCammon, C.A., 1998. Iron crystal chemistry in (Mg, Fe)SiO_3 majorite. Bayeriches Forschunginstitut fur Experimentelle Geochemie und Geopysik. Annual Report, Bayreuth, Germany, 58 pp.

McCammon, C.A., 1997. Perovskite as a possible sink for ferric iron in the lower mantle. Nature 387, 694–696.

Miyajima, N., Fujino, K., Funamori, N., Kondo, T., Yagi, T., 1999. Garnet–perovskite transformation under conditions of

the earth's lower mantle: an analytical transmission electron microscopy study. Phys. Earth Planet. Int. 116, 117–131.
Navrotsky, A., 1999. A lesson from ceramics. Science 284, 1788–1789.
Otten, M.T., 1989. A practical guide to ALCHEMI. Philips Electron. Opt. Bull. 126, 21–28.
Rossouw, C.J., Turner, P.S., White, T.J., 1988a. Axial electron-channelling analysis of perovskite. I. Theory and experiment for $CaTiO_3$. Philos. Mag. B 57, 209–225.
Rossouw, C.J., Turner, P.S., White, T.J., 1988b. Axial electron-channelling analysis of perovskite. II. Site identification of Sr, Zr and U impurities. Philos. Mag. B 57, 227–241.
Spence, J.C.H., Taftø, J., 1983. ALCHEMI: a new technique for locating atoms in small crystals. J. Microsc. 130, 147–154.
Taftø, J., Spence, J.C.H., 1982. Crystal site location of iron and trace elements in a magnesium–iron olivine by a new crystallographic technique. Science 218, 49–51.
van Aken, P.A., Liebscher, B., Styrsa, V.J., 1998. Quantitative determination of iron oxidation states in minerals using Fe-$L_{2,3}$-edge electron energy-loss near-edge structure spectroscopy. Phys. Chem. Miner. 25, 323–327.
van Aken, P.A., Liebscher, B., 2002. Quantification of ferrous/ferric ratios in minerals: new evaluation schemes of Fe-L_{23} electron energy-loss near-edge spectra. Phys. Chem. Miner. 29, 188–200.
van Cappellen, E., 1990. The parameter-less corrections method in X-ray microanalysis. Microsc. Microanal. Microstruct 1, 1–22.
van Cappellen, E., Doukhan, J.C., 1994. Quantitative X-ray microanalysis of ionic compounds. Ultramicroscopy 53, 343–349.
Yagi, T., Akimoto, S., 1982. Rapid X-ray measurements to 100 GPa range and static compression of α-Fe_2O_3. In: Akimoto, S., Manghnani, M.H. (Eds.), High-pressure Research in Geophysics. CAPJ/Reidl, Tokyo, pp. 81–90.

fraction of excited nuclei, and long timescale of relaxation, compared with other spectroscopic techniques at higher frequencies. Secondly, the number of nuclei in the DAC sample is only the order of 10^{-5} of the amount in ordinary test-tube NMR. Thirdly, the metallic gasket tends to shield the rf field. Because of these problems, large numbers of data accumulations are necessary to detect the NMR signal. The rf receiver must be both sensitive and stable for long times, but these features usually conflict with each other in rf circuit design.

The latter two problems relate to the design of the DAC. The DAC should be nonmagnetic so as not to disturb the static field, but this is not easy to achieve. Moreover, if the goal is not only bulk relaxation measurements but also high resolution NMR, every part of the DAC close to the sample should have a magnetic susceptibility similar to that of the sample. This is very difficult to achieve because diamond is diamagnetic but the gasket metal is usually paramagnetic, so that the signs of their susceptibilities are different. Finally, the DAC must fit within the limited working space inside the superconducting magnet. It is therefore required to generate a large force in a compact size, so that a sophisticated force loading mechanism is needed.

These problems make NMR in the DAC very challenging, although progress has been reported since the initial measurements at pressures up to 6.8 GPa (Lee et al., 1987, 1989) with a miniature DAC originally manufactured by Merrill and Bassett (1974). In these studies, a split pair coil was used to induce the rf in the samples. Higher probe efficiency was achieved later (Lee et al., 1992; Pravica and Silvera, 1998a). By development of nonmagnetic lever-arm DACs (Silvera and Wijingaarden, 1985; Bertani et al., 1992), the pressure was extended to 13 GPa (Bertani et al., 1990; Pravica and Silvera, 1998b). High resolution NMR has been performed using gaskets with negative susceptibility, at a limited pressure range to 2 GPa (Yarger et al., 1995).

2. A new nonmagnetic DAC

2.1. Outline

Because the DACs for NMR should be nonmagnetic, beryllium–copper alloy has been adopted for their fabrication (Silvera and Wijingaarden, 1985; Lee et al., 1987; Bertani et al., 1992; Yarger et al., 1995). However, if we intend to apply the DAC for geosciences where high temperature is an important factor, beryllium–copper is no longer suitable. A nonmagnetic DAC made of another material, presumably to work at temperatures to 500 °C, was therefore developed.

Among common nonmagnetic metals, titanium-based alloy is second strongest material behind the beryllium–copper. At room temperature, the former is weaker than the latter. However, there is a substantial difference between their hardening temperatures (Table 1). Metals may lose their strength if they are kept long beyond these temperatures, so that the sample pressure may not be sustained. Thus, the titanium alloy is more suitable for use at high temperatures.

Fig. 1 shows schematic drawings of the DAC fabricated from 6-4 titanium alloy supplied from Daido Steel Co., Ltd., Nagoya, Japan. The cylindrical shape fits the working space 89 mm in diameter in wide-bore superconducting magnets. All parts were designed to be consistent with applicable machining techniques. The DAC has a piston (1) and a cylinder (2), each equipped with a semi-cylindrical rocker (3) on which the diamond anvil is mounted. A couple of upper (4) and lower (5) loading disks are pulled together with three loading bolts (6). Each bolt is supported by a semi-spherical base (7) and laminated Belleville

Table 1
Descriptions of nonmagnetic alloys

Name of alloy	Major component	Minor component (wt.%)	Tensile strength (N/mm^2)		Hardening temperature (°C)
			25 °C	500 °C	
Beryllium–copper 25	Cu	Be 2.0, Co + Ni 0.3	1300	NA	316
6–4 Titanium	Ti	Al 6.0, V 4.2, Fe 0.2	970	∼600	740

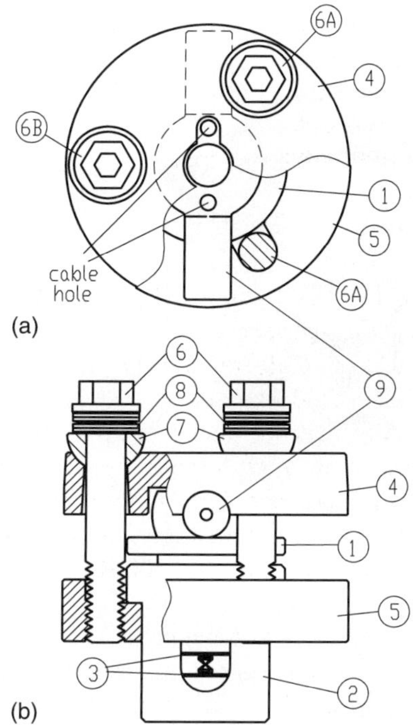

Fig. 1. Schematic drawings of the DAC. (a) Top view and (b) front view. See explanations in the text.

springs (8) which facilitate smooth fastening of the bolt. A center plate with a pivot (9) transfers the force from the upper loading disk (4) to the piston (1), keeping the force aligned in the correct direction and working on the correct point of action.

On the force loading, two loading bolts (6A) at the right of the top view in Fig. 1 are fastened first. The other bolt (6B) at the left is fastened next. Then, as the left bolt fastens, the upper loading disk (4) pushes the center plate (9) with multiplied force. This is because the fastened pair of bolts (6A) works together as the fulcrum of a "lever", which is made up of the pair of the loading disks (4–5). The three loading bolts are accessed from the top of the DAC. They can be fastened or loosened after the DAC is installed into the magnet, so that the pressure is changeable in situ with a handmade nonmagnetic wrench.

2.2. Piston and cylinder

The DAC design of Merrill and Bassett (1974) has been repeatedly used for previous NMR measurements. Because of its simple design, there is sufficient space around the anvils for installation of the rf probe which is the means of excitation and detection of NMR. However, because the anvils are directly fixed on platens, the culets tend to tilt as their platens are pulled together by fastening each bolt. This is a drawback of the simple design which may limit application of this DAC to very high pressure (Jayaraman, 1983, Table II). On the other hand, DACs free from this drawback generally have a tightly fitted piston and cylinder, and several parts to adjust the culets, all of which tightly surround the sample (e.g. Mao and Bell, 1975). The space then is very limited and some of necessary probe parts must be placed away from the sample (Pravica and Silvera, 1998a).

We aimed to combine the advantages of these two DAC designs, while minimizing their drawbacks. Fig. 2 shows a photograph of the piston–cylinder part of our DAC. Inclination and position of the two anvils, one of which is to be mounted on the rocker shown at the lower center, was controlled only by small screws. On the cylinder shown at the right, the rocker is held on two edges by two pan screws. These screws are slotted, not only at their heads but also at their tails. Then the culet inclination is controlled from outside of the cylinder by turning the tails of the screws with a slot driver. Along the two other edges of the rocker, the pair of parts shown at the lower left in Fig. 2 is connected to the cylinder by two countersunk screws. This pair sandwich the rocker to hold it, and by fastening one screw and loosening another, the rocker is pushed along an axis. After the centering is completed and the sample is pressurized, these parts are removed by simply loosening both screws. We have designed another pair of rocker holders that is not removable and these are shown at the lower right.

On the piston shown at the left of Fig. 2, the rocker is held at four corners by pan screws. The culet inclination is controlled by fastening two on one side and loosening the two on the other side. A pair of small holders is fixed between these screws. Set screws in these holders push the rocker along the axis from left to right, which is perpendicular to the direction of movement of the rocker on the cylinder side.

When the piston and the cylinder are coupled, all the screws are along two sides of the pair of rockers, and the other two sides are open. On these open sides, cavities and cable holes are designed in the sidewall of

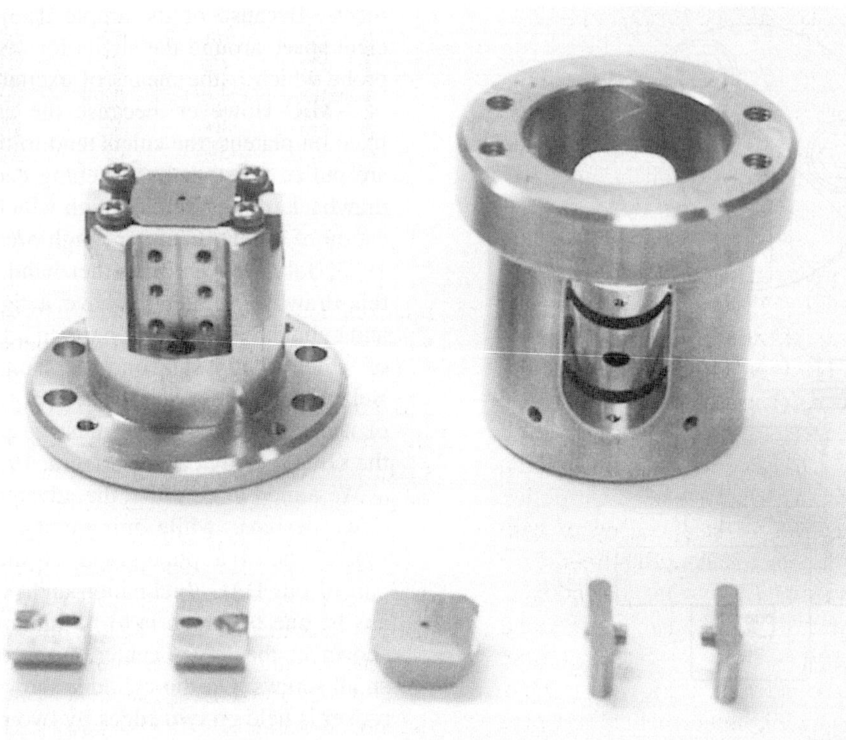

Fig. 2. A photograph of the piston and cylinder. Left, the piston equipped with the rocker; right, the cylinder; lower left, removable rocker holders; lower center, the rocker; lower right, not removable rocker holders.

the piston. Windows were also opened in the sidewall of the cylinder. These cavities and windows are for installation and tuning of the rf probe.

2.3. Rockers

The hardness of the titanium alloy is not sufficient for the rockers, which directly support the diamond anvils. Sintered tungsten carbide is most suitable, but most types contain cobalt as a binder, which shows strong ferromagnetism with a Curie temperature $T_c = 1388\,K$. The only other functional binder is nickel which is also ferromagnetic with $T_c = 627\,K$, with about one-third of the saturated magnetic moment of cobalt. To alleviate this problem, it should be noted that the ferromagnetic moment decreases with temperature once it exceeds half of T_c (e.g., Kittel, 1995, Chapter 15, Fig. 4). The T_c of cobalt is five times higher than room temperature, while that of nickel is two times room temperature. Thus cobalt keeps most of its moment at higher than room temperature, but nickel does not. After this consideration, we adopted the tungsten carbide M009V (Toshiba Tungaloy Co., Ltd., Kawasaki, Japan) which contains a small amount of nickel as a binder. The semi-spherical bases (8) in Fig. 1 were made of the same material.

3. RF probe, transmitter and receiver

A Helmholtz-type rf coil 2.0 mm in inner diameter was prepared from 0.4 mm diameter copper wire. Its inductance was estimated to be ~20 nH. The coil was mounted on a microstripline 3 mm in width with ceramic chip capacitors and trimmers (GRH710 and TZVY2 series, Murata Manufacturing Co., Ltd., Kyoto, Japan), to form a resonating tank circuit. This probe is reusable, robust, and easy to install to the DAC by using microscrews. Because the rf field perpendicular to the conductive metallic gasket cannot penetrate the sample hole, the axis of the Helmholtz coil is set to be parallel to the gasket surface (Lee et al., 1987, 1992).

The S/N ratio of NMR positively correlates with the Q-value and sample filling factor (effective sample/coil volume ratio) of the rf coil (Fukushima and Roeder, 1981). The Q-value of our coil is limited to a few tens, and its diameter is one order of magnitude larger than the sample. Together with other previously described problems regarding the signal amplitude, the estimated S/N should be only about one-tenth per sampling. A narrow-band, low noise superheterodyne receiver was constructed to detect this weak signal. The receiver consists of two cascaded narrow-band GaAs preamplifiers (P207VDG, Gain 22dB, NF 0.5dB, Advanced Receiver Research Corp., USA), a commercial low noise preamplifier, a commercial low input double balanced mixer with matching of local frequency (LO), radio frequency, and intermediate frequency (IF) ports, a handmade phase sensitive detector, two programmable filters, and a two-channel AD converter with 14-bit resolution. Direct digital synthesizers working at 167 and 40 MHz (FSW01 and PCK-50, Digital Signal Technology, Inc., Asaka, Japan) generate LO and IF. Their sum yields 207 MHz rf, which is the Larmor frequency of protons.

The excitation pulse sequence is generated by switching of a continuous output IF synthesizer. A PC-driven digital signal processor (DSP6031, MTT Co., Ltd., Tokyo, Japan) controls the timing. The x, y, $-x$, $-y$ pulses can be programmed. These pulses are amplified by a 50 W transmitter (N146-4799A, Thamway Co., Ltd., Fuji, Japan). A typical $\pi/2$ pulse width is 120 μs. This long width is for compensation of the small transmitter power.

4. Performance

A phosphor bronze gasket 2 mm × 6 mm in size and 0.3 mm in thickness was inserted into the Helmholtz coil in the probe. They were easily installed in the DAC equipped with Type I diamond anvils of 0.45 mm culet diameter. Deionized water was filled in a 0.3 mm diameter sample hole in the gasket, and kept liquid after pressurization so that the applied pressure was less than one gigapascal. The DAC was then installed into a commercial wide bore superconducting magnet operating at 4.86 T. The probe was cabled with a MMCX micro-miniature connector. Access to the tank circuit was easy so that tuning was not bothersome.

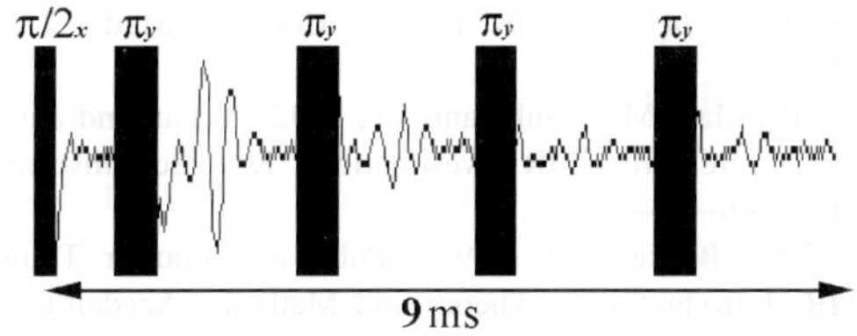

Fig. 3. A spin echo train obtained from H_2O pressurized by the DAC.

The Carr–Purcell–Meiboom–Gill pulse sequence was programmed to detect a spin echo train (Farrar and Becker, 1971). The observing frequency was slightly detuned from resonance, so that oscillating peaks were supposed to be observed in the middle of every two π_y pulses. After $\sim 10^4$ accumulations, it was successfully obtained as shown in Fig. 3. Because this is reliable evidence of an NMR signal, and because it disappeared after the sample was discharged, we conclude that NMR measurement is possible with the DAC described here. The nickel binder apparently did not affect this bulk relaxation detection. While considerable development is still required, we have demonstrated progress towards realizing a practical scheme for high temperature and pressure NMR in the diamond anvil cell.

Acknowledgements

The author thanks T. Yagi at University of Tokyo for advice on DAC design, and K. Suzuki at Nagoya University for advice on mechanical drawing. T. Miura at Ikegami Precision Co., Ltd. is acknowledged for careful machining. The author also thanks K. Kose at University of Tsukuba and T. Haishi at MR Technology Inc. for design and advice on the NMR system. S.K. Lee and S. Gramsch at Geophysical Laboratory are acknowledged for manuscript reading. T. Duffy at Princeton University and two anonymous referees provided critical comments on the manuscript.

References

Bertani, R., Mali, M., Roos, J., Brinkmann, D., 1990. The Knight shift and diffusion in solid lithium and sodium at pressures up

ELSEVIER

Physics of the Earth and Planetary Interiors 143–144 (2004) 619–625

PHYSICS OF THE EARTH AND PLANETARY INTERIORS

www.elsevier.com/locate/pepi

Subject Index of Volume 143–144